MARGARET ROBINSON

DATE DUE			

The Why and How of Home Horticulture

The Why and How of Home Horticulture

D. R. Bienz

Washington State University

W. H. Freeman and Company

San Francisco

Library of Congress Cataloging in Publication Data

Bienz, D R 1926–
The why and how of home horticulture.

Includes bibliographies and index.
1. Gardening. 2. Horticulture. I. Title.
SB453.B49 635 79-19915
ISBN 0-7167-1078-1

Sponsoring Editor: *Gunder Hefta;* Project Editor: *Patricia Brewer;* Copy Editor: *Susan Weisberg;* Designer: *Perry Smith;* Production Coordinator: *William Murdock;* Illustration Coordinator: *Cheryl Nufer;* Artists: *Donna Salmon, Darwen Hennings, Vally Hennings, John Waller,* and *Julia Iltis;* Compositor: *Allservice Phototypesetting Company;* Printer and Binder: *Kingsport Press.*

Printed in the United States of America

3 4 5 6 7 8 9 KP 0 8 9 8 7 6 5 4 3 2

Flower in the crannied wall,
I pluck you out of the crannies;—
Hold you here, root and all, in my hand,
Little flower—but if I could understand
What you are, root and all, and all in all,
I should know what God and man is.

—Alfred, Lord Tennyson

Contents

Preface

When, as a result of the enthusiasm for gardening that developed in the early 1970s, I was asked to teach an elective, introductory horticulture course for nonmajors, I soon discovered that no appropriate textbook existed. The excellent introductory texts available for students of commercial horticulture did not focus on the interests of home gardeners, and the numerous books written about gardening included little of the scientific horticulture necessary for understanding fundamental reasons for horticultural practices. Consequently, I developed my own syllabus, "Horticulture for the Homeowner Who Wants to Know Why."

Although this hurriedly written syllabus was far from polished and was never advertised for sale, I soon began receiving requests for copies from many parts of the country. Former students, instructors of horticulture and biological sciences at several universities and community colleges, and friends who are gardeners encouraged me to expand the syllabus into an illustrated book suitable both as a classroom text and as an informational guide for serious gardeners. What began as a minor revision eventually resulted in a rewriting of the syllabus, augmented with new information, additional chapters, and many illustrations. The outcome is this text, *The Why and How of Home Horticulture*.

The first 13 chapters are organized in a sequence that can be used as the informational basis for a three-hour semester or a five-hour quarter course in home horticulture. The introductory chapters are concerned with reproduction, development, propagation, and planting. These are followed by chapters that deal with the relation of environmental elements to horticultural production and chapters that describe kinds of gardens and individual garden crops.

Although the chapters are sequential, each is a relatively complete unit, and they do not necessarily need to be studied in order. For example, during the fall semester I usually schedule the classes on landscaping after the material in the first 4 chapters has been presented, thus landscape plantings can be observed before they are covered with snow. During warm autumns the chapter on pruning may need to be assigned later so that trees and shrubs will be dormant for concurrent laboratory practice. For the same reason, when spring is early, pruning may need to be scheduled earlier than it would be if the text chapters were assigned in sequence.

The first 13 chapters contain considerable practical information, but most of the instructions for gardening procedures are contained in "The Handbook," Chapter 14. Assembling the how-to-do-it information in one place provides a convenient garden reference and also permits, in the first 13 chapters, a presentation of the sequential pageant of gardening that is more coherent because it is not so frequently interrupted with practical examples.

No specific section is delineated for laboratory work, but a semester or quarter of laboratory exercises, including propagating, planning, planting, storing, processing, flower arranging, and other activities, could be designed based on the descriptive material in "The Handbook."

It is my hope that serious gardeners not involved in

academia will also find this book understandable, interesting, and useful. I would suggest they read the first 13 chapters in sequence, relating the information in each chapter to their own experience. They should then read through "The Handbook" to become acquainted with its contents so it can be used as a reference when specific gardening needs arise.

I thank the many individuals who have contributed assistance and encouragement during my preparation of this manuscript. Naming them all would be impossible, but I would like to recognize especially F. E. Larsen and E. W. Kalin, who contributed parts of "The Handbook" and reviewed sections of the manuscript; R. L. Hausenbuiller, K. N. Nilsen, and K. A. Schekel, who also reviewed sections of the manuscript; Gunder Hefta, who reviewed the entire manuscript; and Patricia Brewer and Susan Weisberg, who spent countless hours correcting and editing.

I also thank Margaret Gurtel, who assembled most of the index and a considerable portion of the glossary and did library research for many of the tables, and my daughter, Marianne, who assisted with library research, glossary organization, preliminary typing, almost all of the final draft typing, and, perhaps most important, enthusiastic encouragement. Above all I thank my wife, Betty, who edited much of the manuscript, in several drafts, for grammar and clarity, who typed preliminary drafts, and who provided patient encouragement during the years of manuscript preparation.

I also thank the many who, over the years, have inspired me with the knowledge and appreciation of plant growing, especially my father, Rudolph Bienz, and Professors J. E. Kraus, Earl New, Leif Verner, and G. W. Woodbury.

August 1979 D. R. Bienz

The Why and How of Home Horticulture

1 Plants and People

With much of the native flora of the world's cities replaced by concrete and brick, urban dwellers no longer live and work in intimate contact with the plant growth that provides the oxygen they breathe; the clothes they wear; the food supply they find so abundantly and conveniently displayed in their supermarkets; and even the rubber, plastic, and upholstery for the cars they drive. Yet, almost without exception, people have the desire to be near green foliage. Thriving flower and plant shops, carefully tended indoor and outdoor home gardens, trim lawns, popular natural parks and walkways, and the use of plastic foliage where live plants cannot be maintained attest to human affection for plants. Some individuals may not be enamoured of the work associated with making plants grow, but almost everyone enjoys a garden environment.

Gardening and Civilization

The desire to associate with plants may stem from the fact that human existence has always depended on the procurement of food and fiber from plants or from animals that feed on plants. From earliest times animals have been hunted and plants have been gathered, activities that persist as interesting hobbies if not necessities on which life depends.

A serious problem associated with living by hunting and gathering was that the success of these activities was uncertain and seasonal, dependent on the vagaries of nature. This problem was gradually alleviated through the process of domestication, which probably began thousands of years ago when observant people discovered that the barley seeds and small pieces of sweet potato discarded to the trash heaps near their shelter would grow into new barley or sweet potato plants. They also found that seeds or root sections gathered from the more desirable wild plants would, in turn, produce better progeny plants when planted near their dwelling. With these discoveries the art of gardening began.

Similarly, the more docile animals were tamed and herded. With their food supply essentially secured by sheep in the paddock and vegetables in the garden, people had time for the civilizing influence of social interaction and the development of art and philosophy. Archeologists have not yet discovered all the pieces to the puzzle of human social development, but they have found enough to make it obvious that gardening played a major role in the rise of civilizations.

From these early beginnings plant domestication and improvement continued with the Egyptians, Assyrians, Babylonians, Dravidians, Chinese, Greeks, Romans, Incas, Aztecs, and countless other civilizations. Even when an empire perished along with its war machines and most of its art and literature, the plants it developed remained

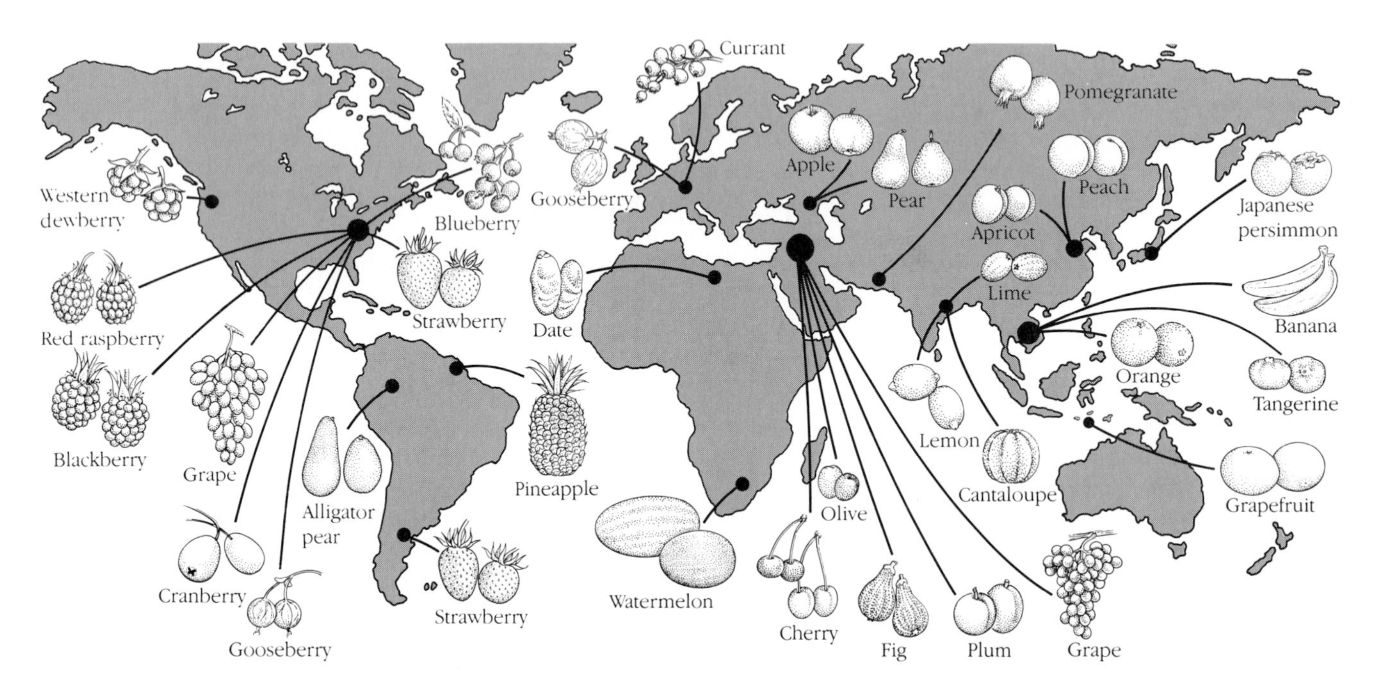

Figure 1–1 Approximate geographical origin of cultivated fruits (above) and cultivated vegetables (opposite). (Map reproduced by permission of Harper & Row, Publishers, Inc., from *Domesticated Plants,* by Bertha Morris Parker and Illa Podendort; copyright 1949 by Harper & Row, Publishers, Inc.)

to enrich the gardens of succeeding civilizations. Explorers have had an important role in carrying garden plants from one part of the world to another. They almost doubled the diversity of domesticated plants when, with the "discovery" of America, they helped combine the plant heritages of the Old and New Worlds in the sixteenth century. Thus the tremendous diversity of domestic plants adapted to the varied climates where people live is a synthesis of the efforts of countless gardeners from many nations over many years (Figure 1–1).

Plant exploration and plant development are being continued today with the impetus of modern science and technology. Each year new types of garden plants and improved techniques for producing them become available. Today's gardeners can utilize not only the garden heritage of all people through all time but also the continuing developments of governmental and commercial research institutes and university plant science departments.

Horticulture in Relation to Other Disciplines

Although the production of garden crops is a horticultural activity, sooner or later most gardeners face problems that require the expertise of specialists in other fields. It seems appropriate, therefore, to devote a few paragraphs to the relationship of horticulture to other agricultural and biological sciences and related disciplines and to explain briefly what the field of horticulture encompasses.

With the very earliest domestication there was a sharp division between the occupation of the animal agriculturist, who was primarily a nomad following flocks and herds from one grassy area to another, and the more or less sedentary plant agriculturist. This division is exemplified in the Biblical story of Cain, the herdsman, and Abel, the gardener. Later, as more animals became domesticated, animal agriculturists were further divided accord-

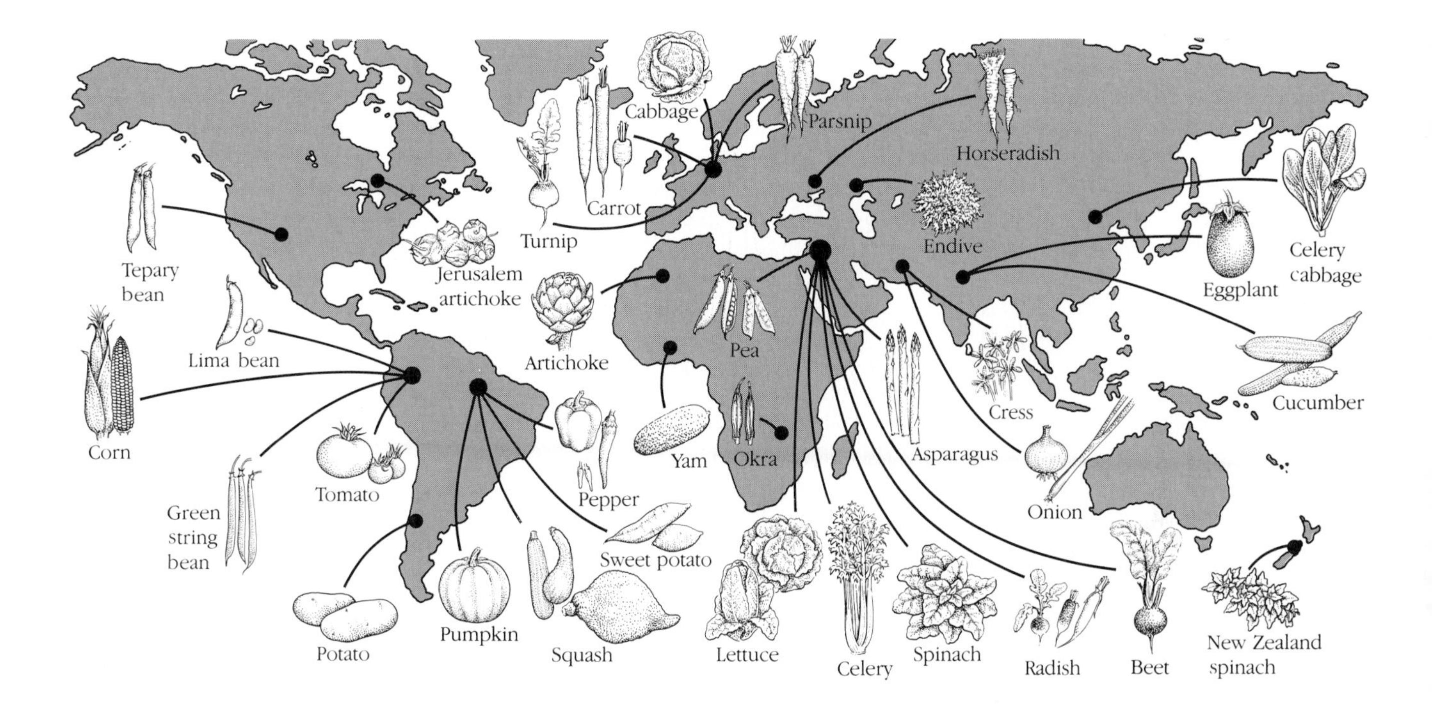

ing to the kind of animal in which they specialized as cattle, sheep, hog, or poultry producers. With the domestication of fowl and swine, the development of planted pastures, the production of hay, and the recognition of the value of manure as fertilizer, there developed a somewhat closer relationship between some plant and animal agriculturists.

As long as planting, sowing, and reaping were accomplished with hand tools and as long as farmers produced food primarily with the help of and for their families, there was little to distinguish growers of one kind of plants from those of others (Figure 1–2). However, with the development of machinery that permitted each grower to produce extensive acreages of certain crop plants, plant agriculturists became divided into two groups: the **agronomists,** who grew field crops on large acreages, and the **horticulturists,** who grew garden or intensively cultivated crops. The distinction between horticultural and agronomic crops persists, even though today some horticultural crops are produced in larger fields than are agronomic crops. The term horticulture is derived from two Latin words that mean, literally, garden cultivation.

Botany and related plant sciences are closely allied to horticulture, and in the past botany and horticulture were often included in the same university department. Horticulture is sometimes referred to as applied botany. Vocational fields related to plant pest control—**entomology** (insects), **plant pathology** (plant diseases), and weed control—as well as soils and agricultural economics are other disciplines closely allied to horticulture.

Divisions of the Field of Horticulture

Horticulture as an educational discipline is both an art and a science and usually includes five different subfields. In some universities these are grouped together in a horti-

Figure 1–2 Harvesting sweet corn during the 1930s. (Photo by C. L. Vincent.)

culture department. At other schools a number of these subfields may be separate departments, or horticulture may be taught with agronomy and plant pathology in an overall plant science department. These five areas are: **pomology,** the culture of fruit; **olericulture,** the culture of vegetables; **ornamental horticulture,** the production and utilization of flowers, shrubs, and trees; **postharvest horticulture,** the processing, preservation, and storage of horticultural products; and **landscape horticulture,** the use of plant materials for beautification.

All or part of the subfield of postharvest horticulture is often part of an interdisciplinary field of food technology dealing with all aspects of storing, preserving, and processing foods. Landscape horticulture is often a part of a much larger discipline that goes by various names and includes different subfields depending on the location. In the past landscape training has been allied with either horticulture or architecture departments; now it is frequently also associated with environmental or outdoor recreation disciplines that may ally it with forestry or physical education. **Viticulture** and **enology,** the culture of grapes and the art of wine-making, respectively, are sometimes treated as fields separate from pomology.

Horticulturists need a knowledge of a number of related fields, including those mentioned above as well as business, climatology, and engineering. People interested in ornamental areas of landscape architecture need a knowledge of design, art, city and recreational facilities planning, and perhaps social science. Horticultural education and garden writing are two other rapidly growing subfields.

Classification of Horticultural Crops

The human mind retains a body of knowledge most easily if the various components can be grouped in a related and orderly fashion. For example, the names of the states of the United States and the provinces of Canada are more easily remembered in groups according to geographical location or in alphabetical order than randomly. Analogously, garden plants are more easily studied if they are grouped in a meaningful relationship. There are many ways of grouping and thus of classifying horticultural crops.

Horticultural Classification

One way of classifying garden crops is referred to as **horticultural classification,** which groups products according to their use:

I. Edible crops
   A. Fruit
      1. Tropical
      2. Subtropical
      3. Temperate
         a. Tree fruit (Figure 1–3)
         b. Small fruit (Figure 1–4)
   B. Vegetables
      1. Cool-season
      2. Warm-season
   C. Drugs, condiments, and beverages
II. Ornamentals
   A. Flowers and foliage plants
      1. Flowers for indoor use
      2. Flowers for outdoor use
   B. Shrubs and trees

Obviously, not all horticultural crops classify neatly into one of the above groups. The term **fruit** as used in a horticultural sense is somewhat different from the term used in a botanical sense. Botanically, a fruit is an enlarged ovary with attached parts. Horticulturally, a fruit is a plant part that can be consumed with little or no preparation as a dessert or snack, and a **vegetable** is a plant part that may or may not require cooking but is usually consumed without much refinement and with the main course of the meal (agronomic crops usually require milling or refining before they are consumed) (Figure 1–5).

Some fruits and vegetables do not readily fit the above classification. For example, tomatoes, peppers, and beans, all botanically fruits, are considered vegetables because they are put to culinary uses we ordinarily associate with vegetables. Rhubarb, usually grown as a vegetable, is normally used as a dessert, and avocado, botanically and horticulturally a fruit, is frequently used in vegetable salads.

Edible plants such as kale, hot peppers, and herbs may sometimes be used as ornamentals. Hot peppers and herbs can also be classified as condiments, a term that

Figure 1–3 Two tree fruits. The orange (*Citrus* sp.) is a subtropical tree fruit (above). The sour cherry (*Prunus cerasus*) is a temperate zone tree fruit (left). (Photo by J. C. Allen & Son.)

Figure 1–4 A berry or small fruit, red currant (*Ribes sativum*).

Figure 1–5 Carrot (*Daucus carota*). Although botanically a biennial vegetable, carrot is grown in the garden as an annual for its root. (Photo by T. W. Whitaker, courtesy USDA.)

describes spices and other products used to enhance the flavor of foods (Figure 1–6). Coffee, tea, and cocoa are the most important beverage plants. Morphine from the opium poppy, caffeine from the coffee tree, and digitalis from the foxglove plant are among the useful drugs produced from horticultural crops.

Figure 1–6 A condiment, *Piper nigrum.* Both black and white pepper come from this plant. Black pepper is made by grinding both seed and dried berry; white pepper is made by grinding the seed after the berry pulp has been removed by fermentation.

Useful or Not Useful. One basis for classifying plants in the garden is whether they are useful or not useful. Plants that are not useful are usually termed weeds. Sometimes a plant is useful in one location but may be a weed in another. Scotch broom, for example, is considered a weed on the Oregon and Washington coast, whereas in parts of the country where it must be planted and nurtured, it is a desirable ornamental. It is conceivable that many plants considered weeds today may someday be found to have useful properties and may be the crop plants of tomorrow. Sunflowers and safflowers are examples of plants once considered to be weeds that have become useful in our own time.

Growth Habit. Another useful method of classifying plants is according to their habit of growth. An **herbaceous** plant is one with nonwoody stems (such as coleus or asparagus) that usually last only one season. The contrasting **woody** plant has woody stems that generally live for several to many years and add new growth each year (Figure 1–7). Herbaceous plants can be classified as upright or vining, and woody plants are divided into **lianas** (woody vines), shrubs, and trees. Plants may also be classified as **deciduous** or **evergreen.** Deciduous plants lose their leaves in the fall; evergreens do not. Evergreen plants may be subdivided into broadleaved and needle evergreens.

Plants are sometimes categorized on the basis of the part consumed or used by humans (Figure 1–8). Examples of this classification would include the leaf (lettuce), stem (asparagus), root (carrot or sweet potato), petiole (celery or rhubarb), bud (broccoli or globe artichoke), fruit (apple or pineapple), and seed (pea or sweet corn).

Length of Life. Plants may also be classified as to the number of seasons they survive. **Annuals** are plants that

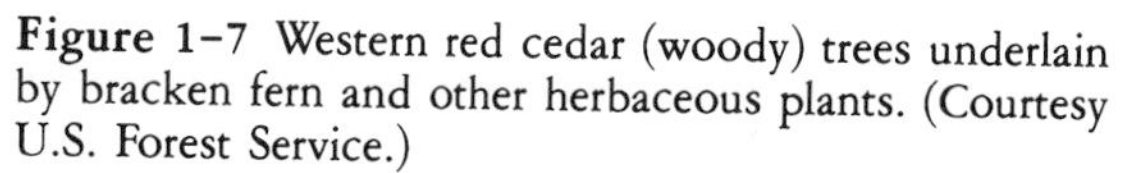

Figure 1–7 Western red cedar (woody) trees underlain by bracken fern and other herbaceous plants. (Courtesy U.S. Forest Service.)

Figure 1–8 Various plant parts useful to humans. (A) Rose, flower; (B) globe artichoke, flower bud; (C) onion, leaf (bulbs are morphologically platelike stems surrounded by fleshy leaves); (D) carrot, root; (E) potato, stem (tubers are the enlarged fleshy tips of underground stems); (F) chard, leaf and petiole; (G) apple, fruit; (H) pea, seed; (I) asparagus, stem and apical bud.

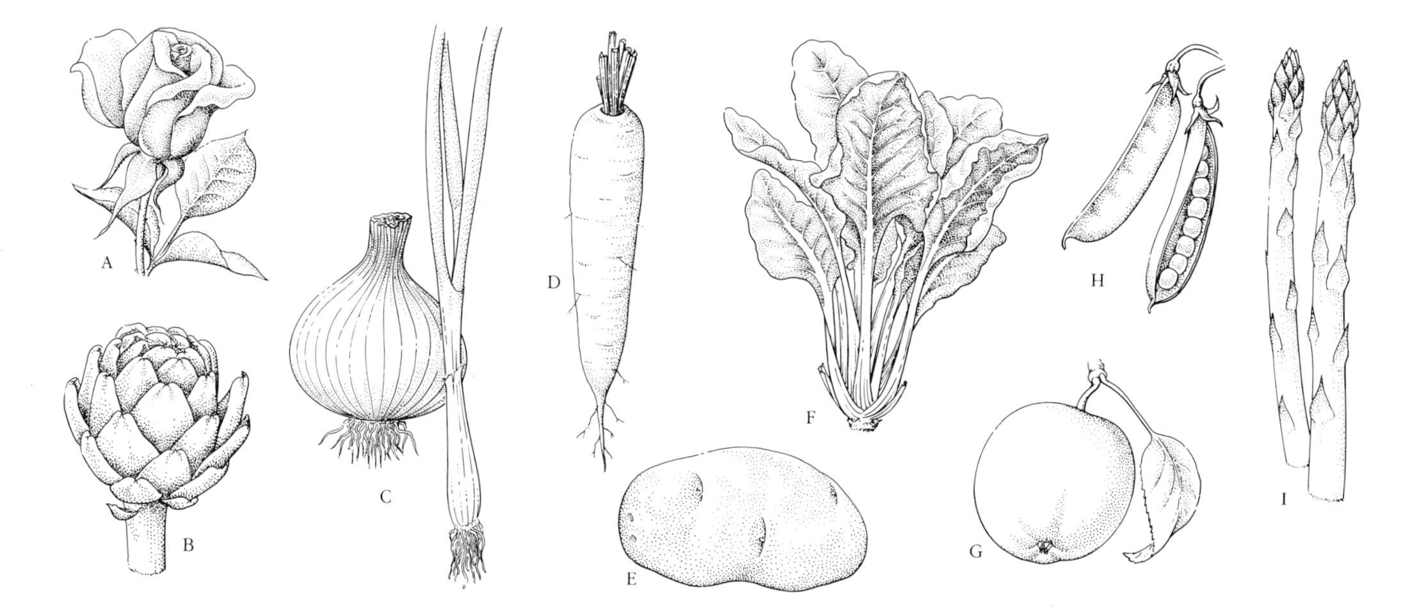

produce vegetative growth, flowers, and seeds, and die during one season. **Biennial** plants produce vegetative growth during the first season. Usually their flowering is triggered by a period of cold weather. They produce flowers, fruit, and seeds the second season and die at the end of their second year. **Perennial** plants are those that survive from three to many seasons.

Plants are usually classified as annual, biennial, or perennial on the basis of the part that is longest lived. Even though a tree has annual leaves, it is considered to be a perennial. Raspberries and other bramble berries have biennial canes, but because the roots are perennial, they are considered perennials. Most perennial flowers have tops that are annual. Carrots are considered biennials because the roots live two seasons even though the tops are annual.

Some crops are grown as annuals, although under certain environmental conditions they may be biennial or perennial. For example, when tomatoes are grown in the topics, they may survive for several seasons; however, in temperate zones they are grown as annuals. Also, when biennial vegetables—celery, carrots, cabbage, beets, turnips, rutabagas—are not being produced for seed, they are grown and harvested during one season.

Temperature Tolerance. Plants are also classified according to their temperature tolerance. **Tropical** crops are those that originated in tropical areas of the earth. Plants such as bananas, pineapple, rubber, cacao, coffee, and even vegetables like watermelon and cantaloupe are of tropical origin and are subject to cold injury at temperatures considerably above the freezing point. **Subtropical** crops will take some freezing temperatures but will not survive in areas with a cold winter climate. Most **temperate zone** plants are able to adapt so that they survive temperatures considerably below the freezing point. Vegetables and flowers that grow in temperate zone gardens may also be classified as cool-season or warm-season crops. **Cool-season** crops, such as radishes, peas, and pansies, are those that withstand some degree of freezing and, as a consequence, can be planted as soon as the ground can be worked in the spring. **Warm-season** crops, such as tomatoes, muskmelon, and sweet potatoes, are mostly of tropical origin and are killed as soon as temperatures drop slightly below freezing. They should not be planted until the danger of frost is over.

Botanical (Binomial) System of Plant Classification

The **botanical system** of plant classification is based largely on the hypothesis that plants evolved from a single, less complex organism and that, with this evolution, plants are all more or less distantly related. The traditional plant categories or **taxa** are kingdom, division, class, order, family, genus, and species. Subgroups are frequently used in this classification system.

Categories of Biological Classification. Most people who studied biology a few years ago learned to classify living organisms as belonging to either the Plant or Animal Kingdom. Recently biological classification has undergone considerable revision, largely as a result of new information. As a consequence, most biologists now recognize five kingdoms: (1) the traditional Kingdom Animalia; (2) Kingdom Monera, or the bacteria and blue-green algae; (3) Kingdom Protista, which includes algae, protozoans and slime molds; (4) Kingdom Fungi; and (5) Kingdom Plantae.

Although horticulturists are concerned with some of the other taxa in dealing with disease-causing organisms and microorganisms utilized in brewing and preserving, the Tracheophyta, or highest division of the Kingdom Plantae contains the higher plants with which we generally work. The division Tracheophyta is grouped into subdivisions, orders, and classes in various ways by various biologists who sometimes disagree with one another. The few orders and classes to which most of the plants commonly used in gardens belong are classified in recent literature as follows:

I. Subdivision Filicophytina (ferns)

II. Subdivision Spermatophytina (seed plants)
   A. Class Cycadinae (cycads)
   B. Class Ginkgoinae (ginkgo)
   C. Class Coniferinae (conifers)
   D. Class Angiospermae (flowering plants)
      1. Subclass Monocotyledonae (leaves with par-

allel veins, flowers with three or six parts, a single seed leaf or **cotyledon,** vascular bundles scattered through the stem)

2. Subclass Dicotyledonae (leaves with netted or branching veins, flowers with four or five parts, two seed leaves or cotyledons, vascular bundles arranged in a ring around the stem or in a vascular ring)

The ferns, the cycads, and the ginkgos are used as ornamentals (Figure 1–9). As horticultural plants, the conifers are also used mainly as ornamentals, although they are of major importance to foresters for lumber, pulpwood, and recreation. Angiosperms are by far the most numerous and most important group of plants used by gardeners. Class Angiospermae is divided into the subclasses Monocotyledonae, or monocots, and Dicotyledonae, or dicots (Figure 1–10); these are subdivided into orders; the orders are subdivided into families, the families into genera, and the genera into species.

The botanical scheme of plant classification is called the **binomial system of nomenclature,** because in it each plant is identified by two italicized Latin names, a capitalized genus name and an uncapitalized species name. These may be followed by the initial or name of the individual who first assigned the botanical name. For example, *Daucus carota* L. is the botanical name of carrot, L. being the initial of Linnaeus, the father of biological nomenclature, who assigned a botanical name to carrot. It is customary to abbreviate, using only the first letter of the generic name, when that name is repeated immediately following its initial use, for instance, *D. carota.*

Plants belonging to the same **genus** (subgroup of a family) have similar structure, appearance, and chromosome makeup (chromosomes are discussed in Chapter 3). Frequently the various subgroups (species) within a genus can be intergrafted one with another, and occasionally hybridization between species within a genus is possible.

The **species** (sp., plural spp.) is the basic unit of classification on which all botanical nomenclature is based. For horticultural plants a species, (for instance, *Spinacia oleracea, Rubus idaeus, Pinus nigra*) often, but not always, is equivalent to a horticultural kind of plant (spinach, red raspberry, Austrian pine). Plants belonging to the same species have numerous **morphological** (structural and

Figure 1–9 Three evolutionarily primitive plants used as ornamentals: (A) sago palm or cycad (*Cycas revoluta*); (B) maidenhair tree (*Ginkgo biloba*); (C) Boston fern (*Nephrolepis exaltata*).

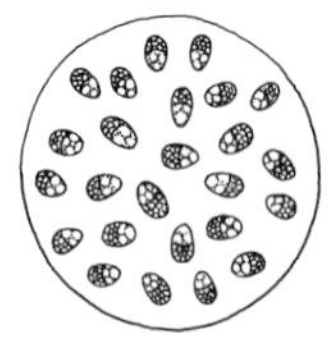

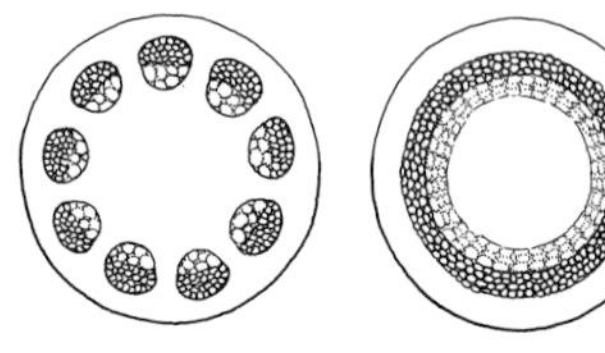

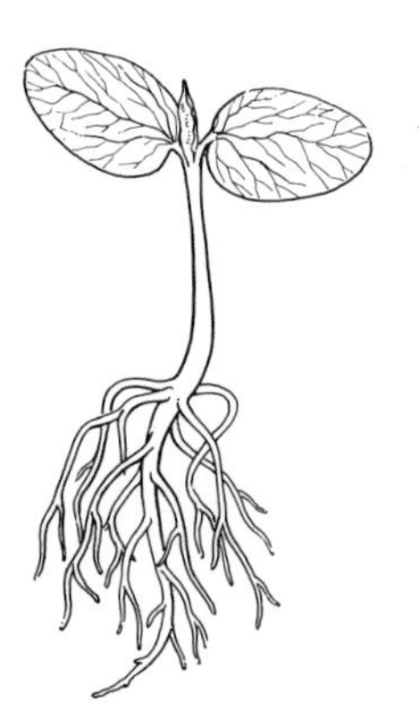

Figure 1–10 A monocotyledonous plant compared with a dicotyledonous plant.

developmental) similarities, and for plant groups that produce seed a species usually constitutes an exclusive interbreeding population.

Subspecies (subsp.), variety (var.), and forma (f.) are three terms that designate subgroups within the traditional species. The use of these terms is sometimes confusing, and even trained taxonomists disagree on the classification into these groupings.

Plants belonging to the same botanical **variety** are generally readily discernible from those of other varieties of the same species. For example, *Brassica oleracea* var. *botrytis* is cauliflower, and *B. oleracea* var. *capitata* is cabbage. Cabbage and cauliflower have the same chromosome number, have similar leaves, and will readily intercross, but they have head characteristics that are distinctly different. Occasionally there are barriers to natural intercrossing of two varieties of the same species. Also in the cabbage group, broccoli, *Brassica oleracea* var. *italica,* is an annual, whereas cabbage is a biennial (Figure 1–11). Although pollen of broccoli will readily pollinate and cause seed to be produced on cabbage flowers, the two varieties normally do not intercross, because broccoli blooms during the fall and cabbage blooms during the spring. (The botanical variety should not be confused with cultivated variety, now referred to as cultivar, which will be mentioned later in this chapter.)

There is no clearly definable distinction between botanical variety and subspecies based on degree of morphological variation. The term subspecies is more often used with natural rather than cultivated populations and is likely to be associated with geographical distribution. Barriers to natural intercrossing are more frequently associated with subspecies than with varieties of the same species.

A **forma** is a group of individuals within a population that differs from the rest of the population in a regular but trivial way. Usually there is less difference between two formae of a population than between two varieties or subspecies. In horticulture the term is often used with ornamentals to distinguish groups with different growth characteristics not reproducible with seed. For example, two important formae of Japanese yew are *Taxus cuspidata* f. *densa,* an upright form, and *T. cuspidata* f. *nana,* a spreading form. New plants are always developed by rooting cuttings of the form desired. Seedlings of

Figure 1–11 Two different kinds of crops that belong to the same species: (A) cabbage, *Brassica oleracea* var. *capitata*, and (B) sprouting broccoli, *Brassica oleracea* var. *italica*. (Photo by Peter C. Kruithof, courtesy Alf. Christianson Seed Co.)

either form would grow into plants of various shapes and sizes but seldom would produce a plant that was typically either *T. cuspidata* f. *nana* or *T. cuspidata* f. *densa*.

Advantages of the Binomial System. There are a number of advantages to the use of the binomial system of plant classification. In the first place, it avoids confusion. Each plant has an internationally recognized name that is the same regardless of language. In fact, North Americans are probably less well acquainted with Latin names of plants than are people from other countries where language localization encourages linguistic diversity. The second big advantage of the binomial system is that is shows relationships. Cultural practices, pesticide tolerances, and soil and climatic requirements are often similar for plants of the same group. The success of grafting one plant to another or of cross-pollinating also parallels botanical relationships.

As a help in classification, a few of the horticulturally more important families are listed in Table 1–1. Table 1–2 provides the botanical classification of three typical crop plants.

Kind, Cultivar, Strain, Clone

Four other classification terms frequently used with horticultural crops are kind, cultivar, strain, and clone. Examples of accepted usage probably provide the best explanations.

Horticultural plants of the same **kind** usually differ from other kinds in several important aspects. For example, cherry, peach, raspberry, and apple are different kinds of fruit. Kind often, but not always, includes all members of one species. An important exception involves the crucifers mentioned earlier—broccoli, cabbage, and cauliflower all belong to the same species but are different kinds of vegetables.

The term **cultivar** is an internationally accepted abbreviation of cultivated variety and is exactly equivalent to the old horticultural term variety. In the *International Code of Nomenclature of Cultivated Plants—1969* cultivar is defined as:

> . . . an assemblage of cultivated plants which is clearly distinguished by any character (morphological, physiological, cytological, chemical or others), and which, when reproduced (sexually or asexually), retains its distinguishing characteristics.

It has been only a few years since the term cultivar was adopted to replace (horticultural) variety in order to avoid confusion with botanical variety. However, the word variety, synonymous with cultivar, is still used by many

Table 1-1 Plant families of importance to horticulture (and some common examples).

A. Horticulturally important conifer families
1. Araucariaceae (araucaria family): Norfolk Island pine, monkey puzzle-tree
2. Cupressaceae (cypress family): cypress, juniper, eastern red cedar
3. Pinaceae (pine family): pine, spruce, fir, true cedars
4. Taxaceae (yew family): yew
5. Taxodiaceae (baldcypress family): bald cypress, redwood, giant sequoia

B. Horticulturally important monocot families
1. Amaryllidaceae (amaryllis family): amaryllis, narcissus
2. Araceae (arum family): philodendron, caladium, dieffenbachia, monstera
3. Bromeliaceae (pineapple family): pineapple, bromeliads
4. Gramineae or Poaceae (grass family): sweet corn and all grasses, including many ornamental grasses
5. Liliaceae (lily family): lily, tulip, crocus, asparagus, Joshua tree, onion
6. Orchidaceae (orchid family): 10,000 typical species; mostly ornamentals, many having aerial roots and living on live or dead organic matter
7. Iridaceae (iris family): iris
8. Palmae or Arecaceae (palm family): coconut, date, and ornamental palms

C. Horticulturally important dicot families
1. Aceraceae (maple family): maple, boxelder
2. Betulaceae (birch family): birch, alder, hazelnut
3. Cactaceae (cactus family): cactuses
4. Caprifoliaceae (honeysuckle family): honeysuckle, elder, viburnum, weigela
5. Caryophyllaceae (pink family): carnations and pinks
6. Chenopodiaceae (goosefoot family): beet, chard, spinach
7. Compositae or Asteraceae (sunflower family): chrysanthemum, sunflower, dahlia, calendula, marigold, zinnia, lettuce, artichoke, dandelion
8. Cruciferae or Brassicaceae (mustard family): radish, cabbage, cauliflower, broccoli, mustard, stock, and many weeds
9. Cucurbitaceae (melon family): watermelon, cantaloupe, squash, cucumber, pumpkin, gourd
10. Ericaceae (heath family): rhododendron, azalea, blueberry, cranberry
11. Fagaceae (beech family): beech, chestnut, oak
12. Juglandaceae (walnut family): walnut, pecan, hickory
13. Labiatae or Lamiaceae (mint family): mint, sage, thyme, lavender
14. Leguminosae or Fabaceae (pea family): clover, lupine, pea, bean, soybean, peanut, lima bean, wisteria, Scotch broom, sweet pea, locust, redbud, mesquite, Kentucky coffee tree
15. Moraceae (mulberry family): mulberry, fig
16. Magnoliaceae (magnolia family): magnolia, tulip-tree
17. Malvaceae (mallow family): okra, hibiscus, flowering maple, Rose of Sharon, hollyhock, cotton
18. Oleaceae (olive family): ash, privet, lilac, forsythia, jasmine, olive
19. Papaveraceae (poppy family): poppy, bloodroot
20. Ranunculaceae (buttercup family): buttercup, larkspur, columbine, peony, anemone
21. Rosaceae (rose family): spirea, ninebark, hawthorn, rose, mountain ash, quince, apple, pear, peach, plum, cherry, apricot, raspberry, blackberry, strawberry
22. Rutaceae (rue family): lemon, orange, grapefruit, lime, citron
23. Salicaceae (willow family): poplar, cottonwood, aspen, willow
24. Saxifragaceae (saxifrage family): mock orange, currant, gooseberry, deutzia, hydrangea
25. Solanaceae (nightshade family): tobacco, petunia, potato, eggplant, pepper, tomato
26. Umbelliferae or Apiaceae (parsley family); carrot, parsley, parsnip, celery
27. Vitaceae (grape family): grape, Boston ivy

horticulturists. The cultivar name is capitalized and written with single quotation marks, for example, 'Hales Best' muskmelon or 'Red Delicious' apple.

For many years horticulturists have designated as **strains** groups of plants within a cultivar selected and cultivated because they differ from other plants of the cultivar. The solid red sports are examples of strains of the 'Red Delicious' apple cultivar. Sometimes one strain

Table 1-2 Botanical classification of pear, Colorado spruce, and cauliflower.

	Pear	Spruce	Cauliflower
Kingdom	Plantae	Plantae	Plantae
Division	Tracheophyta	Tracheophyta	Tracheophyta
Subdivision	Spermatophytina	Spermatophytina	Spermatophytina
Class	Angiospermae	Coniferinae	Angiospermae
Subclass	Dicotyledonae	Coniferophytae	Dicotyledonae
Order	Rosales	Coniferales	Papaverales
Family	Rosaceae	Pinaceae	Cruciferae
Genus	*Pyrus*	*Picea*	*Brassica*
Species	*communis*	*pungens*	*oleracea*
Bot. variety			*botrytis*

will differ from another in having resistance to a disease. For example, 'Hales Best PMR' is a powdery mildew resistant selection of 'Hales Best' muskmelon cultivar. Those responsible for horticultural nomenclature are now recommending that groups of plants having recognizable differences from the cultivar from which they originated be classed as separate cultivars; however, the term strain is still frequently used by those long associated with the industry.

A **clone** is a genetically uniform group of plants derived from a single mother plant by asexual propagation; for example, by cuttings, crown divisions, grafts, or layerage. Many clones are also designated as cultivars. Thus 'Russet Burbank' potato and 'Golden Delicious' apple are clones as well as cultivars. Clones will be discussed further in Chapter 3.

Why Grow a Garden?

The reasons for growing gardens are as numerous and diverse as are the people who grow them. Often the main purpose of the garden is to supplement the family food supply. Latest reports show that 51% of American families now grow some type of vegetable garden. Gardens could become an even more important source of food should there be a national emergency. North Americans take for granted an ample supply of basic and luxury food. But because production, processing, transporting, and marketing have become so complex, the continuing availability of food is extremely vulnerable to such things as strikes, natural disasters, or transportation disruptions. The two weeks' supply stocked by most supermarkets would be quickly depleted if there were an interruption in any phase of food distribution.

Numerous Americans garden because gardening makes them feel better. Unlike many popular sports that are primarily recreations of youth, gardening is a hobby that provides moderate to vigorous exercise for people of all ages. The benefits of gardening in treating emotional stress are becoming widely recognized, and horticulturists cooperating with medical centers at several locations have had spectacular success using gardening to alleviate mental and emotional problems. Gardening can be an especially beneficial hobby for North Americans engaged in indoor occupations that involve considerable stress and a minimum of physical activity.

In most localities today if a family wants to eat fresh peas from the pod, edible podded peas, kohlrabi, kale, cress, currants, dewberries, many other fruits and vegetables, or most herbs; if they want to enjoy the beauty of most kinds of annual flowers; or if they want to savor the garden-fresh flavor of fruits and vegetables, they must grow a garden. With today's mechanized harvest only the kinds and cultivars that can be mass-produced are likely to find their way to the shelves of supermarkets. Moreover, because they must be harvested when still immature and shipped long distances, such products as tomatoes, sweet corn, peas, and strawberries purchased in a supermarket don't have the quality of those grown in the home garden. Often, too, the characteristics that enable a cultivar to withstand the necessary handling and still have eye-appeal when it reaches the market (solid flesh of strawberry and tomato, tough skin on sweet corn kernels) do not provide the ultimate in eating quality.

Gardens are also grown to beautify the surroundings, to give sanctuary to wildlife, and to provide shade and wind protection around the home. For some the assembling of different kinds of garden plants satisfies the urge to collect that is so prevalent in the human race.

Recent nutritional research has resulted in considerable publicity about the effect of diet on health and longevity. Most of the studies have reported that public health could be improved if people would consume more fruits and vegetables.

Fruits and vegetables add flavor, variety, and color to meals. Dinner would be rather bland without the flavor of onions, herbs, or various fruits; the texture of a crisp salad; or the color of carrots, beets, peas, tomatoes, or peaches. Fruits and vegetables are important dietary components because many of them add bulk without calories, promoting digestion and elimination in our sedentary society. Finally, these foods are important sources—in some cases the only source—of vitamins and minerals essential to growth and function of the human body. Table 1–3 lists the amount of nutrients of selected fruits and vegetables.

Table 1–3 Nutrients in common foods in terms of household measures.

Food	Water (%)	Food energy (calories)	Protein (g)	Fat (g)	Total carbohydrate (g)	Calcium (mg)	Iron (mg)	Vitamin A value (International units)	Thiamine (mg)	Riboflavin (mg)	Niacin (mg)	Ascorbic acid (mg)
Mature beans and peas; nuts												
Almonds, shelled; 1 cup	5	850	26	77	28	332	6.7	0	.34	1.31	5.0	Trace
Beans, dry seed:												
Common varieties, as Great Northern, navy, and others, canned; 1 cup:												
Red	76	230	15	1	42	74	4.6	0	.13	.13	1.5	Trace
White, with tomato or molasses:												
With pork	69	330	16	7	54	172	4.4	140	.13	.10	1.3	5
Without pork	69	315	16	1	60	183	5.2	140	.13	.10	1.3	5
Lima, cooked; 1 cup	64	260	16	1	48	56	5.6	Trace	.26	.12	1.3	Trace
Brazil nuts, broken pieces; 1 cup	5	905	20	92	15	260	4.8	Trace	1.21	0	0	0
Cashew nuts, roasted; 1 cup	5	770	25	65	35	51	5.1	0	.49	.46	1.9	0
Coconut; 1 cup:												
Fresh, shredded	50	330	3	31	13	15	1.7	0	.06	.03	.5	4
Dried, shredded (sweetened)	3	345	2	24	33	13	1.6	0	.04	.02	.4	0
Cowpeas or black-eyed peas, dry, cooked; 1 cup	80	190	13	1	34	42	3.2	20	.41	.11	1.1	Trace
Peanuts, roasted, shelled; 1 cup	2	840	39	71	28	104	3.2	0	.47	.19	24.6	0
Peanut butter; 1 tablespoon	2	90	4	8	3	12	.4	0	.02	.02	2.8	0
Peas, split, dry, cooked; 1 cup	70	290	20	1	52	28	4.2	120	.36	.22	2.2	Trace
Pecans, halves; 1 cup	3	740	10	77	16	79	2.6	140	.93	.14	1.0	2
Walnuts, shelled; 1 cup:												
Black or native, chopped	3	790	26	75	19	Trace	7.6	380	.28	.14	.9	0
English or Persian, halves	4	650	15	64	16	99	3.1	30	.33	.13	.9	3
Vegetables												
Asparagus:												
Cooked; 1 cup	92	35	4	Trace	6	33	1.8	1,820	.23	.30	2.1	40
Canned; 6 medium-size spears:												
Green	92	20	2	Trace	3	18	1.8	770	.06	.08	.9	17
Bleached	92	20	2	Trace	4	15	1.0	70	.05	.07	.8	17

Food	Water (%)	Food energy (calories)	Protein (g)	Fat (g)	Total carbohydrate (g)	Calcium (mg)	Iron (mg)	Vitamin A value (International units)	Thiamine (mg)	Riboflavin (mg)	Niacin (mg)	Ascorbic acid (mg)
Beans:												
Lima, immature, cooked; 1 cup	75	150	8	1	29	46	2.7	460	.22	.14	1.8	24
Snap, green:												
Cooked; 1 cup:												
In small amount of water, short time	92	25	2	Trace	6	45	.9	830	.09	.12	.6	18
In large amount of water, long time	92	25	2	Trace	6	45	.9	830	.06	.11	.5	12
Canned:												
Solids and liquid; 1 cup	94	45	2	Trace	10	65	3.3	990	.08	.10	.7	9
Strained or chopped; 1 ounce	93	5	Trace	Trace	1	10	.3	120	.01	.02	.1	1
Beets, cooked, diced; 1 cup	88	70	2	Trace	16	35	1.2	30	.03	.07	.5	11
Broccoli, cooked, flower stalks; 1 cup	90	45	5	Trace	8	195	2.0	5,100	.10	.22	1.2	111
Brussels sprouts, cooked; 1 cup	85	60	6	1	12	44	1.7	520	.05	.16	.6	61
Cabbage; 1 cup:												
Raw, finely shredded	92	25	1	Trace	5	46	.5	80	.06	.05	.3	50
Raw, coleslaw	84	100	2	7	9	47	.5	80	.06	.05	.3	50
Cooked:												
In small amount of water, short time	92	40	2	Trace	9	78	.8	150	.08	.08	.5	53
In large amount of water, long time	92	40	2	Trace	9	78	.8	150	.05	.05	.3	32
Cabbage, celery or Chinese; 1 cup:												
Raw, leaves and stem (1-inch pieces)	95	15	1	Trace	2	43	.9	260	.03	.04	.4	31
Cooked	95	25	2	1	5	82	1.7	490	.04	.06	.6	42
Carrots:												
Raw; 1 carrot (5½ x 1 inch) or 25 thin strips	88	20	1	Trace	5	20	.4	6,000	.03	.03	.3	3
Raw, grated; 1 cup	88	45	1	Trace	10	43	.9	13,200	.06	.06	.7	7
Cooked, diced; 1 cup	92	45	1	1	9	38	.9	18,130	.07	.07	.7	6
Canned, strained or chopped; 1 ounce	92	5	Trace	0	2	7	.2	3,400	.01	.01	.1	1
Cauliflower, cooked, flower buds; 1 cup	92	30	3	Trace	6	26	1.3	110	.07	.10	.6	34
Celery, raw:												
Large stalk, 8 inches long	94	5	1	Trace	1	20	.2	0	.02	.02	.2	3
Diced; 1 cup	94	20	1	Trace	4	50	.5	0	.05	.04	.4	7
Collards, cooked; 1 cup	87	75	7	1	14	473	3.0	14,500	.15	.46	3.2	84
Corn, sweet:												
Cooked; 1 ear, 5 inches long	76	65	2	1	16	4	.5	[a]300	.09	.08	1.1	6
Canned, solids and liquid; 1 cup	80	170	5	1	41	10	1.3	[a]520	.07	.13	2.4	14
Cowpeas, immature seeds, cooked; 1 cup	75	150	11	1	25	59	4.0	620	.46	.13	1.3	32
Cucumbers, raw, pared; 6 slices (⅛ inch thick, center section)	96	5	Trace	Trace	1	5	.2	0	.02	.02	.1	4
Dandelion greens, cooked; 1 cup	86	80	5	1	16	337	5.6	27,310	.23	.22	1.3	29
Endive, curly (including escarole); 2 ounces	93	10	1	Trace	2	45	1.0	1,700	.04	.07	.2	6
Kale, cooked; 1 cup	87	45	4	1	8	248	2.4	9,220	.08	.25	1.9	56
Lettuce, headed, raw:												
2 large or 4 small leaves	95	5	1	Trace	1	11	.2	270	.02	.04	.1	4
1 compact head (4¾-inch diam.)	95	70	5	1	13	100	2.3	2,470	.20	.38	.9	35

(continued)

Table 1-3 *(continued)*

Food	Water (%)	Food energy (calories)	Protein (g)	Fat (g)	Total carbohydrate (g)	Calcium (mg)	Iron (mg)	Vitamin A value (International units)	Thiamine (mg)	Riboflavin (mg)	Niacin (mg)	Ascorbic acid (mg)
Mushrooms, canned, solids and liquid; 1 cup	93	30	3	Trace	9	17	2.0	0	.04	.60	4.8	0
Mustard greens, cooked; 1 cup	92	30	3	Trace	6	308	4.1	10,050	.08	.25	1.0	63
Okra, cooked; 8 pods (3 inches long, ⅝-inch diam.)	90	30	2	Trace	6	70	.6	630	.05	.05	.7	17
Onions:												
Mature:												
Raw; 1 onion (2½-inch diam.)	88	50	2	Trace	11	35	.6	60	.04	.04	.2	10
Cooked; 1 cup	90	80	2	Trace	18	67	1.0	110	.04	.06	.4	13
Young green; 6 small, without tops	88	25	Trace	Trace	5	68	.4	30	.02	.02	.1	12
Parsley, raw; 1 tablespoon chopped	84	1	Trace	Trace	Trace	7	.2	290	Trace	.01	.1	7
Parsnips, cooked; 1 cup	84	95	2	1	22	88	1.1	0	.09	.16	.3	19
Peas, green; 1 cup:												
Cooked	82	110	8	1	19	35	3.0	1,150	.40	.22	3.7	24
Canned, solids and liquid	82	170	8	1	32	62	4.5	1,350	.28	.15	2.6	21
Canned, strained; 1 ounce	86	10	1	Trace	2	5	.3	160	.03	.02	.3	2
Peppers, sweet:												
Green, raw; 1 medium	93	15	1	Trace	3	6	.4	260	.05	.05	.3	79
Red, raw; 1 medium	91	20	1	Trace	4	8	.4	2,670	.05	.05	.3	122
Pimientos, canned; 1 medium	92	10	Trace	Trace	2	3	.6	870	.01	.02	.1	36
Peppers, hot, red, without seeds, dried, ground (chili powder); 1 tablespoon	13	50	2	1	9	20	1.2	11,520	.03	.20	1.6	2
Potatoes:												
Baked or boiled; 1 medium, 2½-inch diam. (weight raw, about 5 ounces):												
Baked in jacket	75	90	3	Trace	21	9	.7	Trace	.10	.04	1.7	20
Boiled; peeled before boiling	80	90	3	Trace	21	9	.7	Trace	.11	.04	1.4	20
Chips; 10 medium (2-inch diam.)	3	110	1	7	10	6	.4	Trace	.04	.02	.6	2
French fried:												
Frozen, ready to be heated for serving; 10 pieces (2 x ½ x ½ inch)	64	95	2	4	15	4	.8	Trace	.08	.01	1.2	10
Ready-to-eat, deep fat for entire process; 10 pieces (2 x ½ x ½ inch)	45	155	2	7	20	9	.7	Trace	.06	.04	1.8	8
Mashed; 1 cup:												
Milk added	80	145	4	1	30	47	1.0	50	.17	.11	.2	17
Milk and butter added	76	230	4	12	28	45	1.0	470	.16	.10	1.6	16
Pumpkin, canned; 1 cup	90	75	2	1	18	46	1.6	7,750	.04	.14	1.2	0
Radishes, raw; 4 small	94	10	Trace	Trace	2	15	.4	10	.01	.01	.1	10
Sauerkraut, canned, drained solids; 1 cup	91	30	2	Trace	7	54	.8	60	.05	.10	.2	24
Spinach:												
Cooked; 1 cup	91	45	6	1	6	223	3.6	21,200	.14	.36	1.1	54
Canned, creamed, strained; 1 ounce	90	10	1	Trace	2	19	.3	750	.01	.03	.1	1

Food	Water (%)	Food energy (calories)	Protein (g)	Fat (g)	Total carbohydrate (g)	Calcium (mg)	Iron (mg)	Vitamin A value (International units)	Thiamine (mg)	Riboflavin (mg)	Niacin (mg)	Ascorbic acid (mg)
Squash:												
Cooked, 1 cup:												
Summer, diced	95	35	1	Trace	8	32	.8	550	.08	.15	1.3	23
Winter, baked, mashed	86	95	4	1	23	49	1.6	12,690	.10	.31	1.2	14
Canned, strained or chopped; 1 ounce	92	10	Trace	Trace	2	7	.1	510	.01	.01	.1	1
Sweet potatoes:												
Baked or boiled; 1 medium, 5 x 2 inches (weight raw, about 6 ounces):												
Baked in jacket	64	155	2	1	36	44	1.0	[b] 8,970	.10	.07	.7	24
Boiled in jacket	71	170	2	1	39	47	1.0	[b]11,610	.13	.09	.9	25
Candied; 1 small, 3½ x 2 inches	60	295	2	6	60	65	1.6	[b]11,030	.10	.08	.8	17
Canned, vacuum or solid pack; 1 cup	72	235	4	Trace	54	54	1.7	17,110	.12	.09	1.1	30
Tomatoes:												
Raw; 1 medium (2 x 2½ inches), about ⅓ pound	94	30	2	Trace	6	16	.9	1,640	.08	.06	.8	35
Canned or cooked; 1 cup	94	45	2	Trace	9	27	1.5	2,540	.14	.08	1.7	40
Tomato juice, canned; 1 cup	94	50	2	Trace	10	17	1.0	2,540	.12	.07	1.8	38
Tomato catsup; 1 tablespoon	70	15	Trace	Trace	4	2	.1	320	.02	.01	.4	2
Turnips, cooked, diced; 1 cup	92	40	1	Trace	9	62	.8	Trace	.06	.09	.6	28
Turnip greens, cooked; 1 cup	90	45	4	1	8	376	3.5	15,370	.09	.59	1.0	87
Fruits												
Apples, raw; 1 medium (2½-inch diam.), about ⅓ pound	85	70	Trace	Trace	18	8	.4	50	.04	.02	.1	3
Apple betty; 1 cup	64	350	4	8	69	41	1.4	270	.13	.10	.9	Trace
Apple juice, fresh or canned; 1 cup	86	125	Trace	0	34	15	1.2	90	.05	.07	Trace	2
Applesauce, canned:												
Sweetened; 1 cup	80	185	Trace	Trace	50	10	1.0	80	.05	.03	.1	3
Unsweetened; 1 cup	88	100	Trace	Trace	26	10	1.0	70	.05	.02	.1	3
Apricots, raw; 3 apricots (about ¼ pound)	85	55	1	Trace	14	18	.5	2,890	.03	.04	.7	10
Apricots, canned:												
Heavy sirup pack, halves and sirup; 1 cup	78	200	1	Trace	54	34	1.0	4,070	.05	.07	1.1	10
Water pack, halves and liquid; 1 cup	90	80	1	Trace	21	27	.7	3,320	.04	.05	.9	8
Apricots, dried:												
Uncooked; 1 cup (40 halves, small)	25	390	8	1	100	100	8.2	16,390	.02	.24	4.9	19
Cooked unsweetened, fruit and liquid; 1 cup	76	240	5	1	62	63	5.1	10,130	.01	.13	2.8	8
Apricots and applesauce, canned, strained or chopped; 1 ounce	80	20	Trace	Trace	5	3	.2	440	.01	.01	.1	Trace
Apricot nectar; 1 cup	85	135	1	Trace	36	22	.5	2,380	.02	.02	.5	7
Avocados, raw, California varieties (mainly Fuerte):												
1 cup (½-inch cubes)	74	260	3	26	9	15	.9	430	.16	.30	2.4	21

(continued)

Table 1–3 *(continued)*

Food	Water (%)	Food energy (calories)	Protein (g)	Fat (g)	Total carbohydrate (g)	Calcium (mg)	Iron (mg)	Vitamin A value (International units)	Thiamine (mg)	Riboflavin (mg)	Niacin (mg)	Ascorbic acid (mg)
Avocados, raw, California varieties (mainly Fuerte):												
½ of a 10-ounce avocado (3½ x 3¼ inches)	74	185	2	18	6	11	.6	310	.12	.21	1.7	15
Avocados, raw, Florida varieties:												
1 cup (½-inch cubes)	78	195	2	17	13	15	.9	430	.16	.30	2.4	21
½ of a 13-ounce avocado (4 x 3 inches)	78	160	2	14	11	12	.7	350	.13	.24	2.0	17
Bananas, raw; 1 medium (6 x 1½ inches), about ⅓ pound	76	85	1	Trace	23	10	.7	170	.05	.06	.7	10
Blackberries, raw; 1 cup	85	80	2	1	18	46	1.3	280	.05	.06	.5	30
Blueberries, raw; 1 cup	83	85	1	1	21	22	1.1	400	.04	.03	.4	23
Cantaloupes, raw; ½ melon (5-inch diam.)	94	40	1	Trace	9	33	.8	[c]6,590	.09	.07	1.0	63
Cherries, sour, sweet, and hybrid, raw; 1 cup	83	65	1	1	15	19	.4	650	.05	.06	.4	9
Cherries, canned:												
Red sour, pitted; 1 cup	87	120	2	1	30	28	.8	1,840	.07	.04	.4	14
Cranberry juice cocktail, canned; 1 cup	85	135	Trace	Trace	36	10	.5	20	.02	.02	.1	5
Cranberry sauce, sweetened; 1 cup	48	550	Trace	1	142	22	.8	80	.06	.06	.3	5
Dates, "fresh" and dried, pitted and cut; 1 cup	20	505	4	1	134	103	5.3	170	.16	.17	3.9	0
Figs:												
Raw; 3 small (1½-inch diam.), about ¼ pound	78	90	2	Trace	22	62	.7	90	.06	.06	.6	2
Dried; 1 large (2 x 1 inch)	23	60	1	Trace	15	43	.3	20	.02	.02	.2	0
Fruit cocktail, canned in heavy sirup, solids and liquid; 1 cup	81	175	1	Trace	47	23	1.0	360	.04	.03	1.1	5
Grapefruit:												
Raw; ½ medium (4¼-inch diam., No. 64s):												
White	89	50	1	Trace	14	21	.5	10	.05	.02	.2	50
Pink or red	89	55	1	Trace	14	21	.5	590	.05	.02	.2	48
Raw, sections, white; 1 cup	89	75	1	Trace	20	31	.8	20	.07	.03	.3	72
Canned:												
Sirup pack, solids and liquid; 1 cup	81	165	1	Trace	44	32	.7	20	.07	.04	.5	75
Water pack, solids and liquid; 1 cup	91	70	1	Trace	18	31	.7	20	.07	.04	.5	72
Grapefruit juice:												
Raw; 1 cup	90	85	1	Trace	23	22	.5	[d]20	.09	.04	.4	92
Canned:												
Unsweetened; 1 cup	89	95	1	Trace	24	20	1.0	20	.07	.04	.4	84
Sweetened; 1 cup	86	120	1	Trace	32	20	1.0	20	.07	.04	.4	78
Frozen concentrate, unsweetened:												
Undiluted; 1 can (6 fluid ounces)	62	280	4	1	72	70	.8	60	.29	.12	1.4	286
Diluted, ready-to-serve; 1 cup	89	95	1	Trace	24	25	.2	20	.10	.04	.5	96
Frozen concentrate, sweetened:												
Undiluted; 1 can (6 fluid ounces)	57	320	3	1	85	59	.6	50	.24	.11	1.2	245
Diluted, ready-to-serve; 1 cup	88	105	1	Trace	28	20	.2	20	.08	.03	.4	82

Food	Water (%)	Food energy (calories)	Protein (g)	Fat (g)	Total carbohydrate (g)	Calcium (mg)	Iron (mg)	Vitamin A value (International units)	Thiamine (mg)	Riboflavin (mg)	Niacin (mg)	Ascorbic acid (mg)
Dehydrated:												
Crystals; 1 can (net weight 4 ounces)	1	400	5	1	103	99	1.1	90	.41	.18	2.0	399
With water added, ready-to-serve; 1 cup	90	90	1	Trace	24	22	.2	20	.10	.05	.5	92
Grapes, raw; 1 cup:												
American type (slip skin)	82	70	1	1	16	13	.4	100	.05	.03	.3	4
European type (adherent skin)	81	100	1	Trace	26	18	.6	150	.08	.04	.4	7
Grape juice, bottled; 1 cup	82	165	1	1	42	25	.8	0	.11	.06	.7	Trace
Lemon juice:												
Raw; 1 cup	91	60	1	Trace	20	27	.5	Trace	.08	.03	.3	129
Canned; 1 cup	91	60	1	Trace	20	27	.5	Trace	.07	.03	.3	102
Lemonade concentrate, frozen, sweetened:												
Undiluted; 1 can (6 fluid ounces)	48	305	1	Trace	113	9	.4	Trace	.05	.06	.7	67
Diluted, ready-to-serve; 1 cup	88	75	Trace	Trace	28	2	.1	Trace	.01	.01	.2	17
Lime juice:												
Raw; 1 cup	90	65	1	Trace	22	22	1.5	Trace	.03	.04	.4	80
Canned; 1 cup	90	65	1	Trace	22	22	1.5	Trace	.02	.04	.4	52
Limeade concentrate, frozen, sweetened:												
Undiluted; 1 can (6 fluid ounces)	50	295	Trace	Trace	109	11	.7	Trace	.01	.02	.2	262
Diluted, ready-to-serve; 1 cup	90	75	Trace	Trace	27	2	.2	Trace	Trace	.01	.1	6
Oranges, raw; 1 large orange (3-inch diam.):												
Navel	86	70	2	Trace	17	48	.3	270	.11	.03	.4	83
Other varieties	86	70	1	Trace	18	63	.3	290	.12	.03	.4	66
Orange juice:												
Raw; 1 cup:												
California (Valencias)	88	105	2	Trace	26	37	.5	500	.20	.05	.6	126
Florida varieties:												
Early and midseason	90	90	1	Trace	23	25	.5	490	.20	.05	.6	127
Late season (Valencias)	88	105	1	Trace	26	25	.5	500	.20	.05	.6	92
Canned, unsweetened; 1 cup	87	110	2	Trace	28	25	1.0	500	.17	.05	.6	100
Frozen concentrate:												
Undiluted; 1 can (6 fluid ounces)	58	305	5	Trace	80	69	.8	1,490	.63	.10	2.4	332
Diluted, ready-to-serve; 1 cup	88	105	2	Trace	27	22	.2	500	.21	.03	.8	112
Dehydrated:												
Crystals; 1 can (net weight 4 ounces)	1	395	6	2	100	95	1.9	1,900	.76	.19	2.5	406
With water added, ready-to-serve; 1 cup	88	105	1	Trace	27	25	.5	500	.20	.05	.6	108
Orange and grapefruit juice, frozen concentrate:												
Undiluted; 1 can (6 fluid ounces)	59	300	4	1	78	61	.8	790	.47	.06	2.3	301
Diluted, ready-to-serve; 1 cup	88	100	1	Trace	26	20	.2	270	.16	.02	.8	102
Peaches:												
Raw:												
1 medium (2½ x 2-inch diam.), about ¼ pound	89	35	1	Trace	10	9	.5	[e]1,320	.02	.05	1.0	7
1 cup, sliced	89	65	1	Trace	16	15	.8	[e]2,230	.03	.08	1.6	12

(continued)

2 Structure and Growth—The Vegetative Phase

From the day a nurse pins on our first cotton diaper until the day the sexton nails the pine cover on our coffin, we use, eat, and enjoy countless plant products that have characteristics of value to us. It is important to realize, however, that these products did not come into existence because a beneficent goddess of plants decreed that they should be produced to please people. Plant stems, roots, leaves, flowers, fruits, and seeds in all their forms and modifications exist because they are required for the survival, growth, and reproduction of the plant producing them. Therefore, the gardener who wants to know the why and how of plant culture and plant propagation must first have some knowledge of the makeup and function of various plant structures.

In existing textbooks plant structure is shown in neatly labeled cross or longitudinal sections with each part in its appointed place. Plant growth is likewise neatly catalogued into verbally standardized phases. This text will be no different in this regard, not because I am happy with this method but because a textbook does not lend itself to a more dynamic approach. The problem with such presentations is that readers often gain the impression that all plants have the same unchanging structure and growth pattern. It would be more appropriate if plant structure and growth could be shown in a three-dimensional movie, complete with sound effects, because the plant is a dynamic, living, changing entity. The structure at any one location will not be the same tomorrow or even in a few minutes, and no two plants are exactly alike. The discussion that follows, therefore, may not describe exactly any plant at any time, but it is useful because it represents the composite of knowledge gained from thousands of scientific observations of plant structure and growth.

Phases of Plant Growth

Dormancy

During the lives of most plants there are periods of active growth, and periods when growth is at a minimum (Figure 2–1). The period of inactivity is referred to as **dormancy.** In annual plants the dormant period may reside entirely within the seed. With biennial and perennial plants the dormant period normally coincides with the winter season or, in the tropics, with the period of drought when growing conditions are adverse. The dormant period of temperate zone perennials and some biennials has two phases. During the first phase, which horticulturists refer to as the **rest period,** growth does not occur even if environmental conditions are favorable. During the second phase the plant remains inactive because of adverse environmental conditions, and growth will resume as soon as temperature and moisture conditions are conducive.

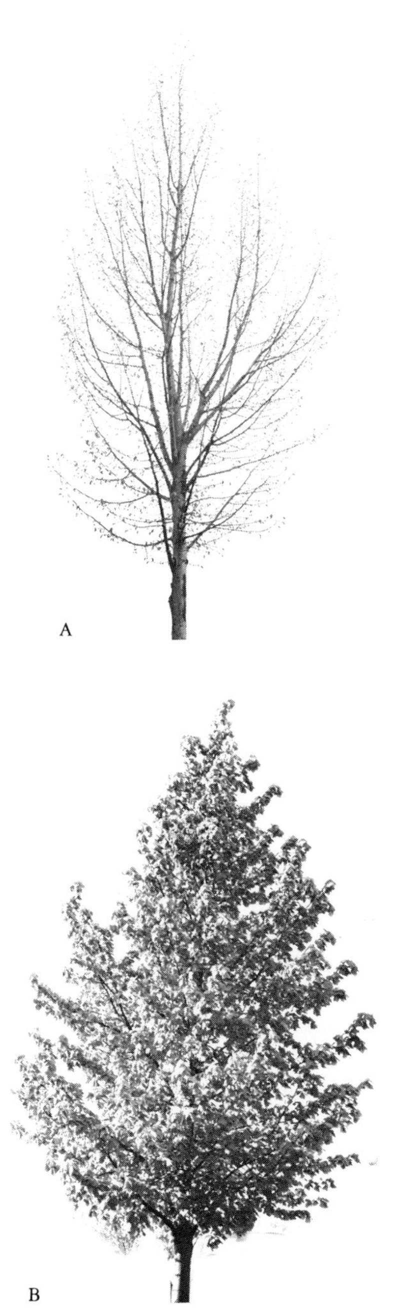

Figure 2-1 Comparison of dormant (A) and actively growing (B) tree. Notice how much easier it is to observe the placement of branches when the tree is dormant.

Figure 2-2 Pear branches forced to bloom in a teaching laboratory. These branches were pruned from the tree on February 25. They were in full bloom on March 11, when the photo was taken. Although its rest period is past, the tree from which this branch was removed is still fully dormant, because outdoor temperatures have remained cool.

For example, in the northern United States the fruit of an apple tree ripens in September or October. The leaves turn yellow and fall from the tree. From October until sometime in January if apple branches are cut and placed in water in a warm room, they will remain dormant. From mid-January, however, if branches from the same tree are brought into a warm room, they form buds that produce flowers and leaves (Figure 2-2). On the tree itself, however, the branches remain dormant until the weather begins to warm in the spring. The rest period of an apple tree lasts from October to January, but the dormant period lasts from October until April.

Vegetative and Reproductive Phases of Growth

Every plant goes through two general phases of active growth, a vegetative phase and a reproductive phase. During the early, **vegetative** phase of the plant, food resources are directed primarily to the growth of leaves, stems, and roots. During their early life woody plants generally will not produce flowers or store food regardless of care or cultural practices employed. This stage, a part of the vegetative phase when reproduction cannot be induced, is referred to as the **juvenile** stage. Some plants, English Ivy and common Juniper, for instance, have different foliage or growth habits during their juvenile stage of growth (Figure 2–3). In the case of long-lived plants, juvenility causes some real problems for orchardists and horticultural scientists, who must bear the expense of caring for the young orchard for many years before they realize any income. Patience is a prime requisite of the tree-fruit breeder, who has to wait 6 to 10 years before the new tree bears fruit.

Later in the life of a plant sugars and starches are stored, and the plant flowers and produces fruit. This later period is known as the **reproductive** phase. In many plants, especially those that survive only 1 year (annuals), reproduction marks an end to the life of the plant. As the reproductive phase progresses, plant tissues begin to degenerate and eventually die. This degeneration and death is referred to as **senescence.** With biennial plants

Figure 2–3 Mature (left) and juvenile (right) form of tamarix juniper (*Juniperus sabina 'Tamariscifolia'*).

the vegetative phase occurs during the first year. The reproductive phase is triggered by cool temperatures, time, a change in day length, or simply a period of dormancy. The second year is the reproductive cycle of the plant, at the end of which the plant becomes senescent and dies.

Only a few tissues of perennial plants normally become senescent with the maturation of the fruit and seeds. In the case of perennial flowers it is the top that matures and dies back, and the roots are capable of sending up new shoots the following year. With deciduous trees only the fruiting body and leaves die with the maturity of the seeds and the end of the growing season.

Sometimes the gardener wants vegetative growth of leaves, stems, and roots, as with cabbage, lettuce, or spinach for the table or grass in establishing a new lawn. At other times reproductive growth of fruits, seeds, or flowers may be the objective, as with raspberries, sweet corn, or annual flowers. Fertilization, irrigation, planting time or cultural practices may have to be varied depending upon whether vegetative or reproductive growth is desirable. With orchard trees, for example, it is important that a balance of vegetative and reproductive growth be maintained. A tree that is overly reproductive will tend to produce too many fruit, all of which will remain small because there is too little leaf area per fruit to manufacture sufficient amounts of carbohydrates necessary for fruit enlargement (Figure 2–4). The vegetative/reproductive balance is correlated with the **carbohydrate/nitrogen balance** in the plant tissue. Carbohydrate is manufactured by plant leaves, and nitrogen is absorbed through the root system. In orchards the carbohydrate/nitrogen balance is regulated mainly by pruning to reduce carbohydrate and fertilizing to increase nitrogen. This use of pruning and fertilization to regulate production will be detailed in later chapters.

The Cell

Whenever plant growth occurs, be it vegetative or reproductive, the growth process takes place within the plant cell. Indeed, the basic unit of each living organism, plant or animal, is the **cell.** Cells are complex factories produc-

Figure 2–4 Diagrammatic sketch showing the effects of fruit thinning. (Leaves have been left out to simplify.) For a thinning experiment a branch of plum (A) with 77 fruits remaining after the "June drop" was selected. Thirty-three of the fruits were removed, and at harvest time 41 fruits remained (B). A similar branch having 83 fruits (C) was left unthinned. At the end of the season 69 fruits still remained on the unthinned branch (D); however, the fruit harvested from the thinned branch weighed 20% more than that harvested from the unthinned branch. Results from thinning are not always so dramatic; however, removing excess fruits when they are still small almost always increases fruit size and improves the quality of the crop. (Experiment reported by V. R. Gardener in *Basic Horticulture,* 2nd ed. Macmillan, New York, 1951.)

ing the chemical reactions basic to the life of the organism. Cells might be compared to the boards, bricks, and stones used as building blocks for houses. Just as building materials vary in shape, size, and function and still have common characteristics, so cells vary in shape, size, and function and still have common structural features. Although science has made considerable progress in elucidating the complexities of cell development, much remains to be learned. Only the briefest outline of cell structure and function can be included here.

Cell Structure

The basic structure of a plant cell is shown in Figure 2–5. The cell is encompassed by a **cell wall** made of cellulose and separated from the cell walls of neighboring cells with a cementing material, the **middle lamella.** Inside the cell wall is the **cell membrane,** which appears to have the function of regulating the flow of nutrients and other materials into and out of the cell.

The living portion of the cell, or **protoplasm,** is divided into two parts: an inner, dense-appearing portion referred to as the **nucleus,** and an outer portion called the **cytoplasm.** Within the cytoplasm are located several different kinds of bodies called **organelles.** One group of these, the **chloroplasts,** are necessary for photosynthesis, the process of utilizing the sun's energy for manufacturing sugar from inorganic elements. Sugar derived by photosynthesis is the basic component of all food consumed by plants and animals of this earth. The chloroplasts are responsible for the green color of plants. Also in the cytoplasm are **mitochondria,** which function to provide energy for cellular activities; and organelles that are responsible for food manufacture, food storage, and other cellular activities. A large portion of the interior of many plant cells is an area of a relatively clear liquid surrounded by a membrane called a **vacuole,** or sometimes vacuoles, as there may be more than one. The vacuole contains dissolved carbohydrates, pigments, organic acids, and other compounds.

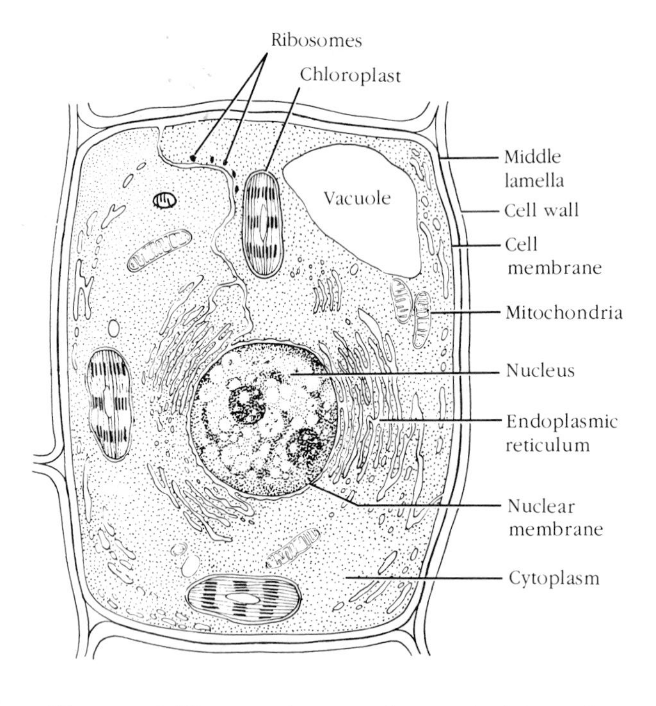

Figure 2–5 Electron microscope image of a living cell.

Running through the cytoplasm is an internal membrane, the **endoplasmic reticulum,** which is now thought to be the wall for channels that permeate all areas of the cytoplasm. The endoplasmic reticulum of a cell is connected to the endoplasmic reticula of other cells and to areas outside the cell by streams of living protoplasm, called **plasmodesmata,** that flow through openings (**pores**) in the cell wall. These intercellular channels greatly increase the opportunity for cellular interchange of gases and liquids. Clustered on the endoplasmic reticulum and other membranes and, to some extent, scattered throughout the cytoplasm are the **ribosomes.** These small spherical entities are associated with **ribonucleic acid** (**RNA**), the substance responsible for relaying genetic information between the nucleus and the cytoplasm.

The nucleus is the regulator of most cellular activities, including reproduction. When a cell is ready to divide, long threadlike structures called **chromosomes** become visible. The electron microscope shows chromosomes to be a double strand of nucleic acid (a major component of proteins) with interconnecting chemical bonds resembling a spiral staircase. This nucleic acid is chemically known as **deoxyribonucleic acid** (**DNA**). DNA, in conjunction with RNA and the ribosomes, controls cell function and, through control of cells in concert, regulates organism structure and function.

DNA is also the hereditary bridge from one generation to the next. Segments of chromosomal DNA called **genes** pass on characteristics of parent to offspring and determine, for example, that the seed of a maple tree will produce another maple tree, or that a baby will look something like its parents. Each plant and animal species has a specific number of chromosomes typical for its species. Human beings have 46, corn has 20, and peas have 14. Usually the chromosomes of an organism are in pairs, one of each pair, or one set, coming from the male parent and one from the female. Corn has 10 pairs and humans have 23 pairs of chromosomes. Plants with the usual two sets are said to be **diploid** and to have $2n$ number of chromosomes. Occasionally plants are developed with more than two sets of chromosomes. Plants with four sets ($4n$) are quite frequent and are called **tetraploids.** Tetraploids are usually larger but less uniform in size and shape than diploids. This may be an advantage, as with large tetraploid snapdragons (see Figure 3–19, Chapter 3), or a disadvantage, as with tetraploid apples, which are larger but always misshapen. Plants with an uneven number of chromosome sets (one, three, five, etc.) usually do not produce seed. Knowledge of this fact is used, for example, in developing seed for growing seedless watermelon. Tetraploid cultivars (four sets) are crossed with diploid cultivars (two sets) to produce watermelon that, because they have three sets of chromosomes (**triploid**), are seedless (see Figure 3–20).

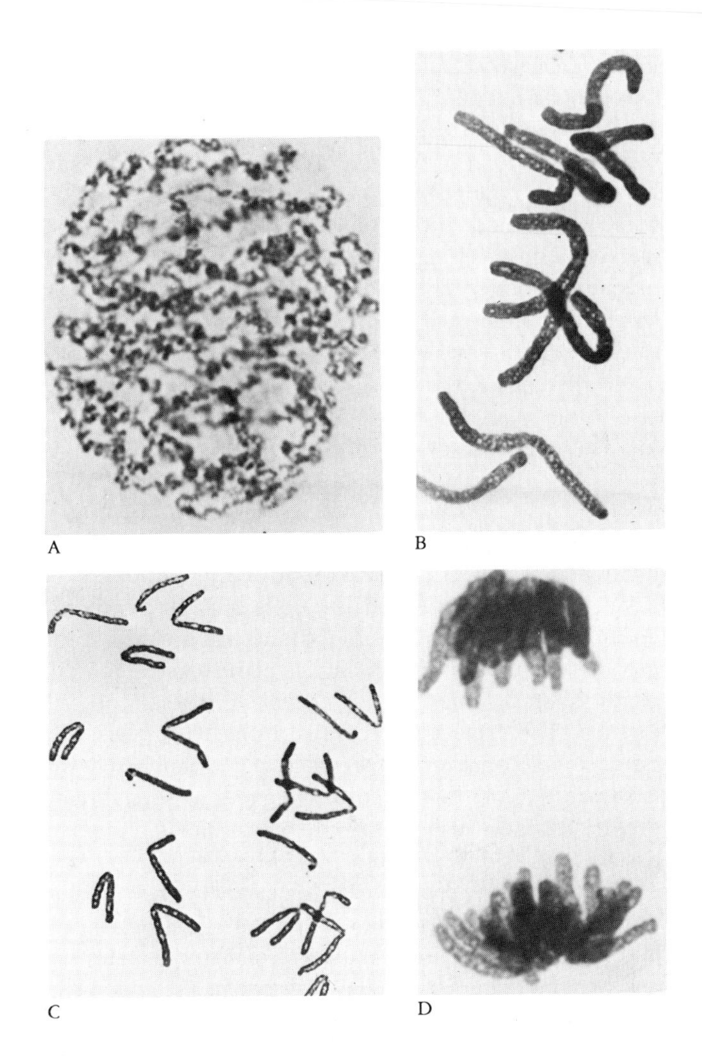

Figure 2–6 Mitosis in the California coastal peony. The vegetative cells of this species have 24 chromosomes ($2n = 24$). Notice how the chromatin material becomes visible as a coiled linear structure (A), how chromosomes line up along the center of the cell (B), how they separate and pull apart toward the sides of the cell (C), and how they finally clump together (D) just prior to a wall being formed between the two groups. At mitosis two daughter cells are formed with the same number and kinds of chromosomes that existed in the original cell. (Courtesy M. S. Walters and S. W. Brown.)

Cell Division

The successful gardener is one who can get plants to grow, and plants grow as a result of cell division or cell enlargement. Two different kinds of cell division occur in plants as well as in most other organisms. The increase in the number of cells in the plant body is the result of a kind of cell division called **mitosis,** in which two "daughter" cells are formed, each having the same chromosome number as the "mother" cell (Figure 2–6). (Another type of cell division, meiosis, which occurs only when an organism reproduces sexually, will be described in Chapter 3.)

Mitosis, and the resulting increase in cell numbers, occurs primarily in the youngest plant tissue, just behind the rapidly growing stem or root tip, at nodes, or in the differentiating fruit, tuber, bulb, or leaf. Because chromosomes and many organelles of newly formed cells are largely protein, and because nitrogen is the major soil-borne element of protein molecules, young plants growing rapidly by cell division must have ample supplies of nitrogenous fertilizer.

The disposition of chromosomes during mitosis is pictured in Figure 2–7. When a cell is ready to divide, the nuclear wall disappears and individual chromosomes become visible and arrange themselves linearly along the mid-axis of the cell. Each splits longitudinally into a pair of chromosomes. The chromosomes of each pair next move away from each other toward their respective sides of the cell, and a cell wall forms to complete the division.

Cell Enlargement

Much of what is seen as plant growth comes about as a result of cell enlargement. Just behind the region of cell division at the tip of stems or roots, or at nodes of plants having growth in the nodal area, is a **region of cell elongation.** Cell elongation accounts for all the length growth and root spread of higher plants except for the relatively small amount that occurs as a result of cell division. The expansion of leaves, fruits, and tubers and other storage organs also occurs largely through cell enlargement (Figure 2–8). Cells can expand to several hundred times their original size in some fruits and storage tissues.

Although protein is utilized for certain differentiating organelles in the elongating cells of stems and roots, cellulose, lignin, fiber, and other structural carbohydrates are relatively more in demand. The increase in size and in the development of sweetness and flavor of fruits, bulbs, tubers, and similar storage tissue also involves the utilization and storage of carbohydrates. The visible and measurable enlargement of cells of storage tissue includes an

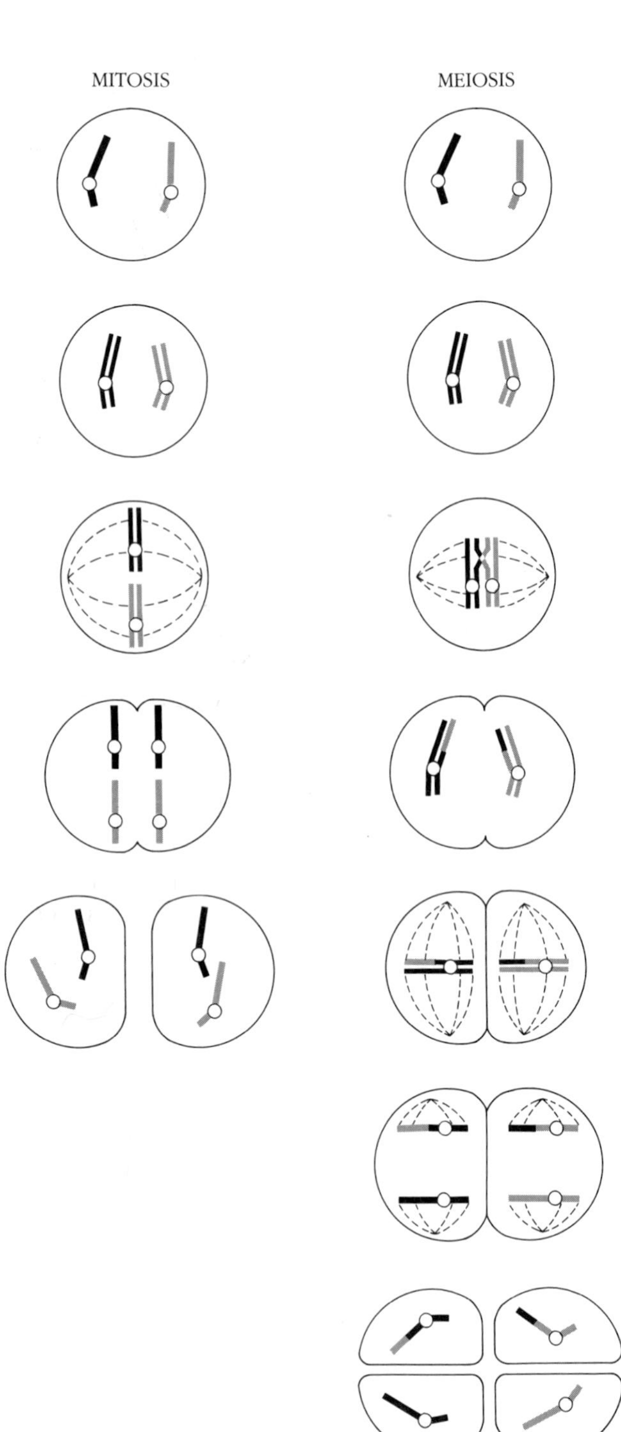

Figure 2–7 A comparison of chromosome disposition with mitosis and meiosis. With mitosis a vegetative body cell divides to form daughter cells with exactly the same number and kinds of chromosomes possessed by the mother cell. With meiosis a cell specialized for reproduction divides to form eventually four daughter cells, each having one half the number of chromosomes possessed by the mother cell.

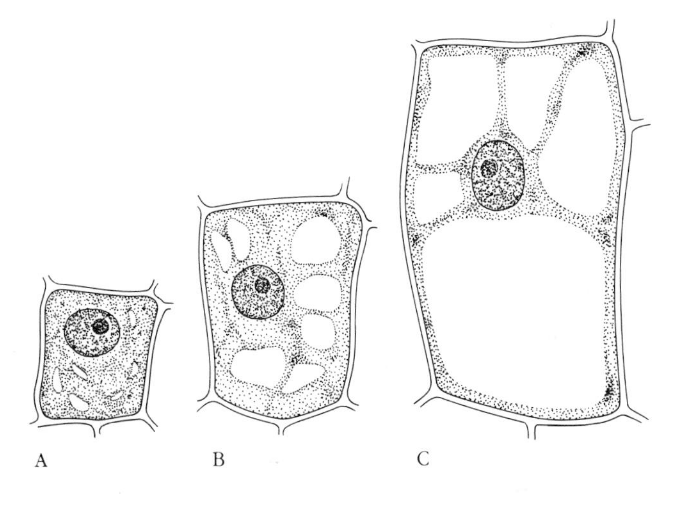

Figure 2–8 Diagrammatic sketch of a newly formed meristematic cell (A), an herbaceous tissue cell of intermediate age (B), and a mature herbaceous cell (C). Although the cytoplasm may increase in size to some extent, the most dramatic change as cells in many parts of the plant mature is their greatly enlarged vacuole.

increase of the size of the vacuole, which contains dissolved sugars and soluble carbohydrates; an increase in starch and insoluble carbohydrate granules in the cytoplasm; and an expansion of the cell wall, which is mostly lignin, cellulose, and fibrous carbohydrate material. Thus while nitrogen absorbed by the roots is the element most needed in tissue containing rapidly dividing cells, an abundant supply of carbohydrates manufactured primarily in the leaves is essential for fruit and tuber development, cell elongation, and other plant growth involving cell expansion. Carbohydrate manufacture by plants (photosynthesis and metabolism) will be mentioned in more detail later in this chapter.

Horticultural Significance of Cell Characteristics

Cell characteristics can have significance in the usefulness of garden products. The juicy edible parts of most fruits and vegetables contain a large proportion of thin-walled living cells capable of growth and differentiation called **parenchyma.** Wood and the tough fibrous plant materials used for string, paper, and clothing have large numbers of elongate thick-walled dead cells called **sclerenchyma fibers.** In the plant sclerenchyma provides support.

There may be significant differences in cells of different cultivars of the same kind of plant. For example, the 'Russet Burbank' cultivar of potato, especially when it is grown in an arid climate with irrigation, is considered ideal for baking, mashing, and processing, because the middle lamellae break down easily when the potato is cooked but the cells remain intact, imparting a mealy dry texture. Potato cultivars such as 'Katahdin' and 'Red Pontiac,' especially if grown in areas of high rainfall, have cells in which the middle lamellae do not break down. Tubers of these cultivars do not "cook to pieces," making them ideal for boiling and frying.

Plant Tissue and Structure

We spoke earlier of the cell as a building block of plants and described briefly the type of cell division, called mitosis, by which plants increase in cell numbers. However, the plant is not simply a helter-skelter collection of cells dividing at random into a plant mass. Rather, each cell is part of a tissue that has a specialized function within the plant. Furthermore, tissues are organized into plant structures, each with a special purpose.

Tissue Differentiation

The differentiation of cells into specialized tissues and structures has been a subject of fascination ever since it was discovered that each living organism develops from a single cell. How is it possible for a single plant cell to multiply itself into a functioning plant, including such diverse structures as leaves, stems, roots, tubers, fruits, and seeds (Figure 2–9)?

The complete answer is not yet clear, but the mystery of tissue differentiation can now be partially explained. Tissue culture, by which single cells or small bits of tissue

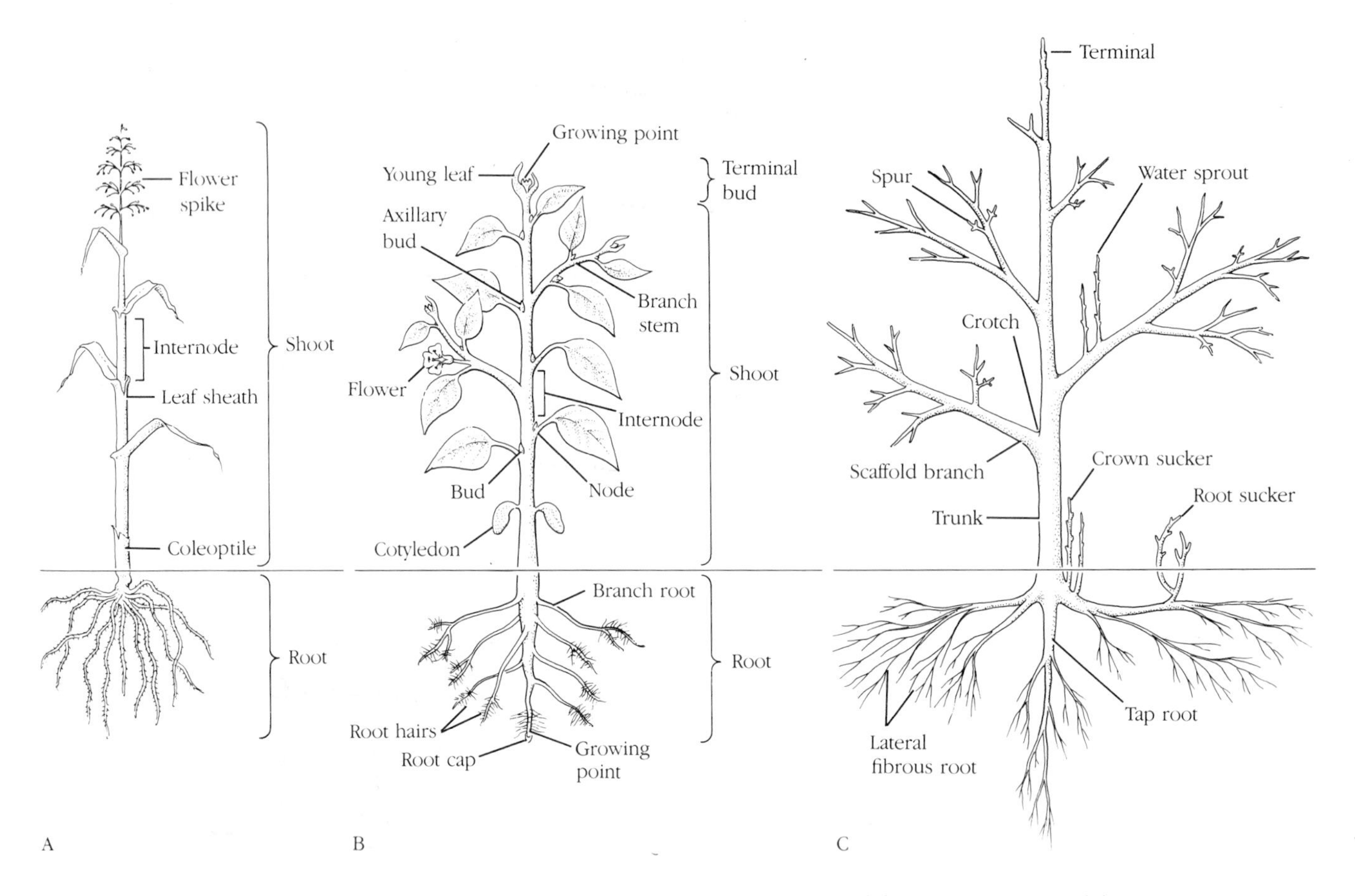

Figure 2-9 Fundamental plant parts: Monocot (A), herbaceous dicot (B), and woody dicot (C). Many of the structures detailed in this figure will be discussed later in the text. (A and B adapted from *Plant Science,* 2nd ed., by Jules Janick et al. W. H. Freeman and Company, San Francisco. Copyright © 1974.)

are grown on special nutrient media, has produced a few answers. For example, a single cell from a carrot root can be placed on a sterile medium containing the elements needed for plant growth plus certain plant growth substances (growth substances are discussed in Chapter 8). If the medium containing the cell is kept circulating, the cell will multiply to produce a mass of tissue in which all parts and all cells are alike. After the tissue mass has reached a certain critical size, which seems to be about the same, regardless of species, the tissue develops into a hollow sphere, or **blastula,** reminiscent of certain stages of lower life forms and also one of the stages of animal embryonic development. From this point if the tissue continues to have all necessary growth requirements and is placed in a stationary location, roots, stems, and other structures will begin to differentiate.

Evidence from these and other experiments suggests that subtle changes in environment—perhaps differences in gravitational pressure or oxygen supply—activate or inactivate segments of DNA in the chromosome. The resulting change in the balance of chemical messages received from its DNA causes the cell to multiply into specialized tissue. For example, reduced pressure at the upper surface and increased pressure at the lower surface of the cellular mass may elicit the chemical message eventually resulting in shoot and root formation, respectively.

The major structures of the plant are the stem, the root, the leaves, and the reproductive organs—flowers, fruits, and seeds (Figure 2–9). Bulbs, corms, tubers, thorns, and the like are modifications of these major structures.

The Stem

Stems provide support for the leaves and the reproductive structures of the plant and contain the tissues that transport water, minerals, and manufactured food throughout the plant. They can also be organs of food manufacture and storage. Stems vary in appearance and structure from plant to plant, but basically there are three types among angiosperms: herbaceous dicot, woody dicot, and monocot.

Dicot Stems. The herbaceous dicot stem has the same structure as that of a first year's growth of a woody dicot; it is shown in Figure 2–10. At the tip of the stem are a series of undifferentiated cells, the **apical meristem.** Meristem is tissue capable of cell division. In a dicot only the cells within a few millimeters of the stem tip are capable of producing plant elongation by dividing or increasing in length. Thus it is only at the tip of a stem that length increases in a dicot plant.

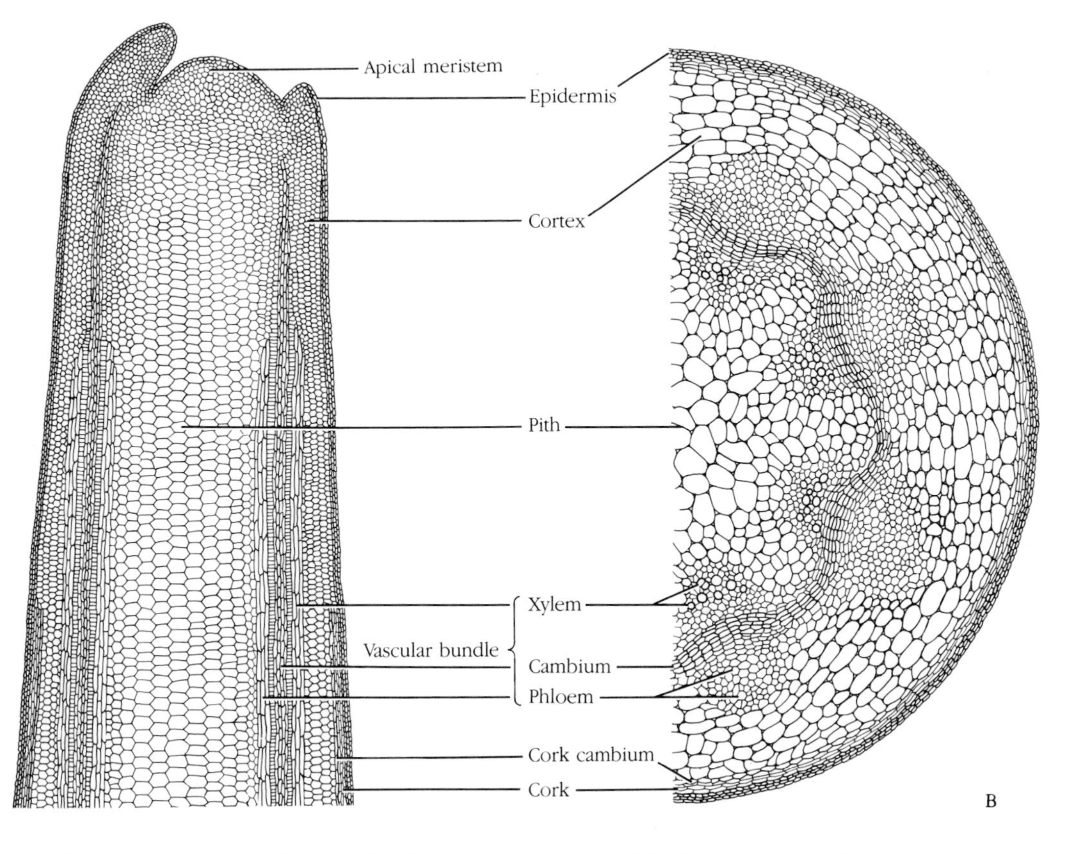

Figure 2–10 Longitudinal section (A) and cross section (B) of the stem of a dicotyledonous plant. This could be either an herbaceous or a woody dicot during its initial stages of growth.

Occasionally a gardener will not prune a low-growing lateral branch of a maple, oak, or other dicot tree, hoping it will rise as the tree grows. Because elongation of a dicot occurs only at the tip of the stem, the limb will never be higher. In fact, it will grow somewhat lower as annual rings add to its girth.

Shortly after the cells in the apical meristem have formed, they begin to differentiate into various stem tissues so that, by the time a stem is a few hours to a few days old, it has cross-sectional structure similar to the stem pictured in Figure 2–10. At the outside and covering the plant to provide protection is a layer of wax-coated cells known as the **epidermis.** Just inside the epidermis is the **cortex.** Cortex cells normally contain considerable chlorophyll and have an important food-manufacturing function in young stems. In the early life of the plant the cortex cells resemble the cells of the apical meristem and remain capable of some degree of differentiation into more specialized tissue.

Just inside the cortex, toward the center of the stem but outward from the pith, are the **vascular bundles.** The vascular tissue is the "circulatory system" of the plant and, like the blood vessels of an animal, permeates all areas of the plant body. A vascular bundle is made up of at least three distinct groups of cells. In its center is a single layer of meristematic tissue, the **cambium.** The cambium is constantly dividing to form phloem tissue toward the outside and xylem toward the center of the plant.

In the young dicot, the vascular bundles lie in a circular pattern between the cortex and pith. **Pith,** the tissue of the center of an herbaceous dicot plant, is made up of more undifferentiated cells. If the pith disintegrates, a hollow-stemmed plant results. The pith is usually replaced by wood fibers from the xylem in older woody dicots. As the vascular bundles enlarge, the cambium of adjacent bundles fuses together until finally a complete ring of cambium is formed. If the plant is of a species in which the stem survives long enough to become woody, some futher differentiation of the vascular tissue in the stem will take place.

Water and mineral elements are transported from the absorbing roots to all parts of the plant through the **xylem.** Although the xylem in a large tree may be many feet in diameter, only the youngest outer layers will be functional in the transport of water and mineral elements. Nonfunctional xylem cells provide support in woody plants. The xylem tissue formed year by year accumulates and gradually pushes the rest of the plant tissue outward. Rapid growth with the subsequent formation of large, thin-walled xylem cells occurs in the spring. Much slower growth, with small thick-walled xylem cells, occurs in the fall; of course, growth ceases during the winter (Figure 2–11). These alternate types of growth produce the annual rings found in the trees of the temperate zone.

Unlike xylem, **phloem** cells do not accumulate from year to year. As they age and are pushed outward by the expanding xylem, they become disorganized and are gradually absorbed by surrounding tissue. Carbohydrates and other foods manufactured or elaborated in the upper parts of the plant are carried primarily in the phloem.

Phloem transport of carbohydrates manufactured in the upper part of the plant is evident whenever the phloem is blocked. Frequently the wire used to tie a young tree to its supporting stake is left too long. Accu-

Figure 2–11 Cross section of a maple tree. The difference in width of annual rings is probably an indication of differences in environment during various growing seasons.

mulating xylem pushes the trunk against the wire, which cuts or girdles the bark, causing a partial blockage of the phloem. The trunk of a girdled or constricted tree will enlarge rapidly just above the constriction as a result of carbohydrates accumulating in that area of the trunk (Figure 2–12). Blockage of phloem may come from root diseases, and in such cases tuber-bearing crops such as potatoes may produce above-ground tubers as a result of accumulation of carbohydrates in the stem (Figure 2–13). Carbohydrate accumulation stimulates reproductive growth, and sometimes phloem transport is purposely blocked by a single trunk-encircling cut through the bark. This practice increases carbohydrates in the stems and leaves, thus inducing fruiting of slow-bearing orchard trees or increasing flowering of sparse-flowering woody ornamentals (see Chapter 8 for more discussion of this practice).

Figure 2–12 Phloem blockage. This woody stem has been constricted by a twining stem of the liana *Clematis columbiana.*

In perennial-stemmed plants and occasionally in annuals and biennials, usually during the first season of growth, one layer of cortex tissue differentiates to form the **cork cambium.** This cork cambium is meristematic tissue that divides to form thick-walled cork cells toward the exterior of the plant. As the layer of cork cells widens, it pushes out to replace the epidermis with a much heavier and more permanent protective layer. Commercial cork used commonly for bottle stoppers is the cork tissue that accumulates on the exterior of cork oak trees (*Quercus suber*).

Figure 2–13 Aerial tubers formed at the nodes of the above-ground stems of potato plants as a result of phloem restriction by a fungus, Rhizoctonia. (From Agriculture Canada Publication 1492, "Diseases and Pests of Potatoes," revised 1974, and reproduced by permission of the Minister of Supply and Services, Canada.)

Monocot Stems. Monocot stems (Figure 2–14) are made up of the same kinds of tissue as dicot stems, but the arrangement is different. In monocots the vascular bundles are scattered at random throughout the stem, and each bundle is surrounded by cortex tissue. In monocots, as in leaves, vascular bundles are partially or wholly surrounded by a sheath of parenchyma cells. Older vascular bundles of monocots do not have a functioning cambium but continue active transport throughout the life of the plant. Monocots do not produce annual rings, even though some, such as palms, grow large and become quite woody. In some monocots growth in length occurs only from a bud at the tip. In others, grasses being a prime example, growth in length occurs from nodes at the lower part of the stem. One reason for grass being a better lawn plant than herbaceous dicots is that mowing does not remove the growing point of the grass plant.

Buds and Nodes. Elongation and branching of stems take place within structures referred to as buds. **Buds** can be described as embryonic stems surrounded by the embryonic leaf and flower tissues that will develop from them. They may be enclosed in protective sheaths that may disappear or remain as bracts or scales when the bud opens. **Nodes** are portions of the stem where visible buds are generally located and where leaf petioles are attached. Areas of the stem between nodes are called **internodes** (see Figure 2–9).

Buds can be actively growing or dormant. Growth in height of dicots commences from terminal buds, and branching commences from lateral buds lower on the stem. Occasionally buds not previously visible grow from internodal areas of stems or from roots of certain species. They are called **adventitious** buds.

Bud placement is an important diagnostic feature of woody ornamentals. Buds are **opposite** when two are located at the same node on opposite sides of the stem, **whorled** when several surround the stem at the same node, or **alternate** when each is at a different node and arranged in a spiral (Figure 2–15). Buds may produce only leaves and stems, only flowers, or both leaves and stems and flowers and are referred to as **vegetative, flower,** or **mixed** buds, respectively.

Buds producing leaves, flowers, and fruit on temperate zone woody plants are usually formed the previous summer and are visible through the winter. Therefore, practices affecting flower and fruit initiation need to be accomplished by early summer of the year preceding the year of production.

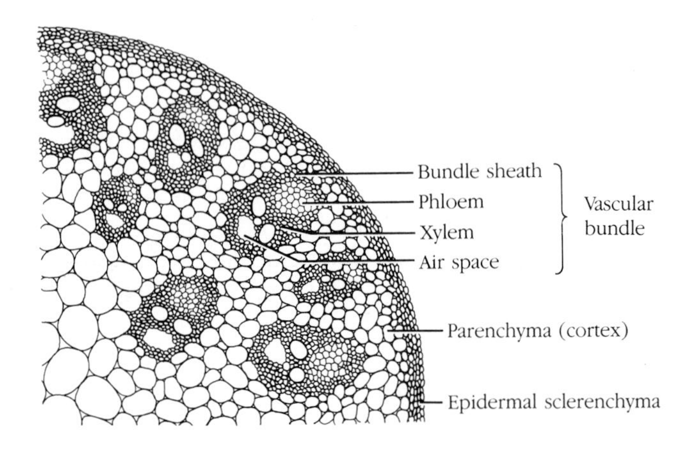

Figure 2–14 Cross section of part of a monocotyledonous plant stem, showing scattered vascular bundles.

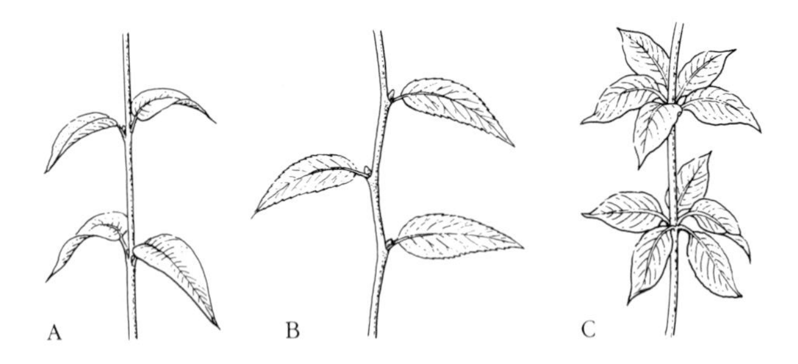

Figure 2–15 Three types of leaf and bud arrangement: (A) opposite, (B) alternate, (C) whorled.

The Root

The primary functions of plant roots are to anchor the plant and to absorb water and mineral elements needed by the plant. The tissues of the stem are connected with those of the root, and most are duplicated in the root. At the tip of a rapidly growing root are several layers of thick-walled cells known as the **root cap.** Just behind the root cap is the region of cell division, behind that is the region of cell elongation, and behind that is the **root hair zone.** It is primarily through these few millimeters at the rapidly growing root tip that the plant is able to take up moisture and minerals. This part of the root system is extremely delicate and subject to injury and desiccation. Almost all of these zones are destroyed with bare-root transplanting, and for this reason a transplanted plant will often wilt severely, even though most of its major roots are still intact. Elongation of roots occurs almost exclusively in the regions of cell division and cell elongation. This elongation is stimulated by favorable growing conditions, which explains why roots become concentrated where moisture and mineral elements are plentiful and why tree roots so easily grow through cracks or joints in moist, nutrient-rich sewer lines.

A cross section near the root hair zone would show a structure similar to Figure 2–16. On the outside of the root is an epidermal layer. In young root tissue this layer is quite permeable to water, and it is from cells of this layer that root hairs are produced. Just inside the epidermal layer is a layer of loosely spaced cortex cells. Water is able to pass quite rapidly through the intercellular spaces of this cortex tissue. Next to the cortex is a single layer of tissue composed of thick-walled, somewhat impermeable cells known as the **endodermis.** In the root-hair zone and younger root tissue the endodermis is permeable enough to permit the passage of water and nutrients. Higher in the root water loss from the vascular system is

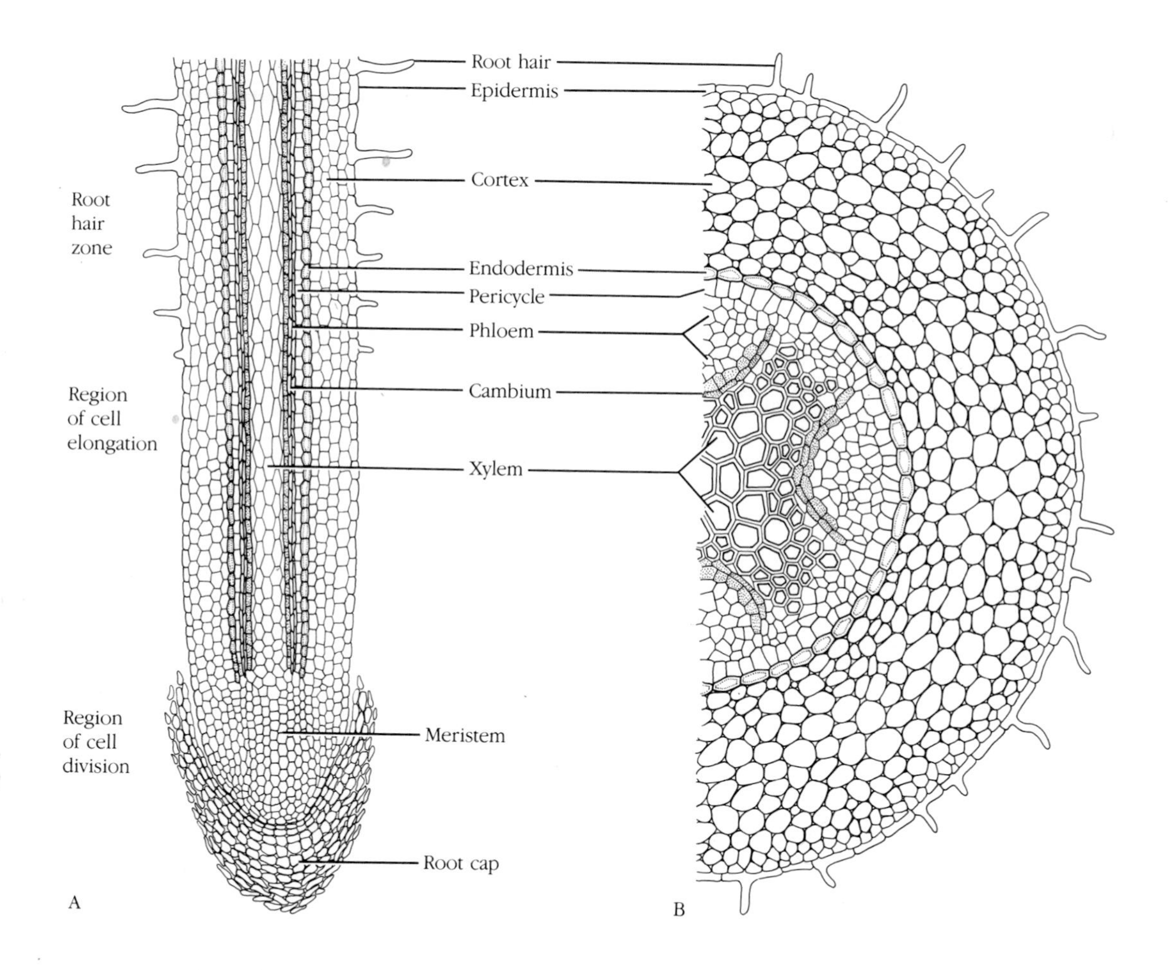

Figure 2–16 Longitudinal section (A) and cross section (B) of a plant root.

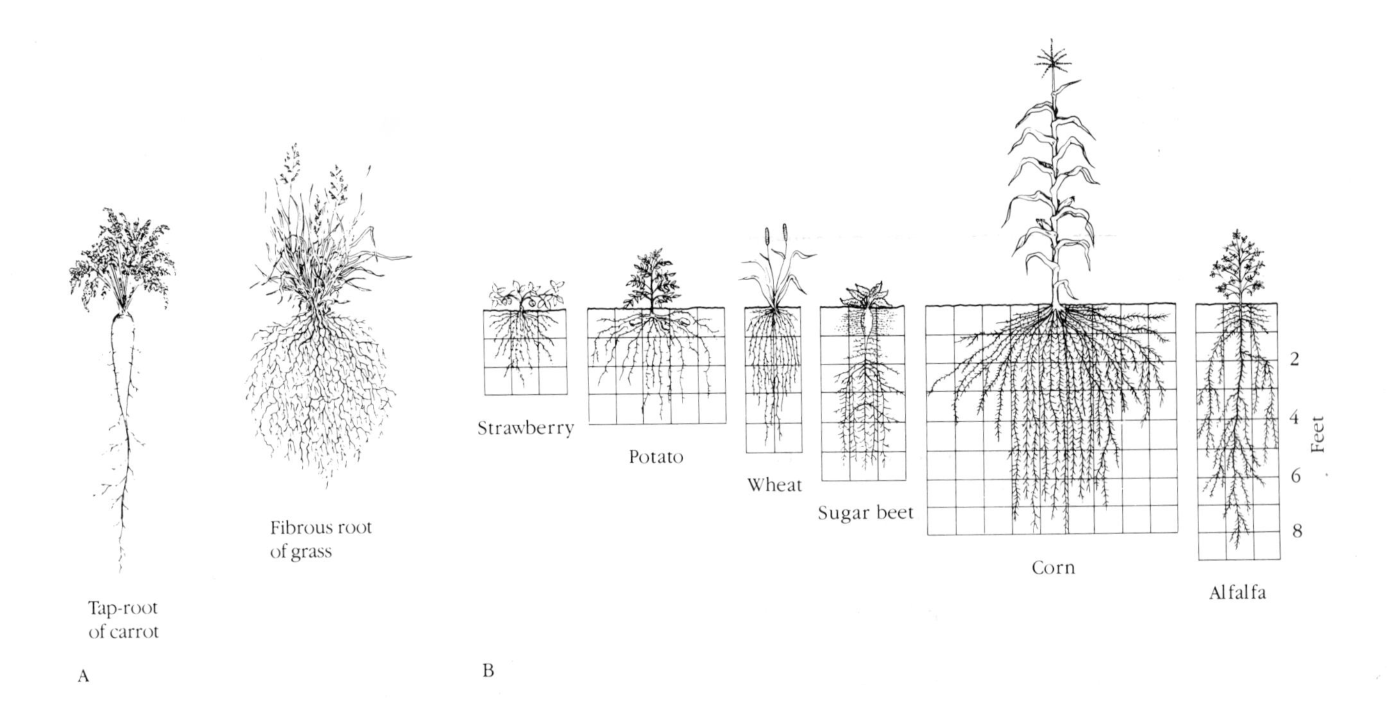

Figure 2–17 (A) Root systems. (B) Comparative root systems of crops in deeply irrigated soils. (From *Plant Science,* 2nd ed., by Jules Janick et al. W. H. Freeman and Company, San Francisco. Copyright © 1974.)

prevented by the impervious nature of the endodermis. Next to the endodermis toward the interior of the plant is a layer of meristematic cells, known as the **pericycle,** from which branch roots form.

The vascular system is at the center of most roots. As is true in the stem, the xylem and phloem are formed from cells cut off by the cambium—the phloem to the outside, the xylem to the inside. Roots of most plants do not have pith.

Root systems vary in appearance and in the extent to which they penetrate the soil. The extensiveness of the root system is often associated with the environment under which the plant evolved. The extensiveness of the root system also determines, to some extent, the frequency and depth of irrigation required for the particular type of plant. Celery, for example, which originated in the swamps of the Middle East where only a small root system was needed because water was plentiful, requires frequent irrigation. Tomatoes and melons have much more extensive root systems and are quite drought tolerant.

Plant root systems can be classified into two general groups on the basis of growth habits (Figure 2–17). When the root system branches into a number of smaller roots near the soil line, the plants are said to have **fibrous roots.** Plants in which a main root grows straight down from the stem and smaller side roots branch from the main root are said to be **tap-rooted.** The carrot is a well-known example of a tap-rooted plant. Because a widely branched root system tends to bind soil particles together, it is easier to keep a ball of earth around a fibrous-rooted plant than a tap-rooted plant. The ball of earth prevents injury and desiccation of the absorbing roots and allows a plant to be transplanted with less disruption in vital functions. To create root branching, nursery workers frequently undercut tap-rooted plants a year or more before they are to be transplanted. Severing the lower part of the tap-root 6 months or a year before transplanting encourages root branching and makes the plant easier to move with the root system and adhering soil particles intact (Figure 2–18).

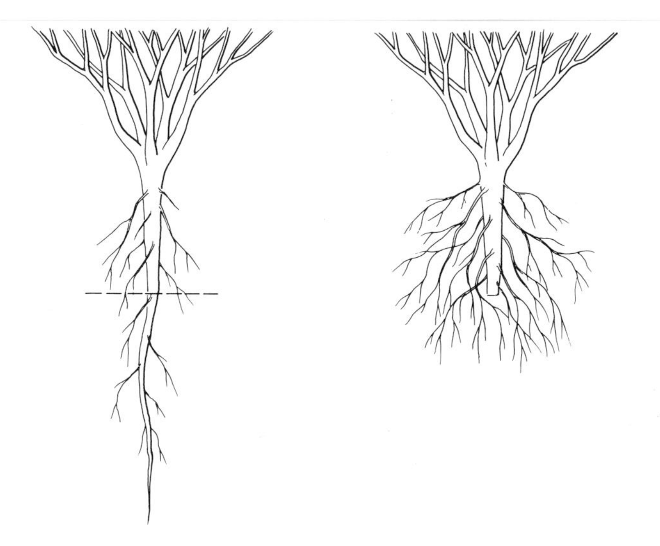

Figure 2–18 Effect of undercutting a tap-rooted plant.

The Leaf

The main function of the leaf is the manufacture of sugar from inorganic compounds. Food for all living creatures on this earth, plants and animals alike, is elaborated from the sugars manufactured by photosynthesis, which is carried out primarily in the leaves and herbaceous stems of green plants.

The structure of the leaf is shown in Figure 2–19. The veins are vascular bundles similar to those found in the stem and the root. They consist of xylem and phloem with cambium in between and are connected to the vascular system of the rest of the plant. Also associated with the vascular bundles of the leaf is a sheath of parenchyma cells and supporting sclerenchyma fibers. The upper surface of the leaf is a layer of epidermal cells coated with a waxy cuticle that makes it impervious to water and other liquids. The impervious nature of the leaf surface effectively prevents water intake through the foliage and explains why special treatments are required for foliar feeding.

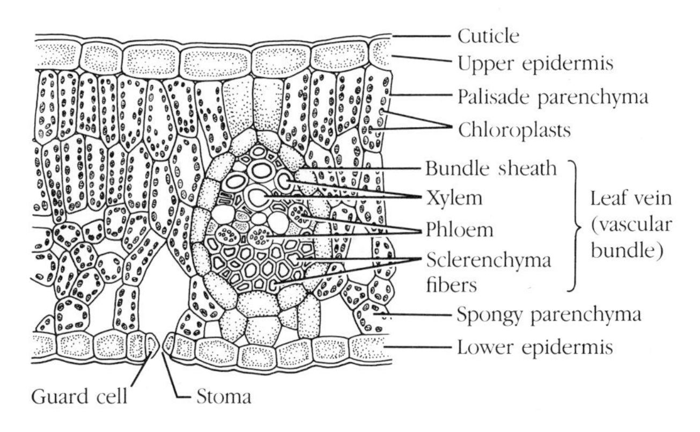

Figure 2–19 Cross section of a leaf.

Just below the upper epidermis is a layer of cells arranged like the marble columns of a Greek temple or like the elongated pillars of basalt sometimes found in areas of ancient volcanic activity. These cells are called the **palisade parenchyma.** Below the palisade parenchyma is an area of rather loosely packed cells called the **spongy parenchyma.** The palisade and spongy parenchyma cells contain numerous chloroplasts in which most photosynthesis occurs.

Below the spongy parenchyma is the lower epidermis, which seldom has the heavy cuticle of the upper epidermis and which is packed with openings called **stomata.** Each stoma is bordered by two **guard cells,** which expand or contract to open or close the stoma in response to environment and to the physiological condition of the plant. Drought conditions or an excess of carbon dioxide, which is likely to occur at night or during cool weather, causes the guard cells to collapse and close the stomata. The stomata are usually open when weather conditions are favorable for photosynthesis.

Modifications of Stem, Root, and Leaf

Earlier in this chapter, when growth by cell enlargement was described, tubers, bulbs, and other fleshy storage organs were mentioned. These as well as other specialized structures have developed in the process of plant evolution. The internal anatomy of these structures resembles the anatomy of a stem, a root, or a leaf, and each is thought to have evolved from one of these three organs. These modified structures are important both as sources of food and as means of propagation for many garden crops. Some of them are shown in Figure 2-20.

Certain plants, the strawberry and strawberry geranium, for instance, produce horizontal above-ground stems called **stolons** or **runners.** Horizontal underground stems, such as those produced by grasses, are called **rhizomes.** Some plants, bearded iris being an example, produce large fleshy rhizomes. **Tubers** are enlarged fleshy sections of rhizomes, their "eyes" being analagous to the buds on a stem. **Corms** are solid thickened underground stems, and **bulbs** are thickened stem plates surmounted by fleshy modified leaves. Bulbs, like those of onion and tulip, having concentric rings, the outer of which form a dry protective cover (a tunica), are called

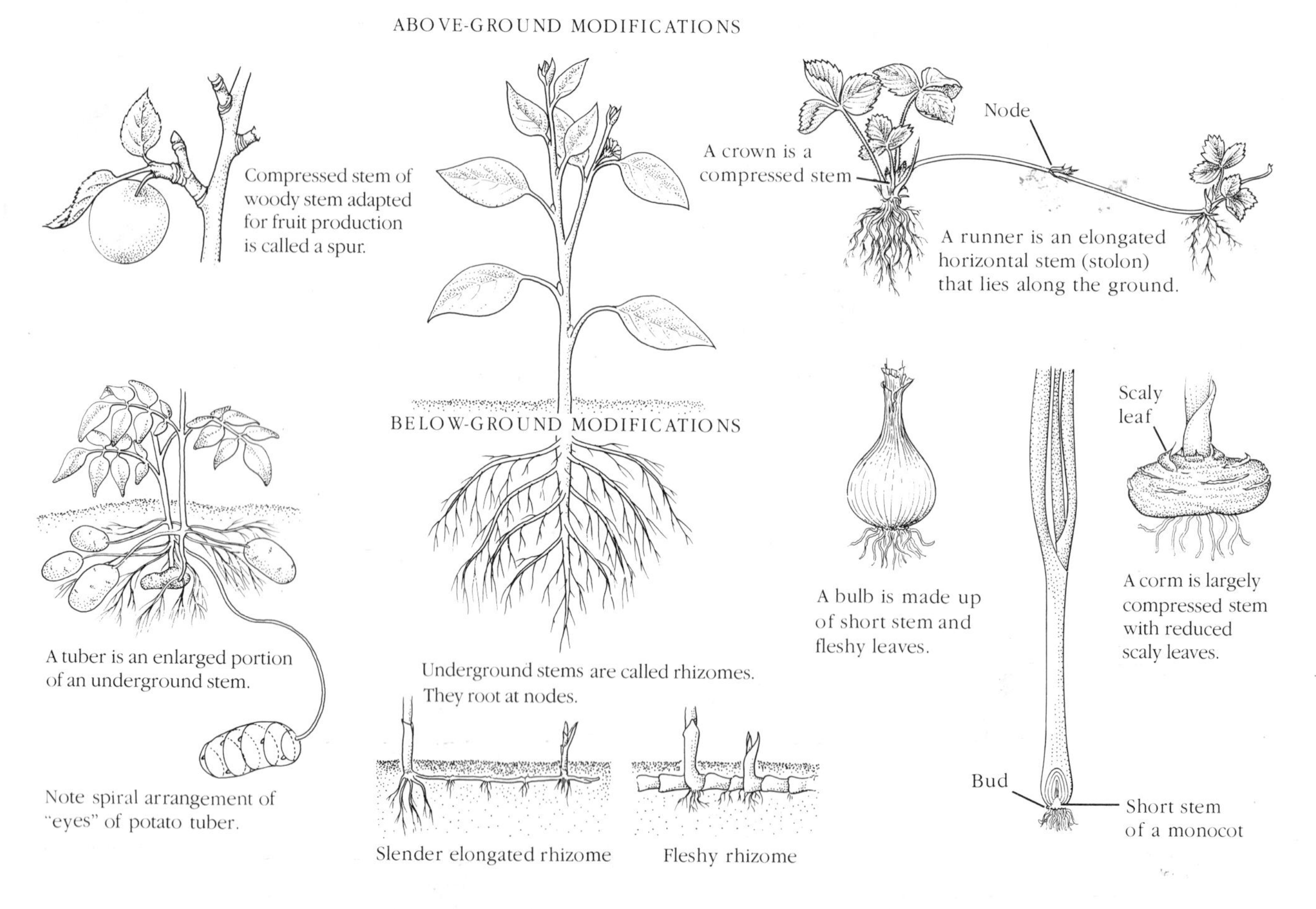

Figure 2-20 Some modifications of stem, root, and leaf.

tunicate bulbs, and those like lily, with fleshy overlapping leaves resembling scales, are called **scaly** bulbs. The first-year stem of most biennial crop plants is a flattened platelike cluster of cells, similar to the stem portion of a nongrowing bulb. The platelike cell cluster elongates to form a typical flowering stem during the second year.

The juncture of stem and root is called the **crown.** Perennial herbaceous plants often have crowns modified into enlarged storage organs that enable the plants to survive through the winter. New shoots, including the edible portion of asparagus and rhubarb and the flower stalk of peony and other perennial flowers, grow from these fleshy crowns each spring (Figure 2–21).

Roots of some species grow into enlarged **storage organs,** the function of which, like that of fleshy stems, is to enable the plant to survive periods of adverse environment. These organs are typical roots having no nodes or leaf scales. Fleshy roots of some plants are capable of initiating buds. Wild morning glory (creeping Jenny) and Canadian thistle are two well-known perennial weeds that unfortunately possess this ability. The large fleshy roots capable of forming buds are more appreciated in sweet potato than in its relative, morning glory. The fleshy storage root of dahlia must remain attached to a piece of the crown having buds if it is to grow into a new plant. Raspberries, some blackberries, and other brambles are capable of initiating buds from their widespread root system and are propagated from shoots or suckers coming from those roots. Staghorn sumac, certain poplars, and some wild roses are not so desirable as ornamentals because of their tendency to produce **root suckers** "all over the lawn" or because suckers from their roots enable them to "take over" an area. Sucker-type shoot growth can also originate from roots near the crown (as with common lilac), from the lower stem, or from **axillary buds,** which originate in the protected angle between the leaf and stem, called the leaf axil (see Figure 2–9). These shoots are called **offshoots, slips, pips,** or other names, and are frequently used for propagation.

Leaves of some plants are thick and fleshy, a modification enabling them to be used for propagation. Leaves may also be modified into bud scales or spines. The spines on cacti are modified leaves, and the thickened fleshy body or flattened fleshy pads are stem adaptations enabling this plant family to withstand extreme drought. Stems of woody plants may be modified to form short, thick, slow-growing **spurs** (see Figure 2–9). Fruit and flowers of woody temperate zone species are frequently borne on spurs.

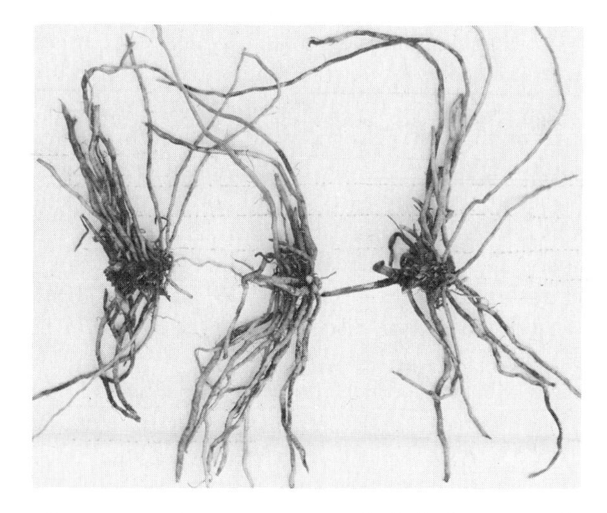

Figure 2–21 Asparagus crown. Carbohydrates stored in the crown of various herbaceous perennials provide the food for early spring growth. These one-year-old asparagus crowns are ready for replanting.

Plant Functions Responsible for Growth

Most of the functions relating to food manufacture and utilization that result in growth take place in the plant's stems, leaves, and roots.

Absorption and Translocation

Water molecules and minerals are absorbed into roots because molecules in a liquid or gas tend to flow to locations where they are least concentrated. Root cells contain membranes that have openings large enough to allow tiny water molecules and mineral particles to pass through but small enough to keep sugar and similar large plant-developed molecules inside. A high concentration of sugar molecules dilutes the water molecules to a lower concentration inside the plant and permits moisture to flow from the soil into the plant root. The flow of a liquid through a semipermeable membrane from a region of higher concentration to one of lower concentration, such as occurs with water absorption into roots, is called **osmosis.** The mechanism of mineral absorption parallels water absorption, but it is not as well understood. Mineral absorption will be discussed in more detail in Chapter 5.

Water will be absorbed into the plant only as long as the concentration outside remains higher than the concentration inside the plant. If water concentration outside is to remain high, soil moisture must be replenished by rain or irrigation. Furthermore, the application of too much fertilizer causes the development of the kinds of molecules that compete with sugar, diluting water concentration in the soil solution, and thus preventing water from entering plant roots. Plant injury resulting from high concentrations of fertilizer is called **fertilizer burn.**

If the absorbed water remained in the root, it would eventually reach an equilibrium with the soil water. This doesn't usually happen, because concentration on the inside is reduced by **translocation** (movement within the plant) of water to other parts of the plant. Translocation occurs as a result of a "pull" created by evaporation from the leaves and by other plant activities.

A high plant-sugar concentration is also necessary for rapid water absorption, and high sugar concentration, in turn, is dependent on a high rate of photosynthesis.

Photosynthesis and Metabolism

The chemical process of **photosynthesis,** never yet duplicated by human beings, utilizes carbon dioxide, water, light energy, and chlorophyll to produce sugar and release oxygen. The release of oxygen is generally considered incidental to the process, but the role of photosynthesis in supplying oxygen to the atmosphere may be as important to life as its role in supplying sugar for food.

The chemical formula for photosynthesis is commonly written:

$$\underset{\text{carbon dioxide}}{6CO_2} + \underset{\text{water}}{6H_2O} + \underset{\text{radiant energy}}{672 \text{ calories}} \rightarrow \underset{\text{glucose}}{C_6H_{12}O_6} + \underset{\text{oxygen}}{6O_2}$$

This equation is greatly simplified; the processes that eventually result in synthesis of sugar (and the other organic compounds now known to be produced during photosynthesis) are a complex set of stepwise reactions dependent on the healthy functioning of all plant organs interacting with a conducive environment. Vigorous roots must be absorbing in warm soil that contains water and mineral nutrients in the proper amounts; young leaves containing chlorophyll must be expanding in an uncrowded, temperate, well-lighted atmosphere that has a plentiful supply of carbon dioxide; the countless systems that manufacture and elaborate food and supply energy to the plant, as well as the transport (vascular) system, must be healthy and functioning at near maximum efficiency. An interruption at any point of the system halts photosynthesis.

In simplest terms maximum production from garden or field requires maximum photosynthesis. The pages of this and succeeding chapters are devoted to plant structure and function, environmental conditions, and pest control largely because these are the things that determine the rate of photosynthesis and, in turn, the growth of the garden.

Just as humans need several kinds of foods, so do plants. The plant forms its proteins, complex carbohydrates, and fats by combining nitrogen and other elements with manufactured sugars. This elaboration of more complex food molecules is one form of **metabolism,** a term used to describe the chemical processes that build up and break down the various food elements in the plant body.

Transpiration

When stomata are open, carbon dioxide enters and water vapor is lost. This water loss through the stomata, called **transpiration,** proceeds slowly or rapidly depending on temperature and humidity whenever humidity is lower than 100% and photosynthesis is occurring. Most of the water lost from a field covered with foliage is through plant transpiration. Several chemical treatments have been devised to reduce transpiration, but except for a few special situations these have not been practical, largely because closing the stomata to reduce transpiration usually reduces carbon dioxide intake and thus slows photosynthesis.

Respiration

The food energy required by plants for the growth processes and metabolic activities throughout their lives is obtained by the breakdown of sugars into carbon dioxide and water, with a resulting release of energy and heat, a process called **respiration.** Since respiration results in the breakdown and utilization by the plant of plant foods, plant growth is negatively correlated with respiration. In fact, plant growth is roughly equal to the food manufactured by photosynthesis minus that used by respiration.

Temperature is a major environmental factor affecting both photosynthesis and respiration. Like other chemical processes, both are speeded by increasing temperature. Each plant has a temperature range over which it will grow, and control of respiration by reducing temperature is important for long storage and quality retention of plant products. The effect of temperature on plant growth, product quality, and storage will be discussed in more detail in Chapter 7.

Selected References

Esau, K. *Plant Anatomy.* 2nd ed. Wiley, New York. 1965.

Heyward, F. B., and Ross, C. *The Structure of Economic Plants.* Macmillan, New York. 1938.

Salisbury, F. B., and Ross, C. *Plant Physiology.* 2nd ed. Wadsworth, Belmont, CA. 1978.

3 Structure and Growth—The Reproductive Phase

The processes of reproduction have long been regarded as mysterious and miraculous. No facet of biological science has received more intensive study during the past few decades and in no area of biology has the increase in knowledge been more rapid. The new understanding has dispelled some of the mystery, but the creation of a new living organism seems as miraculous as ever. Knowledge of the principles underlying plant reproduction can enhance the enjoyment of gardening and is basic to an understanding of plant **propagation**—the production of new plants—and plant growth. Figure 3–1 illustrates the normal reproductive process in flowering plants.

The Flower

A typical flower (Figure 3–2) consists of four major parts, usually situated on a **receptacle** that is the enlarged, terminal portion of a flower stem. Just above and attached to the receptacle is a whorl of connected or separated leaf-like bracts, most often green in color, called (1) **sepals.** Inward from and above the sepals are the frequently brightly colored (2) **petals** of the flower that are collectively known as the **corolla.** Insects, necessary for pollination, are attracted by the corolla. Inward from the corolla are the male organs, or (3) **stamens,** consisting of the **anther,** in which pollen is produced, and the **filament,** the stalk that supports the anther. At the center of a typical flower is the female organ, the (4) **pistil.** The base of the pistil is the **ovary,** which contains the **ovules** or **embryo sacs.** Above and connected to the ovary is the **style,** topped by a usually somewhat broader, sticky **stigma.**

The description above is of a **complete flower,** because it contains the four basic flower parts. A majority of horticulturally important species produce complete flowers, although these flowers vary greatly in size and structure of the various parts (Figure 3–3). If any one of the four major flower parts is missing, the flower is **incomplete.** If the flower contains both male and female structures, it is a **perfect flower.** If either the male or female organs are missing, the flower is **imperfect.**

A plant in which each flower has both male and female organs is referred to as **hermaphroditic.** The majority of garden plants are hermaphroditic; petunia is a common example. A single plant having some flowers with only male organs and some with only female organs is **monoecious.** Sweet corn is probably the best known example of a monoecious horticultural crop. The tassel

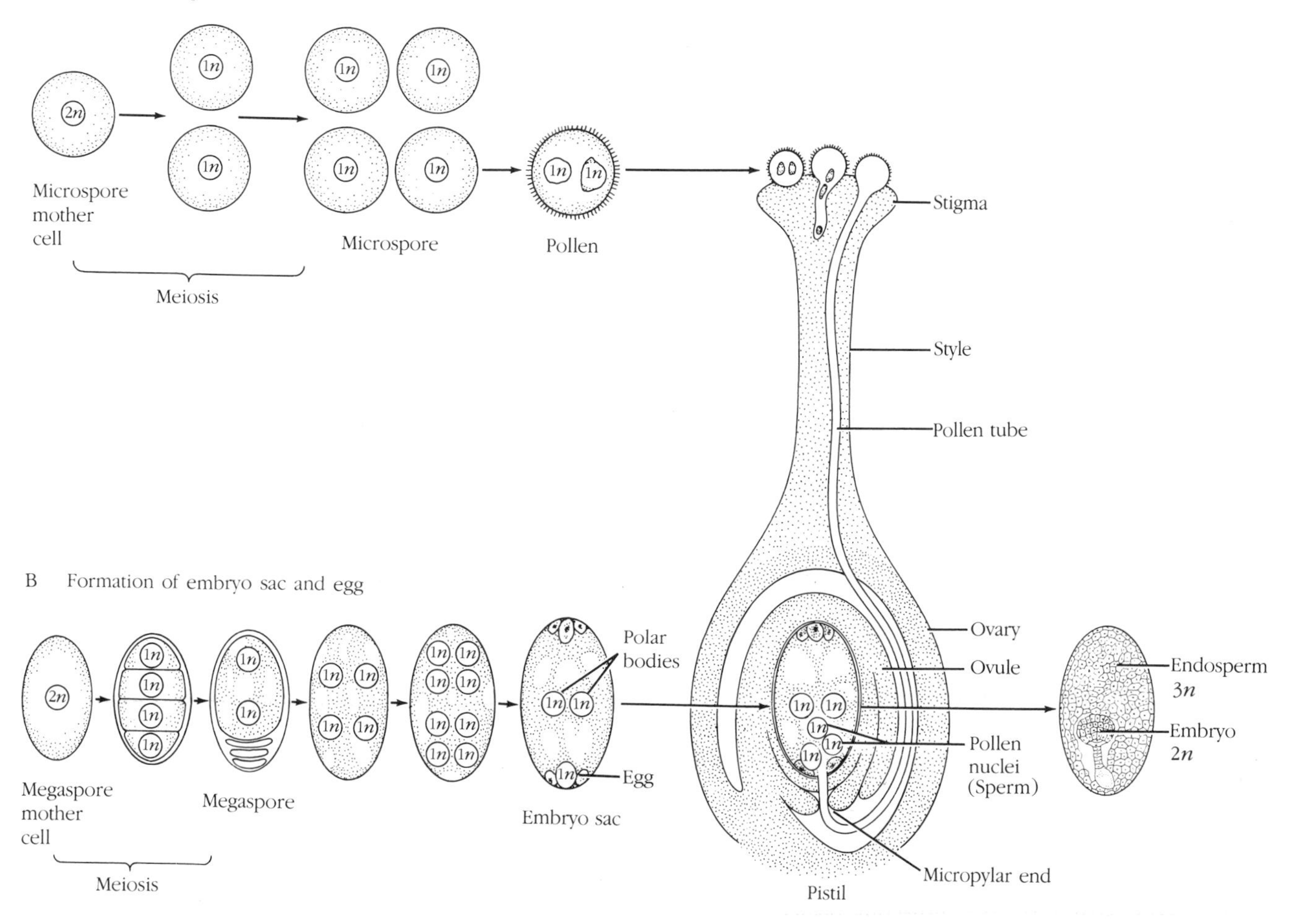

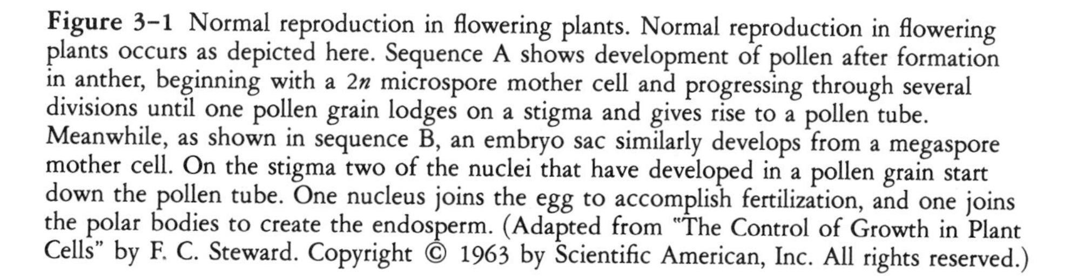

Figure 3–1 Normal reproduction in flowering plants. Normal reproduction in flowering plants occurs as depicted here. Sequence A shows development of pollen after formation in anther, beginning with a $2n$ microspore mother cell and progressing through several divisions until one pollen grain lodges on a stigma and gives rise to a pollen tube. Meanwhile, as shown in sequence B, an embryo sac similarly develops from a megaspore mother cell. On the stigma two of the nuclei that have developed in a pollen grain start down the pollen tube. One nucleus joins the egg to accomplish fertilization, and one joins the polar bodies to create the endosperm. (Adapted from "The Control of Growth in Plant Cells" by F. C. Steward. Copyright © 1963 by Scientific American, Inc. All rights reserved.)

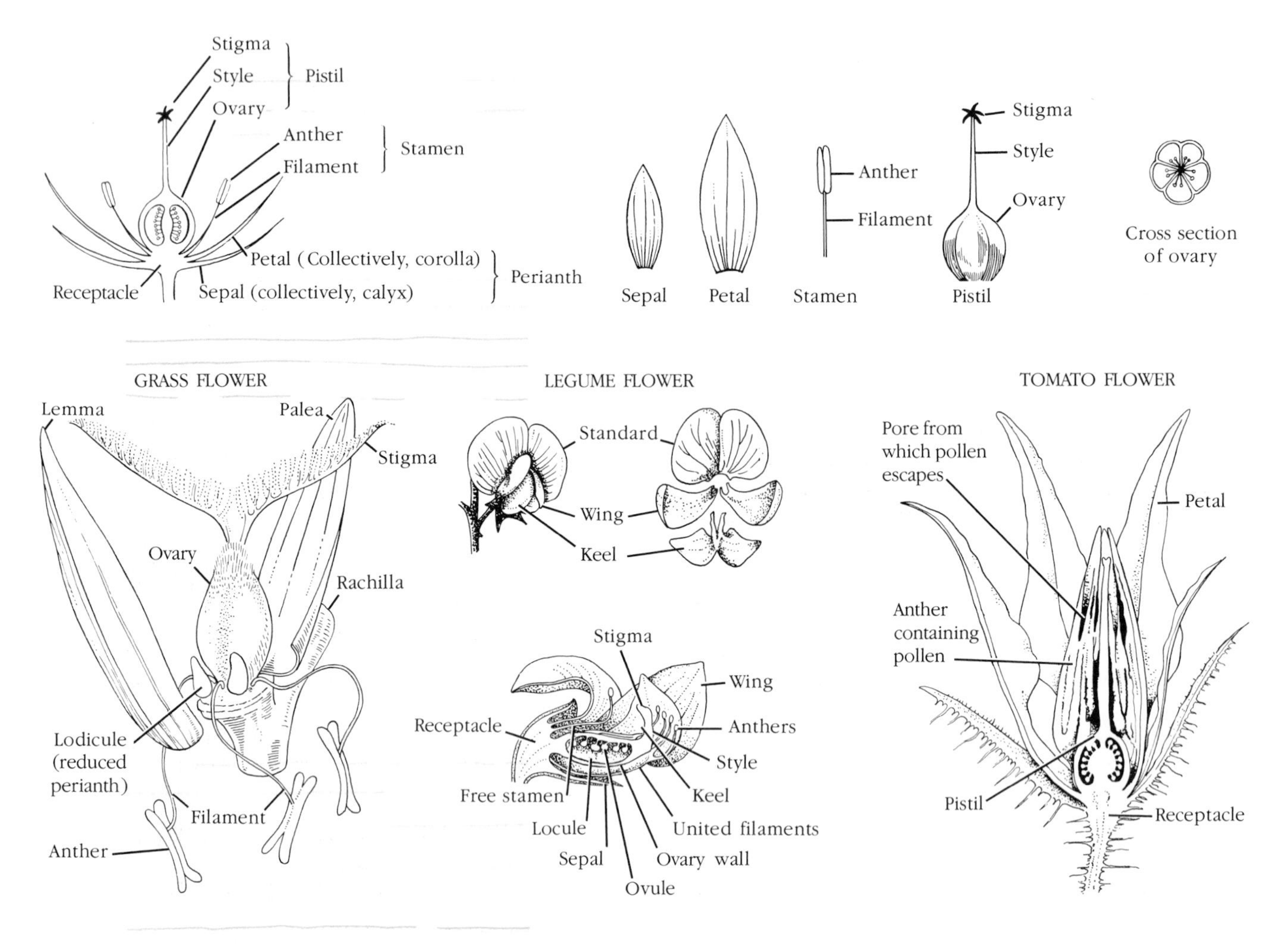

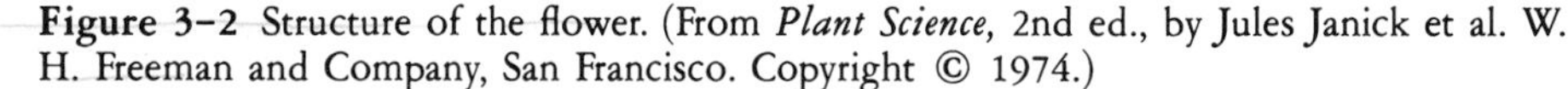

Figure 3–2 Structure of the flower. (From *Plant Science,* 2nd ed., by Jules Janick et al. W. H. Freeman and Company, San Francisco. Copyright © 1974.)

contains the stamens, and the silk and ear contain the pistils. Squash, cucumber, and related plants are also monoecious (Figure 3–3). A species in which the male and female flowers are borne on separate plants is called **dioecious.** Holly, asparagus, and ash are well-known examples of cultivated dioecious plants. Only the female holly plant produces berries, but it must be grown within 100 yards (100 meters) of a male plant if it is to receive sufficient pollen to produce them consistently. The male plant produces only inconspicuous male blossoms. Male asparagus plants produce smaller spears but a higher total yield than do female plants. Male ash trees are more popular for landscaping than are female trees, because the males do not produce seed pods that litter lawns and sprout young trees in flower beds.

Modifications of floral parts are common in cultivars of some species. The so-called double blooms of carnations, stocks, and petunias occur because stamens and sometimes pistils have been modified into petals (see Figure 3–3). As so much plant energy is required for seed production, viable floral parts have been eliminated from some crop species by breeding or selecting. In potatoes seed and fruit formation reduce food reserves available for tuber growth. Because of this plant breeders have developed nonfruiting types, so few potato cultivars produce true seed in the environment in which they are grown. Flowering is also undesirable in coleus, some kinds of begonia, and some other house plants grown for their foliage.

The Seed

A **seed** can be described as a miniature plant with its food supply and a protective cover. Seeds are basic both to the natural world and to the gardener. In nature it is the seed that permits most plant species to survive periods of adverse environment. It is the seed and its attachments that bring about the dispersal of plant species over broad areas. From the gardeners' point of view it is the seed that allows the convenient transport, storage, and propagation of many plants adapted to the garden. It is the combination of chromosomal matter (**germ plasm**) of male and female parents occurring as a result of seed formation that permits the wide variety of individual kinds of plants.

Seeds are useful to people other than as a means for plant reproduction. Much of the world's basic food for both humans and livestock is seed. Fruits of most species do not form, or at least do not grow normally, without the concomitant formation of seeds, and the plant structure from which the seeds develop, the flowers, are an important horticultural commodity. Thus an understanding of seed formation is essential to the "why" of many gardening procedures.

Pollen Formation

In angiosperms and most conifers seed development begins with differentiation of cells within the flower (see Figure 3–1). In the anther, or male portion, of a flower certain cells called **microspore mother cells** begin to

Figure 3–3 Various kinds of flowers. (A) Carnation, double because stamens have been modified to petals; (B) orchid; (C) tomato; (D) squash, female and male. (C, courtesy of Washington State University.)

round up and develop into specialized reproductive cells. These cells lose their nuclear walls, and their chromosomes form into threadlike structures and gradually shorten. The chromosomes line up in pairs (not in a single row, as occurs with mitosis) along the center axis of the cell. If individual chromosomes of a pair touch, they may stick together and exchange equal parts as the two chromosomes of a pair are pulled away from each other toward opposite sides of the cell. After chromosomes are separated, a cell wall is formed through the center axis. This type of cell division by which two daughter cells are formed, each with half the chromosomes possessed by the mother cell, is called **meiosis** (see Figure 2–7).

After the microspore mother cell has undergone its first meiotic division, the chromosomes in each of the two newly formed cells line up in a single line along the center axis of the cell. This time each chromosome splits longitudinally, forming a pair of identical chromosomes. The chromosomes migrate toward their respective sides of the cell, and a wall forms between the two rows of chromosomes. As a consequence of the two divisions, each microspore mother cell forms four **microspores,** each with one half the chromosomes of the original. A heavy, waxy cover having several openings or pores forms around the microspores as they become mature **pollen** grains.

Embryo Sac or Ovule Formation

Meanwhile, a somewhat similar series of events is occurring in the ovary of the pistil. A cell referred to as a **megaspore mother cell** begins to differentiate from other cells, and its chromosomes pair at its center. The megaspore mother cell divides twice to form four daughter cells, each with half the chromosome number of the original mother cell, in a process similar to the divisions that occur in the formation of pollen. However, three of the four daughter cells fail to continue to grow and are eventually absorbed back into the ovary tissue. The fourth, the **megaspore,** enlarges to form an **embryo sac** (ovule). The nucleus within the megaspore gains a thin wall and a slight amount of cytoplasm. It then divides mitotically to form a total of eight cells as the megaspore develops into the embryo sac (Figure 3–1). Three of these cells migrate to the **micropylar end,** the part of the embryo sac that is attached to the remainder of the ovary. This section of the embryo sac has a thin wall, and from this area there is living cellular tissue leading to the main stylar tissue of the flower. One of the three cells near the micropyle becomes the **egg cell.** The other two, which appear to serve no function, are absorbed into the ovular tissue. Three of the five other cells migrate to the other end of the embryo sac, and all three are gradually absorbed. The two remaining cells, called **polar bodies,** stay in the approximate center of the embryo sac.

Chromosomal Behavior at Meiosis

All plants belonging to the same species normally have the same chromosome number, although occasionally a subspecies arises with exactly twice or three times the number typical of the species. Table 3–1 lists chromosome numbers of some representative horticultural plants. The two chromosomes that form each chromosome pair at the beginning of meiosis are essentially the same size, shape, and structure and are called **homologous** chromosomes. They can be identified in all plants of a species. The 10 pairs of chromosomes in corn, for example, have been given the numbers 1 through 10, and scientists, by patiently observing through a microscope the length, thickness, dark and light areas, bulges, and other distinguishing features, can determine which of the 10 chromosomes they are seeing.

Furthermore, the segment of DNA (gene) controlling any given plant characteristic will be at the same location on the same chromosome of all normal plants of the species. As a hypothetical example, if the gene that determines seed color in peas is found to be one-third of the way from the end of the long arm of chromosome 3 in a normal plant, it will be found at that same location in all normal pea plants. Genes at one location will not always produce the same effect however; for example, one pea seed-color gene may cause its plant to produce yellow seed, another may cause its plant to produce purple seed, another, brown, and still another, green; but the gene responsible for seed color will be at the same location in

Table 3–1 Chromosome numbers of representative horticultural plants.[a]

Plant	Chromosome Number	Plant	Chromosome Number
Ornamentals			
Amaryllis (*Amaryllis belladonna*)	22	Jonquil (*Narcissus jonquilla*)	14
Arborvitae (*Thuja occidentalis*)	24, 48	Juniper, creeping (*Juniperus communis*)	22
European birch (*Betula verrucosa*)	28, 42	Kentucky bluegrass (*Poa pratensis*)	62, 100
Calendula (*Calendula officinalis*)	28, 32	Lilac (*Syringa vulgaris*)	46, 47, 48
Carnation (*Dianthus* spp.)	30, 60, 90	*Magnolia* spp.	38, 76, 114
Cherry laurel (*Prunus laurocerasus*)	176	Nasturtium (*Tropaeolum majus*)	64
Chrysanthemum (*Chrysanthemum indicum*)	45, 63	Oak (*Quercus* spp.)	24
Clematis spp.	16, 32, 48	Oregon grape (*Mahonia aquifolium*)	28
Columbine (*Aquilegia vulgaris*)	14, 28	Pansy (*Viola tricolor*)	26
Cyclamen (*Cyclamen persicum*)	48, 96	Peony (*Paeonia* spp.)	10, 20
Dahlia (*Dahlia* spp.)	64	Petunia (*Petunia hybrida*)	21, 28, 35
Delphinium spp.	16, 24, 32, 48	*Philodendron* spp.	30, 32, 34
Dogwood, flowering (*Cornus florida*)	22	*Phlox* spp.	14
Easter lily (*Lilium longiflorum*)	24	Poppy, oriental (*Papaver orientale*)	28, 42
Forsythia spp.	28	*Rhododendron* spp.	26, 52, 78
Geranium (*Pelargonium hortorum*)	18	Rose (*Rosa* spp.)	14, 21, 28
Ginkgo (*Ginkgo biloba*)	24	Norway spruce (*Picea abies*)	24, 48
Gladiolus spp.	30, 45, 60, 75, 90	Sunflower (*Helianthus annuus*)	34
Hemlock (*Tsuga canadensis*)	24	Sweet pea (*Lathyrus odoratus*)	14
Hollyhock (*Althaea rosea*)	41	Tulip, garden (*Tulipa gesneriana*)	24, 36
Hyacinth (*Hyacinthus orientalis*)	16	*Wisteria* spp.	16
Impatiens (*Impatiens* spp.)	14, 16, 18, 20	Yew, Japanese (*Taxus cuspidata*)	24
India rubber tree (*Ficus elastica*)	26	Zinnia (*Zinnia elegans*)	24
Iris, bearded (*Iris* spp. and hybrids)	24, 36, 48, 60		
Fruit			
Avocado (*Persea americana*)	24	Grapefruit (*Citrus paradisi*)	18, 27, 36
Almond (*Prunus amygdalis*)	16	Lemon (*Citrus limon*)	18, 36
Apple (*Malus sylvestris*)	34, 51, 68	Orange, sweet (*Citrus sinensis*)	18, 27, 36, 45
Apricot (*Prunus armeniaca*)	16	Peach (*Prunus persica*)	16
Blackberry (*Rubus* spp.)	28, 35, 42	Pear (*Pyrus communis*)	34, 51
Blueberry, highbush (*Vaccinium corymbosum*)	48	Plum, American (*Prunus americana*)	16
Blueberry, rabbiteye (*Vaccinium ashei*)	72	Plum, European (*Prunus domestica*)	48
Cherry, sour (*Prunus cerasus*)	32	Plum, Japanese (*Prunus salicina*)	16
Cherry, sweet (*Prunus avium*)	16, 24, 32	Quince (*Cydonia oblonga*)	34
Fig (*Ficus carica*)	26	Raspberry (*Rubus idaeus*)	14, 21, 28
Grape, American (*Vitis labrusca*)	38	Strawberry (*Fragaria* x *ananassa*)	56
Grape, European (*Vitis vinifera*)	38, 57, 76	Walnut, English (*Juglans regia*)	32
Grape, muscadine (*Vitis rotundifolia*)	40		

(continued)

Table 3-1 *(continued)*

Plant	Chromosome Number	Plant	Chromosome Number
Vegetables			
Artichoke, Jerusalem (*Helianthus tuberosus*)	102	Muskmelon (*Cucumis melo*)	24, 48
Asparagus (*Asparagus officinalis*)	20	Onion (*Allium cepa*)	16, 32
Bean, snap and dry (*Phaseolus vulgaris*)	22	Pea (*Pisum sativum*)	14
Bean, Lima (*Phaseolus lunatus*)	22	Pepper (*Capsicum annuum*)	24
Broccoli (*Brassica oleracea* var. *italica*)	18	Potato (*Solanum tuberosum*)	48
Cabbage (*Brassica oleracea* var. *capitata*)	18	Rutabaga (*Brassica napobrassica*)	38
Carrot (*Daucus carota*)	18	Squash (*Cucurbita maxima*)	24, 40
Cauliflower (*Brassica oleracea* var. *botrytis*)	18	Squash (*Cucurbita moschata*)	24, 40, 48
Celery (*Apium graveolens*)	22	Squash, summer (*Cucurbita pepo*)	40
Chives (*Allium schoenoprasum*)	16, 24, 32	Sweet potato (*Ipomoea batatas*)	16, 32
Corn (*Zea mays*)	20	Tomato (*Lycopersicon esculentum*)	24
Cucumber (*Cucumis sativus*)	24	Turnip (*Brassica rapa*)	20
Eggplant (*Solanum melongena*)	24	Watercress (*Nasturtium officinale*)	32
Leek (*Allium porrum*)	32	Watermelon (*Citrillus vulgaris*)	22
Lettuce (*Lactuca sativa*)	102		

[a]Counting of chromosomes is difficult, and classification into species is not complete for some genera. Multiple unusual numbers for some species may be the result of an honest counting error, or those species may eventually be reclassified into two or more species.

all pea plants. Genes at the same location on homologous chromosomes are called **alleles**, and the total gene complement, expressed or not, is the **genotype** of the organism.

It should also be emphasized that chromosomes are distributed at random during meiosis. As with the flip of a coin, where chance determines whether heads or tails comes up, chance determines which of each pair of chromosomes goes to a particular pollen grain or egg cell. Assume, for example, that the hypothetical pea plant has a gene for yellow seed on one of its pair of number 3 chromosomes and a gene for green seed on the other. Assume further that on one number 4 chromosome it has a gene for tall vine and on the other a gene for dwarf vine. From this plant will be formed some pollen grains having genes for yellow seed and tall vine, others with genes for green seed and tall vine, others for yellow seed and short vine, and others with green seed and short vine (Figure 3-4). The random distribution of genes to progeny is termed gene **segregation.**

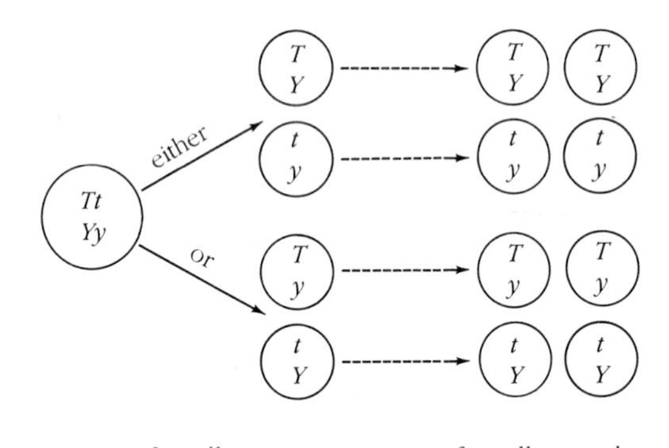

Figure 3-4 The two possible ways genes for tall/dwarf vine and green/yellow seed color in peas can be distributed to the four pollen grains developed from a heterozygous mother cell.

Pollination

Once the pollen grain has matured, it can be transferred to a receptive stigma in several ways. In some species as the style elongates, the receptive stigma is brushed against the ripe anther to pollinate by contact. With certain conifers the pollen drops from male flowers high on the plant to receptive female flowers located farther down on the same or other plants; this is pollination by gravity. Wind currents are instrumental in spreading the pollen of many species. Wind-pollinated species usually produce copious quantities of small light-weight pollen. The whole understory of a pine forest can become coated with a layer of yellow pollen at certain times of the year. A field in which corn is being grown for seed must be at least one-quarter mile (400 meters) from the nearest field of any other corn cultivar to afford reasonable assurance that foreign pollen does contaminate the seed. Insects, especially bees, are probably the most frequent agents of pollination for horticultural crops. Insects are necessary for pollination and production of almost all fruits as well as for melons, squash, and cucumbers. In addition, seed production of a number of other vegetables and flowers is dependent on insects (Figure 3–5). Frequently plant species are pollinated by a combination of agents.

Plant species pollinated by contact are likely to be **self-pollinated;** that is, pollen from a plant usually fertilizes a stigma (normally the stigma of the same flower) of that plant. In nature self-pollination is not as common as cross-pollination; however, many cultivated crops, including peppers, tomatoes, eggplant, peas, beans, and most cereal grains, are self-pollinated.

Crops pollinated by wind, insects, or gravity are likely to be **cross-pollinated;** that is, pollen from one plant normally pollinates the stigma of another plant. It should be mentioned that a small percentage of the plants of many cross-pollinating crops are self-pollinated and that a small percentage of plants of most self-pollinating crops are cross-pollinated. Such variations and their consequences will be discussed later in the chapter.

Aids for Pollination. For growers of crops requiring pollination the season of bloom is a critical time of the year. Blooming usually occurs in the early spring when weather conditions are unstable. The blossoms are especially susceptible to frost. Furthermore, with most early-blooming crops insects, usually active only at warmer temperatures, are required for transfer of pollen. Therefore, if the spring weather is cold, sufficient pollination for good fruit set may not occur.

It is standard practice to place hives of bees in orchards at blossom time (Figure 3–6), but so far there are no good solutions for the fruit tree grower whose orchard

Figure 3–5 Honey bee on apple flower. During the process of collecting nectar and pollen, bees and other insects spread the pollen necessary for seed and fruit production from one flower to another. (Courtesy R. W. Henderson.)

Figure 3–6 Bee hives in an orchard.

environmental conditions are not favorable for bee activity. The tomato grower is more fortunate, because hormone sprays available at garden stores can be used to cause tomato fruits to set when the temperature is warm enough to permit blossom formation but too cold for pollination (day temperature below 60°F [16°C] or night temperature below 45°F [7°C]). Tomatoes caused to set with hormonal sprays will be seedless.

The gardener who attempts to grow fruit-bearing crops such as tomatoes or cucumbers indoors or in a greenhouse must be aware that natural agents of pollination are not present indoors. Even though tomatoes are normally self-pollinating, they require wind movement in order to pollinate effectively. To insure good fruit set indoors, a grower must flip each open flower every day with a fingernail or "buzz" each one with a small electric pollinator designed especially for tomato flower pollination (Figure 3–7). Where large numbers of cucumbers are being grown, one or more hives of bees may be placed in each house to effect pollination. If just a few plants are being grown indoors, flowers can be hand pollinated by brushing the pollen-shedding anther of a male flower against the pistil of each open female flower (Figure 3–8).

Figure 3–7 An aid to pollination. This battery-operated buzzer is used to pollinate greenhouse tomatoes and to collect pollen for potato or tomato breeding programs.

Pollen Tube Growth and Fertilization

As the pollen grain matures, its nucleus divides to form two nuclei. One of these, the tube nucleus, is associated with germination of the pollen grain and growth of the pollen tube. The other, the generative nucleus, is responsible for fertilization of the ovule. When the pollen grain alights on a receptive stigma, it is stimulated to germinate. **Germination** occurs by the **pollen tube** pushing through a pore of the pollen grain and starting growth down into the style (see Figure 3–1). Either just before or just after the pollen tube begins to grow, the generative nucleus divides to form two **sperm.**

The pollen tube grows downward into the style, pushing aside various elongate cells of the style and forcing its way to the micropylar end of the embryo sac. The two sperm follow in the pollen tube behind the tube nucleus and empty from it through the micropyle into the embryo

Figure 3–8 Artificial pollination of cucumber. The corolla has been removed from around the pollen-shedding anther cone of the male flower, and the anther cone is being brushed against the stigma of a female flower. In the foreground the small fruit that is beginning to enlarge was pollinated a few days earlier.

sac. One of these sperm unites with the egg cell and, with this union, the standard (usually diploid) number of chromosomes is again restored. The cell thus formed divides mitotically many times to form the **embryo** of the developing seed. With seed germination, this embryo grows to develop into the new plant. The second male nucleus unites with the two polar nuclei of the embryo sac in what is referred to as triple fusion. The cell from this union will, of course, have three sets of ($3n$) chromosomes. With repeated divisions this cell becomes the **endosperm,** or food portion of the resulting seed (Figure 3–1).

Gene Expression

With fertilization of the egg the new plant regains the gene pairs, one on each homologous chromosome, that are responsible for the expression of each characteristic. If all gene pairs of the parent plant or plants are alike, all pollen grains, all egg cells, and all offspring will be alike. If, however, gene pairs of the parent plants are not alike, random segregation and the interchange of genetic material between chromosomes will insure that no two pollen grains, no two egg cells, and no two offspring are likely to receive chromosome material exactly alike.

If gene pairs are identical, it is easy to predict their expression. If both pea seed-color genes are for green seed color, the seed will be green. However, if the genes of a pair are different, their expression is more complicated. Frequently, the effect of one gene will be expressed and the other supressed. The expressed gene is said to be **dominant** and is usually symbolized by a capital italic letter. The suppressed, or **recessive,** gene is symbolized by a lowercase italic letter. If a pea plant having one gene for yellow seed and one for green seed produces yellow seed, the yellow-seed gene is dominant and the green-seed gene is recessive (Figure 3–9). Genes for many **qualitative** characteristics—color, leaf form, presence of hairs, and so on—are either dominant or recessive. Sometimes when genes are different, both will be partially expressed. In one group of sweet peas, for example, a gene for red color paired with a gene for white color produces pink flowers (Figure 3–10). F_1, widely used in describing hybrid cultivars, is the standard abbreviation of *first filial generation*. It refers to the first generation progeny produced from a cross. F_2, F_3, and so on refer to the second, third, and later generations.

Many plant characteristics, especially those relating to size, shape, yield, and quality, are under the control of large numbers of genes, each of which adds to or modifies the characteristic. Genes that act in this way are called **quantitative** genes or factors. A grossly simplified example of the action of quantitative factors might be illustrated by plant height controlled by three pairs of genes. If three of those genes each add 2 inches (5 centimeters) to the plant, two add 4 inches (10 cm) and one adds 8 inches (20 cm), the plant will be 22 inches (56 cm) high (Figure 3–11).

Expression of genes is also affected by the environment. Russetting, the rough corky skin that is the trademark of the 'Idaho Baking' ('Russet Burbank') potato is reduced if soil nitrogen content is high. The skin color of oranges is much brighter orange when the fruit is grown where the atmosphere is dry and nights are cool. Some high-quality oranges grown in areas or during seasons unfavorable for high skin color are dyed and labeled "color added." Hydrangeas are another dramatic example of the effect of environment and gene expression. Their flowers are blue if the soil in which they are growing is highly acid and pink if it is less acid.

Gene Mutation and Bud Sports. Almost always genes are passed from generation to generation unchanged. The gene for yellow seed color in peas will continue to produce yellow seed for as many generations as it continues to be passed on to progeny. Occasionally, perhaps only once in each hundred thousand or once in each million progeny, a pollen grain or embryo sac will form in which a gene changes. Perhaps a yellow seed gene of a pea changes so that it produces a purple seed. These rare changes are called **mutations,** and the mutated gene will be passed from generation to generation of offspring as faithfully as was the original gene. Many mutations produce plant characteristics already in existence and thus may not be observed. Most of the rest produce changes

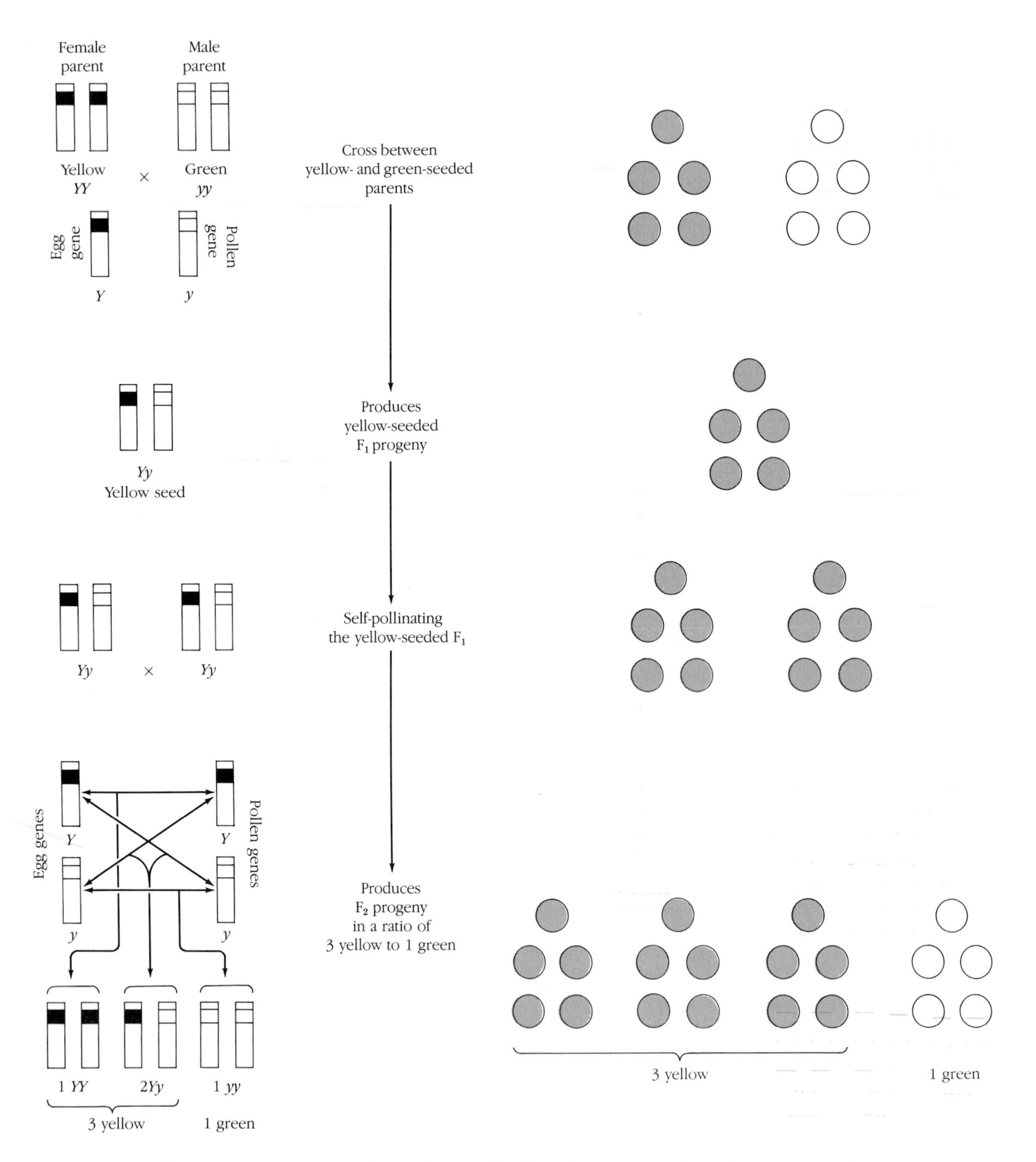

Figure 3-9 Gene segregation and expression through two generations where one gene is completely dominant.

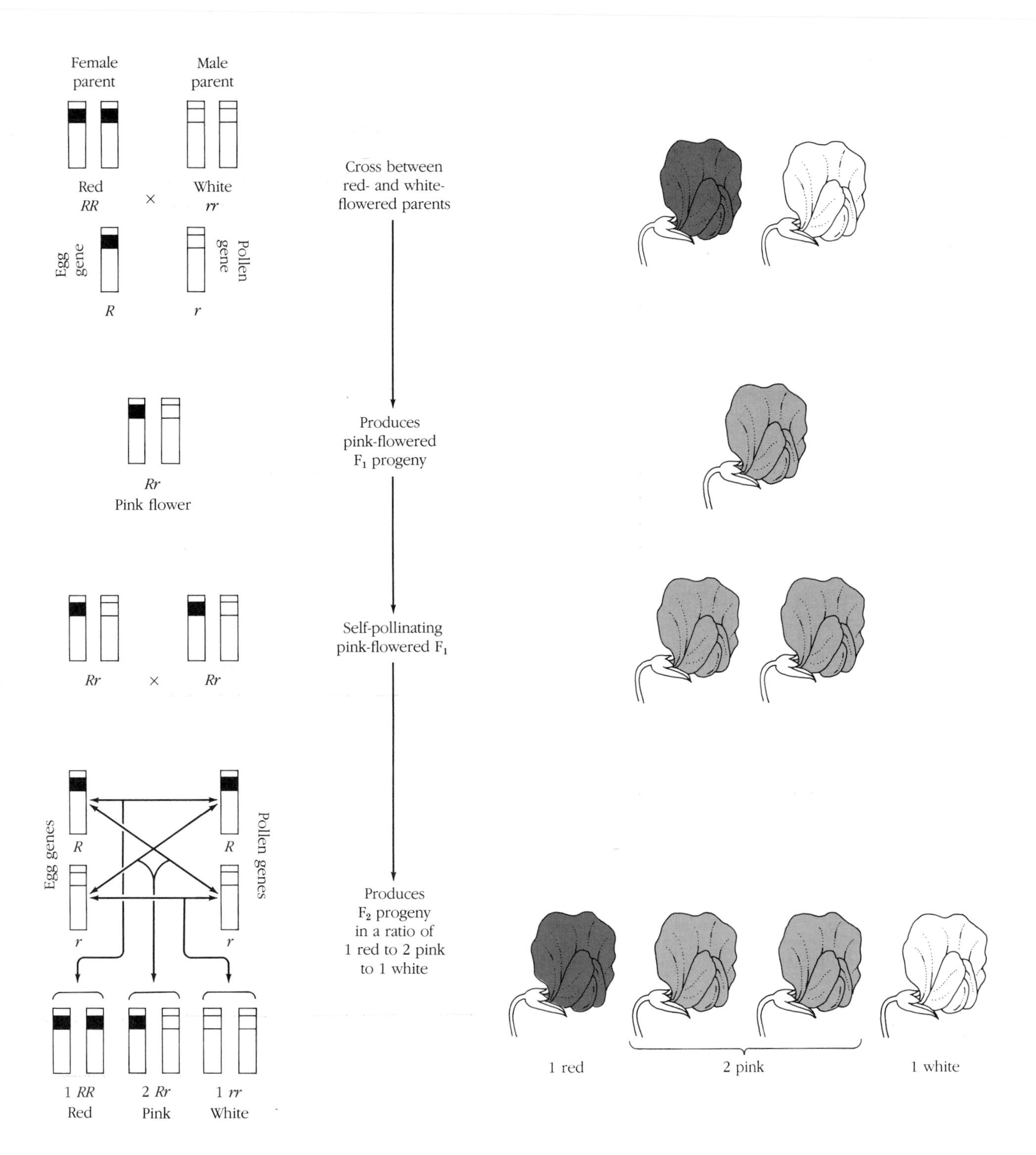

Figure 3–10 Gene segregation and expression through two generations where gene pairs are incompletely dominant.

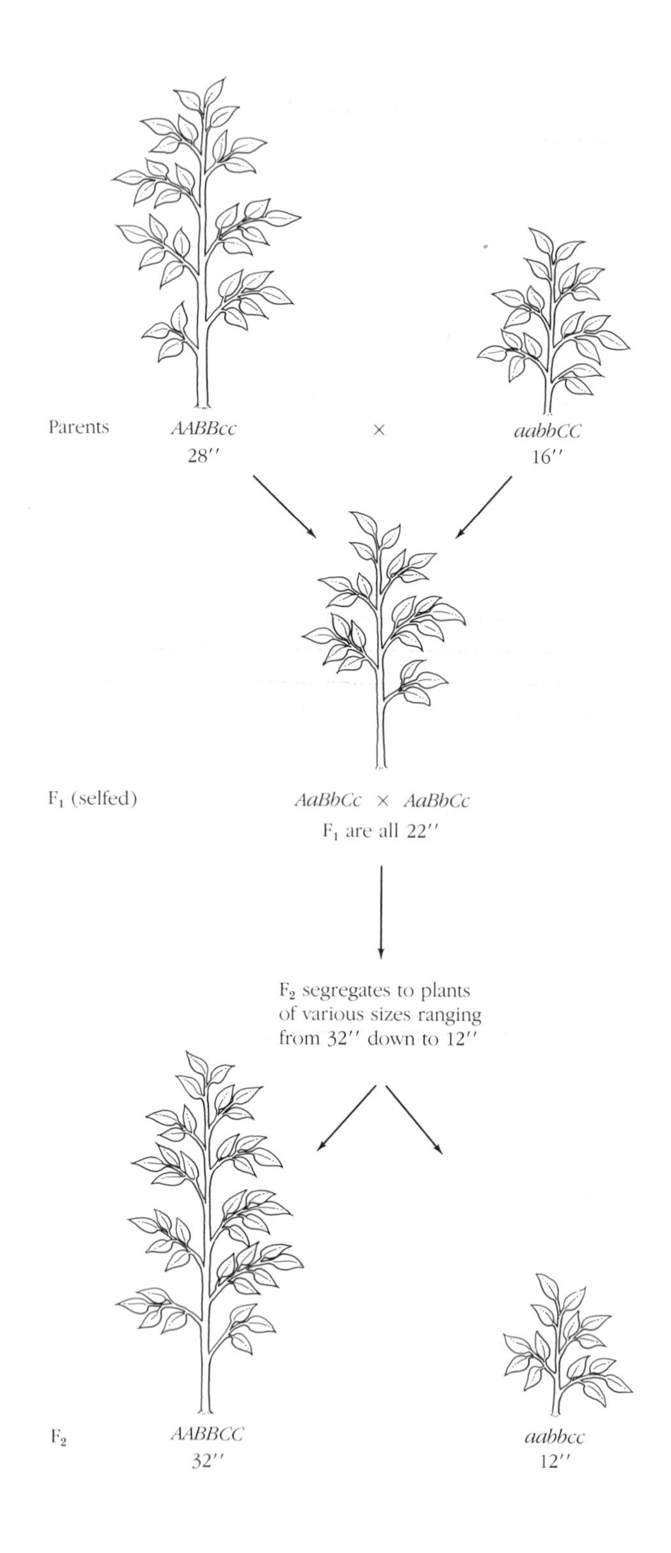

Figure 3–11 Segregation of genes illustrating quantitative inheritance. A theoretical cross is made between two plants in which plant height is controlled by three pairs of genes, each of which contributes to height as follows: $A = 4$ inches; $B = 8$ inches; $C = 4$ inches; and *a, b,* and *c* each contribute 2 inches. Parent plants are 28 inches and 16 inches; the F_1 progeny will be 22 inches, and the F_2 will vary from 32 inches down to 12 inches depending on genotype. Notice that some of the F_2 offspring may be taller or shorter than either parent. Traits such as size, yield, and quality are usually inherited quantitatively.

that cause the plant to be less desirable. Occasionally, though, a change occurs that does increase utility or beauty of a plant. Even though desirable gene changes are rare, mutation is an important tool to plant breeders, who often use chemicals and radiation to speed its rate.

Mutations also occur rarely in the body cells of a plant. These changes are not easily observed unless they are in bud cells from which new plant parts grow. Like mutations of reproductive cells, most body cell mutations cause the development of undesirable characteristics. Even so, a number of important horticultural cultivars have been developed as a result of bud mutation. Cultivars originating from bud mutation are called **bud sports** (Figure 3–12) and include solid red 'Red Delicious' apples from the old striped 'Delicious' and 'Russet Burbank' potato from the old smooth 'Burbank.'

Bud sports can sometimes be profitable for their discoverer. A few years ago a single limb, a bud sport, from a 'Golden Delicious' apple tree sold for over $30,000. Its value was that it had nodes and spurs more closely spaced so that the trees produced from its buds were smaller in size than standard 'Golden Delicious' trees.

New flower colors have often appeared as bud sports. About 70 years ago a double red greenhouse carnation called 'Sim' was introduced. It produced a large, very double flower on a strong stem and possessed other desirable features. Over the years millions of cuttings of this carnation have been propagated. From these occasionally have come bud sports for different colors so that now growers have white, various shades of pink, red and white striped, yellow, and orange carnation cultivars with the

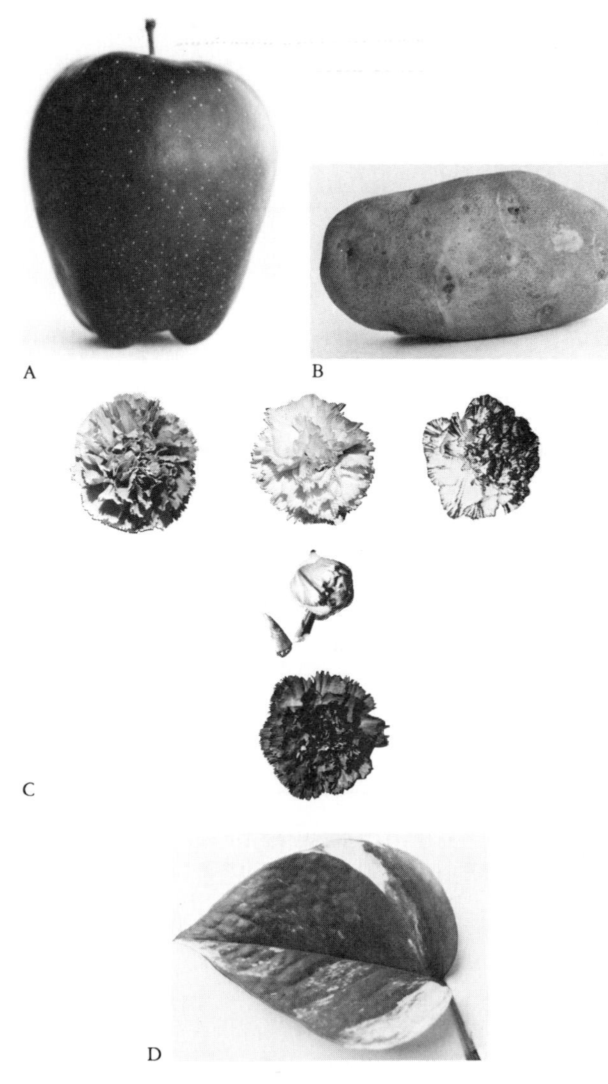

Figure 3–12 Bud sports. The solid 'Red Delicious' apple (A) and the 'Russet Burbank' (Idaho) potato (B) are clonally propagated cultivars of bud mutations of the old striped 'Red Delicious' and the smooth 'Burbank.' Various pink, white, and striped carnation cultivars (C, top) have descended as bud sports from 'Red Sim' (C, bottom). One of the more recently discovered 'Sim' type carnations is 'Tangerine Sim' (center bud). The white markings on the leaves of variegated plants like pothos (D) are usually the result of a somatic mutation in which some of the cells lose their ability to manufacture chlorophyll.

desirable features of the original 'Sim' carnation. 'Tangerine Sim' was found after patenting new plant types became legal, and its discoverer is reported to have received more than a million dollars for propagation rights.

Bud mutation is also often responsible for green and white or green and yellow leaf variegation in certain foliage plants and ornamental shrubs. These come about when a section of a leaf mutates so that it can no longer produce **chlorophyll.** New plants must be propagated from cuttings or divisions containing both green and white or green and yellow portions. If the cutting is all green, a solid green plant will be produced. If it is all white or yellow, the new plant, lacking chlorophyll, cannot manufacture food and will die.

Genetic Consequences of Self- and Cross-Pollination. It is quite possible to be an expert gardener without having a fundamental understanding of genetic segregation. However, several very interesting phenomena likely to be observed by the home gardener and several very practical gardening decisions relating to the choosing of cultivars and to the purchasing or growing of seed can be meaningful only to those having some knowledge of gene and chromosome distribution with self- and cross-pollination. Genetic segregation is based on the laws of chance and can be explained by using an analogy of playing cards.

Assume that the thirteen different kinds of cards (ace, king, queen, etc.) are thirteen different genes scattered along the chromosomes of a species and that the four suits are four different expressions of those genes—for example, the yellow, green, brown, or purple seeds of peas mentioned earlier. Any one diploid plant can have only two of those genes, but all four will be found among plants of the group. Let's assume that the cards are sorted so that kinds are together and laid face down on the table. You are asked to choose two of each kind, that is, two aces, two kings, two queens, and so on. The cards left on the table are discarded. Next you are given other cards that exactly match in suit and kind the cards in your hand. For example, if you had an ace of hearts, an ace of diamonds, a king of diamonds, and a king of clubs, these then are increased to two aces of hearts, two aces of diamonds, two kings of diamonds, and two kings of clubs. Next the cards in your hand are laid face down on the table, and you pick from them at random two of each kind, discarding the rest. Now the new cards in your hand are duplicated as previously, and the cycle of alternate discarding and duplicating is repeated. It is obvious that this procedure carried through several cycles will result in both cards of any one kind that you choose being of the same suit. This is almost exactly analogous to the loss of genetic variability that occurs with self-pollination (Figure 3–13).

Just as the pairs of cards become alike after a few cycles, so the pairs of genes on homologous chromosomes will be alike after 10 to 14 generations of self-pollinating. The site on a chromosome at which a gene is

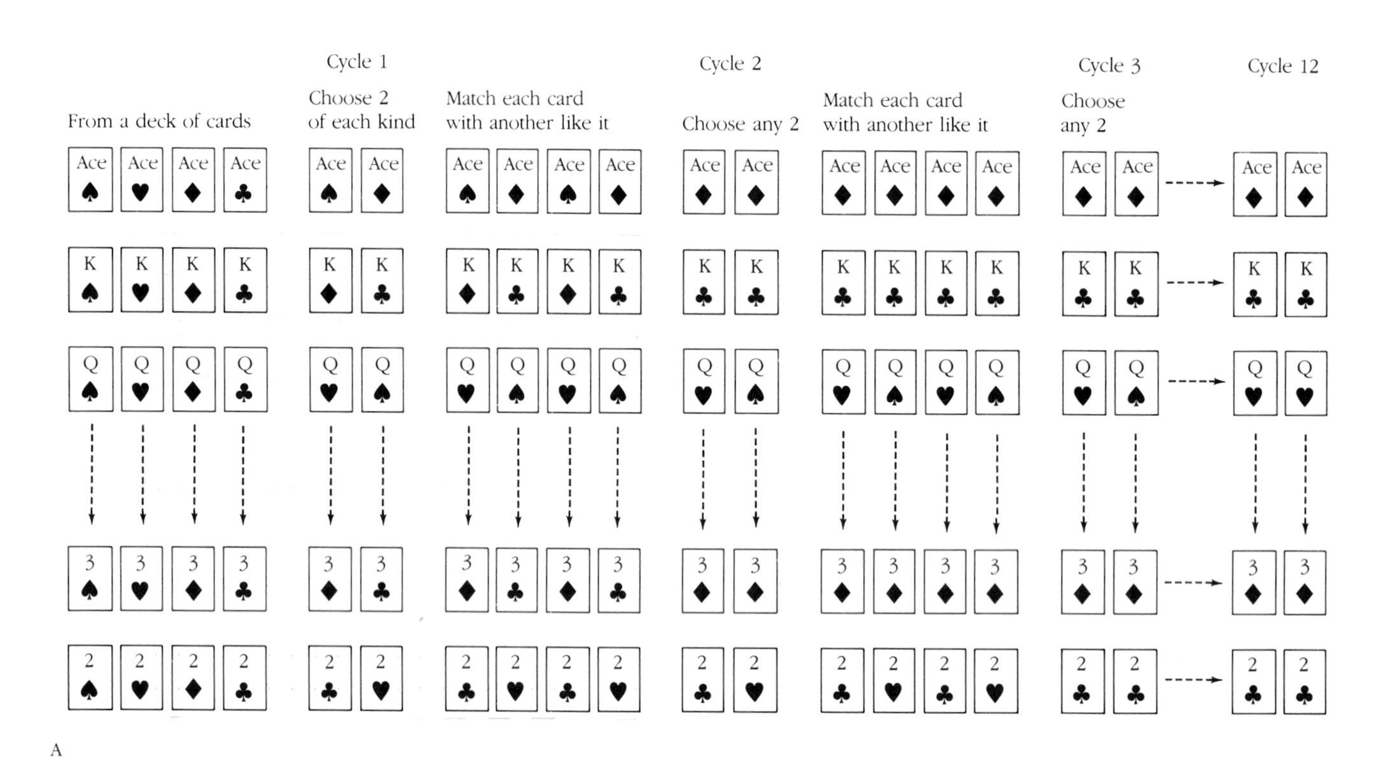

Figure 3–13 Loss of genetic variability with self-pollination.
(A) The loss of genetic variability resulting from self-pollination can be illustrated with a deck of cards. The doubling of number of cards by matching suit and kind after each cycle is analogous to the doubling of chromosomes during meiosis, and the random selection of two cards at each cycle is analogous to the union of sperm and egg (each having a single set of chromosomes) to form the embryo. Approximately half of unlike pairs of cards become alike with each cycle just as about half of the heterozygous gene pairs become homozygous after the passage of each generation.
(B) A hypothetical cross between two genetically unlike pea plants followed by self-pollination. This is the kind of program carried out by breeders attempting to improve a self-pollinated crop. It will result in essentially all gene pairs becoming alike in 10 to 14 generations. Although the two genes of a pair in the parent plants used in such crosses are usually alike (gene pairs I–VII), genetic diversity of F_1 and later generations will not be affected if the genes of a pair are unlike (gene pair VIII and the analogous card example in A).

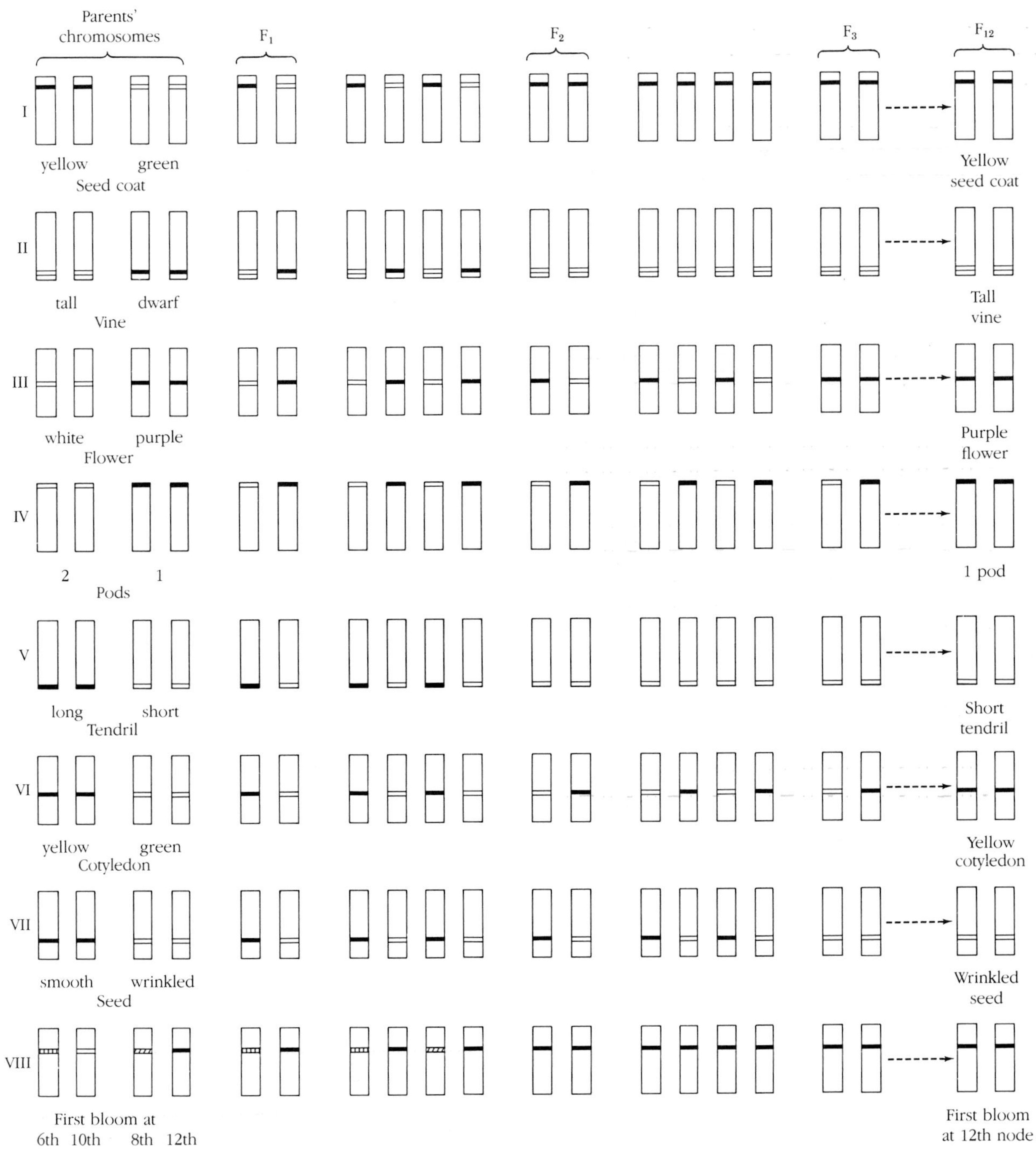

Parents' chromosomes
F1
F2
F3
F12
I
yellow
green
Seed coat
Yellow seed coat
II
tall
dwarf
Vine
Tall vine
III
white
purple
Flower
Purple flower
IV
2
1
Pods
1 pod
V
long
short
Tendril
Short tendril
VI
yellow
green
Cotyledon
Yellow cotyledon
VII
smooth
wrinkled
Seed
Wrinkled seed
VIII
First bloom at
6th 10th
8th 12th
node
First bloom at 12th node
B

located is called its **locus**. When both genes are alike at most loci on all chromosome pairs, the plant is said to be **homozygous.**

For a card analogy of cross-pollination, assume that a large number of players and a number of decks of cards are involved. Each player picks at random two cards of each kind, and these are exactly matched with additional cards of the same suit and kind as in the previous example. This time, however, for the second cycle a player chooses only one card of each kind from his or her own hand and one from the hand of any other player. As the cycle is repeated, cards of all suits will be in the hands of some players and will be distributed to all players from time to time. Thus the two cards of any one kind in a player's hand are likely to be of different suits much of the time (Figure 3–14).

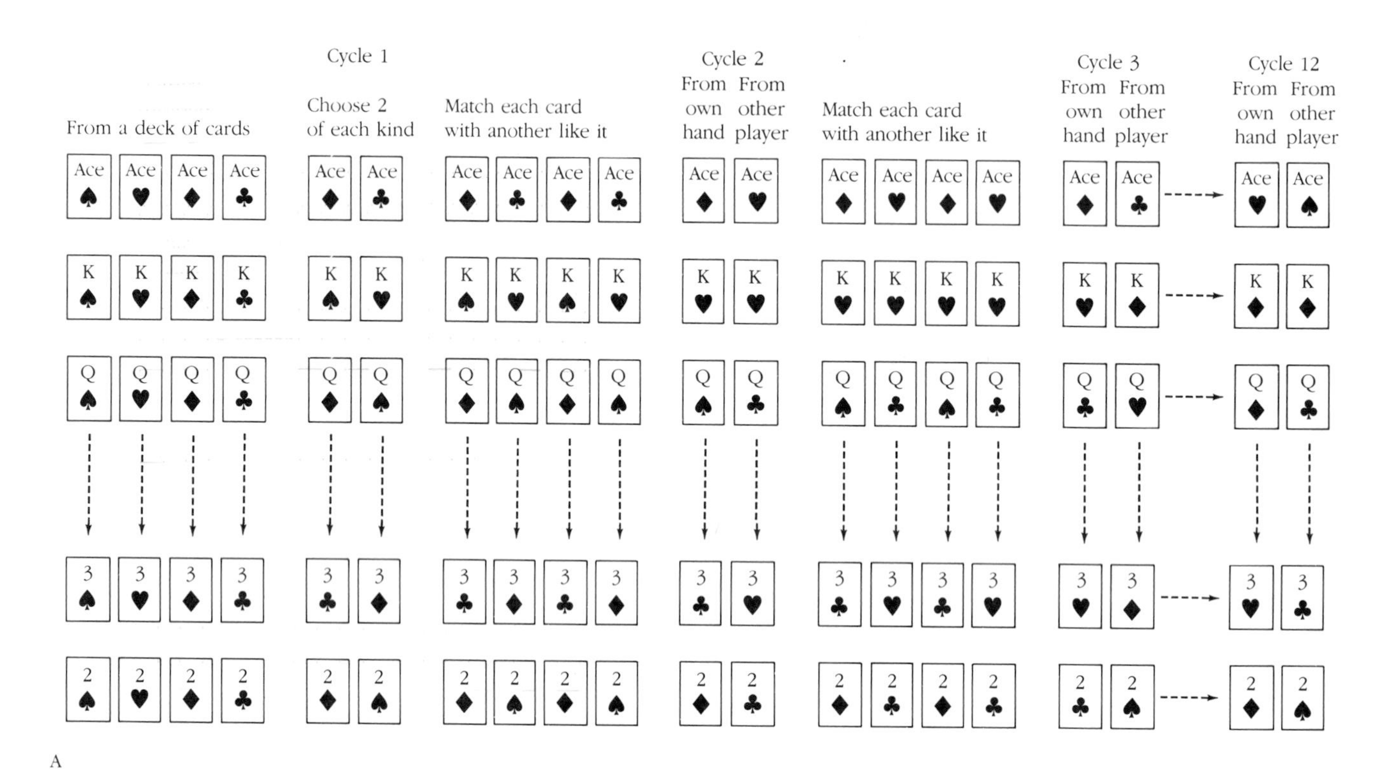

Figure 3–14 Retention of genetic variability with cross-pollination.
(A) The retention of genetic variability with cross-pollination can be demonstrated with a deck of cards. The doubling of the number of cards by matching of suit and kind is analogous to the doubling of chromosomes during meiosis, and the matching of one randomly selected card from the four with one of the same kind from another player's hand is analogous to the kind of fertilization that results from cross-pollination. After 12 cycles the card pairs are as likely to be as different as they were after the first cycle, just as the gene pairs of a cross-pollinated crop plant are as different after 12 generations as they were during the first.
(B) When genes in a plant are unlike, the dominant one (red silk, purple pericarp, red husk, two ears per corn plant) determines the characteristic. In a breeding program involving a cross-pollinated species, selection of similar appearing plants is necessary in order to have a degree of cultivar uniformity after a few generations have passed.

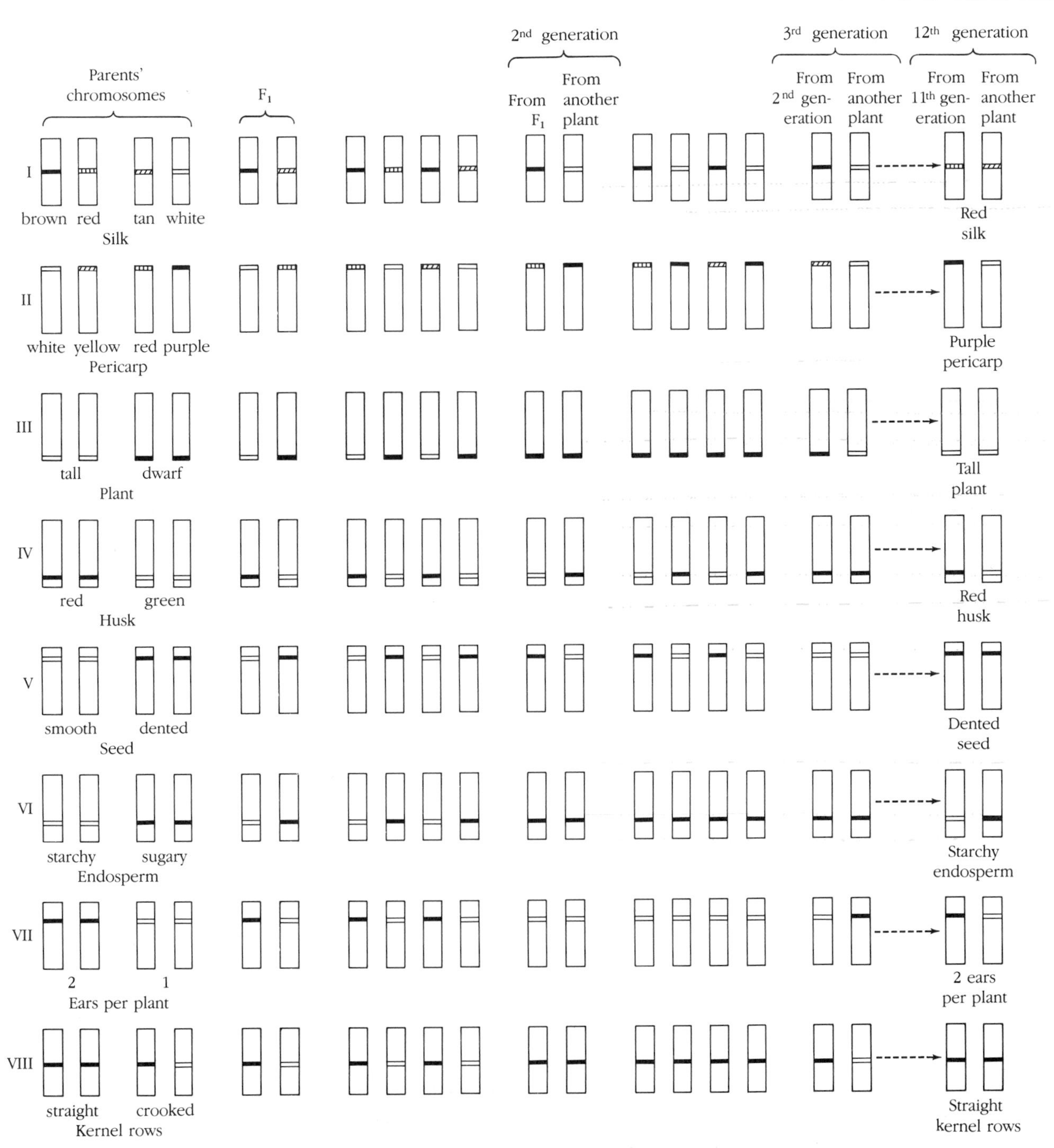
2nd generation
3rd generation
12th generation
Parents' chromosomes
F1
From F1
From another plant
From 2nd generation
From another plant
From 11th generation
From another plant
I
brown red tan white
Silk
Red silk
II
white yellow red purple
Pericarp
Purple pericarp
III
tall dwarf
Plant
Tall plant
IV
red green
Husk
Red husk
V
smooth dented
Seed
Dented seed
VI
starchy sugary
Endosperm
Starchy endosperm
VII
2 1
Ears per plant
2 ears per plant
VIII
straight crooked
Kernel rows
Straight kernel rows

B

This is analogous to what happens with cross-pollination where one set of chromosomes comes from the female parent and one comes from the male parent. Just as the two cards of any one kind are likely to be of different suits, so the two genes at many loci are likely to be different. Plants with most gene pairs not alike are said to be **heterozygous.** Just as cards of all suits will remain in the hands of some player and be distributed to all players from time to time, so the different kinds of genes will remain in the plant population and be recombined into different combinations with each passing generation.

Seed growers find it much easier to develop uniform cultivars of self-pollinating crops than of cross-pollinating ones. By saving seeds for a few generations from single plants having the desired characteristics, they can produce a homozygous uniform cultivar of a self-pollinating crop.

Many cross-pollinating crops can be artificially self-pollinated, of course. However, self-pollinating a species that is normally cross-pollinating leads to death and sterility of some offspring and a drastic reduction in size, vigor, and yield of the surviving plants. Cultivars of cross-pollinated crops (except for F_1 hybrid cultivars discussed later) are developed by selecting plants that appear similar but have considerable genetic diversity. Thus plants of a cultivar of a cross-pollinating crop like carrots are likely to be less uniform than those of a cultivar of a self-pollinating crop like snap beans.

Plant Characteristics Assuring Cross-Pollination

Under natural conditions plant species that cross-pollinate have an evolutionary advantage because of the genetic variation within the group of plants. If the climate changes or if seeds happen to be carried to a different location, variation in a cross-pollinated species is likely to be sufficient so that some individuals will be adapted to the new environment and survive.

In nature there are a number of plant species characteristics that insure the occurrence of cross-pollination. Monoecious species, which have the male and female parts in separate flowers on the same plant, and dioecious species, which have the male and female flowers on different plants, are two such plant adaptations already mentioned. With some crop plants—for instance, the older cultivars of avocado—the stigma of a plant is never receptive at the same time that the pollen of that plant is being shed. If the avocado tree sheds pollen in the morning, its stigmas are receptive only in the afternoon, and vice versa.

Another evolutionary adaptation to insure cross-pollination is the development of self-incompatibility. The ovule of a self-incompatible plant is not fertilized by the pollen from that plant even though the pollen is viable, the stigma is receptive, and the ovule can be readily fertilized with pollen from another plant. Self-incompatibility can be either structural or physiological. With structural incompatibility the stigma and anthers of a plant are located in such a way that self-pollination does not occur. For example, in some tomatoes the style elongates far beyond the anthers so that pollen from a flower does not reach the stigma of the same flower.

Physiological self-incompatibility is more common and causes more problems for horticulturists than structural incompatibility. In such cases self-pollination frequently occurs, but self-fertilization does not. Pollen may fail to germinate on the stigma, or the pollen tube may grow only part way through the style (Figure 3–15).

Self-Incompatibility in Tree Fruits. If fertilization of the ovary does not occur because of self-incompatibility or for any other reason, seeds do not form. Fertilization and usually seed development are essential for development and normal growth of fruit. Thus by preventing fertilization, self-incompatibility will also prevent fruit formation, a fact of great importance to many gardeners.

Recall that, when cells in the body of a plant divide by mitosis, the chromosomes that control their structure and function enlarge and split, providing each new cell with exactly the same genetic material as was possessed by the original mother cell. Thus all plants propagated from the stems, roots, or leaves of a single original plant (the clones mentioned in Chapter 1) will have exactly the same genetic makeup as the original plant from which they were propagated. If the original plant produces red flowers, all the plants of the clone will produce red flowers; if it is self-incompatible, all plants propagated from it will be self-incompatible. All temperate zone tree fruit culti-

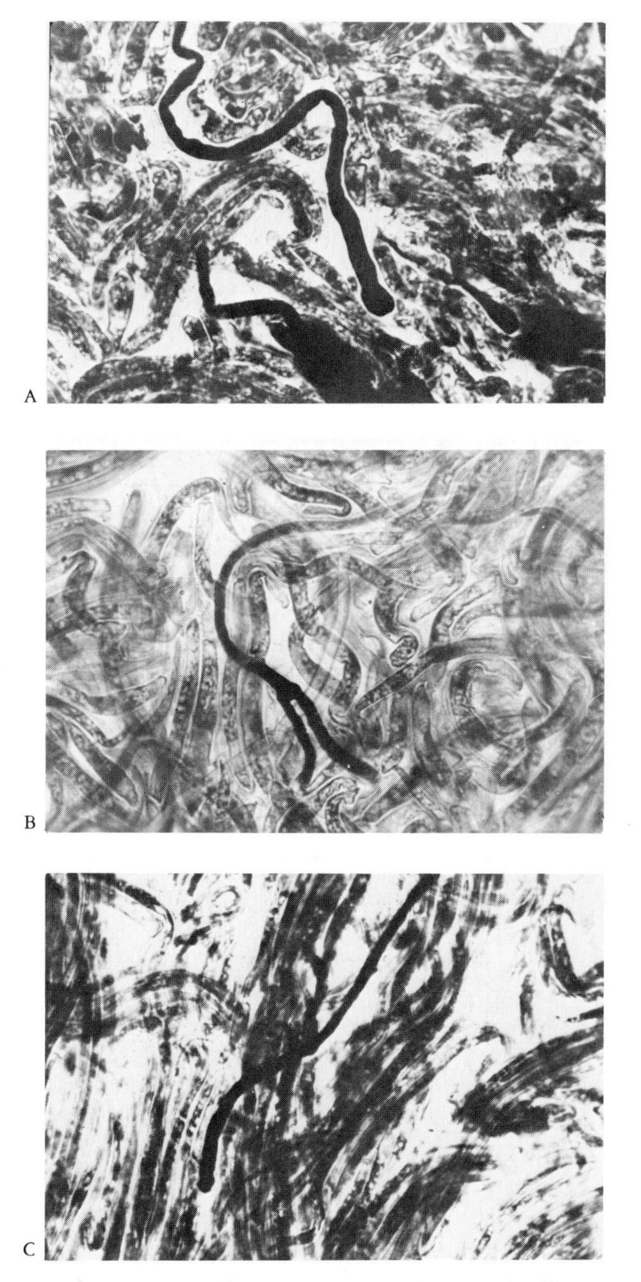

Figure 3–15 Self-incompatibility. The tips of pollen tubes growing through styles in which they are incompatible will frequently become swollen and burst (A). Occasionally they fork (B). A normal tube is shown in C.

vars are clones. If the original tree from which each cultivar developed was self-incompatible, no tree of that cultivar will pollinate any other tree of the same cultivar. Unless they receive pollen from some other source, they will not set fruit.

Self-incompatibility is probably more of a problem with sweet cherries in the western United States and Canada than with any other kind of tree fruit. Not only are sweet cherries self-incompatible, but the three major cultivars grown in the West—'Bing,' 'Lambert,' and 'Royal Ann'—are sister lines that are also cross-incompatible; that is, pollen from any one will not fertilize or produce fruit on the other two. The self- and cross-incompatibility of these three major cultivars of cherries meant, until a few years ago, that Western orchardists had to plant pollinator trees of another cultivar that produced an unsalable crop. Pollinator trees still have to be planted, of course; however, plant breeders have now developed pollinator cultivars that produce marketable fruit even though the quality is not as high as that of the three standard cultivars.

Fruit trees in home gardens frequently fail to set fruit because no pollinator tree has been planted. Since gardeners are usually interested in growing only enough fruit of any one kind for table use, with perhaps a small amount for preserving, a single tree will produce more than enough to satisfy the needs of a family. Unless they are cognizant of pollinator requirements or unless a neighbor has fortuitously planted a pollinator tree, home gardeners are likely to grow a lone tree that blooms profusely but never produces fruit. The pollination requirement can be met each season by placing blooming bouquets of a pollinator cultivar in buckets of water near the nonbearing tree. A more permanent solution where space for planting new trees is lacking is to graft pollinator branches into the fruitless tree. Or two trees, the desired cultivar and a pollinator, can be planted in the same hole. These can be trained to occupy the space a single tree would normally occupy. Figure 3–16 illustrates the various methods of providing pollination.

Cultivars of apple, pear, sweet cherry, and most kinds of nuts are self-incompatible. Except for the three sweet cherries mentioned, any two cultivars will usually cross-pollinate. The 'J. H. Hale' cultivar of peach requires a pollinator. The common apricot and plum cultivars likely to be grown by home gardeners, sour cherry, peach other than, 'J. H. Hale,' and most small fruit cultivars are self-compatible and do not require pollinators. Gardeners should check pollinating requirements before planting rare or unusual plum or apricot cultivars, because a few cultivars of these two fruits are self-incompatible.

Members of the cabbage family, tobacco, petunia, and some other flowers are also self-incompatible. This seldom poses a problem for growers, however, because pollination is not necessary for the formation of a head of cabbage, broccoli, or cauliflower, and failure to set seed is usually an advantage in annual flowers. Even the seed grower is not greatly concerned with self-incompatibility in these crops. As was mentioned, each plant of a cross-pollinated seed-reproduced cultivar has a slightly different genetic makeup (unlike the clonal plants of fruit tree cultivars, which are genetically identical), thus enabling most to pollinate each other.

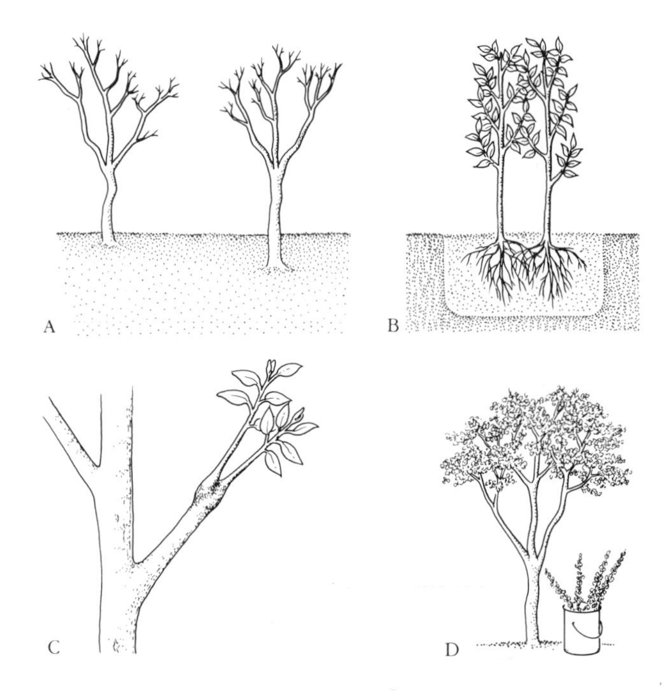

Figure 3–16 Different ways of providing pollination for self-incompatible trees. (A) Planting two separate trees that will cross-pollinate. (B) Planting two trees that will cross-pollinate in the same hole. (C) Grafting pollinator branches into an existing tree. (D) Placing pollinator branches in a bucket of water at the base of the tree during bloom period.

Formation of the Seed

In angiosperms both endosperm and embryo cells grow and divide a number of times. In most monocots, such as the grasses, sweet corn, daffodils, and tulips, these divisions occur until the seed is almost mature. Mature monocot seed will consist of a seed coat formed from the wall of the embryo sac, an embryo formed as a result of the union of a sperm with the egg cell, and an endosperm formed as a result of the union of a sperm with two polar nuclei. The embryo that will develop into the new plant is differentiated into a rudimentary root or **radicle,** shoot or **plumule,** and seed leaf or **coleoptile.** The endosperm, which has three sets of chromosomes, is utilized as food by the embryo as the plant is becoming established (Figure 3–17).

The sequence of events in the formation of dicot seed is somewhat different. Sometime before the dicot seed is mature, the cells of the endosperm cease dividing. Their cell walls break down, and their contents are gradually absorbed by the expanding embryo. The mature dicot seed has a cell wall that is formed from the embryo sac

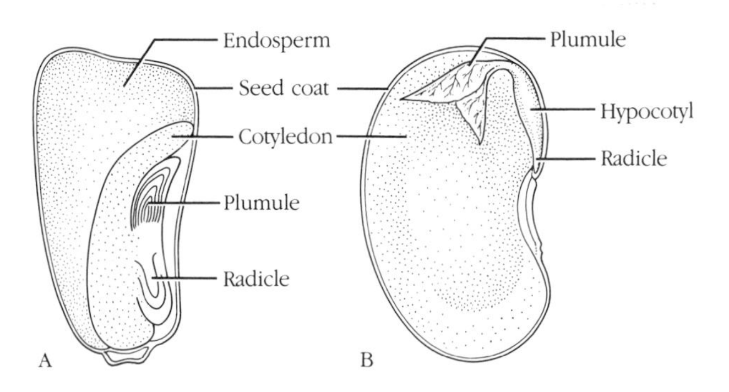

Figure 3–17 Comparison of monocot (corn) seed (A) and dicot (bean) seed (B).

wall and the remnants of the endosperm tissue. The remainder of the dicot seed is embryo and consists of two cotyledons or seed leaves that become the food-supplying part of the developing seedling, a plumule or shoot, a **hypocotyl** or lower stem, and a radicle that becomes the root (see Figure 3–17).

The seed coat of both monocot and dicot seeds consists entirely of mother plant tissue and will have characteristics of the mother plant, but the embryo and endosperm result from the union of male and female gametes and will have characteristics of both the pollen and the mother plant.

F_1 Hybrid Seed

Most growers have willingly paid high prices for hybrid seed since the general acceptance of hybrid corn in the 1930s, often without understanding the basis for hybrid superiority. The term **hybrid** is defined as the progeny of unlike parents, but as generally used in the seed trade, hybrids are those cultivars that come from F_1 hybrid seed obtained by crossing two genetically diverse inbred lines (described below). Seed growers have taken advantage of the reputation of hybrids by introducing and advertising hybrid cultivars of various crop and ornamental plants. Usually hybrids are superior, but sometimes they are no better than the standard or so-called **open-pollinated** cultivars of the species or kind. In a few instances very inferior lots of seed have been sold as F_1 hybrids.

During the past few years some organic and alternative agriculturists have lumped hybrids with hard tomatoes, pesticides, chemical fertilizers, and corporate farming as examples of unnatural and therefore undesirable results of agricultural research. A brief discussion of the scientific bases and techniques for producing hybrid seed may clear some of the confusion.

It has already been stated that self-pollinating and individual plant selecting for a few generations brings about homozygosity, a condition where gene pairs of chromosomes on all plants of a line or cultivar are alike. It has also been pointed out that self-pollinating a species that is naturally cross-pollinated results in death and sterility of some offspring and reduced growth and yield of the remaining progeny. If self-pollination is continued for about 10 generations, the lethality and sterility are eliminated, and a group of plants is developed that remain uniform generation after generation but are much less vigorous and are lower yielding than the original cultivar from which they were developed. Such a group is called an **inbred line** or sometimes just an inbred.

An F_1 hybrid cultivar results from crossing two inbred lines, using one for the male parent and the other for the female parent. As it is homozygous, the male parent will contribute the same set of chromosomes to each progeny plant. The female parent will do the same, although the set it contributes will be quite different from the set contributed by the male parent. Because each F_1 hybrid plant receives exactly the same genetic complement, collectively they will comprise an extremely uniform cultivar (Figure 3–18).

When the right inbreds are crossed, a tremendous stimulation in growth occurs, causing the F_1 hybrid to yield considerably more and mature earlier than the cultivars from which the inbreds were developed. This stimulation in growth is thought to result from the interaction of unlike gene pairs, because the most vigorous F_1 hybrids usually are produced by crossing the most different inbreds. The major advantages of F_1 hybrid cultivars, therefore, are higher yield, earlier maturity, and uniformity.

It is important to emphasize that good F_1 hybrid cultivars cannot be developed from a cross between any two inbreds. Hybrids from many inbred combinations will be no better, or may even be less desirable, than existing standard cultivars. It takes years of patient research on the part of a plant breeder to determine which two inbreds will produce the superior F_1 hybrid. Because gene pairs on the two sets of chromosomes entering the hybrid are so different, their recombination as a result of either selfing or crossing will produce progeny that are extremely variable in maturity, quality, and yield. For this reason seed from F_1 hybrid cultivars should never be saved for replanting.

F_1 hybrid cultivars of self-pollinated species can be easily produced by crossing any two cultivars, because

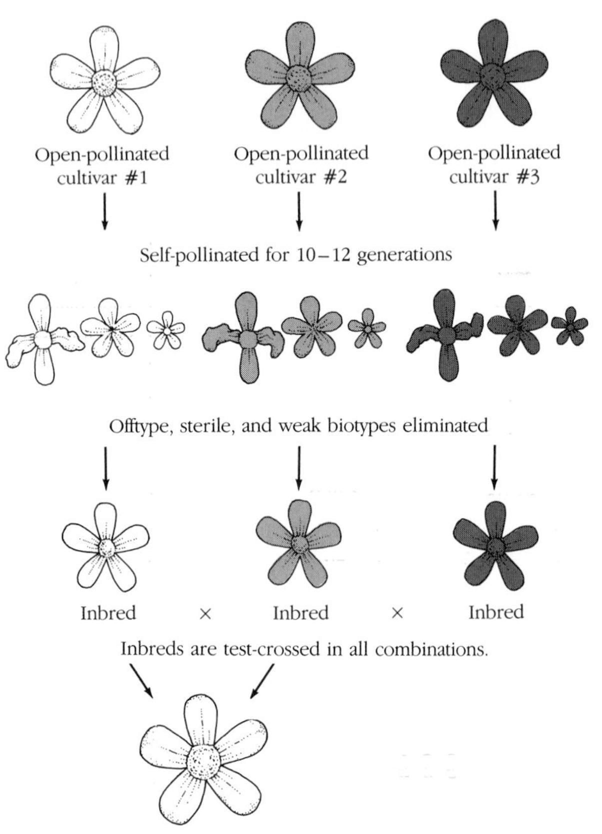

Figure 3–18 Development of an F_1 hybrid cultivar.

self-pollinated cultivars are in reality inbreds. However, F_1 hybrids of self-pollinated crops are not as likely to have a vigor, yield, or earliness advantage over standard cultivars, and standard cultivars of self-pollinated crops are already uniform. As a consequence F_1 hybrids of self-pollinated crops are often not enough superior to standard cultivars to warrant the extra expense of obtaining their seed.

F_2 hybrid seed, which is occasionally advertised, should never be purchased because of the extreme plant variability it will produce.

Each cultivar, hybrid or otherwise, must be judged by its performance in the garden. Despite the adverse publicity given them by some gardening publications, many F_1 hybrid cultivars of both annual flowers and vegetables are unsurpassed in beauty, yield, and quality and certainly deserve consideration by gardeners.

Tetraploid Cultivars

So far in this chapter we have discussed only plants that have two sets of chromosomes, one acquired from the male parent and one from the female parent. As was mentioned in Chapter 2, plants having more than two sets do exist. These include some woody ornamentals, some tree fruit cultivars, potatoes, and certain perennial and annual flowers.

Although there are plants with eight or more sets of chromosomes, the home gardener is not likely to encounter many with more than four (tetraploids). Some species produce tetraploids spontaneously, and in the past tetraploid cultivars were selected from plants or parts of plants growing in field or garden. However, naturally occurring tetraploids are rare and difficult to discover and propagate. Today, with the use of a chemical, **colchicine**, extracted from the autumn crocus (*Colchicum autumnale*), it is not too difficult to produce tetraploid plants of most species. Colchicine prevents cell-wall formation but does not inhibit cell division, so chromosome numbers can be doubled with just enough colchicine to inhibit cell-wall formation during the time necessary for one cell division. In actual practice chromosome doubling involves considerable trial and error to determine the right concentration and timing for a particular species.

Tetraploid cultivars of a number of annual flowers, the most notable being snapdragons, are available. Tetraploid flowers are larger than those of the diploid cultivars from which they are developed (Figure 3–19). They are also likely to produce a smaller percentage of viable pollen and fewer seeds, which is one reason tetraploid seed is expensive. Because they are so large and spectacular, the number of tetraploid flower cultivars available on the market will undoubtedly increase.

Figure 3–19 Comparison of a diploid (left) and tetraploid (right) snapdragon. (Courtesy W. Atlee Burpee Company.)

Seedless Watermelon

In the past seedless watermelons were sometimes mentioned as a joke, because everyone knew that watermelon could not be reproduced without seed. Like so many of the impossibilities of the past, seedless watermelon is now an actuality, and seed for seedless watermelon is available to gardeners. Its production is based on the fact that plants with odd sets of chromosomes do not produce seed. At meiosis chromosomes of cells with even-numbered sets are distributed equally—one set to each pollen grain or embryo sac for diploids and two to each for tetraploids. Equal distribution of chromosomes of cells with three sets is, of course, impossible, and the imbalance of chromatin material results in infertile pollen or inviable embryo sacs or both.

For seed production of seedless watermelon cultivars chromosomes of the cultivar to be used for the female parent are doubled with colchicine to produce a tetraploid (Figure 3–20). This tetraploid line, which has two chromosome sets in each egg cell, is pollinated with

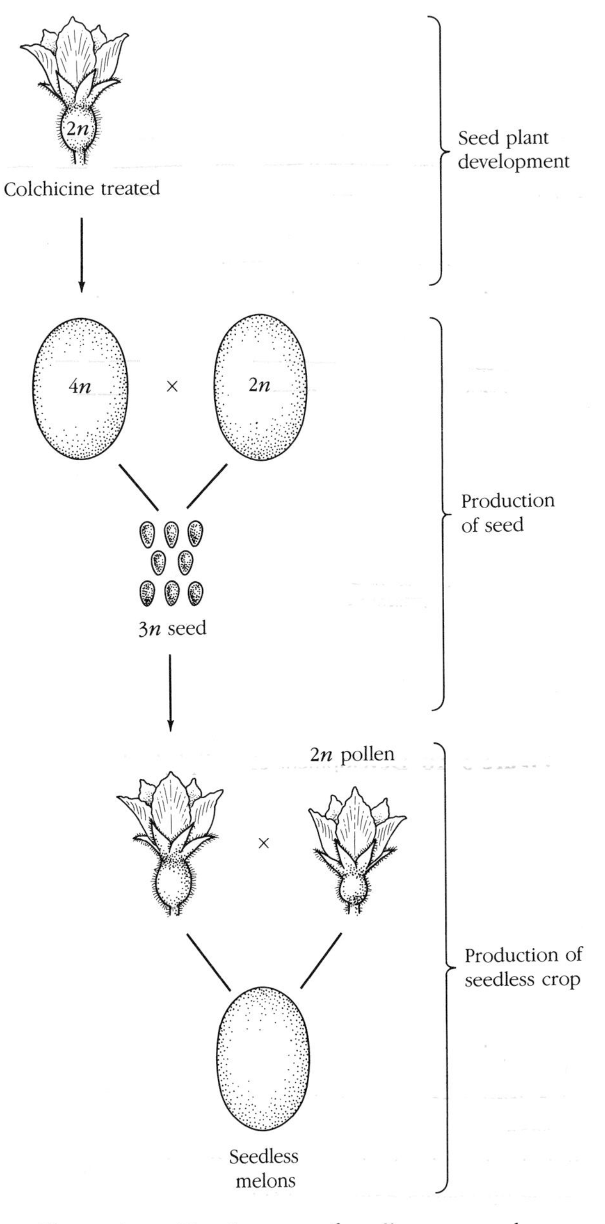

Figure 3–20 Development of seedless watermelon.

pollen from a diploid cultivar, producing embryos with three sets of chromosomes. This triploid seed will produce plants that flower but have no pollen. A regular diploid cultivar is planted with the triploid to produce pollen. Pollination by the diploid stimulates fruit production, but the imbalance of chromosomes in the ovules causes the seed to abort. For production of a seedless crop, a few marked plants of a diploid cultivar are planted to provide pollen. Seedless watermelon seed production requires considerable hand labor and is, therefore, very expensive.

The Fruit

In Chapter 1 it was mentioned that the horticultural definition of fruit was not the same as the botanical definition. Botanically, the fruit is the mature ovary with attached parts. It may include the receptacle, remnants of petals and sepals, pistil, and anthers along with the seeds contained in the ovary.

Fruits can be classified botanically in several ways. One of the most common is whether they are fleshy or dry. Quite a few of the structures normally considered to be seeds are actually dry fruits because they include part or all of the ovary as well as the ovule. For example, the so-called seed of beet or chard is actually a cluster of seeds embedded in dried ovarian tissue. This explains why these crops always emerge as a cluster of plants and why they need to be thinned, regardless of "seed" spacing at planting time. The structure we call carrot seed is also a dry fruit. Many dried seeds, such as peas and beans, are removed or remove themselves from their fruit (in this case, the pod) at maturity. Other types of dry fruits include samaras, which have wings (maple, for instance), nuts, which are one-seeded fruits with a stony wall, and a number of others (Figure 3–21).

Fruits can also be classified according to the number of ovaries incorporated into the fruiting structure as simple, aggregate, or multiple. The **multiple** fruit develops from many separate but closely clustered flowers. Pineapples, figs, and beet seeds are examples of multiple

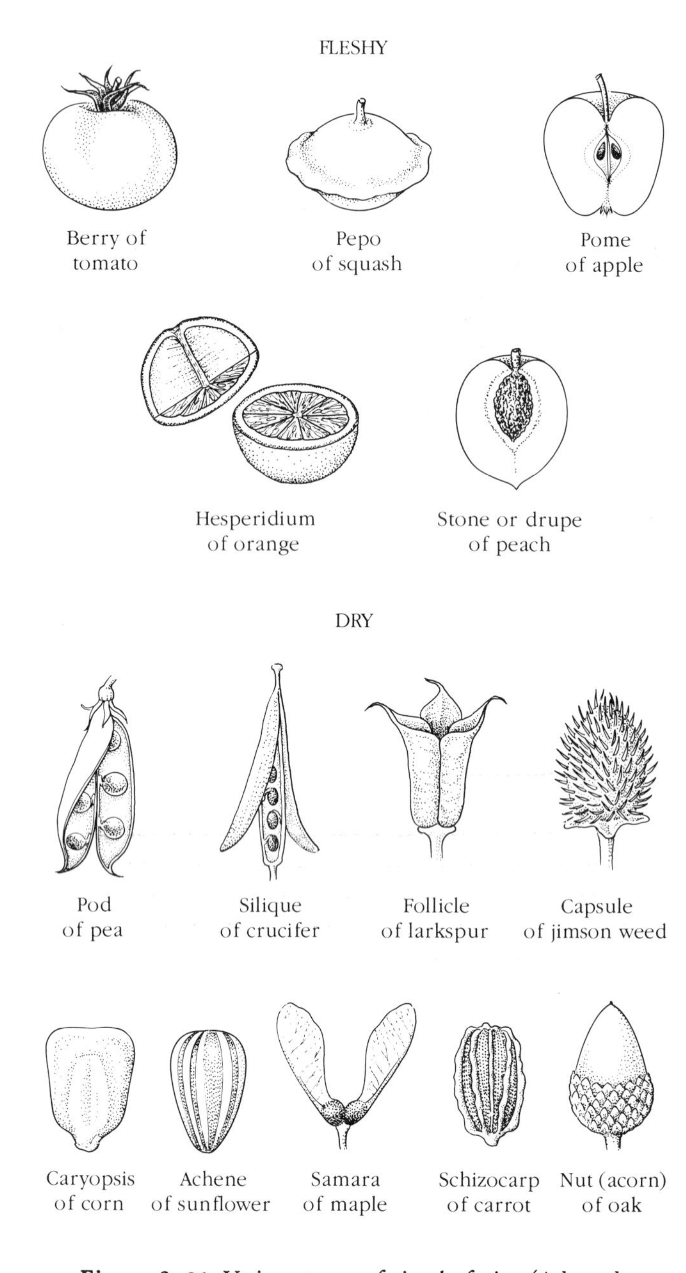

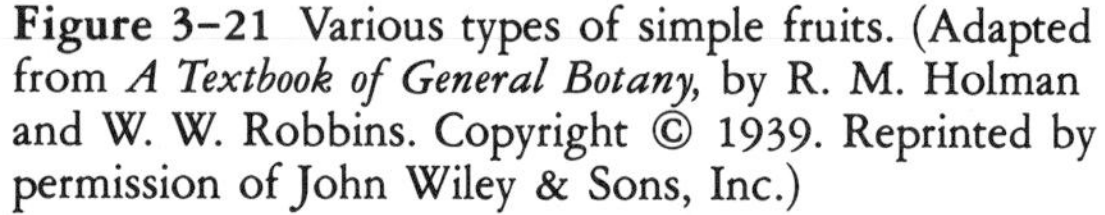

Figure 3–21 Various types of simple fruits. (Adapted from *A Textbook of General Botany,* by R. M. Holman and W. W. Robbins. Copyright © 1939. Reprinted by permission of John Wiley & Sons, Inc.)

fruits. **Aggregate** fruits, for example, raspberries and strawberries, are derived from flowers with many pistils on a common receptacle.

A majority of horticultural plants produce **simple** fruits, those derived from a single ovary. When it becomes part of a fruit, the ovary wall is called the **pericarp** and is divided into three distinct layers—**exocarp** (outer), **mesocarp** (middle), and **endocarp** (inner). When all three layers of the pericarp are fleshy, the fruit is called a **berry.** This usage of the term berry should not be confused with the edible portion of some bush fruits. Tomato, eggplant, and blueberry are berries. So are citrus fruits (called hesperidiums), in which the rind is the exocarp and mesocarp and the pulp is the endocarp; and muskmelons (called pepos), in which the rind is the exocarp, the edible portion is the mesocarp, and the watery portion around the seeds is the endocarp.

Drupe fruits (peach, olive, cherry, plum, for instance) are simple fruits with a stony endocarp. **Pome** fruits (apple, pear, quince) are simple fruits with a papery endocarp.

Alternate Bearing

Flowers and fruits develop from either mixed buds (buds that will develop both flowers and leaves) or blossom buds. Quite a few physiological factors determine whether or not, or how many, blossom buds form on a woody plant, but perhaps the most important plant requirement is a plentiful supply of carbohydrates in its tissues. Blossom buds form during June or July, remain dormant through the winter, and form blooms the following spring and ripe fruit that fall, 15 or 16 months after they are first initiated. June and July are also months of rapid growth of the current season's fruit, and development of excessive fruit may utilize most of the available carbohydrates, leaving little for the formation of blossom buds. When this occurs, the tree or shrub may enter a cycle of biennial or **alternate bearing,** in which a year of heavy fruit production alternates with a year of little or no fruit production (Figure 3–22). Alternate bearing of fruit trees often begins when an entire season's bloom is destroyed by frost. Because there is no competing crop, the set of blossom buds is extremely heavy, followed by too many fruits and no blossom buds the second season after the freeze. Biennial production is most frequent with older apple, pear, and large plum cultivars; it is seldom a problem with peaches, apricots, and cherries.

Trees and shrubs should have high carbohydrate reserves during blossom bud formation if alternate bearing is to be avoided. Heavy nitrogen fertilization and heavy pruning during the spring of the season of light production and early fruit thinning during the year of heavy production (see Chapter 8) will help prevent or overcome alternate bearing of fruit trees.

Alternate-year flowering is common with some ornamentals—old fashioned lilacs, some flowering crabapples, and mountain ash, for example—and can be remedied by removing faded blooms before seed begins to develop or some of the fruit just after petals fall.

Fruit Development

Although a few important kinds of fruits, including bananas, navel oranges, certain cucumbers, and pineapple, develop without ovule fertilization, the vast majority of fruit species require pollination and fertilization for the initiation of fruit development. Moreover, the continuing development of seed is essential for and correlated with the growth of most fruit.

Only a fraction of the blooms of a fruit tree normally produce mature fruit. There are two periods when small fruits are lost from the trees: (1) right after petals fall, the postbloom drop; and (2) four to six weeks later, the June drop, just as the fruit is beginning to enlarge rapidly. A lack of, or faulty, ovule fertilization and localized nutrient imbalance are thought to be the causes, but loss of fruit at these times is normal and should not concern the gardener as more than enough to produce a crop will usually remain.

Shortly after fertilization of the seed there is a rapid increase in numbers of cells in the fruit. By the time an individual apple fruit is a month old, it will have its maximum number of cells, which may be nearly one million.

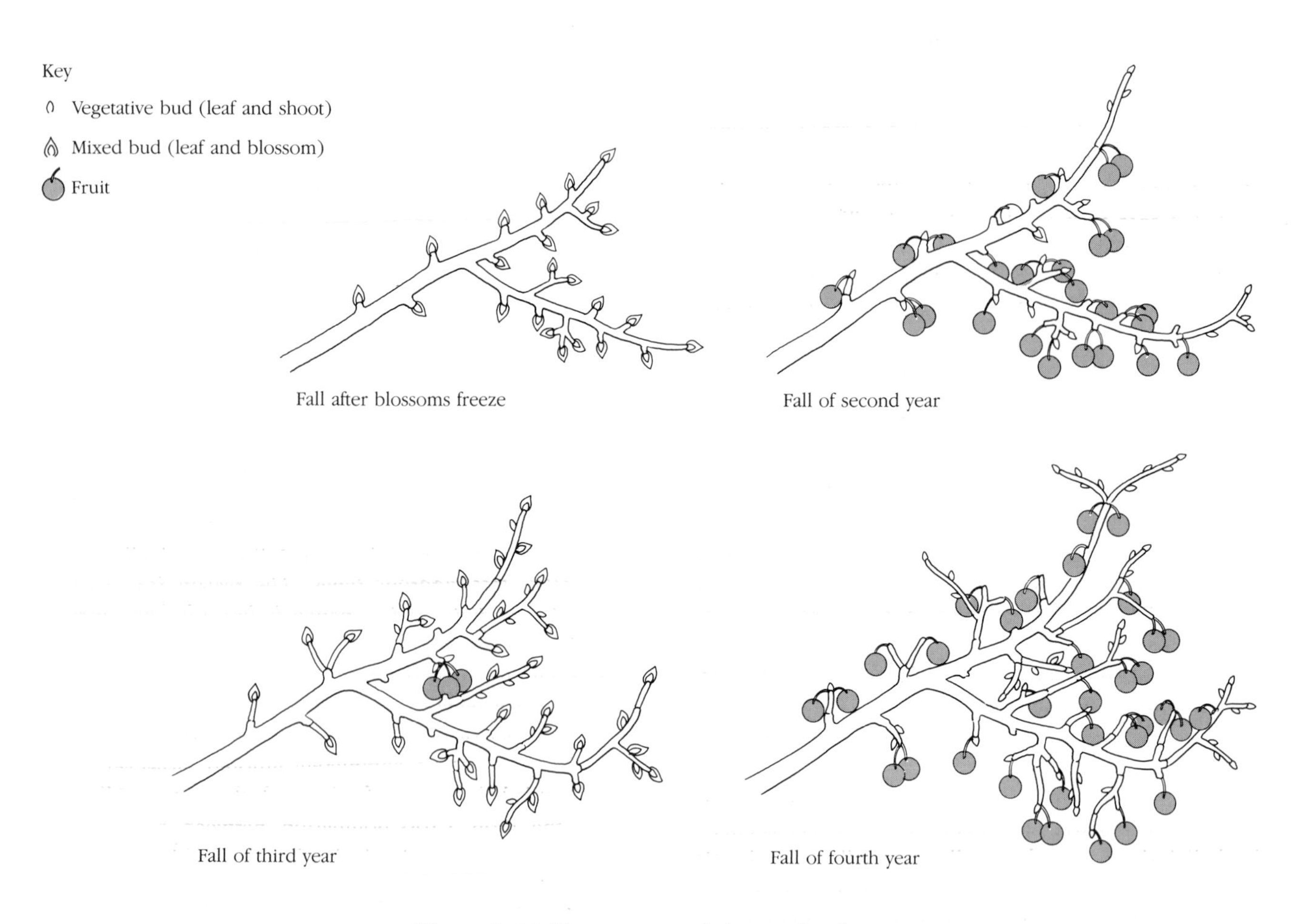

Figure 3–22 The sequence of alternate bearing.

During the second phase of fruit growth the cells enlarge, and most of what is perceived as fruit growth is increase in cell size mainly from an increase in size of the vacuole (see Figure 2–8, Chapter 2). As the fruit begins to mature, sugars and aromatic compounds associated with flavor accumulate.

Accompanying the ripening of many—perhaps all—fruits is a phenomenon called the **climacteric.** With most fruits the climacteric is the time when the fruit can be harvested and possesses, or is able to develop when detached from the plant, its maximum quality. The time of the climacteric is signaled by a rapid rise in the rate of respiration. If the fruit is not harvested, its quality gradually declines as it becomes senescent.

Most fruits will store longest if picked just as the climacteric is reached. Pears are picked shortly after the climacteric, which occurs while they are still hard and green. They do not develop high quality if left on the tree to fully ripen. The climacteric of tomatoes also occurs while the fruit appears immature. Visually, the climacteric in tomato is signaled by the change in color from dark green to lighter green or white as the chlorophyll breaks down; after that the fruit will mature to full color off the vine.

Fruit and Seed Quality and Pollination

"Don't plant muskmelons next to cucumbers" is one of the more persistent warnings that have been passed through generations of gardeners. The idea stems from the belief that fruit resulting from fertilization of muskmelon ovaries with cucumber pollen will taste like cucumber. Some gardeners are reluctant to plant a golden apple tree next to a red one, fearing the effect of cross-pollination on color. Other commonly held notions of the effect of pollination could be cited.

Let's examine the validity of these beliefs in light of the information presented in this and the previous chapters. In the first place, cucumber and muskmelon belong to two different species, *Cucumis sativus* and *Cucumis melo,* respectively. We learned in Chapter 1 that cross-pollination between species almost never occurs, so we must conclude that cucumber pollen cannot affect muskmelon quality, because these two crops do not cross-pollinate.

Secondly, even if cross-pollination were possible, as it is with red and yellow apples, only the embryo and endosperm of the seed would be affected by the foreign pollen. The fruit is the enlarged ovary that develops entirely from tissue of the mother plant, and therefore the source of pollen will not affect its quality. Persistence of the belief that cucumber pollen can affect muskmelon quality stems from the fact that muskmelons taste like cucumbers if they are harvested when immature or if plant disease or unfavorable environmental conditions prevent sugar manufacture by the leaves as the fruit is maturing.

Experienced gardeners sometimes warn beginners that planting white iris next to colored iris will cause the colored iris to turn white, or that planting different colored gladioli together will cause them all to turn one dominant color, such as peach or lavender. The assumption again is that cross-pollination brings about these results. Although cross-pollination and seed production can occur with both iris and gladioli, only rarely will seedlings of either grow to maturity in a home garden. White iris do sometimes become dominant in beds of mixed colors, because the vigorous white iris cultivar gradually crowds out other less vigorous types. The reason for peach or lavender gladioli predominance is less obvious. Perhaps the peach- and lavender-colored gladioli corms multiply more rapidly, or perhaps their corms are selected by the gardener for replanting because they are largest.

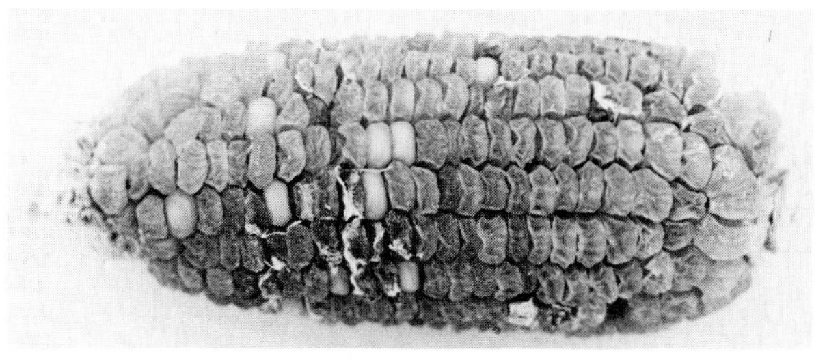

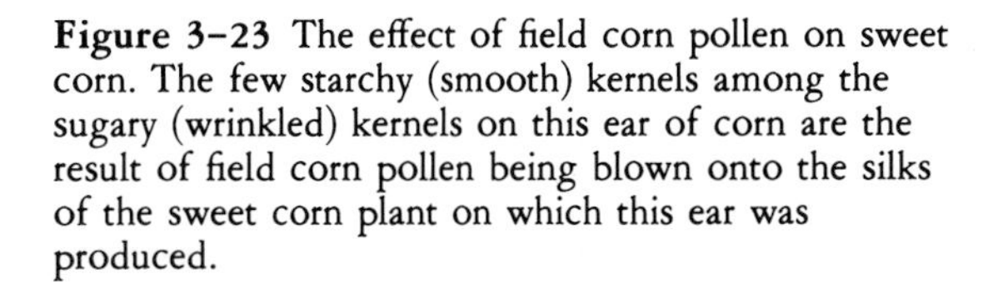

Figure 3–23 The effect of field corn pollen on sweet corn. The few starchy (smooth) kernels among the sugary (wrinkled) kernels on this ear of corn are the result of field corn pollen being blown onto the silks of the sweet corn plant on which this ear was produced.

The effect of pollen source on quality of horticultural products cannot be completely ignored, however. If the product is being grown for its seed, the wrong pollen can ruin the crop. Cross-pollination between popcorn and sweet corn as a result of planting them side by side will cause the popcorn not to pop and the sweet corn to be tough and starchy (compare Figure 3–23). Furthermore, if seed is to be saved for producing a crop the following year, the source of pollen becomes vitally important. Home growing of seed will be discussed in the next chapter.

Selected References

Crocker, W., and Barton, L. V. *Physiology of Seeds.* Chronica Botanica, Waltham, MA. 1953.

Wilson, C. L., Loomis, W. E., and Steeves, T. A. *Botany.* 5th ed. Holt, Rinehart and Winston, New York. 1971.

4 Propagation of Garden Plants

Assuming there is a place to grow a garden, the first step to successful gardening is securing the seed or planting stock. The production of new plants from parts of existing plants has already been defined as propagation. Some species of garden plants, including a majority of vegetables and annual flowers and many herbaceous perennial flowers and woody trees and shrubs are propagated from seed. Seed propagation is referred to as **sexual propagation.** Other plants, including most house plants, many popular landscape materials, most fruit trees, potatoes, sweet potatoes, and rhubarb, are propagated from stems, leaves, or roots; or from bulbs, tubers, rhizomes, fleshy roots, or other specialized structures that nature has modified from basic plant organs. This kind of propagation is referred to as **asexual** or **vegetative propagation.**

Containers and Media for Plant Propagation

Both sexually and asexually propagated plants are frequently started in containers, and they will grow in almost anything that will hold growing media and stay together until the plant parts used for propagating (**propagules**) are ready for transplanting (Figure 4–1). Egg cartons, the bottom third of milk cartons, aluminum foil pans, tin cans, and plastic dishes used to market cottage cheese, liver, and other products are often salvageable from the kitchen and can be used successfully. All must be washed thoroughly with soap and water, and some may require modification. The fibers in some kinds of egg cartons break down before plants become large enough to transplant into the field. Plants grown in cartons made of paper will need extra nitrogen (see "Fertilizing Garden Crops," Chapter 14). Nonporous containers, including plastic and waxed cartons and metal cans, require drainage holes and special care to avoid overwatering.

Commercial flats and pots of various kinds and sizes and small containers for growing individual plants are available at garden shops and stores. Flats are shallow containers designed for starting seedling transplants or small cuttings. Until a few years ago they were always made of wood and were a standard 23 x 14 x 3½ inches in length, width, and depth, respectively. Today they come in many sizes and of various materials, including wood, plastic, metal, fiberglass, and pressed fiber. Small fiber flats each containing 8 to 12 transplants are marketed by the hundreds of thousands around the United States and Canada every spring. Drainage holes should be present in flats made of plastic, fiberglass, metal, or other nonporous materials. Extra nitrogen is required for plants growing in new wood and fiber flats. To provide part or all of the mineral nutrients for growth of transplants, some manufacturers sell fiber flats with

Figure 4–1 Containers for plant propagation. (Courtesy F. E. Larsen and Washington State University Cooperative Extension Service.)

slow-release fertilizer incorporated in the walls and bottom.

Flats are also used to hold individual plant containers, including fiber pots, small plastic pots, plant bands, and peat pellets, all usually 1½ x 1½ inches (4 x 4 cm) or 2 x 2 inches (5 x 5 cm) in size. Plant bands are made from compressed fiber and have no bottom, so they can be folded flat for shipping and storing. Peat pellets are dried compressed peat moss surrounded by coarse netting. They usually contain fertilizer and expand to several times their size when soaked in water.

Pots for plant growing also are available in many shapes and sizes and materials. The standard clay pots have been largely replaced by less expensive, more convenient fiber and plastic pots. Empty gallon (no. 10) and 5-gallon tin and plastic cans in which various foods are packaged for restaurants, hospitals, school cafeterias, and the like, are salvageable to use, especially for large nursery stock. Like other containers, these must have drainage holes when used for propagation.

As with the containers used for propagating, the rooting media for transplants have changed markedly during recent years. Fifty years ago all transplants were grown in a mix of soil, sand, and manure. Today the many materials available for growing plants are more properly called plant-growing media. The two requisites of propagating media are that they retain water and yet drain well so that oxygen can reach plant roots (see "Soil Preparation," Chapter 14).

Propagating Plants from Seed

Even with the recent increases in purchase price the cost of garden seed represents only a small percentage of the total cost of growing a garden. Because the crop can be no better than the seed used to plant it, no gardening activity is more important than securing good seed.

Securing Garden Seed

It is essential that the gardener plant seed that will germinate and grow vigorously; be free of disease-causing organisms, insects, and weed seed; and be the desired cultivar. Obtaining seed with these characteristics is not usually difficult in North America, but as sources for obtaining seed are quite numerous, the inclusion of a few guidelines may help prevent pitfalls.

Purchasing good seed is not as risky today as it was a few years ago when seed was commonly bought from traveling salesmen. Nevertheless, because it is impossible to distinguish by casual observation seed of good or poor viability or seed of different cultivars, a grower must rely on the integrity of the seed dealer. Seed purchased from a reliable local dealer or from a well-known mail-order supplier will usually be satisfactory (Figure 4–2). Many seed companies label their packages with the results of a germination test, including date and percentage of germination and amount of foreign matter. If the test has been made within the past six months, the results will usually be a good estimate of the amount of seed that will germinate if conditions are ideal. A few firms do not include germination test results on their seed packages, relying instead on their reputation of selling only seed of good germinability.

The garden should be planned and the seed ordered as early as possible if the desired cultivars are to be obtained before seed stocks are depleted. Seed fastened to water-soluble tape at the recommended spacing for the crop is available for some garden crops. The tape is laid in the furrow at the proper depth and covered with soil. Taped seed is expensive but convenient when conditions are good for germination (Figure 4–3).

Gardeners often ask if they should grow and save their own seed. This is difficult to answer, because success depends so much on the knowledge and inclinations of the gardener, the time available, and the botany of the desired crops. Seeds of self-pollinating crops can be saved if reasonable care in harvesting and storing are exercised (see Table 14–12, Chapter 14, for a list of self- and cross-pollinating crops). Home-grown seeds of cross-pollinating crops will produce true to type only if the seed came from a cultivar isolated enough that it was not cross-pollinated by other cultivars of the same species. Seeds of F_1 hybrid cultivars should never be saved for the reasons mentioned in Chapter 3.

For the gardener with other pursuits and limited time, gardening activities other than seed growing are probably more satisfying. Because seed represents such a small percentage of the total cost of gardening for the busy gardener, the money saved is not worth the bother and the space required for growing, collecting, threshing, fermenting to separate the seeds from the pulp, drying, storing, and testing. Occasionally, though, it may be neces-

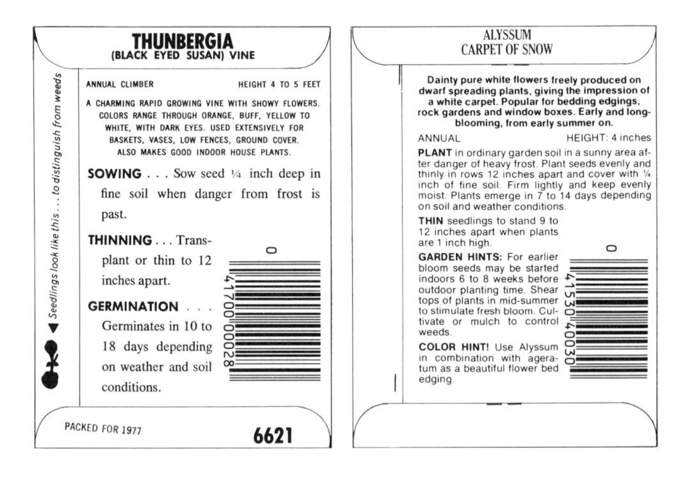

Figure 4–2 Planting information from two seed packets. (Courtesy Northrup King Seeds; W. Atlee Burpee Co.)

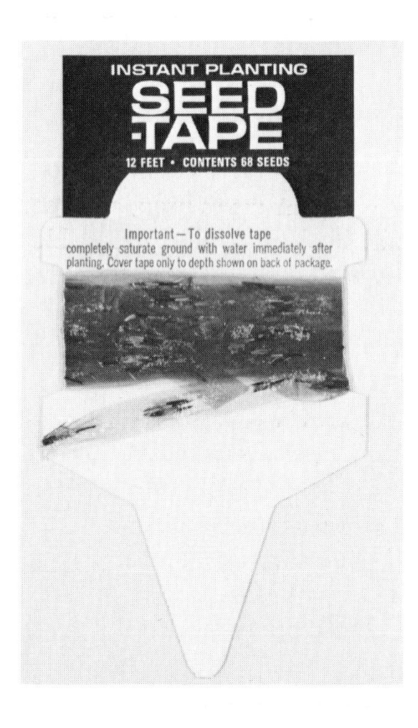

Figure 4–3 Seed tape. (Courtesy Ferry-Morse Seed Co.)

sary to save seeds of a cultivar for which there is no commercial source (see "Growing Garden Seed at Home," Chapter 14).

Seed Dormancy

Before seeds can germinate, they must be subjected to the correct environmental conditions. With some species this involves overcoming physiological or physical seed dormancy. If wild plants of cold winter areas were to germinate immediately after seeds reached the soil in the late summer or fall, the newly germinated, tender young plants would not survive the winter. As a consequence, nature has provided various systems to delay germination until there is a longer period of favorable environment. We refer to these systems of seed germination delay as **seed dormancy** or **seed rest.**

As explained in Chapter 2, a dormant plant is one that is not actively growing, so, technically, all seed could be considered dormant until it begins to germinate. If bud and seed physiology terminology were parallel, seeds that would not germinate even if environmental conditions were favorable would be in a resting period. In reality the term seed rest is almost never used, and seed dormancy is the term usually invoked to describe viable seed that will not germinate even when all environmental conditions are favorable.

Because seed dormancy interferes with cultivation, nondormant cultivars of seed-propagated species have been selected to provide cultivated species that no longer have a dormant period. However, most woody ornamentals and asexually propagated species of garden plants still retain vestiges of either physical or physiological dormancy, and gardeners must know how to overcome this dormancy before they can successfully germinate seeds of these species.

Physical Dormancy. Physical dormancy is common in seeds of legumes such as peas and beans, in seeds of a few other dry-seeded vegetables, and in seeds of many ornamental and perennial flowering plants. The most common manifestation of physical dormancy is the failure of the seed to absorb water because of an impervious seed coat. Such seed, called **hard seed,** is a problem for homemakers as well as for growers, because seeds that vary in degree of hardness cook unevenly. A pot of chili or a lentil casserole with some seeds cooked to pieces while others remain hard enough to crack teeth does nothing to encourage the consumption of beans or lentils.

In nature physical dormancy is overcome by abrading of the seed coat by soil particles and water, by the action of freezing and thawing, by microorganisms, or by chemical action of the soil solution. Hard seed can be artificially opened by abrading the seed coat (**mechanical scarification**), by treating the seed with a strong acid (chemical scarification), by soaking in hot water, by burning in a fire (for a few species), or by manipulating the humidity or other environmental factors. Pea, lentil, and bean seeds grown and stored in areas of low humidity in the western United States are usually put through a mill that subjects them to a light sanding, thus insuring easy water penetration through the seed coat when they are planted or cooked. Seeds of morning glory and New Zealand spinach require a long time to germinate in the garden, primarily because water penetrates through the seed coat at an extremely slow rate. To speed the action, the gardener can make a small notch with a triangular file in the coat of each seed to be planted (Figure 4-4). Nurs-

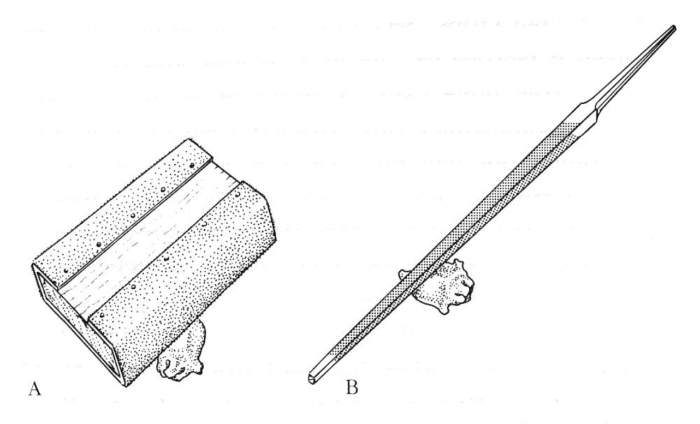

Figure 4-4 Seed scarification. When only a few seeds of morning glory, New Zealand spinach, perennial sweet pea, or others with a seed coat impervious to water are to be planted, germination will be speeded if the seed coat is opened slightly with a piece of sandpaper (A) or a triangular file (B).

eries frequently soak the seed of certain ornamental trees and shrubs for short periods in concentrated sulfuric acid to overcome physical dormancy.

Physiological Dormancy. Physiological dormancy can be manifested in various ways. Sometimes the embryo is immature when the seed is otherwise ready for harvest, a common occurrence with early maturing stone fruit cultivars. The immature embryo can be made to grow if the seed is removed from its shell and placed on a special medium in a controlled environment. Even with this treatment, early seedling growth is likely to be abnormal, with leaves or roots failing to develop completely, and the length of time required for the seedling to develop into a normal plant usually extends to several months.

A more satisfactory treatment is to store the seed from 4 to 14 weeks, depending on species or cultivar, in moist peat moss, sawdust, or similar material in a light plastic bag in the refrigerator. This treatment, called **stratification,** allows the embryo to mature as it would naturally if the seed were to remain over winter in the soil. Treating with activated charcoal or thiourea is also effective with some species.

Physiological dormancy of many fruit and ornamental species is due to chemical inhibitors within or outside the seed. The fruit itself often contains an inhibitor that prevents seeds from sprouting while they are within it. With some crops, tomato being a good example, the seeds will germinate as soon as they are removed from the wet, pulpy material of the fruit. With others stratification is necessary to overcome chemical inhibition of seed germination. Seeds of some woody plants have a combination of physical and physiological dormancy and may require both stratification and scarification before they can be germinated (see "Propagating Plants from Seed," Chapter 14).

Seed Germination

Once dormancy is overcome, or if dormancy does not exist, as with most vegetables and annual flowers, the seed is ready to germinate. The primary requirements for germination are warm temperature, moisture, and oxygen. Recent research conducted by the U.S. Department of Agriculture has revealed light to be much more important in inhibiting or stimulating germination of seeds of many ornamentals than was previously supposed. Even the very low intensities that filter through a shallow layer of soil can affect planted seed. Whether light is stimulatory or inhibitory depends on the species. Its effect is complicated by being interrelated with the age of the seed, temperature, and perhaps other environmental factors. Light usually does not complicate the germination of garden seeds if they are planted at the recommended depth and exposed to light/dark cycles corresponding to normal day/night periods.

The temperature at which a seed will germinate will vary from species to species and, to some extent, from cultivar to cultivar within a species. The temperature required for germination of any species is, however, correlated with the optimum temperature for growth of that species. Seed of 'Hales Best' cultivar of muskmelon, for example, will not germinate until the soil temperature reaches approximately 72°F (22°C), and this cultivar, like all muskmelons, requires high temperatures for normal plant growth.

If the temperature is high enough and if moisture is present, water enters the seed, either through the seed coat or through openings in the seed coat. As water combines with starch and with other materials within the seed, the seed begins to enlarge, and enzymes that control and direct the growth of the developing plant are activated. Growth-promoting substances within the seed direct root growth downward and leaf and stem growth upward. The seed coat and endosperm of monocot seeds usually remain in the soil. With some dicot plants, such as beans, marigolds, and tomatoes, the cotyledons are pushed up through the soil and become active in photosynthesis. With others, such as peas, the cotyledons remain in the soil, and only the plumule and epicotyl emerge (Figure 4–5).

The young seedling is especially vulnerable to adverse environmental conditions during the period when it is becoming established. Until the plumule and cotyledons reach light, begin to develop chlorophyll, and start food manufacture, and until the radicle elongates and develops

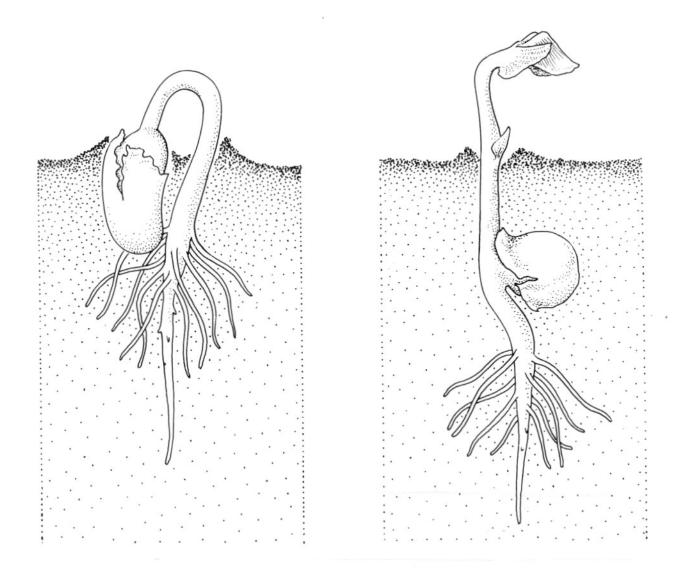

Figure 4–5 Comparative germination of garden bean (A) and pea (B). The cotyledon of the bean is pushed through the surface crust, whereas the cotyledon of the pea remains in the soil.

root hairs, the young seedling is entirely dependent on the food stored in the seed and the moisture that can be soaked into the seed for its survival. If the seed is planted too deeply, its food reserves may be exhausted before the plumule reaches light. If the soil in which the seed has begun to swell becomes dry, the seedling will perish. Young seedlings are also more likely to be killed by frost, high temperature, insects, and diseases than are more mature plants.

Determining Seed Germinability. When more seed has been purchased than can be used during a single year (which frequently happens), the grower must decide whether to use the old or purchase new seed the second or third season. Initial viability, kind of seed, and temperature and humidity of storage will largely determine the length of time seed remains viable. Tables 4–1 and 4–2 show the length of time from harvest to planting during which various kinds of vegetable and flower seed remain fit to plant in most parts of the United States. The years listed in these tables are only approximations and will vary with the climate of the area. Seeds live longest when stored where temperature and humidity are low. Seed of tomato cultivars retained good viability after being stored for 20 years at the high, dry Cheyenne, Wyoming USDA Field Station. Onion seed may lose viability in a few months when kept in open containers in warm, humid locations like the Southeast. Whenever possible, seed should be stored in a cool, dry location.

A germination test should be run on seed of doubtful germinability. For this any technique that supplies heat and moisture can be used (see "Testing Seed Germinability," Chapter 14).

Table 4–1 The approximate length of time during which vegetable seeds of good initial germinating ability stored under proper conditions will germinate satisfactorily.

Vegetable	Years	Vegetable	Years	Vegetable	Years
Asparagus	3	Eggplant	5	Pea	3
Bean	3	Kohlrabi	5	Pepper	4
Beet	4	Leek	1	Pumpkin	4
Broccoli	4	Lettuce	5	Radish	5
Brussels sprouts	4	Muskmelon	5	Spinach	5
Cabbage	4	Mustard	4	Squash	5
Carrot	3	Okra	2	Sweet corn	1
Cauliflower	4	Onion	1	Tomato	4
Celery	5	Parsley	2	Turnip	5
Cucumber	5	Parsnip	1	Watermelon	5

Table 4–2 The approximate length of time during which flower seeds of good initial germinating ability stored under proper conditions will germinate satisfactorily.

Flower	Years	Flower	Years	Flower	Years
African daisy	3	Hollyhock	5	Shasta daisy	3
Aster	2	Marigold	3	Snapdragon	3
Calendula	5	Nasturtium	5	Stock	5
Carnation	5	Pansy	3	Sweet alyssum	4
Chrysanthemum	5	Petunia	3	Sweet pea	2
Cosmos	3	Phlox	2	Sweet sultans	5
Delphinium	1	Salpiglossis	5	Verbena	3
Dusty miller	3	Scabiosa	3	Zinnia	5

(Based partially on information from California Department of Agriculture, *Bulletin*, vol. 26, no. 3, 1937.)

Planting Seeds in the Garden

For direct seeding out-of-doors in the spring, timing is important. Generally, the gardener wants to plant as early as is practical. With some of the cool-season crops, early seeding is essential so that plants can mature before the hot days of mid-summer. With other crops, especially in northern latitudes or at high altitudes, early planting is essential if the crops are to mature during the available growing season. On the other hand, planting before the soil temperature is warm enough for germination will subject seeds and seedlings to a longer period of possible attack by soil pathogens, competition by weeds, and injury by freezing.

Different kinds of vegetables and flowers have slightly different optimum conditions for germination. However, it is possible to divide most of them into two broad groups: the cool-season and the warm-season crops. The cool-season crops include radish, lettuce, spinach, Swiss chard, beet, carrot, onion, cauliflower, cabbage, broccoli, kohlrabi, kale, turnip, rutabaga, pea, snapdragon, pansy, and a number of others. These are crops that germinate at soil temperatures between 40° and 55°F (4.4° and 13°C) and will withstand a certain amount of freezing after they have begun to grow. Although onions, peas, spinach, radishes, and lettuce are usually planted earliest of this group, most of them can be planted as soon as the soil has warmed and dried enough to become workable.

The warm-season crops include tomato, eggplant, pepper, cucumber, squash, watermelon, cantaloupe, snap bean, lima bean, sweet corn, marigold, zinnia, and a number of others. These are crops that require soil temperatures above 60°F (16°C) for rapid germination and are killed by temperatures slightly below freezing. A number of them are injured when subjected to cool temperatures above the freezing point. Although sweet corn, tomatoes, and zinnias can be planted somewhat earlier than muskmelon and watermelon, the general recommendation for time of planting of warm-season crops is after the danger of frost, by which time the soil temperature will usually have warmed sufficiently to promote rapid germination and vigorous seedling emergence.

Seed-bed preparation to level the planting surface and to eliminate large clods and air pockets is important to insure contact between moist soil and the seeds and to obtain uniform depth and spacing of the seeds. Good seed-bed preparation is also necessary for control of weeds, which are easier to eliminate before planting than after. The seed bed should be moist but not saturated with water and should never be worked when excessively wet. If irrigation is required, it should be done before planting. Irrigating after planting should be avoided, if possible, because it can cause crusting, cool the soil, and encourage growth of seed-decaying organisms. Light sprinkling of the planted seed bed may be necessary in areas with low humidity and drying winds and with small-seeded and light-requiring flowers that must be surface-seeded.

Each seed must be planted deeply enough to remain in contact with moisture until it has germinated. However, it cannot be planted so deeply that the plumule is unable to push through the soil before food reserves in the seed are exhausted. The general recommendation for depth of planting is to cover the seed to about three times its width. Small seeds like petunia and snapdragon are scattered on the soil surface and raked in lightly. Seeds should be planted somewhat more shallowly during early spring when temperatures at lower soil depths are likely to be cool and somewhat more deeply during warm, dry summer periods when moisture is likely to be lacking in the upper areas of the soil (see "Propagating Plants from Seed," Chapter 14).

Transplants

Not all plants are seeded directly into the garden. In fact, one of the fastest growing gardening industries is the production of **bedding plants,** which are transplants grown for flower beds or vegetable gardens. The use of transplants is economically justified in situations where each plant occupies a fairly large area and where the crop has a high economic or social value, as is true with most intensively cultivated horticultural plants. The major advantage in using transplants is, of course, earlier production. A plant started in the greenhouse six weeks before it could be seeded out-of-doors may not mature a full six weeks earlier than one directly seeded into the field, but the transplant will probably have a two- to four-week advantage in maturity, which is of real value to the home gardener. Transplanting also enables the grower to germinate seeds under more controlled environmental conditions and thus assures a more economical use of seeds. This is especially important with some double and hybrid flower cultivars, seed of which may sell for several hundred dollars for a fraction of an ounce. Transplanting also permits the gardener to place each plant in its desired location in the garden.

Horticultural crops commonly transplanted include a majority of the annual and perennial flowers and muskmelon, watermelon, celery, tomato, eggplant, pepper, cabbage, cauliflower, broccoli, Brussels sprouts, kale, collards, onion, and asparagus.

Purchasing Transplants. The main consideration in purchasing transplants is obtaining vigorous plants of the proper size and desired cultivar. Transplants should be large enough to be handled easily but not so large that they have become crowded. A compact, bushy seedling will be more likely to survive and produce a better mature plant than one that has grown tall and leggy (Figure 4–6). Plants that have been stunted and those that have a gray-green or yellowish cast from a lack of fertilizer or water should be avoided. A reliable local nursery will generally be the most satisfactory source of supply. Plants purchased from a supermarket or other establishment not specializing in plant production should be carefully inspected for freedom from insects, pathogens, and physiological disorders.

Growing Transplants. Some gardeners like to grow their own transplants. Growing transplants lengthens the season of gardening enjoyment and, because the number of cultivars available for purchase as transplants is often

Figure 4–6 A flat of pansies ready for transplanting. These healthy stocky plants should grow well in the garden.

limited, makes possible the production of cultivars not otherwise obtainable. Even considering these advantages, home gardeners should not attempt to grow their own transplants unless they have adequate facilities. The factor most likely to be limiting in indoor plant production is light. Sufficient sunlight will usually come through an unobstructed south or west window, and good transplants can be grown at such a location if there is not a radiator beneath the window. High-intensity artificial lights are also satisfactory for the growth of transplants (see Chapter 7; see also Chapter 14).

Propagating Plants Asexually

For most species propagation by seed is usually more convenient and less expensive than asexual propagation. With many cultivated crops, however, seed propagation is not feasible. Some of the reasons for propagating asexually are: (1) to perpetuate plants that either never had, or have lost, the ability to produce seeds; (2) to reproduce plants that are heterozygous genetically and, consequently, will not come true from seed; (3) to control the form or size of a plant; and (4) to give the plant a stronger, more pest-resistant or more cold-resistant root system.

Examples of plants that do not produce seed are seedless oranges, seedless grapefruits, seedless grapes, bananas, pineapple, peppermint, French tarragon, and greenhouse carnations. From the horticulturist's and consumer's point of view, the absence of seeds is highly desirable in many crops. Plants that do not come true from seeds include most temperate zone fruits, geraniums, chrysanthemums, roses, junipers, yews, and arborvitae. Seed from a 'Red Delicious' apple, for example, would produce trees having small apples, trees having large apples, and trees with fruit of various colors, shapes, textures and flavors; but not one in 100,000 of the seeds would likely produce another 'Red Delicious' tree. The best known example of propagation to control form or size is the use of size-controlling rootstocks for the development of dwarf apple and pear trees. As another example, the top of the Camperdown elm (*Ulmus glabra* 'Camperdowni') that is so admired for its beautiful weeping growth habit is actually a liana until it is grafted onto an upright rootstock (Figure 4–7). Tree roses are another example of the control of form by grafting.

One of the best known examples of the use of vegetative propagation to control pests is the grafting of European grape (*Vitis vinifera*) onto American grape (*Vitis labrusca*) rootstock. Phylloxera, a grape-root aphid, was introduced into Europe along with the first American grapes taken to that continent. Phylloxera does not extensively injure the roots of American-type grapes; however, European grapes that were propagated with cuttings had no resistance to it whatsoever, and the insect soon started to spread across Europe, threatening to destroy the entire grape industry. Eventually it was found that this insect could be controlled by grafting European grapes onto American grape root systems.

Sour oranges (*Citrus aurantium*) are used to provide a pest-resistant rootstock for plantings of sweet orange (*C. sinensis*). Peach is sometimes used as a drought-resistant root system for plum, and the 'McIntosh' apple cultivar is sometimes used as an **interstock** in developing a cold-resistant framework for trees of such less hardy varieties

Figure 4–7 Changing growth habit by grafting. The unique growth habit of the Camperdown elm is created by grafting an *Ulmus glabra* 'Camperdowni' scion onto a normal Scotch elm (*U. glabra*) rootstock. The graft is made on an older Scotch elm trunk at the height where the tree is to head (form side branches), because the tree "weeps" from the point of the graft union.

as 'Red Delicious' and 'Golden Delicious.' The use of interstocks will be explained in the grafting section of this chapter.

Though not strictly propagation, repairing of certain woody plants is sometimes accomplished by grafting. Bridge grafting is frequently done across areas that have been damaged by freezing weather or crown-girdled by rodents or accident (see Figure 14–47).

Methods of Vegetative Propagation

Vegetative propagation can be categorized into the following groupings: apomictic seed; utilization of specialized stem, leaf, and root structures; cuttage; layerage; and graftage.

Apomictic Seed. In some species, what appears to be seed may be formed without the union of male and female gametes. Such seed, called **apomictic,** may form as a result of the development of a macrospore mother cell into an egg without meiosis, or it may form from a cell produced in the ovule wall. In any event, apomictic seed forms entirely from tissue of the mother plant, and cultivars propagated from such seed are clones. Bluegrasses (*Poa* spp.) are the horticultural crops most likely to be reproduced by apomictic seed. Navel oranges and several other citrus cultivars produce apomictic seed, but they are normally propagated by grafting.

Specialized Stem, Leaf, and Root Structures. Among the stem, leaf, and root modifications used for propagation are offshoots, crown divisions, stolons, rhizomes, tubers, tuberous roots, tuberous stems, bulbs, corms, and fleshy leaves (see Figure 2–20). Among the exotic species pineapple, banana, and sugar cane are propagated by **offshoots.** Perhaps of more interest to the home gardener are the common trees and shrubs, including lilac, flowering almond, mock orange, some of the spireas, and flowering quince, that can be propagated from offshoots.

Plants of some horticultural species can be cut or divided into a number of parts, each of which will grow into a new plant provided it contains some stem and some root tissue. Because the area where the root meets the stem of the plant is called the **crown,** this type of propagation is referred to as **crown division.** In certain cases it is difficult to distinguish propagation by offshoots from propagation by crown division. Rhubarb and some perennial flowers, including chrysanthemum, peony, columbine, lupine, and some perennial poppies, can be propagated by crown division. Ferns also are frequently propagated from crown divisions.

Several important horticultural crops reproduce by above ground trailing horizontal stems called **runners** or **stolons.** Among these, strawberry (*Fragaria* spp.), strawberry geranium (*Saxifraga sarmentosa*), and spider plant (*Chlorophytum elatum*), are three of the better known. Even more numerous are plants that reproduce and can be propagated by underground horizontal stems called **rhizomes.** Lowbush blueberries spread by rhizome-initiated shoots that surface some distance from the mother plant. Wild blueberry clones may spread in this manner to cover several hundred feet in a few years. Grasses spread by rhizomes, and several kinds are propagated from sections of rhizomes. Lily of the valley also spreads by rhizomes, as do a number of troublesome weeds, including quack grass, poison ivy, water hyacinth, and cattails.

Botanically, tubers are enlarged rhizomes. The potato is the best known tuberous crop. It is propagated by cutting the tubers into several pieces, each containing an eye, which is a bud cluster. Seed pieces should weigh from one to two ounces in order to produce a vigorous, healthy plant. The Jerusalem artichoke (*Helianthus tuberosus*) and *Caladium* are also propagated from tubers.

A number of popular flowers, including most of the early spring blooming ones, grow from bulbs or corms (see Table 14–10). Tulip, lily, and narcissus are examples of bulbous crops, and gladiolus, crocus, and water chestnuts are corm-forming. With both bulbs and corms growth of stems, roots, and leaves is initiated and proceeds at the expense of the stored food material in the bulb or corm. After the above-ground portion of the plant has completed its growth cycle, food reserves of the storage organ are replenished from the photosynthetic products manufactured primarily in the leaves. Thus it is important that the leaves of bulb and corm plants not be removed until those leaves have dried. It is also important that the leaves be left with the plant in the garden if the flowers are cut for decorating purposes.

Bulbs enlarge and form new bulbs (**offsets**) at their periphery as food reserves become available. The old corm shrivels, and a new one forms above it as upper parts of corm-producing plants senesce (see Figure 14–35). A large corm may produce two to three stems, and a new corm will be produced at the base of each stem as the plant matures. Thus a single bulb or corm may multiply to form two or three new bulbs or corms during one season. Tiny corms called cormels are also produced around the base of the new corm. With a few years' growth these will become large enough to produce flowers and are a major way of increasing corm-producing crops. With some bulb and corm-forming crops, small bulbils or cormels form in the leaf axils of the plant. Lily bulbs can be separated into individual scales, each of which will produce a new tiny bulb and plant.

A number of horticultural crop plants have the ability to form adventitious buds from the root system. The sweet potato is a true root and is propagated from rooted shoots, called **slips,** produced when the potato is placed in a moist, sterile propagating medium (Figure 4–8). A single sweet potato will produce several crops of slips before its food reserves become exhausted. Most cane berries are capable of producing buds and shoots (suckers) from their root systems. They are propagated either by planting a section of root (root cutting) or, more frequently, by digging a shoot that has already grown from a root-produced adventitious bud. Elderberry, sumac, wild morning glory, and many ornamental shrubs also have the ability to spread by root-initiated shoots (see "Propagating from Fleshy Storage Organs," Chapter 14).

Figure 4–8 Sweet potato with slips beginning to grow from it.

Cuttage. The induction of roots and/or shoots on detached plant parts is called **cuttage.** Cuttage is the most frequently used method of vegetative propagation, because it is convenient and permits large numbers of plants to be produced quite rapidly from a small amount of propagating material. Most indoor foliage plants are propagated from cuttings, as are certain flowering plants, including chrysanthemum, carnation, fuschia, and geranium. Juniper, arborvitae, false cypress, and yew are among the needle evergreens that are propagated by cuttage. Most small fruit species, the broadleaved evergreens, and a majority of deciduous woody ornamental shrubs can be propagated by cuttage. Cuttings may be made of stems or parts of stems, leaves, or roots. Stem cuttings can be made from herbaceous plants. Softwood cuttings are made from the new succulent growth of woody plants that is produced in the spring or early summer. Hardwood cuttings are made from older wood, usually during the dormant period.

Although a cutting with a single bud will produce a new plant, as is evidenced by the successful rooting of leaf bud cuttings of several species, stem cuttings 3 to 5 inches (8 to 13 cm) long containing at least two nodes are more convenient to use. A slanting cut is often made just below the lower node and a straight cut above the upper one. Since there is more meristematic tissue in the vicinity of a node, root formation is more likely to succeed if a node is beneath the rooting medium. The slanting cut at the lower end of the cutting is traditional and is more important in identification of the lower end of the cutting than it is for permitting more surface to be in contact with the rooting medium.

As many leaves as is practical should be left on the stem cutting to provide food for root formation and growth of the new plant. Those leaves that are likely to come in contact with the rooting medium, however, should be removed, as leaves often carry microorganisms that can cause decay when they are in contact with the medium (see Figures 14–19, 14–20, and 14–21, Chapter 14).

New plants can be propagated from the leaves of a few species. Leaf petioles of a number of plants, including coleus, some philodendron cultivars, and other foliage plants, have the ability to form roots but not shoots. With some of these plants the rooted leaf may expand and grow quite large, but it will not produce new shoots unless a stem bud is attached to it (Figure 4–9).

Factors Affecting Rooting Success. Numerous factors affect the success of rooting. Some plants root easily, some root with difficulty, and some cannot be induced to form roots regardless of the treatment applied. The age of the plant part being rooted has an effect on rooting success. Younger plant tissue has more meristematic tissue, which enables it to root more easily. Older tissue, however, contains more reserve carbohydrates, which are necessary for initiation of new roots and root growth. The optimum age of plant tissue for cuttage success varies with the species. With some the easiest to root cuttings may be from stems not quite mature, a compromise between young tissue with abundant meristem and old tissue with ample food reserves. Obviously, the season of the year, the location of the tissue on the plant, the nutritional level, and the general vitality of the plant from which the cutting is taken all affect its ability to form roots.

Figure 4–9 Propagation from leaf cuttings. The detached leaves of some houseplants, for example, pothos (above), most philodendrons, and coleus, will form roots but never initiate bud and stem growth unless an axillary bud is left attached to the leaf petiole.

Chemicals within a cutting partly determine its ability to root. For example, it is difficult to root cuttings from apple, pear, or pine trees, but a willow twig will root if it falls on wet soil. Some chemical differences that are at least partly responsible have been identified—notably auxins (see Chapter 7), carbohydrates, certain nitrogenous compounds, and vitamins. There are probably others of which we are not yet aware.

In addition to *in vivo* plant factors, environmental conditions affect rooting. The ideal temperatures for rooting may vary from plant to plant. Relative humidity normally should be high, and adequate oxygen, moisture, and light are necessary. The rooting medium is also important.

A propagator wants to do everything possible to increase the opportunity for root formation. In large part, this involves the induction of a root system as quickly as possible while preventing the top of the plant from drying. Three special techniques—intermittent mist, hormones, and bottom heat—innovations 30 to 40 years ago, are now routinely used to increase the chance of rooting in nurseries. Intermittent mist prevents the cutting from dehydrating while roots are being initiated and is supplied by a spray nozzle controlled by a solenoid valve and time clock set to permit the spraying of the propagating bench with a fine mist of water for a few seconds every few minutes. Three natural hormones used to stimulate root formation are indole-3-acetic acid (IAA), indole-3-butyric acid (IBA), and α-naphthaleneacetic acid (NAA). Commercial preparations containing one or more of these hormones are available at most garden stores. The lower end of the cutting should receive a very light dusting with the hormone compound, as overapplication will inhibit root formation. Bottom heat is usually supplied with heating cables in propagating benches to keep the bottom of the cutting warmer than the top. Since growth is most rapid where temperature is highest, bottom heat encourages root growth but at the same time allows transpiration to remain low.

Only growers who do considerable propagation will want to go to the expense and effort of setting up and maintaining a bottom-heated mist propagation bed. For the average gardener who wants to start only a few plants now and then, a satisfactory chamber incorporating most of the innovations of the mist chamber can be devised (see "Propagating with Cuttings," Chapter 14).

Layerage. Because **layerage** allows the induction of roots on a part of a stem attached to the mother plant, it is a method of propagation well adapted for the home gardener who needs only a few new plants of any one kind.

The cultivars most frequently propagated by layerage today are dwarf apple and pear rootstocks. Reference was made earlier in the chapter to the extreme genetic variation of apple seedlings. Because uniform rootstocks cannot be propagated from seed and, until recently, apple and pear cuttings were almost impossible to root, a method was devised to propagate them called **mound layerage** (see Figure 4–10 and Chapter 14).

Figure 4–10 Mound layerage being used to root E. M. IX suckers that can be planted for dwarf apple rootstocks.

The homeowner can propagate a number of species, including low-growing evergreens, currants, gooseberries, and many deciduous ornamental shrubs by simply bending down a stem and covering it with soil. With some plants the chance of rooting is increased if a cut is made part way through the stem at the point where the stem is buried. Hormone treatment of the cut portion will sometimes also enhance rooting.

Various large houseplants such as philodendron, ficus (see Figure 14–29, Chapter 14), and dieffenbachia can be propagated by air layerage. Air layerage can also be used to get rid of an unsightly stem when these plants have lost their lower leaves (see "Layering to Renew or Multiply Plants," Chapter 14).

Graftage. In propagation by **graftage** parts of two plants are joined together in such a way that the two parts unite to form a single new plant. Graftage is used to perpetuate plant types that will not come true from seed; to adapt plants to unfavorable soil or climatic conditions; to control or prevent pest damage to roots; to control the size or shape of a plant; to change the top of a mature plant to another cultivar; and occasionally to repair rodent, machinery, or adverse weather damage to trees.

With graftage the part of the plant that will eventually grow to be the above-ground portion is referred to as the **scion,** and the part that will be the root is called the **stock** (Figure 4–11). If the scion consists of a complete cross section of stem with one or more buds, its union with the stock is called **grafting.** If the scion is a single bud, the joining of stock and scion is called **budding.** Grafting is most likely to be successful early in the spring before winter dormancy is overcome. Most budding can be accomplished only while the bark is slipping, usually in July or August in northern states. Apples and pears were formerly propagated primarily by grafting, but with the development of vegetatively propagated rootstocks, they, as well as the stone fruits and roses, are now normally propagated by budding.

Occasionally, because of graft incompatibility or for some other reason, it is necessary to graft a piece of stem of a third cultivar between the stock and scion. This is called double working and, the inserted piece is called an **interstock.** One good example of the use of an interstock is with the development of dwarf 'Bartlett' pear

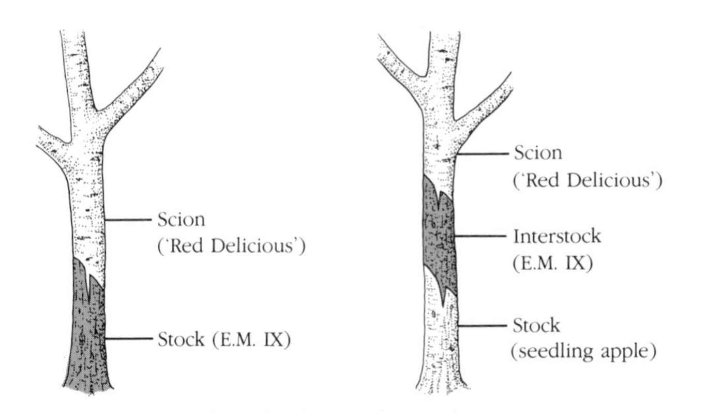

Figure 4–11 Two ways of producing a dwarf apple tree. A dwarfing interstock will have essentially the same effect as a dwarfing rootstock and is sometimes used when the dwarfing stock would produce a weak or inferior root system (compare Figure 8–14).

trees. Pear trees are made dwarf by grafting a scion of the desired cultivar onto quince rootstock. However, the most commonly grown pear cultivar, the 'Bartlett,' will not unite directly with most quince. In order to overcome this problem, propagators use as an interstock a piece of stem of 'Old Home' pear cultivar, which is compatible with both 'Bartlett' pear and quince.

At least three conditions must be met if a graft is to be successful. In the first place, the scion and stock must be graft-compatible. In general, this means that they are relatively closely related botanically. Discovering whether or not two species or two cultivars are graft-compatible is mostly a matter of trial and error. For example, apple and pear are graft-incompatible. However, most of the stone fruits will intergraft, one with another. The second condition that must be met is that the cambium of the stock be in contact with the cambium of the scion. It is the **callus** tissue that grows from the cambium or immature xylem and phloem immediately adjacent to the cambium that permits the union of stock and scion. For the most part, grafting is limited to plants that possess a definite cambium layer with associated layers of vascular tissue. The dicots and the gymnosperms can be grafted, but a union of stock and scion with monocots is not readily developed.

The third condition that must be met is that cut tissue in the graft region must be protected against drying while the graft union is being formed. Usually this is accomplished by coating the union with some type of grafting compound or wrapping it securely with plastic.

It should be borne in mind that grafting is not limited to woody plants. In some of the northern European countries, seedling stems of watermelon and cantaloupe are sometimes grafted onto roots of cold-resistant *Cucurbita* species to enable the melons to grow better and mature earlier. Coleus is another favorite for novel grafting experiments because of the many colors and leaf configurations available.

In addition to being a propagating mechanism, grafting is occasionally used for other horticultural purposes. Twisting together limbs in an approach graft across a weak crotch angle can greatly strengthen the framework of an ornamental or fruit tree. Grafting can also be used to insert pollinator branches into a tree that requires pollination by another cultivar. Already mentioned is the use of bridge grafting for repairing bark-damaged trees. Another use is to change the cultivar of a mature tree, which can be done by gradually cutting off older limbs and replacing them by budding or grafting with branches of the new or desired cultivar. Changing the cultivar of an old tree by graftage is called **topworking** (see Figure 14–54).

Selected References

Brooklyn Botanic Garden. *Propagation.* Handbook 24 (special printing of *Plants and Gardens,* vol. 13, no. 2). Brooklyn, NY. 1956.

Hammet, K. R. W. *Plant Propagation.* Drake, New York. 1973.

Hartmann, H. T., and Kester, D. E. *Plant Propagation: Principles and Practices.* 3rd ed. Prentice-Hall, Englewood Cliffs, NJ. 1975.

U.S. Department of Agriculture. *Seeds* (*Yearbook of Agriculture,* 1961). U.S. Government Printing Office, Washington, DC. 1961.

5 Soil and Soil Fertility

When natural resources are mentioned, most of us think of coal, petroleum, and mineral deposits. However, by far the most important natural resources a nation can possess are abundant areas of good soil and sufficient sources of fresh water. The United States is fortunate to possess a larger area of good, arable soils than any other nation and numerous rivers and lakes. Until a few decades ago most citizens felt and acted as though these resources were inexhaustible, but recently they have come to realize that both soil and water have finite limits and must be conserved.

Physical and Chemical Relationships of Soil

The main functions of soil in plant growth are to support the plant physically and to supply the necessary nutrients and water. It is possible for plants to grow in a medium of almost any kind of solid particles, and artificial media of various kinds are becoming increasingly popular, especially for greenhouse and container-grown plants. Potting soil and artificial potting media can be purchased from most nurseries and garden stores. These products are convenient to use if only a few pots or window boxes are being filled, but they are usually too costly for extensive planting. Despite the commercial importance of artificial media, most gardens are still produced in soil.

Soil Classification

Soil may be classified in various ways, including whether it is mineral or organic, on the basis of particle size, and according to parent rock.

Organic vs. Mineral Soils. The solid particles in soil are of two general types derived from two different sources. One comes initially from the weathering of rocks; it is **inorganic** or mineral in composition and highly resistant to change. The other solid component of soils is **organic**; it comes from decaying plant and animal material. Being relatively unstable, organic matter eventually breaks down, mainly into carbon dioxide, which dissipates into the air, and water, which becomes part of the soil moisture supply. As a consequence, organic matter must be constantly renewed.

When organic matter (O.M.) accumulates in swampy areas, its breakdown may be extremely slow because water excludes the oxygen necessary for decomposition of O.M. by microorganisms. Drained swamps and bogs are the main sources of **organic soils,** which by definition contain over 20% organic matter. These soils are called **muck** if they are well decomposed and **peat** if they are not. Organic soils are excellent for the production of vegetables and small fruits, but as they are generally low-lying they tend to be frosty (see Chapter 7) and, as a consequence, are not used extensively for the production of tree fruits. Organic soils have good water-holding capacity, aeration, permeability to water, and fertility.

If soils are composed of less than 20% organic matter, they are classified as **mineral soils.** The texture and structure of mineral soils determine their suitability for plant growth. Because most gardens are grown in mineral soils, most of the following discussion will relate to these soils.

Soil Texture and Structure. The **texture** of a soil is determined by the proportion of each of the different sized particles it contains. The size range of each kind of particle varies slightly depending on the classification system. The size range of sand, silt, and clay particles in two commonly accepted classification systems is shown in Table 5–1. Virtually all soils are composed of a mixture of particles of different sizes. Those having a textural mix and physical properties favorable for plant growth are called **loams.** Various textural classes are shown in Figure 5–1.

Table 5–1 Classification of soil particles.

Particle	Diameter (mm)
USDA	
Boulders	>256
Cobbles	256–64
Pebbles	64–4
Gravel	4–2
Fine gravel	2–1
Coarse sand	1–0.5
Medium sand	0.50–0.25
Fine sand	0.25–0.10
Very fine sand	0.10–0.05
Silt	0.05–0.002
Clay	<0.002
The International Classification (Atterberg) System refers only to soil particles under 2 mm.	
Coarse sand	2.0–0.2mm
Fine sand	0.2–0.02
Silt	0.02–0.002
Clay	<0.002

(From USDA, 1957.)

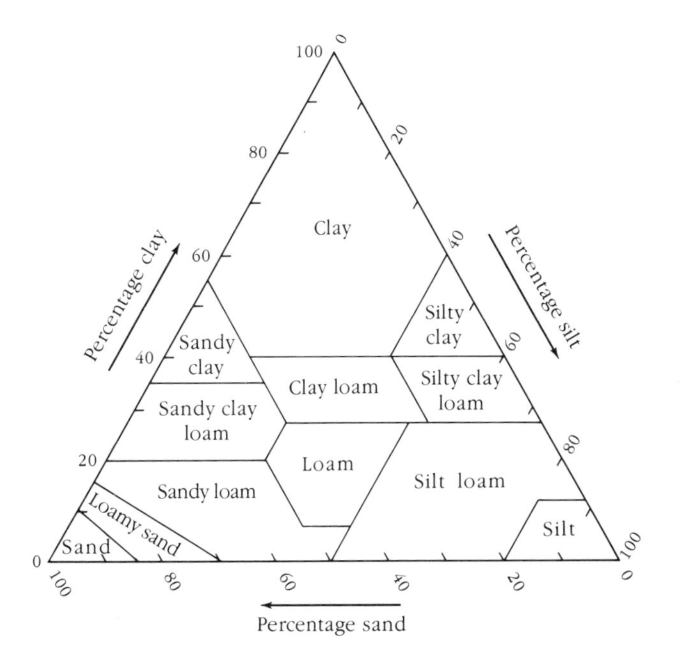

Figure 5–1 Soil texture. The texture triangle shows the relative percentages of sand, silt, and clay in each textural class. In the United States the term loam refers to a soil with more or less equal proportions of sand, silt, and clay. (Courtesy USDA.)

Soil **structure** refers to the way soil particles group into small clumps or **aggregates** (Figure 5–2). Aggregation in soils is dependent on the cohesive nature of the finer particles (clay and organic) and on physical forces that organize them into structural units. Structure is important in fine-textured soils, because it affects pore space. Soil with large structural units has large pores that provide it with aeration and permit water infiltration. Soil with small structural units has small pores that provide it with water-holding capacity.

Soil aggregates can be of several structural forms based largely on shape, including flat or platy, prismlike, blocklike, and spheroidal. **Spheroidal** aggregates are the most desirable, because they provide the best combination of pores for aeration, drainage, and water-holding capacity. The small pores within the structural units retain water, and the large pores between them provide aeration and drainage. Spheroidal aggregates are built around a

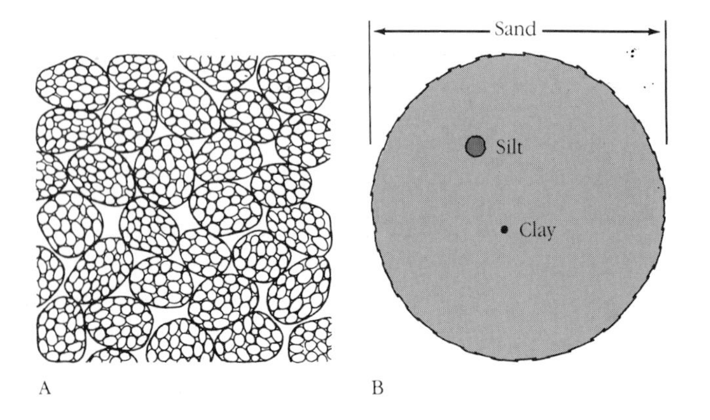

Figure 5-2 Soil structure. (A) Diagrammatic sketch of well-structured soil. Aggregation of soil particles provides the fertility- and moisture-retaining advantages of fine-textured soil and the aeration and workability of coarse-textured soil. (B) relative sizes of sand, silt, and clay soil particles. (B, courtesy USDA.)

central core and have rounded or irregular surfaces. They are called granules if relatively dense and crumbs if relatively porous.

Among the many physical forces that bring about aggregation are alternate freezing and thawing, alternate wetting and drying, activity of earthworms, and growth of roots. Matted roots of grasses are especially conducive to aggregation. Organic matter is important for good soil structure, because it acts as a cementing agent within structural units and because humic acid from its breakdown tends to coat the structural units and preserve their integrity, which, among other benefits, increases the proportion of spheroidal aggregates. For maximum benefit to soil structure, organic matter must be biologically active; that is, it must be in a state of being renewed by additions of manure or crop residues and of being broken down by microorganisms.

Growers who garden in fine-textured soils must be careful not to disturb them when they are wet. Pressure of any kind (cultivation, heavy rain, vehicular traffic, walking) on water-saturated soil causes the particles within an aggregate to slip past one another, destroying structure. If the soil has a high proportion of clay particles, the air spaces will be eliminated, and the soil characteristically becomes a sticky, viscous mass. When fine-textured soils that have lost their structure dry, they form large, hard clods that are difficult or impossible to work (Figure 5-3). Once structure is destroyed, it is reestablished slowly with physical forces mentioned earlier.

Coarse-textured soils do not have much structure. Having large pores, they have good aeration and water infiltration; however, lacking small pores, they lack water-holding capacity.

Soil texture and structure affect many aspects of gardening. Sandy or coarse-textured soils warm rapidly in the early spring and remain friable and amenable to cultivation even when wet. As a consequence, they are especially desirable for early vegetables and flowers and for root, bulb, and tuber crops. Water soaks into sandy soils readily, but such soils have low water- and nutrient-retaining capacities and are likely to be droughty and to lack fertility. Clay or fine-textured soils warm and dry slowly and cannot be worked when wet. Clay soils absorb water slowly but retain nutrients and moisture well, so crops growing on them will not need irrigation or fertil-

Figure 5-3 Soil formed into hard clods because it was worked when excessively moist.

izer as frequently as those growing on sandy soils. Clay soils are best suited for main-season and late-maturing crops where high yields are important.

Soil Profile

A vertical section through the soil, as seen in an excavation, usually shows distinct layers of varying thickness called **horizons.** The series of horizons of a soil is referred to as the **soil profile.** Horizons in each soil vary in thickness and composition, but generally by the time forest or prairie soils have remained undisturbed for a long period, they will have developed three major horizons designated from top to bottom as A, B, and C. The A and B horizons are often subdivided into A_0, A_1, A_2, and A_3; and B_1, B_2 and B_3 horizons (Figure 5–4).

The A horizon is the zone from which soluble substances are dissolved and washed downward (**leached**); the B horizon is the zone in which these substances are accumulated; and the C horizon is made up generally of the weathered bedrock from which the soil was formed. The bedrock itself is often referred to as the D horizon.

From the point of view of the gardener the A horizon is the one that supports most plant roots and from which plants extract most of their mineral nutrients and water. The B and C horizons will have an important effect on water infiltration; drainage; and growth of trees, shrubs, and other deep-rooted plants. In some soils that have been cultivated for long periods the A horizon may have been washed away (**eroded**), and crops will be seeded and growing in the B horizon (Figure 5–5). Such soils are almost always less fertile and less easily cultivated than those with the A horizon intact.

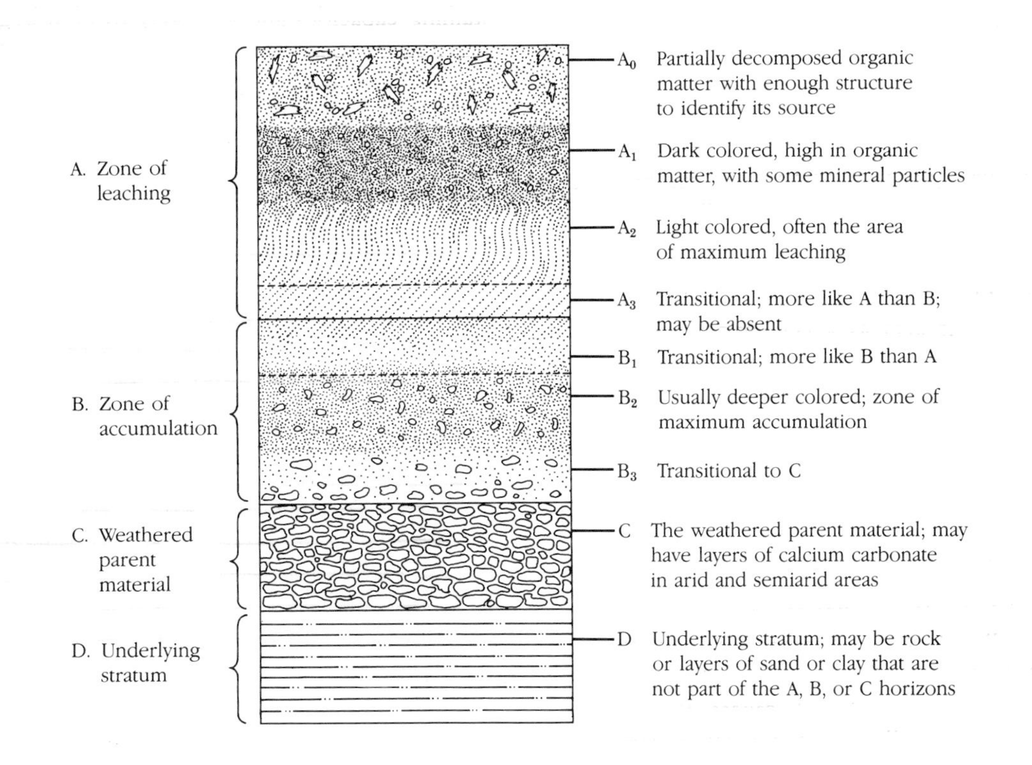

Figure 5–4 A soil profile showing the principal horizons. (Adapted from USDA.)

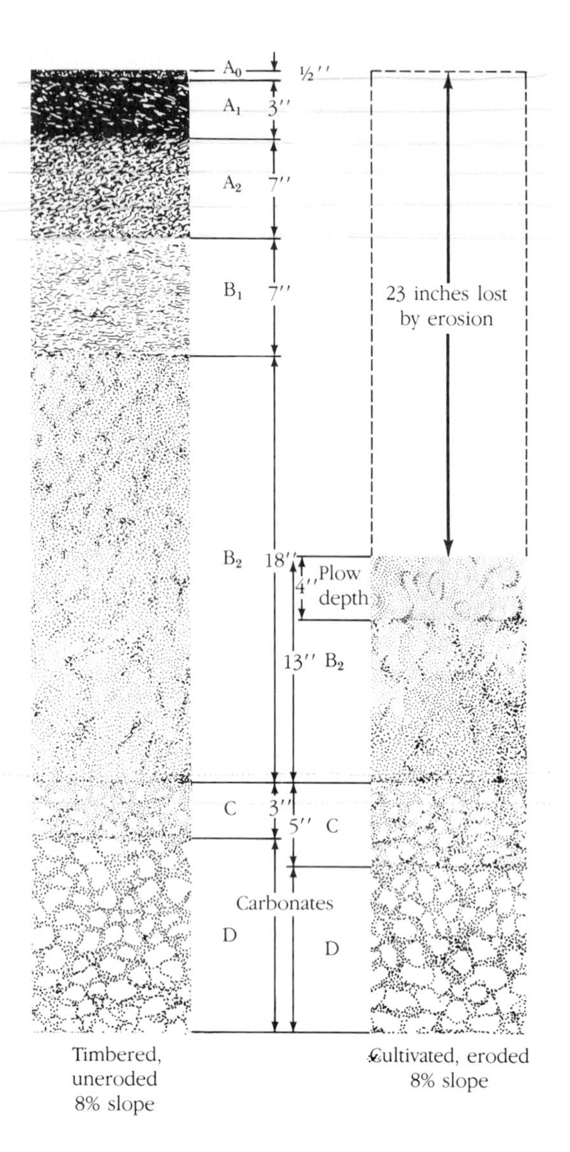

Figure 5–5 Eroded soil. Uncontrolled erosion can result in the loss of productive topsoil, as shown by these profiles taken in Miami silt loam on adjoining fields. Profile at left was taken on a field still covered by virgin timber; profile at right was taken on a field cleared and farmed for 50 to 75 years. (Adapted from USDA.)

Some garden soils that were formed recently by water deposits or by human activity, such as excavation, filling in, and building a new home, do not have well-developed horizons. This is no great disadvantage if the soil has other good plant-growing characteristics.

Organic Matter and Compositing

The organic content of the soil includes living organisms, waste products excreted by animals, and the residues from the decomposition of animal and plant bodies. The extent and importance of living organisms in the soil is sometimes overlooked. An acre-foot (a volume 1 acre in size and 1 foot deep; approximately 1,200 cubic meters) of fertile soil will contain up to three tons of living organisms, including worms, insects, and microorganisms (Table 5–2). Microorganisms are dependent upon organic matter and contribute to the structure formation and maintenance of the soil.

Organic matter is important to soil productivity in many ways. It holds mineral nutrients in forms available for plants and influences soil structure and chemistry. It increases water-holding capacity and permeability of soils. Soils high in organic matter are more amenable to cultivation and less susceptible to erosion.

Sources of Organic Matter. Most mineral soils benefit from added organic matter. Formerly, gardeners obtained organic matter primarily from animal manures, large

Table 5–2 Average weight of organisms in the upper foot of soil (in lb/acre). (Multiply by 1.12 for kg/hectare.)

Organism	Low	High
Bacteria	500	1,000
Fungi	1,500	2,000
Actinomyces	800	1,500
Protozoa	200	400
Algae	200	300
Nematodes	25	50
Other worms and insects	800	1,000
Total	4,025	6,250

(From *Experiments in Soil Bacteriology*, by O. N. Allen, Burgess, Minneapolis, MN. 1957.)

amounts being readily available from barns, stables, and streets. Diversified farming was the rule then, and the market gardener normally also had a small dairy herd, a few beef animals, or a flock of sheep, as well as some chickens. Today, most manure is produced at large feed lots concentrated in certain parts of the country. Growers living near the feed lots can easily obtain manure for their crops, but for the majority of gardeners the source of manure is too distant to make it economical to spread on their land, so they must rely on other sources of organic matter.

Organic matter for today's garden comes mainly from the plowing under of crop residues and the growing of soil-improving crops. Almost any plant can be used as a soil-improving crop. However, two groups are commonly utilized: the legumes and the grasses, including cereal crops. The legumes, plants belonging to the family Leguminosae (Fabaceae), have an advantage in being able, with the aid of nitrogen-fixing bacteria on their roots (Figure 5–6), to change elemental nitrogen from the atmosphere into forms that plants can use. With the decay of leguminous plants, this nitrogen becomes available to other plants growing in the soil. Alfalfa, vetch, sweet clover, clovers, winter peas, and southern peas are some of the legumes grown as soil-improving crops. Because it is a deep-rooted perennial, alfalfa has the additional advantage of opening the subsoil to moisture and aeration. As alfalfa roots die, the channels they have formed remain to conduct water into the lower depths of the soil.

Many grasses have heavy matted root systems that provide considerable organic matter when the grass dies or is plowed under. The cereals, especially rye, are frequently used as soil-improving crops. Because these plants are cool-season annuals, they often can be planted after the garden is harvested. They grow relatively well during cool weather and thus will produce considerable organic matter during the season when garden crops cannot be grown. In areas of intense production organic matter is most frequently added to soil by growing cereal crops during the cooler part of the season.

Sewage **sludge** is available to home gardeners living near larger centers of population. If the sludge has been properly treated, it is clean, safe, and odorless and provides both organic matter and mineral elements to plants.

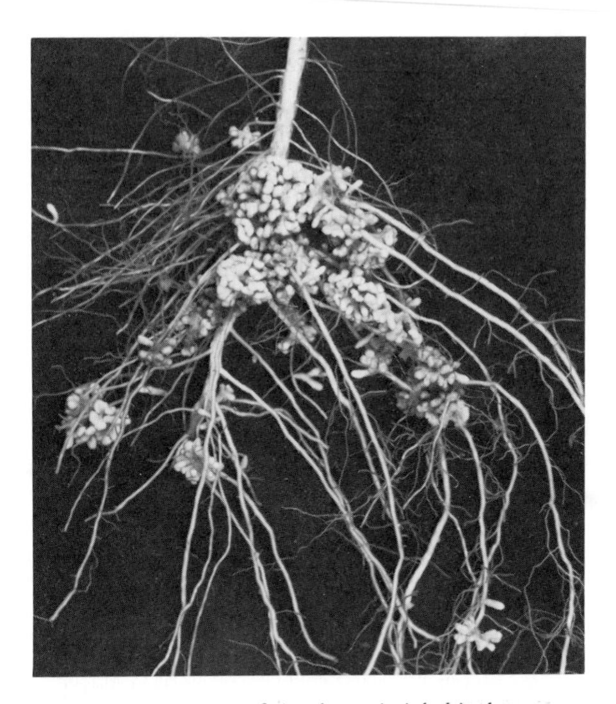

Figure 5–6 Nitrogen-fixing bacteria inhabit the nodules on the root of leguminous plants. Nitrogen from the atmosphere is made available for plant growth in the symbiotic relationship between the bacteria and the higher plant. (Photo by L. J. Klebesadel, courtesy USDA.)

During the course of the year the average home will generate hundreds of pounds of organic waste materials that can be turned into an excellent compost for adding to the garden soil. **Composting** is the process that results in decomposition of organic wastes into innocuous materials that can be used for mulching, fertilizing, and soil conditioning. This kind of decomposition, which takes place whenever organic matter is in contact with soil, occurs as a result of the activity of microorganisms. Artificial composting requires the manipulation of microorganisms, as do a number of other horticultural activities, including food storage and processing and plant disease control. To rapidly decompose organic wastes into compost, microorganisms require moisture, heat, oxygen, and a balanced supply of mineral nutrients (see "Composting," Chapter 14).

Organic Matter and Nitrogen. A quantity of nitrogen in addition to whatever amount is required by the crop must be added if sawdust, straw, fruit and vegetable wastes from the table, leaves, or other nonnitrogenous, undecomposed organic matter is dug directly into the soil or if soil-improving crops or crop residues are plowed or dug under. The reason is that addition of organic matter stimulates reproduction of microorganisms, which can extract nitrogen more easily than can plants. If soil nitrogen is not sufficient to supply both the microorganisms and the plants, it will be utilized by the microorganisms, leaving an insufficient amount for satisfactory plant growth. Extra nitrogen is also needed if new wooden or fiber flats are used for growing transplants; if flats are lined with newspaper to prevent loss of soil through the cracks; or if straw, sawdust, or peat moss mulches are worked into the soil.

Soil Chemistry

Two aspects of soil chemistry, cation exchange and pH, are of sufficient importance to the gardener to deserve some explanation.

Cation Exchange and Absorption. The weathering of rocks, the breakdown of organic matter, and the addition of fertilizer all add mineral salts to soil. When a salt is dissolved by the soil solution, it dissociates into its ionic form. (The **ions** are the negatively and positively charged atoms that join together to form the salt molecule.) For example, calcium chloride, $CaCl_2$, dissociates into Ca^{2+} and two Cl^- ions. Positively charged ions (Ca^{2+}, Mg^{2+}, NH_4^+, K^+, etc.) are called **cations,** and negatively charged ions (OH^-, Cl^-, CO_3^{2-}, SO_4^{2-}) are called **anions.** It is the ions, not the salts, that the plant absorbs and that are the minerals essential for plant growth. Clay and organic matter particles, which are sometimes referred to as soil **colloids** because they are small enough to remain suspended in a solution, possess the ability to attract and exchange cations, and the number of these particles present determines the soil's **cation-exchange** capacity. Organic matter particles will hold 1½ to 30 times as many ions as will clay particles, emphasizing again the importance of organic matter in the soil.

Water molecules and acids dissociate to form H^+ ions, and an equilibrium of H^+ and other cations is maintained in the soil solution. As various cations are absorbed into the plant, their concentration in the soil solution is restored by ions displaced from organic matter and clay particles. The niches on the colloids vacated when the mineral cations move into the soil solution are in turn filled by H^+ ions. Finally, when the concentration of cations is again increased in the soil solution by decomposition of plant residues or addition of fertilizer, the excess of these ions displaces the H^+ ions from the colloids, thus completing the cycle. It should be emphasized again that plants absorb nutrients from soil mainly in the form of ions, a point of importance in deciding the best source of fertilizer for the garden.

Potential fertility of a soil is dependent on its cation-exchange capacity, which in turn is determined by its content of clay and organic matter particles. Soils with a low cation-exchange capacity tend to be low in fertility, because cations added to such soil can readily wash from the soil with excess water. The growing of healthy crops on low cation-exchange soils requires light, frequent fertilization. In contrast, soils with high cation-exchange capacity will tend to retain cations in a form available for plant growth and will require less frequent application. Retention of anions like NO_3^- and SO_4^{2-} will not necessarily be related to cation-exchange capacity.

Soil pH. The second aspect of soil chemistry of importance to gardeners is soil **pH,** which is the relationship of the hydrogen ions (H^+) to the hydroxyl ions (OH^-). These two ions, in turn, determine whether a soil is acid or alkaline. When there are the same number of H^+ as OH^- ions in the soil solution, the soil is neutral and has a pH of 7; when H^+ is higher than OH^-, pH is lowered and the soil becomes acid; and when H^+ is lower than OH^-, the pH is high and the soil is alkaline.

Most exchangeable soil ions affect pH to a degree, but in most soils pH is largely determined by four cations. Two base-forming cations, calcium (Ca^{2+}) and magnesium (Hg^{2+}), are the primary contributors to alkalinity; and two acid-forming ions, hydrogen (H^+) and aluminum

(Al^{3+}), largely determine acidity. Three other base-forming ions, potassium (K^+), sodium (Na^+) and ammonium (NH_4^+) usually are not present in amounts that materially affect pH.

Because of the tendency for the base-forming cations to leach away and be replaced by H^+ and Al^{3+}, soils formed in areas of high rainfall tend to become acid. Desert soils or soils formed in areas of low rainfall are likely to be neutral to alkaline. In North America the Great Plains, the Southwest, the high western plateaus, and many of the Rocky Mountain valleys have neutral or alkaline soils. These are all areas of low rainfall. All areas east of the Great Plains; the coastal valleys of northern California, Oregon, Washington, British Columbia, and Alaska; and a few high-altitude, high-precipitation valleys of the northern Rockies have acid soils.

Most crop plants do best in soils that are neutral or slightly acid. However, a few ornamentals, including rhododendrons, azaleas, and heathers, and cranberries and blueberries, all of which originated in areas of high rainfall, grow best in highly acid soils (Table 5–3). Near their southern limit of adaptability in eastern Idaho and west-

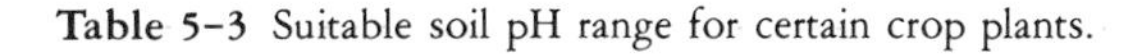

Table 5–3 Suitable soil pH range for certain crop plants.

Crops	pH ranges (4.5 · 5.5 · 6.5 · 7.5)
Alfalfa, Sweet clover, Asparagus	
Beets, Cauliflower, Apples, Onions, Lettuce	
Spinach, Red clovers, Peas, Crucifers, Kentucky bluegrass, White clovers, Carrots	
Juniper, Iris, Squash, Strawberries, Lima beans, Snap beans, Velvet beans, Cucumbers, Many grasses, Tomatoes, Timothy, Barley	
Wheat, Fescue (tall and meadow), Corn, Soybeans, Oats, Alsike clover, Crimson clover, Vetches, Millet, Cowpeas, Lespedeza, Tobacco, Rye, Buckwheat	
Sweet potatoes, Red top, Potatoes, Bentgrass (except creeping), Fescue (red and sheep's)	
Poverty grass, Blueberries, Cranberries, Azalea (native), Camellias, Hydrangea, Gardenias, Rhododendron (native)	

(Adapted with permission of the publisher from *The Nature and Properties of Soils*, 8th ed., by N. C. Brady. Copyright 1974 by the Macmillan Company.)

ern Wyoming, mountain blueberries grow mainly on north slopes of high-altitude areas, most likely because of higher soil acidity at these sites.

The pH relates to many aspects of the soil important to plant growth. The physical condition of the soil may be altered by changes in pH, and soil can lose structure if it becomes strongly acid or alkaline. Soil mineral nutrients are often unavailable to plants when the soil is strongly acid or alkaline—iron, manganese, copper, zinc, and boron are likely to be unavailable to plants in highly alkaline soils; calcium and molybdenum may be unavailable in highly acid soils; and phosphorus may be unavailable in either highly acid or highly alkaline soils. Soil pH can also affect certain plant diseases. For example, potato scab is less severe if the crop is grown in acid soil, and club root of crucifers can be controlled by applying lime to make the soil alkaline.

Lime, the alkaline salt of calcium, can be used quite successfully to raise the pH of soil, and most cultivated soils in high rainfall areas are limed in order to improve crop production (Figure 5–7). Sulfur can be used to lower the pH and make the soil more acid; however, the effect of sulfur is not as predictable, nor is its use as widespread as is the use of lime.

Soil Fertility

Plants obtain nutrients essential for their growth from the soil, but not all nutrients in the soil are utilized in plant growth. Some kinds are not required by plants, and some of the required kinds are combined in chemical forms that plants cannot utilize. Soil fertility is usually judged on the level of available nutrients in the soil and is dependent on a great many complex physical and chemical interrelationships, some of which have already been discussed.

Essential Elements Required by Plants

Sixteen elements are known to be required by plants for proper growth. I have found that the easiest way to remember the chemical symbols for these is by means of the mnemonic device "C Hopkins Cafe managed by mine cousins Mo and Cleo." This stands for C HOPKNS CaFe Mg B Mn CuZn Mo and Cl. The first three elements, carbon (C), hydrogen (H), and oxygen (O), come from the atmosphere; the rest are obtained from the soil, absorbed into the roots as ions, and transported through the

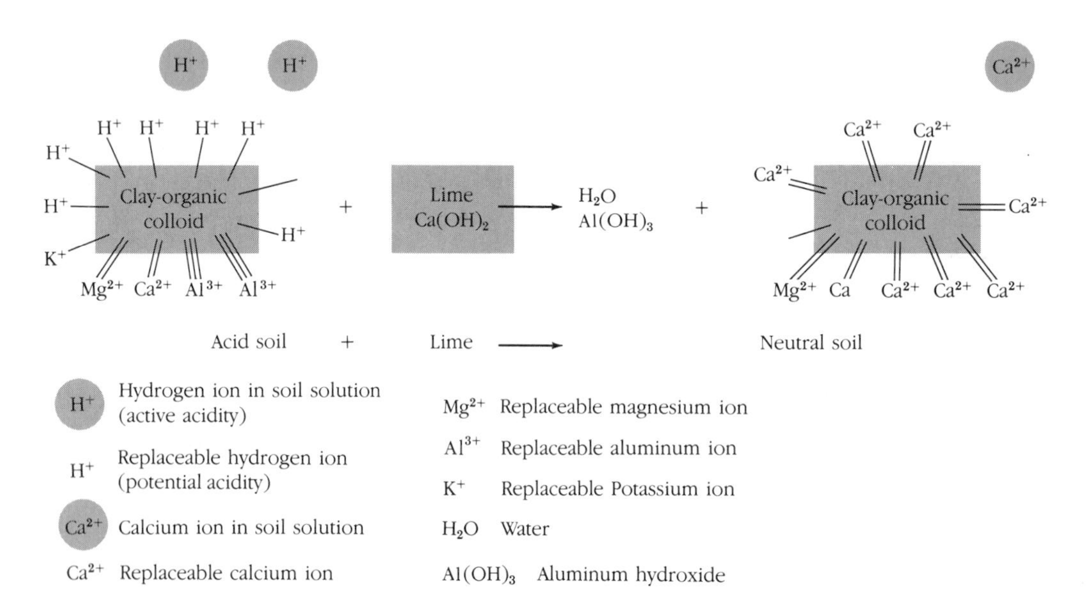

Figure 5–7 Cation-exchange reactions when an acid soil is limed. (Courtesy USDA.)

xylem to various parts of the plant. The next three elements, phosphorus (P), potassium (K), and nitrogen (N), are those most likely to be deficient and most often added to soils. These three, along with sulfur (S), calcium (Ca), and magnesium (Mg), are needed in relatively large quantities to sustain plant growth and are classified as **macronutrients.** The remaining elements, iron (Fe), boron (B), manganese (Mn), copper (Cu), zinc (Zn), molybdenum (Mo), and chlorine (Cl), although essential for growth of plants, are used in minute quantities and are referred to as **micronutrients** (see Table 5–4).

Soil Fertilization

In nature the growth of plants and other organisms feeding from the soil is balanced to utilize the soil nutrients available. If the requirements for a species are not adequate, that species either grows at a reduced rate commensurate with the food available or is eliminated from the ecosystem. The slow growth that would result from letting nature "take its course" is usually not feasible in a garden, so fertilizer must be applied to allow maximum growth of the desired crop (see "Fertilizing Garden Crops," Chapter 14).

Two general types of fertilizers are available: the so-called natural, or organic, fertilizers and the inorganic, or chemical, fertilizers. The organic fertilizers are materials derived from living organisms and include manure, fishmeal, cottonseed meal, bonemeal, and sewage sludge. The chemical fertilizers are commercially prepared from inorganic minerals and include such things as ammonium nitrate, superphosphate, and muriate of potash. Because chemical fertilizers are, on the whole, less expensive to purchase, transport, and apply, the bulk of fertilizers used today are chemical rather than organic.

Growers are using a great deal more commercial fertilizer today than ever before. Partly this is because manure is scarce, but primarily it is because they have found that the addition of fertilizer is the most economical way to increase yields of most crops. An excess of fertilizer, however, can be detrimental. Low color in apples, splitting of carrots, heavy vine growth and delayed ripening of tomatoes, and reduction in flowering and fruiting of ornamentals and fruits can all be the result of heavy nitrogen application or an imbalance between nitrogen and the other elements available to the plant.

Table 5–4 The principal ionic forms of nutrients utilized by plants.

Element	Cations	Anions
	Macronutrients	
Nitrogen	NH_4^+	NO_3^-
Calcium	Ca^{2+}	
Magnesium	Mg^{2+}	
Potassium	K^+	
Phosphorus		HPO_4^{2-}, $H_2PO_4^-$
Sulfur		SO_4^{2-}
	Micronutrients	
Copper	Cu^{2+}	
Iron	Fe^{3+}	
Manganese	Mn^{2+}, Mn^{4+}	
Zinc	Zn^{2+}	
Boron		BO_3^{3-}
Molybdenum		Mo_4^{2-}
Chlorine		Cl^-

(From *Soil Science: Principles and Practices,* 2nd ed., by R. L. Hausenbuiller, Wm. C. Brown, Dubuque, IA. 1978.)

Fertilizer Analysis, Ratio, and Formula. Fertilizer analysis must be understood if fertilizer is to be applied to crops in the correct amounts. Fertilizer **analysis** is the percentage by weight of the available nitrogen (as elemental nitrogen), phosphorus (as P_2O_5), and potassium (as K_2O); by law, these percentages must be listed on every bag of commercial fertilizer sold. (Soil scientists are attempting to standardize the desirable practice of using elemental phosphorus and potassium as they now use elemental nitrogen as the bases for fertilizer analyses, but most commercial companies continue to base their printed analyses on the oxide forms, as we will do in this text.) The percentage of nitrogen (N) is always listed first, phosphorus (P) second, and potassium (K) third. Thus a fertilizer bag with an analysis of 16:20:0 would contain 16% N, 20% P, and no K.

Fertilizer **ratio** is the proportion of the three major elements found in the fertilizers. For example, fertilizer with an analysis of 5:10:5 would have a ratio of 1:2:1, and a 16:20:0 fertilizer would have a ratio of 4:5:0. The balance of various nutrient elements in the soil is as important as is the total amount of each element. Fertilizer **formula** is a statement of the materials and the amounts of each chemical that make up the fertilizer—ammonium nitrate, ammonium sulfate, superphosphate, and so on. Commercial fertilizers are formulated to fulfill varying needs of growers and will often be a mixture of fertilizers of several formulas that may or may not be listed on the fertilizer bag (Table 5–5).

Analysis of different fertilizers varies over a wide range from less than 1% available N, P, and K to more than 80%. Those with more than 30 lb (14 kg) total N, P, K per 100 lb (45 kg) bag are called **high analysis,** and those with less than 30 lb are **low analysis.** Because of the extra weight and consequent cost of handling, low-analysis fertilizers tend to be more expensive per unit of available nutrient. High-analysis fertilizers require more accuracy in application.

Because analyses vary so much, fertilizer recommendations are usually stated as pounds of N, P, or K, and the grower must calculate the amount of fertilizers needed to supply the crop. For example, if 100 lb per acre nitrogen are required, the grower will need to apply 300 lb (140 kg) of fertilizer of an analysis of 33:0:0 or 625 lb (283 kg) of one of 16:20:0 (see "Fertilizing Garden Crops," Chapter 14).

Table 5–5 Comparison of formula, analysis, and ratio of several fertilizers.

Formula	Analysis	Ratio
Ammonium nitrate	33:0:0	1:0:0
Ammonium phosphate sulfate	16:20:0	4:5:0
Sheep manure	2:1:2	2:1:2
Fertilizers synthesized from several formulas	5:10:5	1:2:1
	16:16:16	1:1:1

Determining Fertilizer Requirements. Determining which fertilizer elements are needed is frequently a dilemma for home gardeners. A deficiency of a particular element will produce distinct symptoms in plants, and growers can often determine which elements are lacking by comparing plant symptoms with pictures of nutrient deficiencies in one of the many publications printed on this subject (see Figure 5–8). However, by the time deficiency symptoms appear, it is usually too late to correct

Figure 5–8 Nitrogen deficiency. The small tomato, marigold, and cabbage seedlings on the right show typical light color and stunted woody growth symptoms of acute nitrogen deficiency when compared to more adequately fertilized seedlings of the same age on the left. Seedlings as badly stunted as those on the right will never produce the quality or abundance produced by plants that have not suffered this kind of stress.

them that season, and irreparable deterioration in yield and quality will have already occurred.

There are several ways of determining nutrient deficiencies more satisfactorily than by observing symptoms. Past experience with one's own and neighbors' successes and failures often can be helpful. Most state experiment stations and Canadian Department of Agriculture research stations publish general fertilizer recommendations for most crops in each major area of the states and provinces, and these recommendations are helpful to the home gardener. In recent years gardeners have come to rely more heavily on soil testing. A number of do-it-yourself soil-testing kits are available, and these generally will give a rough (often extremely rough) guide to fertilizer needs. A more accurate recommendation can be obtained by sending samples of the soil to a state or reliable local commercial soil-testing laboratory (Figure 5–9). County agents have information on soil-testing services. Soil-testing laboratories analyze for available nutrients and pH and recommend fertilizer requirements for both major and minor elements.

A new commercial grower service not yet widely available for home gardeners is **foliar analysis,** in which laboratories monitor the nutrient content of plant materials periodically through the season. A slight drop in foliar level of a nutrient indicates a pending shortage that can be corrected by soil or foliar fertilizer application.

Sources of Fertilizer. Home gardeners are often enticed into buying expensive formulations of fertilizers with the idea that the brand contains something a little better than a competitive cheaper brand. In general, the fertilizer that supplies the greatest amount of the element needed per unit of cost is the best buy. There are a few exceptions. One of these might be a formulation where ammonia supplied as a gas would be difficult for the homeowner to apply. Another one might involve the use of special application equipment where the size or shape of the fertilizer particles make a difference in the ease and uniformity of application. Commercial fertilizers containing N, P, and K can be purchased in all areas of the United States and Canada. Other mineral elements are available for purchase at locations where soils require them.

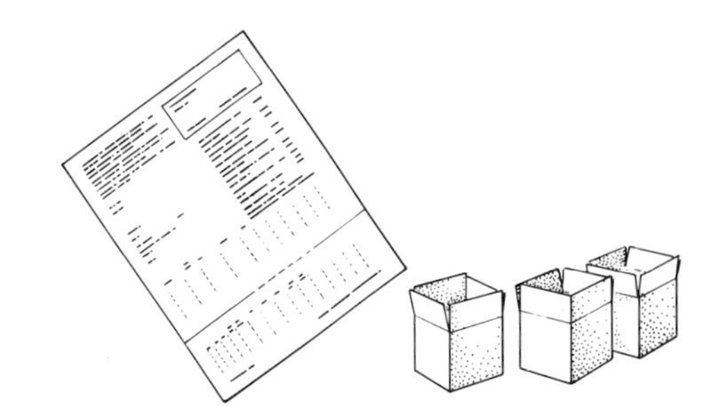

1. Obtain cartons and information sheets from your county agent, the state soil testing laboratory, or other sources.

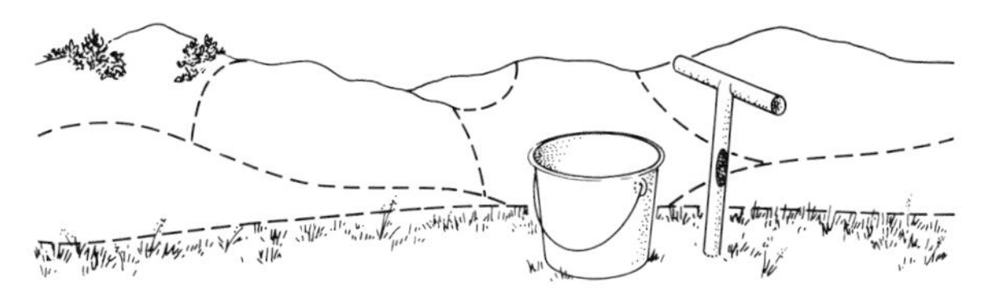

2. Map the different areas within a field—such as hilltops, midslopes, bottomlands, or known areas of different productivity. With a sampling tube take 10 to 15 cores, spaced an equal distance apart. Sample to tillage depth. Place cores from each sampling area in a clean bucket. Mix this composite sample well and fill the soil sample carton (about 1 pint). Repeat this process for each area in the field.

3. A field that is extremely variable, or one where little is known about the variability, requires many samples. Once a field has been intensively sampled and a soil fertility map made, select sites in representative low-, medium-, and high-fertility areas of the field that can be resampled every two to three years. Periodic resampling will show if the general soil fertility level in each area is improving or getting worse.

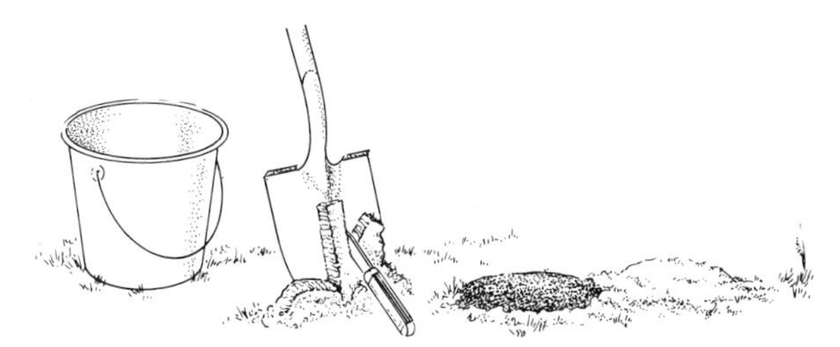

4. It is best to use a sampling tube if that is possible. If you use a spade or shovel, throw away the first shovelful. Then take a 1-inch slice from the back side of the hole (to proper sampling depth) and trim away sides of slice, leaving a 1-inch center core. Place core in a clean bucket, following procedure given in item 2. A garden trowel can be used in place of a spade or shovel.

Figure 5–9 Taking routine soil samples for cultivated crops. (Adapted from Washington State University Extension Circular 387, by A. R. Halvorsen.)

Occasionally the effectiveness of a fertilizer compound will depend on soil conditions. For example, the application of rock phosphate is recommended in some organic gardening texts. The phosphorous in rock phosphate is soluble in acid but not in neutral to alkaline soils. In acid soils rock phosphate decomposes gradually into forms that plants can utilize and provides some fertilizing benefit. In soils of pH 6.5 or higher, however, it never decomposes, and its application is a waste of money and time.

Sometimes one salt of a major element will produce better results than another. Lawns in some areas benefit more from nitrogen fertilization with ammonium sulfate than with ammonium nitrate. In this and other examples of differential response, the more beneficial procedure may be providing a needed element (sulfur, in the lawn example) or perhaps a change in the pH.

Formulation is also important for applications of minor elements in some locations. Soils of semiarid irrigated areas generally contain ample iron, but it is often in precipitated form unavailable for plant growth. Leaves of fruit trees and ornamentals growing in these soils become yellow between the veins, a symptom referred to as lime-induced iron chlorosis. Iron applied to these soils in the form effective on acid soils is precipitated and becomes immediately unavailable for plant use. The most effective treatment for lime-induced iron chlorosis is application of an organic iron formulation called iron chelate developed especially for alkaline soils.

Organic vs. Inorganic Fertilization. Gardeners must also decide whether to use organic or chemical fertilizers. Organics do have some advantages as fertilizers. Many contain slight amounts of trace elements in addition to N, P, and K. They, as well as synthetic organics such as ureaform, undergo a slow change to forms available for plant growth and thus extend the period of availability of an application of fertilizer. This is the basis of the highly advertised slow-release fertilizers.

However, there are disadvantages to the use of organic fertilizers. Slow release of nutrient elements in most situations is beneficial, but it may be undesirable if the crop has immediate need for nutrients. Some manures and sewage sludge from some urban areas contain considerable quantities of heavy metals thought to come primarily from metal sewer pipes and industrial wastes. These could be injurious to crops, although this probably is not a serious problem because of the scattered occurrence of organic materials so contaminated. Perhaps the major disadvantage of organic fertilizers is their bulkiness relative to the amount of nutrients they contain, with the result that they are more expensive to transport and apply. Manures have extremely low analyses, usually less than 3%. See Table 14–19 for an analysis of composition, effect on soil pH, and weight/volume equivalent of some common fertilizers.

Whether its nutrients come from organic or chemical fertilizers makes no difference to the plant. As was pointed out earlier in this chapter, soil nutrients are absorbed into the plant in the form of ions, and ions from organic fertilizers are exactly the same as those from chemical fertilizers. If soil organic matter is maintained with soil-improving crops, crop residues, or compost, plant growth will be as luxuriant with chemical as with natural organic fertilizers.

Managing Soil and Fertilizer for the Garden

Soil for the New Home

The easiest way for a homeowner to be certain of having usable soil is to supervise carefully the excavation and leveling at the time the house is built. In most regions, the top 12 inches (30 cm) of soil will have relatively good structure and will contain the most organic matter. The owner should make certain that this soil is scraped to one side of the lot before any excavating is begun so that it can be spread uniformly over the lot after the building is completed (Figure 5–10). This kind of careful supervision may be impossible at an already built tract or subdivision home, where all too frequently sticky or rock-hard clay subsoil is left on the surface of the yard. From 6 to 10 inches (15 to 25 cm) of good topsoil spread over such a yard would, of course, be desirable, but topsoil is expensive, and a good lawn can be grown in subsoil improved with generous additions of organic matter. If purchased topsoil is limited, it should be reserved for flower beds and other specialized growing areas. Boards, sticks, stones, rubble, plasterboard, and other debris should, of course, be cleaned from the top 12 inches (30 cm) of soil before the area is planted to insure successful root penetration and moisture availability. The frequent complaint of geometric dead or dying patches in a newly planted lawn can almost invariably be traced to pieces of concrete, board, or plasterboard that have been buried shallowly when the yard was leveled.

Figure 5–10 Soil for a new home. Before excavation for a new home is begun, the owner should make certain that the topsoil is scraped from the excavation site and piled to one side so that it can be spread over the surface of the disturbed area.

Managing Soil on Sloping Sites

Sloping sites offer the opportunity for interesting garden effects, but they also present a challenge in erosion prevention. Perhaps the simplest management for a sloping site is a permanent low-growing cover crop like that shown in Figure 5–11. If the topography is not too steep and sufficient moisture is available, grass sod that can be kept mowed may be the simplest, although the least creative, solution. If the site is too steep to mow of if grass is not the preferred cover, low-growing ornamental or small fruit shrubs can be used. Trees, including fruit trees, can be planted with the grass or shrubs if these cover crops are shade-tolerant and the soil is deep enough. A rock garden is another attractive way of managing a slope, although it requires a good deal of labor.

It is possible to use a sloping site for a vegetable or annual flower garden. If the slope is not too steep, erosion can be minimized by cultivating and orienting rows with the contour of the slope. In the early fall seed of rye, Austrian winter peas, or other rapidly growing cold-tolerant crops should be scattered among the remaining garden plants and residues to protect the slope from erosion through the winter.

If the slope is too steep for ordinary cultivation, terracing may enable crops to grow on it (see Figure 5–12). Terraces can be made with soil banks kept in place by shrubs or grasses (or weeds), but upkeep of this kind of terracing is difficult, because soil banks can erode and the cover plantings must be controlled. More permanent terracing can be established with retaining walls built to create a level area, or even several levels when a moderate amount of terracing is required (see "Walks, Drives, Patios, Walls," Chapter 14).

Figure 5–11 Managing soil on sloping sites. A ground cover like this periwinkle in front of a Michigan residence may be the most appropriate management practice for a sloping site. (Photo by R. H. Drullinger, courtesy USDA.)

Cultivation

As the term will be used in this text, **cultivation** includes initial breaking up of the soil in the fall or spring with a moleboard plow, disc, rototiller, digging fork, or other machine or tool; as well as stirring of the soil between crop plants with disc, spike or springtooth harrow, rod-weeder, rototiller, hoe, rake, or other device.

For the initial breaking up of the soil in spring or fall, the moleboard plow has the advantages of being able to reach a somewhat lower depth and of completely covering crop remains, manures, or other residues on top of the soil. Plowing, especially on the contour, produces ridges and an open soil surface, which lessens the likelihood of serious erosion. Discing tends to leave a trashy crop residue on the surface, which lessons erosion but can cause difficulty in seed-bed preparation and planting. The rototiller breaks and mixes the topsoil layer better than the other machines (Figure 5–13), and both plow and rototiller clear the field so that an even seedbed can be prepared and later cultivations can be easily accomplished. Rototilling partially destroys the structure of certain clay soils and leaves fields more subject to erosion than does discing or plowing on the contour.

Figure 5–12 Terracing. Terracing is an art that has been practiced for thousands of years in Southeast Asia. These vegetable fields have been carved from steep slopes in the Cameron Highlands in Malaysia.

Figure 5–13 Soil cultivation. Organic matter may be incorporated and a seedbed prepared by rototilling. (Courtesy Troy-Bilt Roto Tillers.)

Systems of Cultivation. In most of North America cultivation has usually involved clearing away all existing plants before the crop is planted (**clean cultivation**), but complete clearing is not always necessary. In Guatemala and Mexico before Columbus numerous crops were planted together with only enough clearing for seeds to germinate, and the garden grew as a year-round mixture of several kinds of food and fiber plants. As a plant began to age and no longer produced a useful product, it was removed, and one for which there was a need was planted in the spot it had occupied. Similar farming systems are still in existence and are being investigated as alternatives preferable to the slash-and-burn agriculture prevalent in some tropical areas. A similar system called multiple cropping, used with some success by gardeners in the United States and Canada, will be discussed later in this chapter.

Erosion has plagued humanity since cultivation began (Figure 5–14). To prevent it, permanent cover crops of

A

C

Figure 5-14 The problem of erosion. Fifteen hundred years ago ships docked along Harbor Street in Ephesus, Asia Minor (A). Soil, washed from the hillsides above, filled in the harbor. Mosquitoes breeding in the resulting swamps spread disease, which resulted in abandonment of the city. Harbor Street is now almost 10 miles (19 km) from the ocean, and slopes in the mountains above Ephesus that once were cultivated now have no soil whatsoever. (B) Erosion in North America today can be just as devastating and can plague the homeowner (C) as well as the farmer (D). (C and D, courtesy USDA.)

B

D

alfalfa and grass are often the best solution to soil management in commercial orchards and vineyards as well as in orchards planted primarily for home consumption. This is especially true where trees and vines are planted on hillsides. The cover crop is mowed occasionally to permit easier access through the orchard with spray and harvest equipment.

By using herbicides to control weeds, some growers produce field crops without **tillage.** In areas subject to wind and water erosion small grains are being grown in what is known as trashy fallow. With **trashy fallow** the soil is stirred, but dried weeds, stubble, and other debris are left as a mulch to protect fine soil particles from erosion. A farming system called "**minitil**" has resulted in better stand establishment and more rapid early growth of potatoes, sugar beets, and vegetables in areas where spring winds are severe (Figure 5–15). Crops are seeded at their usual row spacing in narrow beds rototilled from small grain stubble or a rye cover crop. Rows are oriented at right angles to the usual wind direction. A strip of stubble or cover crop is left between seeded rows to protect germinating seedlings. Later, as strips of cover crop or weeds begin to compete with the main crop, they can be removed with the use of herbicides or by cultivation. Both trashy fallow and minitil could be adapted for vegetable gardeners on slopes or in areas where spring winds are detrimental to early spring planting.

Figure 5–15 Potato field planted by minitil. (Courtesy R. E. Thornton.)

Benefits of Cultivation. Although other systems of cultivation are sometimes used, the standard cultural practice with vegetables and flowers in the temperate zone is still clean cultivation. A number of benefits are claimed for cultivation of row crops, some of which are truly beneficial and necessary. Others have been shown by modern research to be of little value or, in some instances, even detrimental to crop production. For small seeded vegetables and flowers a smooth, firm seed bed (which can be produced only when crop residues are turned under by cultivation) is highly desirable for uniform planting and good seed germination. However, the main benefit of cultivation has always been and continues to be the control of weeds. Weeds compete with crop plants for space, moisture, light, and soil nutrients, and most horticultural crops cannot compete with the well-adapted, rapidly growing weeds that infest most fields.

One of the supposed benefits of cultivation is the conservation of moisture that allegedly occurs when a dust mulch seals cracks and pores, preventing the upward flow and evaporation of moisture. Research has shown that with a few heavy soils cultivation does indeed conserve moisture by sealing cracks that penetrate deeply when these kinds of soils begin to dry (Figure 5–16). However, with lighter soils, especially where only light rains have fallen, cultivation may actually result in more rapid loss of moisture as the wet soil is stirred to the surface.

Another benefit claimed for cultivation is increased soil aeration. Again, research has shown that with certain kinds of soils cultivation at the proper stage may increase aeration, but that, if the soil has slightly more moisture than is optimum for cultivation, aeration is likely to be reduced. The passage of heavy cultivation equipment almost always reduces aeration and results in compaction in the lower depths of the soil.

Thus the major purposes of cultivation, for the home gardener as well as for the large producer, are to prepare a seed bed, to plow under various crop residues and manures, and, most importantly, to control weeds.

Figure 5–16 Puddled clay soils may form large cracks when they dry out.

Cultivating Equipment. The choice of cultivating equipment for the home gardener will depend primarily on the area that must be cultivated. If the area is small, a spade or digging fork, a hoe, and a rake may be all that is required. If the area is larger, the grower can hire someone with equipment to do the plowing and cultivating or can rent or purchase power equipment. In most regions where gardening is popular students or other individuals do custom rototilling or tractor work on a part-time basis. It is usually possible to rent rototillers and sometimes small garden tractors with plows and/or cultivating equipment from garden and hardware stores. Unless the garden area is large or the gardener can do custom work, renting is usually cheaper, though less convenient, than purchasing equipment when all costs, including interest on the investment, are considered.

Growers contemplating the purchase of power equipment for cultivating a garden should consider carefully the various units available. If possible, they should try out some of the units to see if they will be satisfactory for their specific purposes, and, at the very least, they should study literature available on such equipment and talk to dealers and owners about the advantages of various kinds of equipment. My experience has been that the least expensive units with the minimum power are generally a waste of money. A unit with 4 to 6 horsepower and provision for attachment of digging as well as cultivating equipment is the minimum with which most home gardeners will be satisfied. Often such a unit can power a lawnmower and can be used in the winter for snow removal from walks and drives. For the owner of a somewhat larger acreage, the four-wheeled garden tractor that can be used to power several kinds of farm equipment may be the best buy. A few part-time farmers have purchased a mule or pony broken to the harness as a power source, but anyone contemplating such a purchase needs to investigate the availability of feed and pasture and of horse-drawn hand plows and cultivating equipment as well as local ordinances relating to the keeping of animals.

Rotation, Succession Planting, Intercropping

The year-by-year cropping history of a plot of ground is referred to as its **rotation.** A proper rotation reduces pest problems and permits better utilization of nutrient elements. If sufficient area is available, garden crops should be rotated with climatically adapted field crops on which most insects and diseases of garden crops cannot survive, for instance, grasses, small grains, alfalfa, and other legumes. When the alternate crop is an annual, the plot can be planted to annual flowers or vegetables every other year. Occasionally alfalfa or other deep-rooted perennials should be planted to open the subsoil, in which case a longer rotation will be required. Where space is limited, crops should be rotated within the garden. A garden map showing the location of each crop each year is essential for maintaining a beneficial rotation year after year.

Succession planting is the planting of two or more crops, one after another, during the same season; it is of benefit where garden space is limited. Succession planting is common in areas with a long growing season. In much of North America the area planted in early spring crops—green peas, spinach, lettuce, and radishes—that seldom occupy garden space for more than a fraction of the growing season, can be replanted to summer or fall crops of carrots, lettuce, cauliflower, cabbage, or spinach. The annual flower bed made unsightly by an early freeze can be transformed by transplating a few clumps of hardy chrysanthemums into it. Rye, seeded after harvest of most garden annuals, can provide organic matter and prevent soil erosion during the winter. Numerous other examples of succession cropping could be mentioned.

Intercropping (interplanting) or **multiple cropping** is the growing of two or more kinds of crops on the same area at the same time (Figure 5–17). It is common practice in newly planted orchards where vegetables or small fruits are planted between the rows of trees to utilize space unoccupied while the trees are unproductive. Young orchards thus interplanted often receive better care than those not interplanted, because the grower must care for the plot in order to obtain production from the interplanted crop. The intercrop should never be permitted to compete with the permanent trees, and those likely to spread pests to the orchard should be avoided. For example, the Verticillium wilt organism, which can survive for many years in the soil and can destroy stone-fruit trees, can be spread by plantings of tomatoes, potatoes, and eggplant.

Figure 5–17 Intercropping. Nowhere is the art of intercropping better developed than on the island of Taiwan. As many as seven crops are frequently grown on the same land during a single year. This photo shows a crop of vegetable chrysanthemum being grown in a vineyard while the grape vines are dormant.

Interplanting is also possible with late-planted widely spaced crops such as melons and squash, which can be successfully seeded or transplanted into cleared spaces among early spring plantings of lettuce, radishes, or spinach. By the time the melons or squash plants begin to spread, the earlier planted crops will have been harvested. Annual flowers are often interplanted with spring bulbs to hide the unsightly dying leaves and faded blooms of the bulb plants.

Interplanting should be avoided if it results in plant competition for sunlight, which it often does in northern climates. For example, intercropping of beans, sweet corn, and squash, mentioned earlier and common in Aztec gardens, is not practical in most United States and Canadian gardens. In the north the shaded beans and squash may not mature, caring for a mixed planting is inconvenient, and generally the yield is not significantly greater than it would be if the same area were planted to three separate monoculture plots. Some of the pest control and growth stimulation claims attributed to **companion cropping,** a form of intercropping in which specific kinds of plants are supposed to be mutually benefited by close association in the garden, are probably valid, but I'm skeptical of some. For example, onions and garlic, both highly susceptible to many insects, are often recommended as insect-repellant companion crops. Until there is more precise information on the detrimental effects of chemical interactions among plants (sometimes called **allelopathy**) and between plants and insects, gardeners should not rely too heavily on companion cropping to solve their pest and other garden problems.

Mulching

Mulch is material spread over the surface of soil for controlling weeds, conserving moisture, heating or cooling soil, improving garden aesthetics, improving soil structure, preventing freeze damage, hastening maturity, and preventing erosion. A number of different products are commonly used for mulching garden plants. Among the most popular today are black and clear plastic, pebble-sized pieces of bark, gravel and river rock, sand, sawdust, leaves, straw and other organic trash, and paper (Figure 5–18).

Figure 5–18 Mulches. Plant materials for garden mulches are (A) finished compost, (B) partially composted oak leaves, (C) pruning chips, and (D) shredded oak bark. (Photos by Murray Lemmon, courtesy USDA.)

Plastics by themselves or in combination with other products are the mulching materials most widely used by commercial growers and home gardeners. Although a number of kinds of plastics are used for mulching, polyethylene is by far the most widely used because it comes in varied thicknesses, sizes, and shades and because it is relatively less expensive than most other types. Some of the advantages that polyethylene provides are earlier production, weed control, prevention of moisture loss, and prevention of fruit rot in crops like cucumbers and tomatoes that frequently are attacked by decay organisms when they come in contact with the soil. Polyethylene promotes earlier production with warm-season vegetables and flowers but does not seem to have much effect on the maturity of cool-season crops.

Polyethylene is especially effective in promoting earliness of muskmelons, and research in the Northwest has shown this crop to be 10 to 14 days earlier when grown on plastic that when grown without the aid of this material. For many years it was the consensus that earliness of warm-season crops grown with polyethylene mulch was a result of increased soil temperature under the mulch, but it is now known that black plastic mulch increases soil temperature only slightly and that earliness is promoted even when soil temperature is not increased. When a thin layer of soil is placed on top of black plastic mulch, the temperature under the soil-polyethylene layer remains exactly the same as the temperature at the same depth in unmulched soil. Yet plants growing in the soil-covered polyethylene plot are as early as those grown with exposed polyethylene (Figure 5–19). There is some evidence to show that the increased level of carbon dioxide in the atmosphere around very young plants growing through plastic mulch may be responsible for their earlier maturity and higher yields. The breakdown of organic matter and root respiration in the soil result in the release of CO_2, which escapes through the holes in the plastic sheet and concentrates in the atmosphere surrounding the plants planted through those holes.

The soil under clear polyethylene, unlike that under black polyethylene, is considerably warmer than unmulched soil. As a consequence plants of warm-season vegetables produce even earlier crops (muskmelons in irri-

Figure 5–19 Effects of mulching with polyethylene, (A) Comparative size of muskmelon plants grown with, left to right, no mulch, clear polyethylene, black polyethylene. Plants grown with (B) and without (C) polyethylene mulch.

gated areas of the Pacific Northwest are about seven days earlier) with clear polyethylene than they do with black polyethylene mulch. Unfortunately, clear polyethylene cannot be used as a mulch in many parts of the United States because of weed growth under the plastic. The use of clear polyethylene mulch is practical only in the desert areas where high light intensity day after day keeps the weeds under the plastic burned off until they are shaded out by the growing crop.

Cost is a major factor limiting the commercial use of plastic mulch. Enough .05 mil polyethylene, the least expensive plastic, to cover an acre costs $125, and the machines with which it is applied are expensive. For the home gardener who wishes to mulch only a small area, its use is economically feasible (see "Mulching with Plastic," Chapter 14).

When it is exposed to light in the field, polyethylene lasts for only one season. However, where it is covered with some other material, it may last for many years. As a consequence polyethylene is frequently used under bark, gravel, or river rock to control weeds that would otherwise come through these loose mulches.

A major disadvantage of polyethylene used as a field or garden surface mulch is the necessity of removing it after the growing season has passed. Several kinds of plastics that are biodegradable in the soil are in the developmental stage and, hopefully, will be available within a few years. The use of paper mulch has been suggested as a way of overcoming this disadvantage. However, the very ease with which paper deteriorates in many situations negates its advantages as a mulching material. This is especially true in home gardens where it is likely to become ragged and unsightly long before the season is over.

A number of other materials are used as mulches around the home. As mentioned, gravel, river rocks, and bark are used extensively today in combination with plastic to control weeds and retard moisture loss in ornamental plantings. Straw, sawdust, leaves, peat moss, and other organic mulches can be placed over strawberries and other herbaceous perennials and around roses and other semihardy shrubs in the fall to prevent winter damage. A 2- to 3-inch (5 to 8 cm) layer will provide weed control, prevent moisture loss, and add organic matter in plantings around the home. Loose mulches all tend to keep soil temperatures uniformly cooler, an advantage in the growing of certain ornamentals (Figure 5–20).

Figure 5–20 Loose mulch. Loose mulches discourage weeds, prevent moisture loss from the soil, and help keep the soil uniformly cooler. (SCS photo, courtesy USDA.)

Selected References

Aldrich, D. G. et al. *The Care and Feeding of Garden Plants.* American Society for Horticultural Science and National Fertilizer Association, Washington, DC. 1954.

Buckman, H. O., and Brady, N. C. *The Nature and Properties of Soils.* 7th ed. Macmillan, New York. 1969.

Hausenbuiller, R. L. *Soil Science: Principles and Practices.* 2nd ed. Wm. C. Brown, Dubuque, IA. 1978.

Russell, E. W. *Soil Conditions and Plant Growth.* 10th ed. Longman, New York. 1973.

U.S. Department of Agriculture. *Soils* (*Yearbook of Agriculture,* 1957). U.S. Government Printing Office, Washington, DC. 1957.

6 Water and Irrigation

Water is basic to all life, and adequate fresh water is essential for agricultural and community development. Earliest civilizations were established where rainfall or irrigation water made crop production possible. Most famines, ancient as well as modern, have been caused by too little, or occasionally too much, water, and primary tactics in the present battle against world hunger involve channeling water to arid lands and draining swampy areas.

Supplying the right amount of moisture at the right time is a major factor in gardening, where timeliness of irrigation is as important as amount and where overwatering is often as much of a problem as underwatering. For gardens that rely on rainfall as a basic water supply, planting and cultivating practices may need to be adapted to available moisture. Water also has aesthetic functions in some gardens, where streams, waterfalls, fountains, fishponds, and swimming pools are an integral part of the landscape.

The Moisture Cycle

Water for plant growth condenses and then falls onto the land from the atmosphere as one phase of the earth's **moisture cycle**, which is illustrated in Figure 6–1. The taking up and releasing of moisture by the atmosphere is dependent on the fact that considerably more vaporized moisture can be carried by warm air than by cool air and that the surface of the earth is subjected to differential heating.

Warm air moving away from the equator picks up large quantities of vaporized moisture that is released whenever the air is cooled. Cooling may occur as the warm air meets a mass of colder air moving from the poles, in which case general rains occur; or it may cool when warm air is forced to rise over an area of higher elevation, which accounts for the heavy precipitation on the windward side of mountain ranges. It may also be cooled by what is called a cyclonic disturbance, when warm air near the earth's surface rises rapidly into cool air above, causing thunderstorms.

Moisture falls to the earth as rain, sleet, hail, or snow. Snow in mountainous areas is important for moisture storage, and its slow release keeps streams flowing to lowland areas through the dry summers. Some water from rain or melting snow runs off in streams or rivers, and some soaks into the soil. As mentioned, the amount and rapidity with which moisture soaks into the soil and the amount of moisture the soil will hold depend on soil texture and structure. The infiltration rate is slow in clay soils and rapid in sandy soils and soils high in organic matter. Retention of moisture is high in clay soils and soils high in organic matter and is low in sandy soils.

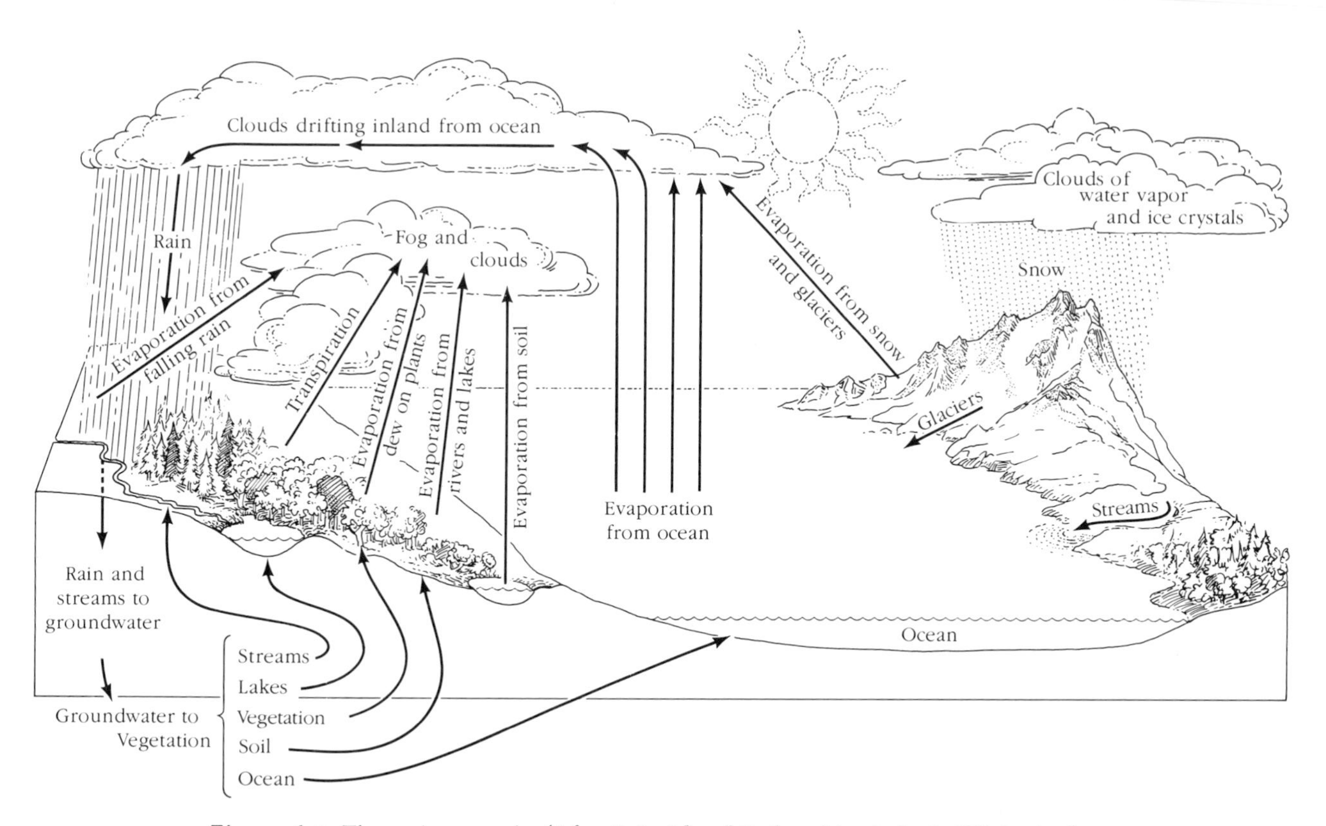

Figure 6–1 The moisture cycle. (After *Principles of Geology,* 4th ed., by J. Gilluly, A. C. Waters, and A. O. Woodford. W. H. Freeman and Company, San Francisco. Copyright © 1975.)

Some of the moisture absorbed into the soil will be pulled down through it by the force of gravity. This **gravitational water** enters the **groundwater** reservoir to become the source of springs and wells. The remainder of the water absorbed into the soil is held in the pore spaces of the soil and on the soil particles. The maximum amount of water a soil will hold against the force of gravity is termed its **field capacity.** The moisture held in the soil is the source of water for plant growth; however, plants can absorb only part of the water contained in a soil at field capacity. Some is held so tightly by soil particles that plants are unable to extract it. The stage in soil-moisture depletion where a plant is unable to take additional moisture from the soil and, as a consequence, becomes wilted is referred to as the **wilting point.** The water held by the soil below the wilting point is **unavailable** or **hygroscopic water.** Water is available for plant growth between the field capacity of a soil and its wilting point, and the amount of water in the soil between these two points is referred to as **available** or **capillary moisture** (Figure 6–2). Clay soils will hold considerable moisture at field capacity but will also have a high percentage of unavailable water. Sandy soils, on the other hand, have a very low moisture percentage at field capacity but may have almost no unavailable water. Table 6–1 lists water-holding capacity of various types of soil.

Moisture is lost from land masses by evaporation directly from the soil and by transpiration of plants, which

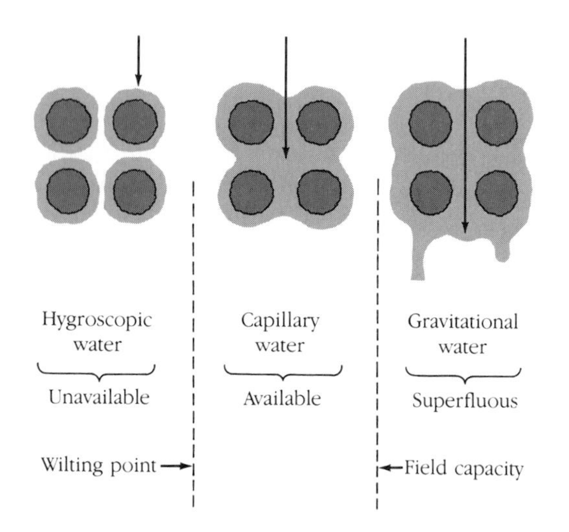

Figure 6–2 The three stages of soil-moisture supply. Gravitational water drains away before it can be used by plants, and hygroscopic water is bound so tightly that plants cannot absorb it. Only capillary water is readily available for plant use. (Adapted from *Principles of Plant Physiology,* by J. Bonner and A. W. Galston. W. H. Freeman and Company, San Francisco. Copyright © 1952.)

Table 6–1 Approximate water-holding capacity of various types of soil.

Soil type	Inches of water/foot (× 3.28 = cm/m)	Inches of available water/foot (× 3.28 = cm/m)
Course sand	0.40–0.75	0.20–0.40
Fine sand, loamy sand	0.75–1.25	0.40–0.70
Sandy loam, fine sandy loam	1.25–1.75	0.70–0.90
Loams and silt loams	1.50–2.30	0.75–1.10
Clay loams	1.75–2.50	0.90–1.25
Clays	1.60–2.50	0.80–1.20
Peat and muck	2.00–3.00	1.00–1.50

Data in this table are averages from several sources.

draw moisture from the subsurface. Wherever a plant cover exists, the amount of moisture lost by transpiration is much greater than the amount lost by evaporation. The transition of large amounts of liquid water to water vapor by transpiration consumes considerable heat energy, partly accounting for the moist coolness of a woody garden. Without a plant cover loss by evaporation from the soil surface is not rapid. This minimal evaporation is the basis of a system employed by farmers in semiarid regions called **summer fallowing,** in which the land is kept completely free of plants every other year in order to store moisture for crop production during the alternate year.

Water vapor transpired by plants or evaporated from soil is an important source of precipitation over large land masses, and the same water molecules may be involved in several precipitation-evaporation cycles as air currents move them across a continent.

Water and Plant Growth

Water plays a vital role in plant growth and development. In some areas all of the time, and in most areas some of the time, lack of water (**water stress**) is the factor most limiting to proper plant functions. Water constitutes the greatest part of the plant body, the leaves of most trees being about 65% water, the roots about 70%, and the fruits about 85%; some leafy vegetables such as cabbage and spinach are about 90%, and melons about 95%, water. Water also serves as a plant nutrient, being one of the chemical constituents required in the photosynthetic process. It serves a fundamental role as the solvent for mineral elements entering the plant and facilitates the transport of minerals and manufactured foods. For many plants and plant parts normal functioning is based on water pressure within their cells. When plant parts are fully distended by being filled with water, they are said to be **turgid,** and most plant functions and plant growth occur only when a plant is turgid. A plant that lacks turgidity is said to be **wilted.** Water also provides the means whereby a plant can expand against external physical forces, as occurs with seed germination, seedling emergence, and most subsequent increase in plant size (Figure 6–3). Most botanists believe that temperature control as a result of evaporative cooling by transpiration is another important function of water in plant development.

Effects of Excess Moisture

Plants growing in soil that is excessively watered will appear unthrifty and may display symptoms similar to those shown by plants lacking moisture. (The term, **unthrifty** is used to describe plants that are not growing as vigorously as they should but don't show symptoms of a specific nature.) An oversupply of moisture will fill soil pore spaces, causing a lack of aeration that will, in turn, restrict the growth and function of absorbing roots. With no absorbing roots an overwatered plant may actually wilt from a lack of moisture in its stems and leaves. Wilting, yellowing, and death of the margins of leaves can all be symptoms indicating that a plant is getting too much water.

Perhaps the group of plants most often overwatered are those growing in glazed pots without drainage holes. If plants are grown in these kinds of containers, care must be taken to apply just enough water to moisten the soil and then to let the soil become fairly dry before watering again. Overwatering is not as much a problem with plants grown in fields and gardens, but it can occur.

At planting time a lack of aeration due to excess moisture may prevent seed germination. Excess moisture may also increase the activity of soil-borne disease organisms, which may lead to damping-off and a reduced stand of seedlings (Figure 6–4). **Damping-off** is the decay or death of a stem at soil level resulting from attack by one

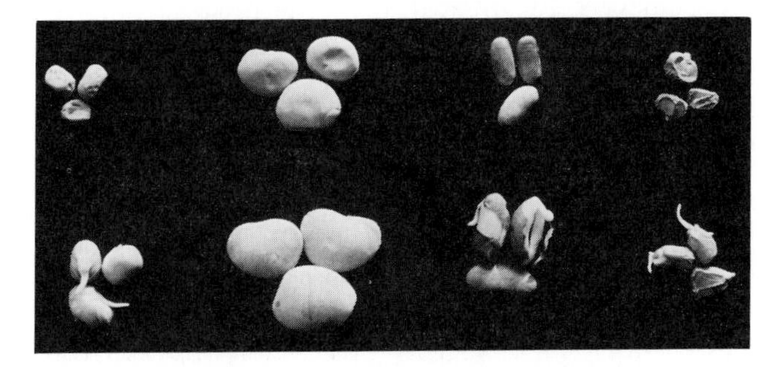

Figure 6–3 Water as a means for plants to exert a force of growth. Peas, lima beans, beans, and sweet corn that have absorbed water as the first stage of germination (lower) have expanded their volume to several times that of comparable dry seed (upper).

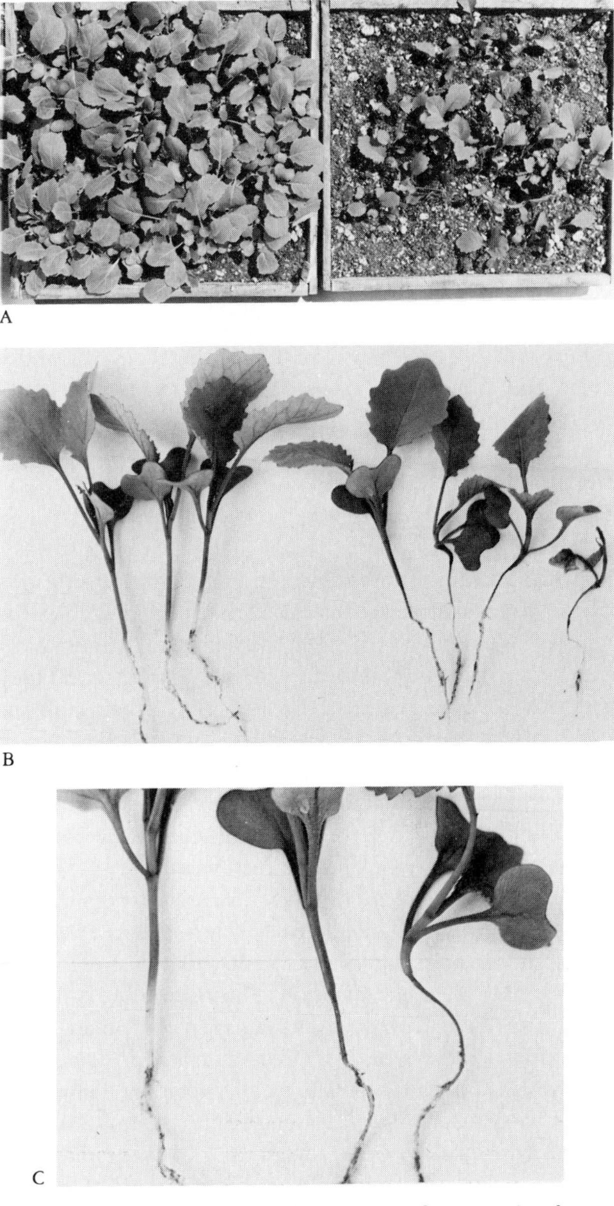

Figure 6–4 Damping-off, a disease often associated with excess moisture. Over half of the cabbage seedlings in the right flat (A) died from damping-off and were removed. Most of the rest are beginning to fall over, wilt, and die because of constriction of their stems near the soil line (B and C). Plants to the left in B and C are healthy.

or more of several different soil pathogens. Seedlings are literally eaten off by microorganisms at the soil line and fall over and die. Usually excess moisture is associated with cold soil, which also delays germination.

Excess soil moisture increases the tendency for splitting of cabbage heads that are left in the garden after they mature. Splitting of cabbage can be delayed by grasping the mature head in both hands and twisting it part way around to sever some of the water-absorbing roots. Excess soil moisture is especially damaging to plants grown for underground parts. With only a slight excess potato tubers will produce enlarged **lenticels,** the small corky areas that allow interchange of gasses between the interior of stems, roots, tubers, and the atmosphere. Enlarged lenticels on potato tubers are unsightly and can be the site for entry of disease-causing organisms (Figure 6–5). Potatoes and root crops are more subject to soft rot in overly moist soil.

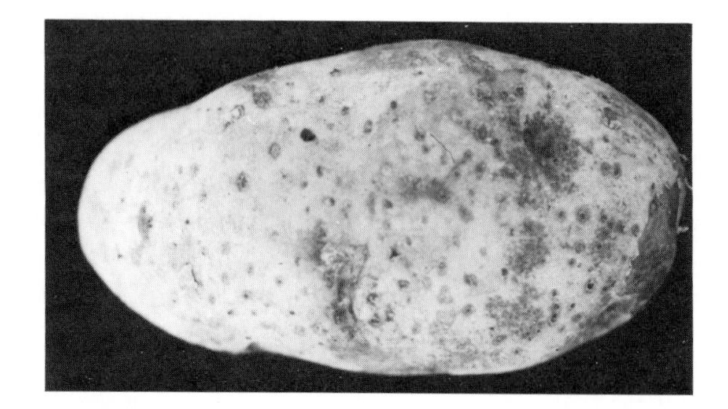

Figure 6–5 Enlarged lenticels on a potato tuber as a result of overirrigation. (Courtesy Robert Kunkel.)

Sometimes moisture falling on above-ground portions of the plant can be detrimental. Fruits with a high sugar content—tomatoes and sweet cherries being prime examples—tend to split if moisture falls on them when they are nearly mature (Figure 6–6). Because of their high sugar content, fruits absorb water by osmosis (see Chapter 2) to the point of bursting. A light rain can destroy a nearly mature cherry crop in a few minutes. Tomatoes irrigated with sprinklers or frequently rained on are more apt to split than those that are furrow irrigated.

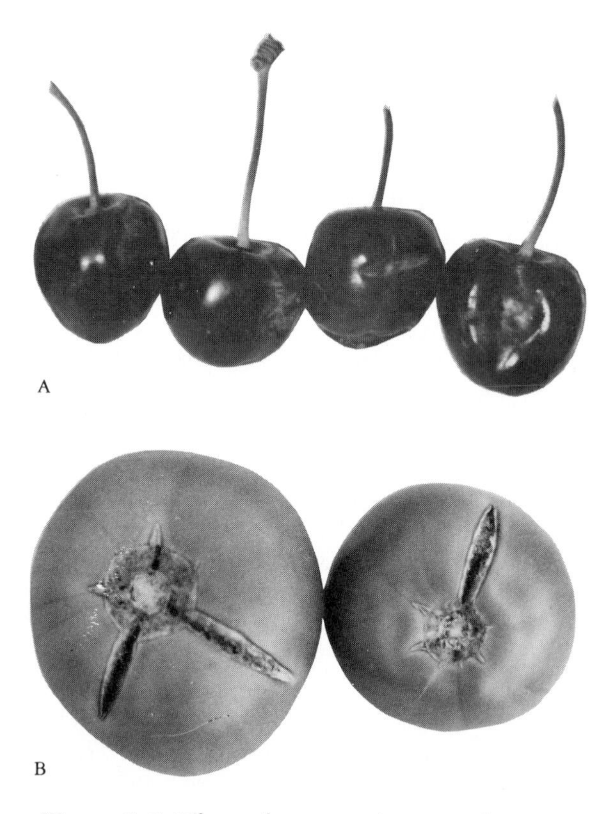

Figure 6–6 Effects of excess moisture on above-ground plant parts. Cracking of sweet cherry (A) and tomato (B) fruits occurs when rain or irrigation water falls on the ripe fruit and is absorbed into the sugar-containing interior by osmosis.

Foliar diseases are more severe in areas of high rainfall and high humidity. The spores of *Phytophthora infestans,* a fungus causing late blight of potatoes, require foliar moisture for their germination. During the years 1845–1851 late blight destroyed or damaged the potato crop over much of the world. In Ireland, where food for the impoverished population was mainly potatoes, famine resulted. Starvation and disease (mainly typhus) killed a million and a half people, almost 20% of the population; another million (or about 13% of the population) left the country, mainly for the United States. Now in areas of high summer rainfall potato plantings are sprayed every week with a fungicide to control this disease.

The seed-borne pathogen that causes halo blight of beans is spread by splashing water. In areas where rain is frequent, halo blight spores can spread from one or two seed-infected plants to destroy many acres of snap or dry beans. It can be controlled by planting seed completely free of the causal organism. This has led to the establishment of the U.S. bean-seed industry in the arid, irrigated Twin Falls area of southern Idaho, where halo blight is rarely found.

Effects of Moisture Deficiency

Injury due to a deficiency of moisture is more common in outdoor gardens than injury due to an excess. The first plant symptom of moisture stress is a slight yellowish or greyish cast of the leaves, a symptom easily recognized by experienced growers. Growth slows, and a reduction in yield may occur even before definite symptoms of drought are present. Wilting, rolling, and finally shedding of leaves and fruit will occur as water stress becomes more acute (Figure 6–7). It is important to distinguish between wilting due to a lack of soil moisture and temporary wilting that may occur during hot days when the plant is transpiring so rapidly that roots are not able to absorb and replace the moisture being lost. With temporary wilting the plant recovers in the evening or when the weather cools.

Water stress lowers the quality of succulent vegetables by causing premature woodiness, stringiness, and toughness. Greens and salad crops, especially, require ample water, because quality of these vegetables depends on their succulence and crispness. Fruit produced under droughty conditions may have a woody texture. If an edible or ornamental plant has insufficient moisture, its flowers blast and its buds drop before they open. Yields of beans and lima beans are greatly reduced by even moderate water stress at blossom time because of the tendency of these crops to lose their buds.

In areas of low humidity during hot weather tomatoes sometimes develop a physiological disorder called **blossom-end rot,** the first symptom of which is a water-soaked area on the blossom end of the tomato. This area then turns brown and finally develops into a dry, shriveled, and sunken lesion that internally may involve most of the fruit (Figure 6–8). The physiology of blossom-end rot is not entirely understood, but it is known to be

Figure 6–7 Effects of moisture deficiency. This coleus plant was not watered for 18 days. It dried to its wilting point in 10 days. It remained without water for one more week (A) and then was watered. Twelve hours later it had recovered (B) except for the loss of a few lower leaves and some slowdown in growth.

correlated with water stress. It is most prevalent on tomato fruits growing on vines that have been produced with optimum growing conditions during the early part of the summer and, after fruit is set, are subjected to either a lack of soil moisture or hot weather and low humidity. It is assumed that, under these conditions, the leaves are able to pull moisture from the fruit, causing dehydration and death of the fruit cells farthest from the moisture source. Since blossom-end rot can also be caused by calcium deficiency, there is probably a physiological relationship between calcium deficiency and water stress.

Potato specialists have long known that the best quality 'Russet Burbank' (Idaho) potatoes are produced with light, frequent irrigation. A high percentage of U.S. #1 potatoes are produced in fields irrigated every 3 to 5 days. Fields irrigated more heavily every 10 to 14 days may produce a comparable total yield, but a high percentage of the potatoes have pointed ends or restricted middles. Potatoes with restricted middles are called "bottlenecks" or "dumbbells" by the trade (Figure 6–9). Research in Idaho has shown that the benefits of frequent irrigation on 'Russet Burbank' tuber quality result from soil cooling rather than from additional water per se. Tuber growth ceases when soil temperatures rise, and the cessation and resumption of growth results in deformed tubers. Round potato cultivars are not nearly so subject to damage by fluctuation of water supply and will usually grow more satisfactorily than long-tubered cultivars in gardens where moisture supply is uncertain.

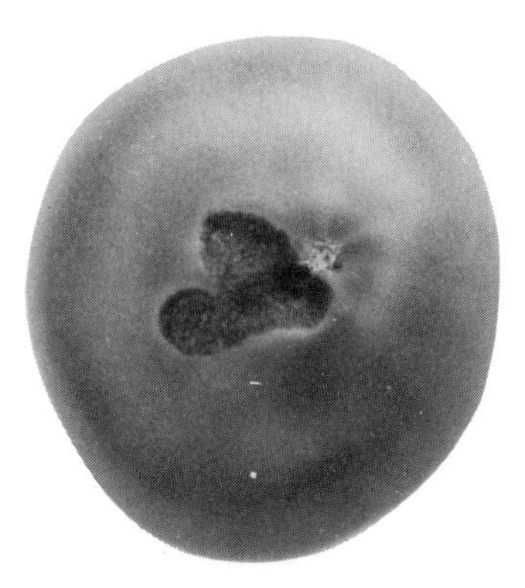

Figure 6–8 Blossom-end rot. This physiological disorder often occurs when tomato vines that have been growing rapidly are subjected to water stress. It can also be related to calcium nutrition.

Figure 6–9 Results of faulty irrigation practices. Many types of misshapen potato tubers can be prevented by more frequent irrigation, which results in more uniform soil temperatures.

Irrigation

Until about 1940 supplemental water was almost never applied to commercially grown vegetables or to vegetable gardens in the more humid areas of North America. An occasional humid area lawn or flower bed was watered during a dry spell, but essentially irrigation was limited to arid areas. This practice has changed, partly because of a better understanding of the benefits of supplemental watering and the introduction of more convenient sprinkler systems and partly because of the developing importance of community beautification and the provision for irrigation water from municipal water supplies. Today, supplemental irrigation is used on garden crops in most parts of North America.

In urban and suburban areas water for irrigation usually comes from the municipal supply. A few communities located in arid areas have two piped systems, one for household use and one for irrigation. In rural areas and some suburban communities where lots are larger, water is often supplied in open ditches, or it may be pumped from an open well.

Three general systems of irrigation are used in commercial production of horticultural crops in the United States—subsurface, surface, and sprinkler. A fourth type, trickle irrigation, which was developed and is used extensively in water-short Israel, is receiving considerable attention, especially for greenhouse irrigation and for field irrigation in areas where water is in short supply or salty.

Because it requires specialized topography and subsoil conditions usually involving large areas such as the drained Everglades of southern Florida, **subsurface** irrigation is seldom feasible for gardens. Most supplemental water applied to gardens comes from either gravity-flow surface irrigation or sprinkler systems; trickle irrigation may soon be added to this list.

Figure 6–10 Furrow irrigation. This commercial field in Arizona is being heavily irrigated to wash out high concentrations of chemicals that sometimes accumulate in soils of arid areas and to provide moisture for planting vegetables. Flooding of this magnitude is not desirable on land that is not almost level or where a crop is already growing. (Photo by W. E. Parsons, courtesy USDA.)

Surface Irrigation

Surface irrigation is used by gardeners to furrow irrigate row crops or occasionally to flood irrigate lawns or other grass-covered plantings. For **furrow irrigation** water is distributed across the garden by small ditches called **rills** or **furrows,** which are 3 to 8 inches (8 to 20 cm) deep. The furrows are usually next to the crop row; for annual crops they will be 20 to 40 inches (50 to 100 cm) apart, with one or two crop rows between them (see Figure 14–12). For **flood irrigation** ridges are usually established along the border of an area to permit a sheet of water to be spread over its surface. Water for surface irrigation can be brought to the garden in open ditches or by pipes or hoses.

The furrow-irrigated garden is planted on raised beds with furrows established at planting time. The garden area must be fairly level, with just enough slope to allow water to pass slowly from the upper to the lower end, or else the rows of garden crops and furrows must be on a contour to permit the same slow, even rate of water flow (Figure 6–10).

Proper furrow irrigation requires patience and considerable skill. Enough water should be channeled into each furrow to flow to the end in a few minutes, and then the amount should be diminished to the point where, theoretically, the last of the flow soaks in just as it reaches the end of the furrow. This achieves maximum infiltration with minimum erosion and leaching of mineral elements. In actual practice a trickle of water usually runs from the lower end of most furrows, because few irrigators are skilled enough to keep the absolute minimum in all rows. Nevertheless, the best furrow irrigator is the one who can spread the available water to irrigate the greatest number of rows simultaneously. The water should be allowed to flow in the garden until it has soaked across each bed a few inches beneath the surface (Figure 6–11).

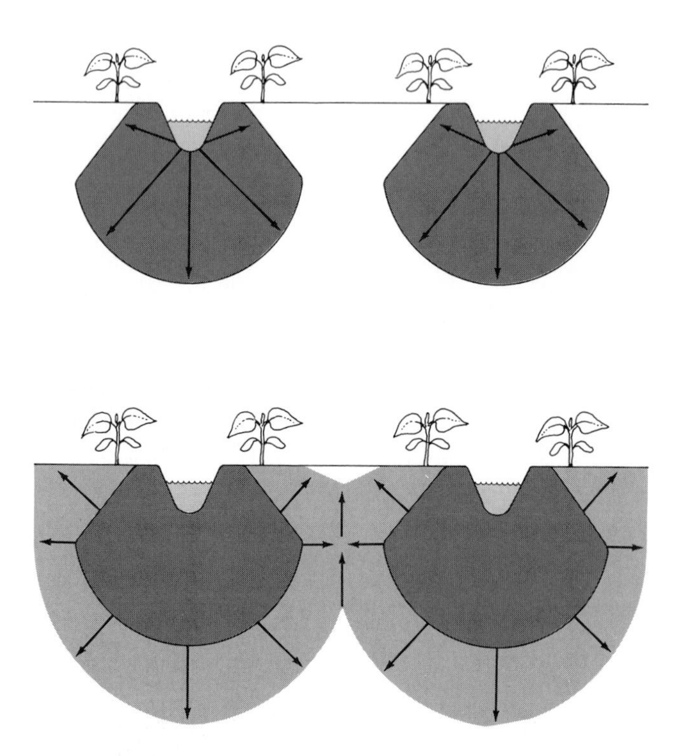

Figure 6–11 Pattern of water infiltration with furrow irrigation. Arrows show the direction of water movement. The darkly shaded area (A) is the area of initial wetting. The lightly shaded area (B) becomes wet later. By the time water has soaked across the bed to within a few inches of the surface, it will have permeated most of the area of root penetration. Although most movement is downward, water also moves toward the center of the furrow and then upward as a result of surface evaporation. In arid areas toxic quantities of salts moving with the water are sometimes deposited as a crust in the center of the bed as the water evaporates.

Lawns and other planted areas to be irrigated by ridging and flooding must be absolutely level. A ridge 4 to 8 inches (10 to 20 cm) high should be established around the periphery, with the slope of the ridge gentle enough that it can be made less obtrusive by continuing the grass or cover crop planting on its surface. With care an area can sometimes be flood irrigated without ridging by distributing water along the upper edge and letting it spread across a sloping area. A flood-irrigated lawn will usually require water less frequently but will require more frequent fertilizer application than one that is sprinkler irrigated.

Disadvantages of surface irrigation are the leveling and contouring required, the attention necessary while water is being applied, and the increased problems with mites and thrips, pests that are partially controlled by wetting of leaves. In addition, more water is required for surface than for sprinkler irrigation. On the plus side, the investment for equipment is usually less, and foliar diseases are less likely with surface irrigation. On a small area where foliar disease is a problem, such as a rose planting, a perforated plastic soaker hose turned upside-down will provide the advantages of surface irrigation.

Sprinkler Irrigation

Sprinkler irrigation is the most widely used system for irrigating small yards and gardens, especially in areas of relatively high rainfall where only occasional supplemental watering during periods of drought is required. Sprinkler irrigation is convenient if the sprinkler can be attached to the pressurized home or municipal water supply. Where irrigation water comes through open canals or ditches, sprinkler irrigation is more costly, because it requires a tank and pump or some other means of providing pressure, piping or hose to deliver water to the garden, and a filter system to eliminate particulate matter and prevent clogging as well as a sprinkler system for water distribution.

Hand-held hose sprinkling is not the best means of sprinkler irrigation (see Figure 6–12). Movable sprinklers are available in a wide range of quality and price. Usually better quality sprinklers are the best buy even though they initially cost more, because they distribute water more uniformly and last longer. Sprinklers vary in their patterns of water distribution, and this should be a consideration when they are being purchased. The sprinkler or sprinkler system selected should be capable of uniformly watering the entire yard with as few moves as possible. Uniformity of application can be checked by setting flat pans at various locations under the sprinkler. Some of the newer kinds of sprinklers distribute water in a rectangular pattern, providing easier watering of garden corners. The height at which water is discharged and the height reached by the discharged water may be important considerations if tall plants are being watered or if wind is likely to affect the pattern of water distribution. The capacity of the water system will determine the maximum number and size of sprinkler heads used. One medium-sized sprinkler may be the maximum if household and irrigation water both must be supplied through a ½-inch (1¼ cm) pipe. The pipe supplying the home from the city water main should be at least ¾ inch (2 cm) if yard irrigation is anticipated.

Permanent-set sprinkler systems with a network of underground pipes have become popular with homeowners. On larger estates these can be connected to a time clock to make sprinkling automatic. The permanent-set system saves considerable time and labor where extensive plantings of lawn and ornamentals must be irrigated. With the use of plastic pipe the cost of permanent-set systems has been reduced, and, with more refined sprinkler heads, their reliability has been increased.

Underground permanent-set systems are, of course, more easily installed before the yard is planted. The grounds should be thoroughly soaked and packed several times to make certain the soil has completely settled before any pipes are installed. If the soil settles after installation, the results will be either a broken pipe at the point of settling or pipes that cannot be completely drained. The installation of an underground sprinkler system requires considerable knowledge and expertise, and the homeowner should hire only well-established, experienced personnel to do the job. Complete water coverage is important and not always easy to achieve. In addition, any low spot that prevents a system from being completely drained will result in broken pipes if water freezes in them during the winter. It is possible for homeowners who are skilled to install their own systems, but, before beginning, they should get the assistance and advice of an expert.

Permanent overhead sprinkler systems (Figure 6–12) are being installed in commercial orchards in some areas and may be feasible for larger garden plantings. They not only supply the amount of water required whenever it is needed but they also help to control mites and small

Figure 6–12 Irrigation by sprinkling. (A) A hand-held hose can be used to wash dust and small insects from shrubbery, but irrigating this way usually results in applying too little water too often. (B) A movable sprinkler that can be adjusted to vary the water distribution pattern usually produces better irrigation results. (C) High overhead sprinklers are frequently used for irrigation, fertilizer and pesticide distribution, and frost protection in vineyards and orchards.

insects, provide a means of applying pesticides and fertilizers, and provide a measure of frost protection (see Chapter 7).

Some of the advantages of sprinkler irrigation are that it can be used on land that is not level, it requires less water than surface irrigation, it removes dust and partially controls certain small insects, and in some situations it may permit application of fertilizers and pesticides with the irrigation water. Commercial growers commonly meter agricultural chemicals into the main water line of their sprinkler systems; however this is not yet common practice for gardeners.

In some arid area gardens sprinkler irrigation may also be advantageous in preventing crop injury from soil surface concentrations of salts. Under natural conditions in arid regions soluble salts accumulate at the depth to which soil moisture soaks each season; this may be a few inches to several feet below the surface. When irrigation water is applied to these kinds of soils, some of this salt layer dissolves and salts are drawn to the surface as a result of surface evaporation. With furrow irrigation salts are carried with the water from both furrows to the center of the bed and then upward with evaporating moisture (see Figure 6–11). In newly irrigated areas these salts often become visible as a maroon or white deposit, and their highest concentration in the center of the bed, which is often exactly where the crop row is growing, may be enough to injure susceptible crops such as strawberries or raspberries. Because water from sprinkler irrigation initially soaks downward over the entire surface, salts are more evenly distributed throughout the upper soil layers, and plant injury is less likely.

Disadvantages of sprinkler irrigation are the cost of equipment for distributing the water, the greater chance of foliar disease, and the likelihood of uneven distribution wherever heavy winds occur.

Trickle Irrigation

During the past few years a new system of irrigation, referred to as **drip** or **trickle irrigation**, has received considerable publicity. Phenomenal crop growth and yield and the saving of one-half to two-thirds of the irrigation water normally used by other systems are claimed for trickle irrigation. Furthermore, because less water is required, salt buildup is reduced and more brackish irrigation water can be used. With this system a trickle of water is discharged continuously near each plant. With fruit trees, small fruits, ornamental trees, and shrubs and greenhouse pots, a small plastic tube attached to a larger feeder line supplies each plant or pot. With row crops perforated plastic pipes are more commonly used. Fertilizers and soil pesticides can be applied through the trickle system as they can through sprinklers. With the trickle system, however, it is possible to keep a constant dilute flow of mineral elements and even of some pesticides, a practice not feasible with sprinklers because of the volume of water a sprinkler applies.

The trickle system has proved successful in greenhouses and other plant-growing structures and shows considerable promise with grapes and tree fruits (Figure 6–13). Trickle irrigation has aroused interest in the American Southwest, where water for irrigation is limited and what is available usually has a high salt content. Because less irrigation water is applied, this concept may become useful for irrigated areas where rising water tables are causing drainage problems. Homeowners have long used the soaker hose, which is one version of drip irrigation. Whether or not other drip irrigation systems become popular with home gardeners will depend on whether or not industry develops equipment that can be used in the garden.

Care of the Irrigation System

Periodic maintenance of the irrigation system is essential if water is to be distributed efficiently. Permanent-set sprinkler systems should be checked for leaks and repaired during the early spring. Hoses, sprinklers, pipes, and all other equipment should also be examined and repaired if necessary before the irrigating season begins. A drop of oil on moving parts will help insure good sprinkler operation. Unlined ditches and canals must be kept free of weeds and trash, because these materials slow water flow, allowing loss of water by seepage through canal banks as well as unwanted moisture in fields adja-

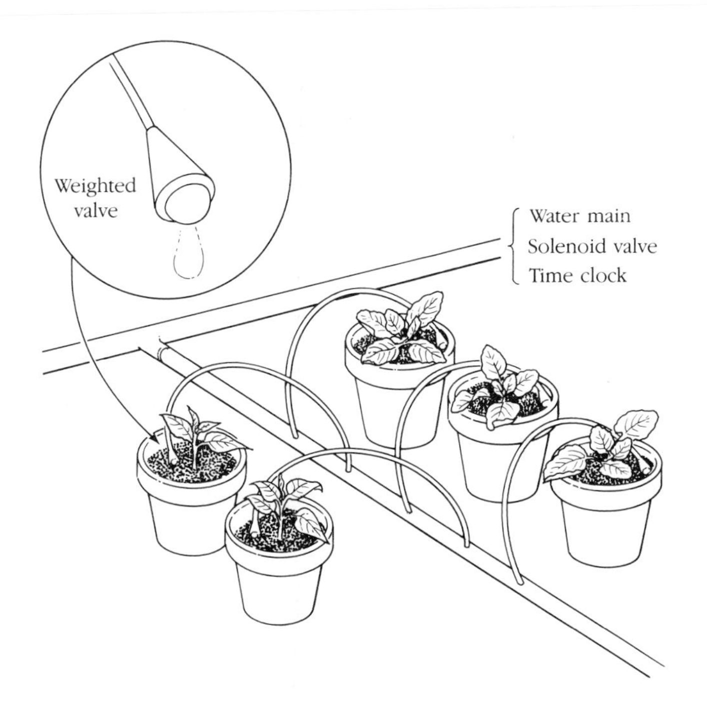

Figure 6–13 Trickle irrigation system for greenhouse watering. The system is supplied from the water main of the home or greenhouse. A solenoid valve controlled by a time clock permits flexibility in timing water application. Weighted valves connected to small plastic tubes that fit into small holes in the plastic distribution pipe are placed in the plant containers to control the amount of water each plant receives. (Adapted from *Horticultural Science,* 3rd ed., by Jules Janick. W. H. Freeman and Company, San Francisco. Copyright © 1979.)

cent to the canal. Weeds along open ditches should be controlled by mowing or with chemicals, and seed screens should be installed in the stream to prevent weed seed and other debris from contaminating the garden or plugging the distribution system.

Irrigation pipes should be moved with care so that dirt and debris do not get into them. All gaskets and moving parts of pipes, hoses, and sprinklers should be examined periodically and worn or damaged parts replaced. Rubber and plastic hoses will last longer if they are stored out of direct sunlight, and hose fittings are less likely to be damaged by being accidently stepped on or driven over with the lawn mower or other vehicles if they are stored when not in use. Damaged fittings on rubber hoses can be easily replaced with a pocket knife and hammer. A supply of male and female replacement couplings and of hose gaskets kept with garden tools enables quick repair of wasteful and annoying leaks. A small cut in a rubber or plastic hose can be repaired with friction tape, but when leaks due to weathering begin to develop, the hose is usually worn out and should be discarded. Small leaks in aluminum pipe can be repaired with "liquid aluminum," a paste material formulated for patching aluminum and available at department and hardware stores.

Before sprinkler pipes or hoses are stored for winter, they should be drained and dried completely. I usually place a drop of light oil on all moving parts of the sprinkler to prevent corrosion, although this is not always recommended by manufacturers because of the dust that sticks to oil. Permanent-set sprinkler systems should be drained completely and pumps winterized according to the manufacturer's instructions.

Frequency and Amount of Irrigation

Most frequently, past experience of the grower or neighbors determines when irrigation water is applied. As mentioned earlier, plants that are in need of moisture change color slightly. With experience, a grower can tell by this color change and by the feel of the soil when irrigation is required.

For lawns and crops covering the soil surface pan evaporation is sometimes used to determine when to irrigate. This method is based on the fact that approximately the same amount of water evaporates from a pan as is transpired from foliage. In some irrigated areas the weather bureau informs the public via radio each morning of the amount of evaporation that has occurred during the previous 24 hours.

There are a number of different meters for measuring soil moisture, most based on the fact that dry soil has more electrical resistance than moist soil. The electric current passing through the soil between two electrodes in a moisture probe or moisture block can be transposed to read directly the percentage of available moisture. Because accurate probes are expensive and difficult to calibrate and interpret, they are not used extensively by gardeners.

The amount of water required by garden plants varies depending upon the plant, the plant's stage of growth, the season, humidity, wind, and other climatic factors. Despite the variation in water requirement from crop to crop, it is possible to recommend general guidelines for garden irrigation.

There is some evidence to indicate that roots are able to absorb moisture equally well from soil with moisture content ranging from field capacity to wilting point. However, since moisture may not move rapidly within the soil, soil particles next to the absorbing rootlets may be at the wilting point long before a soil sample will show complete available moisture depletion. Therefore, it is recommended that most horticultural crops be irrigated when 35 to 50% of the available moisture remains in the soil. In a medium-textured loam soil this is about the time the soil will no longer stick together well when pressed into a firm ball (Figure 6–14).

Many growers irrigate on a regular schedule of every 5, 7, or 10 days. This is a common practice, especially in arid parts of the country where irrigation throughout the growing season is imperative, and irrigation water is often available at fixed intervals. Irrigating on a regular schedule throughout the season is not always to be recommended. Besides wasting water, overirrigation can leach nutrients, injure plants by reducing root aeration, and cause erosion. Irrigation will need to be more frequent on light sandy than on heavy clay soils. Plants use less water during the early part of the season when they are small and the weather is cool than they do during mid-summer. Deep-rooted crops need to be irrigated less frequently than shallow-rooted ones, and, of course, deep-rooted crops should receive more water with each irrigation. As the temperature becomes cooler toward fall and as the crop matures, irrigation can be reduced.

In areas where irrigation supplements rainfall it is important for growers to keep track of the amount of rain. Light showers will often appear to supply more moisture than they actually do, and plants may suffer moisture stress even when showers are frequent. Moreover, plants on the leeward side of a building or under a roof overhang will require supplemental water even during periods of heavy rain (see "Irrigating Garden Crops," Chapter 14).

Gardening with Limited Water

For want of sufficient water, some North American gardeners who dream of growing a shady vista garden with broad green lawns bordered with exotic plants that originated in the rainy forests of Europe and the Orient find

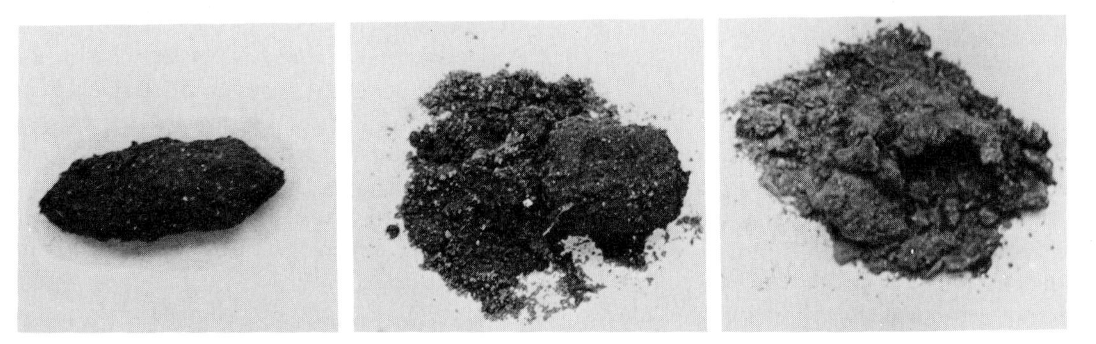

Figure 6–14 When to irrigate. The handful of soil on the left is wet and sticky, and moisture seeps from it onto the paper towel. It is too wet to work. The soil in the center sticks together when compressed but crumbles quite easily. It has about the right amount of moisture for cultivating, potting, and planting. The soil at the right has too little moisture to stick together readily. It is at about the stage of moisture depletion where irrigation is required.

the realization of their dream impossible. Yet even in the areas where rainfall is infrequent and water for irrigation is unavailable, limited, or prohibitively expensive, gardening is still possible and may even be more interesting than traditional gardening because it represents more of a challenge.

Gardeners with limited water should first of all consider using native plants and inanimate materials. Some kind of plant cover grows almost everywhere, and the skillful use of this local flora combined with stone, driftwood, desert-bleached tree trunks, and other materials from the surrounding area can create an aesthetically appealing garden that requires no supplemental water. Cactus and succulent gardens, popular in many areas, are uniquely suitable where water is limited. **Succulents** are a group of plants belonging to several genera that can withstand periods of drought because they are capable of storing water in their greatly thickened leaves. Desert juniper, pinyon, yucca, sagebrush, potentilla, and other desert brush are interesting possibilities for desert gardens.

In regions with slightly more natural moisture Douglas fir, ponderosa pine, Oregon grape, chokecherry, serviceberry, and other plants with localized distribution can be added to the list. These plants should be afforded the space in the garden that they have in nature, where they grow with little competition.

Because a thrifty lawn requires more water than most other plantings, elimination of a lawn should be considered wherever water for gardening is limited (see Figure 6–15; Chapter 11; and "Mulching with Plastic," Chapter 14).

Fruits, vegetables, and flowers grown in arid areas will require supplemental water, but these crops can be grown with much less water than is usually used. Because of the high rate of transpiration of weeds, complete weed control will save more water than any other practice and should be the first and foremost consideration wherever conserving water is essential. Similarly, no cover crop, weedy or otherwise, should be grown around tree or small fruit plantings in minimum moisture conditions. If erosion control is needed, a plastic, straw, sawdust, or bark mulch may be used to cover bare soil. Mulch will also discourage weed growth and reduce evaporation.

Figure 6–15 Eliminating a lawn. A small island planting and a driveway occupy the water-conserving front yard of this home.

Wide plant spacing will permit the production of crops with less water. Sweet corn plants spaced at 3 x 3 or 4 x 4 feet (approximately 1 m x 1 m) will produce good ears with surprisingly little moisture, especially if they are growing in clay loam soil high in organic matter. Short-season annuals, such as lettuce, radishes, spinach, peas, spring flowers produced from bulbs, pansies, bearded iris, asparagus, and rhubarb, can frequently be produced before moisture is depleted in the spring.

Where there is growing space for two gardens, summer fallowing—that is, growing a crop every other year and keeping the ground plant free during alternate years—may be feasible with some vegetables and flowers. During the year it remains fallow the plot can be covered with a mulch. Additional nitrogen fertilizer can then be added to a loose mulch that can be plowed under to provide organic matter during the season of production.

Drainage

Getting rid of excess water is called **drainage,** and good drainage is as essential to successful gardening as is supplemental irrigation (Figure 6–16). Problems with excess water are most easily averted by allowing for drainage at the time the home and landscape are planned. A tile drain

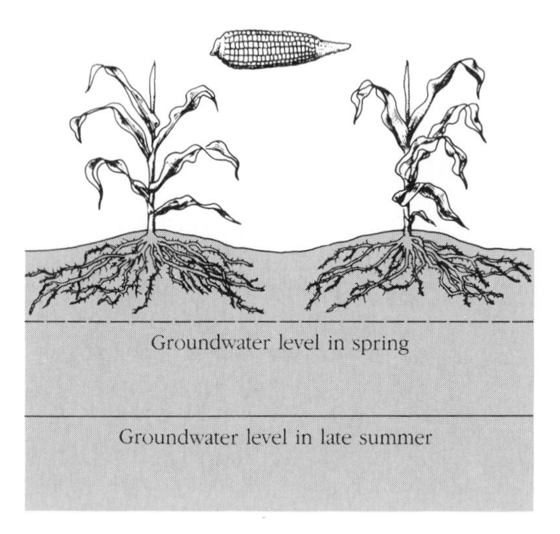

Poorly drained land

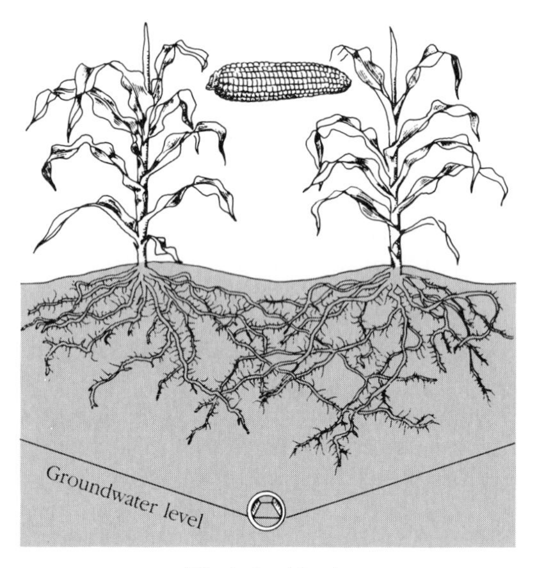

Tile-drained land

Figure 6–16 The importance of good drainage. Plant roots cannot spread in a soil that is wet nearly to the surface. If the water table falls because of summer drought, the shallow-rooted crop "burns up." (Courtesy USDA.)

embedded in gravel or cinders should be placed along the footing of the lowest level of the home to prevent water from accumulating against the foundation and seeping in (Figure 6–17). The tile should drain into an open field, ditch, or storm sewer located below the level of the house footings.

For good drainage of the surface of the yard it is desirable for the home to sit slightly above the level of the street and for the yard to be graded so that it slopes toward the street. If the home is below street level, the yard should be graded so that it slopes away from the foundation for at least 10 feet (3 m) on all sides to keep water from running down against the foundation. Eave troughs should also be installed and channeled into the storm sewer to protect plantings and people from heavy rain deluge from the roof and to lessen the amount of moisture that might otherwise build up against basement walls.

The yard should be graded so that water can drain from every part of its surface. If low spots are unavoidable, they should be tile drained, as should areas that remain damp or boggy for long periods after precipitation. Areas requiring special drainage are prevalent in the regions that have undergone glaciation in the northern United States and Canada.

If there is no way to channel drainage water into a lower elevation storm sewer or field and if the soil strata is suitable, a dry well may get rid of the unwanted water.

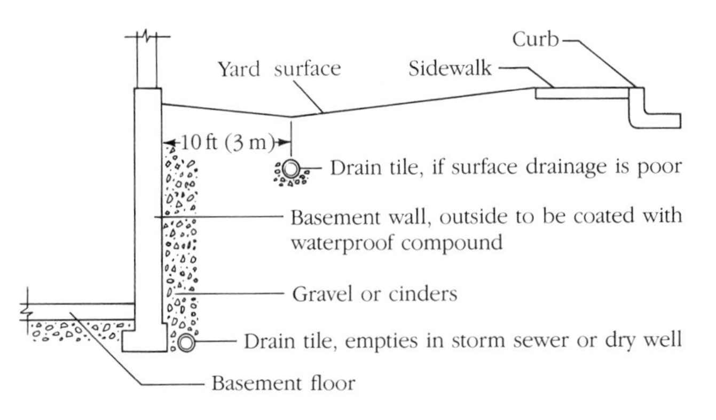

Figure 6–17 Leveling a front yard where the foundation is below street level.

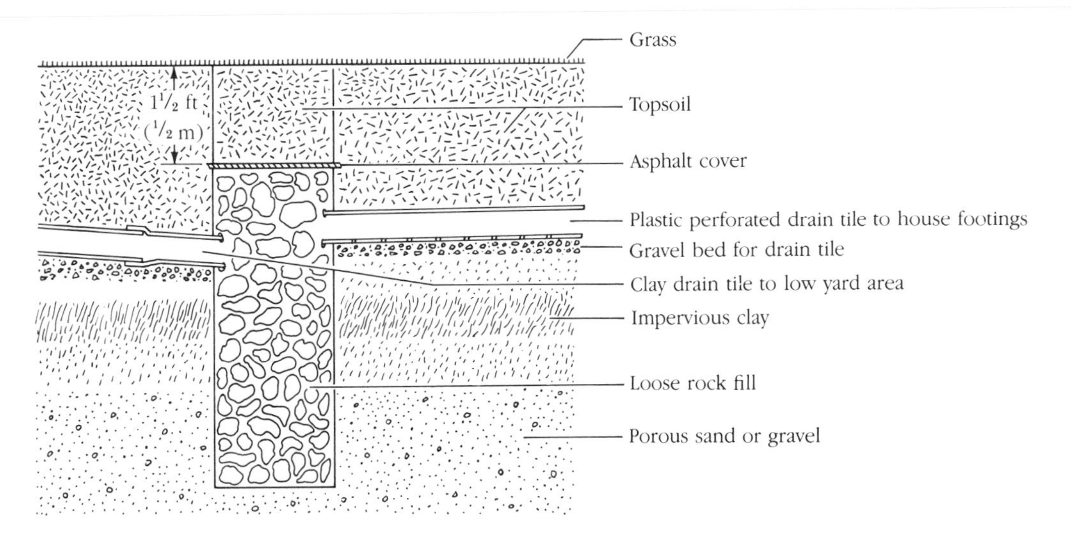

Figure 6–18 Cross section of a dry well located in soil strata ideal for such a structure.

A dry well is a hole several feet in diameter and deep enough to penetrate through the impervious subsoil layer and reach at least 4 or 5 feet (about 1–1½ m) below the lowest level tile drain. Ideally, the lowest end should terminate in a sandy or gravelly soil layer through which moisture can rapidly drain away (see Figure 6–18).

Perforated fiber pipes are usually the most convenient for channeling water from the tile drains to the dry well. A few inches of gravel on the bottom of the trench in which the perforated pipes are laid will further facilitate drainage. The well should be filled with loose solid material to within 18 to 24 inches (45 to 60 cm) of the top, making it a good place to get rid of unwanted rock, pieces of cement, and other solid nonorganic waste that usually accumulate around a building site. Scraps of lumber and other organic material should not be used, as these decompose and cause the surface to settle. A heavy plastic or asphalt cover should be placed over the top of the loose rock fill and topped with soil that can be planted to lawn or other plantings.

Selected References

Israelsen, O. W., and Hansen, V. E. *Irrigation Principles and Practices.* 3rd ed. Wiley, New York. 1962.

Leopold, L. B. *Water: A Primer.* W. H. Freeman and Company, San Francisco, 1974.

Slatyer, R. O. *Plant Water Relations.* Academic Press, New York. 1967.

U.S. Department of Agriculture. *Water* (*Yearbook of Agriculture,* 1955). U.S. Government Printing Office, Washington, DC. 1955.

7 Climate, Temperature, and Light

Tomatoes and cucumbers are more likely to mature at Fort Vermilion in northern Alberta than along the coast of Washington. Summers are cooler in much of central Africa than they are in our own Midwest. Sweet corn will mature in Siberia in about two-thirds the time it requires in Brazil. Sunburn and bright red apples are both more likely in the intermountain valleys of the West than they are along the East Coast. Location within the yard determines survival and earliness of many garden crops.

The importance of heat and light in photosynthesis is understood by most plant growers. Not so well understood are the profound effects of moderate variations of either temperature or light on photosynthesis and on other plant growth processes. These effects, which are illustrated by the above examples, will be discussed in this chapter.

Climate and Horticulture

Temperature and light are two fundamental aspects of climate. Wind, clouds, rain, hail, snow, and humidity, all of which relate to the moisture cycle described in Chapter 6, along with elevation, latitude, and location relative to large bodies of water are some of the many factors that determine climate. Climate of a specific area in the earth's temperate zones varies from day to day, from season to season, and from year to year, and it can vary quite dramatically even in areas only a few miles apart. Nevertheless, climatologists are able to calculate statistical probabilities of certain climatic occurrences likely to affect crop and garden plant performance. Most of these statistics are compiled by and available from the national weather bureaus of the United States and Canada.

Weather Bureau Services

Because the weather affects so many industries and is so important to everyone, most governments maintain a weather monitoring service. Weather stations are in operation in thousands of locations in every state of the United States and every province of Canada. They vary in size and complexity from a covered box housing a thermometer where someone reads daily high and low temperatures to multibuilding complexes where trained climatologists monitor space satellites, track hurricanes, make short- and long-range forecasts, and analyze long-term climatic trends.

Weather bureau information of particular interest to home gardeners includes short-term weather forecasts; frost and foul weather warnings; amount of evaporation; wind velocity; cumulative precipitation; and monthly, annual, and long-term climatic summaries.

The value of short-term forecasts and foul weather and wind warnings is apparent to North Americans. Uses of evaporation data were discussed in Chapter 6. Not so familiar are the monthly, annual, and long-term area weather summaries. These summaries, available at most large libraries and from weather bureau offices, give the average last spring and first fall frost; the average date during spring and fall of other temperatures near the freezing point; high, low, and mean temperatures for each day and month; long-term average daily and monthly precipitation; record high and low temperatures and precipitation; average hours of sunlight; day length; evaporation; wind velocity; and considerably more.

Few garden sites will have exactly the same weather as a weather station; temperatures may be a few degrees colder or warmer and precipitation a little more or less. Nevertheless, these statistics can be useful in determining what to plant, when to plant, when to irrigate, and the timing of other gardening activities.

Macroclimate

The term **climate,** as it is usually used, refers to long-term weather patterns of a fairly broad geographical area; a more correct term is **macroclimate.** Macroclimate is determined by the intensity and quality of solar radiation reaching the surface of the earth; the quantity of the radiation remaining in the surface layers of the atmosphere as heat and light energy; and modifications of the impact of this energy by nearness to large bodies of water, by ocean and wind currents, and by topographical features such as mountains, forests, and irrigation projects. These factors determine not only temperature and light intensity but also precipitation patterns and moisture cycle.

Temperature is inversely correlated with latitude. The farther from the equator, the cooler the average annual temperature becomes, because the sun's rays strike the earth more directly near the equator and, as a consequence, are filtered and scattered by less atmosphere. This concept may be easier to comprehend if one thinks of the earth as similar to a grapefruit with the atmosphere as the peel. A cut directed toward the center of the grapefruit will go through less skin than an oblique cut that slices off one corner (Figure 7–1).

The greatest amount of heat is retained at the earth's surface, where there is the greatest amount of overlying

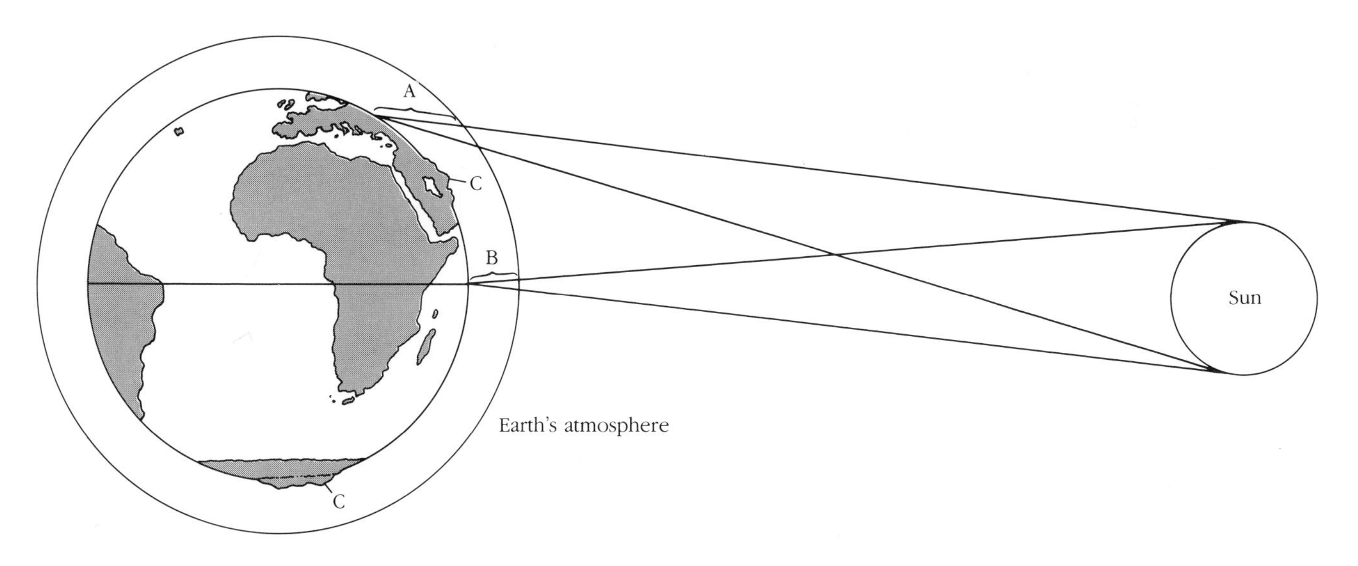

Figure 7–1 The correlation of temperature and latitude. The sun's rays are filtered by more atmosphere when they strike the earth at an oblique angle (A) than when they strike it directly (B). Similarly, there is less atmosphere to filter the sun's rays above mountain ranges (C) but also less atmosphere to absorb and retain heat. (Both mountain range height and atmosphere thickness are greatly exaggerated in this illustration.)

atmosphere to absorb it and to prevent its being radiated back into space. This retention of heat by the atmosphere is called the **greenhouse effect** because it is the same phenomenon that warms greenhouses on sunny days (Figure 7–2). Radiant energy entering through the glass is converted to heat energy, which is prevented from escaping by the glass. Because of the greenhouse effect, temperature is inversely correlated with elevation, each 300 feet rise in elevation usually meaning a 1°F drop in temperature. Differences in elevation account for extreme temperature variations in geographically adjacent mountain-valley locations; this explains, for example, how an hour's drive from Denver or Salt Lake City with their dry, relatively mild climates can end at slopes where virtually year-around skiing is possible.

Altitude also affects the amount and quality of solar energy reaching the earth's surface. At higher elevations fewer of the ultraviolet and short rays are scattered and filtered, because the sunlight passes through less atmosphere. Thinner atmosphere absorbs and retains less heat, which is why at high altitudes it is hot and bright in direct sunlight but cool in the shade, on rainy days, and during the night, and why late spring and early fall frosts are more frequent at high altitudes.

Ocean currents affect nearby land masses. The Gulf Stream is noted mainly for warming the climate of Europe, but it also warms the southeastern United States. The Labrador Current cools eastern Canada and New England, and currents from the Gulf of Alaska cool the Pacific Coast south to Monterey, California. Climatic extremes are also moderated by large bodies of water. Summers are cooler, winters warmer, and late spring freezes less likely near an ocean or a large lake than they are at the same latitude and altitude farther from a body of water.

Water affects freezing in two ways. Many times more heat energy is required to raise the temperature of water than to raise the temperature of air, and water releases large amounts of heat energy when it cools. Thus water acts as a heat or cold buffer, tending to moderate the rise and fall of temperatures of land areas nearby. During early spring days water is likely to be cooler than land, and winds blowing over a body of water will cool the temperature of the land nearby, thereby delaying the bloom of fruit trees and ornamentals. During a frosty night, however, the temperature of water is likely to be considerably warmer than that of land, and plants growing near a body of water are not so likely to freeze as are those growing farther from it (Figure 7–3). This is the reason for the extensive fruit plantings on the leeward sides of the Great Lakes in Ontario, Michigan, Ohio, and New York.

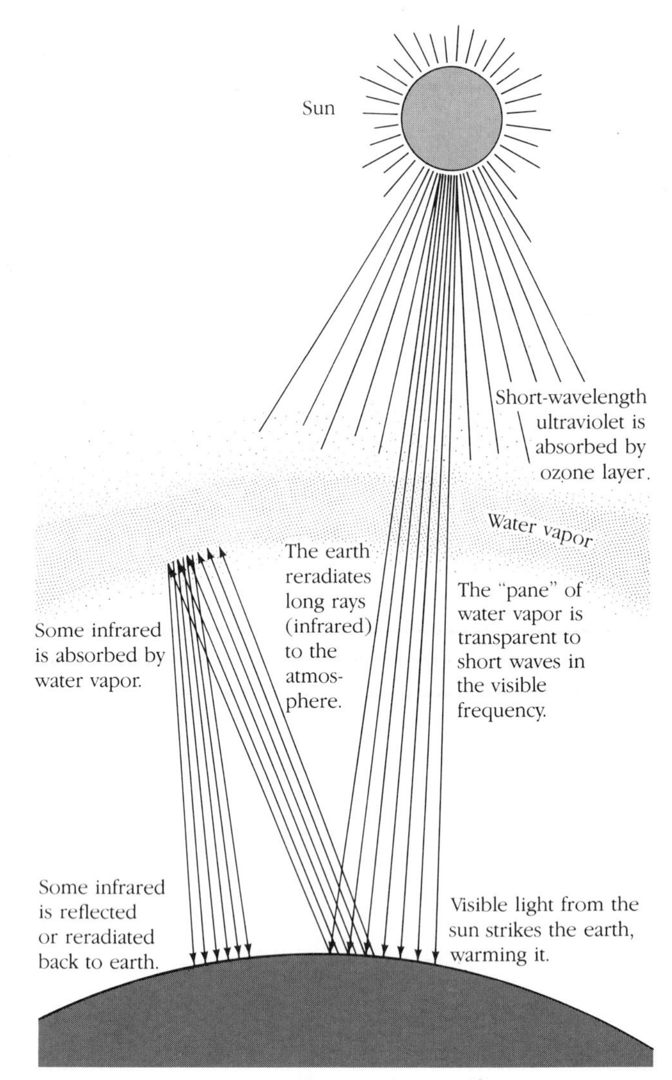

Figure 7–2 The greenhouse effect.

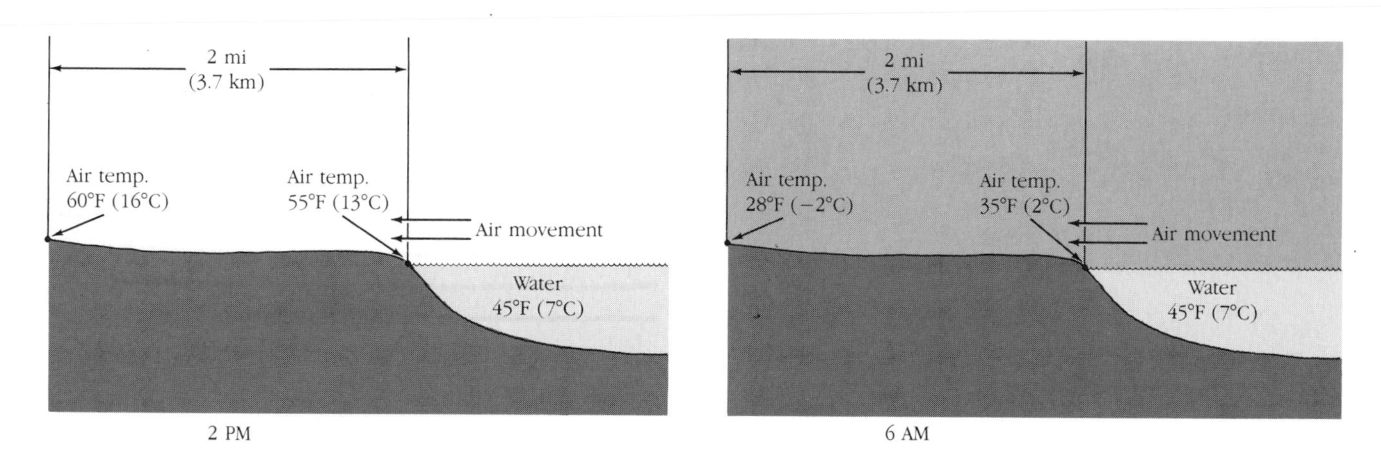

Figure 7–3 Water as a temperature buffer. A body of water cools nearby areas during warm days and warms them during frosty mornings.

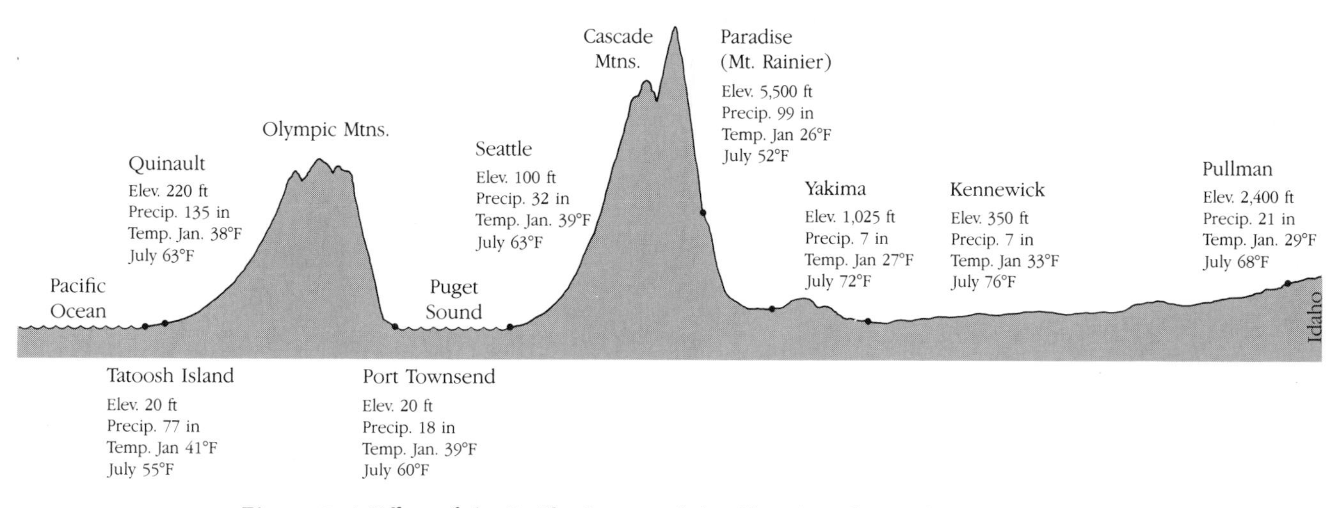

Figure 7–4 Effect of the Pacific Ocean and the Olympic and Cascade Mountain Ranges on the climate of Washington State.

Mountain ranges have a profound effect on moisture and temperature of nearby areas. At the latitude that includes the continental United States and southern Canada, the prevailing westerly winds tend to drop their moisture as they cool in rising over the western slopes and to take up moisture as they warm in dropping down the eastern slopes. Thus some of the wettest and driest areas of the continent are found within a few miles of each other, in Washington and British Columbia, on the windward and leeward sides of the Pacific Coast and Cascade ranges that intercept moisture from Pacific storms. Mountain ranges also prevent the climate-ameliorating effect of the ocean from extending as far inland as it might otherwise, but they frequently divert winds enough to prevent Arctic fronts or severe storms from reaching protected areas (Figure 7–4). Because of the western mountains, moisture and air currents modified by the Pacific do not reach as far inland in North America as moisture and air currents modified by the Atlantic reach in Europe; however, the Sierra Nevada prevent cold air from the Great Basin from spilling into California's Central Valley, and, similarly, the northern Rockies usually protect the valleys of British Columbia and the Pacific Northwest from the bitter winter winds of the Great Plains and the dry summer winds of the southwestern deserts.

Forest and brush cover also affect climate. Forested areas have a higher humidity. Recent reforestation of land denuded for thousands of years in the Mediterranean region is reported to have increased precipitation in nearby areas. Shrub and forest covers also increase water infiltration and reduce runoff, which can be beneficial in preventing erosion or detrimental in reducing the amount of water available for irrigation of nearby lowland areas.

Human activities also tend to modify climate. Extensive reservoirs and irrigation projects increase precipitation along the paths of prevailing winds blowing over them. Metropolitan areas are warmed by combustion and other heat-producing activities, whereas nearby areas may have more cloudy weather and more precipitation because dust and other particulate matter released by auto exhausts and factory smokestacks provide a nucleus around which moisture can condense.

Microclimate

In contrast to macroclimate, which encompasses a fairly large geographical area, **microclimate** relates to climatic variations existing at different locations in a community or even within a single yard as the result of topographic features, direction of slope, or location of buildings or plantings. The location of the yard in relation to surrounding topographic features will affect livability, especially outdoor enjoyment, and also will determine within limits what plants can be successfully grown. A south slope, for example, warms earlier in the spring but may be hotter during the summer; the leeward side of a ridge is less subject to wind and may have reduced precipitation, which may or may not be desirable; and hillside locations are less subject to frosts than the valleys below them.

As mentioned earlier, temperature generally decreases as elevation increases. This is true during the day even for slight increases in elevation. It may not be true, however, for sloping areas during a still night when cold air, which is heavier than warm air, is likely to flow downhill and settle to the bottom of valleys and depressions. Under these conditions there occurs what is known as **temperature inversion,** where slopes may be several degrees warmer than depressions below them. This is the reason that orchards are frequently found on hillsides and not in valleys or "frost pockets" (Figure 7–5). Prospective homeowners may want to consider the reduced likelihood of spring frosts on sloping sites when selecting a homesite.

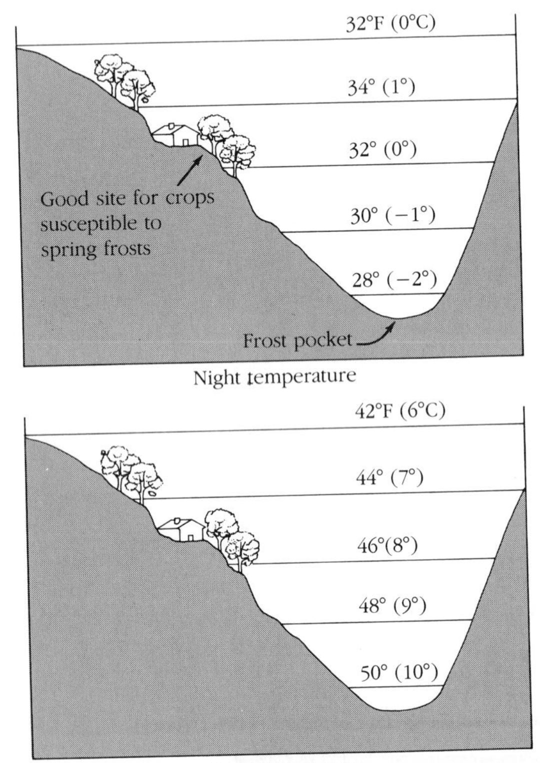

Figure 7–5 Temperature differential with elevation. A homesite on a sloping area above a depression is less susceptible to damage from early spring frosts than is a site in the bottom of a valley.

By the same token, planting frost-susceptible crops at the upper end of an extensive garden may afford them some frost protection.

Buildings modify microclimate, making temperatures near a building warmer during cool weather. Shade-loving plants can be planted on the north side of buildings in the North Temperate Zone. If the climate is cool, melons and tomatoes will mature earlier and grapevines survive better if they are planted on the south side of a building. Special attention to watering must be given plants growing under a roof overhang that interrupts rainfall.

A homeowner can also take steps to modify both indoor and outdoor microclimate. A wide roof overhang will block the hot rays from entering a south window when the summer sun is high in the sky but will permit them to enter during the winter when the sun is lower (Figure 7–6). A deciduous tree planted on the south side of a house will have a similar effect, providing shade during the summer but permitting the sun's rays to filter through during the winter when the leaves have fallen. The home and outdoor living areas can be oriented to take advantage of views, spring sunshine, and cooling breezes or to block cold winds. Planting windbreaks and constructing fences also can provide a more desirable microclimate around the home.

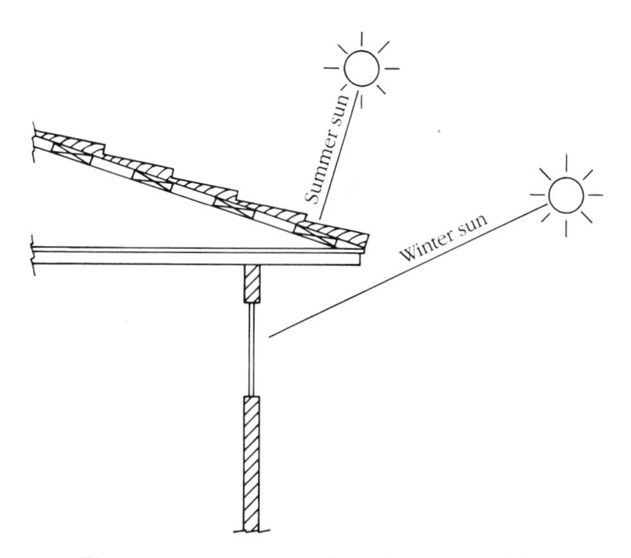

Figure 7–6 Using a roof overhang to modify temperature.

Temperature

Temperature is the climatic factor that, more than any other, determines the kinds of plants that can be grown in particular areas. Temperature affects the quality and maturity rate of garden products and is important to virtually all plant responses, including photosynthesis, transpiration, and respiration, all of which generally increase with a rise in temperature. Temperature also determines the kinds of pests likely to be troublesome; usually the warmer the temperature and the milder the winter, the more severe will be the plant disease and insect problems.

The optimum temperature for growth varies with different plants. Most temperate zone vegetables and annual flowers are classified as cool-season crops, which germinate at temperatures as low as 40°F (4°C), or warm-season crops, most of which will not germinate until the soil temperature has reached about 60°F (16°C). For more detailed information on temperature and germination see Chapter 4 and Table 14–3 (Chapter 14).

Fortunately, most plants grow well over a fairly broad range of temperatures. Though respiration and photosynthesis increase as temperature rises, through the lower three-fourths of the temperature range for any given species the increase in photosynthesis is somewhat more rapid than the increase in respiration. Thus maximum growth rate occurs in the upper part of the temperature range through which the plant will grow.

Effects of Excessively High Temperature

The rate of photosynthesis continues to increase, but the rate of respiration accelerates more dramatically as temperature rises through the upper one-fourth of the temperature growth range, and eventually a point is reached where the food utilized by respiration just equals the amount manufactured by photosynthesis, and growth ceases (see Figure 7–7). This upper temperature limit for

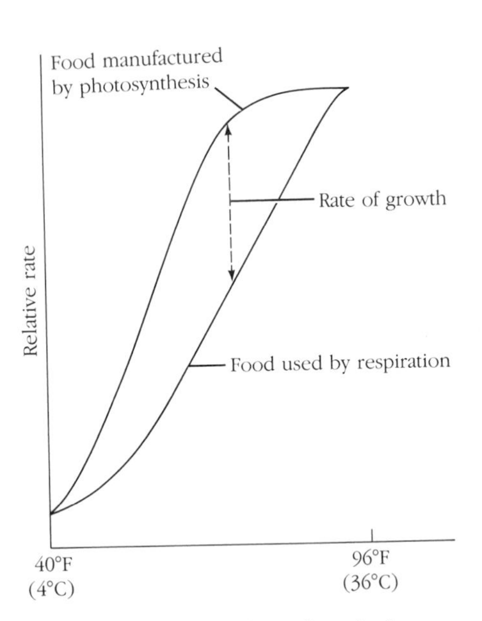

Figure 7–7 Relationship of respiration, photosynthesis, and temperature to plant growth.

growth varies from one species to another; for many crops it is about 96°F (36°C). In some locations temperatures go so high that there may be physiological damage to plants, which, among other things, stops photosynthesis. Thus one detrimental manifestation of excessively high temperatures is the cessation of plant growth.

Temperature and rate of respiration also have important implications in the transporting and storing of horticultural products. Once a plant is harvested, its food production via photosynthesis essentially ceases, while its food breakdown continues to be directly correlated with its rate of respiration. Thus seeds, which have a very low rate of respiration and much stored food, can remain alive for years. Potato tubers can be stored for considerable periods at low temperatures because their respiration rate is low, but leafy vegetables like lettuce and most floral crops, with high rates of respiration and minimal stored food, cannot be kept for long periods under even the most favorable conditions.

As with growing plants, the rate of respiration of stored products is directly correlated with temperature: Heat increases and cold decreases the respiration process. Furthermore, since heat is one of the by-products of respiration, respiration increases temperature and increased temperature brings about more rapid respiration. The heat liberated by respiration of tightly packed fruits or vegetables can have drastic consequences if refrigeration is not adequate or fails in the storage facility. Spoilage resulting from heat-induced deterioration and rapid buildup of microorganisms can result within a few hours. Respiration also results in a loss of stored plant foods such as carbohydrates. Sugars dissolved in cellular liquids, which provide natural sweetness and are the bases of flavor of many fruits and vegetables, are the first carbohydrates to be utilized. On a warm day a noticeable loss of sweetness can be detected only a few minutes after harvesting vegetables such as peas, asparagus, and sweet corn.

The relationships among temperature, respiration, loss of sugar, and storage have implications for the home gardener. Harvesting in the morning when the product is cool will permit longer retention of quality. Harvesting containers made of fibers, cloth, wood, or loosely woven materials are preferable to plastic bags or metal containers, because they permit air to circulate and the heat of respiration to dissipate (Figure 7–8). Floral products will last longer if they are cooled immediately after being gathered from the garden and if the arrangement is kept as cool as possible.

High temperature prevents temperate zone perennials from being grown in tropical and subtropical climates. In warm winter areas deciduous temperate zone woody plants lose their leaves at the normal fall time. There is, however, too little cool weather to permit them to complete their physiological rest period. As a consequence buds remain dormant, and leaves and blossoms form erratically or not at all during the following season. This is why apples and peaches are not grown in southern Florida or in lowland areas of southern California.

Plant diseases and insect problems are likely to be more serious when temperatures are high, partly because plant pathogens and insects reproduce more rapidly during periods of high temperature and partly because a mild winter will permit more than the normal carryover of insects and disease pathogens.

Effects of Low Temperature

Because so many horticultural plants are quite susceptible to frost and cold temperatures, most horticulturists are more concerned about temperatures too low for plant growth than they are about those too high. When temperatures are excessively cool, even though they may not be below freezing, there will be lack of plant growth and failure of seed germination. Most species that originated in the tropics are injured by temperatures slightly above the freezing point. Peppers, tomatoes, African violets, melons, and cucumbers are all injured if the temperature drops much below 40°F (4°C) for any length of time.

Another problem attributable to cool temperature is premature seed stalk formation (**bolting**) of biennial vegetables (Figure 7–9). The natural growth cycle of most biennial plants consists of vegetative growth during the first year, then winter dormancy concomitant with seed stalk initiation, followed by seed stalk formation the second summer. If for some reason cool temperatures occur shortly after the crop is planted, the growth cycle is compressed and seed stalks are likely to be initiated during the first summer. Because biennial vegetables are grown for roots, petioles, or leaves rather than for seed, flowering and seed formation render them useless. Seed bolting often occurs when a crop such as celery is subjected to cool temperatures in order to condition it to withstand the shock of adverse post-transplanting environment. Bolting can also occur if an unseasonably cool spring follows planting and germination of a biennial crop.

The temperature most significant for gardening is undoubtedly 32°F (0°C), the freezing point of water. Length of the growing season, planting time for warm-season annual crops, and harvest time of many garden products are all dependent on the spring or fall occurrence of this temperature.

Just as significant for the garden is the minimum temperature likely to occur during winter. Although most perennial plants that originated in cold climates can withstand some freezing during their dormant period, there is great variation in the conditioning necessary to enable them to tolerate freezing and the minimum temperature

Figure 7–8 Harvest containers. Harvest containers should be constructed so that air can circulate freely to dissipate the heat of respiration.

Figure 7–9 Bolting of onions. These year-old onions have produced a seedstalk as a result of exposure to cold winter temperatures. (Courtesy Washington State University.)

they can be conditioned to withstand. The average minimum winter temperature determines to a large extent which perennial plants can be grown in a given area.

Tissue Freezing and Freezing Damage. Freezing of plant tissue under conditions of slowly lowering temperature typical of most freezes begins with the formation of ice crystals in the spaces between cells. As more of the intercellular water becomes ice, water is drawn from inside the cell by osmosis, causing the membrane around the cytoplasm to shrink away from the cell wall and the intercellular ice crystals to enlarge. If the temperature becomes cold enough, the entire cell eventually freezes. As the temperature warms after a freeze, cells of tissue that was not killed are able to reabsorb water from the intercellular spaces and resume normal activity. Tissue damaged by freezing wilts, because water transport (as well as all other functions) is disrupted. Death of plant tissues as a result of freezing probably results from both physical disruption brought about by ice formation and physiological damage caused by the highly concentrated solution left in the shrunken protoplast.

Occasionally on an extremely still night plants will survive temperatures lower than they normally would because of a phenomenon called **supercooling.** What happens can be demonstrated by simultaneously lowering the temperature of a glass of water and keeping the water absolutely motionless. Under these conditions ice crystals do not form until the temperatue falls several degrees below the freezing point. If, however, the supercooled water is moved slightly, ice immediately forms. Because of the possibility of supercooling, it is advisable to stay out of the garden after a frost until morning temperatures have risen well above the freezing point.

Freezing damage occurs from excessive cold during the winter, from early spring or late fall frosts, or from dehydration of tissues during a time when the water supply cannot be replenished by the frozen root system (physiological drought discussed further later in this chapter). Physical damage caused by the forces of winter also results from plants being pushed from the soil by alternate freezing and thawing (heaving) and from plants being broken by the sheer weight of ice or snow.

Damage from Winter Cold. During an occasional winter in almost every section of the United States and Canada fruit trees and ornamentals are damaged by cold weather. It is often assumed that this damage occurs only when winter temperatures drop below the specific minimum the plant species can tolerate. Actually, the physiological basis of cold temperature damage is complex and dependent on many variables, including the kind of plant, the part of the plant, food and water resources in the plant tissue, the season of the year, temperatures prior to the freeze, rate of temperature drop, temperature during the freeze, temperature after the freeze, amount of air movement, moisture in the soil and in the plant tissue, and perhaps others.

Research has shown that most woody perennials have two distinct levels of cold tolerance: one that develops as a result of dormancy and is not temperature dependent and a second acquired only after several days of freezing temperatures, which condition them to withstand much colder weather. Woody plants gradually become cold-tolerant in the fall after their leaves have been shed. Several weeks after leaf **abscission** (shedding) the plant achieves the maximum cold tolerance it will develop without being subjected to freezing temperature. This innate tolerance enabling it to withstand temperatures considerably below freezing occurs even though the weather remains mild; however, further cold tolerance (**temperature-induced hardiness**) develops only if temperatures remain below freezing. The extent of this temperature-induced hardiness fluctuates with temperature; it is lost if temperatures rise for any length of time but will be regained if temperatures drop again before buds start to swell. Cold conditioning will be further discussed later in this chapter.

Plants are frequently damaged when the weather turns cold before they have developed maximum tolerance. In the Northwest on November 11, 1955, temperatures suddenly dropped to near 0°F (−18°C). Autumn that year had been mild, and leaves were still hanging on many fruit trees and ornamentals. About 80% of the orchards in the region sustained some damage, although a temperature drop to 0°F during the winter would normally cause no problem. Strawberry production was cut in half, and almost every homeowner lost some ornamentals. Injury occurred because plants were not hardened and prepared for such cold weather.

Horticultural crops are often damaged also by excessive cold during mid-winter. Excessive cold may mean a few degrees below freezing in Florida and California or temperatures of −40° to −60°F (−40° to below −46°C) in the upper Midwest, Canadian provinces, and Alaska. During January 1969, also in the Northwest, temperatures dropped to record lows. Some thermometers in the vicinity of Pullman, Washington, recorded −50°F, which was almost 20° lower than any previously recorded minimum. Damage to orchards was extensive, and many species of ornamentals were killed outright. In native stands of ponderosa pine and grand fir, trees over 100 years old segregated for hardiness. Some died immediately, some were damaged and remained unthrifty or died several years later, and others showed no ill effects. Although most plants were at maximum hardiness, they could not withstand the extreme cold that occurred.

The delayed appearance of damage due to cold is not unusual. During the 1950s full grown 'Baldwin' apple trees in some home gardens and in several commercial orchards in the Northeast began to break apart. In the Cornell University orchard in 1953 so many 'Baldwin' trees collapsed under their heavy fruit load that trees of that cultivar were considered too dangerous to harvest. A freeze in the early 1930s had killed enough tissue to permit decay fungi to become established. Although outwardly the trees grew normally, the interior supporting wood was gradually consumed by the fungi, resulting in the trees' sudden collapse 20 years later.

Although there are some variations from cultivar to cultivar, each crop normally has a minimum temperature to which it can be hardened and below which it will be severely injured or killed. Apples, American or European plums, and sour cherries are the most hardy common tree fruits, and some cultivars will withstand temperatures of −30° to −40°F (−34° to −40°C) if conditions for maximum hardiness development have occurred. Peaches are the least hardy of the temperate zone tree fruits, and peach buds are likely to be injured if the temperature drops below about −10° to −12°F (−23° to −24°C), and the trees will be killed if the temperature drops much below −15° to −18°F (−26° to −28°C), even when they have been subjected to ideal conditions for the development of cold tolerance. In the Washington State University orchard at Pullman during the 1969 freeze, 'Montmorency' sour cherry, 'Stanley' plum, and 'McIntosh' and 'Rome Beauty' apple trees sustained little damage. 'Delicious' and 'Golden Delicious' apple, pear, and all sweet cherry trees were injured, but most recovered. All peach and apricot trees were completely killed. Apple and plum trees produced some fruit, but fruit buds of sweet cherries, sour cherries, and pears were killed.

Damage from Spring and Fall Frosts. Probably more damage from cold temperatures occurs in the spring than at any other time of the year. Tender transplants planted out too early are likely to be damaged by late freezes, and even emerging seedlings are frequently frozen. Spring is also the time when the orchardist is most apt to lose a crop as a result of freeze damage, because fruit buds have little resistance to cold temperature once they begin to expand and open into blooms (Figure 7–10). The fruits or blooms of woody ornamentals are frequently damaged by

Figure 7–10 Stage when fruit buds are most likely to be damaged by frost.

spring frosts. The effects of a spring frost may last for several seasons if blossom destruction results in alternate bearing. Late summer or early autumn frosts can damage tender crops before they have matured and also end the riotous color of the late summer flower garden.

Although 32°F is generally considered to be the temperature below which killing frosts occur, most plants do not freeze at 32°F. This is because sugars and other soluble compounds dissolved in the cell solution, acting like antifreeze, lower the freezing point. Tomato transplants will survive temperatures of 31°F (about −1°C); blossoms of temperate zone fruits are not damaged until temperatures reach 28° to 24°F (−2° to −4°C), depending on kind; mature apple fruits will not be seriously damaged to about 24°F; and hardened cabbage will survive a drop to about 20° to 16°F (−7° to −9°C). It is well to remember, however, that, because of temperature inversion on a still night and microclimatic differences, the temperature surrounding a low-growing plant is likely to be lower than that recorded on an eye-level thermometer and often much lower than that of a weather bureau recording device that may be surrounded by the heat producing paraphernalia of a city (see Figure 7–11).

Spring and fall frosts are most likely to happen on a clear, still night when there are no clouds to limit radiation of heat from the atmosphere and no wind to prevent air inversion. The coldest temperature under such conditions is most apt to come just before sunrise, and this is when frost-protective measures may be required.

Physiological Drought. If a plant is permitted to go into the winter with little moisture surrounding its roots, or if a part of the plant dehydrates because the soil or the plant's water-conducting tissue is frozen, injury called **physiological drought** will result. In most northern areas one can observe this kind of water-stress injury on the windward side of exposed ornamental and fruit trees and on the southwest side of trees growing where winter sunshine is prevalent. It is much more likely to occur if the prevailing wind is also from the southwest. Bark injury that exposes the underlying wood and produces a permanent scar is called **catfacing,** and physiological drought is a major cause of catfacing of fruit and ornamental trees (Figure 7–12).

Figure 7–11 Effects of frost on strawberry blossoms. As they bloom early and grow close to the ground where, because of air inversion, temperature is likely to be coldest on a frosty spring morning, strawberry blossoms are especially susceptible to frost. The pistillate cone turns black (right flower) when the blossom freezes. With most plant species the pistil is the tissue most likely to be frost damaged.

Figure 7–12 Catfacing of a young 'Ruby' horse-chestnut tree. This manifestation of physiological drought resulted from drying winter wind coupled with bright sun, which dehydrated the southwest side of the unprotected trunk at a time when moisture could not be replaced because freezing temperature had stopped translocation.

Physiological drought can also occur on evergreens; it will be manifest by the turning brown of needles or by the death of entire leaves or the margins of leaves of broadleaved evergreens.

Physical Damage from Heaving, Ice, Snow, and Freezing. Another kind of cold weather damage is due to **heaving,** alternate freezing and thawing that forces some plants completely out of the soil. Shallow-rooted perennials, including perennial flowers and strawberries, are especially subject to heaving damage.

Breakage by wet snow is frequently the major winter injury to woody plants in areas with a cool, wet climate. Damage can be especially severe when a heavy snowfall occurs early in the autumn before leaves have fallen from deciduous trees and shrubs. Danger of breakage by snow prevents growing broadleaved and semispreading needle evergreens at some locations. The tops of spruce, fir, and pine, especially those having a large crop of cones, are often broken by accumulations of ice and snow. Freak ice storms during which soil and plant temperatures remain below freezing but upper-air temperatures are high enough to permit rain to fall also can be physically damaging. Under these conditions ice accumulates until its weight breaks down the plant (Figure 7–13).

Figure 7–13 Damage to plants from winter storms. (Courtesy USDA.)

Although deep snow that lasts through the winter can insulate plants from freezing, it can also injure them in several ways. Some plants cannot survive being covered for long periods by the exceedingly compact layer of snow that forms as spring thawing commences. Death may result from insufficient aeration or, more commonly, because the combination of plant stress and an ideal environment for the growth of pathogens provides optimum conditions for disease. Compacting of snow as it melts is also responsible for physical damage to plants. Because thawing proceeds throughout the layer of snow rather than from the top, branches trapped in the snow can be pulled downward far beyond their breaking point as the snow settles (Figure 7–14).

Figure 7–14 The weight of snow often breaks trees and shrubs.

Splitting of tree and shrub bark due to differential freezing is a common phenomenon during periods of rapidly falling temperature. The splitting, sometimes so sudden that it sounds like the crack of a rifle, can loosen the bark from the entire circumference of a tree; it will cause

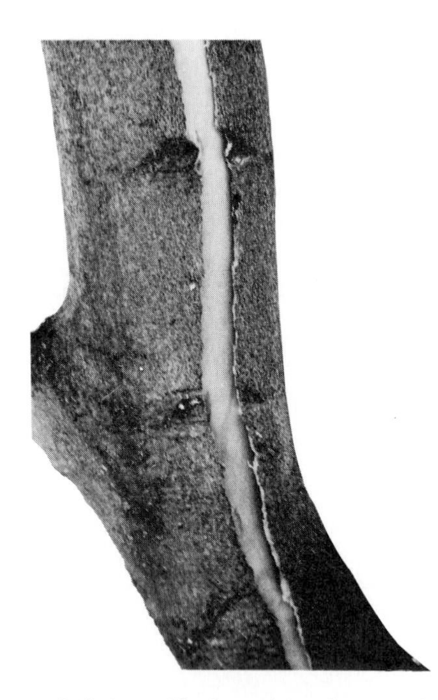

Figure 7–15 Splitting of bark resulting from rapid freezing. This Norway maple died the summer after the photograph was taken. It probably could have been saved had the bark been tacked back in place soon after the damage occurred. The bark loosens at the cambium layer, and fastening it back in place tightly against the wood prevents tissue dehydration and allows the cambial cells to cement the wound with new tissue similar to what occurs in a graft union.

the tree to die unless the damage is repaired (Figure 7–15). Not only is breakage due to winter weather physically damaging; it also provides pathways for the entrance of insect and disease organisms. Pest injury may be seen immediately, or its effects may not be noticeable for several years, as was the case with the 'Baldwin' apple trees mentioned earlier in this chapter.

Cold Temperature Survival

Whether a plant is able to survive a period of cold weather depends on its genetic constitution, the physiological conditioning it has undergone, and the environment to which it is subjected before, during, and after the cold. Gardeners can reduce the possibility of damage from cold by plant and site selection, by conditioning the plant with proper planting and cultural practices, and by the use of mulches and other frost protective devices.

Plant and Site Selection. Gardeners should select plants known to be hardy in their area. An occasional choice specimen that is only half-hardy to the region can be brought through the winter with special care, but it takes dedication that most of us do not have to nurture very tender plants through winters in climates not suited for them. The U.S. Department of Agriculture has compiled a Plant Hardiness Zone Map based primarily on minimum winter temperatures (see Figure 14–107, Chapter 14). Plant lists found in gardening texts generally cite the zones from this map to which each kind of woody ornamental is adapted. Because they involve a smaller area, plant lists published by state experimental stations and regional publishers contain more precise adaptation lists. These maps and lists are good general guides to the probable cold survival of ornamental species. However, climatic adaptation depends on many factors, and gardeners should seek advice from local authorities before making extensive plantings of materials listed as having borderline hardiness for the area.

Choice of location even within a yard may determine whether or not a particular plant will be injured by cold temperature. The more hardy rhododendron and azalea cultivars, for example, are more subject to wind damage and fluctuations in temperature than to cold temperature per se. Planting them on the north side of a building in a location where they are protected from the wind will often enable them to grow in climates where they would not otherwise survive. On the other hand, when grapes are grown in some northern regions, they fail to manufacture and store sufficient carbohydrates. Trellising against a south wall will sometimes provide them with sufficient heat to permit manufacture of the sugars needed for survival at locations where they otherwise would not be hardy. South slopes and sandy soil will warm more rapidly in the spring, and crops can be planted on them earlier and will mature more quickly than if they are planted on a colder north slope or on clay soil. The importance of slope and nearness to a body of water for frost prevention have been discussed.

Cold Conditioning and Cultural Practices. It goes without saying that gardeners will be less likely to have a garden freeze if they wait to plant until after danger of cold injury is past. As has been mentioned, the general rule for planting vegetables and annual flowers is to plant cool-season crops as soon as the ground can be worked and to plant warm-season crops after the danger of frost is over. Where winter damage is frequent, lawns and woody plants should be planted during the spring rather than during the fall so they can become established before winter commences.

Whether or not a plant survives a fall or winter freeze is greatly influenced by the temperature before that freeze. Plants will reach their maximum tolerance only when temperatures have been fairly cold for a period of time and when there has been a gradual reduction in temperature over that period. The intensity and duration of the cold and the rate of drop or rise in temperature will affect the extent of freeze injury. If the drop in temperature is gradual over a period of several days and if the cold period is not too long, damage will be less than if temperature falls rapidly or the cold period is prolonged. Air movement will intensify the injurious effect of cold temperatures. A freeze is likely to cause considerably more damage if there is a wind blowing when it occurs.

The level and type of stored food has a profound effect on how cold-tolerant a plant is likely to be. Plant tissues well supplied with carbohydrates and water will withstand considerably more cold than those that are not. Reducing the amount of nitrogen fertilizer and water supply as autumn approaches and refraining from heavy summer pruning are practices that slow vegetative growth and increase carbohydrate accumulation. Trees and shrubs with their carbohydrate supply depleted by heavy crops of fruit or seed become extremely susceptible to winter injury. Limiting production by removing some or all of the faded blooms or immature fruits and seeds increases the chance for woody plants to escape freeze damage the following winter.

The importance of maintaining abundant soil and plant moisture throughout the winter to prevent physiological drought has been emphasized. Training a branch to shade the southwest exposure of trunks of fruit and ornamental trees or wrapping the trunks with commercial tree wrap to prevent drying and to reflect solar radiation will often prevent catfacing caused by physiological drought.

Conditioning a plant to withstand an adverse environment is called **hardening**, and the cold conditioning of woody perennials described above is a way of hardening. More commonly, the term hardening refers to the conditioning of transplants before they are planted into the garden or field. This has generally been done by subjecting them to cooler temperatures for a period of time. Reduced temperature hardening is satisfactory for transplants of annual crops but should not be used for celery, cabbage, and related biennials as it can sometimes induce seed instead of petiole or head formation because it replaces winter cold (see Figure 7–9). Transplants of biennial crops should be hardened by reducing the fertility of the growing media slightly or cutting down somewhat on the amount of water supplied. Care should be taken not to overharden, because overhardened transplants will require a long time to resume growth after they have been transplanted and will never produce as high yields as those that have not been severely hardened. Many horticulturists now recommend that transplants not be hardened at all.

Modifying Temperature with Plant Covers and Mulches. Various kinds of plastic and waxed paper covers can be used to start crops earlier than normally (Figure 7–16). Night temperatures under a plastic tent will be 4° to 6°F warmer than those on the outside, and the use of plant protectors may enable the planting of tomatoes and melons as much as a month earlier than without them. Daytime temperatures under a plastic tent can be as much as 20° to 25°F warmer than outside temperatures, and on warm, sunny days it may be necessary to remove the tent or at least provide ventilation for the plants under it. Other kinds of plant-growing structures used to alter temperatures will be discussed in Chapter 10.

The relative merits of mulching materials that can also affect time of crop maturity were discussed in Chapter 5. It should be mentioned here that most mulches do affect

Figure 7–16 Installing plastic and waxed paper plant protectors.

growth and maturity of plants growing through them. Black and clear polyethylene have been used extensively to hasten maturity of some warm-season vegetables. For example, muskmelons growing through a plastic mulch will be as much as 10 to 14 days earlier in certain northern areas than will similar plants growing without the benefit of mulch. A loose mulch of peat moss, straw, sawdust, shavings, or leaves is often raked over the crowns of strawberries, roses, and perennial flowers to prevent winter damage from freezing and heaving or to reduce soil temperature fluctuations (Figure 7–17). In areas where climbing roses, grapes, and trailing blackberry hybrids are not reliably winter-hardy, their canes can be taken from their trellis, placed along the ground, and covered with a loose mulch for winter protection. Special pruning of grapes is necessary to produce a low branching trunk if the trunk is to be easily covered.

Plants should not be mulched for winter protection until after the first hard freezes, middle to late November being early enough in most areas of North America. Loose mulches should not be used around woody plants if there are rodents in the yard. Mice find straw- or shaving-mulched shrubs especially attractive for home building, with tasty, protected bark of such shrubs providing a readily available food supply. Where rodents are a problem in rose plantings, soil from between the bushes can be hilled over the crowns in the late fall to protect the lower stems from winter damage.

Snow is an excellent insulating mulch and many semitender shrubs survive the intense cold of mountain winters under an annual protective cover of snow. Alpine flowers that routinely survive winters where temperatures reach −50°F (−46°C) in their native high mountain habitat often winterkill when transplanted to lower elevations because their new, warmer environment does not provide a protective cover of snow. In areas where snow almost always falls, it can be used to protect cold-susceptible plants by being piled over low-growing shrubs when extremely cold temperatures are forecast. Taller shrubs can be afforded the same protection with heavy plastic or cloth tubes constructed around them and filled with snow. Nails can be used to pin the tube onto and around the shrub (Figure 7–18). A snow cover is especially effective in protecting blossom buds, which are usually the most cold-susceptible structures of early spring-blooming fruit and ornamental shrubs. All materials needed for protecting plants against a mid-winter freeze should be assembled during the good weather of autumn; otherwise, saving the plant may not seem worth the effort when it comes time to face the storm.

Preventing Physical Damage from Snow and Cold Weather. At locations where heavy snows accumulate, plants may require special winter attention. Securing easily broken plants with sturdy stakes, tying upright branches of evergreens together with string (or string and burlap) so that snow cannot accumulate on them, and occasionally shaking the snow from plants during heavy snowstorms will reduce breakage. Pruning should be de-

Figure 7–17 Mulching to provide protection. Mulching of strawberries not only provides winter protection but also keeps the fruit clean and conserves moisture after the mulch is raked from the plants in early spring. If frost is predicted, the mulch can be raked back over the plants until the hazard passes.

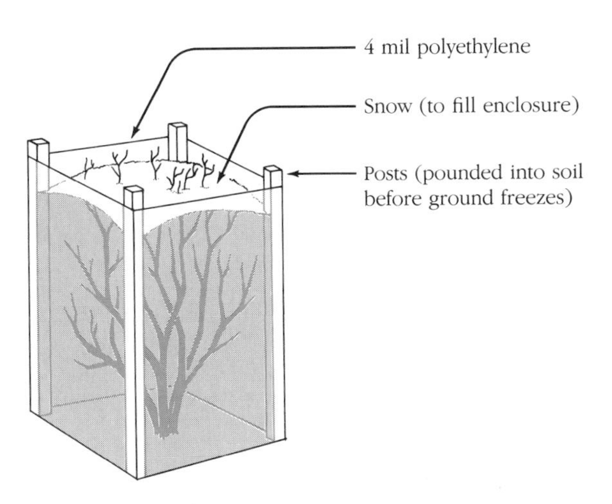

Figure 7–18 Snow as a protection from cold. A semihardy shrub can be protected from mid-winter freezing by creating a plastic tube around it and filling the tube with snow.

layed until severe winter weather is past so that the optimum number of undamaged branches can be left and so that winter-damaged wood can be removed. In addition, the extra branches remaining through the winter may provide some protection for the rest of the plant. Snow-covered plants should be observed frequently while snow is melting during early spring so that branches in danger of being torn off by settling snow can be released. During extremely cold periods trees should be examined frequently for split bark.

Preventing Spring Freezing. The fruit crop of an entire season is often destroyed by late spring frosts that occur during only a few hours on one or two nights. In most orchard areas petroleum-fueled heaters are used to protect the commercial crop during these occasional cold periods. Some of these heaters would be suitable for home plantings, but gardeners contemplating their use should check to assure that the heaters comply with local antipollution ordinances. If only one or two trees near the home are to be protected, heating lamps or a gasoline lantern hung in the lower branches (because hot air rises) may provide the protection required. (Don't permit heating devices to contact bark or other parts of the living plant.) A sheet of polyethylene held above the tree by long poles at each corner can be used as a canopy to keep the warm air from rising too high. Covering small plants with a long-lasting, spray-on foam and preventing air inversion with wind machines are two additional methods commercial growers sometimes use to protect against early spring freezes.

Irrigation water also is used to prevent early spring freezing. As was mentioned earlier, considerable heat energy is released as water cools, and even more (80 calories for each gram) is lost as it freezes. To prevent the freezing of crops, sprinklers should be turned on during the early morning when the temperature drops to 34°F (2°C) and left flowing until the sun has warmed the area to above the freezing point. Ice will accumulate when the temperature drops to freezing, but as long as some water remains liquid, the temperature in the vicinity will not drop below the freezing point. Frost protection with sprinklers is feasible for light freezes only, because the weight of accumulating ice can physically damage plants.

Wind Protection. Wind damages plants by desiccation, by physically breaking or abrading branches and leaves, and by amplifying the effects of cold temperature. In areas where wind is frequent, seedling growth can be greatly improved by planting on the leeward side of a high bed or by providing some kind of artificial barrier on the windward side of the plant row (Figure 7–19). In areas of high spring winds where a cereal is grown as a winter cover crop, early growth of vegetable seedlings and transplants has been markedly improved by leaving strips of the cover crop for wind protection. Planting strips are prepared for seeding with a narrow rototiller so that each seeded row alternates with a strip of cover crop. The remaining strips of cover crop are removed by cultivation or with herbicides before they begin to crowd the seedlings.

Windbreaks and shelterbelts are the more traditional and more permanent means of wind protection for both the home and garden; these will be discussed in Chapter 11. Where space for a windbreak is limited or nonexistent, an ornamental border, a row of bush fruits, or a row of espaliered fruit trees (see Chapter 8) along the windward side of the garden will provide some permanent wind protection. Limited temporary protection can be achieved with rows of sweet corn, sunflowers, or other tall annuals.

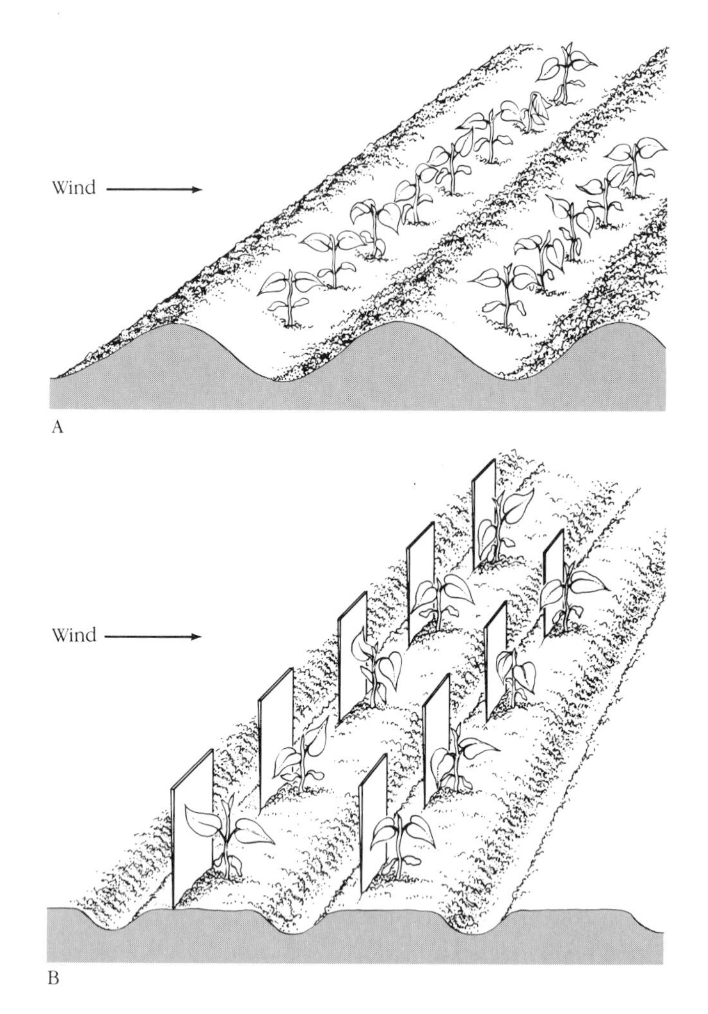

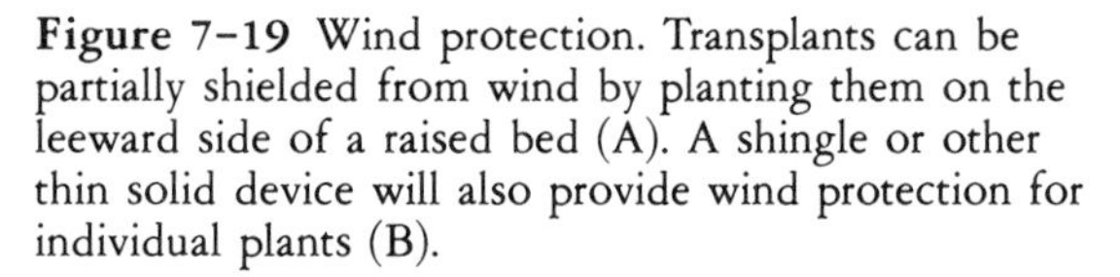

Figure 7–19 Wind protection. Transplants can be partially shielded from wind by planting them on the leeward side of a raised bed (A). A shingle or other thin solid device will also provide wind protection for individual plants (B).

Heat Units

Heat units are a measure of growing efficiency of the weather based on the fact that temperature is the climatic factor best correlated with the length of time required for a crop to mature. Heat units for any one day are calculated by subtracting a base temperature (usually 40° or 50°F; 5° or 10°C where Celsius measure is being used) from the average temperature for the day. Bases of 40° or 50°F are predicted on the fact that most cool-season crops produce some growth whenever the temperature is above 40°F, and warm-season crops don't grow unless the temperature is above 50°F. Estimating time of maturity, latest feasible date for fall planting, and whether long-season fruit cultivars will mature in a specific locality are common uses of accumulative heat units (see "Climate, Hardiness, and Maturity," Chapter 14).

Light and the Growth of Garden Plants

Light is that part of the sun's energy visible to the human eye, but visible light is only a part of the total spectrum of solar radiation that reaches the earth. Although it is now known that the short- and long-wavelength radiation on either side of the visible spectrum is utilized by some plants, it is visible light that is of primary importance for plant growth. Consequently, this discussion will be limited to that part of the solar spectrum detectable by the human eye and will be concerned primarily with light as a growth requirement of plants.

Kind or Quality of Light

At one time or another most people have observed the use of a prism to separate light into its various wavelength components that the human eye interprets as color. The rainbow comprises the same separation. The human eye can see solar radiation from about 390 to about 760 nanometers in length (see Figure 7–20). The shortest visible rays are violet and blue, the longest are red, and green and yellow rays are intermediate in length. Some of the shorter blue and ultraviolet rays are filtered out by atmospheric gases and airborne particulate matter. The ultraviolet rays, which cause sunburn, are, along with the red rays, important in the formation of **anthocyanin** pigments that provide the red color to many plant products. This is the reason that orchards in the dry, clear mountain valleys of the arid West produce highly colored apples (Figure 7–21) and also the reason that a Western suntan is likely to be darker than an Eastern suntan. The relationship between light and anthocyanin production also accounts for the faded color of coleus and other red-leaved houseplants when they are grown in reduced indoor light.

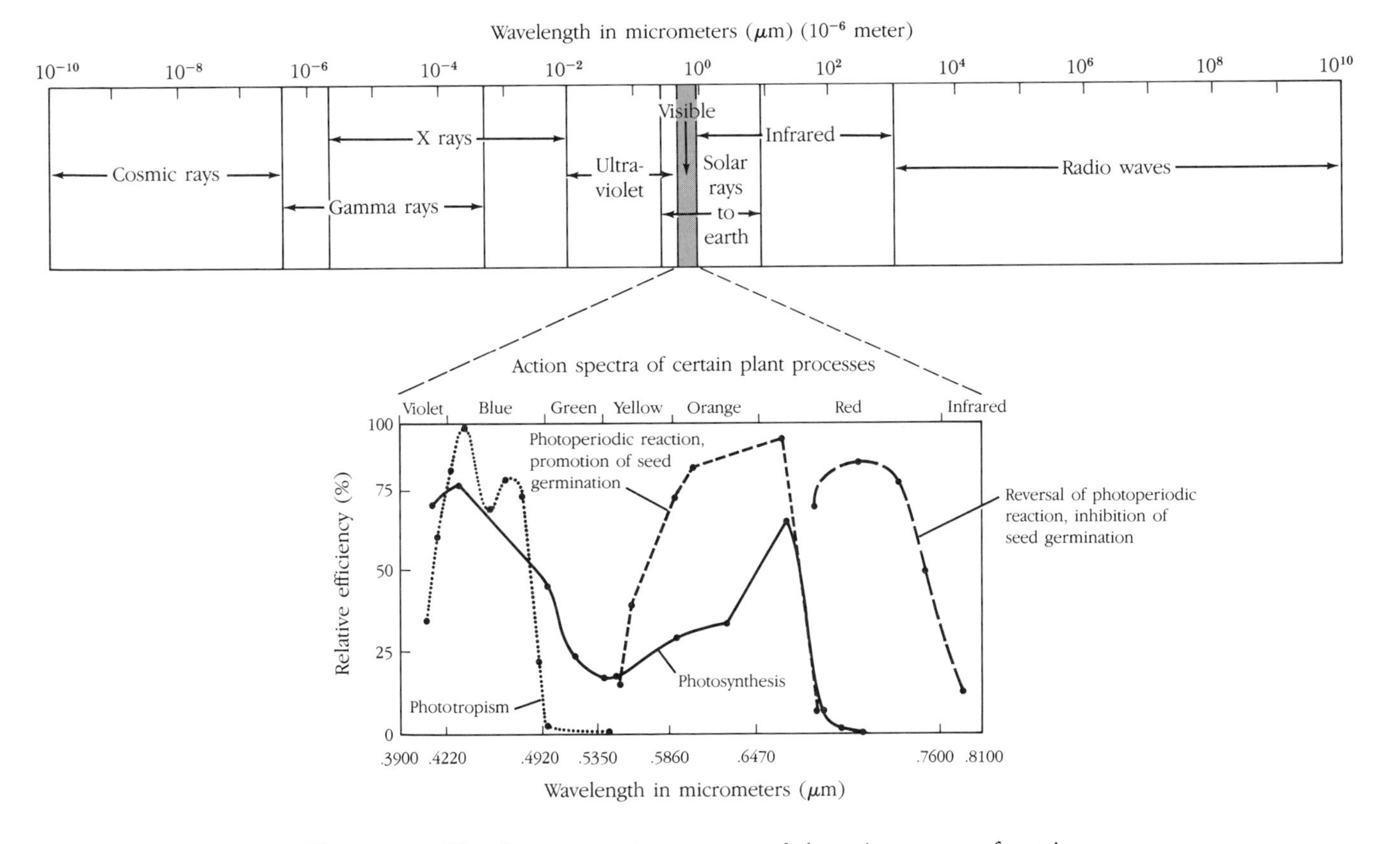

Figure 7–20 The electromagnetic spectrum and the action spectra of certain plant processes. (Adapted from *Plants in Action,* by L. Machlis and J. G. Torrey. W. H. Freeman and Company, San Francisco. Copyright © 1959.)

The bending of flower heads and some other plant parts toward the sun is a phenomenon of which most gardeners are aware. The bending of a plant part in response to light, called **phototropism,** is mediated by the blue areas of the spectrum. Although the physiological basis of phototropism is not completely understood, scientific evidence suggests that light either destroys the growth-promoting plant auxin IAA or else causes it to move from the area where light intensity is greatest. (An **auxin** is a plant growth substance that promotes cell elongation.) Regardless of the exact mechanism, the result is more auxin on the side of the stem away from the source of light, and, as a consequence, that side grows more rapidly than does the side on which intense light rays are falling (Figure 7–22). This enables the stem to orient the flower head toward the sun or other source of light. Partly because of phototropism potted plants growing near windows should be turned frequently to permit balanced growth.

Figure 7–21 The influence of light on pigmentation. Light is necessary for the formation of anthocyanin pigments, which give the red color to apples. These labeled fruits were produced by sticking tapes on the fruit before the natural formation of pigment. (From *Horticultural Science,* 3rd ed., by Jules Janick. W. H. Freeman and Company, San Francisco. Copyright © 1979.)

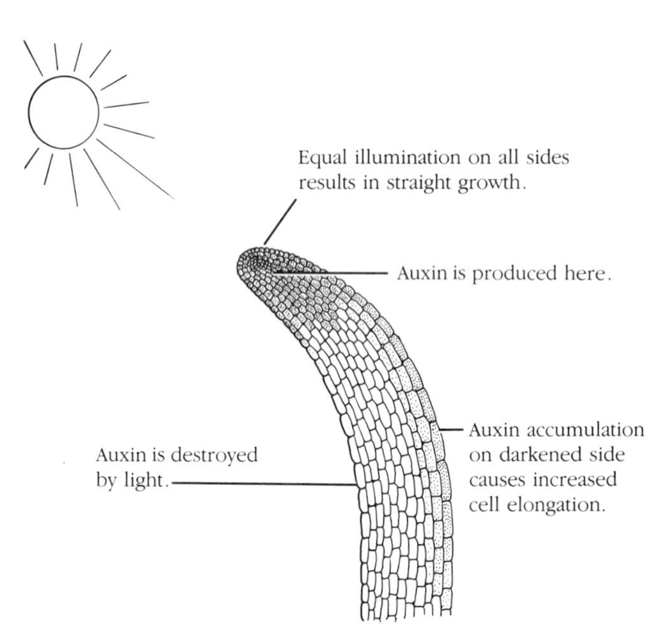

Figure 7–22 Phototropism. (From *Horticultural Science,* 3rd ed., by Jules Janick. W. H. Freeman and Company, San Francisco. Copyright © 1979.)

Light energy is absorbed by plants primarily from the blue-violet and orange-red parts of the spectrum. In fact, plants appear green because they are reflecting or transmitting considerable green light rather than absorbing it. This plant characteristic of not utilizing green light has suggested that lamps be designed for plant growth that produce light mainly in the blue and red areas of the spectrum. The earliest fluorescent lamps of this type, which gave off red and blue light almost to the exclusion of other wavelengths, were called "Gro-Lux." Later, the same manufacturer introduced "Gro-Lux Wide Spectrum," which emitted a high proportion of red and blue light but also provided light from other areas of the spectrum. These lamps deliver light shown by recent research to be more closely correlated with the light spectra utilized in photosynthesis. The lavender glow emanating from many greenhouses each morning and evening during the winter months comes from these special plant-growing lamps. The advantage of such specialized lamps is that they provide plants with more usable light in proportion to the electrical energy consumed. For plant growing under artificial light, a mixture of lamps is often most desirable (Figure 7–23). Cool white, the most common type of fluorescent lamp, plus incandescent lamps (the latter to supplement the far-red area of the spectrum), can be combined to provide another useful light source that may or may not be used in conjunction with the "Gro-Lux" lamps.

Photoperiodism, or response of plants to day length, which will be discussed later in this chapter, is triggered by light from the orange-red and far-red areas of the light spectrum.

Light Intensity

Light intensity has traditionally been measured in units called **foot-candles** (F.C.). Although the foot-candle is no longer the universally accepted light measuring unit to which physiologists relate plant growth, it is still the one most commonly found in gardening literature. In the eastern and midwestern United States light intensity in full sunlight at midday will measure approximately 10,000 F.C. The intensity is much less early in the morning or later in the evening when solar rays are reaching the earth at an oblique angle and thus are being filtered by more layers of atmosphere. For the same reason intensity is much less in the winter. Full sunlight in desert areas can surpass 12,000 F.C. at noon on a summer day. Light intensity is also high on clear days in the tropics. With heavy winter overcast the light intensity outdoors at noon may be as low as 600 to 900 F.C. The interior of lighted homes will measure from 50 to 300 F.C.

Plants vary in their ability to utilize light. Because philodendrons, African violets, begonias, and many other houseplants evolved under a jungle canopy, they grow well at low light intensities, making them valuable for indoor decoration. An individual leaf of most crop plants can utilize only about 1,200 F.C. of light, but because of shading of lower leaves plants adapted to full sun will generally respond to several times that intensity. Increased growth of some plants will parallel increased light intensity up to about 4,000 F.C. The intensities mentioned earlier are far higher than plants are capable of utilizing; these were maximum intensities at noon on a clear midsummer day. When reduced intensities during spring and fall, morning and evening, and cloudy and overcast days are considered, the light intensity is probably not appreciably higher than the maximum that can be utilized by

Figure 7–23 Growing plants by artificial light. Artificially lighted tables, benches, and shelves are commercially available in many shapes and sizes. They can also be constructed. The table shown in photo A has eight 8-foot (2.4 m) fluorescent tubes with a measured light output of 750 F.C. (foot-candles) at the height of the lower plants and a timeclock for day/night regulation. In photo B an under-counter single-tube kitchen light that emits about 150 F.C. at plant height is shown. The under-counter light has been left on continuously, partly to compensate for low intensity and partly because it would be inconvenient to remember to turn it off and on. The plants produced under it—African violet, jade plant, rex begonia, and peperomia—have all grown slowly but have remained reasonably healthy.

most crop plants during most of the growing hours at most locations.

Plants that receive less than the required amount of light respond with increased elongation. Nodes will be far apart, leaves will be broad and thin, and the plants will have a loose, open structure (Figure 7–24). Reduced light intensity also induces succulence. Tobacco leaves used for cigar wrappers are artificially shaded so they will produce a broad, thin leaf. Where light intensity is low, plant growth will be more normal, although extremely slow, if temperatures are also kept low. Lengthening the daily period during which plants receive light will compensate to a degree for low light intensity.

Plants adapted for growing under low light intensity or plants that have been growing with low intensity will sunburn, wither, and die if they are placed where light intensity is high. This frequently happens when potted plants are placed outdoors for the summer or when certain shade-loving plants, such as African violets, philodendrons, spurges, and ferns are placed in direct sunlight.

There is some evidence that variation of light intensity above the optimum for plant growth may have considerable impact on the growth of plants. Most crop plants grow somewhat taller in the East and the Midwest than do plants of the same cultivar in high altitude areas of the West. The upper layers of cucurbit leaves often show damage when those crops are grown in the western desert, although yields do not seem to be affected. Both dwarfing and leaf damage can be corrected by shading. The relationship of variations in light intensity and light quality to plant growth are not yet well understood, and alert gardeners are in a position to make interesting observations in this area.

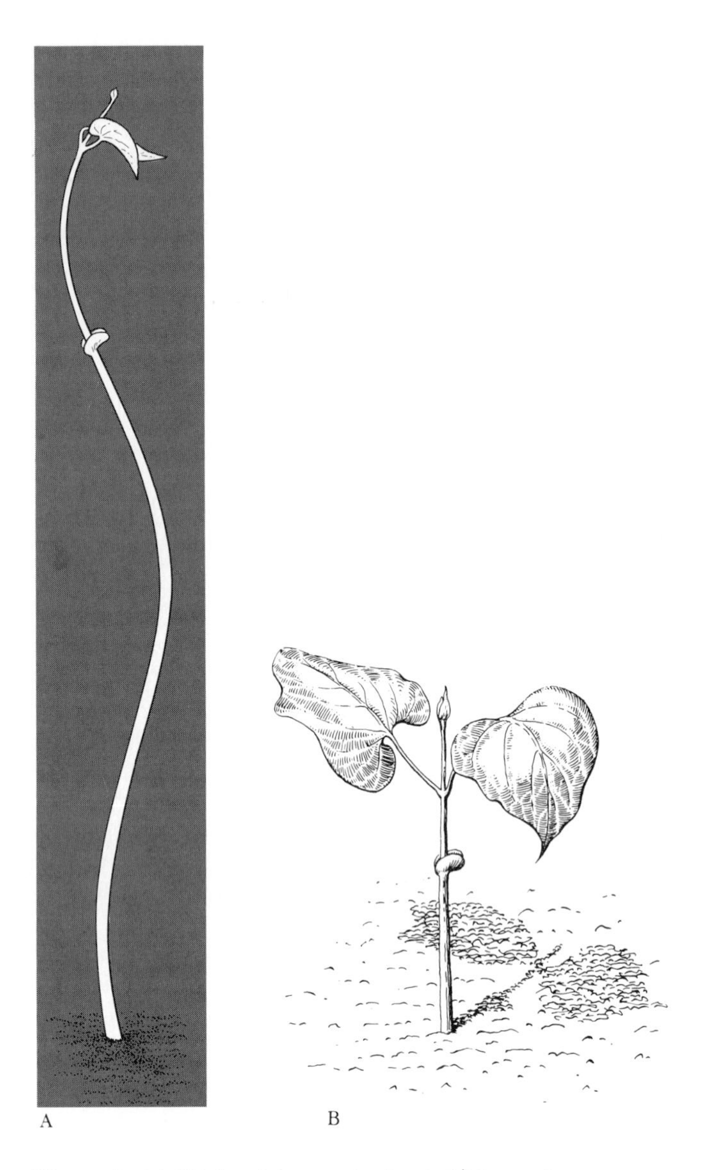

Figure 7–24 Etiolated (grown in the dark) bean plant (A) and normal (grown in light) plant (B). (From *Principles of Plant Physiology,* by J. Bonner and A. W. Galston. W. H. Freeman and Company, San Francisco. Copyright © 1952.)

Duration of Light

During 1920 two U.S. Department of Agriculture scientists working at Beltsville, Maryland, W. W. Garner and H. A. Allard, reported that the time of flowering of a certain tobacco cultivar was determined by the length of the light period during each 24 hours. It was soon discovered that many responses in certain plants—flowering, bulbing, tuberization, and the like—are brought about as

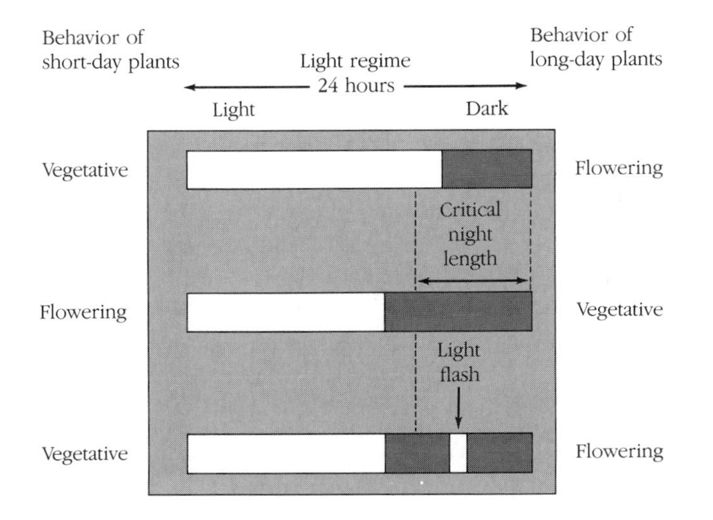

Figure 7–25 Response of long-day and short-day plants to different light regimes. Interrupting the long dark period with a few minutes of light prevents flowering in short-day plants and permits flowering in long-day plants.

a result of the plants' exposure to a particular day length; this phenomenon is called **photoperiodism** (Figure 7–25). Some plants produce a response only at day lengths longer than a certain minimum; these plants are referred to as **long-day plants.** Plants responding to a day length shorter than a certain maximum are referred to as **short-day plants.** Plants that show no visible response to day length are referred to as **day-neutral.**

A number of very important horticultural crops respond to day length. Chrysanthemums, for example, are short-day plants; however, there are cultivars that are induced to bloom by day lengths ranging from 16 hours down to 7 hours or less. Because chrysanthemums are a popular greenhouse crop, cultivars that bloom during each period between late summer and early winter at all latitudes have been developed. However, growers can have any cultivar bloom at just about any season of the year by covering it to shorten its day length for a period of time if they wish to have earlier blooming or by increasing the day length with artificial light if they wish to delay bloom. The light intensity required to delay blooming of chrysanthemums is very low, and the blooming of outdoor mums is sometimes delayed by porch or street light.

Spinach is a long-day plant, and if planted late in the spring, it produces seed stalks before it produces edible leaves. Present spinach cultivars require relatively long days to initiate flower buds, and just as soon as that day length arrives, the spinach plant blooms regardless of its size. Blooming of spinach is also hastened by high temperatures, so spinach is produced either early in the spring or late in the fall when temperatures are cool and days are short.

Onion bulbing is also a long-day response, and cultivars that produce bulbs at various latitudes have been developed. For example, the Bermuda-type onion has been developed to produce during the winter in Texas; it initiates a bulb whenever the day length reaches 11 to 12 hours or more. If a cultivar of this type is planted in New York or Minnesota where climate decrees that onion planting cannot occur much before mid-March, it produces a rather interesting response. When the leaves come through the soil, the day length in those northern climates is longer than that necessary for bulbing to occur. As a consequence, the seedlings immediately initiate bulbs. However, because little food will have been manufactured by the extremely small leaves, the bulbs will be about the size of a pea. Cultivars such as 'Sweet Spanish' and 'Yellow Globe' have been developed for northern areas and do not bulb until day length has reached 15 to 16 hours. These cultivars germinate in March or April, producing considerable vegetative growth by mid-June, at which time bulbs are initiated. Bulbs will continue to grow and will not be fully developed until late August; because there is plenty of food material stored, they will be relatively large. If, on the other hand, the 'Yellow Sweet Spanish' cultivar is grown in Texas where the day length, even during mid-summer, never reaches 16 hours, growth will continue without formation of bulbs, and the grower will have a crop of exceptionally large scallions. Figure 7–26 illustrates photoperiodism in onions.

Many other plants respond to day length. The poinsettia produces its red or white bracts under short-day conditions. Commercial poinsettia growers have to know exactly when to reduce the day length so that their plants

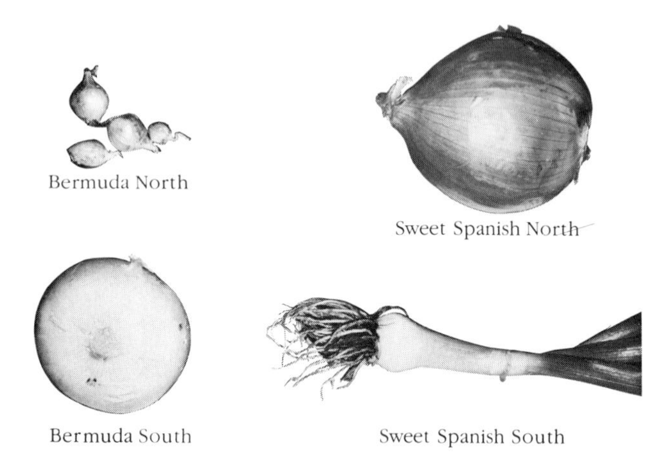

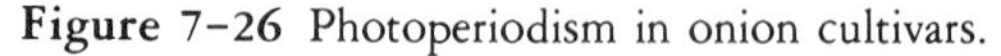

Figure 7–26 Photoperiodism in onion cultivars.

will be in bloom for the Christmas season; poinsettias are not of much value on December 26. Poinsettias can be kept as houseplants through the summer, but they require special lighting if they are to bloom a second Christmas. This is because artificial lighting in most homes increases the day length to a point where blooms are never induced. To cause blooming, start in early November to keep the plant in a room where only natural daylight enters. All artificial lights should be removed, because only a few minutes of interrupted darkness are required to delay or eliminate the flowering response. Flowering can also be induced by covering the poinsettia at sunset with a light-proof box and removing the box at sunrise. The special treatment can be stopped as soon as the bracts show color.

Like the response of spinach, the response of most plants to day length is modified by temperature and may be speeded or delayed by cold or warm weather. This fact becomes of major importance when growers are attempting to produce plants to flower on specific dates—large mums for homecoming, poinsettias for Christmas, or lilies for Easter.

Selected References

Bickford, E. D., and Dunn, S. *Lighting for Plant Growth.* Kent State University Press, Kent, OH. 1972.

Brooklyn Botanic Garden. *Gardening under Artificial Light.* Handbook 62 (special printing of *Plants and Gardens,* vol. 26, no. 1). Brooklyn, NY. 1970.

Chang, J. *Climate and Agriculture: An Ecological Survey.* Aldine, Chicago. 1968.

Neiburger, M., Edingen, J. G., and Bonner, W. D. *Understanding Our Atmospheric Environment.* W. H. Freeman and Company, San Francisco. 1973.

Rabinowitch, E., and Govindjee. *Photosynthesis.* Wiley, New York. 1969.

U.S. Department of Agriculture. *Climate and Man (Yearbook of Agriculture,* 1941). U.S. Government Printing Office, Washington, DC. 1941.

8 Regulating Plant Growth

Although it is not often expressed in these terms, the major goal of gardening is to utilize the finite plant growth requirements—mineral nutrients, light, and water—of the garden space in the way that will best produce the results desired. Optimum utilization of minerals, water, and light is achieved by regulating inter- and intraplant competition by spacing of garden plants; by regulating plant growth with pruning, fruit thinning, size-controlling rootstocks, growth-regulating chemicals, phloem disruption, and plant breeding (discussed in this chapter); and by controlling garden pests (discussed in Chapter 9).

Plant Spacing

Perhaps the most obvious way of regulating interplant competition is by optimum spacing of each plant, but determining the optimum is not always easy. The close spacing necessary for maximum yield of some crop plants may reduce the size or quality of their product and increase their susceptibility to foliage diseases. Furthermore, with ornamentals the spatial organization for aesthetic effect is more important than total utilization of plant growth resources of the area.

Considerable publicity has been given to plant spacing research that started in England a few decades ago and demonstrated that optimum yield and size of vegetables usually result when plants are spaced approximately the same distance in each direction rather than in the customary wide between-row and narrow within-row planting distances. These results were achieved, however, only with complete chemical weed control, which, as will be explained later, is not usually practical in the home garden. Also essential in the home vegetable garden is space for walking between rows to weed, harvest, and perform other necessary activities. Thus except for the tiniest plots where not even a hoe is needed for maintenance and where plants can be reached from the garden borders, the garden arrangement is most convenient if vegetables are planted in rows. Extensive plantings of annual and perennial flowers and small fruits are also easier to care for when they are in rows.

Thinning of direct-seeded annuals is essential. Annual flowering plants produce larger blooms and blossom more profusely when plants have room to grow to full size. Lettuce, cauliflower, and cabbage do not produce heads if plants are crowded, and root vegetables remain small and become misshapen if too many are left to grow in the row. As explained in Chapter 6, plants should be widely spaced if there is likelihood that either water or minerals will be insufficient. Recommended spacing for vegetables is listed in Table 14–2 (Chapter 14).

Size-controlling rootstocks permit growing fruit trees in a space as small as 10 x 10 feet (3 x 3 m), and new training systems mentioned later in this chapter make it possible to produce tree fruits in even less space and to use them as dual-purpose ornamental and food plants in hedges and borders.

Planting woody ornamentals too close together, which restricts their uniform growth to full size and beauty, is the mistake most commonly made with these plants (Figure 8–1). Because trees and shrubs increase in size so tremendously as they develop, wise gardeners will draw a scale model of the yard or garden showing planting materials at full-scale size before they do any planting. This will enable them to resist the temptation to overplant. Small temporary shrubs and annual and perennial flowers can be used as filler plants to provide a pleasing effect until the more permanent plants grow to fill their allotted space. Trees and shrubs that grow to the correct size should be selected for each location. Keeping a woody plant pruned to fit a location too small for it, although possible, requires a great deal of effort. Besides, a heavily pruned plant normally will not be as attractive as one permitted to grow naturally (Figure 8–2).

Figure 8–2 Excessive pruning. This flowering crabapple has been pruned drastically (and ruinously) in an attempt to make it fit its location. Such drastic pruning would not have been necessary had a smaller tree or this cultivar grafted onto a dwarfing rootstock been planted.

Figure 8–1 Overplanting. Crowding by the 'Cutleaf Weeping' birch caused this Douglas fir to become lopsided and ruined its appearance as a specimen.

Growth Control by Pruning

Pruning, the removal of parts of a plant, is the most common method of opening the plant's canopy to allow sunlight to reach leaves, flowers, and fruits and of regulating other intraplant competition. Pruning to restrict plant size is also a common method of reducing interplant competition. Roots are sometimes pruned, but the term pruning, when used alone, usually means removal of above-ground portions of the plant, generally stems. The main purposes of pruning are to modify plant growth and to influence the amount and character of flower and fruit production.

A term commonly used in conjunction with pruning is **training**, the modification of plant shape or size to suit the needs of the horticulturist. Plant training by pruning is one of the less exact, more controversial horticultural practices, perhaps because satisfactory results are possible with considerable variation in pruning techniques. There is some truth in the cliche, "most plants grow relatively well in spite of the pruning they receive," and, indeed, some good gardeners who have not overplanted confine their pruning to the removal of dead branches and branches that are obstructing an activity or view. For producing fruit and for attractive maintenance of most ornamental trees and shrubs, however, more extensive pruning is essential.

Figure 8–3 The dwarfing effect of pruning. About one-third of the wood of the Nanking cherry (*Prunus tomentosa*) shrubs on the left was removed 8 years before this photo was taken. The pruned shrubs are still noticeably smaller than the shrubs on the right, which were never pruned.

Effects of Pruning Cuts

Because it is impossible to prescribe exact pruning procedures for each plant in each situation, intelligent pruning requires a knowledge of the impact of various amounts and kinds of pruning cuts on the physiology and growth of the plant. Basically, pruning dwarfs the plant, changes its carbohydrate/nitrogen ratio, and affects growth through auxin imbalance.

Dwarfing. Removal of branches or roots has a dwarfing effect on the entire plant, both top and root system (Figure 8–3). The dwarfing effect is not always apparent, because pruning tends to stimulate vegetative growth, especially in the vicinity of the cut. Practical use is made of this localized effect in developing the framework system of fruit trees or in directing the growth of ornamentals. Cutting tip growth back to an outward-facing bud will tend to direct new growth outward and give the plant a more open growth habit. However, although length growth may be stimulated, even greatly stimulated, with excessive top pruning, the total size of pruned plants will be less than that of unpruned plants.

Altering Carbohydrate/Nitrogen Balance. Pruning is one of the most important tools a gardener has for regulating the carbohydrate/nitrogen balance and, concomitantly, the vegetative/reproductive growth ratio of woody plants. Pruning of above-ground plant parts removes some of the stored carbohydrates as well as reduces the carbohydrate manufacturing organs and thus increases the amount of nitrogen in relation to carbohydrates. This, in turn, stimulates vegetative growth relative to reproductive growth. Root pruning, on the other hand, increases the carbohydrate balance and causes the plant to become more reproductive.

Largely because it reduces the carbohydrate ratio, heavy pruning during the early life of a plant may delay flower and fruit production by several years, a delay especially disconcerting to growers of fruit trees that require many years of growing before much fruit is produced even in optimum circumstances. As a general rule, if early flowering is desired, young fruit and ornamental trees and shrubs should be pruned as little as possible consistent with the development of a desirable framework and shape. Commercial apple orchards grown without pruning between their second and sixth years produce a crop 1 to 2 years earlier than conventionally pruned trees. Few orchardists follow this practice completely, however, because young trees grown with no pruning are often structurally weak.

The importance of maintaining a balance between vegetative and reproductive growth of woody plants has been stressed several times in previous chapters. To be an effective tool for regulating the amount and quality of production, pruning must be done at the right time and be correlated with mineral fertilization, thinning, and other management practices. In general, heavy pruning increases the production of stems and leaves and reduces the number of fruits and flowers. Young, vigorously vegetative trees and shrubs and those growing in fertile soils will require little pruning. Older trees and shrubs and those growing on less fertile soils will tend to be overly reproductive and will produce larger flowers and fruits if they are pruned quite heavily. Excessive pruning can completely eliminate flower and fruit production for several years, and heavy pruning during late summer may increase freezing susceptibility because it reduces carbohydrate reserves.

Figure 8–4 Apical dominance in conifers. (A) Rapidly growing apex and reduced growth of upper laterals are typical of young trees with strong apical dominance. (B) Rounding of the crown is typical of maturity and reduced apical dominance.

Auxin and Apical Dominance. Pruning also affects growth habit and production by its effect on **apical dominance**, the tendency for the **apex** (uppermost bud) to grow more rapidly than lower buds. Apical dominance is thought to be due to growth stimulation by an auxin that is manufactured in the highest part of the plant. Plant and plant parts show varying amounts of apical dominance. Young conifer trees, for example, develop their typical "Christmas tree" shape because of strong apical dominance (Figure 8–4). On the other hand, spreading plants such as Pfitzer junipers show little tendency for a single apical bud to be dominant. Normally it is the uppermost branch that shows the greatest apical dominance, but suckers, the rapidly growing, nonbranching stems that shoot from adventitious buds anywhere in a tree or shrub, are examples of extreme apical dominance lower on the tree.

Apical dominance affects plant growth in several ways besides stimulating growth of the apical bud. Buds near but below the apex either fail to grow, or the shoots they produce remain short. In addition, the crotch angle at which side branches meet the main trunk is wider as long as the terminal bud is exerting the influence of apical dominance. If the terminal bud is removed, the shoots from buds just below the terminal begin to grow rapidly and turn upward, causing their crotch angles to become narrow. Solid wood often fails to form in the vicinity of narrow crotch angles, resulting in a weak framework for the tree and often in breakage when the tree has to carry a heavy load of fruit or snow (see Figure 14–73, Chapter 14).

Heavy pruning and high nitrogen fertilization enhance apical dominance, suggesting that auxin stimulation is correlated with a low carbohydrate/nitrogen ratio, especially as plant parts subject to strong apical dominance do not blossom or bear fruit. Apical dominance and vegetative/reproductive growth are also affected by branch orientation. Bending a branch downward reduces its apical dominance and enables it to bloom and fruit. The vegetative/reproductive balance of trees and shrubs trained to the espaliered or pillar system (see "Pruning and Training," Chapter 14) is maintained by orienting their branches in a horizontal or downward position. If the branches were allowed to grow in the normal upright position, the heavy pruning required for these training systems would completely inhibit flowering and fruiting.

Apical dominance is most intense in young plants and lessens as a plant ages. This lessening is related to the greater reproductive tendencies and the open growth habit of older trees. A forester knows that a conifer is mature when it loses its Christmas tree shape and its top begins to spread (Figure 8–4).

Plant parts other than branches—potato tubers, for example—show a form of apical dominance. In late December or early January, when the tuber first becomes able to produce growth after its rest period, only a single sprout will develop at the bud end of the tuber. As is true with trees, aging reduces tuber apical dominance, and by May several buds will germinate from each eye on the potato. Another interesting aspect of the sprouting of potato tubers is that, if the tuber is cut into several pieces, apical dominance is overcome. However, there will be a tendency for each seed piece to produce its first sprout from the terminal bud of the eye that was closest to the bud end of the potato (Figure 8–5).

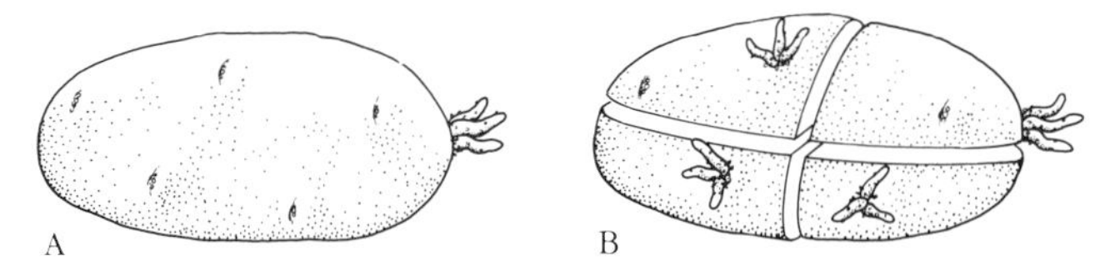

Figure 8–5 Apical dominance in the potato. When "tuber dormancy" first begins to break (in December or January in northern areas), only the eye at the bud end will initiate sprouts (A). If the tuber is cut into sections, apical dominance is overcome and other eyes will initiate sprouts (B).

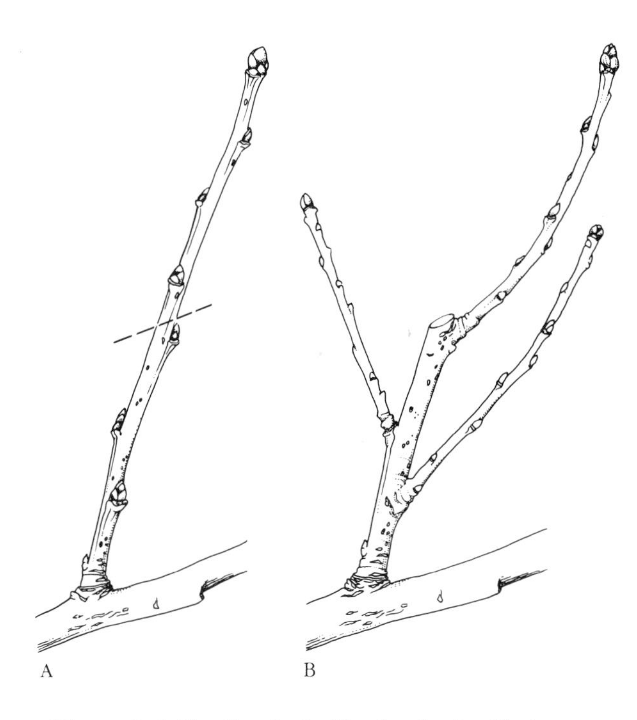

Figure 8–6 Heading back. Heading back will change growth toward the direction of the highest remaining bud (A). It also frequently stimulates growth of lower buds, resulting in a more compact growth habit (B).

Heading Back vs. Thinning Out

The two general kinds of pruning cuts are heading back and thinning out. **Heading back** is cutting off a part of a limb or branch. When a branch is headed back, the buds nearest the cut are those most likely to be stimulated into shoot production (Figure 8–6). If only a small part of the tip of the branch has been removed, frequently only the terminal bud that remains with the plant will produce shoot growth. Therefore, pruning lightly to an outside bud is one way to force the growth of the branch in a slightly more outward direction. With may shrubs, especially where considerable portions of a branch are removed, a number of latent buds will be stimulated to produce shoots. Thus heading back generally tends to produce a more compact or bushy plant. The extreme example of heading back is shearing plants to produce hedges or topiaries (Figure 8–7). The art of **topiary** is the shaping of woody plants to resemble objects other than plants, such as animals or geometric designs.

Thinning out is the removal of an entire shoot or branch. Although growth of branches near the one removed will be stimulated and sometimes latent buds near the base of the removed branch will be forced into growth, thinning out generally tends to produce a plant with a more open growth habit (Figure 8–8).

Figure 8–7 Hedge and topiary. A hedge (A) and a topiary (B) represent the ultimate in compact growth stimulated by shearing, one method of heading back.

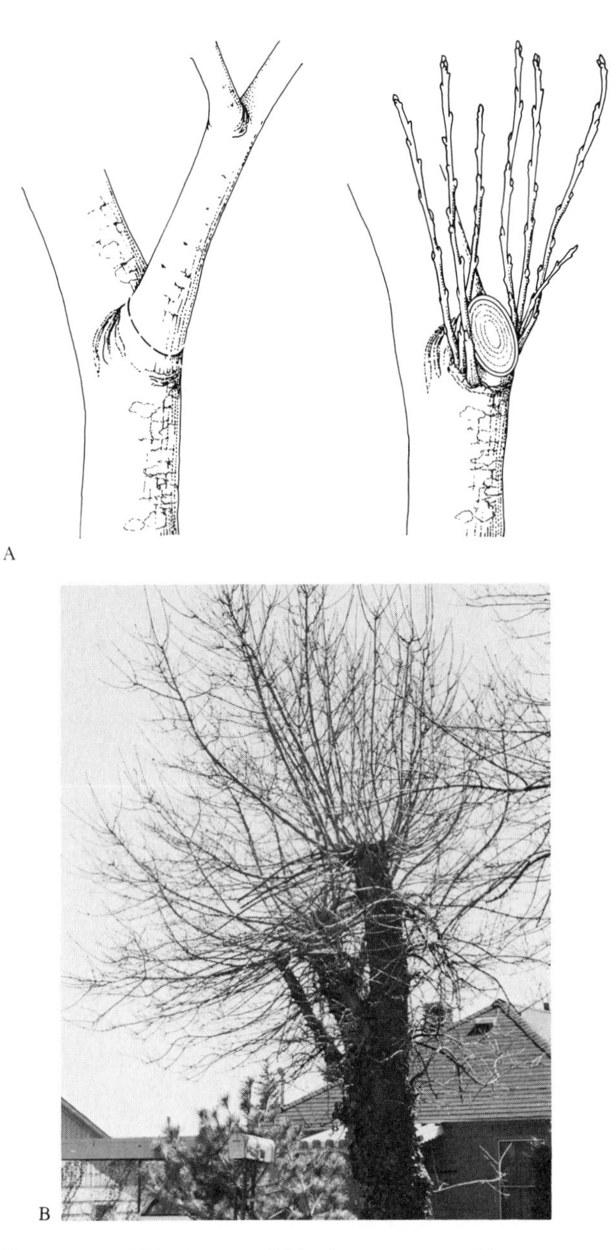

Figure 8–8 Thinning out. Thinning out cuts will usually cause the plant to be more open; however, removing a large limb will sometimes stimulate adventituous buds to produce sucker growth (A). Note the large number of suckers that have developed near the ends of the stubs on the excessively pruned tree skeleton (B).

Time of Pruning

The timing of pruning depends on a number of factors, including climate, species, and the results desired. Generally, in northern latitudes fruit trees should be pruned in the late winter or early spring. Pruning when a plant is actively growing has a more retarding effect than does pruning during the dormant season. Moreover, it is easier to see the framework of the tree when there are no leaves (see Figure 2–1, Chapter 2). At locations where woody plants are subject to winter damage, pruning is usually delayed until after the coldest winter weather is past, as pruning somewhat increases susceptibility to winter damage. Small fruits can be pruned right after harvest except in areas of extremely cold winters, where pruning is generally delayed until early spring. Pruning in the spring permits the removal of winter-damaged parts, and the extra branches left through the winter may provide some support to prevent breakage from heavy snow or strong winds.

There is an old adage that pertains to pruning of ornamentals: "Prune when the knife is sharp." Since earliness of maturity and amount of production are not primary concerns of those who produce ornamentals, gardeners should prune landscape plantings when they have the inclination to do so and observe that pruning is needed. There are, of course, a few precautions and guidelines for the time to prune ornamentals. One already mentioned is that heavy pruning of any woody plant late in the summer is likely to stimulate late vegetative growth and make the plant more susceptible to winter damage. Furthermore, cut surfaces are more quickly hidden by new growth if the plant is pruned just prior to or during its season of rapid growth. It is recommended that flowering trees and shrubs be pruned after they have bloomed. This permits the removal of dead blooms that are unsightly and also prevents a shrub or tree from forming seed or fruit that would use food reserves that otherwise could be utilized for production of leaves and blooms for the following season. Those shrubs that bloom in the spring and early summer are generally pruned after they have bloomed; those that bloom in late summer or autumn are generally pruned the following spring. The recommendation to prune right after blooms have faded does not apply, of course, to flowering shrubs and trees that also produce attractive fruit or seed pods. Hedges will need to be pruned several times during the growing season whenever they begin to look shaggy.

Pines and rhododendrons normally are pruned in the spring as their new shoots elongate, but most other evergreens can be pruned at any season of the year. Spring pruning of evergreens does have an advantage in that cut ends are soon covered by new growth.

Training Systems for Trees

There are a number of training systems for trees, and the choice of which to use depends on growth habit, length of time the tree normally lives, and purpose for which the tree is being grown. If an ornamental or shade tree has been properly selected for the location in which it is planted, it can usually be permitted to assume its natural growth habit which is generally similar to the growth habit developed by the **central leader** system of training, best illustrated by the shape of conifer trees. With this system, the trunk is encouraged to form a central axis with branches distributed laterally around it. Strong trees suitable for shade or timber but too tall for fruit production develop from this system.

In the open center or **vase** system of training, the central leader is cut off 18 to 30 inches (45 to 75 cm) from the ground, and two or three side branches become the scaffolds and spread to form the framework of the tree (see Figure 14–76, Chapter 14). This system is satisfactory for peaches and a few other stone fruits; however, with plants where the tendency for apical dominance is strong, the side branches tend to grow too upright if the central leader is removed.

With apples, pears, and some stone fruits, the **modified central leader**, or delayed open center, system of training is extensively utilized (see Figure 14–72, Chapter 14). In this system the central leader is headed back slightly but not completely removed until after the tree is 5 or 6 years old and has borne a crop of fruit, at which time the main framework will have been established.

Several other systems of training can be used when there is limited space. When trees are trained flat in one plane so that they grow against a trellis or wall, the French term, **espalier**, is used to describe them (see Figure 14–74, Chapter 14). With this type of pruning the branches are brought out from the trunk in geometric patterns such as the U, double U, and oblique. In some sections of Europe where lack of sufficient heat units often prevents proper ripening of fruit, this system of pruning has been used for pears and peaches. Grown against the sunny sides of walls, the trees may receive sufficient reflected heat to mature the fruit. This system can also be used to grow fruit along a fence or border and to grow ornamental shrubs against a wall (Figure 8–9).

Several of the many new training systems developed by orchardists could probably be utilized to the benefit of gardeners. With the **mold-and-hold** system trees are permitted to grow to any convenient size desired and then are kept at that size by heavy pruning. The downward orientation of side branches overcomes the inhibition of fruiting that would otherwise occur as a result of the heavy pruning. With the **pillar** system a permanent trunk is developed, and new side branches are permitted to grow each year (see Figure 14–75, Chapter 14); the following year's crop comes from newer side branches. As with the mold-and-hold system side branches oriented downward are kept to encourage fruiting. The **hedgerow** system, where dwarf trees are supported by and trained to fences and trellises, can also be adapted to small home plantings.

Figure 8–9 Espalier. Training a fruit tree to two dimensions (espalier) along a fence or wall permits growing fruits in minimal space.

General Pruning Practices

In the final analysis, pruning is an art. There are basic principles that should be followed, but once these are mastered the pruner learns mainly by experience (see Figure 8–10). Good pruners are observant people who have had considerable experience. Pruning practices for various groups of plants are briefly outlined in Chapter 14. Although each plant provides a unique pruning challenge, a few practices apply to all plants:

1. Dead and diseased tissue is always removed. This is important for improving plant appearance as well as for limiting spread of disease. If there is the slightest suspicion that a branch is diseased, that is, if the branch has not obviously died from being overcrowded or from a lack of sunlight or if it is dying back from the tip, the pruning cut should be at least 6 inches (15 cm) behind the farthest advance of dead tissue to insure removal of disease-causing organisms. Where disease is present, all prunings should be burned, and with bacterial diseases, such as fireblight of apples and pears, pruning equipment should be sterilized between cuts.

2. Old wood, weak wood, branches that cross, and branches that form balanced crotches are removed or cut back. A balanced crotch, which is formed by two branches growing at about the same rate from the same point on a tree, is undesirable because it tends to split easily (see Figure 8–11). Balanced crotches are prevented by pruning back one of the branches to keep it subordinate to the other.

3. Plants should be pruned heavily to encourage vegetative growth and fewer and larger flowers or fruit, and lightly or not at all to encourage reproductive growth and numerous but smaller flowers and fruit.

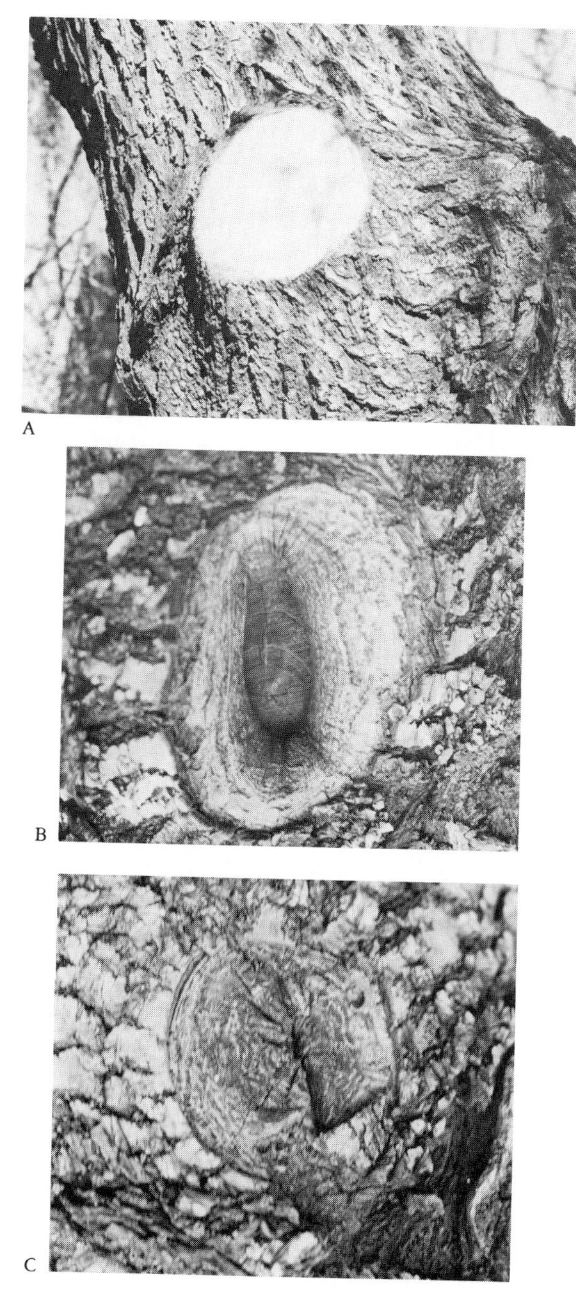

Figure 8–10 Healing of a pruning wound. A pruning cut made smooth and flush with the remaining wood (A) will begin to callus over (B) and eventually become completely covered with bark (C).

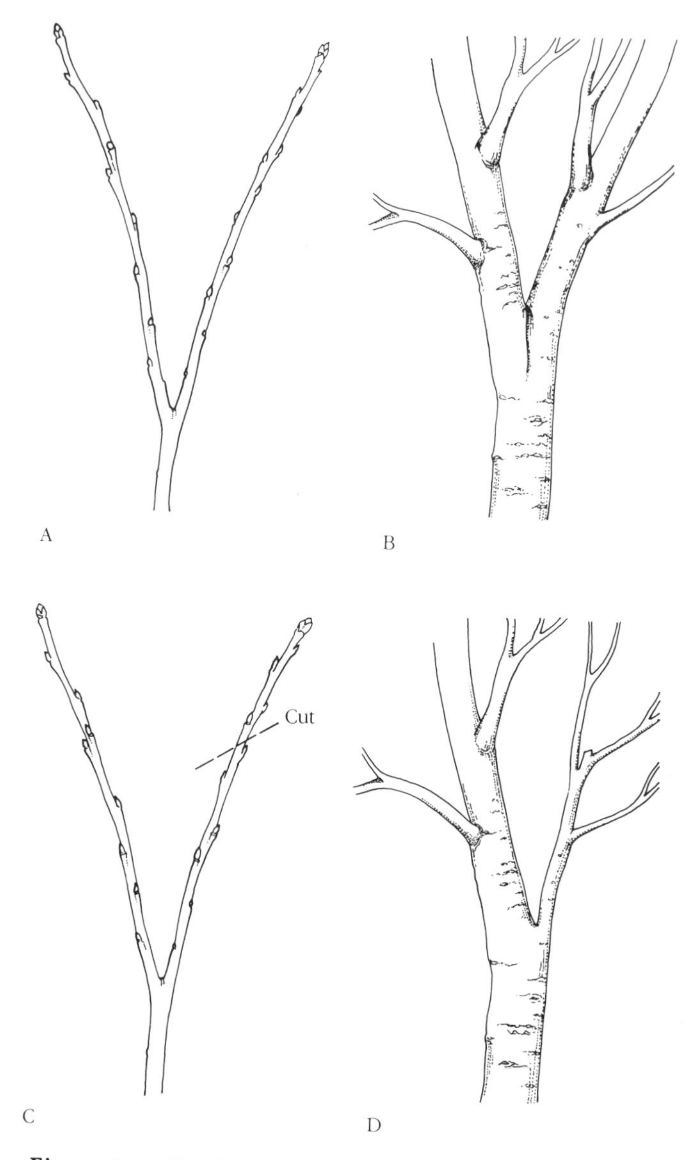

Figure 8–11 Pruning a balanced crotch. If the two branches of a balanced crotch (A) are left to grow naturally, they will produce a tree with a weak framework (B) as they grow larger. If one of the branches is removed or subordinated by pruning (C), the tree will develop a much stronger framework (D) as the branches enlarge.

4. The direction of plant growth can be determined to some extent by the kind of pruning cut. Heading to an outward-facing bud will cause new growth to be somewhat more spreading. In general, heading back produces plants with a compact growth habit, and thinning out, by removing an entire branch, produces plants with an open growth habit.

5. Excessively heavy pruning, especially on trees, is undesirable and unnecessary. The tree or shrub should be selected to fit the location. Where pruning is necessary to keep a plant in bounds, light to moderate pruning each year is infinitely better than excessively heavy pruning that becomes necessary when the plant has grown too large for its location.

Fruit Thinning

Side buds are occasionally pinched from roses, peonies, chrysanthemums, and carnations to promote the development of larger terminal flowers. Although this is a thinning operation, it is called **disbudding** (see Figure 14–85, Chapter 14). To a horticulturist thinning generally means removal of some of the small immature fruits from fruit trees (or removing excess plants from a seeded planting, discussed earlier). Because developing fruits compete primarily for plant carbohydrates, the regulation of competition for mineral nutrients, light, and water by fruit thinning is indirect but just as effective as other methods of regulation.

The main purpose of thinning is to improve fruit size and quality by providing more leaf-manufacturing area for each fruit. If thinning is done early enough, it may also increase blossoming for the following season and reduce the tendency for alternate-year bearing common with some cultivars. Thinning also allows the gardener to discard imperfect fruits—those with stings, rub damage, hail marks, and the like—permitting utilization of plant resources by the better fruit. Thinning may also improve winter hardening of fruit trees by reducing the drain of carbohydrates.

Timing of fruit thinning depends largely on the propensity of the tree for alternate bearing. For stone fruits and the apple and pear cultivars that normally produce a crop each year, thinning should be delayed until after the June drop (see Chapter 3) to avoid the labor of removing fruit that would fall anyway and the possibility of thinning too heavily. Normally, thinning after the June drop will be too late to release carbohydrates for blossom bud formation for the following year's crop, so some of the older apple cultivars such as 'Baldwin,' 'Yellow Transparent,' 'Wealthy,' and 'Duchess,' which all tend toward alternate bearing, as well as other apple, pear, and plum trees that have entered an alternate-bearing cycle as a result of a spring freeze, should be thinned shortly after petals have dropped and fruits begin to enlarge.

Considerable fruit thinning, especially of stone fruits, is done during the pruning operation. With peach trees at least half the blossom buds should be removed with the wood pruned in late winter. Commercially, fruits are thinned with chemicals, but because timing is critical and application procedures vary with environment, home gardeners should not use them without local advice. The number of fruits to be left will depend on the kind of tree and its vigor. A healthy, medium-sized peach tree can mature about 800 quality peaches. Each apple of large-fruited cultivars such as 'Delicious' requires about 40 leaves to develop maximum size and quality (Figure 8–12). Only a single fruit should be left at each spur or node. Because measuring distance between fruits on a branch is usually easier than counting leaves, the following spacings are recommended by several authors, the wider ones being for larger cultivars:

Apples and pears	6 to 8 inches (15 to 20 cm)
Peaches and nectarines	4 to 5 inches (10 to 13 cm)
Large plums and apricots	3 to 4 inches (8 to 10 cm)

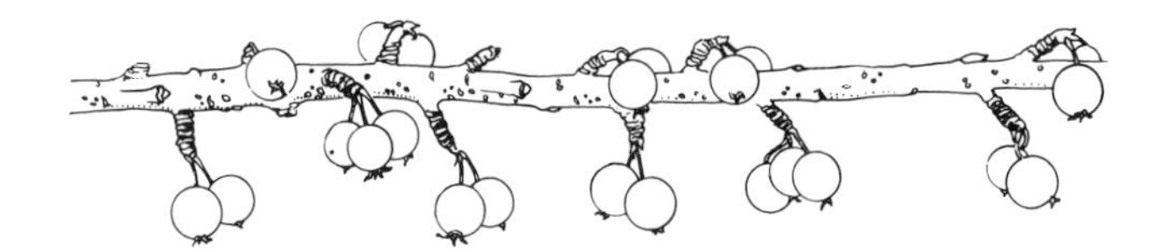

Figure 8–12 Fruit thinning. The 5 to 8 leaves produced by each spur and bud of this 15-inch (38 cm) segment of an apple tree branch will provide optimum nutrition for only 2 or 3 of the 23 fruits that have set. With set this heavy many fruits will fall with the June drop, but judicious fruit thinning will still be required.

Crabapples, cherries, and nuts are not usually thinned. More fruit should be left on trees thinned prior to their June drop. For optimum spacing of fruit on early-thinned trees, the gardener should examine the trees and remove any surplus after the June drop.

Other Ways of Modifying Plant Growth

Besides pruning and thinning, plant growth is modified by grafting onto size-controlling rootstocks, by scoring or girdling, by use of chemical growth regulators, and by breeding for plant size.

Grafting

Grafting (see Chapter 4) to control size is an ancient practice. With the scientific development, study, and classification of apple rootstocks by the East Malling Research Station in England, nursery workers now have the potential to produce uniform-sized apple trees ranging all the way from 8 to 10 feet (2 to 3 m) in width and height up to standard height, which may be 30 to 40 feet (9 to 12 m). Dwarfing rootstock development has been more extensive with apples than with other fruits; however, quince is usually used to dwarf pear trees, and rootstocks that slightly dwarf plum and orange trees are available. Dwarf cultivars of peach and sour cherry are available, and dwarf sweet cherries should be available soon.

Dwarf fruit trees have been a real boon to home gardeners not only because they are small enough to grow on a city lot but also because they produce sooner after being planted and are much easier to prune, spray, harvest, and otherwise take care of. Moreover, the quantity of fruit produced by a dwarf tree is more in line with the amount that a family can consume.

Dwarfing apple rootstocks are designated by the letters E.M. (East Malling) or M.M. (Malling Merton) followed by a Roman numeral. The dwarf rootstock most popular with home gardeners, probably because it produces the smallest sized tree of any widely available rootstock, is E.M. IX. Trees on E.M. IX rootstocks are not well anchored and must be staked to prevent their being blown over or broken off by the wind. They are small, however, seldom growing over 10 feet (3 m) in height, and they often start producing fruit the second year after being planted, which accounts for their popularity. A better anchored, fully dwarf apple tree can be produced by using a standard seedling rootstock with E.M. IX used as an interstock, but these are more expensive and their superiority is questioned by many authorities.

An interstock must be used for dwarfing of many pear cultivars, including the popular 'Bartlett.' 'Bartlett' and some other cultivars are graft-incompatible with quince (*Cydonia*), and an interstock from a cultivar such as 'Old Home,' which is compatible with both 'Bartlett' pear and quince, is used as a bridge between the two (Figure 8–13).

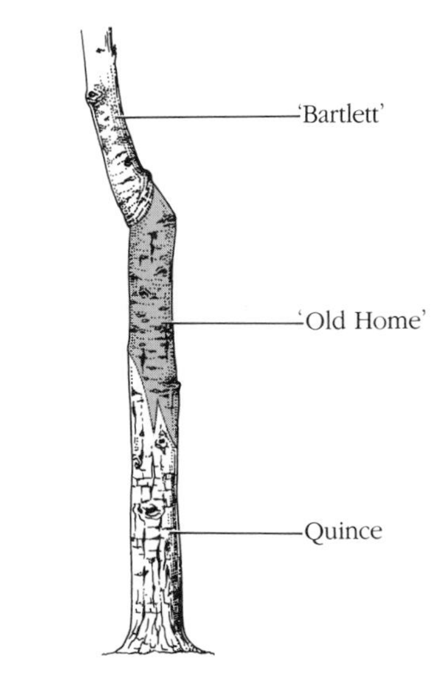

Figure 8–13 Using an interstock to overcome 'Bartlett'-quince graft incompatibility. During early spring an 'Old Home' scion is whip-grafted onto quince (*Cydonia*), a dwarfing rootstock of pears. Later during the summer a 'Bartlett' bud is budded onto the 'Old Home' portion of the new tree. 'Old Home,' a pear cultivar no longer extensively grown for its fruit, is used as a rootstock and an interstock because it has resistance to root diseases and is graft-compatible with quince and also with most pear cultivars.

Homeowners frequently ask if quality is affected by the rootstock. In general, dwarf fruit trees produce fruit of just as high quality as do normal-sized trees. In some instances fruit quality of the dwarfs will be higher, either because they are easier to thin out by pruning or because they do not tend to produce as much shade as do normal trees, and abundant light increases color and flavor. In a few instances quality of the scion may be affected by the root. For example, most cultivars of orange grafted to rough lemon rootstock are lower in sugar content than are the same cultivars of orange grafted on other rootstocks. This is probably due to the effect of the rootstock on the carbohydrate/nitrogen ratio rather than directly on the quality.

Grafting may also be used to stimulate growth or at least provide a better root system for plants that normally produce a poor one. Hybrid tea and floribunda roses are grafted onto multiflora rose rootstocks. A multiflora root improves the vigor of the scion and also is resistant to a number of root-borne diseases. Upright junipers, *Juniperus virginiana* cultivars, produce poor root systems and are easier to propagate and better anchored if grafted onto the sturdy root system of *Juniperus chinensis* 'Hetzii.'

Dwarf trees should be planted so the graft union is above the soil surface (Figure 8–14), otherwise the scion will form a root system of its own and the effect of the rootstock will be lost. Branch growth originating below the graft union should be removed, because if will not be of the desired cultivar and because it often outgrows the scion if allowed to remain.

Figure 8–14 Depth of planting grafted plants. The 'Ruby' horsechestnut (A) has been planted with the graft union above the soil surface, an important detail when the rootstock has an effect on the growth of the tree or shrub. In areas where winter damage is frequent, rose bushes (B) are often planted with the graft union below the soil surface. This permits the growth of new shoots from undamaged below-ground scion wood in case the top growth is frozen back to ground level.

Phloem Disruption

Juvenility or the delay of fruiting, most common in apples and pears but also prevalent with plantings of other kinds of fruits, has been mentioned several times. A similar problem sometimes occurs with woody ornamentals, some cultivars of wisteria that fail to bloom year after year being a notable example. Reproduction of a nonfruiting tree or shrub can frequently be hastened by disrupting the phloem, which temporarily halts the downward transport of carbohydrates and increases the carbohydrate/nitrogen ratio in the top of the plant. It

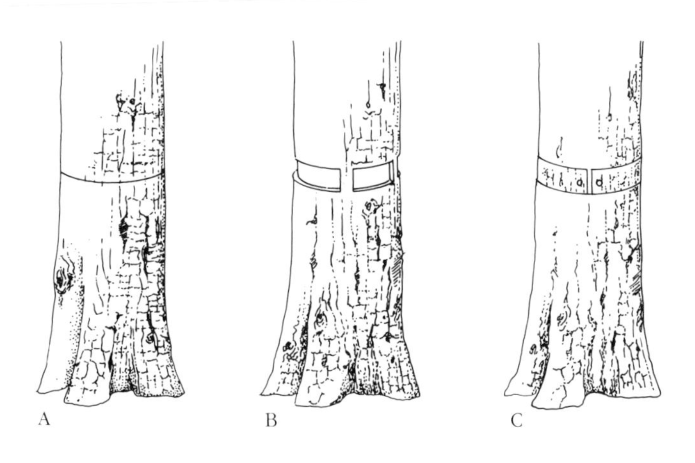

Figure 8–15 Phloem disruption. Three methods of temporarily interrupting the downward flow of solutes through the phloem are scoring (A), girdling (B), and bark inversion (C).

may also alter auxin and other growth-regulator relationships. The timing of **phloem disruption** is important. It must occur just prior to the time when blossom buds are being formed. For most temperate zone tree fruits and spring blooming ornamental shrubs this would be during late May of the season before blossoms and fruit are produced to affect blossom buds that are formed a month or two later in June or July.

Phloem disruption is accomplished by scoring, girdling, or bark inversion (Figure 8–15). **Scoring**, the least drastic of the three practices, consists of running a knife blade around the tree or branch to cut through the phloem. With scoring the effect is of short duration because the wound heals quickly. With **girdling** a strip of bark about ½ inch (1 cm) wide is removed. Generally the strip removed is not continuous; several undamaged sections of bark are left intact across the girdled area so that some phloem transport can continue. With **bark inversion** a strip of bark is removed, turned upside down, and tacked back to the area from which it was removed. Bark inversion has a less drastic but more long-lasting effect than girdling. The disruption of phloem by any of these methods can kill a tree or shrub if it is not done carefully, and it is probably a practice of last resort where other efforts to induce reproduction have failed.

Chemical Modification

Beginning during the early decades of this century and continuing to the present, plant physiologists have identified a number of chemicals that affect plant growth out of proportion to their concentration. Some of these diverse substances occur naturally in plants, and some have never been found in plant tissue. Some of the purely artificial ones are horticulturally most lastingly useful, because the plant does not have "antimechanisms" to break them down. These chemicals have been known by various names. Although all of them do not strictly affect plant size, they are known most commonly as plant **growth substances.**

The use of plant growth substances for rooting cuttings and the relation of indole-3-acetic acid (IAA), an *in vitro* growth substance, to various plant growth responses and pruning have been discussed earlier. Commercially, growth substances are used extensively by horticulturists and promise to be even more useful as our understanding of their action increases. Other than for rooting cuttings, they are used to inhibit or promote sprouting of potatoes (Figure 8–16) and dormancy of other crops, to thin flowers and fruit, to initiate flowering, to modify sex expression, to promote fruit set, to promote or delay ripening, to prevent preharvest fruit drop, to promote leaf and fruit abscission, to increase or

Figure 8–16 Effects of chemical growth substances. The potatoes on the left were harvested from vines that had been sprayed in early September with the sprout inhibitor maleic hydrazide. Those on the right were grown in the same location but were untreated. The photo was taken in April.

decrease plant size, and to control weeds. Probably the most extensive use of these substances at present is for controlling weeds; this will be discussed in Chapter 9.

Several growth substances besides those that enhance rooting of cuttings and assist in weed control may be useful in gardening. Compounds that cause fruit to set on tomato plants during cool periods when flowers would normally abscise (drop) are available at nurseries and are widely used by gardeners in cooler areas of the country. Also available are "stop-drop" sprays, which prevent fruit from being blown from the tree before it is fully ripe. Chemicals for blossom or fruit thinning are also available, but the timing, distribution, and concentration of their use is so critical that the home gardener who has only a few trees is wise to thin the fruit by hand. Chemicals that have a dwarfing effect on plants and encourage branching may occasionally be useful to home gardeners.

Some growth-inhibiting chemicals are being used experimentally to keep grass from growing and consequently to reduce or eliminate the need for mowing. This and countless other potential uses of growth substances give promise of easier and more satisfying gardening. As is true with all chemicals applied to the garden, directions on the labels of growth substances should be followed carefully.

Breeding to Modify Growth

A number of years ago wheat breeders at Washington State University imported some unthrifty-appearing dwarf wheat lines from Japan. By crossing them with some of the tall wheat cultivars that had been grown for years in the Pacific Northwest, Orville Vogel and his co-workers were able to introduce short, stiff-strawed wheat cultivars that produced large heads and were adapted to that region. The major advantage of those short cultivars was that, in areas where moisture was plentiful, they could be heavily fertilized without danger of **lodging** (falling over). Yields of wheat in the Northwest increased dramatically. The same dwarf germ plasm was used by Norman Borlaug and his co-workers to develop dwarf wheats adapted to Mexico and other subtropical regions, and a similar technology was used by plant breeders in Southeast Asia to develop rice cultivars. These introductions brought about the so-called Green Revolution and earned the Nobel Prize for Borlaug. Dwarf cereals have increased yield potential of grains in many developing nations and are credited with saving millions of people from starvation.

Though not strictly a horticultural development, the above illustrates what can be achieved by modifying plant growth by breeding. The development of dwarf determinant tomato plants (small-branched plants that bear fruit clusters at branch tips as well as on the stem) has made available earlier maturing cultivars and also concentrated the maturity of this crop, in turn making mechanized harvest feasible. Bush cultivars of winter squash and pumpkins make these crops better adapted for gardens with limited space.

Plant improvement is not always a matter of cross-pollinating two varieties. Often it consists of selecting a promising individual plant that occurs in nature and propagating a new cultivar from it. During recent years orchardists have discovered occasional limbs of apples that have shortened internodes and consequently have buds and spurs much closer together. Trees propagated from scion-wood from these branches, called **spur-type** trees, grow to only about two-thirds the height of normal trees and produce excellent yields. We now have spur-types of many of the most popular apple cultivars. These trees possess the advantages of dwarf trees without the disadvantages of dwarf rootstocks.

Good dwarf rootstocks of sweet cherries are not available, and naturally occurring spur-type limbs have not been found on sweet cherry trees; however, scientists in British Columbia irradiated branches to induce a genetic mutation that brought about spur-type growth in several sweet cherry cultivars, and from these they have developed spur-types of several of the more popular sweet cherry cultivars.

The potential for directing plant growth by breeding is just now being recognized. Most horticultural practices, including planting, cultivation, pest control, pruning, and especially harvest, could be facilitated by modifying the size or shape of certain plants. The most permanent and practical method of size modification of many plants is by genetic improvement.

Selected References

Baumgardt, J. P. *How to Prune Almost Everything.* Morrow, New York. 1968.

Brooklyn Botanic Garden. *Pruning Handbook.* Handbook 28 (special printing of *Plants and Gardens*, vol. 14, no. 3). Brooklyn, NY. 1957.

Free, Montague. *Plant Pruning in Pictures.* Doubleday, Garden City, NY. 1961.

Galston, A. W., and Davies, P. J. *Control Mechanisms in Plant Development.* Prentice-Hall, Englewood Cliffs, NJ. 1970.

Sunset magazine. *Pruning Handbook.* 2nd ed. Lane, Menlo Park, CA. 1972.

Tukey, H. B. *Dwarfed Fruit Trees.* Macmillan, New York. 1964.

9 Garden Pests

Losses to commercial agriculture due to insects, diseases, and weeds are estimated to be between 10 and 15 billion dollars annually in the United States and Canada. Birds, rats, squirrels, mice, and other animals add greatly to these extensive crop losses. Control of plant pests is perhaps the major problem of commercial horticulture because of the ever-present yet ever-changing nature of pests. Consumers eventually pay most of the cost of agricultural pest control in higher prices for food and fiber.

Control of pests is frequently more of a problem for gardeners than for commercial horticulturists, because gardeners are usually concerned with more kinds of crops and consequently a greater number of actual or potential pests. In addition, many of the more effective and labor-saving pest control measures are not adaptable to the small multicrop garden.

To many gardeners pest control means getting out the chemical pesticide and spray gun. Chemical pesticides have received a great deal of research emphasis during the past four decades and are, at present, the most rapid and complete control measure for most controllable crop pests. Because some are dangerous to animals and humans when misused and because some can pollute the environment, pest-control chemicals are among the most maligned of agricultural and garden products. Undoubtedly, chemical pest control has been overemphasized in some situations, and, fortunately, other control measures are now receiving more attention. A welcome trend is the careful monitoring of pest prevalence coupled with the integration of several approaches against each of the more devastating crop pests. It should be emphasized, however, that for many pests other approaches are either not available or not as reliable as chemical control, and considerably more research technology will be needed before other methods can replace chemical control of most pests. Although chemicals are by no means the only, or for many purposes the best, control measure, it is doubtful if chemical pest control ever will be or even should be entirely replaced.

Pesticide Regulations

Chemicals used in agriculture are subject to strict federal controls. Before a chemical is released, the company producing it must show that, when the chemical is applied at the recommended dosage and time, either it will leave no residue or the residue left cannot possibly cause injury to humans or warm-blooded animals. Each chemical must be cleared for each individual crop. For example, even though a company has proven there is no residue of a particular chemical on apples when it is applied in a specified way at a specified concentration and stage of the seasonal growth cycle, it must also prove that the same amount, method, and time of application of the chemical to peaches will produce no residue.

Most states have pesticide regulations also, and both state and federal regulations require that pesticides be cleared for individual states and sometimes for sections of a state. Every container of pesticide sold must be labeled to show the crops and pests for which it is cleared and the specifications to be followed in applying it. Growers who apply the chemical in higher concentrations or later in the season than recommended or to crops other than those for which it is cleared are subject to arrest and can have their entire crop confiscated and destroyed.

Regulations and enforcement of those regulations are, of course, necessary for public protection, but they pose special problems for growers of crops consumed in small quantities and therefore grown on few acres. As a consequence of stringent regulations it costs 4 to 8 million dollars or more to obtain government clearance for a single chemical on a single crop. Obviously, commercial companies are not interested in developing pesticides for crops not planted to extensive acreages. Since hand weeding is prohibitively expensive, the lack of suitable herbicides is tending to eliminate a number of minor horticultural crops from commercial production or greatly increase their price. Similarly, elimination by government regulation of insecticides traditionally used on garden crops and failure to clear replacements for them are at least partly responsible for the scarcity and high cost of small fruits and some vegetables, and they pose a serious threat to the entire floriculture (flower-growing) industry.

To protect the applicator and farm workers, there are also strict laws regulating the amount of time that must elapse before a field can be reentered after pesticide application and, in most areas, even stricter regulations on when and how the chemical can be applied to reduce the chances of injury to nontarget organisms. Those who apply pesticides to commercial farms must pass an examination and be licensed. Proposals to prohibit pesticides from being applied in incorporated population centers as well as to prohibit unlicensed individuals from applying pesticides, even around their own homes, have been considered by federal, state, and local governments.

Undoubtedly, the confusion surrounding the use of pesticides on small acreages will diminish as various laws are interpreted by the courts. In the meantime, gardeners can usually comply with the law by following explicitly the labels on containers of pesticides newly purchased from reliable dealers. Before a pesticide purchased during a previous year is applied, or if there is any question concerning the legality or wisdom of a proposed application of pesticides, gardeners should consult the county extension office or other local authority. For their own and their neighbors' safety, gardeners should use only those pesticides formulated for gardens.

Most states issue new pesticide application guides annually to aid growers in keeping abreast of new materials and changing regulations.

Weeds

A **weed** has been defined as a plant out of place, and almost any plant under some circumstances can be classified as out of place. However, there are some plants that are out of place under most conditions, and these are the ones that we usually categorize as weeds.

The most pestiferous weeds have certain characteristics that enable them to compete vigorously in cultivated fields. Generally, they grow more rapidly than do most crop plants. They frequently produce numerous seeds that remain viable for long periods in the soil. Usually some of the seeds will display some type of dormancy and may require light, scarification, or stratification in order to germinate. In a Midwest experiment in which seeds of crop plants and wild species were stored underground, three wild species—curly dock (*Rumex crispus*), evening primrose (*Oenothera biennis*), and moth mullein (*Verbascum blattaria*) still germinated after 80 years, and moth mullein still germinated after 90 years. Seed of all crop plants had aged into inviability many years before.

The longevity and dormancy of their seeds partially explains why weeds seem to grow year after year even when they are never permitted to go to seed. Some seeds may remain physically dormant in the soil for 1 to 20 years or longer, until abrasion and soil chemicals eventually wear a hole in their seed coats, permitting them to germinate. Seeds requiring light for germination remain dormant until they are brought to the surface, often by cultivating equipment, one good reason for only shallow cultivation of row crops. Some of the most pestiferous weeds can reproduce vegetatively as well as with seed. Bindweed (creeping Jenny, morning glory), quack grass, and Canadian thistle, three of the most notorious perennial weeds, reproduce readily from small pieces of root or rhizome (Figure 9–1).

Because weeds grow so rapidly, the damage they cause results primarily from their competition for light, water, carbon dioxide, nutrients, and space. Sometimes certain weeds inhibit germination or prevent growth of other species by exuding compounds selectively toxic to other plants (the allelopathy mentioned in Chapter 5). For example, many kinds of seeds do not germinate well in soils containing large amounts of quack grass, and research has shown that a solution made by soaking dead or living quack grass rhizomes in water will prevent the germination of seeds of many species.

Figure 9–1 Three notorious weeds. Quack grass, *Agropyron repens* (top), Canadian thistle, *Cirsium arvense* (lower left), and bindweed, *Convolvulus arvensis* (lower right), are three noxious weeds that reproduce from underground rhizomes or roots.

Weed-control methods for today's gardens can be classified primarily as either cultural or chemical, although biological control is presently used on pastures and rangelands and may be a major control method in cultivated fields in the future.

Cultural Weed Control

Despite agricultural advances during the past 100 years most garden weeds are still controlled by hand pulling, by use of various hand tools and power equipment, and by manipulating growing procedures to reduce the weed population—all methods of **cultural** control.

Long-term weed control can be made easier by keeping weeds and weed seeds from the garden, causing the reservoir of weed seeds in the garden soil to decline gradually. To prevent planting weed seed, growers should purchase garden seed from a dealer of known reputation. Such seed may initially cost more, but it will almost always be free of weed seed. Seed picked up at a bargain from an unknown source can be even more costly in the long run if using it results in the contamination of the garden with a new kind of weed.

Perennial weeds are often introduced with a tree or shrub transplanted from the yard of a neighbor. The area around such a plant should be examined weekly for a period of time after transplanting so that any introduced weeds can be destroyed before they spread to the rest of the garden. Even those gardeners who keep their gardens free of weeds all summer often allow the weed seeds in their soil to increase by neglecting after-harvest weed control. Many kinds of annual weeds are frost-hardy, short-day plants that can produce a crop of seeds when they are only a few inches high and hardly noticeable among the frozen skeletons of the abandoned garden.

Weed control must be a community project to be successful. Even complete control will not reduce weed potential in the garden very much if seed is blowing in or being scattered from an unmowed roadside, an abandoned yard, or a neighbor's weedy garden (Figure 9–2). Community effort is also necessary to reduce weed seed in irrigation systems, a major source of weeds spread to gardens watered from open ditches. Ditch banks should be mowed or chemically sprayed and screens installed to remove seeds before the water reaches cultivated areas. If there are no weed screens in the main canal system, growers should install their own in the waterway leading to the garden.

Weed control for annual crops is much easier if planting is delayed until the soil has warmed sufficiently to permit crop seeds to germinate and seedlings to grow rapidly. The planting bed should be cultivated lightly just prior to planting. If no large weeds are present, light raking with a garden rake may disturb the soil sufficiently to dry out and destroy germinating weed seeds and small seedlings.

The hoe, various kinds of cultivating equipment described in Chapter 5, and hand weeding are the most common methods of ridding the yard and growing garden of weeds. One very important aspect of cultivating for weed control, whether it be tractor cultivation or hand weeding, is that weeds are much easier to kill when they are small. A few hours each week spent keeping the garden free of weeds will result in a much more economical use of time than will several days of work once the weeds have become large.

Small weeds growing within the crop row can frequently be smothered by using a hoe or cultivator to mound enough soil onto the crop row to cover the weeds. Crop plants must not, of course, be covered. **Mounding** also helps keep irrigation furrows open in rill-irrigated fields. Later in the season cultivation should be shallow, both to prevent damage to the expanding crop roots and to avoid bringing a fresh supply of weed seeds to the surface, where they can germinate. When larger weeds are pulled from the plant row, one hand should be placed on the soil surrounding small nearby crop plants to prevent their being pulled with the weeds (Figure 9–3).

Figure 9–2 The difficulty of weed control. Seeds blowing from these uncontrolled dandelions can spread dandelion seedlings over many square blocks of lawns and gardens.

Figure 9–3 Weeding by hand. Small crop plants should be held firmly against the earth to prevent their being uprooted when large nearby weeds are being pulled.

Chemical Weed Control

With the advent of **herbicides** (plant killers) in the 1930s the science of weed control began an advance that has continued at an ever-increasing pace. Today there are hundreds of compounds designed to control weeds under one condition or another. Chemical weed control, like chemical control of all pests, has received a great deal of criticism and has become subject to ever-increasing governmental regulation during the past few years. It should be pointed out that, compared to most insecticides, herbicides normally are less toxic to humans or warm-blooded animals; however, a few are quite toxic. In addition, most can destroy nontarget crops when carried to them by wind or water. The main long-term detrimental effect of herbicides is in altering the flora of a particular area treated with a single herbicide repeatedly over a span of many years. An example of this result would be the elimination of a number of broadleaved species in certain wheat areas of the Great Plains and Pacific Northwest where phenoxy herbicides have been used year after year over extensive areas.

Classification of Herbicides. There are several terms with which plant growers must be acquainted if they are to apply herbicides intelligently. First of all, herbicides can be selective or nonselective. A **selective** herbicide is one that kills some kinds of plants with little or no injury to others; a **nonselective** herbicide is one that kills indiscriminately all plant growth to which it is applied (Figure 9–4).

Herbicides can also be classified as **contact** or **translocated,** depending on whether they injure only those parts of the plant to which they are applied or are translocated through the vascular system of the plant so that roots and other organs to which the spray was not applied are also killed. Herbicides are classified as **residual** or **nonresidual,** depending on whether they kill only at the time of application or whether they remain active in the soil for a longer period. The period of time during which an herbicide remains active will depend on many factors and may range from a few days to a number of years. I recall a neighbor who used sodium chlorate when it first became available to control wild morning glory in

Figure 9–4 Effect of herbicides. A well-chosen, selective herbicide will keep a crop free of weeds most of the growing season (A). Compare the untreated plot (B) and the plot on which the herbicide destroyed the crop as well as the weeds (C).

his garden. The morning glory was completely eradicated, but it was 7 or 8 years before anything else would grow on that plot of ground!

Soil texture, amount of moisture, exposure to light, and other environmental factors will affect the residual life of an herbicide. Many herbicides tend to remain active longer in sandy soils or soils low in organic matter, perhaps as a consequence of low soil microorganism activity and low moisture in these kinds of soil. For example, atrazine used on sweet corn in the Midwest always breaks down during the season it is applied, but in some irrigated areas of the Northwest atrazine can be injurious to crops planted one year or even two years after it has been applied to a field. These kinds of variation in residual activity create serious problems for commercial companies who supply these herbicides and for growers who utilize them.

Herbicides are also classified according to the stage of growth of the crop when they are applied. A **preplant** herbicide is one that is applied before the crop is seeded, a **preemergence** herbicide is applied after the crop has been seeded but before it comes through the ground, and a **postemergence** herbicide is applied after the crop has emerged.

Use of Herbicides in Home Gardens. Because so many different kinds of plants grow around most homes, and because weeding with a hoe or by hand is not completely prohibitive, herbicides are not as essential for the home garden as they are for larger acreages. Nevertheless, considerable quantities of herbicide, mostly 2,4-D (2,4-dichlorophenoxyacetic acid) for weed control in lawns, are used around American homes. Although 2,4-D was one of the earliest herbicides, it is still one of the most widely used. It is one of the group of phenoxy compounds that kill broadleaved plants but do not damage most kinds of lawn grasses unless applied at excessive rates. Homeowners who are contemplating the use of 2,4-D should make sure there are no regulations in their community prohibiting its use. This material should not be applied when there is a wind or even a slight breeze, and nearby broadleaved plants should be protected by covering or otherwise making sure there is no drift toward them. As is true for application of any pesticide, labeled directions and local recommendations should be carefully followed. Usually an amine salt of 2,4-D is the only type supplied by nurseries and garden stores. The ester forms should never be applied around the home as they are very volatile and are almost certain to drift and injure nearby susceptible plants.

Nonselective residual herbicides can be used to control unwanted weeds in a gravel pathway, along a fence row, or beneath a loose brick, cement block, or graveled patio area. Anyone contemplating the application of long-lasting residual herbicides should be certain that the area treated will not be needed for growing plants in the foreseeable future and that the herbicide chosen cannot be transported to other locations by either water or vaporization.

Use of herbicide formulations available for controlling herbaceous weeds around woody plants is sometimes practical for the yard or garden. In some vegetable gardens a preplant or preemergence treatment to control annual weeds is possible. An example is the use of the herbicide trifluralin, which prevents germination and seedling growth primarily of plants in a single plant family, Chenopodiaceae. When the crop plants to be seeded—beans, lima beans, tomatoes, and so on—are not susceptible, and most of the weeds—redroot pigweed, lambsquarters, tumble pigweed, and so on—belong to the Chenopodiaceae family, the preemergence spray will control the weeds for at least half the summer. Postplant herbicides are available whenever relatively large plots of a single crop are grown.

Garden Plant Injury from Herbicides. Herbicide damage to home plantings, often unrecognized, is common in many sections of North America, and, unfortunately, very little can be done to alleviate the problem once damage has occurred. If the injury is not too severe, affected plants may ultimately recover, but prevention is the only sure solution.

Symptoms of herbicide damage vary with kind of herbicide, concentration of dosage, species of plant, and environment. Herbicide damage is sometimes difficult to distinguish from disease or nutrient deficiency symptoms. The type of herbicide damage most common in home gardens is caused by phenoxy compounds such as 2,4-D, mainly because phenoxys are frequently used in suburban areas and because several garden crops are susceptible to injury by such compounds. Tomatoes and grapes are extremely sensitive, lilacs and maples are easily damaged, and most broadleaved plants can be killed by moderate concentrations. Symptoms of phenoxy damage occur commonly on parts of the plant that are growing rapidly. Expanding leaves become elongated, and their veins become lighter in color and contrast sharply with the remainder of the leaf (Figure 9–5). The growing point is usually abnormal and is often killed, and the plant appears generally stiff and stunted.

Herbicide mixtures containing amino triazole prevent formation of chlorophyll, causing growing parts of plants injured by this herbicide to be white instead of green. Nonlethal doses of several other herbicides cause severe upward cupping of leaves. **Fasciation,** the growing together of plant parts that are normally separated, is another symptom common with herbicide damage. In fact, herbicide damage is suspected whenever there is abnormal plant growth with no other obvious cause.

Herbicide damage frequently results from a grower's own carelessness. Protecting susceptible plantings may require more than preventing spray materials from falling on them. Herbicides should *never* be applied when the air is moving, and label precautions concerning application at high temperature should be followed, because most herbicides are volatile at high temperatures. Some nonselective translocated herbicides can enter roots of woody plants in treated fencerows or ditchbanks and be translocated to kill trees and shrubs some distance away. During the past few years some puzzling herbicide damage to woody plants has been traced to the use of fertilizer-herbicide mixtures formulated for spring lawn weeding and feeding. Gardeners should be cautious in using these materials around trees and shrubs.

Figure 9–5 Herbicide damage. The squash leaf on the right has been damaged by a phenoxy herbicide. Compare it with the normal leaf (left).

Herbicide contamination of equipment used for application of other pesticides is a common cause of plant damage. Because some herbicides affect sensitive plants at concentrations ranging down to parts per billion, and because it is almost impossible to cleanse application equipment of every trace of a chemical, sprayers used for herbicides normally should not be used to apply insecticides or other garden sprays. Numerous instances of suspected plant damage by insecticide sprays have been traced to herbicide contamination.

Growers who live in areas where weeds along roadsides and in vacant lots are controlled are fortunate, but the spraying of such waste places by inexperienced or careless public employees frequently results in herbicide damage to sensitive plants in nearby yards. If herbicides can be smelled, sensitive plants growing near that spot will be damaged.

Community cooperation, zoning for home building, and regulation of pesticide application are essential in areas where gardening and extensive field monocropping exist side by side. Gardeners must realize that there may sometimes be unintentional drift of herbicides such as 2,4-D from commercial agricultural areas. Every effort is usually made to avoid such problems, but unusual weather factors can cause unexpected drift. Herbicides are as necessary to present-day production of some commercial crops as are light and water. Anyone contemplating the purchase of a few acres for garden crop production, or even for a landscaped home in a rural setting, should consider the herbicides being used on crops in the area. A person who decides to grow grapes in an area surrounded by wheat monoculture where phenoxy herbicides are aerially applied to every field several times each year will not be a happy neighbor. Because most herbicides are heavier than air, low-lying areas are more subject to herbicide damage than are slopes.

Damage to houseplants from herbicides or other chemicals accidently introduced in potting soil mixes is not uncommon and is difficult to trace. Soil washed from fields treated with herbicides and manure from stables where strong disinfectants have been used or from feedlots treated to prevent plant growth have occasionally been the source of chemical contaminants in potting mixes. Most of the recent instances of houseplant damage from herbicidelike chemicals to come to my attention were ultimately traced to sand purchased from sand, gravel, and cement companies. It has not yet been determined whether the contamination came from nonselective herbicides used to control weeds around the sand pit or from some material added to sand used for building purposes, or whether this problem has more than local significance.

Relative Toxicity of Herbicides. Recent publicity concerning the toxic nature of some herbicides has frightened many people into believing that all herbicides are health hazards. Unlike insecticides, however, most herbicides, including 2,4-D, are not highly toxic to humans or animals. People sometimes mistake 2,4-D for 2,4,5-T, another phenoxy compound that is an effective brush killer and was a main ingredient of Agent Orange, used in large quantities to kill jungle growth during the Vietnam War. Some batches of 2,4,5-T have contained a contaminant that develops into dioxin, a compound highly toxic to humans. Because of the high incidence of miscarriages and illnesses in some areas where 2,4,5-T had been used, the Environmental Protection Agency suspended its registration in March 1979.

Those working with herbicides should, of course, exercise the same precautions they would with any chemical in applying, handling, and storing the product.

Biological Weed Control

Biological control, which has received considerable publicity in recent years, is the utilization of one living organism to control another. There are two major categories of biological control of weeds—crop competition and introduction of insects or diseases to which the weed is susceptible. Weed control is much easier where a good stand of the crop is produced with plenty of light, moisture, and fertilizer (Figure 9–6). With many horticultural crops weed control ceases to be a problem as soon as the leaves

Figure 9–6 Crops for weed control. A good stand of a vigorously growing crop controls weeds by crowding them out of existence. (Courtesy Washington State University.)

of the crop plant have spread to the point where most of the ground beneath them is shaded. Under these conditions light and perhaps other factors necessary for plant growth are so limiting that weeds cannot become established (Figure 9–7). Crop plants may also control weeds to some extent by chemical competition. The fact that some weeds give off substances that are toxic to crop plants has already been mentioned. Conversely, some (perhaps all) crop plants exude substances that tend to inhibit or prevent the germination of seeds or the growth of other competing plants in the immediate vicinity. The study of chemical "warfare" among plants is just beginning, but an understanding of this phenomenon would certainly provide ammunition for growers in their age-old struggle to control weeds.

Another facet of weed control that has received renewed interest in recent years but so far has had little impact on horticulture is the introduction of diseases to which they are susceptible or of insects that feed exclusively on a weed or weeds. This method of control has been used with some success in the control of certain range and pasture weeds. For example, the beetle *Chrysalina gemellata,* for which the Klamath weed, *Hypericum perforatum,* is the only known host, has been introduced into areas infested by this weed and has provided considerable control of it. This weed is a serious pest of western rangelands because it is toxic to animals.

The use of pathogenic organisms and insects to control weeds has tremendous potential. Recent focusing of attention on detrimental effects of chemical control and widespread research interest in biological control should result in new and interesting weed-control techniques.

Figure 9–7 Biological weed control. Creeping junipers of various kinds are often planted to control weeds; however, many weed species can grow through the prostrate branches of these shrubs. Weeds are better suppressed by spreading conifers such as the tamarisk juniper (*Juniperus sabina* 'Tamariscifolia') that covers this steep bank.

Plant Disease and Insect Pests

Plant diseases and insects will be discussed together, because their control measures are similar and can be categorized into the same general topics. A **disease** is often defined as any kind of injurious abnormality; it can be a physiological disorder due to environmental influence as well as an abnormality caused by living organisms. A **pathogen** is a biological agent that incites an injurious abnormality and so, in the broadest sense, insects and even rodents, deer, and elk can be regarded as pathogens. Generally though, when we speak of pathogens we refer to four different groups of organisms—viruses, bacteria, fungi, and nematodes. Several other terms used during the remainder of this chapter should be defined. A **parasite** is an organism that derives nourishment from another living organism; the **host** is the organism that provides the nourishment. An **obligate** parasite is one that can survive only on the living host. The **host range** is the group of organisms on which a pathogen or insect normally survives. A pathogen or insect with a narrow host range, that is, able to survive on only a few plants, is generally easier to control than one having a wide host range.

Most people are aware of the different kinds of pathogens as all of them are associated with human and animal disorders as well as with plant problems. A **virus** is an infectious particle made up of a nucleic acid surrounded by a protein sheath. Known viruses are obligate parasites. **Bacteria** are one-celled plants that enter the host through wounds and most frequently cause a rotting of the tissue they infect. **Fungi** are multicelled plants containing no chlorophyll. They are responsible for a majority of those disorders referred to as rusts, smuts, molds, mildews, and blights. **Nematodes** are unsegmented worms that, under optimum viewing conditions, are barely visible to the naked eye. In addition to countless diseases known to be caused directly by nematodes, many others are caused by bacteria and fungi that enter into the plant through wounds made initially by nematodes (Figure 9–8). A number of root diseases occur only when a fungus and a nematode are both present in the soil.

The class of organisms most gardeners mean when they use the term **insect** is the larger group to which insects belong—the Arthropods. This group includes ticks, mites, slugs, snails, and various other similar organisms, but common usage lumps all of these organisms as "insects" (Figure 9–9).

The variety of forms through which insects pass during their life cycles poses problems in insect identification and control. Most start as an egg, pass through several **larval** stages into a **pupa** (resting stage) and finally an adult (Figure 9–10). Usually insects are most damaging in the larval or feeding stage. It is sometimes difficult to correlate the larvae with adult flies, moths, or butterflies because the organisms look so different at each stage.

Figure 9–8 Root knot lesions, a result of nematode infection.

How Pathogens and Insects Damage Plants

Crop-infesting insects and plant pathogens are troublesome primarily because they reproduce and feed. They continue to reproduce in prodigious numbers until one of three things occurs: (1) Their food supply becomes exhausted; (2) the environment becomes unfavorable for their continued multiplication; or (3) they are destroyed by a predator.

The food supply of an insect or plant pathogen becomes exhausted only when its entire host plant population is destroyed. This occurs most frequently when an

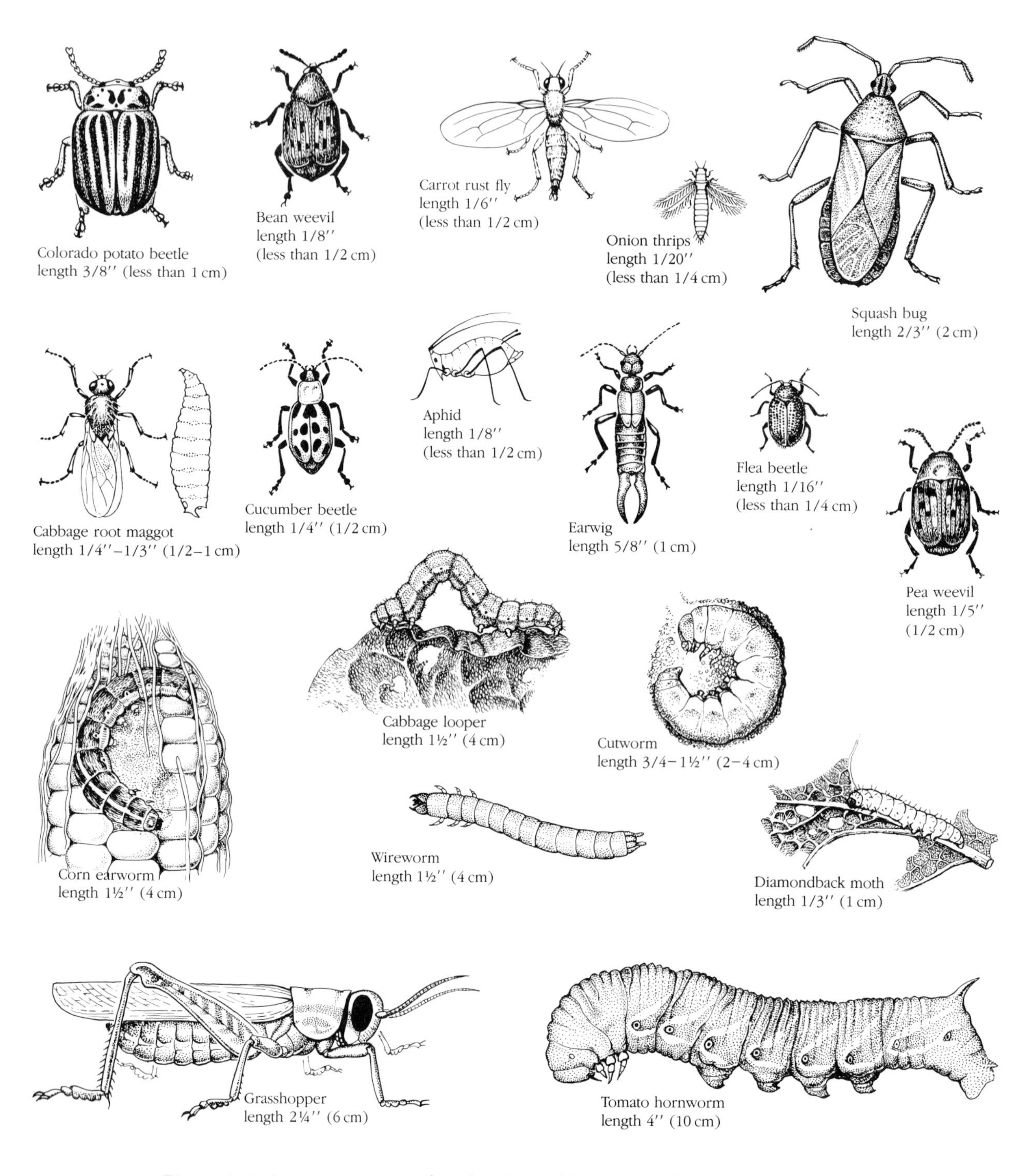

Figure 9-9 Some insect pests of garden plants. (Courtesy Washington State University.)

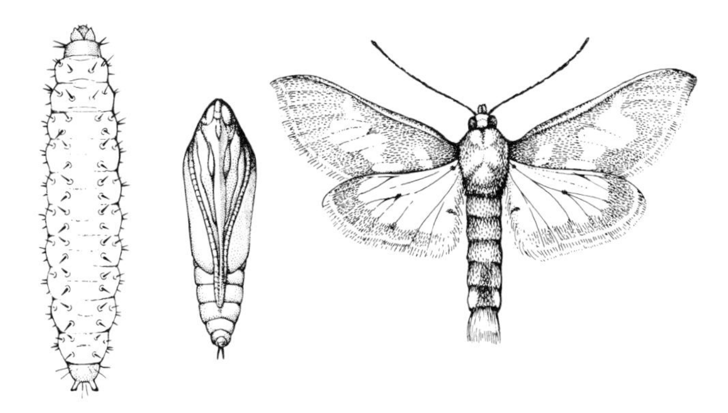

Figure 9–10 The larva, pupa, and adult of pickleworm. (Courtesy USDA.)

entirely new pest is introduced into a region, as has occurred or is occurring with several North American species. The destruction of the American chestnut by chestnut blight introduced from the Old World is one example of a species totally annihilated by a newly introduced pathogen, and the relentless advance of Dutch elm disease across the continent is an example of a plant species in the process of being destroyed by a pathogen. Sometimes a host species is saved from a virulent new pathogen when genetic resistance appears in a few plants, from which a new population less vulnerable to the insect or disease can be developed. This appears to be occurring with white pine in its relationship to white pine blister rust, a disease introduced from Europe that threatens the existence of all five-needled pines in North America.

The changing of the seasons is the most common environmental phenomenon that helps to control insects and plant pathogens in temperate climates. Even a few degrees drop or rise in average daily temperature or a drop or rise in humidity can mean the difference between serious loss to a pest and no damage at all. Pesticide action can be viewed as another example of environmental change, because most pesticides control by changing the chemical environment so that it is unfavorable for the pest.

Under natural conditions virtually all living organisms have other organisms that prey upon them, keeping their population in check. This predation, coupled with genetic changes in the resistance or susceptibility of host and pest, brings about the so-called balance of nature that is nearly universal where host and pest have been associated for long periods. Because almost all garden plants were introduced into North America from elsewhere and because many popular cultivars are new developments, natural balance has not yet become established with most cultivated species. Furthermore, even where a natural balance among an insect or pathogen, its predators, and the host crop does exist, the pest control it affords is usually only partial.

Damage to a host crop by fungi, bacteria, nematodes, and insects comes about in at least three ways—(1) from direct action on the plant, usually through loss of tissues or juices by direct feeding; (2) from toxic substances secreted into the plant; and (3) from secondary organisms that enter the plant as a result of activities of the insect or pathogen. Viruses bring about plant injury by interfering with translocation, growth, and photosynthesis and by directly affecting the genetic mechanisms of cells.

Symptoms and Signs of Insect and Disease Infestation

Symptoms (plant expression of or response to attack of an organism) and **signs** (structures produced by the causal organism) of insects and diseases are many and varied, ranging from a measurable yield decrease with no visible symptoms to complete consumption or sudden death of the plant. Rarely, an infectious organism will actually stimulate plant growth, an example being the root-rotting organism, *Gibberella zea,* which also produces growth-promoting gibberellic acid. The nature of insect damage usually depends on how the insect feeds. Because chewing insects—beetles, caterpillars, and the like—eat away parts of the plant, the results of their feeding can easily be seen (Figure 9–11). Some chewing insects, such as the tent caterpillar, produce a web. Rasping insects like thrips and mites remove only the surface layers of a leaf or petal, causing affected plants to have uneven coloration or brown spots; a stunted, unthrifty appearance; abnormal growth; and sometimes (for example, thrips on gladiolus) flower bud failure. Mites may also leave a webbing on the underside of affected leaves, which often have a dusty appearance.

Figure 9–11 Insect damage. (A) This cabbage plant has been extensively damaged by the cabbage looper. (B) The leaf miner is difficult to control because it lives and feeds beneath the epidermis.

Leaf miners tunnel under the surface of the leaf, causing white areas or streaks where the interior leaf cells are destroyed. Plants infested with sucking insects such as aphids, scale, and leafhoppers grow slowly, partly because the insects are utilizing their manufactured food. Sucking insects often cause abnormal growth of meristematic tissues either by injecting digestive toxins into the phloem or by transmitting viruses. Secretion of enzymes that affect cells in the vicinity of their feeding is common among sucking insects, but a few affect tissues a considerable distance from where they are feeding. For example, a single squash bug feeding at the base of a squash plant can cause its entire host plant to go into shock and wilt. Like other sucking insects, scale insects cause slow growth and unthrifty appearance of their host plant. The diagnostic feature for most scale insects, however, is the rough, scaly shell or cottonlike cover they build for their own protection and for securely anchoring themselves to their host.

The most apparent signs of fungus diseases that attack above-ground plant tissues are the **mycelium** (threadlike vegetative structures) and the fruiting bodies—rust pustules, smut bodies, leaf spots, spore capsules, sclerotia, and toadstools (Figure 9–12). **Necrotic** (dead) areas on leaves, stems, or other plant parts are common symptoms of fungus diseases. The first noticeable symptoms of fungus attacking roots or the vascular system may be sudden wilting or death as a result of root destruction or xylem blockage. Phloem blockage by fungi may result in enlarged nodules on the upper part of the plant where carbohydrates are accumulating.

Bacterial diseases are characterized by rapid spread. In herbaceous tissues they often result in soft rot and frequently are secondary invaders after initial infection by fungi. In woody tissues attacked by bacteria the bark at first appears dark and water soaked and later becomes blistered and peels. Some bacteria may cause abnormal growth of plant tissue—galls, root enlargement, and the like.

Viruses cause death of the plant, stunting and yellowing of the plant body (Figure 9–13), **mosaic** (light and dark areas) of stem and leaf, abnormal growth, or no obvious symptoms at all. The presence of a virus in a virus-infected plant showing no symptoms (a symptomless carrier) can be detected only by transmission to an-

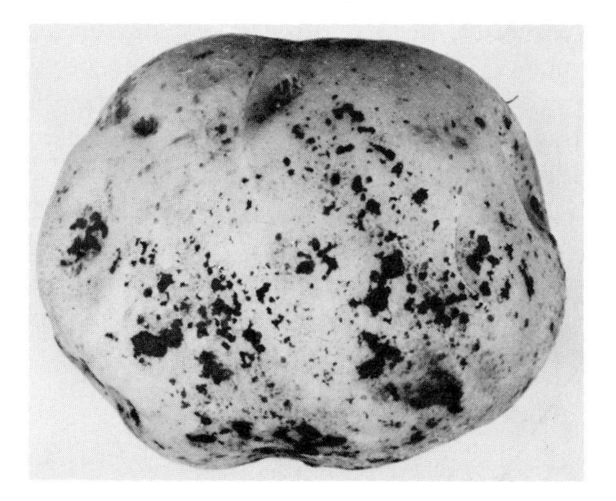

Figure 9–12 *Rhizoctonia* (black scurf on potatoes). The black specks on the tuber are the hard fruiting bodies (sclerotia) of the fungus. The sclerotia send out mycelia that attack and kill the developing potato sprouts when infected tubers are planted. (From Agriculture Canada Publication 1492, "Diseases and Pests of Potatoes," revised 1974, and reproduced by permission of the Minister of Supply and Services, Canada.)

Figure 9–13 Virus leafroll of potatoes. Like many viruses, leafroll causes distortion of plant tissues, lack of growth, loss of chlorophyll, and reduction of yield. (Courtesy Washington State University.)

other plant, on which it does produce symptoms. Often two viruses, neither of which produce symptoms when alone, together will severely injure infected plants.

The presence of many species of nematodes is first signaled by areas in the garden or field where susceptible plants do not grow well or sometimes do not grow at all. These areas enlarge year by year whenever susceptible crops are produced. One of the most common nematodes, the root-knot nematode, causes odd shaped, fleshy enlargements along lateral roots and raised lumps on carrots or potato tubers (see Figure 9–8).

Transmission of Plant-Infesting Insects and Diseases

Many, perhaps a majority of, plant pests rely wholly or partly on the movement of water and wind to spread from one plant to another. Some, for example, the corn earworm and certain rusts, survive through the winter only in mild-winter areas and spread north with the south and southwest winds as the season progresses. Others, such as the beet leafhopper, winter over on plants in desert areas or along fence rows and spread into cultivated fields during the summer. For some, including most wind-disseminated fungi, wind is the only means of long distance spread. For others wind merely increases the distance and rate of their travel. For example, although leafhoppers are quite capable of flying long distances, they are always more prevalent and spread greater distances in the direction of the prevailing winds. Winged aphids, too, rely on wind to speed their movement and increase their travel distance.

Several of the more serious fungus diseases, including late blight and early blight of potatoes and tomatoes, can spread only if free water is present for a specified period of time, and most fungal and bacterial diseases are more severe in areas with wet climates than they are in drier areas. Free water is necessary sometimes for growth of the pest, sometimes for its spread, and sometimes for both.

Temperature influences the spread of insects and diseases by affecting both survival and reproduction of the pest and the susceptibility of the host. Generally, the higher the temperature, the more rapidly pests multiply and spread. Freezing temperatures greatly reduce the numbers of insects and disease organisms and may eradicate some species from the area. Plants are more susceptible to pest attack when temperatures are either above or below the range optimum for plant growth.

Fungus spores, bacteria, and certain viruses can be spread by birds, animals, or people on feathers, fur, clothes, claws, beaks, or hands. Some viruses are transmitted only by insects and frequently by a single species of sucking insect; for instance, aphids, leafhoppers, and white flies transmit many kinds of virus diseases and are responsible for the spread of most kinds that attack primarily the phloem.

Human carelessness is often responsible for spreading garden pests. Viruses and some diseases caused by other organisms can be spread with plant parts used for asexual propagation. Reducing the dissemination of such diseases is the reason for plant certification programs for seed potatoes, small fruit planting stock, and bulbs and corms.

Pathogens can also be spread with seed. Most seeds destined for gardens are treated with chemical dusts, partly to eliminate pathogens that might be on the exterior of the seed and partly to protect the seed from soil-borne pathogens. Seed of a few garden crops can carry virus or bacteria within itself. Use of disease-free seed is usually the only control for these kinds of diseases, but with a few, blackleg of cabbage being one, it is possible to heat the seed to a temperature high enough to destroy the pathogen but not so high that it damages the viability of the seed. The seed packet of cabbage and related crops will usually carry the label "hot water treated."

The grower who transplants from a friend's garden runs the risk not only of bringing in disease or insects on the above-ground plant parts but also of introducing nematodes and other soil-borne pathogens that are almost impossible to control. Soil-borne pests can be introduced on soil particles that adhere to equipment and tools used previously in an infested garden or even from mud tracked into the garden on dirty shoes. Some virus and bacterial diseases can be spread from plant to plant with unwashed hands. Plants that are handled frequently, such as staked tomatoes trained to a single stem or pinched chrysanthemums, are especially prone to the spread of these diseases. The transmission by smokers of tobacco mosaic virus to such relatives of tobacco as tomato, eggplant, pepper, petunia, or potato is probably not common, but, because live viruses can be found in cigars and cigarettes, smokers should wash their hands as a precaution before handling any of these solonaceous garden crops.

Unsanitary garden conditions, failure to dispose of insect- or disease-infested plant material promptly by burning, use of diseased plant materials in compost piles, failure to prune away diseased plant tissue, and any practice that reduces garden plant vigor will increase the likelihood of disease and insect damage.

Control of Garden Insects and Diseases

The Pest Cycle. Although gardeners can control insects and diseases by following directions from a publication or from the label of a pesticide container, they should realize that the recommended control is based on a knowledge of the **pest cycle,** the sequential changes and interactions that occur in the host/pathogen relationship through their respective life cycles. Researchers who search for and eventually devise controls for various insects and pathogens must have a thorough knowledge of the pest cycle because control is easiest, and sometimes only possible, during one stage of the life of a crop or of insects or disease pests.

Legal Control. The most effective and least costly way to combat a pest is to keep it out of the area. Because this involves laws and their enforcement, it is referred to as **legal control,** and it is usually administered by federal or state plant quarantine offices. If you have been overseas recently, you may be aware that customs officials do not permit you to bring certain plants and animal products into the United States or Canada. They do this in an attempt to prevent the entry of insects and disease pests that are not found here but are serious problems in other parts of the world. Some of the most severe disease and insect epidemics (as well as almost all of our serious weed infestations) have resulted from importation of a pathogen or insect (or weed seed) on products from other countries. Chestnut blight, Dutch elm disease, white pine blister rust, Japanese beetle (Figure 9–14), earwigs, fireants, and late blight of potatoes are but a few of the unwelcome plant-pest immigrants of the past 150 years.

Legal control cannot be effective without the cooperation of all citizens. The number of American citizens traveling abroad is so great and there are so many ways of sneaking in illegal plant products that it becomes impossible for police action to control all illegal plant introduction. It must be the responsibility of each citizen, and especially those interested in plants and gardens, to understand and educate others to the serious consequences of illegal plant and animal introduction.

Besides quarantine at the national level, there is some legal control on interstate and interprovince shipments of plant products. The grape root louse, phylloxera, and the corn borer have so far been effectively kept out of some of the western states partly as a result of inspection and control of interstate shipment of grape vines and corn seed. For many years all people driving into California have been stopped at the border and subjected to an inspection in order to prevent the importation of certain citrus fruit pests into that state. Control of shipment of plant products to and from Hawaii is also quite stringent.

Plant quarantine officers are not trying to restrict the importation of all plant products, and importation will be allowed if there is reason for it and if the product is routed through the plant quarantine center in Washington, DC, where it can be properly inspected. It cannot be too strongly emphasized, however, that the responsibility for legal control must rest with individual citizens, and that only through education to the dangers of plant importation can legal control keep out of this country the many serious diseases and insect pests that are not presently found here.

Cultural and Physical Control of Insects and Diseases

Cultural and **physical** control of insects and diseases includes a broad spectrum of practices relating to cultivation, physical removal of infested or infesting entities, and barriers against infestation. **Roguing,** the removal from the garden of diseased or insect-infested plants, is quite effective with diseases or insects that

Figure 9–14 Japanese beetles. These common garden pests introduced to the eastern United States and Canada from Europe are spreading westward. (A) Adult beetle feeds on blackberry leaf. (B) Japanese beetle grubs (larvae) destroy lawn grass and other plants by feeding on their roots. (Courtesy USDA.)

spread slowly from plant to plant. Pruning provides a way of eliminating diseased plant parts or those likely to become diseased. Fireblight, a serious bacterial disease of apple and pear trees, is controlled by keeping all diseased wood pruned from the orchard and by periodic antibiotic spraying. Plowing under infested plant refuse, stirring soil to destroy insect egg masses, and other kinds of cultivation practices are effective in controlling some pests. Clubroot of crucifers and scab on potatoes can be partially controlled by regulating soil pH. Effective disease control may include various sanitation practices, such as burning diseased or insect-infested plant residues, cleaning away weeds that might harbor disease organisms or insects, washing hands frequently when garden plants are being handled, and refraining from smoking when handling tomatoes and cucurbits subject to the tobacco mosaic virus. Draining or filling low-lying land, planting windbreaks, and land leveling are effective controls for certain insect and disease problems.

Some insects and diseases can be physically excluded by construction of physical barriers, such as screens placed over the openings of a greenhouse to prevent their entrance or large, deep water-filled excavations to intercept crickets and related insects before they enter cultivated areas. The hot-water treatment of seed to destroy a seed-borne pathogen is a physical method of control. Mites, thrips, and aphids can often be controlled by frequent foliar sprinkling with water. If they are not too numerous, large insects, such as mealybugs on houseplants, tomato hornworms, and Colorado potato beetles, can be controlled by hand picking them from plants. Insect traps featuring such attractants as light, food, or sex-related chemicals are frequently used to determine the presence of damaging insects (Figure 9–15). Traps coupled with attractants and poison baits or other killing mechanisms are being used in a few places for insect control and have promising future potential.

Figure 9–15 Gypsy moth trap. (Courtesy USDA.)

Chemical Control of Insects and Diseases

The most common method for controlling diseases and insects during the past forty years has been chemical, primarily because it is rapid and effective. Chemicals used to control insects and diseases are called **pesticides** and can be classified as repellents, protectants, insecticides, miticides, fungicides, bactericides or antibiotics and nematicides. **Ovicides** (egg killers) and **larvicides** (larva killers) are other terms sometimes used to classify chemical pesticides.

Repellents are used to protect clothing from moths and to drive away blood-sucking insects, but their possible use against plant pests has not been adequately researched. Insect repellents presently available are nontoxic to animals and humans, and they can be applied so that no residue is left on plants. These two characteristics tremendously enhance their acceptability over other forms of chemical control.

Protectants are used in largest quantities as seed treatments to prevent attack of seedlings by soil-borne pathogens. Protective dips are used to delay spoilage of some kinds of fresh fruits and vegetables and to lengthen storage life of processed foods. Copper and zinc compounds, elemental sulfur, and some antibiotics are often used to protect plants from infection by fungi or bacteria.

Insecticides, fungicides, bactericides, and **nematicides** are by definition killers of various microorganisms. They are the compounds usually envisioned when chemical pesticides are denounced or defended and, of course, the ones most useful and available for farm or garden. The fact that viricides are not mentioned among the chemicals used for disease and insect control is indicative of the fact that viruses are not normally controlled by chemicals. Indeed, only a few of the plant bacterial diseases are subject to control by chemical means. Nematicides are relatively new.

Application of Pesticides. Chemicals can be applied as dusts, sprays, soil fumigants, or volatile granules. Dust formulations are most effective if applied in the early morning when there is some dew or residual moisture from rain on the plants that are to be dusted. Various kinds of pump dusters are available for the home gardener, and special dust formulations that come in disposable cardboard shake or pump packages are available for application on a few plants or small areas. The application of highly toxic chemicals in dust form should be avoided.

Application by spray generally will provide more complete coverage and adherence of the pesticide to the plant than will dusting. A hand-pump pressure-tank sprayer with a capacity from 2½ to 4 gallons (11 to 18 liters) is the most satisfactory application equipment for most home garden insecticide and fungicide sprays (Figure 9–16). A small trombone-type sprayer with a screened inlet hose that fits into a bucket is sometimes used; a sprayer of this sort has two handles that can be pumped back and forth to create pressure at the spray nozzle. This type of sprayer is fairly effective, but the application rate and coverage is not as uniform as it is with a tank sprayer. Another good sprayer available to home gardeners consists of a nozzle and small container in which is placed a concentrated solution of the chemical to be applied. This sprayer is designed to be connected to a garden hose and depends on water and water pressure for correct dilution and application of the spray material. This is a good device for spraying large areas or moderate-sized trees.

Figure 9–16 A hand-pump pressure tank sprayer.

Effective soil fumigation usually requires specialized equipment. Furthermore, many soil fumigants will destroy nearby plants that happen to be rooting into the fumigated area. Consequently, fumigation is generally not very practical for the home gardener. Possible uses of soil fumigation by the amateur might be in the pasteurization of potting soil for the hobby greenhouse or the fumigation of an area on which a plant nursery is to be established.

Pesticide Safety. Pesticide formulations packaged for sale for the home gardener or amateur horticulturist are relatively safe to use if instructions printed on the label are carefully read and followed. Pesticide formulations packaged for the large-scale grower or commercial applicator often presuppose that the user will have had considerable experience with pesticides and safety equipment. Even though pesticides packaged and sold in bulk by agricultural dealers are less expensive and often more effective than those sold at garden stores, they should not be used by people who are not thoroughly familiar with their toxicity or who do not have the protective clothing and special equipment designed to be used when they are being applied (Figure 9–17). Even garden formulations can cause injury to animals, humans, and plants if they are not applied as directed. All pesticides should be labeled and stored where they cannot be reached by children or others not acquainted with their dangerous nature. Pesticides should always be kept in their original containers. Small children who take a drink from the pesticide-filled soda-pop bottle set on a shelf or who roll each other around in a discarded parathion drum have been the most frequent victims of pesticide tragedies.

Figure 9–17 Pesticide safety. Garden chemicals may cause harm if they get on the skin. When spraying, wear full-length clothing, a wide-brimmed hat, and gloves. (Adapted from USDA.)

Problems Associated with Chemical Pesticides. There is not space in this text to analyze the many charges leveled against the use of chemical pesticides, but a few problems that might affect gardeners as both potential pesticide users and food consumers should be mentioned. The first is the possibility of harmful residues on edible products. Residues on food products entering market channels are strictly monitored so they remain well below the amount that might pose a hazard to human health. Garden produce will be safe to eat if gardeners follow directions on pesticide labels. Edible products growing along unfamiliar roadsides and cultivated fields, as well as flowers and other products sold primarily for decoration, should not be consumed, although the possibility of serious poisoning from such sources is remote.

The second problem is the technical difficulty of application. Whether application is with dust or spray, by hand sprayer, by a 500-gallon mobile ground rig, or by airplane, placing the right concentration of the right chemical in the right place at the right time poses numerous problems for growers and applicators. In my experience the most common cause of faulty pesticide application is a mistake in measurement. Pesticide users should read all labels and check all measurements at least twice, be careful and accurate in applying the material, and use only clean application equipment that is in good repair.

A third problem is the possibility of injury to the host plant or to a crop plant that may be in the vicinity or may be planted at a later date. Many pesticides must be applied at specified conditions of temperature and humidity, and problems may result if there is a rapid change in the weather after application or if the grower gambles at a time when the application is necessary but environmental conditions are not ideal.

In recent years the development of genetic resistance by pests has been appearing more and more frequently. In Chapter 3 we spoke of mutation, the slight change in genetic determiners occasionally occurring in all living species. Although mutation may occur in only a minute percentage of the population of any species, the total number of mutated individuals will be large in species where numbers within a small area may equal trillions, as is the case with most pathogenic organisms. If a mutation permits an individual pathogenic organism to survive in the presence of a pesticide that destroys all of its sister organisms, the resistant individual can multiply astronomically because it has no competition from other organisms of its kind. In a short time the progeny of the resistant individual will number in the billions, and a new race with genetic resistance to the formerly effective pesticide will have been developed. Most readers are aware of housefly resistance to DDT and several other fly-spray materials and of the constant need for developing new pesticides to replace those that no longer control the pests they were once effective against.

Chemical control of plant pests can disturb biological balance, sometimes to the detriment of the grower. This may result from the killing of beneficial organisms. Mites, for example, were not a serious problem in most orchards until DDT was introduced as an orchard spray. In addition to killing the codling moth and other apple pests, DDT was also a very effective killer of mite predators, and its use permitted the buildup of injurious mite populations. Summer spraying to rid cities of flies and mosquitoes, which was common a few years ago, frequently resulted in an almost immediate increase of thrips and aphids in the sprayed area. Apparently the fly and mosquito spray killed ladybird beetles and other predators but had little effect on aphids and thrips, perhaps because it did not reach them on the undersurfaces of leaves and in the other protected places where they reside.

Honeybees and other pollinating insects are frequently the victims of pesticide poisoning. Unfortunately, they are more susceptible to the newer insecticides than they were to DDT. Insecticide applications for fruit trees are usually needed more after the bloom period is past, and, with proper planning, gardeners should seldom need to spray a crop while it is in full bloom and most attractive to bees. Bee losses can also be reduced by waiting to apply insecticides until late evening so that their most lethal period occurs during the night when bees are not active.

How Much Insecticide and Fungicide for the Garden? Despite the problems that accompany their use, the benefits derived from chemical pesticides outweigh their disadvantages. Gardening in many areas would be difficult or impossible without them, and they are essential to the abundant harvests prevalent in well-fed North America.

My advice to gardeners regarding the controversial subject of chemical pesticides would be to use them, but only when necessary. Often gardens can be grown easily with fewer pesticide treatments than are recommended by local extension people, because gardens are more isolated than commercial fields and because the presence of one or two insects does not prevent a garden crop from being utilized. In my own garden if a few earworms are acceptable on the latest planting of sweet corn, if hornworms and potato beetles are handpicked from tomatoes and potatoes, and if mildew-resistant rose cultivars are selected for planting, only two or three pesticide treatments on apple, pear, sweet cherry, gladiolus, and crucifers are required annually for satisfactory pest control. Not all areas are equally pest-free, but with experience most gardeners can grow a variety of crops with minimal use of pesticides.

Biological Control of Insects and Diseases

The first successful scientific use of biological pest control occurred in California in 1890 with the use of the vedalia beetle introduced from Australia to control cottony cushion scale of citrus trees. With the realization of the shortcomings of chemical disease and insect control and the environmental problems sometimes arising as a result, there has been a renewed interest in more natural biological systems of control. As advocates of biological control are quick to point out, these control methods are the ones used by nature, and, fortunately for the home gardener, the natural system of checks and balances will often bring about control of plant pests where chemical control may not be feasible (Figure 9–18).

Most people are aware that birds eat many kinds of insects and that ladybird beetles, if prevalent in a garden, will provide control of aphids and some other plant-feeding pests. Fewer are aware that yellowjackets feed on and destroy cabbage loopers and similar caterpillars or

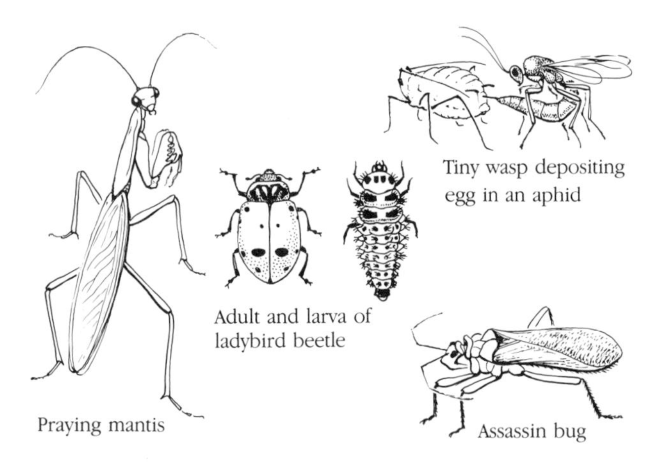

Figure 9-18 Beneficial insect predators that provide biological control in the garden. (Courtesy USDA.)

that all living organisms are subject to destruction by other living organisms. During the past 5 years much of the pest-control research has been on biocontrol. The benefits of this research are new pest-control methods based on an understanding of the relationship among crop plants, their pests, and the parasites and pathogens of those pests.

Another intriguing aspect of biological control just beginning to receive scientific recognition is the nature of evolved and evolving plant and pest chemistry, which appears to be the basis of much pest resistance and susceptibility. This area includes the inhibition of pests by plant **volatiles** (substances given off as vapors, generally aromatic). We know, for example, that a cedar closet repels moths, that some gardeners rely on marigold and garlic to repel certain kinds of insects from their gardens, and that scientists accidently discovered insecticidal properties in a species of balsam fir when insect larvae failed to survive in cages covered with paper manufactured from that fir.

Although proven biological methods for controlling diseases and insects are limited, a few that are available to the home gardener should be mentioned. Both ladybird beetles and praying mantises are available commercially and can be purchased to spread around the home garden. The major problem with purchasing these two predators is that both are extremely mobile, and consequently the purchaser may not be the one who benefits from their activities. At least one disease organism of a serious insect pest is formulated commercially for controlling that pest: *Bacillus thuringiensis,* a natural pathogen of the cabbage looper, has been available since about 1965 and is sold in spore formulations under several trade names. If utilized as a spray within a few months of its formulation, it does a very effective job of controlling cabbage loopers and larvae of certain other moths and butterflies.

Certain of the pest-control measures advocated by organic gardeners probably should receive more scientific investigation. Some undoubtedly are beneficial in controlling particular pests; others are probably worthless. One recommendation that will almost always reduce pest damage is to do everything possible to keep plants growing vigorously.

The use of plant extracts that have pesticidal properties, although not strictly biological control, was one of the earliest methods of killing insects with a natural product. At least two, pyrethrum from a relative of chrysanthemum and rotenone from various species of the tropical genera *Derris* and *Lonchocarpus,* have been used for many years to control certain insects. Both are relatively nontoxic to humans and provide fair control for a rather broad group of plant pests. Nicotine sulfate, formulated from an extract of tobacco, was formerly a popular insecticide, but it is no longer used extensively by home gardeners because it is extremely toxic to humans.

Breeding for Control of Insects and Disease

Home gardeners should have some knowledge concerning the development of disease- and insect-resistant cultivars by plant breeding. Most gardeners recognize the value of disease-resistant cultivars and have had experience with at least a few. One mistake made by some home gardeners is assuming that a cultivar listed as disease-resistant will show general resistance to all diseases; this is not generally true. Any one cultivar usually will be resistant to only one or two diseases, and unless those diseases are prevalent in the area, that cultivar will produce no better crop than will a nonresistant one.

Because development of resistant cultivars has been the only reliable method of control for many virus, bacterial, and soil-borne diseases, breeding for resistance to disease has received considerable research effort. Breeding for insect resistance received only minor attention until the last 2 or 3 years, perhaps because chemicals have provided satisfactory control of most insect pests. Because developing and testing a new cultivar requires from 12 to 15 years, it will be some time before insect-resistant cultivars become generally available to the public.

Few people realize the dependence of our food supply on breeding for disease resistance. Many crops can be produced today because cultivars with improved disease resistance are introduced every few years. Pathogens and predators can mutate so that they are able to infest cultivars that formerly were resistant to them.

Wheat is an example of a major crop where profitable production is entirely dependent on new cultivars being introduced at frequent intervals. Because of the rapid mutation of organisms pathogenic to the wheat plant, mainly the rusts and smuts, a wheat cultivar remains profitable for only about 7 years. Because the breeding and testing necessary for the introduction of a new wheat cultivar require from 12 to 15 years, a crash program cannot develop a new resistant cultivar the year after people start to become hungry. It is essential that wheat breeding, as well as other agricultural research, be continuing programs that can anticipate needs and count on public support and funding on a long-term basis. If funds for breeding programs were to be eliminated, it would be only a few years before this country and the world would be faced with a general famine.

Most horticultural crops are not as vulnerable to disease and insect ravages as wheat; some cultivars have remained resistant to a particular disease or insect for 20 or 30 years. But resistance is seldom permanent, and research personnel must be ready with new cultivars if the commodity is to be produced when the old cultivar fails.

Few commercial agricultural firms are economically able to support research programs having the uncertainty and requiring the amount of time necessary for new cultivar development. Agricultural research is but an infinitesimal part of our federal and state budgets, but much of the world food supply is dependent on that research budget. The public must be aware of the disaster of hunger that would result if that budget were eliminated.

Integrated Control of Insects and Diseases

One other interesting development in pest-control techniques is the relatively new concept of integrated control. Often in the past pest control has consisted of the use of a pesticide at predetermined intervals; for example, an apple tree was sprayed every 10 to 14 days in order to prevent infestation of codling moths. As has been mentioned, this kind of program destroyed predators and hastened the development of pesticide resistance. With integrated control the pest/host relationship is monitored carefully, and every conceivable control measure, including cultural practices, environmental manipulation, and the use of natural predators, is utilized to control the pest. Chemicals generally are used as the last resort and then with as low concentration and as minimal coverage as possible.

Birds and Animals

Because they are often located near uncultivated areas and because they usually contain only a small planting of any one crop, gardens are especially vulnerable to the feeding of birds and animals. Because birds and animals such as toads, frogs, snakes, and lizards benefit garden crops by feeding on insects, and because many people are reluctant to harm birds and mammals, their control poses special problems.

Gardens and orchards located where deer or elk are abundant should be surrounded with a fence high enough to keep out those animals. Rodents can be controlled by trapping or by poisoning. Rodent poisons should be placed underground if possible to lessen the likelihood that seed-eating birds, such as pheasants, doves, and grouse, or meat-consuming birds and animals, such as hawks, owls, eagles, coyotes, and weasels, will be poisoned by eating the poisons or the bodies of poisoned rodents.

Although poisoning has been used for many years to control rodents, there is a dearth of information of the impact of such control measures on the total ecology of an area. It is possible that growers who use strychnine-treated oats for controlling ground squirrels are defeating their own purpose by destroying predator animals and birds. Pellets or special dispensers that release poisonous gases are quite effective against burrowing rodents and pose little threat to other birds and animals, but they require extreme caution to prevent injury to the user. Steel traps have been designed for most kinds of rodents and are effective for small areas where animals are not numerous. Cage-type traps using bait to lure the animals are more humane.

Rodents can be discouraged by eliminating nesting sites and litter around the garden area. House cats cannot be relied on to keep rodents under control; those that are good hunters are as likely to control song and game birds as they are rodents.

My own observations lead me to believe that the best way to control rodents is to encourage a good population of hawks, owls, weasels, skunks, foxes, or coyotes, even though these predators undoubtedly destroy some song birds and their presence may necessitate special protection for farm poultry and small animals. City dwellers benefit from a predator population because control of rodents on farm lands surrounding a community keeps them from multiplying and spreading into the urban area. Encouraging a predator population requires community or area-wide education and cooperation. Reliable, well-published information on the feeding habits of various predators would also be a valuable research contribution to farmers and gardeners.

Bird pests are most destructive to emerging seedlings, ripening seed crops, and ripening fruit, especially small fruit and sweet cherries. Pheasants and crows will systematically dig every germinating corn seedling from a row, pick off the sprout, and eat out the embryo and endosperm, leaving the seed coat almost intact. Because of this method of feeding, seed treatment with bird repellents is ineffective in many areas. Small devices such as metal discs that whirl in the wind, firecrackers, recorded bird calls of distress, and scarecrows designed to frighten birds are usually effective for a short time until the birds become used to them. State experimental stations distribute publications containing plans for cage-type traps designed to help control starlings and similar bird pests. One effective (though expensive and bothersome) bird control for the small home orchard and garden is a mesh cover. Cheesecloth can be used over a row or two of strawberries, and specially designed covers are available for small fruit trees. A narrow strip of screen or chicken wire can be bent into a convex cover over germinating seeds until seedlings are large enough to resist bird damage (Figure 9–19). Plastic covers may serve dual purposes protecting seedlings from weather and birds. Covers are,

Figure 9–19 Netting spread over a frame to keep birds from eating berries. (Courtesy USDA.)

of course, impractical if more than a few rows or trees are being grown. Harvesting fruit and seed crops as soon as they are mature will reduce the chance of birds finding and damaging the crop.

Selected References

Brooklyn Botanic Garden. *Garden Pests.* Handbook 50 (special printing of *Plants and Gardens,* vol. 22, no. 1). Brooklyn, NY. 1966.

Brooklyn Botanic Garden. *Weed Control.* Handbook 73 (special printing of *Plants and Gardens,* vol. 30, no. 1). Brooklyn, NY. 1974.

Chupp, C., and Sherf, A. F. *Vegetable Diseases and Their Control.* Ronald Press, New York. 1960.

Crafts, A. S., and Robbins, W. S. *Weed Control.* 3rd ed. McGraw-Hill, New York. 1962.

Roberts, D. A., and Boothroyd, C. W. *Fundamentals of Plant Pathology.* W. H. Freeman and Company, San Francisco. 1975.

U.S. Department of Agriculture. *Insects* (*Yearbook of Agriculture,* 1952). U.S. Government Printing Office, Washington, DC. 1952.

U.S. Department of Agriculture. *Insects and Diseases of Vegetables in the Home Garden.* Home and Garden Bulletin 46. U.S. Government Printing Office, Washington, DC. Revised periodically.

U.S. Department of Agriculture. *Insects and Related Pests of House Plants: How to Control Them.* Home and Garden Bulletin 67. U.S. Government Printing Office, Washington, DC. Revised periodically.

U.S. Department of Agriculture. *Insects on Deciduous Fruits and Tree Nuts in the Home Orchard.* Home and Garden Bulletin 190. U.S. Government Printing Office, Washington, DC. Revised periodically.

U.S. Department of Agriculture. *Plant Diseases* (*Yearbook of Agriculture,* 1953). U.S. Government Printing Office, Washington, DC. 1953.

Walker, J. C. *Plant Pathology.* 3rd ed. McGraw-Hill, New York. 1969.

Ware, G. W. *The Pesticide Book.* W. H. Freeman and Company, San Francisco. 1978.

Ware, G. W. *Pesticides: An Auto-Tutorial Approach.* W. H. Freeman and Company, San Francisco. 1975.

Watson, T. F., Moore, L., and Ware, G. W. *Practical Insect Pest Management: A Self-Instruction Manual.* W. H. Freeman and Company, San Francisco. 1975.

10 Indoor and Container Gardening

In our urbanized society not everyone has a plot of ground on which to grow a garden. However, almost everyone does have a room with a window or an electric light, and almost everyone can obtain a tin can or plastic carton and some soil or potting medium, all that are required for growing an indoor or container garden. The fact that container gardening is possible in apartments, dormitories, trailer houses, condominiums, offices, school rooms, and public buildings where outdoor gardening is not feasible probably accounts for the increasing popularity of this form of gardening (Figure 10–1).

Environmental Limitations to Growth of Plants Indoors

As anyone who has tried to grow them knows, most food-crop plants (vegetables, fruits, small grains, etc.) do not grow well in the interior of homes. The reasons for this can be summarized: (1) too much heat, (2) too little light, (3) too low humidity, (4) atmospheric pollution, and (5) restricted rooting area. Plants designated as houseplants are able to withstand these adverse conditions to some extent, because most of them were developed from ancestors native to the shade of tropical forests. Moreover, it is often possible to choose plants adapted to special home conditions—for example, cacti for an extremely dry atmosphere, or ferns for a cool, moist corner. Even so, growth of any plant indoors is likely to be limited by one or more of the five conditions mentioned.

Temperature

Homeowners are often surprised to learn that temperatures in their homes are too high for the optimum growing of plants. An outdoor temperature of 70° to 80°F (21° to 27°C) is certainly not detrimental to most plants; however, there is a difference between outdoor and indoor conditions. Temperatures are high out-of-doors usually in conjunction with high light intensity, and the combination of high light and high temperature permits a high rate of photosynthesis. Indoors, high temperatures frequently occur with low light intensity, and plants suffer because all of their activities except photosynthesis are speeded up. In addition, home temperatures are often kept relatively higher at night than would occur in the plant's native climate. High night temperature speeds respiration, resulting in rapid utilization of photosynthates. Turning the thermostat down to 60°F (16°C) at bedtime will benefit most houseplants.

Choosing the right location may be the secret of success with certain temperature-sensitive plants. Plants from the jungles do relatively well in the 70°-plus temperatures of modern homes. However, geraniums, coleus, wax-leaved begonias, impatiens, and some of the other

Figure 10–1 Four long-lived easy-to-grow houseplants. (A) Jade plant (*Crassula argentea*); (B) Dumbcane (*Dieffenbachia*); (C) Boston fern (*Nephrolepis exaltata*); (D) Snake plant (*Sansevieria trifasciata*).

window-box plants of grandmother's generation do poorly simply because they are adapted to the cool night temperatures and relatively high light intensities that existed in south and west windows of grandmother's wood-stove heated house. They will grow vigorously only if there is a well-lighted, cool location available.

2. Light

Perhaps the major factor limiting growth of plants within a home is a lack of light. Most foliage plants sold for indoor growing do best with the amount of light coming from an east or north window. Philodendron, peperomia, ficus, begonia, and other similar jungle plants can be grown with fair success toward the interior of rooms that are well lighted with floor-to-ceiling windows, but plant growing should not be attempted in the poorly lighted interior of rooms unless supplemental light can be provided. The relative amount of light required by some of the more popular houseplants is mentioned in Table 10–1, and a brief discussion of artificial light sources for indoor plant growing is given later in this chapter.

3. Humidity and Atmospheric Pollution

Humidity in homes, especially when they are being artificially heated during the winter, is likely to be lower than humidity out-of-doors. Humidity can be raised somewhat

Table 10–1 Selected easy-to-grow houseplants

Name	Growing requirements and characteristics
	Moderate-sized specimen plants
Aglaonema (*Aglaonema* spp.)	A group of plants that tolerate fairly dark conditions; easy to root.
Aluminum plant (*Pilea cadierei*)	Shiny variegated leaves need spraying occasionally; easily damaged by overwatering.
Boston fern (*Nephrolepis exaltata*)	Rich soil; high humidity; avoid direct sunlight; fronds become almost trailing.
Dracena (*Dracaena deremensis*)	Diffused light to shade; water freely; grasslike leaves.
Dumb cane (*Dieffenbachia picta*)	Diffused light to shade; when spindly, it can be cut off and rooted in moist soil.
Philodendron (split-leaf) (*Monstera deliciosa*)	Diffused light; spray leaves; can be pinched to keep low or trained to climb.
Piggyback plant (*Tolmiea menziesii*)	Easy to grow; tolerant of gas and relatively low light intensity.
Screw pine (*Pandanus veitchii*)	Grasslike ornamental leaves; tolerant of varying environment.
Snake plant (*Sansevieria trifasciata*)	Long, straight, stiff leaves; slow-growing; tolerant of drought.
	Low-growing specimen plants for moist, shaded situations
African violet (*Saintpaulia ionantha*)	Does best in a north window; warm temperature; keep water from leaves.
Fittonia (*Fittonia verschaffeltii*)	Diffused light; good small plants for terrariums.
Maidenhair fern (*Adiantum cuneatum*)	Humid atmosphere; soil high in humus; terrariums.
Peperomias (*Peperomia* spp.)	Diffused light; easy to propagate; some appropriate for terrariums.
Polkadot plant (*Hypoestes sanguinolenta*)	Diffused light; needs pinching; easy to root.
Prayer plant (*Maranta leuconeura*)	Diffused light; leaves require occasional spraying.
Sansevieria, birdsnest (*Sansevieria trifasciata* 'Hahnii')	Will grow in a terrarium if not overwatered.
	Low-growing specimen plants for dry, light situations
Aloe (*Aloe variegata*)	Sun, sandy soil; slow-growing succulent.
Fish-hook cactus (*Mammillaria* spp.)	Group of low-growing, spiny, pot-sized cacti; useful for hot, dry locations.
Hen and chickens (house leek) (*Sempervivum* spp.)	Sandy soil; light frequent watering.
Maternity plant (*Kalanchoe daigremontiana*)	Propagated from plantlets on leaf margins; keep slightly moist.
	Trailing or climbing plants
Artillery plant (*Pilea microphylla*)	Diffused light to shade; don't overwater; terrariums or hanging baskets.
Baby's tears (*Helxine soleirolii*)	Water from bottom; likes high humidity; good for terrariums.
English ivy (*Hedera helix*)	Tolerant of a wide range of soil and light conditions; can be used in terrariums.
Grape ivy (*Cissus rhombifolia*)	Pinch tips to encourage branching; needs diffused sunlight.
Nephthytis (*Syngonium podophyllum*)	Fast-growing; diffused light to shade.
Pothos (*Scindapsus aureus*)	Variegated creepers with waxy leaves; tolerant.
Spider plant (*Chlorophytum comosum*)	Easy to grow; best where new plantlets can trail.
Swedish ivy (*Plectranthus australis*)	Ground cover or hanging basket; will grow in water or soil.
Wandering Jew (*Tradescantia* and *Zebrina* spp.)	Diffused light; humid atmosphere; will grow in water.
Wax plant (*Hoya* spp.)	A group of waxy-leaved vines tolerant of indoor conditions.
	Plants that grow into small trees
Fiddle-leaf fig (*Ficus lyrata*)	Grows to 5 ft (1½ m) or more; quite tolerant of shade.
Kentia palm (*Howeia forsteriana*)	Said to be one of the easiest palms to grow; good light but not necessarily sun.
Norfolk Island pine (*Araucaria excelsa*)	Diffused light to full sun; beautiful specimen tree.
Rubber plant (*Ficus elastica*)	Diffused light; will grow to 10 ft (3 m); broad leaves should be washed with water.
Weeping fig (*Ficus benjamina*)	Grows 2–5 ft; small leaves with graceful branches.

around plants by placing their containers on gravel in trays containing water (Figure 10–2). The growth of ferns and other humidity-loving plants will be aided by a weekly sprinkling. Their containers can be placed in a bathtub or sink or on a waterproof surface and their leaves lightly sprinkled with a small hand sprayer such as is used for rinsing dishes or for spraying laundry starch or washing windows (Figure 10–3). Although most other houseplants are not as adversely affected by a dry atmosphere as ferns, many will benefit by having their foliage washed off a few times a year. Exceptions would be African violet, gloxinia, and other similar hairy-leaved plants, which are damaged by sprinkling with cold water.

Even better than gravel beds or sprinkling for increasing humidity is the installation of a home humidifier. This would benefit both plants and humans, especially during the winter when humidity is likely to be extremely low.

Most houseplants are sensitive to all sorts of atmospheric pollutants. The minute quantities of propane and other manufactured gases that escape in lighting stoves and other appliances make it very difficult to grow houseplants in a building where this kind of gas is being used. Cooking fumes and fumes from most common household cleaners are not noticeably damaging to plants in the amounts normally released in the home; however, fumes from certain adhesives and from special cleaners can be injurious if they become highly concentrated.

In industrial and congested urban areas pollution that is damaging to outdoor plants can seep indoors and also damage houseplants. In rural areas chemicals used for weed control on farms may injure houseplants. Both types of pollutant are lessened by air conditioning units that utilize air filters.

Figure 10–2 African violets in a tray of moistened perlite.

Figure 10–3 Spraying houseplants. Spraying occasionally with lukewarm water increases humidity; washes away dust; and helps control aphids, mealy bugs, thrips, and spider mites.

Managing the Growing Medium

Because the amount of growing medium is so restricted for plants growing indoors, it must be especially suitable for supplying them with support, water, and mineral elements. Soil and growing media were discussed in Chapter 5. Potted plants, of course, must receive adequate moisture; however, overwatering is much more of a problem than is underwatering. The grower of indoor plants must be especially careful when attempting to grow them in containers that do not have adequate drainage. Container-grown plants should be watered just enough to bring the soil to field capacity, never so much that water stands in the container. The soil should be permitted to become rather dry before they are watered again. If containers are of adequate size and if humidity is not too low, plants can often go a week between waterings. Container-grown plants should be fertilized approximately every four weeks with a commercially formulated or home-prepared fertilizer mixture. Cacti and some of the succulents will not require either as much water or as much fertilizer as do most other kinds of plants.

Insects and Diseases

The insects and insectlike pests that are most troublesome on houseplants are aphids, mealybugs, scale, spider mites, thrips, and whiteflies. The diseases most likely to infect houseplants are mildews, damping-off, nematodes, and viruses. All of these with their symptoms or the damage they are likely to cause are briefly described in Table 14–24 (Chapter 14). Ants, cutworms and other grubs, earthworms, millipedes, slugs, sowbugs, and springtails are sometimes troublesome in greenhouses but seldom require more than handpicking to remove them from houseplants. Fungus gnats, delicate, dark-grey, flylike insects about 1/8 inch (1/4 cm) long that are attracted to light and tend to swarm over windows, are sometimes a problem. Their larva, whitish 1/4 inch (1/2 cm) maggots, feed on roots and burrow into the crown. Injured plants make little growth, appear dull colored, and may lose leaves.

The best control of insect and disease pests of houseplants is to prevent infestation. Cut flowers and new plants should be examined to determine that they are free of pests before they are brought into the home. New plants should be isolated for three or four weeks before they are placed next to other plants. Using sterilized soil for potting will prevent the introduction of such soil-borne problems as root rot, damping-off, nematodes, slugs, centipedes, and earthworms.

It is important to examine houseplants frequently so that insects or disease infestations can be controlled while numbers are low. A few aphids, mealybugs, thrips, spider mites, or scale can be washed from one or two infested plants by spraying with a light spray of lukewarm water or by washing the plants with a soft cloth and soapy water made with two teaspoons of mild detergent in a gallon of water (10 ml in 4 liters). The washing may need to be repeated several times as eggs hatch or nymphs mature. It is also possible to handpick and kill larger insects if only a few plants are infested. A swab dipped in alcohol can be used to eradicate a minor infestation of mealybugs or aphids. Destroying severely infested plants may be the most logical solution if the plants are not too valuable.

Mildew is most easily controlled by reducing moisture. Moving an infected plant to a less humid location and keeping water from contacting the foliage will also help.

When more than a few plants have insects or diseases or where infestations are severe, chemical control will be necessary. Only formulations prepared for houseplants should be used. These will contain chemicals such as malathion, rotenone, and the pyrethrins, all of which are relatively nontoxic to people and animals. The formulations contained in spray cans are convenient when only a few plants are to be treated. Concentrated solutions or wettable powders from which growers can mix their own dips or sprays are available. These more concentrated formulations are also more dangerous to both plants and the applicator, so labels should, of course, be read and followed religiously.

Systemic insecticides are being used on potted plants in greenhouses and may soon be available for houseplants. These are applied to the soil in the pot, absorbed into the root system and translocated to all parts of the plant with the water and mineral nutrients in the xylem. These insecticides are often spectacularly effective because they control both sucking and chewing insects at all stages of growth.

Kinds of Plants for Indoor Gardening

A wide selection of traditional and unique plants are available for growing under varying indoor conditions. Growers may include selections from many different groups or specialize in certain special kinds, such as ferns or cacti and succulents.

Traditional Indoor Plants

A selected list of some of the easier-to-grow indoor plants is included in Table 10–1. This is by no means a comprehensive list, and many excellent houseplants are not included, but it does include plants that tolerate low light intensity and relatively warm night temperatures. Other plants, not listed, may do better if temperatures are cooler or a south window is available. For a more complete list of houseplants with photographs, descriptions, and directions for individual care, the reader should consult one of the references listed at the end of this chapter.

Unusual Indoor Plants

Cacti and Succulents. Cacti and succulents (a special group of plants with thick, drought-tolerant leaves) thrive with little care and often do well in the hot, dry interiors of homes (Figure 10–4). They should have a sunny window, if possible, but many will survive through a winter in a well-lighted room. If they cannot have a sunny window, they should be placed outside to rejuvenate during the summer. For plants growing without sun during the winter the transition should be gradual, just a few hours of sun each day until they become accustomed to the higher light intensity. They require a well-aerated, coarse growing medium and should not be watered or fertilized as often as most houseplants.

Plants from Bulbs and Corms. Several kinds of bulbous plants can be forced indoors. Amaryllis bulbs, for example, can be potted in January with the bulb above the soil. If the planted bulb is kept at room temperature, the flower stalk will begin to elongate almost at once and will produce large pink, red, or white lilylike flowers before the leaves are fully grown. The bulb should be permitted to continue growth after the flowers have faded. If the plant is kept growing through the next summer, its leaves will die down during the fall and it will produce another flower during the next winter or early spring. The amaryllis will bloom even if kept in a location away from a window or artificial light, but it should have full sun through the summer while its leaves are manufacturing carbohydrates.

Figure 10–4 Cacti as houseplants. These photos show several of the interesting types of cacti and succulents suitable for dry conditions. Cacti will tolerate environmental adversities other than drought, for instance, low light intensity.

April is the best time to plant gloxinia, although corms of this plant are available from nurseries during other times of the year. Gloxinia corms, which produce plants having beautiful velvety leaves and trumpet-shaped flowers, do best if grown in a north or east window. Gloxinias can also be grown from seed or propagated from leaf cuttings.

Tulips, narcissi, crocus, snowdrops—in fact most of the bulbous flowers that normally bloom in the spring—can be forced indoors during the winter. These bulbs require about three months of temperatures under 45°F (7°C) before they will commence growth. They can be stored in a container in peat moss or sawdust in the refrigerator or another cool location. Tulips and daffodils will bloom about two months after they are transferred from cool into warm temperatures.

Garden Annuals as Indoor Plants. The bloom period of several garden annuals, including ageratum, browallia, nicotiana, *Phlox drummondii*, and petunia, can be considerably extended if they are brought indoors just before the first fall frost. Plants that are just beginning to bloom will remain attractive longest; sometimes these will be available from seedlings self-sown during the summer. If self-sown seedlings are not available, seeds can be sown in early August to produce potted annuals for indoor bloom during autumn and early winter. These plants should be transplanted with a ball of soil around their roots so that, as much as possible, the roots are not disturbed. As most are sun-loving, they should be placed where there is ample light, in a south or west window if possible.

Fruits and Vegetables Indoors. Most vegetables require a fairly high light intensity, but if a sunny location is available, several kinds can be grown in containers. The 'Patio' cultivar of tomato, for instance, was developed especially for container growing. It produces medium-sized fruit on an upright bush and makes a very attractive addition to any south window. The 'Tiny Tim' tomato cultivar produces a profusion of cherry-type fruits on plants barely 6 inches (15 cm) high. Tomatoes will not yield well even in a south window from about December 15 to February 15 in most northern areas of the United States; thus the timing of their planting is important. Onions or scallions, butterhead and leaf lettuce, kale, and most kinds of herbs grow quite well in window containers.

Several products from the supermarket also can be grown into attractive and interesting indoor plants (Figure 10–5). Seeds from citrus fruits can be germinated quite easily and will grow into small trees. A sweet potato or part of a sweet potato placed in moist sand will produce slips that can be removed and planted in soil to grow into attractive and relatively long-lived vines. Avocado seeds placed in moist sand will frequently produce small trees. An unusual plant can be grown from a fresh pineapple. The top should be cut with about 1 inch (2½ cm) of fruit attached. Trim away the soft flesh of the fruit and permit the cut surface to air dry for a few days; then plant the cutting to the base of the lower leaves in potting soil. The plant should receive diffused light, and the soil should be permitted to dry between waterings.

Terrariums and Bottle Gardens

Terrariums and bottle gardens make interesting conversation pieces and attractive gifts. Bottle gardens can be made from any size bottles. Plants tend to grow quite well in lightly tinted bottles, but heavily tinted ones should not be used as the colored glass blocks the rays of light from entering. Planting in narrow-necked bottles can be accomplished in one of two ways. The bottle can be carefully cut near its base with a glass cutter and glued together with waterproof cement after it has been planted. A more common method is to use plants with leaves pliant enough to force through the bottle opening without breaking.

A layer of fine gravel (in amount, about one-third of the growing medium) is placed on the bottom of the bottle for drainage. On top of this is added enough good, properly moistened growing medium to fill the bottle about one-third full. Many growers mix a handful of charcoal in the growing medium to absorb plant-damaging gases that occasionally form. In order to prevent its sides from becoming soiled, the bottle should be

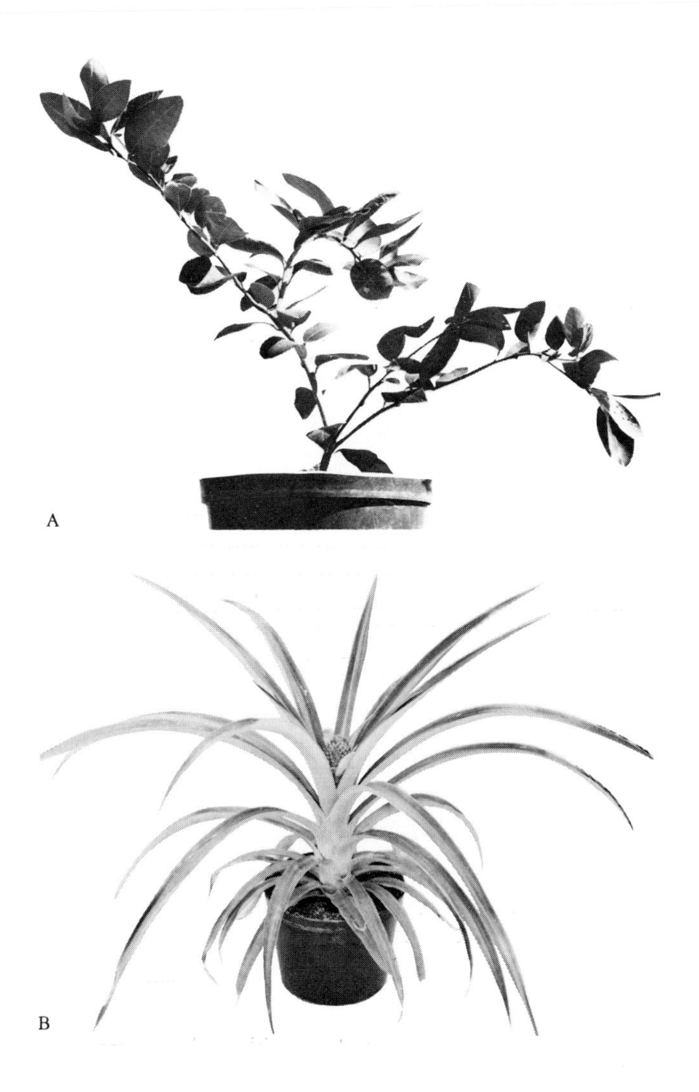

Figure 10–5 Three plants from supermarket products. The lemon tree (A) was started from seed. The pineapple (B) grew from the upper part of a pineapple fruit. The sweet potato vine (C) grew from shoots produced from a sweet potato root.

completely dry. Using a rolled sheet of paper to funnel the growing medium to the lower part of the container will also help keep the sides of the bottle clean.

Arrangement of plants should be planned on paper before any planting begins. Soil should be carefully washed from the roots of each plant. A pair of forceps long enough to reach the bottom of the container and thin enough to fit through the opening is needed to maneuver plants and soil. Other tools can be constructed from the stiff wire of an old coat hanger. A straight piece of wire can be used as a probe to dig holes and position plants and a piece with the end bent 90° and fashioned into a small loop makes a convenient tamper for firming the soil around plants (Figure 10–6).

After holes are scraped in the appropriate locations, each plant is gently pushed (with its roots down) through the container opening. The bottle can be tilted slightly so that the plant drops into the hole prepared for it. The growing medium is tamped firmly around the roots. The growing medium should be moistened to just below field

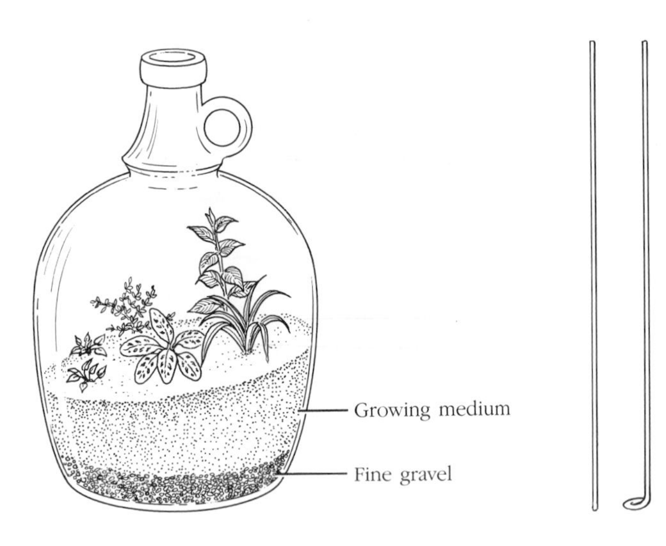

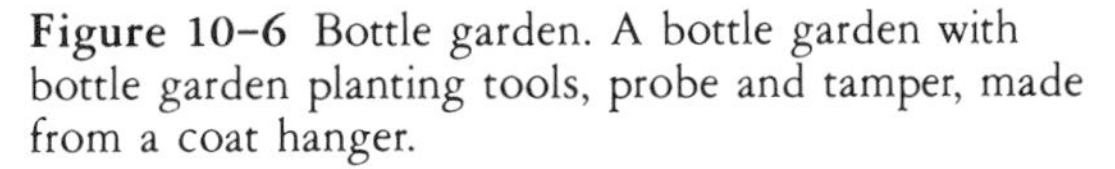

Figure 10–6 Bottle garden. A bottle garden with bottle garden planting tools, probe and tamper, made from a coat hanger.

Figure 10–7 Terrariums.

capacity. The water added can serve a dual purpose of irrigating and simultaneously washing off any particles that might be clinging to the inside walls of the container.

Terrariums are similar to bottle gardens, except the container has a somewhat wider opening. Both terrariums and bottle gardens can be covered with a sheet of glass or some plastic wrap. If disease-free plants are properly planted in a pasteurized growing medium and watered to just below field capacity, these kinds of container gardens should not require much attention for several months.

Miniaturization of Plants

The art of **bonsai** has become popular in this country in the last few years. This ancient art, which was developed in Japan, is the miniaturization of what would normally be large woody plants. Miniaturization is brought about primarily by judicious root and top pruning, and the plant is shaped by pruning, staking, or tying (Figure 10–8). In Japan bonsai plants several hundred years old and worth hundreds of dollars are prized and jealously guarded by the families who own them. Many of the Japanese bonsai were originally collected from rocky crevasses where they had been gnarled and twisted by years of adverse environment. Periodic pruning and shaping have kept them in their natural form.

Several students in one of my horticulture classes produced instant bonsai by transplanting small, twisted sagebrush plants into containers in their rooms (Figure 10–9). The addition of a piece of igneous rock containing a bit of lichen produces a very attractive miniature landscape. The person who makes a bonsai out of sagebrush would, of course, have to enjoy and not be allergic to the smell of sage. I suspect shrubs in other parts of the country would be just as suitable.

Plants with small leaves or needles are most appropriate for bonsai as small leaves or needles are more in scale with miniaturization. Most bonsai fanciers keep their plants out-of-doors in a greenhouse or in a sunny window during most of the year and bring them in only for special occasions. If a temperate zone plant is being used, it must receive its usual alternation of seasons, with enough cold to overcome its rest period.

Figure 10–8 Bonsai. The art of bonsai is a rewarding garden hobby that requires little space. This Scotch pine is about 15 years old.

Figure 10–9 Sagebrush bonsai.

Living Christmas Trees

Needle evergreens can be enjoyed as temporary houseplants. Each fall my family transplant a 2- or 3-foot Engelmann spruce from some land we own into a 2-gallon container, keep it indoors through the winter into the following summer, and then transplant it back to the out-of-doors (Figure 10–10). Because its rest period is never overcome, it produces little growth, but it remains green and attractive even in a dark corner where the original was brought to provide a little color. A plant so treated will not survive a second winter without cold to break its rest. Although we have used only Engelmann spruce, other needle evergreens would do as well if purchased from a nursery after their cold requirement had been satisfied. They could be used as Christmas trees and then be kept growing indoors until the following summer, when they could be added to the outdoor landscape.

Figure 10–10 Evergreens as houseplants. This small Engelmann spruce dug from the woods during late fall will provide indoor greenery throughout the winter.

Taking Care of Cut Flowers and Gift Plants

Flowers from the garden for an arrangement should, if possible, be gathered in the afternoon. The reason—that they will have more accumulated sugar at this time of day—may not be physiologically valid, but flowers cut in the afternoon do seem to have a slightly longer vase life than those cut in the morning. A sharp knife should be used so that the cut is clean. The stems of most herbaceous flowering plants should be placed in lukewarm (100° to 110°F; 38° to 43°C) water as soon after cutting as possible. If there is a delay in getting the cut stems into water, a second cut should be made so that the conducting tissues will not have sealed over when the stems are placed in water. The container with water and flowers should be placed in a cool location (40°F [4°C] if possible) for a few hours. Warm water is more easily taken into the vascular conducting channels than cold, and cool temperatures reduce transpiration and respiration so that the flower becomes fully distended with water and utilizes a minimum of its stored carbohydrates. Treatments such as the above that result in prolonging the life of plants or plant materials are called **conditioning.**

Flowers that produce a milky exudate from cut surfaces—poinsettias, poppies, and dahlias, for instance—should have the end of the stem plunged into boiling water for 30 seconds before they are placed in lukewarm water. Some woody-stemmed flowers are better able to take up water if the ends of their stems are crushed with a hammer. Containers, "frogs," and accessories should always be washed well with soap and water and rinsed in hot water to destroy decay-causing and xylem-plugging organisms before plant materials are placed in them.

Although flowers purchased from a florist will probably already have been conditioned, cutting off a few centimeters of their lower stems and placing their cut stems in lukewarm water in a cool room for a few hours may also increase their vase life somewhat. Keeping flower arrangements in a cool location when they are not being observed will prolong their vase life.

Homeowners often question how they should treat potted flowering plants such as chrysanthemums, cyclamens, poinsettias, or azaleas. Most of these plants were grown by the florist to be a temporary bouquet. Many of them require special growing conditions, and the cultivars sold by florists are often those that are not suitable for growing around the home. They should be placed where they can be viewed to the best advantage. They will last longer if, like cut flowers, they are placed in a cool location when not being observed. When their beauty has faded, the homeowner can with a clear conscience throw them in the garbage can.

Those who want to attempt to continue growing a gift plant after it has finished blooming should remove faded blooms and move the plant to an inconspicuous corner while it undergoes its after-bloom dormancy. If the weather is favorable, mums can be planted out-of-doors, though they may not be hardy nor programmed to bloom during the growing season. Many of the newer cultivars of poinsettia will continue to grow and produce new leaves, but they will probably not bloom again unless they are placed in a room that is dark for 14 or 15 hours through the night (see Chapter 7). These potted plants, of course, will need water just as any other potted plant does.

Figure 10–11 Container-grown rose. Miniature roses, such as 'Opal Jewel' shown here, are ideally suited to grow on small patios. They can be brought indoors for special occasions. (Courtesy USDA.)

Container Planting Outdoors

Container plants adapted to growing outdoors are popular with homeowners who have small yards and with apartment dwellers who have only a patio (Figures 10–11, 10–12). Containers that can be removed to a protected location during cold periods are also used to grow nonhardy plants in cold winter climates. Temperature,

Figure 10–12 Container planting outdoors. (Courtesy USDA.)

light, and humidity are not as limiting with outdoor as with indoor container-grown plants, but close attention to growing media, water, and fertilizer are necessary because root systems are restricted. As was mentioned in Chapter 7, roots are the plant organs most susceptible to freezing, and special insulation may be required to protect the root system when it is no longer insulated by the soil mantle of the earth.

Theoretically, any plant can be grown in a container; but from a practical standpoint trees larger than about 10 feet (3 m) in height are too bulky for most home container plantings. A major advantage of container gardens is that plantings can be rearranged. Containers for large shrubs and small trees should be on some kind of dolly if it is necessary to move them for winter protection or other reasons. In some cities it is possible to rent plants growing in containers. Renting is perhaps justified for special occasions, but it would be prohibitively expensive on a long-term basis.

Specialized Structures for Plant Growing

Shortly after glass came into common usage in the sixteenth and seventeenth centuries, which was about the time of the extensive explorations by European navigators, there developed in Europe an interest in exotic plants, and the collecting and growing of plants introduced from the Far East, Africa, and the Americas became the fascinating hobby of the European nobility. Perhaps the favorite of the many introduced exotic foods were oranges, and since oranges could not be grown out-of-doors in northern Europe, special glass houses were constructed in which orange trees were planted. These special buildings, known as *orangeries,* were the forerunner of our modern glass greenhouses.

For several centuries glass was the universal covering for greenhouses, hotbeds, and cold frames, which were the only kinds of plant-growing structures in common use. These were used primarily to produce flowers and vegetables (mostly tomatoes, cucumbers, and lettuce) on a commercial basis during the winter when these products could not be grown outdoors. The introduction of fiberglass and plastic materials and the development of fluorescent lights, coupled with the relative prosperity that has existed in this country during the last 30 years, have made possible the development of greenhouses and other plant-growing structures for the hobby grower and homeowner. Such structures range from a plastic or wax paper tent that fits over a single plant to protect it from early spring frosts, to a light fixture suspended over a few plants in a corner of the living room, to plant growth chambers and greenhouses where temperature, humidity, and sometimes light can be carefully controlled.

Purposes of Plant-Growing Structures

Gardeners use plant-growing structures for a variety of reasons. One of the major reasons is to grow transplants for earlier crops. Because produce prices are high early in the season, home-grown early vegetables save the most money. Growing from transplants also makes a home garden more profitable by spreading production over a longer season. It is also frequently impossible to purchase transplants of the vegetable and flower cultivars that are best adapted to a particular location, and growers can produce them early only if they grow them as transplants.

The hobby greenhouse provides the satisfaction of gardening, the beauty of flowers, and the flavor of fresh vegetables when outdoor conditions are unfavorable. The gardening season can be extended with structures other than greenhouses: Plastic and waxed paper covers are used to permit the seeding or transplanting of plants out-of-doors a month to six weeks earlier than they can be started without protection. Such an early start extends the season when vegetables, small fruits, and flowers are available from the garden.

Plant-growing structures can shade or protect certain plants during a part of their life. Many of the most beautiful houseplants, including coleus, philodendron, and most begonias, will not grow well in the brilliant sunlight that occurs during the summer in much of the United States. Nursery stock often needs protection from the intense rays of the sun. Plant-growing houses covered with enough lathe to filter out one-half to three-fourths of the sun's rays can be used to provide this protection for shade-loving plants otherwise adapted to the area. Various semiopaque compounds are sprayed on glass greenhouses to cut down the intense heat and provide some shade to plants grown in conventional greenhouses during the summer. A dark, fairly heavy nylon netting is sometimes hung over shade-loving plants grown through the summer, out-of-doors or in a conventional greenhouse.

Plant-growing structures also protect crops from early fall frosts. It is frequently possible to have tomatoes into late November in the northern United States by constructing a plastic tent that can be opened on sunny days over the tomato vines about the time the first frost is likely to occur. Geraniums, coleus, and other potted plants can be brought through the winter indoors. They can be moved into a greenhouse if one is available, but many growers winter them near a window in an unheated basement.

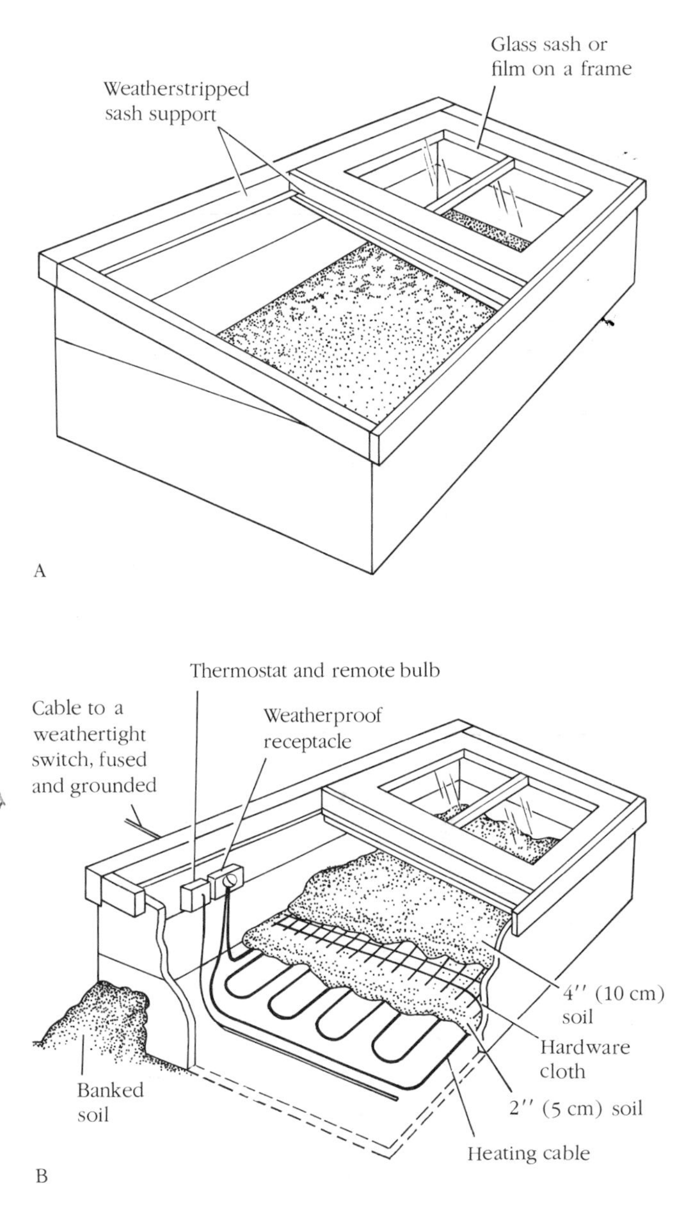

Figure 10–13 Cold frame (A) and layout of heating cable for hotbed (B). (Adapted from USDA.)

Types of Plant-Growing Structures

Plans, do-it-yourself kits, and preassembled plant-growing structures in a wide variety of kinds, shapes, and sizes made from a wide variety of materials are available to the home gardener. They can be categorized into four groups: hotbeds and cold frames, plant protectors, indoor growing structures, and greenhouses.

Hotbeds and Cold Frames. Typically, hotbeds and cold frames consist of four well-insulated sides and a transparent top cover that slopes south in order to capture the rays of the sun (Figure 10–13).

If no heat is added to the structure, it is called a **cold frame.** The main purpose for cold frames in cold winter areas is the hardening off of transplants before they are transplanted out-of-doors. In milder climates cold frames are used for holding plants through the winter. Most homeowners will wish to add heat to this frame; then the structure is called a **hotbed.**

The sides of the structure can be made from whatever material is available and convenient. Brick, poured concrete, and pumice blocks are sometimes used, but more commonly the structure will be made of wood. Where the structure is to be temporary, soil, straw, leaves, or sawdust are sometimes piled around the wooden frame to provide insulation. If fiberglass or other insulation is used in a more permanent structure, the insulation should be at least 4 inches (10 cm) thick.

Normally, the north wall of a hotbed or cold frame will be built to stand 18 to 20 inches (45 to 50 cm) above ground level. Side walls should slope toward the front at least 1 inch per foot (2½ cm each 30 cm) so that the front or south side will be 10 to 14 inches (25 to 35 cm) above the soil if a standard 3 x 6 foot (1 x 2 m) sash is to be used as a cover. With the standard sash the width of beds is 6 feet from the center of the front wall to the center of the back wall, providing approximately 5 feet 8 inches (2 m) inside growing space. The tops of the back and front walls should have the same slope as the sides, giving the sash 2 inches (5 cm) of firm supporting surface on top of each wall. With standard sash the distance between cross-supports will be 3 feet (1 m), and the length of the hotbed will be some multiple of 3 feet.

Hotbeds and cold frames should always be located on well-drained land that is free from depressions or danger of flooding during heavy rains. The south side of a building is an ideal location, because the building helps to trap some of the warmth from the sun's rays. Obviously the area should not be shaded by large trees or buildings. Plants growing in hotbeds or cold frames require frequent irrigation, so a source of water should be available nearby.

Hotbeds can be heated in many different ways. Steam, hot water, or hot air can be used if one of these sources of heat is available. The most convenient source of heat for small hotbeds if they are in an area where the cost of electricity is reasonable is an electric heating cable. Heating cables come in various lengths and wattage ratings. The 60-foot (18-m), 400 watt cable is common and adapted for use with ordinary service current of 110–120 volts. (A 400 watt cable will use about the same amount of current required by four 100 watt light bulbs.) USDA leaflet 445 recommends 10 to 12 watts of cable heating capacity for each square foot of bed (105–125 watts/m^2). One 60-foot cable is required for a 6 x 6 foot (4 m^2) bed and two 60-foot cables for a 6 x 12 (8 m^2) bed for late winter and early spring heating if fairly high temperatures are required. Where temperatures are extremely cold, three cables may be required to heat each 6 x 12 bed. Where only moderate heat is required, as for spring growing in the South, one 60-foot cable for a four-sash 6 x 12 foot bed may be sufficient.

If electricity is to be used, a hotbed must be well insulated and joints well fitted, electric cables should be buried in approximately 6 inches (15 cm) of growing media and, of course, a thermostat should be installed. All electrical connections should be watertight, and all wiring should be designed for outdoor use. The major advantage of an electrically heated hotbed is that the amount of heat can be automatically controlled by the use of thermostats, and thus extremes can be avoided.

Before the advent of electricity and the development of modern heating systems, manure was the common source of heat for hotbeds, and where manure is available, it can still be used. Temporary hotbeds can be constructed by placing a board frame on or around a flat pile of manure 12 to 24 inches (30 to 60 cm) in depth. Sash covering is placed on the frame, and about 5 to 6 inches (13 to 15 cm) of good soil is spread on top of the manure. The more usual practice for the construction of a manure-heated hotbed is to dig a pit beneath the frame of the hotbed. Where the temperature does not get below 12°F (−11°C) during the plant-growing period, a layer of manure 12 to 15 inches (30 to 38 cm) thick will be sufficient, provided the bed is well banked on the outside with soil or manure and some form of covering is used over the sash during periods of low temperature. The depth of the layer should be increased 1 inch for each degree of lower temperature. Thus, about 24 to 28 inches (60 to 70 cm) of manure should be used under conditions of 0°F (−18°C) temperature. The pit should be dug deep enough so that about 5 or 6 inches of screened garden loam can be spread evenly over the manure.

Directions for preparing the manure for use in a hotbed are from USDA Farmer's Bulletin 1743, which recommends that only good quality straw-bedded horse or mule manure be used for hotbeds; the directions for preparation are based on this kind of manure. I know of successful hotbeds being heated with cow and chicken manure, but the length of time the manure needs to be piled and turned and the time that elapses before seeds or plants can be placed in the hotbed will probably be somewhat different. The amount of straw or shavings mixed with the manure will also cause variation in the rapidity and amount of heating produced.

The manure should be placed along the side of the bed and turned occasionally during the period of 2 to 3 days until it begins to heat. If the manure becomes dry, moisture should be added as it is being turned. It is forked uniformly into the pit below the hotbed as soon as it has begun to heat uniformly. The temperature in manure-heated hotbeds is likely to rise as high as 90° or 100°F (32°or 38°C) during the first few days. Seeds should not be sown in the hotbeds until after the temperature has dropped to 85°F (30°C) or below.

Hotbeds are inexpensive to construct, fairly easy to maintain, and quite satisfactorily produce salad greens through the winter and bedding plants for spring transplants. They deserve more attention from home gardeners.

Figure 10–14 Plant protectors. In the experiment pictured, watermelon transplanted under plastic tubes in early April matured 40 days earlier than those direct-seeded at the usual planting time in late May.

Plant Protectors. Plant protectors are almost any kind of covering that can be placed over tender young seedlings or transplants to prevent freezing damage on cold nights during the early spring. During the early part of this century several companies manufactured plant covers of translucent reinforced paper attached to a heavy cardboard rim. The completed protector had approximately the shape of a Mexican sombrero. These structures were known collectively by the brand name of the most widely used protector—Hot-Kap. Hot-Kaps are still available in some areas. More recently polyethylene has replaced most of the other kinds of plant protectors (Figure 10–14). A square of polyethylene and two wire hoops placed crosswise are sometimes used to protect individual plants. Plastic structures for plant row protection are pictured in Figure 7–16 (Chapter 7).

Plastic can also be used to bring a nonhardy specimen shrub through a few extremely cold winter days. In many areas only the two or three subzero days each three to four winters prevent the growing of saucer magnolia, dwarf holly, pyracantha, rhododendron, or certain other choice shrubs or trees. If these plants are kept small by pruning, the homeowner can quickly erect a plastic protector around them whenever injuriously low temperatures threaten.

The potential for using polyethylene tent protectors is probably only beginning to be realized by home gardeners. In Greece commercial strawberries are being produced 35 to 40 days earlier than normal when the plants are covered in early February with a perforated plastic tent. In the Columbia Basin of Washington, watermelon have matured 40 days earlier than normal when transplants were placed under plastic tents in mid-April. In the northern United States, Alaska, and Canada, where the accumulation of heat units is insufficient for the production of such warm-season crops as melons, tomatoes, sweet potatoes, and eggplant, there would seem to be real promise for the use of large plastic tents with some sort of variable ventilation to prevent temperatures from rising too high on extremely bright days.

Indoor Growing Structures. The requirements for plant growth are available within the walls of most homes. Temperature, oxygen, and carbon dioxide are all a part of a normal home environment; a growing medium, water, and fertilizer can be easily supplied. There are some plants that do relatively well with the 250 to 400 foot-candles of light normally found in a living room.

Those requiring more light can be grown next to a south window. However, many plants do require extra light, and frequently there is no convenient place next to a south window where plant containers can be located. Thus the growing of plants with artificial light has become popular. Artificial light sometimes is also used to supplement winter sunshine in the hobby greenhouse.

Because of homeowner interest in artificial light for plant growing a number of companies are manufacturing various kinds of light tables (Figure 10–15). These will vary from a simple fluorescent fixture with a couple of tubes or a sunlamp with a plant holder beneath, to floor-to-ceiling light shelves, which have several layers of light fixtures, each fitted above a plant-growing tray. There are also fixtures with sliding glass doors, which aid in controlling humidity as well as temperature. Most of the more expensive units are fitted with a time clock, and some room dividers and other furniture are made up partly or wholly of plant-growing trays and fluorescent lights.

Homeowners can save considerable money by building their own light tables and further have the benefit of a unit custom built to fit into the space available. Custom-made light tables may vary from a fluorescent tube placed under an upper kitchen cabinet to light a few plants growing on the counter to very complex, carefully controlled units that may light a number of square feet (Figure 7–23, Chapter 7).

People contemplating purchasing or building an artificial plant-growing table should keep a few facts in mind: First of all, the intensity of light is much higher close to the lamp than at a distance from it. Longer fluorescent tubes are more efficient in light output than are shorter tubes. A reflector placed above the tubes will increase the intensity of light reaching the plants below. Furthermore, plants utilize about the same spectrum of irradiance as can be seen by the human eye, except that much of the light from the green area of the spectrum is reflected; thus most fixtures used for home lighting will also be satisfactory for plant growing. Fluorescent tubes are used more extensively than other light fixtures for growing indoor plants because they are relatively efficient in the use of electrical current and because they put out less heat in proportion to the intensity of light they develop.

Figure 10–15 A lighted plant-growing structure. (Courtesy Earth Way Products, Inc.)

Of all the fluorescent tubes that can be bought at local stores, the cool-white type is probably most satisfactory for plant growth. As mentioned in Chapter 7, specially formulated tubes with trade names such as Wide-Spectrum, Gro-Lux, and Plant Gro have been manufactured specifically for plant growing. Most of these provide maximum output in the blue and red ranges of the spectrum and a minimum of green light. Incandescent bulbs are also frequently used in conjunction with fluorescent tubes to provide light in the far-red area of the spectrum, but for growing most foliage plants fluorescent tubes alone are probably sufficient. Mercury vapor lamps, which provide high light intensity, have also proven satisfactory for plant growth under most conditions.

The question is often asked as to what kinds of plants are suitable for growing under artificial light. Unfortunately, if light tubes are placed too close together, the light impinging on one tube from those next to it is converted into heat. Therefore, without special fan and cooling equipment light tubes should be spaced no closer than 4 to 6 inches (10 to 15 cm) apart. As a consequence most reasonably priced artificial light structures do not provide intensities high enough to be optimum for most crop plants. Most ornamentals grown for their foliage, as well as gloxinia, African violet, some orchids, coleus, Christmas cactus, poinsettia, and begonias, all grow very well under artificial light. When seedling transplants are to be grown, the light should be placed within a foot of the container in which seedlings are being established in order to provide them with the highest intensity possible.

Some compensation for low intensity of artificial light-growing tables is accomplished by having the lights on for a longer period of time. Unless short-day plants are being grown, artificial lights are kept on for 16 hours. A time clock should be installed between the electrical outlet and the light source as a dark period is necessary for normal growth of most plants.

The Hobby Greenhouse

Greenhouses are becoming almost as popular with American homeowners as are swimming pools. They extend the gardening season so that gardening becomes a year-around hobby, and they provide the pleasure of flowers and growing plants throughout the year (Figure 10-16).

Planning the Greenhouse. The choice of greenhouse, its size, and the material used for its construction will depend on the location available, the plants to be grown, and the finances available for its construction and maintenance. Hobby greenhouses come in a variety of shapes and sizes. Small reach-in units that attach to and replace an existing window are the simplest. Where a south wall is available, the attached lean-to greenhouse may be the choice. Most common is the even-span or standard type, which most people visualize when they think of a greenhouse. The greenhouse must be located on a well-drained site where it will receive sun throughout the day, and the

Figure 10-16 Inside a hobby greenhouse. (Courtesy Lord & Burnham.)

heating bill be considerably reduced if the location is protected from stong winds.

The width of the greenhouse should be divisible into convenient work and traffic areas. Side benches should be kept 3 inches (8 cm) away from the side wall and should be no wider than the owner can reach across, 2 to 3 feet (up to 1 m). Center benches that can be serviced from both sides can be as wide as 6 feet (2 m). Walks will need to be at least 18 inches (45 cm) if one is to just squeeze between benches or wider if wheelbarrows are to be used. Normally the height at the eaves should be a minimum of 5 feet (1½ m), and the pitch of the roof should be between 25° and 30° for a glass house and somewhat steeper for one covered with fiberglass (Figure 10–17).

If a glass greenhouse is the choice, its size will often depend on the size of the units commercially available. Glass greenhouses can be constructed from a pattern, glass, sash bars, and other building materials. If construction is from basic materials, size should be a multiple of the basic structural unit. Usually, however, glass greenhouses are purchased as a complete unit, either designed to be set up by the company handling them or prefabricated with do-it-yourself instructions. These are available in a variety of models and sizes (Figure 10–18).

Hobby greenhouses are also constructed of fiberglass and plastics. Fiberglass does not break as easily as glass, but it has the disadvantage of being somewhat less transparent. The outer layer of some older types of fiberglass tends to weather, which makes them even less transparent. Fiberglass and glass are comparable in cost, but glass is more costly to install and more difficult for the do-it-yourself structure.

Greenhouses made of plastic are much less expensive than fiberglass or glass greenhouses, both because the plastic is less costly and because lightweight plastic requires only a minimum of framing materials. The major disadvantage of plastic greenhouses is that they must be recovered each year. Recovering a plastic greenhouse is not a difficult undertaking, but the fact that the greenhouse is likely to remain uncovered during a part of each summer precludes the growing of many kinds of plants that require a permanent greenhouse environment. Greenhouse frames are often covered with two layers of plastic to reduce the loss of heat.

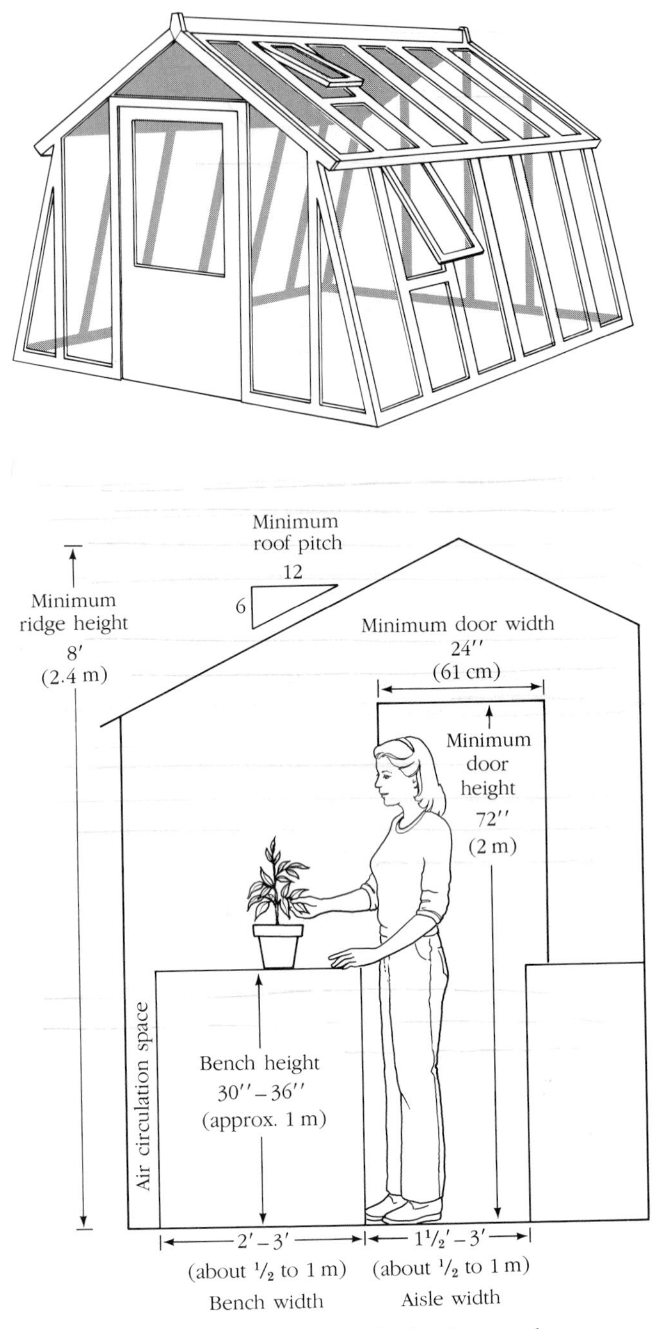

Figure 10–17 Specifications for hobby greenhouses. (Courtesy USDA.)

Figure 10–18 Various kinds of hobby greenhouses. (Courtesy Lord & Burnham.)

Greenhouse Plant-Growing Requirements. Steam, hot water, and hot air furnaces; electric cables; and various kinds of space heaters can be used to heat greenhouses (Figure 10–19). The capacity of the heating system will depend on the climate, the size of the greenhouse, and the temperature required by the plants to be produced. Often the home heating system will have enough capacity to heat a small greenhouse, or perhaps the home system with an electric or a gas- or oil-fired space heater for occasional cold weather emergencies will be sufficient to provide needed heat. During the winter night temperatures of 55° to 60°F (13° to 16°C) and day temperatures of 60° to 70°F (16° to 21°C) are sufficient for most plants. When space heaters are used, ample ventilation should be provided. If oxygen becomes limiting, incomplete combustion may result in the formation of carbon monoxide.

Ventilation and cooling are most often provided by vents and fans. Evaporative cooling systems, which in arid regions can reduce temperatures by up to 25°F (14°C) during warm weather, are available and not too expensive. Automatic controls are available for cooling systems. Greenhouses without an air-conditioning unit are shaded during the summer by application of a special shading compound to the outside of the house. The compound is washed off by fall rains.

Supplemental light is quite expensive and not necessary for most ornamental plants likely to be grown in the greenhouse. Temperature should be kept low during the winter when light is limited to maintain a balance of plant-growing factors. In northern areas vegetables may require supplemental light during mid-winter.

The moisture supply for greenhouse crops must be carefully monitored. During cool weather plants may not require water oftener than once or twice a week, whereas during hot weather water may be needed twice or three times a day. Soil and fertilizer recommendations for potted plants are discussed in Chapter 5.

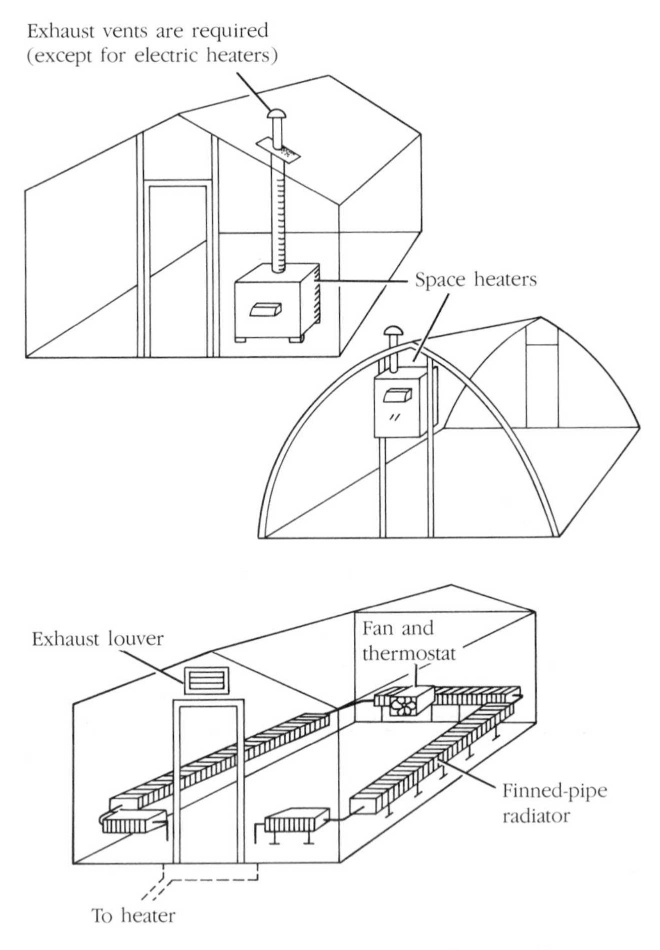

Figure 10–19 Some heating systems for greenhouses. (Courtesy USDA.)

Plants for Greenhouse Growing. Most vegetables, strawberries, and even some tropical and subtropical tree fruits have been grown in greenhouses; however, flowers and houseplants are by far the most popular hobby greenhouse products. Houseplants; specialty items such as orchids, carnivorous plants, and exotic ferns; and cuttings, seeds, corms, and bulbs of traditional greenhouse plants are available from nurseries that cater exclusively or partially to needs of greenhouse growers. Containers, pesticides, stakes, and supplies of various kinds are also available from these dealers.

Greenhouse Pest Control. Weeds in greenhouses should be controlled even if they are under benches where they do not crowd other plants. Weeds tend to harbor and nourish insects and insect eggs and often carry diseases that can spread to other plants. Greenhouse weeds should be pulled or hoed as herbicides pose too much of a threat to other plants in a closed environment.

The major insects and diseases that are likely to be a problem in greenhouses as well as controls for minor infestations are the same as those that affect houseplants discussed earlier in this chapter. Pesticides for major disease or insect problems are most frequently applied as sprays, and formulations for greenhouses are likely to be much more toxic to people than are formulations for houseplants. Whenever pesticides are applied in a greenhouse, it is important that signs listing materials used and times of application be posted and that doors be locked so that children or others don't inadvertently enter.

For greenhouses and large isolated plant-growing rooms, fumigation with smoke bombs is sometimes the most practical insect control. Smoke bombs contain an insecticide that is spewed into the room in the form of smoke when the wick is lighted. Their active ingredient is usually highly toxic, and people using them must be extremely careful. In fact, only a licensed applicator is permitted to apply many of these kinds of materials.

As is true with houseplants, strict sanitation and daily examination of plants for insects and diseases will minimize pest problems and the amount of pesticide required for the home greenhouse.

Selected References

Acme Engineering and Manufacturing Corp. *The Greenhouse Climate Control Handbook: Principles and Design Procedures.* Acme Engineering and Manufacturing Corp., Muskogee, OK. 1970.

Carlson, R. M., guest ed. *Gardening under Artificial Light.* Brooklyn Botanic Garden Record, (special printing of *Plants and Gardens,* vol. 26, no. 1). Brooklyn, NY. 1970.

Cathey, H. M., and Campbell, L. E. *Indoor Gardening.* USDA Home and Garden Bulletin 220. U.S. Government Printing Office, Washington, DC. 1978.

Crockett, J. U., and Editors of Time-Life. *Flowering House Plants.* Time Inc., New York. 1971.

Crockett, J. U., and Editors of Time-Life. *Foliage House Plants.* Time Inc., New York. 1972.

Crockett, J. U., guest ed. *Greenhouse Handbook for the Amateur.* Brooklyn Botanic Garden Record (special printing of *Plants and Gardens,* vol. 19, no. 2). Brooklyn, NY. 1963.

Edison Electric Institute. *Electric Gardening.* Edison Electric Institute, New York. 1970.

Lord & Burnham. *Your Gateway to Year-Around Gardening Pleasure.* Lord & Burnham Corp., Irvington, NY. 1971.

McGourty, Frederick, ed. *A Houseplant Primer.* Brooklyn Botanic Garden Record (special printing of *Plants and Gardens,* vol. 28, no. 3). Brooklyn, NY. 1972.

Perkins, H. O., guest ed. *House Plants.* Brooklyn Botanic Garden Record (special printing of *Plants and Gardens,* vol. 18, no. 3). Brooklyn, NY. 1962.

Pfahl, P. B., and Kalin, E. W. *American Style of Flower Arranging.* Prentice-Hall, Englewood Cliffs, NJ. 1980.

United States Department of Agriculture. *Building Hobby Greenhouses.* Agriculture Information Bulletin 357. U.S. Government Printing Office, Washington, DC. 1973.

United States Department of Agriculture. *Electric Heating of Hotbeds.* Leaflet 445. U.S. Government Printing Office, Washington, DC. 1956 (out of print).

11 The Ornamental Garden

The ornamental garden has always been an important form of artistic expression. Much has been written about ornamental gardens of the world, which are as varied in design and purpose as are the people who have created them and the landforms on which they were created. This text is concerned primarily with gardens around the home.

The use of plants and inanimate materials to enhance the utility and beauty of the area around a home is called home **landscaping.** There may be several different gardens along with nongarden utilitarian areas within the landscape (Figure 11–1). What is important is that their combination have beauty and harmony as well as practicality.

Figure 11–1 Lush home landscaping, entrance planting.

Figure 11–2 Placement of trees. This tree was planted too close to the house; despite a belated effort to reduce its size, the tree appears to be pushing the house from its foundation.

The Home Landscape

Not uncommon are the home that is almost being pushed from its foundation by the "cute little sapling" that someone planted next to it 20 years ago and the living room that must be artificially lighted all day because three gorgeous rhododendrons exclude light from its windows (Figure 11–2). There are many homes with an outdoor eating area that is seldom used because food, table service, and all the incidentals have to be carried from the kitchen, down a flight of stairs, through the basement, and across 100 yards of back lawn. Only by careful planning can these kinds of mistakes be avoided.

The outdoor area around a home, whether it is a 10-acre estate or a porch with a few planter boxes, can be just as important to the enjoyment and well-being of a family as the indoor area. A house is not a home until it is landscaped in a manner that pleases the homeowner and provides for the outdoor needs and comforts of the family. Landscaping is a functional part of the home, and it should be planned at the time the house is planned.

Considerations During Construction

The family fortunate enough to build their own home should plan for the use of outdoor as well as indoor areas. First and foremost, the house should be planned to fit the lot. Enjoyable views should be preserved and enhanced or created with landscaping if none presently exist. The house should be placed on the lot so that a maximum amount of yard space can be used for family activities (Figure 11–3). This usually means building as

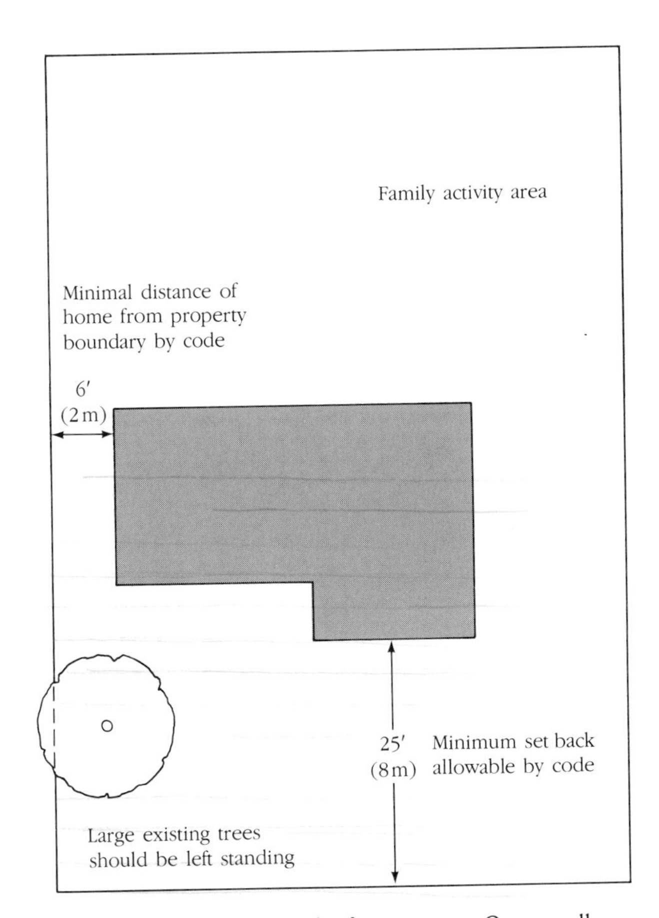

Figure 11–3 Allotting the landscape space. On a small urban or suburban lot it is generally desirable to place the home as near to one side and the front of the lot as zoning restrictions will allow. This placement permits maximum family activity area.

Figure 11–4 Using an existing tree. The tree allowed to remain on the lot of this newly constructed home is already an asset to the landscape.

close to the street and as near to one edge of the property as city codes will allow. Such placement provides maximum usable area with privacy behind the house and, with most lots, some usable space at one side. For suggestions on grading, topsoil, and drainage, see Chapter 5.

If at all possible, large trees and shrubs already on the lot should be saved (Figure 11–4). These plants grow slowly, and the pleasure they will provide is difficult to measure. From a purely economic standpoint a good, large tree will often add hundreds, sometimes thousands, of dollars to the value of a new home; for example, 'Live Oak' trees growing along St. Charles Avenue in New Orleans are insured for $25,000 to $50,000 each!

Excavation and land leveling must be carefully done if existing plants are to be saved. If the soil level around the tree is to be raised, provision for aeration of the roots must be provided. This is usually done by installing drain tile and leaving an area a few feet in diameter next to the trunk unfilled. If soil level is to be lowered, a wall to protect the soil around tree roots will be necessary. (See "Landscape Construction," Chapter 14.)

Economic and Sequential Considerations in Landscaping

Most people want a landscape that provides maximum use and beauty at minimum cost and upkeep. It is not wise to sacrifice quality to save a little money. It is estimated the cost of landscaping should be 10 to 30% of the cost of the home. Much of this cost is in labor, an expense that can be saved if owners are able to do the work themselves. Having professional assistance in planning the landscape is, of course, highly desirable and often not very expensive. Large landscape offices are sometimes reluctant to work on small lots, but in many communities part-time or full-time independent professionals may have offices where they consult with homeowners. Large nurseries employ personnel experienced in the art of landscaping to provide assistance for those who purchase plant materials from the nursery. In some urban areas the county extension staff has one or more individuals who assist with home landscape problems, and there are numerous government bulletins, commercially published books, and other publications designed to help homeowners who must do their own planning.

In spite of its desirability, complete landscaping immediately after a home is built is often not possible. A young family is likely to build as soon as they are able to scrape together enough money for a down payment on the construction, a situation that leaves little cash for landscaping. However, even if there isn't money to complete it, the landscape should be planned when the house is planned. The various elements that go to make up the landscape can then be budgeted over a period of years.

The first essential is probably a front sidewalk and lawn so family and visitors are not constantly tracking mud into the house. The owner may also wish to plant a few trees. Purchasing the major trees as inexpensive, small whips the first year will provide the same results as purchasing much larger trees during succeeding years and in the long run will save considerable money and planting effort. A few dollars worth of annual flower seed will provide color and filler until the major shrubs and trees can be planted or grow larger. In many areas it may be necessary at least to gravel the driveway before the first winter (Figure 11–5).

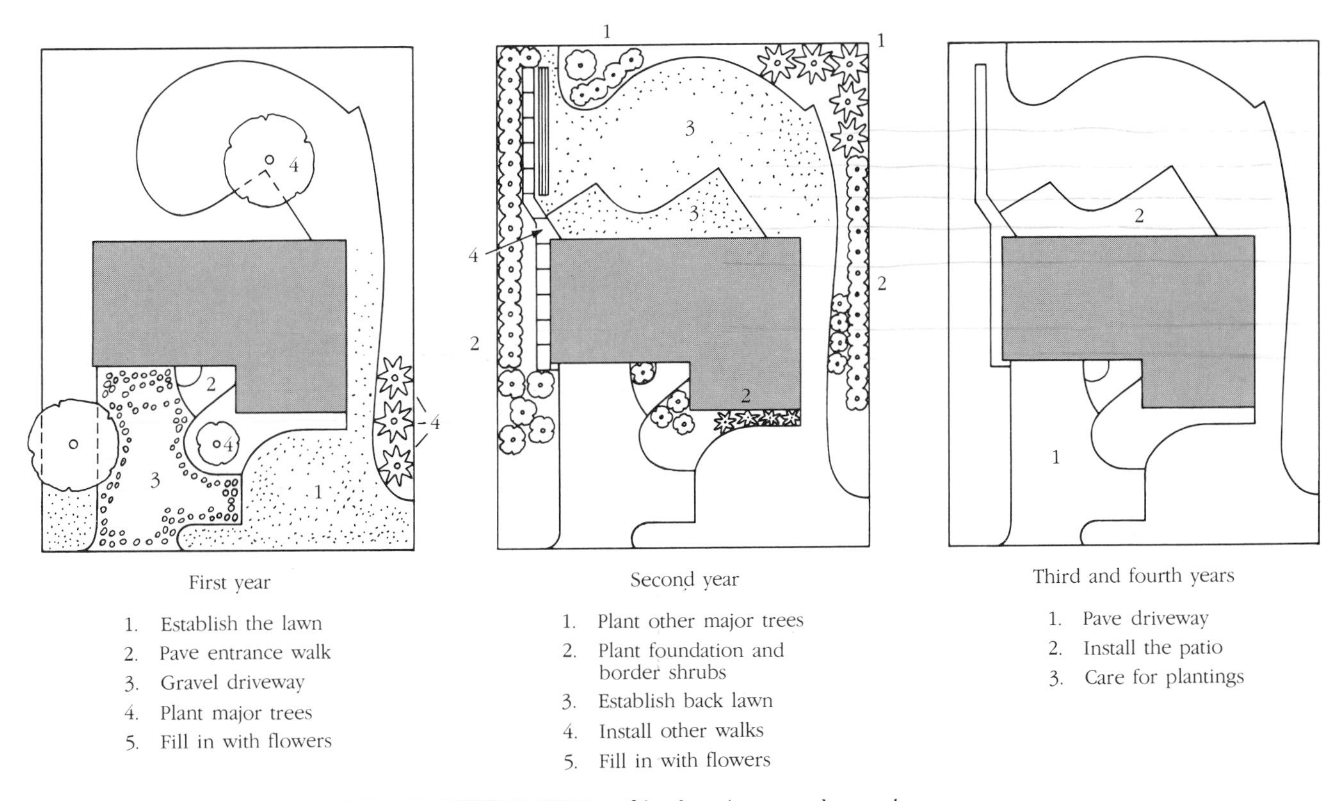

Figure 11–5 Sequence of landscaping spread over three years.

In the second year the back lawn and the rest of the major trees with some foundation shrubs can be planted. Retaining walls and rock gardens may not cost very much if the family can furnish the labor. If money is scarce, it may be necessary to delay until the third or fourth years the building of benches, fountains, and pools or the purchasing of shrubs for specimen and border plantings.

Many, perhaps most, individuals will not have the privilege of building their own home. If they do become homeowners, they will purchase either a newly built tract home or an older home that already has some landscaping. A carefully planned landscape is probably even more important for a tract home than for an owner-built home, because the landscape can provide uniqueness for what would otherwise be one of a monotonous row of look-alike houses. Purchasers of tract homes will not have the opportunity to plan for the coordination of outdoor-indoor areas, but they should make certain that the property has the potential to provide the outdoor amenities essential to the lifestyle the family desires. The topsoil left around a tract home is frequently poor. For soil improvement suggestions see Chapter 5.

Occasionally an older home will have a well-planned landscape that fits the needs and delights the aesthetic senses of the purchaser; more often minor or major renovation will be necessary. Seldom is it necessary to relandscape completely. If the initial landscape plan was sound, judicious pruning and trimming and a little fertilizer can do wonders. Even if the new owners decide that the landscape is hopeless, they may find that careful planning will enable them to retain many of the existing trees and larger shrubs.

Planning the Landscape

Planning the landscaping of a home can be divided into three parts: (1) choosing design elements or materials; (2) determining limitations or considerations that make your landscape different from all others; and (3) actually planning on paper the complete landscape around the home.

Design Elements. Probably the most important design element is that of cubic space. All of us are familiar with the number of square feet it requires to place a chair or a kitchen cabinet or for convenient movement from area to area in our homes, but we sometimes must be reminded that the cubic space that human beings move about in and occupy as they perform the acts of daily living is as important to emotional well-being as floor space is to furniture placement or to traffic patterns. For example, the ceiling of a room is usually higher than necessary for a human to squeeze into because architectural builders know that it takes more space than that occupied by a person's body to provide adequate ventilation and the feeling of atmosphere in the room. Outdoors, ceilings are not of so much concern because the ceiling is the sky or in some cases the canopy of tree branches overhead. Nevertheless, the cubic space designed into the garden should fit the dimensions of the body, including the eye, and should provide for maximum utility.

Land **contour,** or the shape of the land, whether it is a hill, a valley, or a perfectly level site, is an important design element. When structures are built to accommodate a family's needs, the land around those structures must also be sculptured to fit the particular needs. A sloping site may be well adapted to an artistic design, but it may require considerable alteration if the family needs a level game or play area.

Plants, the element that some people think to be the only element in landscaping, are not to be considered lightly. The plants used outside the house, inside, and for the transition in between are most certainly the element that gives character to the landscape scheme. However, plants are a building material, and unless the landscape has been designed, even the most beautiful plants may not make the home grounds attractive or usable.

Rock adds interest and variety and often helps to make the landscape appear natural. Not only are the broad surfaces of rock used for walls and patios, but rock provides texture and color when it is used as a gravel path or drive. Rock in walls is used to retain soil and in rock gardens to simulate a natural rock outcrop. Plants that require deep, cool, moist rock pockets filled with broken stone and porous decayed vegetable matter are grown in rock gardens. Still finer particles of rock mixed with humus become the soil that is essential for the growth of plants.

Water as a landscaping element is used for contrast, variety, movement, and sound. Consider the beauty of the reflective surfaces of a pool or lake, the movement and sound of a fountain or a brook, or the enjoyment of swimming in a backyard pool. Water that falls as rain or that is used to irrigate or to supply nourishment to plants is a design factor that must not be overlooked. Water in the atmosphere or relative humidity may limit or enhance certain types of garden development.

A long list of manufactured building materials used as design elements, including brick, dimensioned lumber, concrete block, pipe, plastic, glass, canvas, fiberglass, and other fabrics, add much to the convenience and comfort of the landscape.

Limitations and Considerations. Most of us could not afford and would not want a "Renaissance estate." Such gardens were designed to satisfy the desires of a special class of people who had relatively large incomes and the leisure time to enjoy them. Today, gardens should be designed to satisfy the physical and emotional needs and fit the pocketbook of the person or family using them.

An athletic family may need a volleyball court, a swimming pool, or a games area. Another family might want an area set aside for the cultivation of vegetables. Others may want facilities for eating outdoors or a secluded place in which to sit and read. Such needs should be seriously considered and planned for (Figure 11–6).

The major physical limitation in gardens of today is the size and shape of the property, which is often determined by the amount of money the family can spend for a living site as well as by the neighborhood they choose. A very small lot in an exclusive neighborhood may cost more than several acres in a less prestigious area. One should not, however, consider only the price quoted by the realtor in judging the relative cost of two lots. Public utilities, such as paved streets, sidewalks, storm sewers, sewer lines, and water mains; as well as cost of commuting to work, school, church, and shopping centers are all important considerations in the expense of a homesite. A small lot with parks and playgrounds nearby may be more desirable than a larger lot so far from these amenities that a playground and picnic area must be provided in the yard. Regardless of cost, the lot—that piece of land with definite geometrical borders imposed by civilization—is the factor that, in many cases, decides the kind of house that is built, and it always dictates landscape patterns.

The orientation of the house is important for comfort and convenience. With many homes the location of the lot will determine house orientation. Wherever possible, the house should be placed so that maximum sunlight reaches all sides and certain rooms at certain times of the day. Sunny kitchens and living areas, especially where the climate is cool, are more important than sunny bedrooms. Those areas around the home that are always shaded are a problem to landscape because few plants thrive in perpetual shade. If the house is oriented at a 45° angle to the north, it will have the smallest amount of perpetually shaded area. Windows and glass doors of the living areas should, if possible, face the garden or a scenic view rather than a busy street. Where wind velocity is high, doorways and large windows should face away from the direction of prevailing wind or be protected from its force by a wall or planting.

The style or type of architecture is another definite consideration. A two-story Cape Cod house requires a different landscape than a low, ranch-style house; a box style, different than a cottage.

Climatic influence is one of the most important considerations in landscaping. House orientation is based on climate. Prevailing winds must be utilized and controlled for pleasant living both inside and outside. The factor of most concern as far as the livability of plants is concerned is temperature. Cold temperature during the winter determines which plants survive in the landscape, and spring frosts often damage early blooms or limit early planting of annuals. High summer temperatures in some areas preclude the utilization of many fine cool-climate species and increase the need for shade-providing trees. It is also important that the landscape be attractive during the entire year, not just in summer (Figure 11–7).

Figure 11–6 Planning for the family's needs. An area for outdoor eating need not be large, but it should be close to the kitchen.

Figure 11–7 The winter landscape. The landscape should be attractive during all seasons.

Figure 11–8 Framing a view. Note how this view is framed and limited by the foreground trees.

Figure 11–9 Creating a view. The owners of this property wanted a day and evening garden view yet did not want to entirely hide the rolling landscape behind their home. This semiformal low, lighted garden is their solution.

Views are an important consideration. Some people buy a lot with a view at a premium price and maintain long lines of utilities and roads in order to have that view. If your lot has a good view, take advantage of it by framing it with trees and locating outdoor living areas so that you can use it (Figure 11–8). If you have a view that is undesirable, screen it with a fence or a living plant screen. Views can be created within the confines of the property with imaginative design and placement of gardens (Figure 11–9).

Maintenance is important. An elaborate place poorly maintained is less attractive than a well-maintained place with little landscape. Most people have a minimum of time to devote to the art of gardening and hiring a gardener is expensive, so a simple, easy-to-maintain landscape is the most satisfactory solution for most of us.

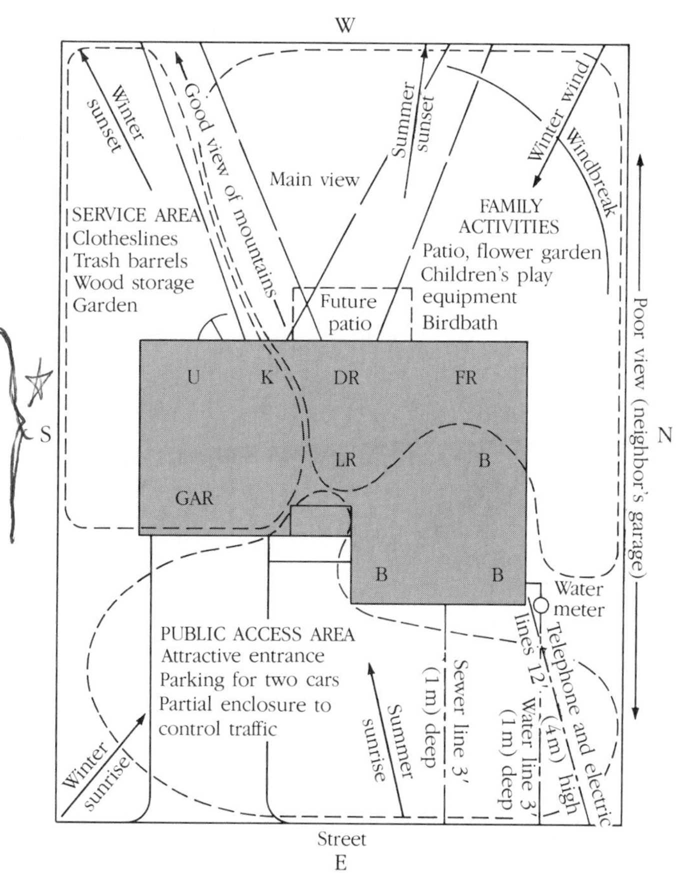

Figure 11–10 A completed analysis plan. (Courtesy E. B. Adams, from *Homescaping,* Intermountain Regional Publication 4. Extension Service, University of Wyoming, Laramie, WY. Reprinted and/or revised periodically.)

The Landscape Plan

In order to develop a coordinated and pleasing landscape, a person must make a guide or plan. In fact, several rough sketches and several preliminary plans will generally be needed. It is easiest for most individuals inexperienced in the art of drafting to use a large piece of graph paper to show to scale the yard area and the location of buildings and plants that are to be saved. The initial drawing should be constructed so that a square on the graph paper is equal to an appropriate number of square feet of yard space. The graph paper plan can be traced to provide the basic house-yard relationship for the analysis and other preliminary plans. The graph paper or a neater traced copy can be used for the finished plan.

Analysis Plan. The **analysis plan** (see Figure 11–10) should include views to be preserved or enhanced; unsightly views or objects that need screening; the general direction of wind during various seasons; the direction of summer and winter sun; and the locations of utility poles, overhead lines, underground cables, sewer pipes, and water pipes. It should also include general locations of rooms within the house. The kitchen, laundry, and garage are service areas; the entrance way or room where guests first enter the home is the public area; and the family room, dining room, and living room are areas of family activity or privacy. Logically these indoor divisions should extend to the outdoors, and the next step is to divide the yard into public, service, and family activities or private areas.

Each of these areas should be considered separately, and several trial sketches showing walks, drives, walls, plantings, and other design elements should be drawn. While making these sketches, the designer should keep

Figure 11–11 Landscaping the public area. A mountain ash (left) and a beech (right) help to frame this home.

note of the analysis plan so that sun and wind effects are planned for, tall-growing trees are not planted under utility lines or walks placed directly over water or sewer pipes, and willows or similar trees are not planted where their wide-ranging roots will enter and plug sewer lines. Separate sketches drawn to a large scale may be necessary for detailed explanation of such elements as rock gardens, planter boxes, pools, and walls.

The Public Area. The public area, or front yard with its drive, walks, lawn, and trees, should be arranged to serve the purpose of attractively displaying the home to the owner and to those who pass by. The public area is traditionally landscaped with a broad expanse of lawn in front of the house, a couple of trees and/or some large shrubs along the sides of the yard to frame the house, and some foundation shrubs to help tie the house to its surroundings (Figure 11–11). If the house is balanced and appears to have approximately the same weight on either side, the framing trees should be about equal in size. If the house has unequal balance, a larger tree should be planted on the smaller-appearing side. A large existing tree that heads above the home is a desirable asset even if it is directly in front of the house. Large trees normally should not be planted in front of a home, however, because they hide so much of it for such a long period while they are growing.

Foundation shrubs should be planted at the corners of the house and on both sides of the doorway. If the house has a wide front, a cluster of shrubs at a few other strategic locations and/or a small flowering tree 10–12 feet (3–4 m) from the foundation and a few feet toward the wide side of the house from its entrance will help tie the house to its surroundings. The foundations of most modern homes are not so unattractive that they must be completely hidden with shrubbery. As a general rule, foundation shrubs look best in clusters of two or more rather than as single specimens (Figure 11–12). They should be planted far enough from the foundation so that they do not crowd against it when they are mature.

Although in North America front yards have traditionally been planted to lawns, there are other attractive ways of landscaping the public area. In England the area in front of the home is frequently planted to a flower gar-

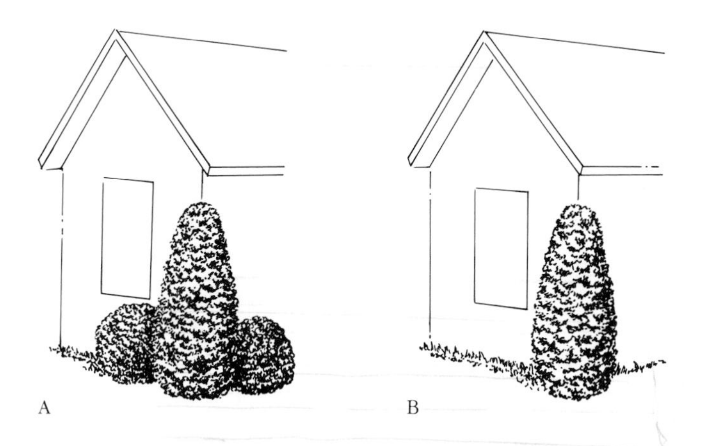

Figure 11–12 Foundation shrubs. If there is room, a cluster of two or more shrubs (A) is usually more attractive than a single specimen (B) as a foundation planting.

Figure 11–13 Alternatives to a lawn. (A) Much of the space usually planted to lawn is covered with cedar bark underlain with 4 mil polyethylene. (B) The public area here is planted entirely to drought-tolerant shrubs.

den. In the American Southwest, where water is scarce, cacti gardens often occupy that area. With impending energy, water, and land shortages more and more homeowners throughout the United States and Canada are using something other than grass for the public area (Figure 11–13). For some homes a semicircular driveway with an island of rock and low-growing shrubs is the practical solution. Where the yard is sloping, a rock garden planted to native or drought-tolerant, slow-growing shrubs may offer a low-maintenance alternative. Low-growing shrubs planted through a mulch of black plastic covered with river rock, cedar bark, or cedar rounds is another way of keeping down maintenance labor and cost. In some situations the front yard may be covered with a masonry or asphalt paving and decorated with border plantings, raised bed plantings, and/or potted plants. Such a yard may be left open for public view, or it can be screened with a fence, wall, or hedge to provide a family living area.

Regardless of the choice of landscaping, the public area planting should enhance the appearance of the home, should not interfere with convenience, and should be easy to maintain. Perhaps the most common mistakes in landscaping the public area are selecting foundation shrubs that grow too large, thus hiding the home and/or covering the windows, and planting too much. It is difficult to picture shrubs at their mature size when all one has to look at are small specimens from the nursery; a planting plan with shrubs and trees scaled to mature size helps to overcome the temptation to overplant.

The Service Area. The service area should be located convenient to the kitchen and driveway and should be screened from the public and private areas by plantings or fences, as it is here that clotheslines, wood pile, garbage cans, and other such objects of domestic need are located. The service area should be as compact and well planned as the modern kitchen to increase its convenience and usefulness. Vegetable and cut-flower gardens are often included in the service area; their presence or absence and size are determined by the time and effort the owner wishes to spend in caring for them.

The Private Area. The private area, or the area for outdoor living, is usually at the rear of the lot and should be connected with the living side of the house. It should be screened from public view. Privacy can be secured with informal shrub groups or formal hedges, trellises, fences, or walls. Outdoor patios or living terraces, outdoor fireplaces, space for athletic events, pools, fountains, flower gardens, shrub borders, and shade trees are located in this area. Fences, walls, walks, statuary, and other inanimate elements can be important parts of the private area.

Walks and Drives. After the use areas are identified, they must be interconnected with a circulatory system of walks, drives, paths, patios, or terraces. Walks and drives should be direct and useful, durable, and built on good foundations, with sufficient slope for drainage. Walks should be at least 2 feet (⅔ m) wide for each person likely to walk abreast. Those surfaces that are constantly used, such as entrance walks and patios, should be surfaced with concrete, brick, flagstone, or other durable material; paths used only occasionally may be surfaced with stepping stones, cedar rounds, rolled gravel or thick turf (Figure 11–14). Drives should be built for service. They should be easily accessible from the street, have convenient turns, include adequate off-street parking, be built at least 10 feet (3 m) wide with a crown and gutter, and

Figure 11–14 Surfaced walks. Cedar rounds (A), clam shells (B), bricks (C), and river rock (D) are some of the varied materials used for walks and patios.

be surfaced with gravel, asphalt, or concrete. (See "Landscape Construction," Chapter 14.)

Suggestions for landscaping sloping sites, building terraces, and constructing retaining walls can be found in Chapters 5 and 14.

Locating the Plantings. Following the designation of walks, drives, patios, retaining walls, and other hard-surfaced areas, the location of large trees should be determined. The character of the landscape picture is dependent on trees. Trees are used to frame views, for shade, for windbreaks, for background, and as specimen plants.

After trees are located, plant walls or borders can be established. A location from which the outdoor living area will be viewed frequently is selected as the vantage point. This may be a sliding door or window opening onto the outdoor living area, or it may be the patio that serves as the transition between indoor and outdoor living areas. Next an axis line should be established between the vantage point and a focal point at the far end of the outdoor living area. The focal point may be a natural view beyond the confines of the yard, or it may be a garden house, garden statue, specimen tree or shrub, or other special object within the yard to which the eye is drawn. The most effective location for the plant walls will be at about the visual periphery as one looks from the vantage point along the major axis toward the focal object. A secondary axis is naturally located at a right angle to the first at the widest point of clear visual perception, which will usually be about two-thirds to three-fourths of the distance toward the focal point (Figure 11–15).

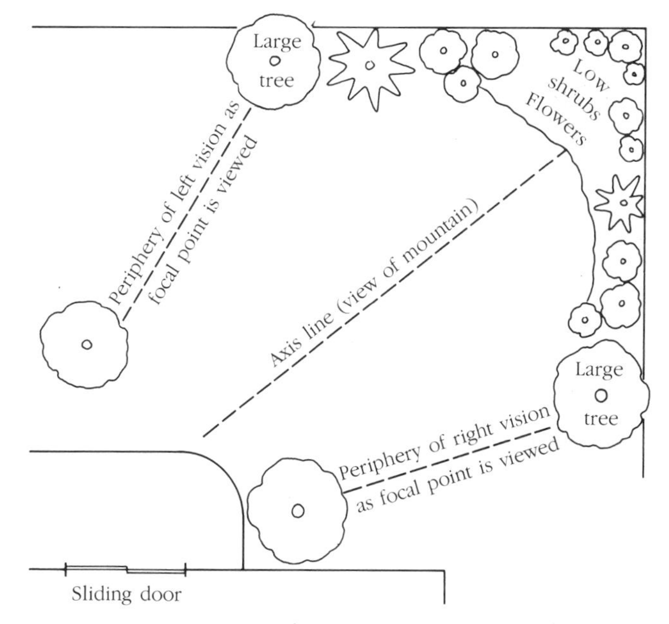

Figure 11–15 Method of determining visual walls for framing and enhancing a view.

The outdoor wall can be a solid mass of several rows of shrubs and trees. If open space is desired, the wall can be limited to a couple of trees planted relatively close to the vantage point that serve the dual purpose of providing shade and forming the focal point. In this case a fence with a few espaliered shrubs or small trees or a formal hedge may serve as the ends of the secondary axis and the background of the focal area as well as provide privacy and conserve maximum area for athletics or other purposes.

Shrubs should be located on the plan according to size and height before their foliage, texture, color, or flowering habit are considered. With the traditional naturalistic planting masses of shrubs and trees grown as plant walls are referred to as **border masses.** Border masses for naturalistic plantings should be planted in staggered groups and be at least two rows wide. Generally they are planted in groups of three or more of a single kind. An interesting skyline is created with rhythm and repetition in plant form.

Once the basic size and shape of plants are established, specific kinds can be identified. Nursery catalogues, garden books, local nurseries, and extension publications will help in selecting the best plant for each location (see also the plants listed on tables in this text).

The Finished Plan. After the design elements have been located on the preliminary sketches, a finished plan should be constructed. Plants and other design elements are numbered on the landscape plan and specifically identified on an accompanying key (see Figure 11–16).

Suggested symbols to be used in constructing the home landscape plan are shown in Figure 11–17.

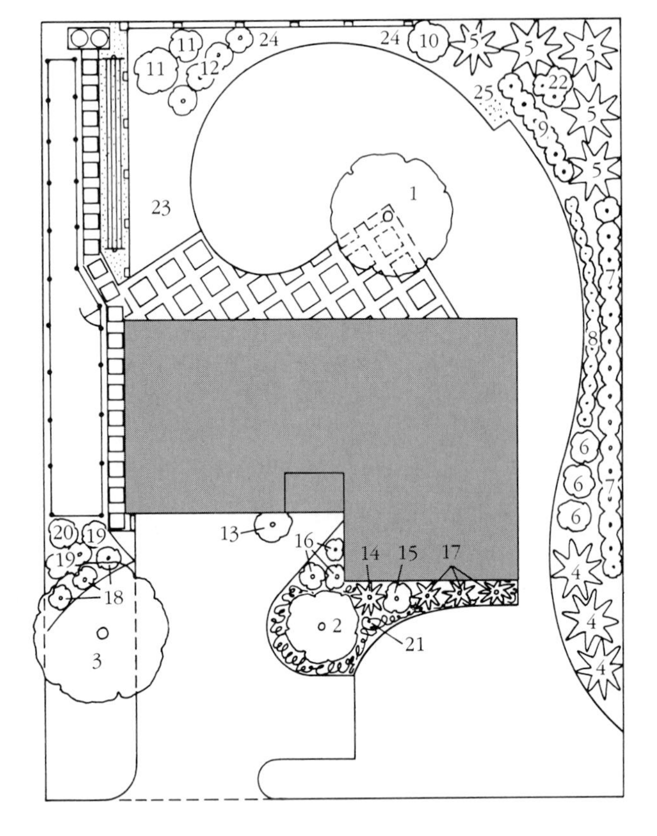

Planting key

1. Thornless honeylocust
2. Staghorn sumac
3. Cutleaf weeping birch
4. Colorado blue spruce (3)
5. Pondorosa pine (5)
6. Dwarf apple trees (3)
7. Lilac hedge (17)
8. Blue mist (Caryopteris) (16)
9. Spreading cotoneaster (6)
10. Snowball bush
11. Russian olive (2)
12. Redleaf barberry (5)
13. Mugo pine
14. Pfitzer juniper
15. Mentor barberry
16. Silver sage (3)
17. Tamarisk juniper (3)
18. Dwarf winged euonymous (3)
19. Apple serviceberry (3)
20. Caragana
21. Ground cover (Kinnikinnick)
22. Washington hawthorne
23. Salad garden
24. Annual and perennial flowers
25. Peony bed

Figure 11-16 A completed planting plan keyed for plant identification. (Courtesy E. B. Adams, from *Homescaping,* Intermountain Regional Publication 4. Extension Service, University of Wyoming, Laramie, WY. Reprinted and/or revised periodically.)

North

Slope down

View

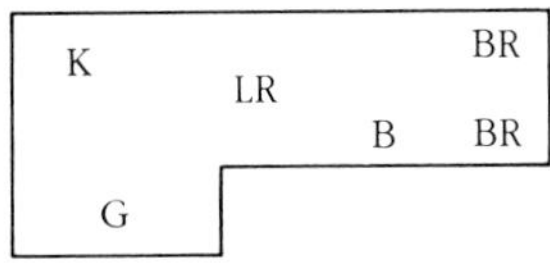

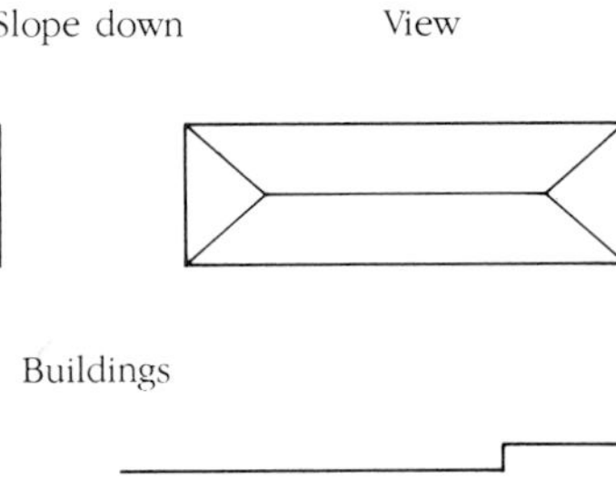

Buildings

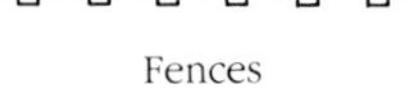

Fences

Drives

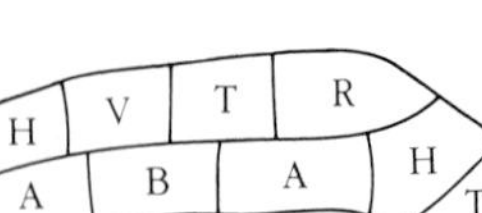

Flowers

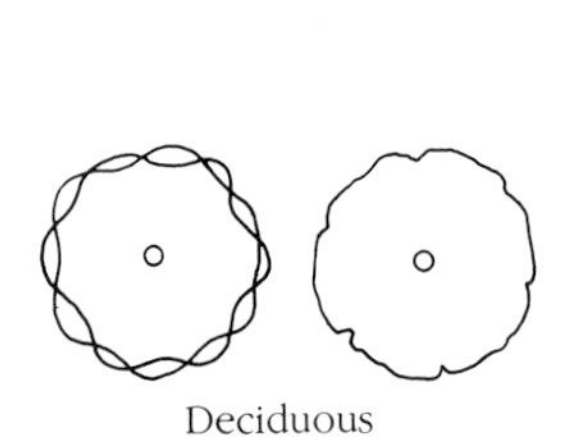

Deciduous

Individual

Clump

Deciduous

Evergreen

Trees

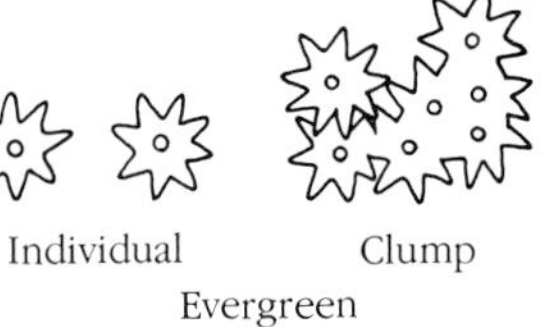

Individual

Clump

Evergreen

Shrubs

Figure 11-17 Standard symbols used on the landscape plan.

Woody Plants and Plantings

Much has already been said about the importance of selecting appropriate plants for each location in the landscape. There are many factors to consider in choosing plants, but perhaps the two most important are (1) whether, during most of its life span, the plant will be of a size suitable for the location where it is to be planted and (2) whether the plant is adapted to grow in the climate (including microclimate) and the soil.

Woody plants, including trees, shrubs, and lianas, are the screening walls of the garden and the larger permanent fixtures that accent the landscape.

In addition to their climatic adaptation and size, woody plants should be selected for the garden on the basis of texture, color (foliage and bark as well as flower color), bloom date, rate of growth, and how well they relate to other elements of the design.

Woody plants should be attractive in the landscape during all seasons of the year. The color contrast of snow on the dark green foliage of a fir, the yellow-green branches and buds of the golden willow that are reminiscent of the spring even in January and February, the velvety red fruit clusters and twisted shape of the staghorn sumac, the lacy branches and winter catkins of the birch, and the unusual branch pattern of the saucer magnolia make these plants as interesting in winter as they are in summer. Woody plants that are messy should be avoided unless the mess they make is outweighed by other desirable characteristics. For example, the gardener must determine whether the beauty, hardiness, and dense shade of the Norway maple compensates for its constant drip of honeydew, or whether the colorful berries and their attractiveness to birds of European mountain ash or certain flowering crabapples are worth the mess they make when their fruit falls to the ground and rots. Trees and shrubs that constantly drop dead branches, leaves, or seed pods should be avoided.

Colors of the planting should be complementary during each season of the year. For instance, planted alternately in the parking area around a church occupying an entire block were Dolga crabapples and scarlet maple. For me the clash of the carmine blossoms of the crabapple with the yellow blooms and maroon unfolding leaves of the maple spoiled the effect of both during the spring season, when either type of tree alone would have been magnificent.

As for hardiness, it is possible for the dedicated gardener to provide winter protection for a few exotic plants, but most of us are too involved with other activities to spend cold winter evenings and mornings covering and uncovering tender shrubs, especially when there are so many good plant materials adapted to even the coldest regions.

Woody plants should be free of disease and insect pests. Often the experience of neighbors may be the best—perhaps the only—criterion on which to judge the susceptibility of a certain species to local plant pests. Certain kinds of willows and some viburnum may be impossible to grow without extensive spraying in some regions because of their susceptibility to aphids. In many areas the various kinds of ash are hardy and pest free, but in other locations they may be subject to scale or aphids or both. It is futile to plant an American elm if Dutch elm disease has ravaged mature elms in the town in which you live.

Trees

Among their many beneficial attributes, trees provide shade and beauty, are a sanctuary for birds, control wind, and help prevent floods and erosion (Figure 11–18). It

Figure 11–18 Trees as a sanctuary for birds. (Photo by R. Bennett, courtesy USDA.)

should be emphasized again that each tree planted should fit the location where it is to grow (Figure 11–19).

The age of family members and the length of time the family plan to live in the home will at least partially determine whether to plant a fast- or slow-growing tree (Figure 11–20). Fast-growing trees such as willow and poplar are almost always short-lived; drop branches, fruit, and leaves; and frequently have suckers and shallow roots that can crack sidewalks, plug sewers, and interfere with the growing of lawn. They do, however, provide shade and bird sanctuaries much sooner than do oak, hickory, and

Figure 11–19 Inadequate planning. Look up when planting a tree! The beauty of this conifer has been destroyed because one side has been trimmed away from utility lines.

Figure 11–20 Differences in growth rate of trees. The trees in this photo are the same age. The ponderosa pines (*Pinus ponderosa*) in the background are much better adapted to their situation than is the concolor fir (*Abies concolor*) in the foreground. In a different environment the growth of the two species would be more nearly equal.

other slow-growing trees. Owners will have to decide whether they want a tree that produces light shade or one that provides heavy shade. Grass grows much better under the canopy of trees such as birch and honeylocust, which permit some sunlight to filter through, than under maple, oak, or sycamore. A number of factors will determine whether a deciduous or evergreen tree should be planted in a particular location. On the south side of a home, for example, a deciduous tree will provide shade during the summer and permit the sunshine to come through during the winter. It is usually a mistake to plant a conifer in front of a house. Because they are large growing, conifers can completely hide a home and create dense shade winter and summer; they belong in the border or as specimens at the side or behind the house (Figure 11–21). They also make excellent year-round windbreaks.

Characteristics and adaptability of various trees are provided in Table 14–25 (Chapter 14).

Shrubs

Shrubs come in all shapes and sizes, and there is some disagreement as to whether certain plants are small trees or large shrubs. Generally, plants that grow with multiple stems at the base are considered to be **shrubs;** those that grow from a single stem are trees. Shrubs grow to full size more quickly than do trees, so planting the wrong shrub is not quite as serious as planting the wrong tree. Because they are smaller, shrubs are more easily protected from freezes and other adverse weather conditions than trees.

Most of the guidelines listed previously for selecting woody plants apply to shrubs. It is usually considered more artistically pleasing to have a mass of several of the same kind of shrubs than to have one each of a number of different kinds; however, as with other aspects of landscaping, the preference of the owner takes precedence. For example, if space is limited, a gardener may wish to plant single specimens in order to enjoy as many different kinds as possible.

Lists of shrubs suitable for various purposes along with their characteristics and climatic adaptability are given in Table 14–25 (Chapter 14).

Figure 11–21 Poorly placed conifers. These large conifers hide the home and create dense shade winter and summer.

Lianas

Lianas, or woody vines, serve many purposes in the modern garden. They are especially valuable for small yards where they use much less space to provide suitable plant cover for certain locations than do trees or shrubs. For example, a screen of ivy (Figure 11–22), Virginia creeper, or clematis trained to a fence or a flat wall will take only a few inches of growing room along a property boundary, leaving the remainder of the yard for other activities; a shrub or tree border, on the other hand, will require several feet of yard width to provide the same amount of screen.

Lianas are used to cover or soften banks or unsightly walls or fences. They can be trained to screen locations where screening with other plant materials would be difficult or impossible. A number of lianas such as clematis or wisteria are grown as dramatic specimens providing spectacular color to the garden during certain seasons of the year. Characteristics of the more popular lianas are listed in Table 14–25 (Chapter 14).

Figure 11–22 Ivy cover for walls and banks.

Insect and Disease Control for Woody Ornamentals

It is impossible in a text with limited space to make an annotated list of the ubiquitous insect and disease pests that infest the hundreds of woody plant species. The sheer size of most shade trees, the extensiveness of woody plantings, and the numerous related wild-growing and uncared-for neighborhood specimens that become reservoirs of pest infection often render chemical control of the pests of woody ornamentals by the homeowner unfeasible or ineffective.

The best way to limit insect and disease damage to woody ornamentals is to plant kinds that are not subject to pests found in the region and to keep them growing vigorously. Proper fertilization and irrigation, along with pruning and thinning so that light can reach the branches and air can circulate through them, will do much to prevent pest damage to woody ornamentals.

Cultural or chemical control of occasional infestations that occur on specific species, for example, sprinkling with water to reduce mite infestation or spraying with insecticides to control the periodic buildup of army worms and tent caterpillars, are often necessary. A number of other pests, including bark beetles, spruce budworms, and tussock moths, may be present in large numbers only during an occasional growing season when specific environmental sequences occur. When these insects multiply to significant numbers, immediate control measures are essential to save the infected plants. Sometimes control of insects is necessary for reasons other than the welfare of the tree; for example, control of aphids reduces the honeydew falling from maple trees. Occasionally other disorders will be mistaken for insect damage. Defoliation of the fir in Figure 11–23 was initially attributed to mites, which were causing similar damage to spruce in the neighborhood.

A description of insects that are general feeders and control suggestions for them are listed in Table 14–24 (Chapter 14). Local pest-control experts or literature should be consulted if it becomes necessary to treat a particular species for an unusual disease or insect pest.

Figure 11–23 The defoliated appearance of this concolor fir is identical to the appearance of a conifer attacked by one of several needle-destroying insects. This tree lost its foliage as a result of a severe drought. Now that moisture is again available, limited new growth is beginning at the tips of the branches.

Windbreaks and Woody Plant Borders

Windbreaks and woody plant borders are discussed together because they have much in common. Both are usually made up of one to several rows of woody plants of various shapes and sizes. Although the motive for planting windbreaks is generally utilitarian and the primary purpose of woody plant borders is usually ornamental, both reduce wind velocity, stop erosion, shelter birds and wildlife, and at the same time beautify.

Windbreaks are usually unnecessary in the East and those sections of the Northwest where trees and woody shrubs are the native cover. The purposeful planting of windbreaks is also not important in most urban areas where closely spaced buildings and landscape plantings provide the protection usually afforded by windbreaks. However, windbreaks are almost mandatory for rural comfort in the plains states and provinces and the semiarid intermountain valleys. Even in newer city subdivisions and in suburban areas a windbreak can often make the difference between a yard that can be utilized for family outdoor activities and one that cannot.

The right planting of trees can essentially eliminate heavy wind from around the home while still permitting light breezes to enter the yard. Because an increase in velocity of wind of one mile an hour (2 kph) has the same effect as dropping the temperature 1°F (½°C), a good windbreak not only makes winter, spring, and fall outdoor living more pleasant, but it also can reduce home heating bills considerably.

Specifications for a Windbreak. Windbreaks consist of from one to approximately eight rows of trees and shrubs. Where there is space available, several rows of shrubs and trees of various sizes and growth habits, both deciduous and evergreen, should be utilized to provide the most effective protection against wind and snow (Figure 11–24). The planting should be at right angles to the prevailing winds and, where there is space, should extend at each end 50 feet (15 m) beyond the boundaries of the area to be protected in order to prevent winds from whipping around the windbreak. Where drifting snow is a problem, the home should be located at least 60 feet (18

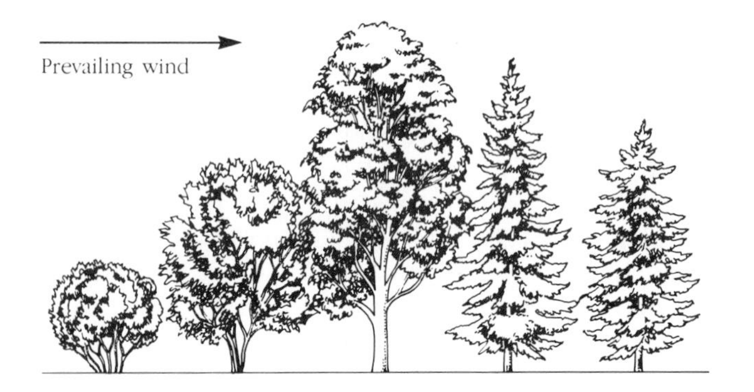

Figure 11–24 Wind protection from trees and shrubs. (From *Trees Against the Wind,* Pacific Northwest Bulletin 5. Extension Service, University of Idaho, Moscow, ID. 1962.)

m) beyond the last row of the windbreak. A five-row planting, such as that shown in Figure 11–25, gives the greatest amount of protection per row of windbreak. Plants in the windbreak should have adequate spacing (Table 11–1). Trees and shrubs for windbreak planting are available at production cost or less from the state departments of forestry or resource management in most areas.

Where wind is a problem around urban and suburban homes, space for windbreak planting may be limited. In such a situation a single row of common or upright juniper or arborvitae can make a windy backyard more livable and reduce heat loss from the home.

Of course, windbreaks need to be carefully tended if they are to become effective. They should be kept weeded and cultivated. Some pruning may be necessary and the area should be kept free of livestock, as livestock can ruin a windbreak by chewing the lower leaves and bark and compacting the soil around the trees.

The woody plant border may have many purposes, including those mentioned for windbreaks. It can screen unsightly views, provide privacy for the family, be a sanctuary for birds and wildlife, provide beauty and shade, subdue the sound of traffic and other noise, and to some extent modify atmospheric pollutants, as well as protect from wind and drifting snow.

If the woody border is to be made up of several rows of trees and shrubs, the taller ones should be planted toward the outside and the shorter ones toward the center lawn or open area of the yard. Each row should consist of several different types of shrubs or trees to provide contrast in silhouette, in color and texture of the foliage, and in flower and leaf display through the season.

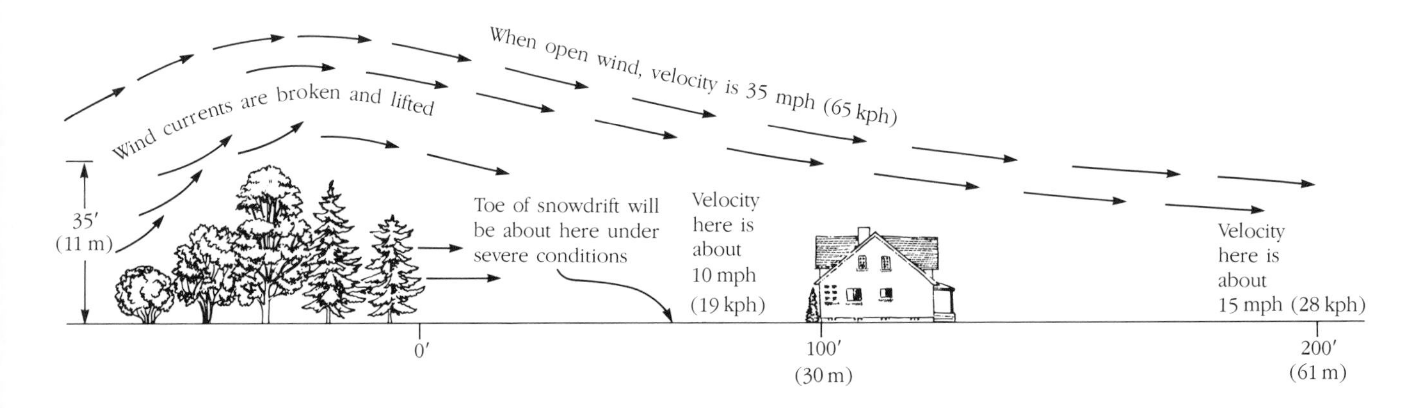

Figure 11–25 Typical windbreak. (From *Trees Against the Wind,* Pacific Northwest Bulletin 5. Extension Service, University of Idaho, Moscow, ID. 1962.)

Table 11–1 Recommended spacings for windbreak plantings

Situation	Tree types	Spacings to use: On irrigated land (feet)	Spacings to use: On dry land (feet)
Farmstead windbreaks, main field windbreaks, and supplemental field windbreaks with more than one row:			
Between rows	All species	16	20
Between trees in the rows	Dense shrubs (row 1)[a]	3	3
	Medium-sized deciduous (row 2)	6–10	8–10
	Tall deciduous (row 3)	8–12	10–12
	Tall evergreens (row 4)	8–12	10–12
	Dense medium-height evergreens (row 5)	6–10	8–12
Single-row supplemental field windbreaks and living snow fences:			
Between trees	Tall deciduous or tall evergreen interplanted with dense shrub	6[b]	6– 8
	Dense, medium-height evergreens	6	8
	Medium-sized deciduous	6– 8	8–10
	Dense shrub	1.5– 3	3

(From *Trees Against the Wind.* Pacific Northwest Bulletin 5. Extension Service, University of Idaho, Moscow, ID. 1962).
[a]Refer to Figure 11–24 for row number.
[b]The tall trees should be 12 feet apart with a dense shrub planted between them.

The Rose Garden

In many respects roses are in a category by themselves. They grow on a woody plant but are the most popular of all garden flowers. Some kinds of roses can be grown in almost all parts of North America, including ones available for growing on arbors or trellises for borders, as hedges, as bedding plants, as specimens (e.g., tree roses), and as a source of cut flowers. Rose cultivars in many shapes, colors, and sizes are available and new ones are being developed by plant breeders each year.

Purchasing Roses. Rose bushes are available from several sources. Older cultivars that are no longer patented may be purchased inexpensively as packaged, bare-rooted specimens from grocery and department stores. Before purchasing packaged plants, examine them carefully to make certain the stems are alive and healthy. Both patented and unpatented roses are also available from mail-order nurseries.

Container-grown rose bushes are available throughout the growing season from local nurseries. Although they cost considerably more than packaged rose bushes, they are the only satisfactory planting stock for late spring and summer planting. Both older standard cultivars and the newer more expensive cultivars that are still under patent are available in containers. Cultivar names will be wired to each bush, and often the nursery will have a catalog showing colored pictures of the cultivar in bloom. Table 11–2 lists characteristics of selected types of roses.

Planting Roses. If winter temperatures do not go below 10°F (−12°C), bare-rooted roses can be planted any time they are fully dormant. If winter temperatures do not

Table 11-2 Characteristics of selected types of roses

Type	Height and spacing (feet)	Bloom season	Hardiness zone[a]	Remarks
Hybrid tea	2–6 x 3–4	All summer in cool-season areas; spring and fall in South	7	Most popular rose; flowers large, one per stem or clusters of 3–5; many cultivars are fragrant; wide range of colors.
Floribunda and Polyantha	2–6 x 3–4	Same as hybrid tea	5	Will tolerate neglect; flowers produced in clusters, smaller but more numerous than hybrid tea; often not distinguished from hybrid tea in nursery; used for border or mass plantings.
Grandiflora	2–6 x 3–4	Same as hybrid tea	7	Flowers smaller but more numerous than hybrid tea, one per stem; flowers good for cutting.
Hybrid perpetuals	4–8 x 3–6	Late spring or early summer	3–5	June roses of our great-grandmothers' day; large flowers, somewhat coarse in appearance; hardy.
Shrub roses	Dense bushes, variable	Spring and early summer	2–3	Miscellaneous group; flowers small, single, numerous; used in borders and hedges.
Old fashioned	6–8 x 5	Spring and early summer	3–4	One of the first roses to be cultivated; flowers are less attractive than some modern types but extremely fragrant.
Tree roses	5–8 x 3	Same as hybrid tea	7	Made by grafting standard rose on an upright trunk.
Miniature roses	½ x ½	Late spring or early summer	6	Used for planter boxes, rock gardens, edging beds; flowers and plants are small.
Climbing roses	Canes 8–15	Variable	3–7	Developed from all the first five bush roses on this chart; each cultivar resembles its bush counterpart in bloom season and hardiness.
Ramblers	Canes to 20	Spring and early summer	4–5	Flowers small and clustered; bloom on previous season canes; subject to mildew.
Trailing roses	Canes 8–15	Spring and early summer	4–5	Climbers with numerous single flowers; adapted to covering walls and banks.

[a]Hardiness zones are illustrated in Figure 14–107 (Chapter 14). The hardiness zone listed is for unprotected plants. With winter protection roses can be grown in zones two to three numbers lower.

go below 0°F (−18°C), roses can be planted either in fall or spring; but if winter temperatures go regularly below 0°F, roses should be planted in the spring only. Container-grown plants can, of course, be planted at any time through the summer. Bush roses should be spaced 2 to 3 feet (⅔ to 1 m) apart and climbers about 8 to 10 feet (2 to 3 m) apart.

Roses grow and bloom best where they have full sunshine all day (Figure 11–26); however, they will produce satisfactory growth if they have at least 6 hours of sun a day (see "Transplanting Woody Plants," Chapter 14).

Winter Protection. Many kinds of roses will need winter protection in areas where the temperature drops below 0°F (−18°C). Bush roses can be protected by soil heaped over the crowns after the early frosts but just before hard freezes are expected (mid- to late November in most areas). For winter protection the canes of climbing roses should be unfastened from their trellis; held on the ground with heavy weights, notched stakes, or wire pins; and covered with several inches of soil. The soil should be removed in the early spring after danger of severe frost is past. Tree roses can be protected from

Figure 11–26 An All-American selection of the hybrid tea type of rose. Roses grow best where they receive sunlight all day but will grow satisfactorily if they receive at least six hours of sunshine daily. (Courtesy George E. Rose, Shenandoah, Iowa.)

winter cold by wrapping them with straw and covering them with burlap if the temperature doesn't drop below −10°F. Where temperatures drop to 10° to 15°F below zero (−23° to −26°C), tree roses should also be covered with soil. This can be done by digging carefully under the roots on one side of the plant until the plant can be pulled over on the ground without breaking all root connections. In the spring after the soil thaws soil cover should be removed and the plant set upright again.

Roses also require heavy pruning (see "Pruning and Training," Chapter 14). Suggestions for controlling the major insect and disease pests on roses are given in Table 11–3.

Herbaceous Plants and Plantings

Herbaceous plants provide temporary or permanent cover and filler and add variety to the landscape. Because they are less permanent than woody plants, herbaceous plantings can be changed from year to year to suit the changing needs and whims of the owner.

Flowers

Flowers are enjoyed in many different ways by almost all societies. The front yard of the English cottage is planted to flower beds instead of lawn; service stations and public buildings in Europe are commonly landscaped with flower beds; and vendors throughout the Middle East sell bouquets of wild and cultivated types of flowers to people who take them home once or twice a week. In Japan flower arranging is a time-honored activity that commands as much respect and interest as athletics. On the island of Bali no successful businessperson would think of starting the day without a floral offering to the gods.

Although in the United States flowers are traditional for funerals, weddings, and other special occasions, their use in and around the home has not been as universal as in some other cultures. Nevertheless, flower growing as a hobby is becoming more popular each year with U.S. and Canadian gardeners.

Flowering plants come in all shapes and sizes, but the group usually classified as flowers are herbaceous flowering plants. They can be classified as spring-flowering bulbs and corms, summer-flowering bulbs and corms, annual-flowering plants, and perennial-flowering plants.

Spring-Flowering Bulbs and Corms. Some of the easiest to grow and most satisfying flowers are produced from bulbs and corms. Bulbous flowers are either large and attractive or else they bloom early before there is much else to show in the garden. Most spring-flowering bulbs and corms should be planted in the fall to enable them to bloom in the spring, but some that originate in the tropics are not winter-hardy. These nonhardy types can be forced as potted plants or stored during the winter

Table 11–3 Insect and disease pests of roses

Insect pests	Symptoms and damage	Control
Japanese beetle	The beetle consumes flowers, buds, and foliage during July and August. At present its damage is limited to the East and Midwest.	With moderate infestation insecticides can be used. In heavily infested areas flowers may have to be protected with cheesecloth cages.
Rose chafer	Yellowish-brown beetles appear suddenly on rose petals; may destroy the entire flower.	An insecticide.
Rose leaf beetle	The small, metallic green beetle feeds on buds and flowers, often riddling them with holes. Most numerous in suburban gardens near uncultivated fields.	An insecticide into the open flower.
Rose leafhopper	The tiny greenish-yellow jumping insect sucks the contents of leaf cells from the underside, causing stippling of the leaves that resembles injury by spider mites.	An insecticide to the lower side of the rose leaf.
Rose slugs	The larvae of three species of sawflies, slugs cause the leaf to become skeletonized by feeding on them.	An insecticide as required.
Thrips	Flower petals, especially those of white varieties, become brown. The tiny yellow or brown thrip can be seen if an infested flower is shaken over a sheet of white paper.	No form of satisfactory control is available, because the opening petals cannot be completely covered with dust. Cheesecloth cages around prize blooms may protect them.
Aphids	Aphids may accumulate in large numbers on the buds, sometimes causing them to fail to open. Sticky honeydew accumulates on foliage.	Insecticide as needed.
Rose scale	Scale become encrusted on canes, where they suck the sap from plants.	Biweekly insecticide treatment will kill the young rose scale crawlers. Infested stems should be pruned away during the dormant season and the remainder of stems sprayed with a summer oil emulsion.
Rose midge	A tiny yellowish fly lays its eggs in the growing tips of rose stems. The maggots that hatch destroy the tender tissue, killing the tips and deforming the rose buds.	Cut off and burn the infested tips daily for one month to eliminate the maggots before they complete their growth and drop to the ground. No good insecticide control is available.
Spider mites	Various species of spider mites suck juices from rose leaves, which soon become spotted, turn brown, curl, and drop off.	Avoid excessive use of insecticides, such as carbaryl, that control mites' predators but not the mites. Where mildew and blackspot are not a problem, sprinkling the underside of the leaves daily will help control mites.
Rose stem borers	Larvae burrow through the pith of the stem and sometimes cause breakage or death of the stem.	Infected parts should be trimmed off and burned.
Diseases	**Symptoms and damage**	**Control**
Black spot	Circular black spots frequently surrounded by a yellow halo appear on the leaves. Infested leaves turn yellow and die prematurely. Plant may become almost defoliated.	Black spot is spread by water, which must remain on the leaves for at least 6 hours before the infection takes place. Heavy pruning in the spring helps remove diseased branches. Spray or dust with a fungicide weekly throughout the growing season.
Powdery mildew	White powdery masses of spores appear on the young leaf shoots and buds. Buds and foliage are stunted or distorted.	The disease is spread by wind and winters over in fallen leaves. Dust with fungicide.
Rust	Yellow or orange pustules appear on the leaves. Plant may become defoliated. Rust winters over in fallen leaves and is spread by the wind. Troublesome mostly along the Pacific coast.	Fungicide.
Cankers	Small reddish spots start on weakened plants and enlarge to girdle the stem, causing it to die.	Keep bushes growing vigorously and free of black spot. Provide winter protection. Prune all cankered canes out and disinfect pruning tools with alcohol after use on a cankered shoot.
Crown gall	Galls usually begin at ground level but sometimes higher up. They increase in size, and infected plants become stunted.	Prevent crown gall by planting roses in soil that has been free of crown gall–infected plants for at least two years. Remove any infected plants and burn them.
Virus diseases	Viruses usually are spread by propagation. They cause small angular colorless spots on the foliage. Infected plants are usually dwarfed.	Purchase plants that are free of the symptoms of viruses.

to be planted in the early spring. Except for those noted as needing special protection, most of the genera listed in Tables 11–4 and 11–5 will survive without protection in areas along the Pacific Coast and south of the Ohio River. Farther north, bulbs and corms not definitely known to be winter-hardy should be protected with a 2–5 inch (5–13 cm) mulch of straw, leaves, or shavings.

Spring-blooming bulbs must have time to develop a root system before winter and should not be planted later than about the last of September in a northern part of the country and in late October in the South. Because they are also subject to injury from excess winter moisture, they should be planted only in well-drained locations. If soil is fertile and plants are deep green and healthy, spring fertilization is probably not necessary. It is best to fertilize in the autumn by spreading about ¼ pound (about 100 g) of ammonium nitrate or an equivalent of nitrogen from another source over each 50 square feet (4½ m^2) of bulb garden.

Faded flowers should be removed before they go to seed and rob the bulb of nutrients. However, leaves should be left on the plants until they have dried and their nutrients have been translocated into the bulbs. To camouflage the untidiness of dying leaves of spring-flowering bulbs, a gardener can plant fast-growing annuals such as verbena among the bulbous flowers.

Most spring-flowering bulbs can be left to multiply for several years, but when they become crowded and the flowers start to become smaller, they should be dug and separated. Tunicate bulbs (those with concentric rings, such as tulip and onion) and corms should not be dug until their plant has died and their brown protective cover has formed. After being dug, they should be kept out of direct sunlight. Daffodil and iris bulbs should be dried as quickly as possible to prevent rotting. Tunicate bulbs should be kept where the temperature is high (80°F; 27°C) for two weeks after being dug. Then, if possible, they should be stored where the temperature is about 50°F (10°C) and the humidity 75%. In the early fall the clumps should be separated and individual bulbs replanted into the garden.

Scaly bulbs of lily should be separated and replanted immediately after being dug because even moderate dehydration can kill them. They are usually separated and replanted as bulbs consisting of a cluster of scales. Individual scales will produce new plants, but plants thus produced will be small and require several seasons of growth before they produce blooms.

Table 11–4 Characteristics of selected types of tulips, narcissus, and bulbous iris

Kind	Bloom date[a]	Height (inches)	Flower shape	Color	Remarks
		Tulips (planting depth 4–7 inches; bulb spacing 2–8 inches; full sun to partial shade)			
Single early	Mid-April	6–12	Egg to cup	Various	Some, such as 'Duc van Tol,' are good for early forcing.
Double early	Mid-April	8–12	Peony	Various	Forced in great quantities.
Darwin hybrids	Late April	24–28	Cup	Various	Hybrids between *T. fosteriana* and Darwin; like Darwin but flowers are larger. Mendel and Trumpet types are crosses between Darwin and Single early.
Darwin	May	24–32	Cup	Various, bright	Most popular of outdoor tulips; strong stems.
Cottage	Late April to June 1	to 36	Cup	Various, bright	Cottage, Breeder, and Darwin have been intercrossed to produce intermediate cultivars.
Breeder	May	24–32	Cup	Muted	Flower opens during day; colors bronzed; not as popular as formerly.
Lily flowered	May	20	Star	Various, bright	Flower petals pointed and folded out.

(continued)

Table 11-4 *(continued)*

Kind	Bloom date[a]	Height (inches)	Flower shape	Color	Remarks
Late double	May	24	Large peony	Various	Flowers won't stay upright with rain or sprinkler irrigation.
Parrot	May	24–32	Serrated and twisted	Various	Most are sports of other types, primarily Darwin.
T. kaufmanniana	Early April	8–16	Star open	White, yellow, red, multicolor	Attractive for early bloom in rock gardens or mass plantings.
T. fosteriana	Early April	9–12	Large open	Red (mostly), yellow, white	'Red Emperor' is the most famous cultivar.
T. greigii	Early April	9–12	Bell; pointed petals	Red, yellow	Striped leaves; good for rock gardens.
Rembrandt or broken	May	24–32	Cup	Multicolored	The unique coloration occurs because of virus infection; don't plant these near other tulips.
T. praestans	Late March	8	Open	Red	Good for rock gardens; multiflowered.
T. clusiana	Early April	12	Star	White with outer red stripe	Good for rock gardens.
T. tarda	April	6	Star	Yellow	Up to five flowers per stem.
Narcissus (planting depth 2–6 inches; plant spacing 3–8 inches; full sun to partial shade)					
Trumpet	April	8–19	Single	Yellow, white	Trumpet as long or longer than petals; one flower per stem.
Cupped	April	14–20	Single	Orange to white	Trumpet ⅓ petal length; one per stem.
Double	April	8–20	Double	Orange to white	One or more flowers per stem; fall down in rainy weather.
Triandrus hybrids	Late April	8–15	Pendant	White, yellow	Two to five flowers per stem; petals folded back.
Cyclamineus hybrids	February to March	8–15	Single	Orange, yellow	One flower per stem.
Jonquilla hybrids	Early May	11–17	Single	Orange-white	Three to five flowers per stem; strong pleasant fragrance.
Tazetta hybrids	April	15–17	Single	White, yellow, orange	Up to fifteen flowers per stem; not hardy; mostly used for forcing indoors.
Bulbous iris (planting depth 2–4 inches; plant spacing 3–10 inches; sun to partial shade)					
Dutch	June	20		Blue, white, yellow	After stratification can also be planted during spring to bloom in September.
Spanish	Late June	20		Various; yellow most popular	Needs winter protection.
English	July	20		Blue, purple, white	Naturalizes in milder regions.

[a]Bloom time will be correct most seasons for many locations in hardiness zones 5, 6, and 7. Blooms will be earlier in warmer regions and during seasons with above-normal temperatures and later in cooler regions and during seasons of below-normal temperatures. (Hardiness zones are illustrated in Figure 14–107.)

Table 11–5 Characteristics of selected less common spring-flowering bulbous plants

Kind	Planting depth (inches)	Spacing (inches)	Bloom date[a]	Height (inches)	Flower color	Light exposure	Remarks
Allium	2–6	2–12	June–July	12–18	Pink, purple	Sun to part shade	Ornamental onions; many kinds.
Anemone	2–3	3–8	March–April	6–12	Various	Part shade	Many kinds and colors; wood anemones grow in heavy shade.
Brodiaea (triteleia)	3	1½	May–June	18–48	Various	Sun to part shade	Alpine gardens; many species; likes sandy soil; dig and store during summer if soil is wet.
Bulbocodium	2	3	February–March	4	Lavender	Sun	Needs well-drained soil and frequent replanting; closely related to crocus.
Camassia (spring meadow saffron)	3–4	5–9	April–May	24–30	Blue, white	Sun to part shade	Native to North America. *C. quamash* used for food by Indians of Northwest.
Chionodoxa (glory of the snow)	2	2–3	March–April	4–9	Blue, pink, white	Sun to part shade	Native to alpine meadows; short tubular open flowers in spikes.
Convallaria (lily of the valley)	1	6–12	May	9	White	Shade	Grows from rhizomes but is similar in culture to bulbous plants; grown as much for its fragrance as for its ornamentation.
Crocus	2	1–2	February–March	6–8	Yellow, purple, white	Sun to part shade	One of the first flowers of spring; some will also bloom in autumn.
Eranthis (winter aconite)	1–2	2–3	February–March	6	Yellow	Part shade	Even earlier than crocus; looks like buttercup.
Erythronium (adder's tongue)	1–2	2	April	8–18	Yellow, pink, purple	Part shade	Several species; good for woodland gardens.
Fritillaria	3–6	4–15	April–May	6–12	Various	Sun to part shade	Many species of various types.
Galanthus (snowdrop)	1–2	2–3	February–March	6–12	White	Part shade	One of the harbingers of spring.
Hyacinth	2–4	3–10	April	6–12	Various	Sun to part shade	Showy spikes of very fragrant blooms; in moist climates bulbs should be lifted and stored in a cool, dry place through the summer.
Ixia (cornlily)	2–3	1–2	June	18	Various	Sun	Needs heavy mulch for winter protection north of southern states and Pacific Coast. Frequently forced in pots.
Muscari (grape hyacinth)	2–3	5	April	8	Blue spikes	Sun	Easy to grow and propagate.
Oxalis	2–3	2–4	May–July	3–12	Various	Sun to part shade	A large, varied genus; many are bulbous and hardy; good in rock gardens.
Ranunculus (buttercup)	1	3	April–May	6–8	Yellow	Sun to part shade	Tolerant.
Scilla	1	3	March–April	6–20	Blue	Part shade	Likes moderately moist soil; good for interplanting with tulips and daffodils.
Trillium	3–4	4–6	March–April	6–10	White, purple	Part shade	Needs moist soil; good for ground cover in shady places.

[a]Bloom time will be correct most seasons for many locations in hardiness zones 5, 6, and 7. Blooms will be earlier in warmer regions and during seasons with above-normal temperatures and later in cooler regions and during seasons of below-normal temperatures. (Hardiness zones are illustrated in Figure 14–107.)

Summer-Flowering Bulbs. Among the summer flowers that grow from bulbs, corms, tubers, or similar underground storage organs are begonia (tuberous rooted), caladium, calla, canna, dahlia, gladiolus, hemerocallis, and lily (Figure 11–27). With the exception of some of the lilies, these are flowers of tropical origin. The storage organs from which they grow are not frost-hardy, and in regions where winters are cold, they must be dug and stored through the winter. Peony and bearded iris, which are sometimes classed with summer-flowering bulb plants, will be included in the section on perennial flowers.

Figure 11–27 A few of the many kinds of flowers that grow from summer-flowering bulbs: (A) dahlia, (B) gladiolus, (C) calla.

Tuberous-rooted begonias are adapted to cool summer climate areas with acid soil, such as England, the U.S. Northeast, and the North Pacific Coast. They require special treatment to grow successfully in parts of the country with a warmer growing season or neutral or alkaline soils. In warmer areas they can be grown during the spring or fall, or, if the temperature doesn't get too hot, during the summer in a location that is shaded. Soil should be kept continually moist (but not saturated) to maintain humidity around the foliage. Some gardeners successfully grow begonias by modifying soil of neutral or alkaline pH with acid peat and sulfur. In areas with this kind of soil it is better to grow begonias in planter boxes or pots containing a special acid growing medium that can be purchased at nurseries or garden stores. Another satisfactory growing medium for begonias is decomposed organic matter from coniferous forests. Where this material can be conveniently obtained, it can be used to fill containers or to replace the top 8 inches (20 cm) of soil in the begonia bed.

Begonia tubers can be planted directly into the garden, but they bloom much earlier if started indoors four to ten weeks before they are to be planted out-of-doors. They should be started in a friable growing medium containing considerable peat. They should be planted with the concave (bud) side up and can be grown relatively close together until sprouts are an inch long. Care should be taken not to damage roots when the plants are transplanted. They should not be planted into the garden until about two weeks after danger of the last frost, or about two weeks after it is safe to transplant tomatoes. They should be grown in a cool area that has light shade throughout the day. In the right location with proper care begonias will continue to bloom until frost, at which time they can be dug and allowed to dry for a few days with the plant left attached to the tuber. The tuber should be detached from the dried plant and stored in a cool but frost-free location with a humidity of 60–70%.

Because caladium is grown for its colorful leaves, flower buds should be removed as soon as they appear. Many varieties and types are available, ranging in size from the dwarfs that are less than 9 inches (23 cm) tall to the elephant's ear, which will grow up to 6 feet (2 m) in height. Caladium also do best in an acid soil; they will not tolerate extreme heat and do not do well in areas where night temperatures drop below 55°F (13°C). Thus in the north they are grown mainly indoors.

Caladium are usually started in pots in a peat-sand mix six to ten weeks before they can be planted out-of-doors. Frequently they are left in pots throughout the summer. In the fall, before killing frosts, plants are allowed to become dry, then dried tops are cut off and tubers are stored in unwatered pots, preferably in a location where the temperature remains about 60°F (16°C).

Callas are tall, showy plants grown as much for their attractive leaves as for their white, yellow, pink, or red blooms. The so-called flower is actually a cluster of florets on a spadix surrounded by a colorful bract or sheath. In areas where frosts seldom or never occur, callas can remain in the same location for many years without replanting. In other areas they are generally grown as potted plants.

Cannas are tall flowers with showy red or yellow blooms. They do best where the season is fairly long and the summer quite warm. Their propagation is similar to that of large dahlias.

Dahlias are grown in two ways. The small 'Unwin' hybrids are grown from seed and treated as annuals; the large, showy dahlias, which may have blooms up to 12 inches (30 cm) across, are grown from what are called tubers. Actually, the so-called dahlia tuber is thickened root and must be attached to a small piece of the lower stem containing an "eye," or bud, in order to grow, because the root is unable to initiate buds. Dahlias do best where summer temperatures are relatively cool. Like gladioli, they should be planted in loose, friable soil so that they emerge after danger of frost is past. By the end of the growing season the single root that was planted in the spring will have developed into a clump of roots. These should be dug, dried off somewhat, and stored with the lower stem attached in a cool but above-freezing location with 60 to 70% humidity. As spring approaches, the clump can be divided. Exceptional plants of the dwarf type produced originally from seed can be propagated a second season by lifting the roots and storing them through the winter.

Figure 11–28 Gladioli corms.

The gladiolus is a popular garden flower because it is relatively easy to grow, is extremely showy, and lasts well as a cut flower. Gladioli corms can be planted any time from a couple of weeks before the last killing frost until as late as late August in some of the warmer parts of the country (Figure 11–28). Biweekly planting of gladioli corms will insure blooms through most of the summer. They need full sun and well-drained soil. For longest life as cut flowers, they should be cut when one to four florets have opened. Leaves should be left on the plant to permit food production for the corm.

For winter storage gladioli corms should be dug when plants are mature but prior to heavy freezes, and placed in a cool, dry location with good air circulation. The tops should be removed when they have completely dried, the scales loosened, the old corm and scales thrown away, and the new corms treated with an insecticide and a fungicide. The pest most likely to be a problem with gladiolus is the thrip. These insects feed and multiply on the flowers and foliage and, if they are serious enough, may cause the flower buds to fail to open or be deformed. Thrips can be controlled by dusting of the corms and spraying of the plants, following practices recommended locally.

Hemerocallis, or daylily, is hardy and easy to grow throughout most of the United States and Canada. Many types of various heights and blooming times are available, with red, pink, orange, yellow, or cream flowers. The leaves remain green and produce an attractive border even after the flowers have faded. Daylilies can be started from seeds or tubers. Seeds can be planted in late fall or early spring. Tubers, which are edible, are planted just below the surface of the soil usually during the spring, but they can be planted almost any time of year. Hemerocallis is tolerant of most kinds of soil and will survive some drought. Seedpods should be removed as flowers fade. Plants will need to be dug and separated when they become crowded, approximately every four to five years. In many parts of the country hemerocallis has escaped cultivation and can be found along fence rows and stream banks.

Lilies, a symbol of purity for centuries, are among the most popular, most varied, and most widely grown of all garden flowers. The native habitat of various species of lily ranges from the Arctic to the tropics. Among the dozens of popular hybrids, species and cultivars are:

Lilium candidum (white madonna lily) grows 3–4 feet (about 1 m) in height and blooms in June.

Lilium longiflorum (Easter lily) is grown widely as a potted plant to be sold at Easter time. In climates where average January temperatures are above 35°F (2°C), potted Easter lilies can be planted out-of-doors after their blooms have faded. They will grow and bloom in July or August year after year in these climates.

Lilium regale blooms in July and grows 3–5 feet (1–1½ m) tall; produces white or yellow flowers.

Lilium speciosum and *L. auratum* bloom in August or September; plants grow to 4–6 feet (about 1–2 m). There are many hybrids between these two types.

Lilium testaceum is one of the oldest and best of the hybrids; produces large apricot flowers on 5–6 foot (1½–2 m) plants in June.

Lily bulbs should be planted in the fall at a depth three times the height of the bulb. They will require well-drained soil and do best in full sun.

Flowering Perennials. There is a large group of flowering plants that die back each year after they have bloomed but grow each succeeding season from the crowns remaining in the ground. These plants are propagated from crown division or from seed and generally do not bloom until the second season after they are planted. Most perennials will brighten the garden year after year with minimal care. Mainly they need to be lightly fertilized and irrigated, have their faded blossoms removed, and be dug and divided whenever they become crowded. Perennials should be selected for the area where they are to be grown. Gardeners should notice what grows well in local gardens, consult nurseries, and check with the state experiment station. They should be cognizant of the flowering times and plant a mixture of annuals and perennials that will provide color in the garden over a long period. In areas where winters are severe and snow cover is lacking, some perennials will need a loose mulch protection. Characteristics of selected perennials are listed in Table 11–6; a more extensive listing of flowering perennials can be found in the references at the end of this chapter.

Flowering Annuals. No other group of plants adds so much color to American gardens as annual flowers, and no other group is as easy to grow (Figure 11–29). As a consequence annuals are enjoying unprecedented popularity with gardeners and have received a great deal of recent attention from those engaged in the seed and nursery business as well as from amateur and professional plant breeders. This attention from plant specialists has resulted in hundreds of new cultivars with larger, more colorful flowers and more profuse flowering habits, and techniques that make it easier to produce annual flowers.

Frederick McGourty, editor of the Brooklyn Botanic Garden's *Plants and Gardens,* suggests in the Garden's *Annuals* handbook (see Selected References at end of this chapter) that a major advantage of annuals, especially for

Table 11–6 Selected perennials for the flower garden

Kind	How propagated	Bloom date[a]	Height	Flower color	Remarks
Chrysanthemum	Cuttings (crown division in spring)	July–November	½–3 ft	Various	The most popular autumn flower; blooms in response to day length; many types of flowers available; also popular as a forced flower; cultivars vary in winter hardiness.
Columbine	Seed	May–June	2–4 ft	Various	Open growth habit.
Delphinium	Seed	June	4–5 ft	Pink, blue, white	Will bloom later during season if faded blossoms are cut back; flower head is a tall spike.
Hollyhock	Seed	June–July	5–7 ft	Various	Comes in both double and single flowers; needs staking where storms or winds are severe.
Iris (bearded)	Crown division after bloom	May–June	3 ft	Various	There are also dwarf species that bloom very early in spring; needs well-drained soil and a minimum of water.
Lupine	Direct seed	May–June	3 ft	Various	Drought tolerant; large spikes; doesn't transplant easily.
Penstemon	Seed	June	1½–2 ft	Various	Drought tolerant; new cultivars are being developed.
Peony	Crown division	June	2–4 ft	Red, white	Can be kept in same location for many years; foliage is attractive all season.
Phlox (summer)	Crown division	July	3 ft	Various	Moss phlox blooms in early spring and is low-growing; can be propagated from seed, but plants are variable.
Poppy (Iceland)	Seed	July	1½ ft	Scarlet to white	Poppies often seed themselves; can be grown as an annual in warm winter areas.
Poppy (Oriental)	Seed	July	3 ft	Scarlet to white	Brilliant flowers on rapidly growing plants. Short-lived in warm climates.
Primrose	Seed	April	6–9 in	Various	Rock garden plant; seed requires freezing before it will germinate.
Violet	Seed	April	3–4 in	Purple, yellow, white	Cool climate flower; some are scented; low-growing, petite.

[a]Bloom time will be correct most seasons for many locations in hardiness zones 5, 6, and 7. Blooms will be earlier in warmer regions and during seasons with above-normal temperatures and later in cooler regions and during seasons of below-normal temperatures. (Hardiness zones are illustrated in Figure 14–107.)

Figure 11–29 A bed of annual flowers. Tall hollyhocks growing along the fence provide a background for intermediate-sized snapdragons, which, in turn, are fronted by low-growing petunias. (Courtesy USDA.)

the novice gardener is that "mistakes—the occasional grouping of tall plants in front of low-growing ones or the red salvia sizzling next to the pinkish-purple petunias—don't haunt forever, as does the misplacement of trees and shrubs." People with gardening experience can also find pleasure and challenge in the tremendous variety of fine annual flowering plants available today. Annuals provide gardening enjoyment for a low initial investment, an especially attractive feature for those who have invested their capital in a new home and lot and have little money left for needed plantings.

Annual flowers are mostly propagated from seed. The methods, advantages, and disadvantages of growing or purchasing transplants or direct seeding into the open soil were discussed in Chapter 4. I emphasize again the importance of not planting until the soil is warm enough to permit seed germination. In addition, seedling beds should be kept moist until seedlings are well-established; after that annuals should be watered enough to soak the soil well, no oftener than once each week. Annual flowers require enough fertilizer to keep them healthy, but they should not be overfertilized, especially with nitrogen. Heavy nitrogen fertilization will reduce or even prevent flowering. Removing dead or faded flowers not only keeps the flower bed more tidy but also will direct plant nutrient resources into blooms instead of seeds.

Most annual flowers can be grown in containers, a special advantage for city gardens, for areas where soil is poor or nonexistent, or where frost or other adverse environmental conditions preclude growing flowers in outdoor beds. Special protection is more easily provided for container-grown plants than for plants growing in the open; however, container-grown plants require more careful attention to watering and fertilization.

Characteristics and suggestions for growing selected annuals are given in Table 11–7.

Table 11–7 Characteristics of selected garden annuals

Plant	Height (inches)	Exposure	Spacing (inches)	Color	Start from seed	Trans-plants	Remarks
Ageratum	6–10	Sun or partial shade	10–12	Blue, violet, pink, white		X	Pinch tip of plants to encourage branching; remove dead blooms; good for edging.
Aster	6–30	Sun	10–15	Blue, white, red, pink		X	Good cut flower.
Calendula	14–18	Sun	10–15	Yellow	X	X	Good for window gardens.
Celosia (cockscomb)	16–40	Sun	10–15	Red, yellow, pink, rose	X	X	Cut flowers, plants for drying.
Cosmos	30–48	Sun	12–18	Pink, rose, white, orange	X	X	Cut flowers, background.
Dahlia (unwin)	18–20	Sun or partial shade	12–14	Many		X	Blooms early; good cut flower.
Dusty miller	6–12	Sun	10–15	Gray foliage		X	Good for edging.
Four o'clock	20–24	Sun	12–24	Rose, yellow, white	X	X	Temporary hedge; easy to grow.
Impatiens	6–12	Partial or deep shade	12–14	White, red, pink, orange		X	Does well in containers.
Larkspur	18–48	Sun	6–8	White, pink, blue, violet	X		Hard to transplant; grow in pots.
Lobelia	6–12	Partial shade to shade	12	Blue, white		X	Good for edging.
Marigold	6–30	Sun	4–30	Yellow, gold	X	X	Cut flowers, bedding, containers.
Nasturtium	10–12	Sun	8–12	Yellow, red, orange	X		Does well in poor soil; needs drained soil.
Pansy	6–10	Sun or shade	6–8	Various	X	X	Good cut flower; does best in cool season; pick off seed pods.

(continued)

Table 11-7 *(continued)*

Plant	Height (inches)	Exposure	Spacing (inches)	Color	Start from seed	Trans-plants	Remarks
Petunia	8–24	Sun	12–24	Many		X	Long blooming period; does well in containers.
Pinks	6–16	Sun or partial shade	8–12	Pink, white		X	Remove dead flowers; good cut flower.
Poppy	12–16	Sun or partial shade	8–12	White, pink, red, orange	X	X	Hard to transplant.
Portulaca	6–9	Sun	10–12	Many	X		Does well in poor, dry soil; stands heat.
Snapdragon	10–36	Sun	6–10	Pink, white, yellow, red		X	Pinch tips to encourage branching; good cut flower.
Spider plant	30–36	Sun or shade	12–14	Pink, purple, white	X	X	Long blooming period; grows in dry, poor soil; good background plant.
Sweet alyssum	6–10	Sun or shade	10–12	White, purple	X	X	Long blooming period; grows in dry, poor soil but needs good drainage.
Sweet pea	8 in–6 ft	Sun	6–10	All but yellow	X		Bush or vining; needs cool temperature.
Verbena	9–12	Sun	10–15	Many	X	X	Edging, ground cover, cut flowers.
Vinca rosea	15–18	Shade or partial shade	18	White, pink, rose		X	Grow in beds; perennial grown as annual.
Zinnia	6–18	Sun	10–12	Many	X	X	Grow in rows for cutting.

(Adapted from *Growing Annual Bedding Plants*. Washington State University Extension Bulletin 611. Pullman, WA. 1973.)

Rock Gardens

Rock gardens originated as an attempt to duplicate the wildflower beds of Alpine regions. Because most of the plants suitable for rock gardens originated in those regions, rock gardens can be grown best in areas where the growing season is relatively cool. The extent and design of the rock garden will depend on the climate, the area available, the topography, the kind of stone available, and the means for transporting and placing the stone (Figure 11–30).

The rock garden should be carefully planned and sketched on paper before any stones are moved. Areas for walking, plant materials to be used, and the general contour of the garden, including a rough sketch of the size and shape of stones required, should be included on the plan. Rock garden plants generally do best in soil of intermediate fertility, high in organic matter. In many areas the most successful rock gardens are grown in soil

Figure 11–30 A rock garden on a difficult-to-manage slope.

brought from a forested area and placed among the rocks. Rock gardens are most beautiful during the early spring, because that is the time of year when the climate of populated areas most closely duplicates the high light intensity and cool nights of high altitude regions. Selected rock garden plants are listed in Table 11–8.

Sometimes the rock garden is built around a natural rock outcrop; more often, rocks must be brought from other locations. In some regions sponge lava or other lightweight rock is available. In most situations native rock is more natural and, consequently, more pleasing than exotic rock brought in from a distant area no matter how colorful the latter may be. Using rock with moss and/or lichen already growing on it will give a newly planted rock garden an established appearance.

Flower Gardens

Flowers are grown in three types of plantings: (1) the flower border, (2) the flower bed, and (3) the cut-flower garden. Like any other planting, the flower garden is most attractive only when it is well planned. Because flowers are grown primarily for their color, color and time of bloom are two of the most important attributes to consider in planning a flower garden, though height of the plant, texture of the foliage and flowers, and climatic adaptability are also important. Landscape planners feel that a solid mass of one color of flower is more effective than a planting of mixed colors. Nevertheless, I often buy mixed cultivar seed packets because I enjoy the anticipation of something different with each opening bloom.

Flower beds are frequently planted in the center of large areas of lawn in parks and open public areas. Home flower plantings are easier to care for, however, if they are planted along one edge of the yard or against the foundation of the house, because less labor is necessary to keep border grass from encroaching into the flower planting. Both bed and border planting can satisfactorily consist of a mixture of annual and perennial flowers. Bulbs and other perennial flowering plants bloom during early to late spring, and annual flowers can be planted to camouflage the stems of the fading perennials and provide color in the flower garden through summer and fall. Late fall

Table 11–8 Selected easy-to-grow rock garden plants

Name	Hardiness zone[a]	Growth habit	Height x spread (inches)[b]	Soil and exposure[c]	Remarks
Achillea Yarrow	All	Perennial	8 x 8	A	Several species of traditional herbs.
Adiantum pedatum Maidenhair fern	4–9	Fern	24 x 36	B	Good for deep shade and cool climate.
Ajuga reptans Carpet bugle	5–9	Perennial	4 x S	A	Glossy green ground cover, blue flowers.
Alyssum saxatile Golden tuft	4–9	Perennial	12 x 12	A	Cut back after bloom.
Aquilegia Columbine	All	Perennial	24 x 24	A	Blooms May and June.
Arctostaphylos uva-ursi Kinnikinnick	4–9	Liana	4 x S	A	Evergreen leaves; red berries.
Artemisia schmidtiana Silvermound	3–9	Shrub	9 x 12	D	Several cultivars of differing size.
Aubrieta deltoidea Aubrieta	3–8	Perennial	8 x 8	A	Purple flowers, May, June.
Blechnum spicant Deer fern	4–8	Fern	24 x 24	B	Evergreen for shade.

(continued)

Table 11–8 *(continued)*

Name	Hardiness zone[a]	Growth habit	Height x spread (inches)[b]	Soil and exposure[c]	Remarks
Cerastium tomentosum Snow-in-summer	All	Perennial	6 x S	A	White flower, June, July; spreads.
Convallaria majalis Lily of the valley	4–9	Perennial	8 x S	A	Dainty fragrant white flowers, April; tolerates shade.
Cotoneaster dammeri Bearberry cotoneaster	6–9	Shrub	24 x 48	A	Slow growing; red, persistent berries.
Daphne cneorum Rose daphne	6–10	Shrub	10 x S	A	Pink flowers in early spring.
Echeveria imbricata Hen and chickens	8–10	Succulent	4 x S	A	Most common hen and chickens for California.
Erica carnea 'King George' King George heath	6–9	Shrub	12 x S	B	Blooms February to April; tolerates less acid soil than most heaths.
Festuca ovina 'Glauca' Blue fescue	All	Tuft	18 x 18	A	Ground cover or edging; blue-grey grass.
Iberis sempervirens Evergreen candytuft	5–10	Tuft	8 x 8	A	Several cultivars.
Iris cristata Crested iris	All	Clump	4 x S	D	Swordlike leaves; blue flowers in May.
Juniperus spp.	All	Spreading	S	A	Several low-growing types (see Table 14–25, Ch. 14).
Linum alpinum Alpine flax	All	Perennial herb	6 x 6	A	Blue flowers, May to September.
Nepeta mussinii Persian nepeta	4–10	Perennial herb	20 x 20	A	Violet flowers, May to September; grey foliage.
Oscularia deltoides Ice plant	9–10	Shrub	12 x 12	D	Tolerates drought; purple, fragrant flowers.
Penstemon glaber Blue penstemon	All	Perennial herb	8 x 10	A	Blue flowers, July to October.
Phlox subulata Moss phlox	2–8	Tuft	4 x 15	A	Mound of pink or blue flowers in April.
Potentilla verna Spring cinquefoil	All	Tuft	4 x 15	D	Yellow flowers, summer.
Sarcococca humilis Himalaya sarcococca	7–10	Shrub	15 x S	B	Evergreen; white flowers, October to March.
Sedum lineare	3–10	Succulent	5 x S	A	Various shapes.
Sempervivum spp. Houseleek	3–10	Succulent	4 x 10	A	Several types of interesting clustered plants.
Teucrium chamaedrys Germander	All	Shrub	12 x 18	A	Clip flower heads; good also for edging or low hedge.
Thymus Thyme	All	Ground cover	3 x S	A	Types with grey and variegated leaves.
Viola cornuta Tufted pansy	All	Perennial	3 x 5	A	Likes cool conditions.

[a]See Figure 14–107.
[b]S designates shrubs that have a spreading growth habit; spread may vary depending on age, pruning practices, and environment.
[c]A—General garden loam; sun to light shade tolerant.
B—Needs acid, well-drained soil; usually shade tolerant.
C—Tolerates wet marshy situations.
D—Needs perfect drainage, full sun; usually alkaline tolerant.

perennials, such as autumn crocus or chrysanthemum, can provide color through the late fall and early winter.

Tall flowers in the border should be planted toward the back, shorter ones in front. The borders should be carefully planned so there will be bloom through the entire season and so the colors of flowers in bloom at any one time will complement each other.

A popular and simple combination succession is an early spring mixture of tulips or daffodils and pansies, followed by a single cultivar of a long-lasting annual, such as petunias or geraniums, to provide a mass of color through the rest of the summer.

Because the dying remains of cut flowers may not be attractive through the late summer months, the cut-flower garden is usually relegated to the utility area. It may be a large, separate area with several rows each of a number of flowers, or it may be a couple of rows of a favorite cut flower planted with the vegetable garden or integrated into the flower border.

Spring and summer flowering bulbs, such as tulips, daffodils, iris, gladiolus, and lilies; and tall-growing annuals, such as snapdragons, carnations, upright zinnias, tall marigolds, and cosmos, are the kinds of flowers usually grown in the cutting garden. Perennials such as chrysanthemums, tall phlox, daisies, bearded iris, peonies, penstemon, delphiniums, and hybrid tea roses are excellent cutting flowers that must be grown in a permanent location.

Figure 11–31 Ornamental vegetation. This ornamental kale is growing with wire hoops for protection on a strip of soil between an apartment house and sidewalk along a busy street in Tokyo.

Herbaceous Plants Grown for Their Vegetation

The advantage of growing plants that have attractive or unique vegetative structures is that their beauty lasts through the growing season. A few, notably ornamental kales, coleus, and some recently introduced impatiens cultivars, are grown for their colorful foliage. In Japan, ornamental kale is an important component of almost all winter gardens (Figure 11–31). Most plants in this category are noted for their picturesque form, although many do produce foliage having unusual shades of grey or green. Some, such as the cacti, provide spectacular blooms during a short season, but the blooms are a secondary bonus not the primary purpose for which they are planted.

Unlike flowers, most of these plants are not adapted to being displayed en masse. Their location must be carefully chosen so that their unique characteristics can be viewed in an appropriate setting. Most should be planted only where there is ample space for them to fully expand to their natural form, and many require a specific microclimate. Certain crops usually grown as vegetables can add interesting foliage and form to the ornamental garden. These include leaf lettuce, several crucifers, globe artichoke, pepper, the cucurbits, chard, and carrots. Most herbs also can enhance ornamental plantings.

Herbaceous plants that can "climb" by either twining around or attaching to other objects are called **vines.** Vines are similar to and are used for the same purposes as lianas. Although ivy and some other perennial vining plants are herbaceous in nature, they are usually classed with the lianas; only annual vining plants are considered as vines. Vines are used primarily for short-term, rapid-growing screens. A selected list of herbaceous plants, including vines, grown for their vegetation is contained in Table 11–9.

Table 11-9 Selected herbaceous plants with ornamental stems and leaves

Name	Hardiness zone[a]	Use	Height	Soil and exposure[b]	Remarks
			Plants grown as annuals		
Anethum graveolens Dill	2–9	Background	3 ft	A	Herb; airy-open growth habit.
Brassica oleraceae acephala Kale	2–9	Specimen	1 ft	A	Colorful ornamental as well as interesting vegetable types.
Capsicum spp.	5–10	Border; specimen	1–2 ft	A	Colorful fruit; ornamental and vegetable types suitable for landscape.
Coleus blumei Coleus	5–9	Color	1–2 ft	B	Many cultivars; colorful leaves; injured by cold nights.
Cucurbita pepo Bush summer squash	4–9	Specimen	3 ft	A	Bold leaf form; several cultivars.
Geranium spp.	3–10	Specimen; border	1–4 ft	A	Many types; grown for foliage form and fragrance.
Lycopersicon esculentum 'Tiny Tim' tomato	4–10	Border	1 ft	A	Attractive mound of foliage; marble-size fruit; early; other determinant cultivars also decorative.
Ricinus communis Castor bean	4–10	Specimen	8 ft	A	Grows into a huge tropical-appearing plant in a few weeks.
Senecio cineraria Dusty miller	3–9	Ground cover	6 in	A	Grey foliage; a perennial if given winter protection.
			Annual vines		
Cobaea scandens Cup and saucer vine	All	Screen	25 ft	A	2 in purple to greenish cuplike flowers in saucerlike calyx.
Cucurbita pepo ovifera Ornamental gourd	4–10	Ground cover	12 ft	A	Multicolored ornamental fruit; picturesque leaves.
Dolichos lablab Hyacinth bean	5–10	Screen	10 ft	A	Sweetpealike flowers; velvet pods.
Ipomoea Morning glory	All	Screen	15 ft	A	Several cultivars with various flower colors.
Phaseolus coccineus Scarlet runner bean	All	Screen	8 ft	A	Clusters of scarlet flowers; pod with beans edible when young.
			Herbaceous perennials		
Aegopodium podagraria Bishop's weed	3–6	Ground cover	6 in	A	Prefers partial shade; can become invasive.
Allium schoenoprasum Chive	3–9	Border	1 ft	A	Herb; small grasslike mounds; evergreen if winter is mild.
Asparagus officinalis Asparagus	4–9	Background	4 ft	A	Edible, spring; airy hedge, summer; golden with red berries, fall.
Cortaderia selloana Pampas grass	8–10	Specimen	20 ft	A	May grow 8 ft in one season; fountain of saw-toothed leaves.
Cynara scolymus Globe artichoke	8–9	Specimen	3 ft	A	Cool frost-free climate; colder zones with winter protection; edible flower buds.

Name	Hardiness zone[a]	Use	Height	Soil and exposure[b]	Remarks
Festuca ovina 'Glauca' Blue fescue	3–10	Specimen	6 in	A	Tufts of blue grass; doesn't spread; needs drainage.
Hosta spp. Plantain lilies	4–8	Ground cover	6 in–3 ft	A	Large heart-shaped, veined leaves; dies back in fall.
Iris spp. Bearded iris	3–9	Accent	6 in–3 ft	A	Swordlike leaves attractive all seasons; various sizes.
Lamium galeobdolon variegatum Silver nettle vine	4–9	Ground cover	6 in	B	Leaves mottled with white.
Lysimachia nummularia 'Aurea' Moneywort	3–9	Ground cover	6 in	C	Bright yellow foliage; needs shade.
Miscanthus sinensis gracilimus Eulalia grass	4–9	Specimen	6 ft	A	Tall feathery grass; attractive winter and summer.
Nepeta cataria Catnip	2–9	Specimen	2 ft	A	Mint herb; attractive to cats; grey-green leaves.
Pachysandra terminalis Japanese spurge	4–9	Ground cover	6 in	B	Leaves clustered atop evergreen herbaceous stems.
Paeonia spp. Herbaceous peony	3–9	Specimen	2 ft	A	Perennial flower; leaves shrublike all summer.
Salvia officinalis Garden sage	3–9	Ground cover	6 in	A/D	Grey-leaved herb; many other Salvia.
Thymus lanuginosus Wooly thyme	3–9	Ground cover	3 in	A	Flat, matted, grey moss; several other good Thymus.
			Selected ferns		
Adiantum pedatum Maidenhair fern	4–8		2 ft	B	Deciduous; fine, lacy; shade loving; spreads by rootstocks.
Athyrium pycnocarpon Narrowleaf spleenwort	5–9		2–3 ft	B	Deciduous; native to rich woodland.
Blechnum brasiliense Dwarf tree fern	9–10		4 ft	B	Evergreen; nearly erect fronds; compact clusters.
Blechnum spicant Deer fern	7–9		2 ft	B	Evergreen; native to Pacific Northwest; deep shade; woodsy soil.
Botrychium virginianum Rattlesnake fern	5–9		1–2 ft	B	Deciduous; native to open woods.
Cystopteris bulbifera Bulblet fern	5–9		1–2 ft	C	Deciduous; limestone cliffs; small bulbs form on upper fronds.
Dryopteris erythrosora Oriental wood fern	4–9		2 ft	B	Deciduous; young fronds reddish, turning deep green.
Dryopteris marginalis Leather wood fern	5–9		2–3 ft	B	Evergreen; native of eastern rocky woods; adaptable.
Lygodium palmatum American climbing fern	6–10		4 ft	B	Deciduous; climbs; difficult to grow; highly acid, moist soil.

(continued)

Table 11–9 *(continued)*

Name	Hardiness zone[a]	Use	Height	Soil and exposure[b]	Remarks
Osmunda cinnamomea Cinnamon fern	6–10		4 ft	B	Deciduous; moist, acid soil.
Osmunda regalis Royal fern	4–10		4 ft	B	Deciduous; massive; sun or shade.
Platycerium bifurcatum Staghorn fern	9–10		3 ft	B	Evergreen; the hardiest of epiphytic ferns; to 22°F.
Polystichum acrostichoides Christmas fern	4–9		1–2 ft	B	Evergreen; common eastern native; tolerant; shade.
Polystichum munitum Western sword fern	4–10		3 ft	B	Evergreen; large, coarse native from California to Alaska to Montana; shade.
		Cacti and succulents (see also Table 11–8)			
Cephalocereus senilis Old man cactus	10	Specimen or pot	40 ft	D	Slow growing; Mexican native; can be kept small; long white hairs.
Coryphantha vivipara (sold as *Mammillaria*)	3–10	Specimen	6 in	D	Hardy; 2 in knob-covered bodies; showy purple flowers.
Echinocactus grusonii Golden barrel cactus	9–10	Specimen	4 ft	A/D	Best known barrel cactus; showy 3 in spines; yellow 2 in flowers, April.
Echinocerus triglochidiatus	4–9	Specimen	1 ft	A/D	Round ribbed bodies; intense profuse scarlet flowers; Great Plains native.
Ferocactus wislizenii Fishook barrel cactus	7–10	Specimen	8 ft	D	Curves toward sun; 3 in yellow to red flowers, late summer.
Lemaireocereus thurberi Organ-pipe cactus	9–10	Specimen	15 ft	D	Arizona native; columnar, branching; 3 in purple blooms at night.
Opuntia compressa Eastern prickly pear	4–9	Ground cover	10 in	A	Native to Eastern America; flat pads; spring bloom.
Opuntia ficus-indica Indian fig cactus	9–10	Specimen	15 ft	D	Large treelike succulent; 4 in yellow flowers, spring.
Opuntia fragilis	3–9	Ground cover	8 in	A/D	Great Plains native; hardy to central Alberta.
Opuntia polyacantha	4–9	Ground cover	8 in	D	Needs sandy soil; 3 in yellow or carmine blooms, spring.
Sedum acre Goldmoss sedum	3–9	Ground cover	4 in	A	Spreading evergreen; yellow spring flowers; can be weedy.
Sedum lineare	3–9	Ground cover	to 1 ft	A	Spreading, trailing, rooting stems; yellow flowers, late spring.
Sedum spectabile	3–9	Specimen	18 in	A	4 in rose to carmine flower clusters; 3 in blue-green leaves.

[a]See Figure 14–107 (Chapter 14).
[b]A—General garden loam; sun to light shade tolerant.
B—Needs acid, well-drained soil; usually shade tolerant.
C—Tolerates wet marshy situations.
D—Needs perfect drainage, full sun; usually alkaline tolerant.

Lawns and Other Ground Covers

Trees, shrubs, vines, and flowers are all necessary to the garden environment; however, it is the open spaces that add depth to the vistas, accommodate movement of people and vehicles, and provide space for relaxation and recreation. Because unprotected soil turns to either dust or mud, these open areas must have some kind of cover. The cover may be hard-surface paving or mulch or a lawn or other low-growing plants.

Establishing a Lawn

Although there is sentiment for reducing the size of lawns or eliminating them altogether, the majority of homes still have lawns, and most individuals who become homeowners will at some time be concerned with the growing and care of a lawn. As a recent bulletin on home lawns states, "A high quality lawn pleases the eye and increases the value of the property, but beautiful lawns don't just happen. They are the result of wise planning, hard work and proper care" (from *Home Lawns*. Washington State University Extension Bulletin 482, revised. Pullman, WA. 1977).

Perhaps the most important requisite for a satisfactory lawn is proper leveling and grading of the soil around the home. Once the lawn is established, it is difficult to fill in holes or to change the grade. Suggestions for grading, soil, and soil preparation for lawns have been given in Chapter 5. If the soil is clay or clay loam, or if deep excavations have occurred in the yard, special precautions will be necessary to make certain that the soil has completely settled before the lawn is planted. If possible the yard should be left to settle over a winter. After the lawn area is leveled and graded, it should be soaked to the depth of any excavation. The area should then be allowed to dry and any resulting depressions filled with soil. If considerable settling occurs with the initial soaking, the area should be thoroughly wet and allowed to settle a second time.

Plowing, rototilling, or spading, followed by leveling with hand or power equipment, are the best methods of preparing the soil for planting (Figure 11–32). Hand rak-

Figure 11–32 Grading. The yard can be rough graded with power equipment (A), but a hand rake is needed for the final leveling (B). (A, courtesy Ford Tractor Operations.)

ing is a desirable final leveling procedure to fill depressions where water might concentrate and to level high spots.

Lawns can be planted by sodding, by seeding, or by vegetative planting. **Sodding** is the covering of the soil surface with a layer of established lawn that has been grown and brought from elsewhere for that purpose (Figure 11–33). Areas around commercial buildings where an immediate cover is needed and steep banks where erosion is likely to be a problem are frequently sodded. Sodding is expensive, and the homeowner who can afford it probably will be able to afford to have a commercial company lay the sod.

Most kinds of grasses recommended for lawns in the central and northern parts of the United States are seeded. Generally, seeds are sold as lawn mixtures of several cultivars. Analysis tags on seed containers should be examined carefully before seed is purchased. State and federal laws require that tags list the percentage of each kind of grass seed in the container, germination percentage, and date of germination test. Homeowners should consult the county agent or state agricultural experiment station, a seed dealer, or a nursery before deciding on the mixture required for their lawns. Figure 11–34 illustrates regions of the United States suitable for growing various grasses. Grass seed should be purchased on the basis of quality not quantity, because inexpensive mixtures often contain a high percentage of large-seeded annual grass, which is of very little value in establishing a permanent lawn.

Seed can be planted with a seeder or scattered by hand. If it is to be hand scattered, the amount required should be weighed and divided into two portions. The first portion should be scattered in one direction and the second crosswise to the first sowing. Seeds should be covered by hand raking, not more than one-fourth inch deep (Figure 11–35). The seeded area should then be firmed by rolling with a light roller or cultipacker.

Grasses that reproduce asexually can be planted by plug sodding, strip sodding, sprigging, or stolonizing.

Figure 11–33 Sodding. (Photo by E. B. Trovillion, courtesy USDA.)

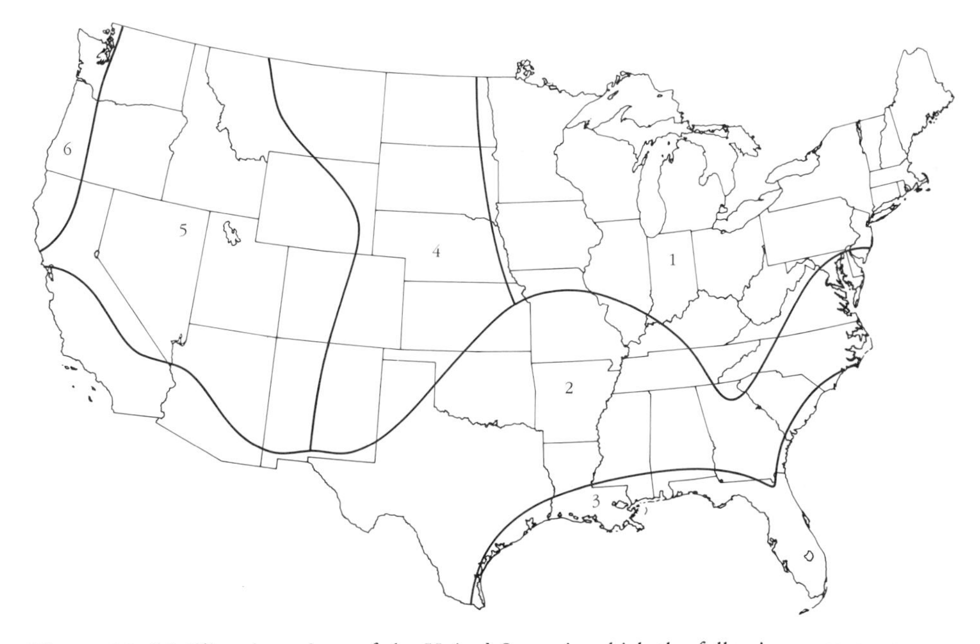

Figure 11–34 Climatic regions of the United States in which the following grasses are suitable for lawns: Region 1. Common Kentucky bluegrass, Merion Kentucky bluegrass, red fescue, and colonial bent. Tall fescue, Bermuda grass, and zoysia in southern portion of the region. Region 2. Bermuda grass and zoysia. Centipede grass, carpet grass, and St. Augustine grass in southern portion of the region; tall fescue and Kentucky bluegrass in some northern areas. Region 3. St. Augustine grass, Bermuda grass, zoysia, carpet grass, and bahia grass. Region 4. Nonirrigated areas: Crested wheat grass, buffalo grass, and blue grama. Irrigated areas: Kentucky bluegrass and red fescue. Region 5. Nonirrigated areas: Crested wheat grass. Irrigated areas: Kentucky bluegrass and red fescue. Region 6. Colonial bent and Kentucky bluegrass. (From *Better Lawns.* USDA Home and Garden Bulletin 51. U.S. Government Printing Office, Washington, DC. Revised periodically.)

Figure 11–35 Covering hand-scattered lawn seed. (Courtesy USDA.)

Grasses that must be planted by vegetative means include zoysia, improved strains of Bermuda grass, St. Augustine grass, centipede grass, creeping bent, and velvet bent. With **plug sodding** small plugs generally are set 1 foot (30 cm) apart, but they may be set closer together if more rapid coverage is desired. With **strip sodding** strips of sod 3 to 4 inches (8 to 10 cm) wide are planted end to end in rows that are about 1 foot apart. **Sprigging** is the planting of individual plants, cuttings, or stolons obtained by tearing apart established lawns. Sprigs can be planted end to end in rows or at spaced intervals. Bermuda grass may be established by spreading shredded stolons and raking lightly to firm them into the soil.

In many climatic regions lawns can be established at almost any season of the year if common-sense precautions are observed, but they are easiest to establish during wet periods when the temperature is moderately cool. For fall seeding in cold weather areas grass should be started at least 45 days before hard freezes are expected. In areas where extremely cold weather is likely, spring seeding is best. Whether a lawn is seeded, vegetatively propagated, or sodded, it must be kept moist until it becomes established. Lawns planted when the weather is hot may need to be mulched and/or sprinkled several times a day. Further lawn planting information is listed in Tables 11–10 and 11–11.

Table 11–10 Characteristics of lawn plants

	Seed, pounds per 1,000 square feet	Time of seeding	Mowing height (inches)	Adaptations	Special characteristics
			Grass		
Bahia	2–3	Spring	1	Warm, humid areas.	Coarse; mostly ground cover.
Bermuda (common)	2–3	Spring	½	Warm areas.	Will grow in low fertility, highly acid soil.
Blue grama	1–1½	Spring	1½	Cool, dry areas, Great Plains.	Drought resistant; won't tolerate heavy traffic.
Buffalo	½–1	Spring	1½	Well-drained, heavy soils; Central Plains.	Drought resistant.
Canada blue	2–3	Fall	1½	Gravelly soils of low fertility.	Resists wear; thin open turf.
Carpet	3–4	Spring	1	Fertile, sandy soils, moist year-around.	Resists wear.
Centipede	2–3	Spring	1	Best low-maintenance lawn grass for the South; spreads rapidly.	Can destroy grazing value of pastures so should not be planted in farm lawns.
Chewings fescue	3–5	Fall, spring	1½	Shade-tolerant grass for cool regions.	Nonspreading, difficult to establish a good stand.
Colonial bent (Highland and Astoria)	1–2	Anytime ground isn't frozen	½	Cool, moist; New England and Oregon, Washington coast.	Forms highest quality, fine-textured lawn; requires special care.
Crested wheat	1–2	Spring	2	Spring and fall lawn for unirrigated cool regions; dormant during midsummer.	Tough, coarse; very drought resistant.
Kentucky blue (Common, Delta)	2–3	Fall, spring	1½	Most common lawn grass in U.S.; cool humid and cool irrigated areas.	Withstands wear; drought resistant.
Kentucky blue—improved (Merion, Newport, Cougar, Windsor)	1–2	Fall, spring	¾	More heat-tolerant and leaf spot resistant than common bluegrass.	Requires more fertilizer than common bluegrass; Merion is rust susceptible.

(continued)

	Seed, pounds per 1,000 square feet	Time of seeding	Mowing height (inches)	Adaptations	Special characteristics
Red fescue (Pennlawn, Ranier)	3	Fall, spring	1½	Good for cool, humid, or shady areas; tolerates acid soils.	Common in mixes with Kentucky bluegrass for cool areas.
Red top	1–2	Fall	1½	To establish quick cover for temporary lawn or in mixtures with other grasses; overseeded with Bermuda in winter to provide year-round green.	Seldom lasts more than 2 years with heavy mowing; tolerates poorly drained acid soils.
Rye grass (annual, perennial)	4–6	Fall, spring	1½	Good for temporary cover on sloping sites and other places.	Seed is large, so rye grass is a common component of inexpensive lawn mixtures.
St. Augustine grass	(Vegetative only)	Spring, summer	1	No. 1 shade-tolerant grass for area south of Augusta, GA, Birmingham, AL to coastal Texas; good for Florida muck soils.	Withstands salt spray; subject to damage by chinch bugs and several diseases.
Tall fescue (Kentucky 31, Alta)	4–6	Fall	1½	Used for playing fields and other locations where wear is more important than beauty; somewhat shade tolerant.	Extremely wear resistant; coarse textured.
Velvet bent	1–2	Fall	½	Cool humid regions; New England and Pacific Northwest coast.	Finest textured of lawn grasses; requires close mowing, regular watering, fertilizing, and disease control.
Zoysia (Japanese lawngrass, Manila grass)	1–2	Spring, summer	1	South of a line from Philadelphia to San Francisco; Manila grass is the best shade grass for the mid-South.	Turns straw yellow with first frost; resistant to wear; stands close clipping.
75% bluegrass 25% red fescue	2–4	Fall, spring	1½	Mixture for sunny locations of Canada and north and central U.S.	All-purpose mixture.
25% bluegrass 75% red fescue	2–4	Fall, spring	1½	Mixture for shady locations of Canada and north and central U.S.	All-purpose mixture.
			Other seeded lawn plants		
Dichondra	Seed or vegetative plantings	Cool season	1	Central and southern California; not recommended for other areas.	Requires large amounts of water; becomes stemmy in many locations.
White clover	1–2	Fall, spring	1½	Cool areas; will not persist during hot weather.	Provides nitrogen for lawn if uniformly mixed with grass; killed out or becomes patchy if sprayed with weed killers.

(Adapted from *Better Lawns*. USDA Home and Garden Bulletin 51. U.S. Government Printing Office, Washington, DC. Revised periodically.)

Table 11–11 Vegetative grasses—rate and time of planting

Grass	Amount of planting material per 1,000 square feet	Time of planting
Bermuda grass	10 square feet of nursery sod or 1 bushel of stolons	spring–summer
Buffalo grass	25–50 square feet of sod	spring
Carpet grass	8–10 square feet of sod	spring–summer
Centipede grass	8–10 square feet of sod	spring–summer
Creeping bentgrass	80–100 square feet of nursery sod or 10 bushels of stolons	fall
Velvet bentgrass	80–100 square feet of nursery sod or 10 bushels of stolons	fall
Zoysia	30 square feet of sod when plugging; 6 square feet of sod when sprigging	spring–summer

(Adapted from *Better Lawns*. USDA Home and Garden Bulletin 51. U.S. Government Printing Office, Washington, DC. Revised periodically.)

Renovating and Maintaining a Lawn

Whenever the decision to improve a poor lawn is made, a choice between renovating the old lawn or establishing a new one becomes necessary. If the lawn is poorly drained or there is little grass of desirable species, it is usually best to tear up the lawn with a plow or spade and establish a new one.

If there is a high percentage of desirable grasses scattered throughout the area, the lawn can usually be renovated by the following steps: (1) Apply weed-control measures as necessary to rid the turf of as many weeds as possible; (2) clip the old grass close to the ground and rake away leaves, grass clippings, and any other foreign matter that has accumulated; (3) rake vigorously to loosen the surface and remove thatch; (4) cultivate and reseed bare spots; and (5) apply nitrogen fertilizer and water enough to keep the top 12 inches (30 cm) of soil moist.

Several herbicide treatments at 10- to 20-day intervals may be necessary in order to completely eliminate broad-leaved weeds from a long-neglected lawn area.

Timely and constant maintenance is the key to successful lawns. Suggestions for maintenance fertilization are given in Chapter 5 and "Fertilizing Garden Crops," Chapter 14 (see also Figure 11–36). Suggestions for irrigation are in Chapter 6. Lawns should be mowed frequently, even though little of the top growth is removed (Figure

A

B

Figure 11–36 Fertilizer spreaders: (A) Typical "drop-type" suited for nonpelleted fertilizer; (B) typical broadcast spreader for pelleted fertilizer. (Courtesy Washington State University.)

11–37). With frequent mowing and good fertilizer practices it is not usually necessary to remove the clippings. Bentgrass and Cougar and Merion bluegrass will look best if cut to a height of between ¾ and 1 inch (2–2½ cm). Most other cool-season grasses should be kept at a height of about 1½ to 2 inches (4–5 cm), especially during hot weather. Crabgrass is reduced by the shading effect of taller permanent grasses.

Warm-season grasses, particularly Bermuda grass, will require closer mowing than do most cool-season grasses. Bermuda grass should be cut frequently to a height of ⅝ inch (about 1½ cm) to maintain a fine quality turf. Other warm-season grasses, such as zoysia, centipede grass, carpet grass, and St. Augustine grass, should be mowed to a height of about 1 inch. The mower should be kept sharp so the grass will be cut cleanly without bruising or tearing.

Proper mowing, irrigating, and fertilizing will usually minimize the build-up of **thatch,** the accumulation of a dry layer of clippings at the soil surface, which is most likely to be a problem with fine-leaved grasses. If thatch does accumulate to a depth of ½ inch, it should be removed, as it prevents penetration of air, water, and plant nutrients. There are several types of machines available for thatch removal that can be rented from hardware or garden stores. Early spring is generally the best time to remove thatch if it becomes necessary.

Pest and Weed Control

The best pest control for any lawn is good management. Weeds, insects, and diseases are all likely to be less of a problem in lawns that have received a balanced fertilizer application, timely irrigation, and frequent mowing to the recommended height. Pest control in well-cared-for lawns growing in cooler regions may require only a single spring application of herbicide. In warmer regions even well-managed lawns may require more attention to keep pests from damaging or destroying them.

Even with good maintenance, weed seeds may get established in a lawn, having spread from a neighbor's yard or nearby weedy roadside, and herbicides may be necessary to help control them. Three different types of

A

B

Figure 11–37 Reel (A) and rotary (B) lawn mowers. Use of these lawn mowers requires extreme caution. Feet can slip into the blade, and the blade can propel rocks and other solid objects at extreme speed. (A by Jack Schneider, courtesy USDA.)

weeds are the usual offenders—crabgrass, other weedy grasses, and broadleaved weeds.

Crabgrass is an annual that develops from seeds produced the previous year. New plants can continue to establish from late spring until the first fall frost. Crabgrass is most serious in the warmer parts of the Midwest, the South, and the Southwest. It will not tolerate shade, and a thick, dense turf that is cut no shorter than 1½ inches retards its growth. Raking the lawn just before mowing to raise the seed heads within reach of the mower blades will help reduce the amount of crabgrass seed in the soil. Herbicides are available that prevent germination of crabgrass seed for several months but do not seriously damage established perennial grass. However, these materials also prevent the germination of lawn grass seed, so replanting a bare spot is not possible for some time after treatment to control crabgrass. There are also herbicides suitable for controlling crabgrass as it germinates and after it has established some growth. Crabgrass killers are sold under various trade names at garden and farm stores. Directions on the label should be followed carefully to avoid damage to the lawn and nearby plants.

The control of perennial grasses such as orchard grass, timothy, and others is usually accomplished with dalapon (2,2-dichloropropionic acid) at ¼ pound dissolved in 1 gallon of water (150 g/l) and applied to individual plants or infested areas. All grasses in the treated area will be killed. Dalapon usually will dissipate from warm, moist soil within 3 to 6 weeks; it may persist longer in dry, cool soil. The treated area will then have to be reseeded or resodded with sod from a less conspicuous part of the yard. Single clumps of coarse perennial grass can be dug from a lawn. All rhizomes must be removed, and the area should be reseeded or resodded immediately.

Broadleaved weeds in lawns are generally controlled with 2,4-D (2,4-dichlorophenoxyacetic acid). The sodium salt or amine forms of 2,4-D are best for home use. The more volatile ester forms are likely to drift onto and damage nearby shrubs and trees. Sprays are more effective and less expensive than dust. 2,4-D should be applied in the early spring at about the time dandelions begin to bloom or in the early fall during peak periods of weed germination and growth. Herbicides should be applied when rain is not expected for at least 12 hours, and the treated lawn should not be mowed for at least 3 or 4 days. The chemical can be applied in a coarse spray from a sprayer or from a sprinkling can. Applying fertilizer just before or just after treating with 2,4-D will stimulate grass growth so that it crowds into bare spaces left by killed weeds (Figure 11–38). Many of the commercially available herbicide formulations for lawns will contain small amounts of silvex (2-[2,4,5-trichlorophenoxy]-propionic acid) and perhaps some other herbicide. For some weeds these are more effective than 2,4-D alone. For further discussion of herbicides, see Chapter 9.

A number of commercial companies are formulating mixtures of herbicide and fertilizer for application to turf. These mixtures have given excellent results where large areas of turf, such as golf courses, college campuses, or parks, are to be treated. Some homeowners have had problems with these materials. Accidental application to trees, shrubs, or gardens will, of course, kill the treated plants; and one of the persistent herbicides sometimes used in these formulations has occasionally accumulated to concentrations high enough to damage trees and shrubs with roots in the treated turf.

Figure 11–38 Results of treatment with 2,4-D. On a mixed clover and grass area 2,4-D will kill some of the clover, resulting in a patchy-appearing turf.

Lawn grasses vary in their susceptibility to insects and diseases, but all kinds are susceptible to some pests. Lawns are susceptible to a number of soil-inhabiting insects, which can be best controlled during the spring by applying an insecticide and immediately afterward sprinkling the lawn thoroughly. One application may control the pest for several years. Control of soil insects is slow and it may be some time before the insecticide becomes fully effective. Insecticides for above-ground insects should not be applied unless the insect is present and causing damage. When they are applied, the lawn should be sprinkled very lightly to wash some of the insecticide to the base of the plants. Plants should not be watered again for several days.

A description of some of the more troublesome lawn disease and insect pests and controls for a few of them are listed in Table 11–12.

Other Lawn Problems

Injury from other causes is occasionally mistaken for disease or insect injury. Burning with chemical fertilizer; damage from dog urine; scorching caused by placing rugs, mats, or plastic on the lawn in hot weather; injury by weed killers and drought all may be mistaken for disease damage. Lawns are sometimes damaged by moles, pocket gophers, and field mice. Traps in mole and gopher tunnels are effective; poison bait will control pocket gophers and field mice but not moles.

Moss growing in the lawn is generally the result of lack of fertility. It can also be caused by poor drainage, high soil acidity, excess shade, and soil compaction. Moss can be removed by hand raking, burning with ammonium sulphate, or spraying with copper sulphate.

Shade often prevents growing an attractive lawn. Under trees grass receives insufficient light and is likely to be robbed of nutrients. The problem can be partially overcome by planting trees such as honeylocust or birch that cast only light shade. Cutting away lower branches and heavy pruning to reduce the amount of shade will also help. Fertilizing trees so their roots can feed below the level of the grass roots will allow grass to utilize more of the surface-applied fertilizer. This can be done by placing fertilizer for trees in holes 18 inches (45 cm) deep punched with a metal rod at intervals beneath the tree canopy. Watering thoroughly through the root zone of both the grass and the trees will help keep the tree roots from concentrating on the surface. Tree roots that do protrude above the soil surface can usually be pruned off without damaging the tree. Shade-tolerant grasses, such as St. Augustine grass and Manila grass for the South and fescue for the North should be planted in heavily shaded areas. Finally, if it is impossible to get turf grass to grow under trees, a shade-tolerant ground cover such as periwinkle, pachysandra, or ivy could be used.

Plants Used as Ground Covers

Technically any plant is a ground cover; however, the term is usually reserved for low-growing spreading ornamentals. Some or most kinds of many plant groups already discussed are classed as ground covers, including low-growing shrubs, lianas and vines, and herbaceous plants grown for both their flowers and their vegetation.

Ground covers are the ideal planting for banks and slopes too steep to mow; small, isolated areas that would be awkward to mow; shady areas where lawns will not grow well; and areas not designated for activities that require paving or turf and where the owner does not want the responsibility of continual maintenance of turf.

Even more than other plantings, ground covers must be ecologically adapted to their location. Unless they grow vigorously, their growing site will require constant hand weeding to eliminate competition from other species. Ground covers that grow 12 or more inches (30 cm) in height will shade out the unwanted growth of weedy locations better than will fully prostrate materials (see Figure 9–7, Chapter 9). On the other hand, plants 1 to 2 feet tall may not provide the desired effect. A favorite springtime scene is a bed of daffodils that have elongated through low-growing, purple-flowered vinca. Similarly, the carpetlike effect of the ground-hugging Bar Harbor juniper (*Juniperus horizontalis* 'Bar Harbor'), kinnikinnick (*Arctostaphylos uva-ursi*), or some wooly thyme (*Thymus*

Table 11–12 Major insect and disease pests of home lawns

Insect pests	Symptoms and damage	Control
Grubs and wireworms	Grubs are larvae of several kinds of beetles. They burrow around and kill grass roots, causing the lawn to be patchy and unthrifty. Wireworms, the larvae of the click beetle, are dark brown, hard, smooth, slender, and ½ to 1½ inches (1¼ to 4 cm) long. The grub of the Japanese beetle is probably the most destructive grub in the eastern states.	Count grubs by cutting from several places in the lawn 1 foot square (90 cm^2) strips of sod on three sides, undercutting 2 to 3 inches (5 to 8 cm) beneath the surface, and laying them back using uncut side as hinge. If there is an average of three or more grubs per square foot, an insecticide should be applied. Milky disease spores (see billbug control) are effective for some grubs. Don't apply both spores and insecticide.
Ants, wasps, and wild bees	Grass is buried by their mounds. Grass seed and roots are often damaged by their nesting.	Apply insecticide to soil in vicinity of mounds.
Mole crickets	Light brown cricketlike insects with lower surface lighter than upper. The burrowing and feeding of one mole cricket can damage several yards of newly seeded lawn in a night. Most numerous in south Atlantic and Gulf states.	Where these insects are numerous, treat with a soil insecticide before planting the lawn.
Billbugs	Grubs feed on roots. Adults are tan to reddish-brown beetles ⅕ to ¾ inches (less than 2 cm) long with long snouts and strong pincers. They burrow into the grass stem near soil surface and feed on leaves. Zoysia grass is especially susceptible.	Billbug grubs are susceptible to a bacterial disease, milky disease. Spore formulations can be purchased, but it may require several years for complete control. Once the disease is established, control is permanent.
Earthworms	Make mounds or castings that can interfere with mowing and give lawn an uneven surface.	Rake down castings with a fine-toothed rake. Increase mowing height to hide castings. Earthworms improve aeration and water infiltration and help decompose thatch. Can be controlled with insecticide.
Sod webworms	Light brown, hair-covered, ¾ inch (2 cm) long larvae of lawn moth. They rest in silken webs during the day and feed at night. As they grow older, they build grass and silk-lined tunnels near the surface of soil. Most damaging in areas south and east of a line from Kansas to Maryland. Irregular brown spots in lawn are first sign of damage.	Break apart drying sod. If more than three or four webworms are in a 6 inch square (38 cm^2) section, insecticide should be applied.
Army worms	Black-striped green larvae that usually are found in clusters of several hundred. Can eat grass down to the roots in a very short time.	Isolated groups can be destroyed by squashing them. Use insecticides for heavier infestations.
Cutworms	Brown or gray caterpillars that hide in the soil during the day and cut off plants by feeding on their bases at night.	Apply insecticide.
Chinch bugs	Yellowish spots that turn rapidly into brown dead areas are the major symptoms. Young nymphs are about the size of a pinhead, bright red with a white band across their backs. Adult nymphs are black and have a white spot on back between wing pads. Adults are about ⅙ inch (less than ½ cm) long with black and white markings. Damage is most likely in the East and South.	Identify infestation by sinking a can open at both ends halfway into the turf. Fill can with water and watch for about 5 minutes for chinch bugs to float to surface. Insecticide application most effective in early spring.
Scale insects	Scale insects attach themselves to the crown or roots of grass and secrete a hard waxy or white cottony covering. They damage the plant by sucking juices from it. Plants attacked usually turn brown and die, leaving patchy, dead places in the lawn. Scale insects in lawns are most serious in the South and Southwest.	Fertilize and irrigate to keep grass growing rapidly; dispose of infested clippings. Several weekly treatments with an insecticide effective against scale during spring season may be necessary.
Leafhoppers	These ½ inch (1¼ cm) long, speckled insects suck sap from the leaves and stems of grass, killing newly seeded lawns and causing whitened patches that may be mistaken for drought damage in older lawns.	Apply insecticide when leafhoppers become numerous enough to do damage.

Insect pests	Symptoms and damage	Control
Insect mites	Tiny, barely visible insects that suck sap from grasses, causing leaves to be blotched and stippled and can eventually kill the plant. A fine webbing on the leaves is usually associated with mite infestation. When mites are numerous in the yard, they sometimes enter homes in large numbers.	Periodic application of miticides. Mite damage often occurs because insecticide application for other pests kills their predators.
Earwigs, ticks, chiggers, slugs, snails, fleas	These insects do not damage the lawn but often live in grass and cause annoyance or injury to people on the lawn. Some of them may enter the home from the lawn area.	Insecticides applied directly to the lawn or, in the case of ticks and fleas, insecticide treatment of animals that carry the insects onto the lawn.

Diseases	Symptoms and damage	Control
Brown patch	Fungus disease that attacks practically all kinds of turf grass and produces irregularly shaped brown spots 1 inch (2½ cm) to several feet in diameter. The spots have a dark, smoke-green effect around the outer edges where the fungus is active. Most serious during the period of high humidity when daytime temperatures drop to develop dew or fog at night.	Avoid overstimulation with nitrogen. Apply a turf fungicide.
Dollar spot	A fungus disease most severe on Kentucky bluegrass, bentgrass, rye grass, centipede grass, and St. Augustine grass, causing straw-colored spots about the size of a silver dollar. Most serious during spring and fall periods of cool nights and warm humid days.	Turf fungicide.
Leaf spot	A disease causing reddish-brown to purple-black spots on the leaves of Kentucky bluegrass. It may spread to the crown, causing considerable damage.	Mow no shorter than 1¾ to 2 inches (4 to 5 cm). Use adequate but not too much fertilizer. Merion and other improved Kentucky bluegrasses have some resistance to leaf spot. Fungicide application every 7 to 10 days starting with the first symptoms will also control the disease.
Snow mold	Commonly found in the northern United States and Canada during winter and early spring, usually known as white mold on dead brown patches of old grass as the snow melts in the spring.	Avoid nitrogenous fertilizers late in the fall. Rake away heavy matted grass and leaves before the snow comes. If severe, mercury fungicides through the winter will help control the disease.
Leaf and stem rust	Reddish-brown to black powdery spots in leaf and stems. Most serious on Merion bluegrass. Forms as a result of heavy dews during warm weather.	Increase nitrogen fertilization; water during dry periods and mow frequently.
Red thread	A fungus disease that occurs most frequently in fescues and bentgrasses during cool, wet periods. Most serious in the coastal area of the Pacific Northwest. Diseased areas vary in size and shape; appear scorched. Grass changes color from green to pink to brown and finally light tan. The fungus produces characteristic red threads or strands over the dead blades of grass.	Maintain balanced fertility and apply a cadmium fungicide in spring and fall.
Damping-off	Young seedlings in newly seeded lawns decay at the soil line and fall over. The disease is favored by cool, damp weather.	Plant the lawn when growing conditions are favorable; prepare a good seed bed and avoid overwatering. Treat the seed with a protectant fungicide.
Powdery mildew	Most severe on bluegrass; mildew appears as grey-white powdery masses on leaves and stems. Severely affected leaves may turn yellow and die.	Keep turf well fertilized and watered to maintain high vigor.
Fairy ring	These rings are caused by several fungi. Rings or circles of dark green grass appear and gradually grow larger. Just inside this ring there is often a second ring of dying or dead grass, although the grass in the center of the circle may be normal. Sometimes mushrooms grow on the ring's edge.	Keep the turf well fertilized; daily soaking of the affected area with water for a month will help control fairy ring.

Figure 11–39 Wooly thyme as ground cover.

lanuginosus) may provide just the right contrast to accentuate the color and form of semidwarf evergreen shrubs or perennial flowers (Figure 11–39). The low spreading shrubs and many of the lianas described in Table 11–9 and Table 14–25 (Chapter 14) would make good ground cover plants.

Selected References

Adams, E. B. *Homescaping.* Intermountain Regional Publication 4. Published by the University of Wyoming, Laramie, for the combined extension services of various Rocky Mountain, Great Plains, and Pacific Coast states. Revised and reprinted periodically.

Brooklyn Botanic Garden. *Annuals.* Handbook 74 (special printing of *Plants and Gardens,* vol. 30, no. 2), 1974. *Gardening in the Shade.* Handbook 61 (special printing of *Plants and Gardens,* vol. 25, no. 3), 1969. *Nursery Source Guide, 1200 Trees and Shrubs.* Handbook 83 (special printing of *Plants and Gardens,* vol. 33, no. 2), 1977. *Rock Gardens.* Handbook 10 (special printing of *Plants and Gardens,* vol. 8, no. 3), 1952. *Small Gardens for Small Spaces.* Handbook 84 (special printing of *Plants and Gardens,* vol. 33, no. 3), 1977. *Succulents.* Handbook 43 (special printing of *Plants and Gardens,* vol. 19, no. 3), 1963. All published by the Brooklyn Botanic Garden, Brooklyn, NY.

Bush-Brown, Louise, and Bush-Brown, James. *America's Garden Book.* rev. ed. Scribner's, New York. 1966.

Carpenter, P. L., Walker, T. O., and Lanphear, F. O. *Plants in the Landscape.* W. H. Freeman and Company, San Francisco. 1975.

Laurie, Alex, Kiplinger, D. C., and Nelson, K. S. *Commercial Flower Forcing.* 7th ed. McGraw-Hill, New York. 1968.

Nelson, W. R. *Landscaping Your Home.* rev. ed. University of Illinois, College of Agriculture Cooperative Extension Service Circular 1111, Champagne-Urbana, IL. 1975.

Sunset magazine. *The New Western Garden Book.* Lane, Menlo Park, CA. 1979.

U.S. Department of Agriculture. *Better Lawns.* USDA Home and Garden Bulletin 51. U.S. Government Printing Office, Washington, DC. Revised periodically.

U.S. Department of Agriculture. *Landscape for Living.* (*Yearbook of Agriculture,* 1972). U.S. Government Printing Office, Washington, DC. 1972.

12 The Vegetable Garden

The headlines of a recent Associated Press news release read: "Farmers lose over $14 billion in vegetable sales." The article went on to explain that, during the past year, home gardeners had grown about half the vegetables consumed in this country, valued at $14 billion, a sum that farmers supposedly would have earned had there been no home gardens. The amount of the $14 billion remaining after expenses are subtracted is a virtual tax-free savings enjoyed by the estimated 50% of U.S. families who grow vegetable gardens each year. The percentage of Canadians who are vegetable gardeners is about the same.

Other news releases periodically remind us of the dietary value of the high fiber, vitamin, and mineral, and low fat and calorie content of vegetables in combating degenerative diseases such as obesity, hypertension, heart disease, and cancer. In addition, those who grow their own vegetables benefit from the physical exercise involved with gardening.

To many, saving money and having healthier bodies may not be the most important reasons for growing vegetables. They enjoy the thrill of watching plants grow, the primal satisfaction of supplying food for the body in a direct way, or the gustatory pleasure derived from consuming freshly harvested vegetables.

Planning the Vegetable Garden

Planning the vegetable garden is essential if optimum production is to be obtained from the area available. The arrival of seed catalogues in early January is often the catalyst that starts those living in cold winter climates planning their gardens.

Like the landscape, the vegetable garden plan should begin with an accurate measurement of the site. The plot with each crop row properly spaced should be drawn to scale. Vegetables that are to be planted at the same time should be grouped together so that the areas they occupy can be weeded, cultivated, and, if feasible, replanted together. Perennials should be along one side so that they will not be disturbed when the rest of the garden is worked. The plot plan showing final location of each crop should be saved from year to year so that rotation is possible. Unless enough land is available so that a field or cover crop can be produced at least every other year, the gardener should alternate the location of individual vegetables. This will permit better utilization of plant nutrient reserves and prevent the buildup of insects and soil microorganisms detrimental to a specific crop. The growth habit of each kind of plant should be considered in planning so that, for example, a sun-loving crop isn't planted in the shade on the north side of a tall-growing crop such as sweet corn or so that vine crops aren't located where they will spread to choke out nearby low-growing, less vigorous vegetables. A sample garden plan is shown in Figure 12–1.

If more than one site is available, the garden should be as near the kitchen as possible but should not be near large trees as these will shade the site and deplete it of mineral nutrients. All problems and interesting observations that pertain to the garden should be recorded so

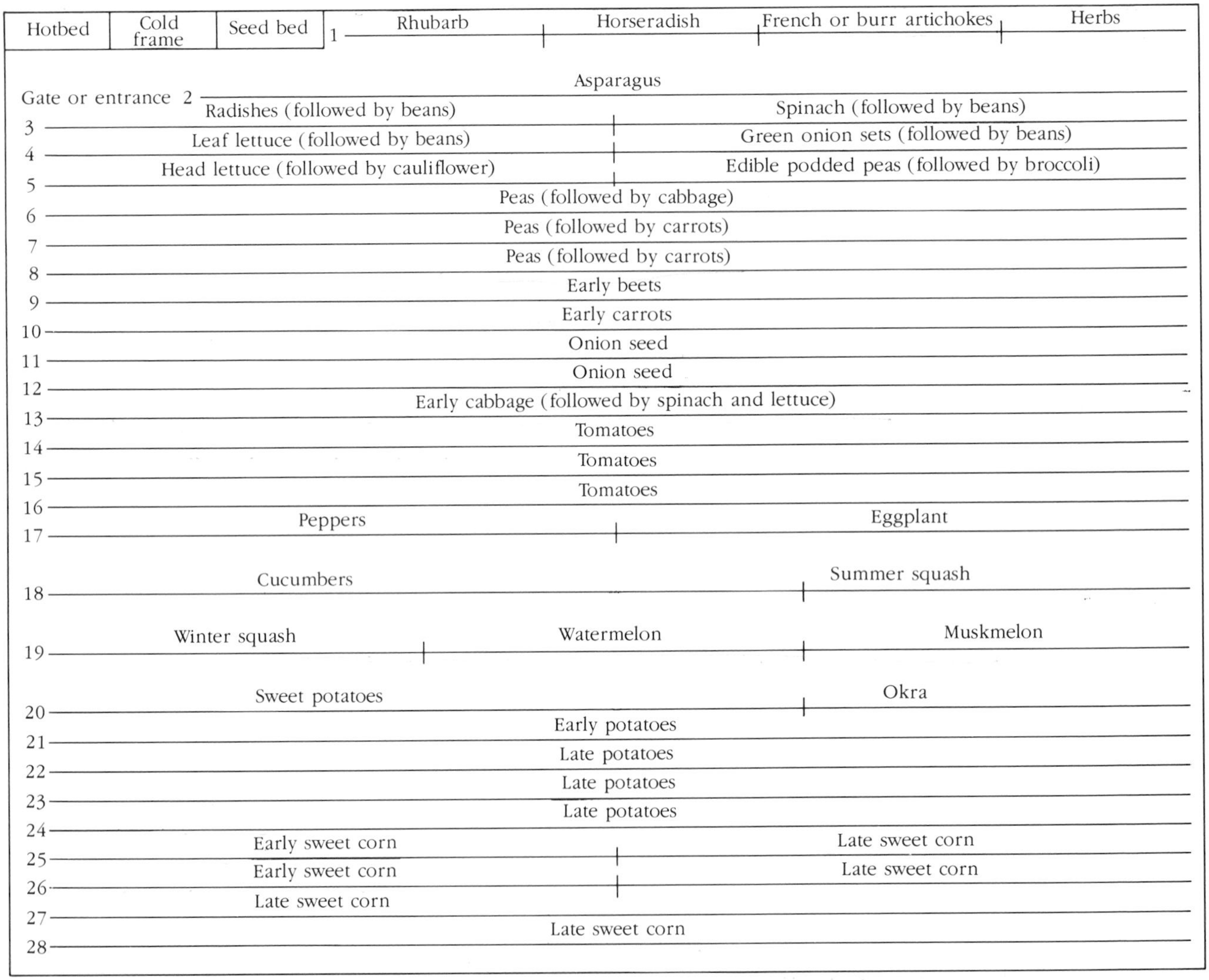

Row	Planting		
	Hotbed	Cold frame	Seed bed
1	Rhubarb	Horseradish	French or burr artichokes / Herbs
2 (Gate or entrance)	Asparagus		
3	Radishes (followed by beans)	Spinach (followed by beans)	
4	Leaf lettuce (followed by beans)	Green onion sets (followed by beans)	
5	Head lettuce (followed by cauliflower)	Edible podded peas (followed by broccoli)	
6	Peas (followed by cabbage)		
7	Peas (followed by carrots)		
8	Peas (followed by carrots)		
9	Early beets		
10	Early carrots		
11	Onion seed		
12	Onion seed		
13	Early cabbage (followed by spinach and lettuce)		
14	Tomatoes		
15	Tomatoes		
16	Tomatoes		
17	Peppers	Eggplant	
18	Cucumbers	Summer squash	
19	Winter squash	Watermelon	Muskmelon
20	Sweet potatoes	Okra	
21	Early potatoes		
22	Late potatoes		
23	Late potatoes		
24	Late potatoes		
25	Early sweet corn	Late sweet corn	
26	Early sweet corn	Late sweet corn	
27	Late sweet corn		
28	Late sweet corn		

Figure 12–1 The vegetable garden plan. Such a garden—length 200 feet (60 m), width 100 feet (30 m)—should produce all the vegetables a large family can use throughout the growing season and a surplus for canning, storing, and drying. (From *Suburban and Farm Vegetable Gardens,* USDA Home and Garden Bulletin 9. U.S. Government Printing Office, Washington, DC. 1967.)

that they will be remembered and avoided (or capitalized on) during subsequent seasons.

Size of the Garden

Don't give up the idea of a garden if you have only a small plot. A friend of ours grew so many tomatoes and beans on a 14 x 22 foot (4 x 7 m) plot that he had to give some away, and he also grew lettuce, spinach, radishes, strawberries, and cucumbers on this plot. He staked his tomatoes and trained them to a single stem. Planted one foot (30 cm) apart in rows 24 inches (60 cm) apart, they yielded a bushel to the picking from only a few vines. Cucumbers also can be staked and grown on a trellis or fence.

If the garden is extremely small, don't grow low-yielding kinds of vegetables such as corn, watermelon, cantaloupe, or peas. Use plenty of fertilizer and water and practice succession cropping. For instance, cabbage can be planted following spinach, or carrots can be planted in the space where radishes or lettuce were harvested.

If the garden plot is larger, of course more can be planted. In this case it may not pay to stake plants, because gardening time rather than space is likely to be limiting. Moreover, the large garden often can be arranged so that it can be cultivated by machinery. Don't overplant. A small garden well cared for will yield a great deal more than a large plot poorly managed.

Climate

The climate will also determine kinds and cultivars that can be grown. In cooler areas radishes, lettuce, beets, carrots, turnips, peas, potatoes, cauliflower, cabbage, green onions, and generally beans will grow to perfection. In warmer areas corn, tomatoes, beans, cucumbers, and melons do well, and some of the cool-season crops cannot be grown except during early spring or late fall. If your home is located on a south slope, you may be able to grow crops that you would be unable to grow on a north slope in the same area. Planting on the south side of a building will also hasten maturity. Crops grown in a sandy soil are earlier than those grown in a heavier clay soil; but heavy soils can be conditioned by adding humus. Growing from transplants will allow crops to mature earlier and also permit production of kinds that would not normally mature if they were direct seeded in some short-season areas. Because of their shorter growing time in the garden, transplants also free space for other crops. Short-season crops such as radishes or spinach often can be matured before transplants are set out, and the earlier maturity of transplanted crops may allow time for growing a succession crop.

Quantity to Plant

The amount of each vegetable planted will be determined by the length of time it can be harvested from the garden, whether or not it can be stored or processed, the space available, the number of people using the garden produce and their likes and dislikes, and the yield of the vegetable. For example, planting a few feet of row at any one time will usually produce all the lettuce a family can consume during the time it retains good quality in the garden. Planting a small amount of this crop each 2 weeks will provide lettuce of the right maturity throughout the growing season for lettuce in an area. A short row of Swiss chard or New Zealand spinach and three hills of summer squash will usually provide as much of these vegetables as a family can consume.

Growing the Vegetable Garden

The ideal soil for all-around vegetable production is one that warms early, has good drainage but is retentive of moisture, works easily, and does not pack. Sandy loam soils probably come closest to this ideal, especially if irrigation water is available. Both sandy and clay soils can be improved by the addition of manure. Where manure is not available, vegetable growers must use commercial fertilizers and soil-improving crops to maintain soil structure and fertility (see Chapter 5 and "Fertilizing Garden Crops," Chapter 14). Although clay soils may not be ideal for gardens, tomatoes, sweet corn, beans, squash, and cucumbers will produce excellent yields in them, and even root crops can be produced in carefully managed clay loams.

Vegetables are heavy users of fertilizers; however, gardens are as frequently overfertilized as underfertilized. A soil test carried out by a state or reliable commercial laboratory is the best way to determine fertilizer needs. A number of factors are involved in developing a fertilizer program that will produce high yields. One is the type of soil being used. For example, acid soils frequently need phosphorus, sandy soils may be low in nitrogen and potassium, and muck or peat soils often are low in phosphorus and potassium. Another factor is the type of plant to be grown. Leafy vegetables are heavy feeders and need large and continuous supplies of nitrogen, root and tuber crops often need additional potassium, and fruit crops such as tomatoes respond to phosphorus. Ordinarily, most vegetable crops grow best on soils with a pH around 6 to 7. However, potatoes are freer of the disease, scab, when grown on soils either more acid or more alkaline.

An adequate, uniform water supply is needed for the production of quality vegetables. Succulents and most root crops lose quality rapidly under conditions of insufficient moisture. The entire root zone should be brought up to field capacity with each irrigation. In areas and during periods where rainfall does not occur, irrigation should usually be provided each 7 to 10 days during the warm summer. If irrigation water is unavailable and rainfall may be limited, a gardener can produce spring and fall crops. Some compensation for a lack of moisture can also be provided by wide spacing of the vegetable plants, by control of weeds, and by mulching. For seeding time and procedures for choosing, producing, and planting transplants see Chapter 4 and "Propagating Plants from Seed," Chapter 14.

Winter Vegetable Gardening

In warmer parts of Florida, Texas, California, and Arizona vegetable gardening can be a year-round hobby, cool-season vegetables being grown during the fall, winter, and spring and warm-season vegetables during the summer. As soon as one crop is harvested, another can be planted in its place. Because plant pests are not killed by cold temperatures in these areas, rotation among unlike vegetables or with field crops is especially important.

Slightly further north and along the Pacific coast where heavy freezes do not often occur, vegetables such as lettuce, broccoli, spinach, onions, and peas can be planted in the fall to winter over and mature during the early spring. Some of the biennial crops can also be wintered over in these areas if they are very small during the period of cold temperatures. If the plants have produced a second or third set of leaves before their growth is halted by winter cold, seedstalks will be initiated and will form as soon as the weather warms in the spring.

Even further north in many areas of the United States and Canada such crops as spinach, lettuce, and some of the crucifers can be fall seeded if the small seedlings can have a snow cover or can be provided with a loose mulch of straw or shavings to protect them from hard winter freezes. Also, in these regions many gardeners tend to utilize only part of the growing season. For example, in eastern Washington state in mid-November we are still harvesting head and leaf lettuce, broccoli, carrots, beets, turnips, onions, chard, cabbage, radishes, and potatoes from the garden even though the first light frosts occurred almost two months ago. We have often been able to keep a continual supply of tomatoes until after Thanksgiving by harvesting them at various stages of maturity just after the first light frost and keeping them in a room where temperatures are 55° to 60°F (13° to 16°C). The autumn here is somewhat milder than in some other areas at our latitude (47° north), but in many sections of the northern United States and southern Canada gardens could produce vegetables until late October and early November (see "Estimating Planting Date," Chapter 14).

Several vegetables will produce food for the table indoors during the winter. If there is a corner of the basement where the temperature ranges between 50° and 70°F (10° and 21°C), rhubarb can be forced. For forcing, 2-year-old or older rhubarb crowns should be dug before the ground freezes. To overcome their rest period, they must be left for at least 6 weeks where the average diurnal temperature is below 40°F (4°C). In most northern areas they can be brought into the building in late December or early January. They should be covered with peat moss or shavings kept moist by frequent watering. Within a few weeks they will produce light pink, delicately flavored rhubarb petioles. Witloof chicory, sometimes called French endive, as well as the common weed dandelion can be forced in the same way. They produce tender, light-colored shoots that are excellent additions to winter salads.

In areas that have winter sunshine a number of kinds of vegetables can be grown in containers or window boxes next to a south or west window. Some of the better vegetables for window boxes are tomatoes, leaf lettuce, and onions. 'Patio' and 'Tiny Tim,' small cherry tomatoes, and 'Bibb,' a butterhead lettuce, are cultivars adapted to container growing. Herbs also can be grown in containers.

Storage of vegetables is discussed in "Storing Horticultural Products," Chapter 14. The home gardener should store a winter's supply of carrots, beets, potatoes, onions, and squash and should also consider canning or freezing surplus vegetables for winter use (see "Home Processing," Chapter 14).

Figure 12–2 Vegetables as ornamentals. Rhubarb, one of the first vegetables to become available in northern gardens, can also be used as an ornamental.

Vegetables in the Landscape

Where there is no room or no particular desire to have a separate vegetable garden, vegetables can be produced in the form of landscape plants (Figure 12–2). Early in the season a border of 'Ruby,' 'Black-seeded Simpson,' 'Salad Bowl,' or some other leaf lettuce will be attractive. The 'Tiny Tim' cherry tomato will produce an ever-changing, interesting border throughout the summer and, from July until frost, will supply salad tomatoes. Peppers also are ornamental and make excellent border plants. Some years ago I observed an interesting ornamental display bed made up of 'Savoy' and 'Red Rock' cabbage. Alternate plants of red- and white-stalked Swiss chard provide beauty and greens throughout the summer if only the outer leaves are removed for table use. New Zealand spinach and kale are frequently grown for their ornamental, textural effects. Squash, pumpkin, and cucumber provide very interesting ornamentals, and their vines can be trained to form an arbor or trellis, provided large fruit is given some type of support. Herbs can be used as rock garden plants, chives and sage being two that provide attractive contrast in any planting. A bed of asparagus can be a welcome vegetable in the spring when greenery is abundant and fresh vegetables are scarce and a green hedge through the summer and autumn when other vegetables are abundant and green foliage less plentiful. 'Red Emperor' tulips and daffodils can be planted with asparagus to provide color in the spring. Later the asparagus ferns will hide the dying leaves while the bulbs are maturing. These are only a few examples of the numerous possible uses for vegetables in the ornamental landscape.

Growing Hints for Various Vegetables

In the following discussion vegetables will be divided into three groups: perennials, cool-season annuals, and warm-season annuals. Individual annual vegetables will be discussed primarily in the order of their ability to withstand cold temperatures, essentially the order in which they can be planted in the spring.

Perennial Vegetables

Perennial vegetables include asparagus, rhubarb, globe artichoke, horseradish, and Jerusalem artichoke. These crops all respond to large amounts of organic matter worked into the soil before they are planted. Land that is to grow horseradish should be manured the autumn prior to being planted, because fresh manure causes misshapen roots (see "Fertilizing Garden Crops," Chapter 14).

Asparagus *(Asparagus officinalis).* Asparagus has been used as a food in the Mideast and Europe for over 3,000 years, and it was grown in this country in colonial days. It is one of the first vegetable crops to reach edible maturity in the spring. It is easily grown in most areas of the United States and is an excellent ornamental after the cutting season. Because there is considerable labor involved in its commercial harvest, it tends to be expensive when purchased; consequently, it is an ideal crop to produce in the garden or yard. Asparagus will grow in almost any soil, including those that are relatively alkaline, but it is best adapted to well-drained sandy loams. Crowns do poorly if planted where asparagus has previously grown, probably because of soil-borne diseases. Growth and development are retarded by excess moisture.

Asparagus is usually grown from 1-year-old crowns, although it can also be produced by direct seeding. For either direct seeding or producing crowns for transplanting, seed is sown as early in the spring as the soil can be worked. Radish seed is often mixed with the asparagus seed to mark the rows for initial cultivation. Radishes can be harvested shortly after the slow-germinating asparagus emerges from the soil. Plants are thinned to 3 to 6 inches (8 to 15 cm) in the row if crowns for transplanting are being grown and 8 to 12 inches (20 to 30 cm) if the bed is being direct seeded.

Under average conditions crowns will be large enough to set in the field the following spring. They should be dug before growth commences. The smallest crowns should be discarded, as they are genetically inferior and will produce a lower total yield during the life of the planting. One-year-old asparagus crowns can also be purchased.

Asparagus is planted by placing crowns on loose soil in the bottom of a 6- to 8-inch trench in rows 4 to 6 feet (1 to 2 m) apart with 1 to 2 feet (30 to 60 cm) between crowns within the row. The trench should be filled gradually as the spears grow up through the soil. Weeds, of course, should be controlled, and care must be exercised in hoeing so that the asparagus shoots are not cut off.

Asparagus shoots or spears are not usually cut for the table until 2 full years after the crowns are set out (3 years after seeding), although it may be possible to harvest a meal or two the second year. Asparagus spears will begin growing in late March to late April depending on the climate, and all spears should be cut as they reach 8 to 10 inches (20 to 25 cm). The cutting season should not be longer than 8 to 10 weeks (Figures 12–3, 12–4).

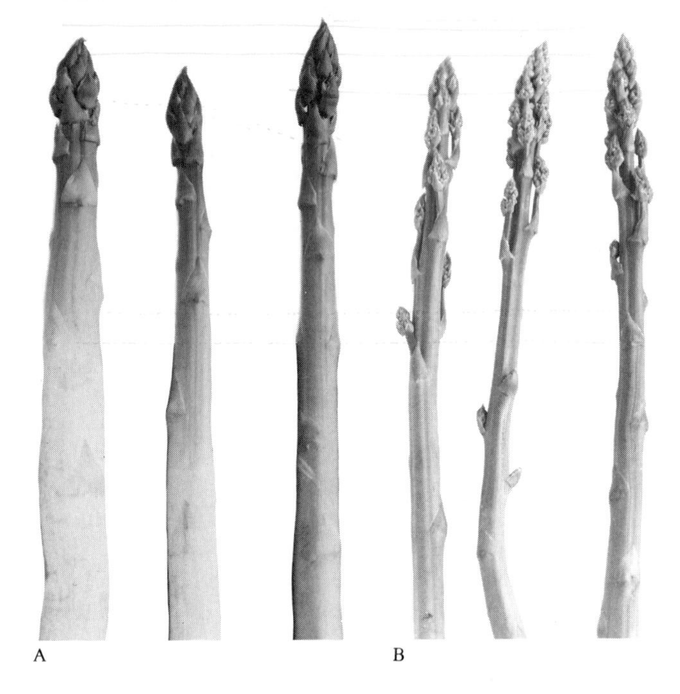

Figure 12–3 Asparagus. The tight asparagus spears (A) are at optimum harvest maturity. The loose, seedy spears (B) either are overmature or were formed when the temperature was high. (Courtesy USDA.)

Figure 12–4 Harvesting asparagus. An asparagus spear usually will break off just above the section that is too tough to consume. Spears that have been broken off do not have the tough butts that are left on spears harvested the usual way, by cutting below the surface.

At the end of the cutting season asparagus beds should receive a thorough cultivation, and the ferns should be permitted to grow through the rest of the season without being cut. Asparagus requires only light fertilization, and once a stand is established it is also quite drought tolerant. A cultivar resistant to rust called 'Mary Washington' was developed in the early 1900s; the principal cultivars grown today are improved selections from 'Mary Washington.' These are rust resistant, produce good spears, and are not too sensitive to hot weather. Once established, asparagus plantings last an average of 18 to 22 years, and some plantings as old as 40 to 50 years are still in existence. Fifteen to 25 plants should produce enough to supply fresh asparagus for a family of four to six plus some left over for freezing.

Rhubarb *(Rheum rhaponticum).* Rhubarb is grown for its thick leaf stalks or petioles, which are cooked into a fruit sauce or made into pies. It is one of the most acid of all vegetables, the juices having a pH of about 3.2. Like asparagus, the edible crop, which is produced early in the spring, is grown from food stored in the crown and roots during the previous growing season. Rhubarb is a hardy, easily grown, attractive plant and deserves a place in the yard of families who live in the cooler parts of North America and who enjoy rhubarb sauce or rhubarb pie.

Rhubarb is usually propagated by crown divisions, although it can be grown from seed. Each division must have at least one bud or eye with as much root or rhizome as possible (Figure 12–5). Plants from seed take at least three years to mature, and they will not be true to the parent type.

In harvesting, which occurs over a 6- to 8-week period, the stalks should be pulled not cut. The petioles will not

Figure 12–5 A rhubarb crown division that has begun to grow.

wilt as quickly if the broad leaves are removed from them immediately after they are pulled. Like asparagus, rhubarb needs an annual light application of fertilizer and occasional good irrigations during seasons and in areas where rainfall is limited. 'MacDonald' and 'Cherry Red' are two highly colored cultivars; 'Victoria,' 'Linnaeus,' and 'German Wine' are older standard sorts. Four good-sized plants will supply an average family.

Globe Artichoke ***(Cynara scolymus).*** Globe artichoke is another perennial grown commercially in this country in only a small area along the California coast near Monterey. It requires a cool growing season but will not withstand freezing of the crown or roots. Its area of adaptation is relatively restricted. It grows quite well in the cooler parts of northern California and in coastal valleys of Oregon and Washington. With some protection of the crown and roots during the winter, it can be grown in home gardens in other relatively cool areas of the country.

It is best propagated from crown divisions, which must contain an eye or bud and a piece of crown. The home gardener is not likely to find crown divisions for sale and may have to produce globe artichoke plants from seed, which is entirely feasible.

Plants should be spaced 3 to 4 feet (about 1 m) apart with 4 to 6 feet between rows. The artichoke plant is a thistle and the artichokes of commerce are flower buds (Figure 12–6). To be edible they must be harvested before they open. If this plant is being produced in areas having cold winter weather, a 6- to 8-inch layer of leaves, straw, or sawdust should be placed over the crowns as soon as the first frosts have occurred in order to protect the crowns through the winter.

Horseradish ***(Armoracia rusticana).*** Horseradish is a perennial grown for its pungent root, which is used as a relish for meat and certain other foods. It is propagated from side roots, which can be purchased. Gardeners who wish to procure their own planting stock can cut side roots 8 to 14 inches (20 to 35 cm) long, the diameter of a lead pencil or slightly larger, from established plants. Traditionally, the bottom of the planting stock is cut slanted and the top straight across, because top and bottom can-

Figure 12–6 The globe artichoke. (A) leaves, (B) fleshy flower head. (B, courtesy J. C. Allen & Son.)

not otherwise be distinguished. Root cuttings are planted 2 x 3 feet apart in the bottom of 3- to 5-inch (8 to 13 cm) deep furrows. The basal end of the cutting should be pressed down so that it is slightly lower than the top.

For commercial production horseradish roots are pruned twice during the season, initially when tops are 8 to 10 inches high and again 6 weeks later. The top of the root is carefully uncovered and lifted slightly without disturbing the lower roots. All but the main crown of leaves and all upper branch roots are rubbed off before the root is recovered with soil. This procedure is necessary only if the main root is to be harvested each year. Commercially, the whole root is dug just before the soil freezes in late October or November. When horseradishes are being grown in the home garden, enough side roots can be removed to supply the family and the main plant is left to grow for many years. The roots are hardy so can remain in the garden in most of North America. For convenient accessibility roots can be dug and stored through the winter (see "Harvesting" and "Storing Horticultural Products," Chapter 14).

Horseradish is prepared for the table by grating it into 4½ or 5% distilled or white wine vinegar. (It will turn dark if grated into ordinary cider vinegar.) The pickled product will remain "hot" under refrigeration for a few weeks.

Jerusalem Artichoke *(Helianthus tuberosus).* Although the Jerusalem artichoke, a member of the sunflower family grown for its edible tuber, is propagated like the potato, it is usually considered a perennial (Figure 12–7). The starch it contains is inulin, and formerly it was recommended as a food for diabetics. The tubers are relatively rough and do not store well out of the soil. It is propagated by planting tubers 18 to 20 inches (45 to 50 cm) apart in rows about 3 feet apart. Jerusalem artichokes will grow in relatively poor soil. Tubers are dug with a spade as needed, and where the soil doesn't freeze they can be harvested through the winter. They can be prepared for the table in the same ways potatoes are prepared. Tubers normally do not freeze, and the Jerusalem artichoke can become a weed if small tubers are scattered through the garden with cultivating equipment.

Figure 12–7 Jerusalem artichoke.

Cool-Season Annual Vegetables

The cool-season annuals can be grown throughout the summer in the northern states and along the Pacific Coast from central California northward. Many of them are spring, fall, or winter crops in other parts of the country. A number of them are botanically biennials, but as the edible portion is produced the first year, they are grown in the garden as annuals.

Even though all of these vegetables are classified as cool-season crops, their response to temperature varies somewhat. For example, radishes, lettuce, peas, and spinach will produce an acceptable crop only if the weather is cool during their entire growing period. Onions must be planted early if a large bulb is to be produced, but they grow quite well in areas where the summer climate is relatively hot. Carrots and beets produce the best quality crop where temperatures during the growing season average between 60° and 70°F (16° and 21°C) but will grow under a wide range of environmental conditions.

All of the cool-season crops except potatoes will withstand some frost; potatoes are quite subject to injury by temperatures even slightly below freezing. The general recommendation is to plant cool-season crops as soon in the spring as the ground can be worked.

Peas *(Pisum sativum).* Peas are grown extensively in the home garden, but few are sold fresh on the market anymore. They are one of the more important processing crops but produce a quality product only when the weather is cool. In most areas they do not grow well as a fall crop.

Pea plants are vines in growth habit. The bush types grow to a height of 18 to 24 inches (40 to 60 cm) and the pole or climbing peas will reach 8 or 9 feet (2½ to 3 m) in height. The shorter vine type are the most popular today. Because in most areas peas must be planted as early in the spring as possible to mature before hot weather occurs, it is desirable to treat the seed with fungicide to protect it against decay. Usually pea seed is treated by the dealer before it is sold. Peas can be grown on soil types ranging from light sandy loam to heavy clay. For early crops lighter soils are desired. Well-drained clay loam soils will give larger but later maturing crops. Pea seeds are relatively large and fairly easy to plant. For home use they are spaced in rows 2 to 3 feet (⅔ to 1 m) apart and 1½ to 2 inches (4 to 5 cm) apart in the row; they should be planted 1 to 2 inches (2½ to 5 cm) deep. Peas should be harvested while green and tender and, if possible, during the cool early morning hours. They lose quality rapidly either as they become overmature on the vine or if they are left unrefrigerated for any period of time after they are harvested (Figure 12–8).

Edible-podded or sugar peas and the similar snap peas do not have a tough membrane at the inner surface of the pod. They have been used in this country by immigrants of German or Oriental extraction and are becoming popular with people who enjoy exotic dishes. They are an excellent cooked vegetable and are easily grown and prepared for the table. Their growth requirements are almost identical to those of other peas. They can be harvested as soon as pods form and must be harvested before the seeds are fully developed. To prepare them for cooking, pull from the pods both ends along with any strings that may be attached to the ends. Peas and edible-podded peas are adapted to processing by freezing.

Spinach *(Spinacia oleracea)*. Spinach is the most important of the greens or potherb crops. There are two types, the cool-season or broadleaf, and the warm-season or New Zealand spinach (*Tetragonia expansa*).

Broadleaf spinach is seeded where it is to mature and is grown during the cool spring and fall months in the north and during fall, winter, and spring in southern areas where winters are not severe. It develops seedstalks when days become long. In colder areas where winter temperatures remain above 0°F (−18°C), spinach can be seeded in the late fall to winter over and develop an earlier spring crop. The earliest harvest of spinach can be a thinning operation with well-placed plants left to grow larger. With later harvesting the plants are cut close to the surface of the ground (Figure 12–9).

New Zealand spinach may be seeded in place, or the plants may be transplanted. The seed tends to be impervi-

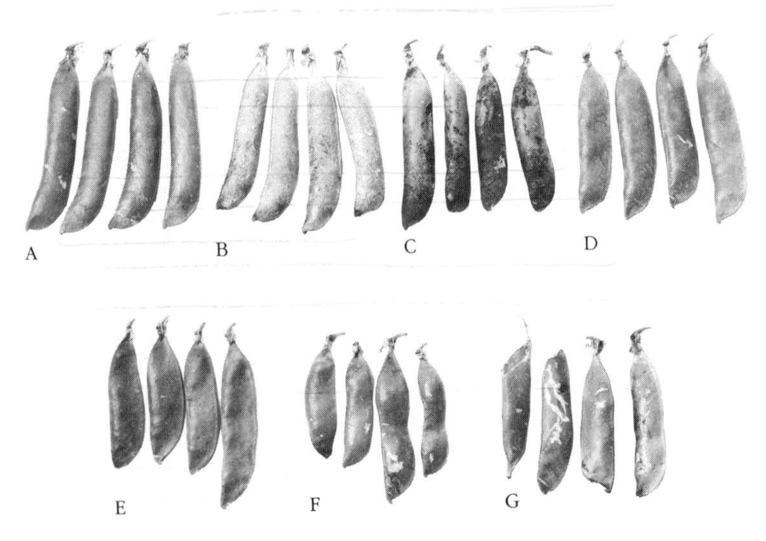

Figure 12–8 Peas at various stages of maturity. (A) Well-filled pods at optimum maturity; (B) overmature pods; (C) pods damaged by pressure or friction, peas still edible; (D) undermature pods; (E) unfilled flat pods; (F) partially filled pods, two seeds or less; (G) broken pods. F and G result from faulty fertilization.

Figure 12–9 Spinach. Spinach bolts to seed as a result of long days and high temperature (A). A healthy planting of spinach and lettuce (B).

ous to water, and filing a notch in it will produce more rapid germination. Only the tender leaves and young side shoots are used. These can be harvested and cooked like regular spinach at any time during the summer. New Zealand spinach is injured at temperatures below freezing.

Chard ***(Beta vulgaris).*** Chard is a leaf beet. Its planting and care are similar to that of spinach, and the leaves and petioles can be cooked to provide a potherb similar to spinach. When harvested, it should be cut so that the growing point is not destroyed. This will allow the plant

Figure 12–10 Chard about a foot high with outer leaves ready for harvest. (Courtesy USDA.)

to continue to produce all summer. An even better harvest method is to remove the outer leaves, leaving the inner ones to expand for later harvests (Figure 12–10).

Root Crops. Root crops include radishes, turnips, rutabagas, beets, carrots, and sweet potatoes. All except the sweet potato are cool-season crops that can be planted very early in the spring. All are herbaceous biennials except the radish, which is an annual, and the sweet potato, which is a perennial grown as an annual. Cultural practices are similar for all except for the sweet potato, which will be discussed under warm-season vegetable crops. All but the sweet potato are grown from true seed, and all are popular home vegetable garden crops. All of the cool-season root crops can be grown in any good loamy soil but will produce better shaped roots if the soil is loose and friable. They can be planted in rows spaced as close as 12 inches (30 cm) apart. Rutabagas, turnips, beets, and carrots all form undesirable seedstalks if they are subjected to several weeks of cool temperatures after the roots have begun to enlarge.

Two types of radishes (*Rhaphanus sativus*) are grown. The ones available at grocery stores and commonly grown by home gardeners are annuals that grow only during cool weather and mature in 4 to 6 weeks. The biennial, winter, or storage radishes are planted in the summer, harvested in the fall, and stored for winter use. These are popular in the Orient but are not grown extensively in this country.

The first plantings can be made early in the spring in the north and during the winter in the south. For succession harvests radish seed should be planted each 10 to 14 days. Seed of fall or winter radishes can be sown from June through July. Fall season plantings of the annual short-season cultivars can be made from August into September. The short-season types can be thinned by pulling those that are large enough to use. The larger summer and winter types should be thinned to 2 to 4 inches (5 to 10 cm) apart. Harvesting should begin as soon as the roots reach edible size. The quick-maturing cultivars must be used at once, because they soon age to become tough and pithy. The winter and fall cultivars remain edible much longer, and winter radishes are usable for several months with proper storage.

Insects and diseases that attack radishes will vary with locality, and local recommendations for pest control should be followed. The cabbage maggot is an almost universal pest of all vegetables that, like radishes, belong to the mustard family, and the row in which radishes, turnips, and rutabagas are seeded will usually have to be treated with an insecticide if they are to grow edible roots.

Turnips (*Brassica rapa*) and rutabagas (*Brassica napobrassica*) are both essentially cool-season crops, although rutabagas can withstand some heat and drought. Because turnips are relatively quick maturing, they are grown as spring and fall crops in the north and as a winter crop in the south. Rutabagas require 4 to 6 weeks longer to mature than turnips. Both do best in deep rich loams. Both crops are grown from seed and planted where they are to mature. After the plants are well established, they should be thinned to stand 3 to 6 inches (8 to 15 cm) apart in a row. (See information on root maggots in the discussion on radishes.)

Turnips are used as greens or potherbs or as roots. For greens they are pulled before much storage root development has taken place. If their roots are to be used, young turnips are harvested when these roots reach a diameter of 2 to 3 inches (5 to 8 cm); if the roots grow too large, they become strong flavored and stringy. The best rutabagas for table use are 4 or 5 inches (10 to 13 cm) in diameter. When either of these crops is to be stored, the tops should be removed. Storage at 32°F (0°C) and about 95% relative humidity is best. Rutabagas can be stored through the winter.

Beets (*Beta vulgaris*) are planted where they are to mature. Thinning is always necessary no matter how evenly the seed has been spaced, because the so-called seeds are actually fruits or seed balls containing more than one embryo. Thinnings may be cooked as a potherb, beet greens, which many people prefer to spinach. The final spacing in the row should be from 3 to 4 inches. Final thinning can be delayed until small beets are large enough for table use.

Beets are of best quality (i.e., with uniform color and a sweet flavor) when they are grown with an average daily temperature that does not exceed 70° to 72°F (21° to 22°C). They are not normally subject to many disease or insect pests. In some irrigated areas of the West they cannot be grown because of curly top, a virus disease spread by the beet leafhopper that affects many kinds of vegetables.

The roots should be harvested when they reach a diameter of about 1½ inches. The leaves are cut off approximately a half inch (1¼ cm) above the crown. The tap root and peeling are not removed until after the beet is cooked. Any cut or injury to the root will permit excessive juicing out in the cooking water. Beets can be wrapped in foil and baked like potatoes. They are also popular as pickles.

Beets can be stored for a period of time at 32°F at 95% relative humidity. 'Detroit Dark-Red' and 'Crosby Egyptian' are two of the more popular cultivars. Orange-colored cultivars have recently been developed and are quite popular with home gardeners. They have a delicious flavor and do not juice out or stain, which to some is an objectionable feature of normal red beets.

The per-capita consumption of carrots (*Daucus carota*) has increased steadily over the past 50 years since the discovery of their dietary importance. They are an especially good source of vitamin A and also contain appreciable amounts of several of the B vitamins. Of all the root crops carrots are probably the most susceptible to misshapen roots. Heavy soils can be made more amenable for growing carrots by adding organic matter to them. An ample supply of moisture is required while the seed is germinating and the seedlings are starting their growth. It is difficult to obtain a good stand during hot weather. Carrot seed is difficult to plant by hand, and a salt shaker makes a convenient planting assist if no regular seeder is available (Figure 12–11).

Figure 12–11 An aid to seeding carrots. A salt shaker (or pepper shaker if holes are large enough) can be used to plant small-seeded crops such as carrots.

The most convenient way to thin garden carrots is to pull them for table use as soon as they become large enough to eat. Carrots have the best shape and quality when grown at temperatures between about 55° and 75°F (13° to 24°C). At lower temperatures the shape tends to be longer, and at higher temperatures it is more blunt. Carrot roots will develop less color and consequently less vitamin A at either high or low temperatures. Carrots can be held in storage for as long as 6 months if the humidity is high and the temperature is under 40°F (4°C) (see "Storing Horticultural Products," Chapter 14).

Many cultivars of carrots are available; however, they belong mostly to three different types (Figure 12–12). The extremely long carrots are the 'Imperator' type popular in grocery stores. 'Chantenay' and 'Danvers Half Long' are pointed, half-long types of high quality. Commercially these cultivars are used for processing. 'Nantes' is one of the highest quality cultivars; it is too brittle to market well and too small for processing but is excellent for the home garden. 'Oxheart' is a very short, stubby type fairly low in quality. It should be grown only where soils are too heavy to produce good-quality roots of longer cultivars.

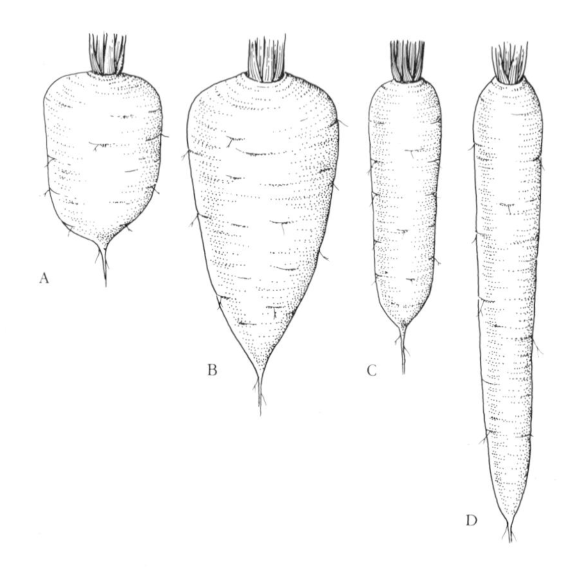

Figure 12–12 Carrot cultivars: (A) 'Oxheart,' (B) 'Chantenay' or 'Danvers,' (C) 'Nantes,' (D) 'Imperator.'

Salad Crops. A number of vegetable crops, grown primarily for the consumption of their uncooked leaves or petioles, are classed as salad crops. Only two, celery and lettuce, are important commercially; and only one, lettuce, is grown extensively in home gardens. Other salad crops that could have a place in the home garden are cress, endive, and French endive, sometimes called Witloof chicory.

Cress (*Nasturtium officinale*), sometimes referred to as watercress, is frequently gathered from the wild in this country, but it is a common garden crop in England, where it is grown on extensive acreages. It requires a steady supply of water and probably has a place in only a few specialized home gardens.

Endive (*Cichorium endivia*) and Witloof chicory (*Cichorium intybus*) both require about the same growing conditions as lettuce. Plants of both crops should be thinned to about 3 to 6 inches apart. Endive is sometimes covered with two boards arranged as an inverted trough over the row, which blanches the green color from it and reduces the amount of bitterness. Although Witloof chicory leaves can be used for salads, the crop is more often grown to maturity in the fall, when roots are dried and ground and can be used as a coffee substitute. More frequently, Witloof chicory is used for winter forcing (described earlier in this chapter). The roots are dug late in the fall before the ground freezes and stored under cool, moist conditions until time for forcing.

Lettuce (*Lactuca sativa*) is the most extensively grown and most important of all salad crops. All cultivars thrive best in cool weather with an abundant supply of moisture. Most lettuce cultivars become bitter and go to seed in hot weather. Lettuce can be grown in all kinds of soil, but it requires high amounts of nitrogen. Lettuce is an excellent home garden crop because large amounts can be grown on a few square feet of ground. Plantings should be made approximately two weeks apart in the early spring for late spring and early summer harvest and starting again in July or August for autumn harvest. Summer crops of lettuce are possible in some cool coastal areas and mountain valleys of the West and in the far north.

There are two types of lettuce: leaf and head (Figure 12–13). Leaf lettuce forms a loose, open cluster of leaves; head lettuce forms a solid mass of tightly compressed leaves. Because it is easier to grow, leaf lettuce is generally preferred for home gardens, although there is no reason why gardeners cannot produce high-quality head lettuce. For earliest maturity head lettuce can be seeded

Figure 12–13 Three types of lettuce. 'Greenhart' leaf (A) and immature heads of 'Buttercrunch' butterhead (B) and New York head (C) lettuce. (A, courtesy Burpee Seeds.)

indoors about 6 weeks before it is to be transplanted. Later crops and leaf types are usually seeded where the plants are to mature. In the South the seed may be planted in the fall. With direct field seeding, the seeds are drilled in rows from 12 to 18 inches (30 to 45 cm) apart. Lettuce can be thinned by pulling surplus plants to use in salads. Head lettuce will not form good heads unless it is thinned to 8 to 15 inches (20 to 38 cm) apart.

Leaf lettuce may be harvested at any stage of growth before the plants send up seedstalks. Head lettuce is harvested when the heads are solid but before any sign of seedstalk development is seen. If a surplus of head lettuce is produced, it can be held in the refrigerator for as long as three weeks.

Head lettuce can be classified into three groups. The crisp types are most important commercially, and strains of the 'Great Lakes' and 'New York' cultivars are most widely adapted to home gardens. Butterhead cultivars used to be available only in gourmet restaurants because the main cultivar, 'Bibb,' was best adapted to greenhouse growing and, as a consequence, was very expensive. Today butterhead cultivars such as 'Buttercrunch,' 'Summer Bibb,' and 'Big Boston' can be grown easily in home gardens. The cos or romaine lettuce forms loose elongated heads. It is somewhat resistant to heat. 'Paris White Cos' is a popular cultivar.

Because it requires somewhat specialized growing conditions, celery (*Apium graveolens*) has never been an important home garden crop. Celery needs a uniform moisture supply to develop good quality; hollow and stringy stalks may develop after sudden dry periods. It is best adapted to areas that have a relatively long, fairly cool, damp growing season. It can withstand some freezing temperature if properly matured.

Celery seed is very small, so it is planted under glass or in specially prepared beds. The seeds should be covered lightly with sand or sandy soil. Four to 5 weeks after seeding the plants are transplanted or thinned to stand 1½ to 2 inches apart. Five to 7 weeks later the plants can be set into the garden; thus 10 to 16 weeks are required from seeding to transplanting. Rows in the garden should be from 3 to 6 feet apart and the plants spaced 6 to 8 inches in the row.

Because its root system is shallow, celery requires heavy fertilization and frequent irrigation. It should have a light to medium irrigation approximately each 3 to 4 days during periods of warm dry weather. Celery is subject to a deficiency of several minor elements, especially boron, which causes cracking of the petioles. This condition can be prevented or cured by soil amendments of borax or boric acid.

In harvesting of celery the roots are cut at or just below the crown, and the plants are lifted, trimmed, and washed. When necessary, celery can be held for several months in cold storage. Celery is a biennial that will initiate premature seedstalk development if the young plants are subjected to long periods of cool temperatures.

Crucifer or Cole Crops. The cole crops include cabbage, cauliflower, and broccoli, which are well known in this country; and Brussels sprouts, kohlrabi, kale, and collards, which are not so well known. Collards and kale are essentially leaf cabbages, and their culture is similar to the culture of cabbage. Both are frequently grown to add texture to an ornamental planting, and there are a number of highly colored ornamental kale cultivars.

The culture of Brussels sprouts (*Brassica oleracea* var. *gemmifera*) is similar to that of cabbage except that Brussels sprouts require a relatively long growing season. They are frequently grown from transplants. The sprouts, which resemble small cabbages, form in the leaf nodes on an elongated stem (Figure 12–14). Sprouts mature from the bottom up and can be harvested for table use or freezing whenever they become solid. 'Jade Cross' is earlier than most standard Brussels sprouts cultivars and should be an excellent choice for the home gardener.

Of all vegetables the cole crops are perhaps the ones most subject to attack by insects and diseases. They should not be grown in soils that have a history of nematodes or club root. Club root is a soil-borne disease characterized by extreme swelling of the roots into which the pathogen has entered. It can be controlled to some extent by raising the pH of the soil. In most sections of the country a grower of cole crops must be prepared to treat the soil for cabbage maggots and periodically to dust or spray to control the cabbage looper, cabbageworm, and aphids.

Cabbage (*Brassica oleracea* var. *capitata*) can be grown on almost any soil, but sandy loams are most desirable for both early and late crops. This crop requires an ample

Figure 12–14 Brussels sprouts. Sprouts are harvested at this stage when they are 2½ to 3 inches (5 to 8 cm) in diameter and have become solid. (Courtesy Ferry-Morse Seed Co.)

but not excessive supply of moisture for maximum development; lack of water will cause premature heading. Cabbage is a cool-season crop but is not as susceptible to damage by high temperatures as are cauliflower, peas, or lettuce. It can be produced through the growing season in areas where the average July temperature is under 74°F (23°C). In the South it can be grown from late fall through the winter and early spring. For a continuous supply three crops should be grown. Seed for the first crop is planted under glass 6 or 8 weeks before the plants are to be set in the field. The second crop is direct seeded during early spring to mature in July and August. Seed for the late crop, planted in open ground from the middle of June to early July, will produce a crop that should mature in October and can be stored.

Cabbage should not be harvested until the heads are solid as soft heads are inferior in quality. For storage the plants are pulled and held at temperatures about 40°F in a fairly humid atmosphere.

Cabbage cultivars vary widely. Some have pointed heads and others have round or flat heads. Leaf color may be green to red or purple. The leaves themselves may be smooth or wrinkled (the savoy cabbage, for instance). They mature at different seasons of the year. The most important early types are 'Golden Acre' and 'Copenhagen Market.' 'Marion Market' is a round-headed, early cultivar that is resistant to a root disease called cabbage yellows. Mid-season cultivars include 'Glory of Enkhuizen,' 'All Seasons,' and two yellows-resistant varieties—'Wisconsin Globe' and 'Wisconsin All Seasons.' Two late cultivars are 'Danish Ballhead' and 'Wisconsin Hollander' (yellows-resistant).

The culture of cauliflower (*Brassica oleracea* var. *botrytis*) is similar to that of cabbage except that cauliflower will not withstand warm temperatures. In warm-season areas the spring crop should be transplanted or direct seeded as early as possible. The fall crop can be started in July to mature during the cool days of late fall. During hot weather the heads or curds may be discolored or fail to form. When pure white heads are desired, blanching can be accomplished by tying the outside leaves over the head as it begins to form to exclude light from the curd (Figure 12–15).

Depending on weather and cultivar, the heads should be ready to harvest one or two weeks after they start to develop. The head is cut off at the juncture of the head and leaves. 'Imperial' and 'Snowball' are two of the most popular cultivars. However, there are dozens of strains, and the strain of seed used may be more important than the cultivar. Follow local recommendations in selecting cauliflower cultivars. 'Early Purple Head,' a cultivar that

Figure 12–15 Cauliflower. This head of cauliflower has been darkened by exposure to sunlight. White cauliflower is produced by tying the leaves together to shade the head. Notice also the holes in the leaves and the black droppings on the head as a result of looper damage.

resembles a cross between broccoli and cauliflower, will produce good heads at somewhat warmer temperatures without blanching.

Broccoli (*Brassica oleracea* var. *italica*) is grown in much the same manner as cauliflower, except that the numerous heads are not blanched and broccoli will withstand a little more summer heat than cauliflower. Heads are formed both terminally and laterally. The first or terminal crop should be cut when the heads have reached maximum diameter but before they begin to separate. From 4 to 6 inches of stem is included with the head. Succeeding cuts come from small individual heads that develop on shoots from leaf axils after the first head has been cut. 'Italian Green Sprouting' is a spring cultivar widely grown and 'Waltham 29' is one of many cultivars planted in summer for fall harvest. Some recently introduced early hybrids, many of which were developed in Japan, appear to be excellent for the home garden; 'Green Comet' is one that seems widely adapted.

Kohlrabi (*Brassica oleracea* var. *caulorapa*) is a vegetable popular in Europe that deserves more attention from home gardeners in this country. The edible portion is an above-ground enlargement of the stem that resembles a very mild turnip in shape and flavor. Raw, it is an excellent addition to a relish plate; it also can be boiled or stuffed and baked. Its planting, culture, and harvest are similar to those of turnips. Because the edible portion is above the ground and also because it seems to have some resistance, kohlrabi is not as susceptible to cabbage maggot damage as are the other crops of the Cruciferae family.

Bulbs. The onion is the most widely grown of the bulb crops. The others—garlic, chives, shallots, and leeks—are produced in smaller amounts. Garlic (*Allium sativum*) is grown from thickened bulb scales called cloves; it should be planted as early in the spring as possible and harvested when the top has died down and the plant is completely mature. Chives (*Allium schoenoprasum*) are perennials grown for their tender leaves, which have an onionlike flavor and can be harvested for salads and seasoning from early spring to late fall. Leeks (*Allium porrum*) and shallots (*Allium ascalonicum*) produce succulent, thick-necked plants with a mild onionlike flavor. They are used as cooked vegetables, in salads, and as flavoring for soups and stews. Planting early in the spring enables both to be produced in most parts of the United States and Canada.

The onion (*Allium cepa*) is one of the hardiest of all vegetable crops. Commercially onions are usually grown from seed sown as soon as the frost is out of the ground. In the South they can be planted in the fall. For home gardens onions may be grown from seed, sets, or transplants. Sets are very small onions produced by thick seeding on poor ground late in the season the previous year. Crowding and late planting checks the growth of the sets, and they ripen prematurely. Planted out in the spring, these sets complete their growth and develop into larger bulbs.

Onions are grown in rows spaced from 12 to 24 inches apart. If large bulbs are desired, they should be thinned to 3 to 6 inches apart in the row. The thinnings can be used to flavor salads or as scallions. Onion sets and transplants are spaced at planting time. Onion tops do not produce much shade and onions have shallow root systems, so this vegetable cannot compete with weeds and requires heavy fertilization as well as frequent irrigation during dry periods. Onions grown to be used as green or bunch onions, also called scallions, are pulled as soon as they are large enough for use. Onions for storage must be ripe when harvested. Maturity is indicated by softening of the neck tissues and falling over of the tops and dying of the roots. At this time the bulbs enlarge rapidly. They are ready to harvest when about 70% of the tops have fallen over. Onions should then be pulled and, if the weather is sunny, left in the garden until the tops and outer scales are completely dry. This procedure is called **curing**. The tops can be spread over the bulbs to prevent sunburn. After the bulbs have cured, they are topped. Ideal storage conditions are 32°F with 70% humidity, but onions keep relatively well in a cool, dry basement or attic.

There are three distinct colors of onions: yellow, white, and red. Onions also differ in shape, some being round, some flat, and some long. Some have a very pungent flavor, and others are mild. Many of the onions grown today are hybrids. Because bulbing is initiated by day length, most onion cultivars are not widely adapted (see Figure 7–26, Chapter 7). Almost any cultivar will produce satisfactory green bunch onions, but the grower who wishes to produce large bulbs should seek local cultivar recommendations.

Potato *(Solanum tuberosum)*. All of the vegetable crops discussed so far can be hardened to withstand some freezing and, as a consequence, can be planted about as soon as the ground can be worked. By contrast, the potato, although it is a cool-season crop, cannot withstand frost. Nevertheless, because the potato seed piece contains more stored food than the seed of most crops, the potato can regenerate top growth in case initial sprouts are damaged by frost or other adverse conditions. Furthermore, because potatoes require about 2 weeks from time of planting to emergence, they can be planted 2 to 3 weeks before the last frost is expected in the area.

The potato is grown as an annual and is propagated vegetatively by means of whole or cut tubers or seed pieces. Some cultivars may produce true seed in tomatolike fruits in northern areas, but this seed is not used to grow commercial crops. The tuber that is the edible part of the plant is a thickened underground storage stem with scalelike leaves and buds or eyes. Potatoes are subject to numerous diseases, many of which can be carried in the seed piece. For this reason gardeners should use certified seed that has been grown in cool regions where virus disease spread is minimal and under carefully regulated conditions enforced by inspection. Potatoes sold in the spring for table use should never be used for seed. Not only are they likely to harbor disease organisms, they also are likely to have been treated to prevent them from sprouting.

In preparation of planting stock seed tubers can be left whole or cut into two or more pieces (sets), depending on their size. Sets of 1 to 1½ ounces (30 to 40 g) will produce better plants than will smaller sets, especially if growing conditions are unfavorable. Commercial growers sometimes cut their seed several weeks in advance and then permit them to **suberize** (form a protective layer resembling a new skin over the cut surface). Suberization requires high humidity, oxygen, and a temperature of 60° to 70°F. For the home garden it is probably best to cut the tubers just before they are to be planted and allow them to suberize in the soil, where conditions for suberization are usually good.

Planting depths vary. For the early crop the seed is covered with 2 to 3 inches of soil. The late crop on heavy soils should be planted about 3 inches deep. On lighter soils the depth of planting should be increased enough to bring the seed piece down to moisture. Most potatoes are planted in rows 2½ to 4 feet (⅔ to 1¼ m) apart with 9 to 14 inches (23 to 35 cm) between seed pieces in the rows (Figure 12–16). Planting may be done with a hoe or shovel or with special planting equipment. Optimum summer growing temperatures are around 60° to 70°F. Potatoes do well in areas where the nights are relatively cool, and tuber shape and quality of long-tubered cultivars are best when soil temperatures are cool and uniform. As mentioned in Chapter 6, frequent light irrigation is recommended, especially during warm periods. Under these conditions irrigation is more important for cooling the soil than for providing moisture. Fertilizer needs vary with different soils. Potatoes use heavy amounts of nitrogen and potassium (see "Fertilizing Gar-

Figure 12–16 Potato sets.

den Crops," Chapter 14). Cultivation should be shallow but sufficient to control weeds and keep the soil loose and open. Soil should be worked up around the plants when they are 6 to 8 inches high; this helps keep tubers covered and prevents their becoming exposed and sunburned.

Early potatoes are frequently dug before the tubers are mature. Sometimes home gardeners can space plants close enough so that every other plant can be removed to provide small tubers for early eating. Plants that appear diseased can also be dug for early tubers as tubers on these kinds of plants never grow very large anyway. For winter storage potatoes should be dug when they are fully mature and after the vines have died and the skin no longer scuffs from the tuber. If tubers need to be harvested for storage before vines die, they can be matured by cutting or pulling the vines 10 days to 2 weeks before tubers are dug. Potatoes should be stored at a temperature of 38° to 40°F (3° to 4°C) if possible, with a relative humidity of about 90%. If potatoes are stored at temperatures below 38°F, the starches in the tuber are changed to sugar and the potato becomes unappetizingly sweet. If storage temperature is above 42° to 45°F (6° to 7°C), potatoes are likely to sprout as soon as their rest period is broken in early January.

Warm-Season Vegetables Grown as Annuals

Vegetables described in this section require warm temperatures for germination and grow best in areas with a relatively warm summer. Melons and sweet potatoes, especially, cannot be matured without a long, hot growing season. The introduction of new, early cultivars of most of the other warm-season vegetables permits them to mature in most parts of the United States and many areas of Canada. None of these crops should be planted until danger of frost is past. Earlier production is possible with transplanting. Solanaceous crops transplant easily with bare roots, but the cucurbits and sweet corn will not survive if their roots are disturbed, and therefore they must be grown in some kind of transplantable container if they are to be transplanted.

Beans and Sweet Corn. Snap beans (*Phaseolus vulgaris*) and lima beans (*Phaseolus lunatus*) are among the most popular home garden vegetables. They grow well in most parts of the country and mature in 55 to 70 days. They can be planted so that they are ready to eat during most of the summer and fall and can be enjoyed fresh from the garden or as a canned or frozen product. Beans vary in growth habit from dwarf bush to long, viny pole types, and the pods can be green, yellow (wax), or purple. Dry beans of various kinds and most green shell beans, which are stripped from the pods like peas while still tender, belong to the same species as snap beans but require a somewhat longer growing season.

Because the seed must push large cotyledons through the soil (Figure 4–5, Chapter 4), planting should be no deeper than necessary to get seed to moisture. In most areas this will be from 1 to 3 inches (2½ to 8 cm). Pole beans can be planted in rows and trained on wires or strings or planted in hills approximately 18 inches (45 cm) apart. Rows should be 2 to 4 feet (about ⅔ to 1½ m) apart depending on the method of cultivation. A pole is placed in each hill for the plant to climb, and the tops of four poles are often tied together for stability. Where cultivation is by hand or with a garden tractor, rows of bush beans may be spaced as close as 18 to 20 inches (45 to 50 cm), but where field equipment is used row spacing should be at least 30 inches (75 cm). Plants should be spaced 2 to 3 inches (5 to 8 cm) in the row. In many areas only disease-resistant cultivars can be produced, and cultivars resistant to rust, mosaic, curly top and bacterial blight are available. Although beans are legumes, their capability for nitrogen fixation is not great so all types generally require fertilization. They are susceptible to moisture stress and will not form pods if moisture is deficient at the time of bloom (Figure 12–17).

Lima beans require a somewhat longer growing season than snap beans, their seed requires slightly warmer soil for germination, and they are even more susceptible to blossom abscission if hot, dry weather occurs when they are blooming. Otherwise, growing requirements are similar to those of snap beans. Lima beans also grow on either bush or vining plant. Lima bean cultivars are either large seeded or small seeded.

Snap beans should be picked before the pods reach full size and when the seeds are beginning to fill. Plants should not be picked when they have moisture on the leaves, as handling the plants when they are moist tends to spread rust, a fungus disease. Green shell and lima beans should be picked as soon as the bean seeds have attained full size but before they turn white or harden. Shelling is easier if the pods are allowed to wilt for a few hours. Dry beans are harvested as soon as most of the pods are fully mature and have turned yellow. They must be harvested before the lower pods begin to shatter. Small plantings of dry beans are pulled, but bean harvesters that cut the vines just below the ground are used for large quantities. Where weather is dry, beans are allowed to dry in the field. Small quantities can be threshed in the age-old way by tromping the beans from the vines on a canvas and tossing them in the air to permit wind to blow away the chaff. Various-sized combines and threshing machines are available for larger plantings.

A

B

Figure 12–17 (A) A healthy snap bean plant starting to form pods. (B) Snap beans growing in a research trial to test the effect of different spacings.

Sweet corn (*Zea mays*) is a warm-season crop native to the Western Hemisphere. It is a vegetable in which highest quality can be appreciated only when it is cooked freshly harvested from the home garden. Sweet corn will succeed on any soil that is suitable for general crop production. Where early maturity is wanted, a warm, sandy soil is desirable. Higher yields are produced on heavier soils more retentive of moisture and minerals. Corn needs more nitrogen that most other vegetables (see "Fertilizing Garden Crops," Chapter 14).

Corn is planted in hills spaced 2 to 3½ feet (about ⅔ to 1 m) apart or in drills 10 to 12 inches (25 to 30 cm) apart in rows spaced 2½ to 4 feet apart. In most soils the seed should be planted 1 to 2 inches deep. Cultivation should be shallow but sufficient to control weeds. Adequate moisture is essential and critically necessary during and shortly after silking. If either fertilizer or moisture is likely to be limiting, plants should be spaced no closer than 3 x 3 feet (1 x 1 m).

Open-pollinated cultivars of sweet corn have almost all been replaced by F_1 hybrids. Hybrid cultivars that mature at different times can be planted to provide sweet corn for a long season, thus eliminating the need for succession plantings and making cultivation and weed control easier.

Sweet corn is a highly perishable crop. After the sugar content in the corn reaches its maximum, it quickly changes to starch. Thus it is necessary that the corn be picked at the right stage of maturity for highest quality. For best fresh-from-the garden eating quality or for freezing, this is the "milk" stage, when a thin milky juice exudes from kernels pressed with a thumbnail. For canning sweet corn should be harvested at the slightly more mature "cream" stage.

Because corn loses its quality rapidly after harvest, especially when temperatures are high, it should be picked and husked as close to meal time as possible. For home processing it should, if possible, be harvested while temperatures are cool early in the morning and canned or frozen immediately. Sweet corn should always be refrigerated if storage is necessary.

The larva of the corn earworm and the European corn borer are the most obvious pests of sweet corn. The corn earworm (Figure 12–18) is found almost universally in gardens and fields in North America. The female moth lays her eggs at the top of the developing ear and the black-striped, tan, or green caterpillars that hatch from those eggs eat their way among the upper kernels, leaving feces and other debris (see Table 14–24, Chapter 14). The European corn borer hatches in the soil and burrows its way into the interior of the stalk all the way to the tassel and ear. These burrows, usually invisible from outside, frequently result in breakage of the stalk and/or damage to the kernels.

Figure 12–18 The corn earworm. This insect is difficult to control and most gardeners choose simply to clip off the damaged part of the ear tip. (Courtesy USDA.)

Solanaceous Fruit Crops. In this group are tomatoes, peppers, and eggplant. Tomatoes and peppers are of tropical American origin; eggplant has been used as a food in India and Southeast Asia for millennia. All three of these vegetables are warm-season crops that should not be planted in the field until all danger of frost is past. Often the seed is planted under glass or in specially prepared beds 6 to 8 weeks before field planting (see "Propagating Plants from Seeds," Chapter 14). In regions with short growing seasons the seed for home gardens can be sown 10 to 12 weeks before the plants are set out, and plants may be blooming before they are transplanted.

The tomato (*Lycopersicon esculentum*) ranks second among the vegetable crops grown in the United States, surpassed only by the potato. Its fruits are high in vitamins A, B_1, B_2, and C, in spite of the fact that they contain over 94% water. Tomatoes are grown by all types of vegetable gardeners on all types of soil for both fresh consumption and canning. For early maturity and in areas where the frost-free season is short the lighter, warmer soils are desirable. More fertile loam, silt loam, and clay loam soils produce greater yields.

In growth habit tomato plants may be indeterminant or determinant. **Indeterminant** cultivars can continue growth indefinitely. Normally a blossom cluster is produced at every third node, and the terminal bud continues vegetative growth and elongation. Until recently the principal commercial cultivars were indeterminant. In contrast, growth of **determinant** cultivars eventually terminates in a flower cluster, and shoot elongation stops. This gives a "self-pruning" or "self-topping" habit of growth. Determinant cultivars are usually earlier than indeterminant ones and are especially desirable where the growing season is cool or short. Determinant cultivars also tend to ripen their fruit more nearly at one time and are the type produced commercially for mechanized harvest. F_1 hybrid cultivars are available, but seed is expensive, and F_1 hybrid tomatoes are usually only slightly superior to standard cultivars.

A field spacing of 2 to 4 x 4 feet is common for most indeterminant tomato cultivars when they are set out as transplants. Determinant cultivars can be transplanted as close as 12 inches in the row. Where the growing season is long, tomatoes, especially determinant cultivars, may be direct seeded. With direct seeding plants may be left as close as 6 inches apart with little effect on fruit size. Direct seeding and close spacing are practiced in irrigated areas of the West where curly top (western yellow blight) prevents the growing of transplanted tomatoes. Direct-seeded tomatoes, especially if they are left closely spaced, do not seem to be as susceptible as are transplanted ones to insect-spread diseases.

Staking is practical mainly with indeterminant cultivars. With staking, plants are usually spaced 1 to 2 feet apart in the row. A stout stake 5 to 7 feet (1½ to 2 m) long is placed next to each plant, and the plant is allowed to grow to a single stem by pinching of the vegetative shoots from the axils of the leaves. The plant must be tied to the stake with string or specially purchased ties (Figure 12–19). Staked tomatoes usually mature almost a

Figure 12–19 Staked tomatoes. (Courtesy Washington State University.)

week before those not staked, and staking also reduces fruit rot, which frequently occurs when fruit comes in contact with wet soil. However, staking will sometimes increase the amount of sunburn and cracking.

The tomato is a deep-rooted crop that can withstand considerable drought. Where irrigation is necessary, the entire root zone, which may extend to a depth of 6 to 10 feet (2 to 3 m), should be thoroughly soaked.

Tomato fruits are tender and highly perishable, so they must be picked and handled carefully. If necessary they may be picked at the mature green stage, when cream-colored streaks show in the green ground color at the blossom end, the seeds have become firm, and the skin is still tough. Characteristic color will develop in fruits picked at the mature green stage, and they will ripen to fairly good quality. Fruits picked after some red color is evident will color well and will develop flavor almost equal to that of tomatoes picked when fully ripe. To ripen with full color, the fruit, whether on or off the vine, must be kept at a temperature between about 55° and 85°F (13° to 30°C). For home use tomatoes should be picked at the firm ripe or ripe stage, that is, when they are fully mature but before they have begun to soften.

A number of diseases and insects affect tomatoes; however, there are many areas of the country where a tomato crop can be grown with no more pest control than the picking off of an occasional hornworm. In humid areas and areas with high rainfall, late blight (*Phytophthora infestans*) and early blight (*Alternaria solani*) may require weekly preventive sprayings with a copper or zinc fungicide. The tomato hornworm and corn earworms occasionally may be serious enough to require spraying. Attending carefully to sanitation, washing hands frequently, and refraining from smoking when handling tomato plants will help to reduce the incidence of tobacco mosaic virus, which can be serious especially with greenhouse and staked tomato plantings. Close spacing and direct seeding will be necessary in some areas of the West to prevent the spread of curly top.

Several distinct types of garden peppers (*Capsicum frutescens*) are grown. The large-fruited or bell types usually are mild and sweet; the small-fruited cultivars are pungent and distinctly hot. A small-fruited type is grown for decorative purposes. The 'Cayenne' or red pepper of commerce is the ground dried fruit of a small pungent cultivar. Paprika is made from the dried fruits of certain mild, sweet cultivars. Both sweet and hot peppers are distinct from commercial black pepper, which is made from the fruit of a tropical vine, *Piper nigrum* (Figure 1–6, Chapter 1).

Culture of peppers is similar to that of tomatoes. Plants are spaced 1 to 2 feet in rows 3 feet apart. Peppers thrive best in warm climates with long growing seasons. They are definitely a warm-season crop and should not be planted in the open until all danger of frost is past. Peppers require at least 8 weeks from seeding for the production of optimum-sized transplants. Peppers can also be direct seeded after the soil has warmed to 65°F (18°C).

Stage of maturity at which the fruit is harvested depends on use. Sweet bell peppers usually are picked when of good size but still green in color, although red fruits are in demand for some uses. Hot peppers are harvested when ripe and then dried or pickled. Local recommendations should be followed in selecting cultivars. Curly top and other virus diseases can be serious, but in many areas peppers are free of pests. Peppers have the highest amount of ascorbic acid per pound of any common food, several times as much as citrus fruit. New early cultivars that will grow in most sections of the country are becoming available, and they deserve a place in the garden of anyone who likes them.

The eggplant (*Solanum melongena*) is an important food crop in Asia and the Middle East but is not grown extensively in North America. Yet almost everyone who has tried Oriental or Middle Eastern eggplant casseroles is enthusiastic about this most delicious and versatile vegetable. It requires a warm season, and plants should not be set in the field until after the last spring frosts have occurred. The fruits generally are large and purple or white in color. The skin is smooth and shiny, and the seeds are imbedded in the flesh. Soil for eggplant production should be fertile, and a moderate but continuous supply of moisture is essential. The Colorado potato beetle is a common pest of eggplant (Figure 12–20).

The fruits are harvested as soon as the grower feels they are large enough but before they are fully mature. They are cut with the calyx and a short piece of stem left intact. Fruit of 'Black Beauty,' the most commonly

Figure 12–20 The Colorado potato beetle. This serious pest of all solanaceous crops seems to prefer eggplant when it is available.

grown cultivar, will grow large, and if left on the vine too long, it becomes tough and bitter. Some of the newer F_1 hybrids are of better quality and earlier maturity.

Vine Crops or Cucurbits. Included in the vine crops are watermelons, muskmelons and related melons, pumpkins, squash, and cucumbers. All of them are warm-season crops that cannot be hardened to withstand frosts. Vine crops require bees for pollination. If bees are not present because of weather or other factors, hand pollination is possible by rubbing the anthers of a male flower onto the pistil of a female one. The female flower will have an enlarged ovary (small fruit) under the open flower (Figure 3–3, p. 45).

Cucumber (*Cucumis sativus*) plants generally are monoecious in flowering habit; that is, the flowers are either pistillate or staminate, with both being borne on the same plant. They are killed by even light freezes, but, because their crop matures in 60 to 70 days, they can be grown in nearly all sections of the United States and southern Canada. They are adapted to most soils.

Cucumbers are planted either in hills spaced 3 to 5 feet apart or in drills with the rows 4 to 6 feet apart. Six or seven seeds per hill are planted, and the plants eventually are thinned to two or three per hill. When seeded in drills or rows, the plants are thinned to stand 1½ to 2 feet apart.

Cucumbers are harvested on the basis of size rather than age. For slicing they are picked when 6 to 10 inches (15 to 25 cm) long; for pickling when they reach 2½ to 6 inches in length. Because healthy vines will continue to produce fruit for several months, cucumber pickers should be careful not to injure the vines. Spraying may be necessary to control striped and spotted cucumber beetles. Planting resistant cultivars is the best control for mildew.

The muskmelons, or cantaloupes, and honeydew melons (*Cucumis melo*) all are grouped under the general term melon. For satisfactory production a frost-free season of at least 120 warm or hot days is needed. A few early cultivars such as 'Minnesota Midget' have been developed for northern gardens, but even these require bright sunlight and high temperatures for high sugar content and flavor. Use of plastic mulch and transplanting will considerably shorten the growing period required (Chapter 5). Melons can be grown on a wide variety of soils. Because they are quite drought resistant and require high temperatures, they are especially adapted to sandy loams. Melon plants are tender, and the seeds will not germinate at soil temperatures lower than 70°F (21°C). Planting times, spacing, and seeding methods are the same as those for cucumbers.

To develop maximum quality, melons must ripen on the vine. However, for commercial handling and shipment the stage of maturity at which they are picked will depend on how they are to be marketed. Most melon cultivars form an abscission layer at maturity, and the stem loosens from the fruit. When the stem is nearly loose, the melon is said to be at the "full slip" stage (Figure 12–21). For shipment the fruits are picked less mature, or at the "half slip" stage. Other indexes of maturity are softening at the blossom end, a change in color of the base of the pedicel from green to waxy, aroma, taste, and a change in ground color. At the grocery or vegetable stand melons with a piece of stem attached should not be purchased, as they were harvested when too green and will be of inferior quality. A solid melon greenish-yellow in color with a strong cantaloupe aroma will usually have the best flavor.

Figure 12–21 Muskmelons at the "full slip" stage. (Courtesy USDA.)

'Hales Best' and its various strains are important as shipping melons. Other good cultivars are 'Iroquois,' 'Honey Rock,' and 'Hearts of Gold.' 'Honey Dew,' 'Honey Ball,' and 'Casaba' are grown in the West. 'Minnesota Midget' and 'Chipman Lake Champlain' will often ripen where the season is too cold for other cultivars. 'Burpee Early Hybrid Crenshaw' is a delicious home garden cultivar that deserves a trial in areas where standard melon cultivars mature.

Watermelons (*Citrullus vulgaris*) require a long, hot growing season and are most in demand during midsummer. For this reason they are grown mainly in the South, although short-season cultivars can be grown in many areas of the country. Soil selection, planting, and cultural methods are similar to those used for muskmelons and cucumbers. Watermelons are usually planted five or six seeds per hill and thinned to two or three plants. Each hill requires 80 to 100 square feet (7½ to 9 m^2) of space.

There are no completely reliable maturity guides for watermelons. Size of fruit and color of rind are not reliable indexes (Figure 12–22). Some of the more useful

Figure 12–22 Taste—the best test of a watermelon's quality.

indications are the sound the melon makes when rapped or thumped, the color of the spot where the melon rested on the ground, and the condition of the tendril where the fruit stem is attached. Ripe melons, when cool, give a dull sound when thumped, the ground spot takes on a yellowish tinge as the fruit matures; and the tendril begins to dry up.

Some of the leading cultivars are 'Tom Watson,' 'Klondike,' 'Dixie Queen,' 'Kleckley Sweet,' 'Charleston Gray,' and 'Stone Mountain.' Short-season cultivars, including 'New Hampshire Midget' and 'Rhode Island Red,' are usually of low quality. 'Summer Festival' is a recently developed early melon, slightly smaller than standard size, with relatively good quality.

Pumpkins (*Cucurbita pepo*) and squash (various *Cucurbita* species), both native to the Americas, have similar cultural requirements. They are sensitive to frost but are also somewhat more tolerant of cool, moist environments than are melons or cucumbers. They need a warm but not extremely hot growing season. Summer types, which bear early and are used immature, can be grown in about 60 days; the winter types need a long growing season. Most cultivars will tolerate partial shade and are sometimes interplanted with corn.

Any fertile soil rich in humus but not excessively acid or alkaline can be used. The seeds do not germinate in cold soils but do not require as much heat as melons, so they should not be planted until the soil temperature has reached 60°F. The seeds usually are planted six or seven to a hill and later are thinned to a stand of two or three plants. The flowers are monoecious. Hills of vining types should have about 60 square feet (6 m^2) of space and bush types need 30 to 40 square feet (3 to 4 m^2).

Cultivation should be shallow and frequent enough to control weeds. When plants cover the ground, they will shade out most weeds and cultivation can stop. Summer squash are harvested at any stage before the seeds begin to mature and the rind to harden—small for slicing, a little larger if they are to be stuffed. Winter types are picked when fully ripe and after the rind has hardened. Winter squash and pumpkins can be stored for 2 or 3 months at temperatures around 45° to 50°F (7° to 10°C) and a relative humidity below 75%. They should be harvested before they are injured by frost.

Summer squash cultivars include 'Crookneck,' 'Straightneck,' 'Scallop' or 'Patty Pan,' 'Cocozelle,' 'Zucchini,' and 'Vegetable Marrow.' The leading winter squash cultivars are 'Hubbard,' 'Table Queen' or 'Acorn,' 'Golden Delicious,' 'Buttercup,' and 'Butternut.' Because squash have better cooking quality and even make better "pumpkin pie" than pumpkins do, pumpkins are grown more for fall decoration than for eating. 'Jack-O-Lantern' and 'Connecticut Field' are two decorative pumpkin cultivars; 'New England Sugar' is a small pie pumpkin.

Okra *(Hibiscus esculentus)*. Okra is a favorite vegetable of the old South, where it is cooked in many ways and is an ingredient of many southern dishes. Most northerners have tasted it in commercially canned gumbo soups, however, and have not developed a liking for okra prepared in other ways. Okra requires about the same growing temperature as winter squash. In northern areas plants tend to be relatively small, but they do bloom and produce fruit. The edible portion is a fruit or pod that must be harvested while immature. Plants tend to be spiny, and gloves are needed to harvest the pods. 'Clemson Spineless' and 'Dwarf Green Long Pod' are two commonly grown cultivars.

Sweet Potatoes *(Ipomoea batatas).* Of all the vegetables commonly grown in the United States sweet potatoes require the longest, hottest summer growing conditions. The use of plastic tents to permit earlier setting out of plants may enable growers in a few selected northern areas to produce sweet potatoes, but gardeners ordinarily will not plant sweet potatoes unless the July/August temperature in their area averages over 77°F (25°C).

The sweet potato is related to morning glory, and the sweet potato of commerce is an enlarged fleshy storage root capable of initiating leaf and shoot buds. It is propagated by placing fleshy roots in moist sand or similar media. The sprouts that grow from the fleshy roots are removed when they reach 6 to 7 inches in height (see Figure 4–8, Chapter 4) and are used as transplants. Several crops of sprouts can be produced by a single sweet potato if the season is long enough to permit several plantings. In the South sweet potatoes are often grown from vine cuttings secured from an early planting of slips. About half the butt end of a 15-inch (38 cm) piece of vine is buried. Plants are spaced at 4 to 12 inches in rows 3 to 5 feet in width.

Sweet potatoes are of two types—the moist, orange fleshed and the dry, yellow fleshed. 'Nemagold,' 'Red Nancy,' and 'Porto Rico' are three of the moist-fleshed cultivars; 'Big Stem Jersey' and 'Yellow Jersey' are two dry-fleshed cultivars.

Sweet potatoes should be dug when soil is dry. Slight frost will not damage the roots if they are well covered with soil, but they should not be left in the soil after frosts begin to occur frequently as they are subject to chill damage. Bruising should be avoided. If they are to be stored, sweet potatoes should be cured for a few weeks at a relatively high temperature and humidity; 85°F (29°C) and 85% humidity are ideal. They should be stored in a warm, dry room.

Selected References

Doty, W. L. *All About Vegetables.* Ortho Division, Chevron Chemical Company, San Francisco. 1973.

Heiser, C. B., Jr. *Nightshades, The Paradoxical Plants.* W. H. Freeman and Company, San Francisco. 1969.

Jones, H. A., and Mann, L. K. *Onions and Their Allies: Botany, Cultivation, and Utilization.* Interscience (Wiley), New York. 1963.

Nierwhof, M. *Cole Crops.* Leonard Hill, London. 1969.

Salaman, R. N. *The History and Social Influence of the Potato.* Cambridge University Press, Cambridge, England. 1949.

Ware, G. W., and McCollum, J. P. *Producing Vegetable Crops.* Interstate, Danville, IL. 1968.

Whitaker, T. W., and Davis, G. N. *Cucurbits: Botany, Cultivation, and Utilization.* Interscience (Wiley), New York. 1962.

13 The Fruit Planting

The appeal of high-quality fresh fruit is universal, and few foods are as popular or as nutritionally beneficial. Even so, the per-capita consumption of most kinds of fruit has steadily declined in the United States during the last 30 years. A major reason for this decline is that, at its present price, fruit is considered by many people to be a luxury. No economical way has yet been found to mechanize the planting, the pruning, and especially the harvesting of most kinds of fruit grown for the fresh market. The packing and shipping of fruit require more hand labor than the packing and shipping of other food products, and fruit requires special types of storage to insure prime condition. As a consequence, fruit is costly to produce and expensive to purchase. Growing fruit as a hobby is an excellent way to improve one's diet and at the same time reduce the food bill.

Home gardeners who produce their own flowers and vegetables may not consider growing fruit because they think it requires too much space and too much specialized knowledge and equipment. Such reasoning is not valid today. The many kinds of fruits that grow on herbaceous plants, lianas, and bushes require little space. By replacing small flowering trees and ornamental shrubs with dwarf fruit trees like the 'Stanley' plum or pear in Figure 13–1 and fruit-producing shrubs, gardeners can make their landscape provide both food and beauty. The 'Lodi' cultivar of apple on a dwarf rootstock displays almost the same show of blooms as a flowering crabapple and can produce a bonus of at least two boxes of apples in late July when this fruit is most welcome. Low-growing ornamentals or ground cover can be replaced by currant bushes that have attractive blooms during early spring, bright red berries and juice for jelly during summer, and, if all are not harvested, dried berries to attract birds during late fall and early winter. Brambles, such as the various kinds of raspberries and blackberries, make a pleasing hedge or, if the right types are used, an almost impenetrable fence. Grapes can be trained over an arbor to provide summer shade, and pears or other kinds of tree

Figure 13–1 Dual purpose trees. This semi-dwarf 'Stanley' plum is part of the corner planting of this home. An espaliered 'Bartlett' pear is at the right.

fruits can be espaliered along a fence line or against a wall to hide "architectural mistakes" or soften geometric patterns.

Like all garden plants, fruit trees and shrubs need timely care but not much more than ornamentals do. The pruning of fruit trees, for example, often viewed as a slightly mysterious and extremely precise operation, requires only a little more expertise than does the pruning of ornamentals. Furthermore, fruit pests are generally easier to control than mites on junipers, thrips on gladiolus, or the various pests of roses. In fact, at most locations strawberries, raspberries, blackberries, grapes, sour cherries, and, away from commercial orchards, even peaches and apricots require only minimal pest control. Pest management measures recommending relatively innocuous materials for various sections of the country are available with easily followed directions from state experiment stations, as are bulletins and pamphlets suggesting kinds of cultivars, planting procedures, and cultural practices adapted to local conditions.

Fruit-producing plants are classified according to the climate in which they are produced as tropical, subtropical, or temperate zone fruits. In the United States tropical fruits can be grown only at the southern tip of Florida, and consequently these will not be discussed in this book. We will mention briefly the production of subtropical types such as citrus, figs, and avocado. Most of the discussion that follows, however, will be concerned with the temperate zone fruits classified as **small fruits**—those produced on bushes, lianas, and herbaceous plants—or as **tree fruits**. Temperate zone tree fruits are divided into two groups—the **pome** fruits, including apple, pear, and quince; and the **stone** fruits, including peaches, nectarines, apricots, sweet and sour cherries, and plums.

Small Fruit Production in the Home Garden

Small fruits are widely adapted, are relatively easy to produce, and do not require a great deal of space. With some winter protection some cultivars of raspberries, strawberries, gooseberries, and currants can be produced in even the coldest regions. Although blueberries are not adapted to the excessively high summer temperatures and alkaline soils of the inland West and Southwest, they are a profitable home garden possibility in many areas of the country. One or more cultivars of grapes are also adapted to most areas of the United States, except for the upper Midwest and the northern Rocky Mountains, and to some warmer sections of Canada.

Grapes

Commercially, grapes are by far the most important of the small fruits. The grapes produced in North America are primarily of three species—*Vitis vinifera, V. labrusca,* and *V. rotundifolia*—and of two interspecific hybrids, French and American. The European or California grape, *V. vinifera,* is grown for fresh eating and for the production of raisins and wines. Most of the other commercial grapes are of the species *V. labrusca*, the only one of several native American grape species that is extensively cultivated. The muscadine grape, *V. rotundifolia*, is produced mainly in home gardens in the southeastern part of the country. French and American hybrids were developed in an attempt to combine the quality of *V. vinifera* with the disease tolerance and cold hardiness of various American species.

All grapes have a liana habit of growth, but there are differences in other plant characteristics. European grapes have fleshy, succulent roots, whereas the American and muscadine types have roots that are hard and woody. The skin of the European grape adheres tightly to the flesh, which is high in sugar content. In contrast, American grapes have a "slip" skin, watery juice, and lower sugar content. Muscadine grapes separate easily from the cluster when mature and are usually harvested as individual berries rather than bunches.

Climatic Adaptation. Most cultivars of both *Vitis vinifera* and *V. rotundifolia* are injured when temperatures drop below 0°F (−18°C). *V. rotundifolia* does best in the Southeast where summers are long, warm, and humid, and *V. vinifera* is best adapted where summers are long with hot, dry days but relatively cool nights. As a consequence, commercial production of *V. vinifera* grapes is limited to California and a few warm valleys of other western states, although with winter protection and special pest controls they are produced in home gardens in other areas of the United States and in the warmer areas of British Columbia and Ontario. American-type grapes are more cold resistant. They need fairly cool temperatures in the early part of the growing season, with warm weather and bright sunshine to mature the fruit properly. Some French and American hybrid cultivars resemble European grapes but are able to withstand temperatures as low as $-15°$ to $-17°$F ($-26°$ to $-27°$C).

In areas where winter damage is likely to occur, European and other less hardy grapes can be grown by keeping the permanent trunk no higher than 8 inches (20 cm) and laying vines on the ground and raking leaves, straw, or peat moss over the entire plant for winter protection (Figure 13–2). In some locations certain grape cultivars may be hardy in that they will winter over without damage, but they still may not be adapted because the summer climate does not provide enough heat units for their fruit to mature. In the cool-season areas maturity can be hastened by planting the grapes along a south or west wall. Sometimes vines can be trained to grow where there is ample sunlight even when roots are shaded. Furthermore, where summers are cool late cultivars will not develop maximum resistance to winter cold, apparently because insufficient carbohydrates are accumulated in the plant, and cultivars advertised as hardy to $-15°$F may be damaged by much less severe winter cold in such areas.

Characteristics of a few grape cultivars are listed in Table 13–1, but to be certain of planting types that are climatically adapted, gardeners should obtain recommendations locally.

Grapes have an extensive root system and require deep, well-drained soil. Although they grow on both coarse- and fine-textured soils, highest yields in both the Northeast and the Northwest are produced on sandy loams. In the South and the Southwest nematodes are sometimes a problem in sandy soils.

Propagation and Planting. Grapes are propagated by cuttage, layerage, or graftage. They do not grow true to type from seed. European-type grapes are usually grafted

Figure 13–2 Grapes, dual-purpose lianas. This semi-hardy cultivar, which helps soften an otherwise monotonous expanse of wall, is trained so that its canes can be lain on the ground and covered with leaves for winter protection. In this photo its recently uncovered canes, on which growth is just beginning, are being fastened to the wall.

Table 13-1 Characteristics of a few grape cultivars popular in North America

Type and cultivar	Heat units	Hardiness[a]	Uses[b]	Color
		American and American Hybrid		
Beta	1,450	VH	J	Blue
Himrod (seedless)	1,600	M	T,J,W	White
Interlaken (seedless)	1,600	M	T,J	White
Van Buren	1,600	VH	T,J,W	Blue
Schuyler	1,700	M	T,J,W	Blue
Seneca	1,800	H	T,J,W	White
Niagra	2,100	H	T,J,W	White
Concord	2,300	VH	T,J,W	Blue
Delaware	2,300	H	T,J,W	Red
Catawba	2,500	H	T,J,W	Red
		French Hybrid		
Aurora	1,700	M	W	White
Foch	1,700	VH	W	Blue
Cascade (Improved Seibel)	1,800	M	W	Blue
		European		
Early Muscat	1,700	M	T,J	White
Csaba	1,700	M	T, W	White
Thompson Seedless	2,100	M	T,J,W	White
Zinfadel	2,600	LH	W	Blue
Emperor	3,000	LH	T	Red
Ribier	3,000	LH	T	Blue
		Muscadine		
Dearing	2,700	LH	J,W	White
Scuppernong[c]	2,700	LH	J,W	Red

[a]VH=very hardy; H=hardy; M=moderately hardy; L=least hardy.
[b]T=table; J=juice; W=wine.
[c]Requires a pollinator; others listed are self-fertile.

onto American-type rootstocks, but more often American, Muscadine, and French hybrid grapes are propagated by rooting cuttings or by layering (see Chapter 14).

Grape plants can be purchased from nurseries either bare-rooted or in containers as 1- or 2-year-old planting stock. If 2-year-old plants are to be purchased, they should be examined carefully to make certain they are not the culls that were too weak or unthrifty to be set out as 1-year-old plants the previous season. (**Culls** are products that don't reach the standard of lowest marketable grade.)

The tops of 1-year-old grape transplants should be pruned to a single cane having two or three buds, and their root system should be divested of excessively long roots before the plants are planted. Two-year-old transplants should be cut back to a single stem and should have their damaged or excessively long roots removed (see "Transplanting Woody Plants," Chapter 14).

The standard spacing for grape vines is 7 to 10 x 7 to 10 feet (2 to 3 m). If trained in two dimensions on a fence or wall, each plant should have approximately 60 square feet (6 m^2) of space. Grape plants should be placed where they will receive full sunlight for at least half of each day.

Culture and Management. Fertilizer should be applied in the early spring, two to three weeks before the plants leaf out. In some areas a split application, part in the early spring and part just after blossom formation, may be desirable. Late summer or fall applications should be avoided, because they stimulate growth at a time when plants should be storing carbohydrates in preparation for the onset of cold weather (see "Fertilizing Garden Crops, Chapter 14).

Grapes have a high requirement for potassium. Deficiency of this element is manifested by interveinal scorching along the margins of the leaves. Phosphorus deficiency is not as common as deficiencies of nitrogen and potassium, but extra amounts will be needed in some areas. Magnesium deficiency is common in the East where soil pH is low, and zinc and iron deficiencies are frequent in vineyards growing in soils of high pH in the West.

Because of their extensive root systems grapes do not require as frequent irrigation as many garden crops. The soil around their roots should be filled to field capacity at the start of the dormant season. In areas where irrigation is required the vineyard should be soaked thoroughly and then allowed to dry before the next irrigation. Even in arid areas permitting 3 weeks to elapse between irrigations will generally not be harmful if the area between vines is cultivated. Weekly irrigation during dry periods will be needed if grapes are growing in a lawn or with another heavy cover crop. Where humidity is high, sprinkler irrigation should be avoided to reduce the tendency for mildew. In dry areas where mildew is not a problem thoroughly wetting the leaves twice a week will help reduce mite infestation. Irrigation should be reduced or eliminated 6 to 8 weeks before harvest to speed maturity and assist the onset of dormancy.

Vitis vinifera, V. labrusca, and the hybrids are self-fruitful (able to produce fruit without another cultivar as pollinator), as are many of the newer cultivars of *V. rotundifolia*. Local nurseries or garden advisors should be consulted concerning pollinators if muscadine grapes are to be planted, however, as most older muscadine cultivars produce only male or female, not perfect, flowers.

Fruit is produced on current-season wood, and some cultivars tend to be overproductive and require fruit thinning. When initial buds are damaged, by frost or otherwise, grape vines can initiate secondary fruiting buds that will produce a partial crop.

The grape makes its best growth and fruit production when heavily pruned to reduce the number of fruit buds and keep fruiting wood near the trunk and root system. Because fruit develops primarily from shoots produced from buds on 1-year-old wood, replacement canes for the following season's crop must be developed each year. Various pruning systems, each best adapted to specific areas and/or cultivars, have been developed. Regardless of whether the vines are trained on wires, on a trellis over an arbor, on a wall, or as short canes attached to a trunk, the same principles apply. After the plants have grown to maturity, they should be pruned so that only enough previous season wood to contain 40 to 80 buds remains (see "Pruning and Training," Chapter 14).

Pests and Related Problems. Grape vines are extremely susceptible to injury by 2,4-D and related phenoxy herbicides, and growers are warned never to use volatile forms of these compounds close to grape vines. If 2,4-D must be used near grapes, the home lawn formulations should be sprayed with extreme caution and only when the air is still and the temperature is lower than 85°F (29°C). Symptoms of 2,4-D damage are manifested by distorted younger leaves with light-colored raised veins and midribs along with distorted or nongrowing shoot terminals.

In many regions grapes will not require pesticides. This is true especially of American and Muscadine grapes. European grapes will need a weekly application of fungicide for control of mildew in cool humid regions. Other pests of grapes are listed in Table 13–2.

Strawberries

Strawberries (*Fragaria* spp.) are divided into two types on the basis of flowering and fruiting habits—the everbearing, which form flower buds throughout the growing season without regard to length of day, and the single-crop or June-bearing cultivars, which form flower buds only during short days in the fall and produce fruit in early summer. The total season production for good June-bearing cultivars is about the same as for ever-

Table 13–2 Major insect and disease pests of grapes

Insect pests	Symptoms and damage	Control
Phylloxera	Root mite fatal to European-type (*V. vinifera*) grapes.	Plant only plants grafted to resistant rootstocks.
Mites (various species)	Green and cream mottling of upper leaf surface and fine webbing underneath. Most serious in arid areas. Plants are weakened.	Thorough wetting of both leaf surfaces twice a week will help. Limit use of insecticides, which destroy mite predators. Miticides may be required.
Grape-berry moth	Small, brown worms that develop in fruit, causing it to ripen and drop prematurely. Not serious in most gardens.	Insecticides, if needed.
Cutworms	Damage shoots and buds by night feeding.	Cutworm baits.
Leafhoppers	Small, elongate, pale green insects with yellow and red markings; jump from leaves when disturbed. Cause rusty appearance on upper leaf surface as they suck sap from underside of leaf.	Insecticides.
Scale	Insects that cover themselves with a hard shell and remain attached to branches during the dormant season and emerge in crawler stage from May to July. Weaken plants by sucking juices.	Dormant spray.
Grapevine flea beetle	Small, steel-blue jumping beetles that eat opening buds and destroy new canes and fruit. Larvae feed on leaves in summer.	Insecticides.
Diseases	**Symptoms and damage**	**Control**
Powdery mildew	Characterized by white spots that make leaves appear as if dusted with flour; most limiting disease of *Vinifera* grapes.	Recommended fungicide, preventive sprays. Keep foliage as dry as possible.
Downy mildew	Yellow patches on upper leaf surface, followed by white cottony mildew below. Serious during cool, humid periods.	Sulfur dust as a preventive when temperatures are cool. Plants are injured by sulfur when temperature is above 85°F (29°C).
Dead arm	Caused by a fungus that enters the plant through wounds and causes loss of vigor and eventually death by girdling of a part of the plant.	Disinfect pruning equipment between cuts in vineyards where disease is present. Fungicides.
Black rot	The fruit rots, blackens, and shrivels, and is covered with tiny black pimples.	Fungicides.
Viruses	Cause various plant distortions, leaf mosaics, and stunted growth. Don't confuse with 2,4-D damage.	Plant certified virus-free plants. Prune off and burn diseased canes or destroy diseased plants.

bearing cultivars, so the choice of type to plant depends on whether the gardener wants production concentrated or would prefer to have fewer strawberries over the length of the entire summer season.

Climatic Adaptation. With some winter protection strawberries will grow in most parts of the United States and Canada where gardening is possible. However, each cultivar is generally adapted to a relatively restricted area. Highest quality berries are produced where the temperature is relatively cool during the maturing and ripening periods. Runners do not form on strawberry plants grown in the tropics, and fruit fails to set where temperatures are excessively hot.

Strawberries require well-drained soil with the water table at least 2 feet (60 cm) below the surface at all times, and they grow best in soils of moderate fertility that contain a good supply of organic matter. Although some older cultivars of strawberries require pollination, all cultivars readily available today are self-fruitful and do not require the planting of another cultivar for cross-pollination.

Commercial plantings of strawberries are not considered profitable for more than 2 production seasons or 3 years altogether. Home plantings, if they are well cared for and if plants are thinned occasionally, may produce profitably for 5 or 6 years or even longer in some instances. Where lack of winter cold or disease prevent the normal strawberry growth cycle, everbearing cultivars can be used to produce a single crop. If they are planted in the early spring and their blossoms are removed until July, they will build up sufficient carbohydrates to allow them to produce through the fall and early winter. They can be plowed out when production starts to diminish.

Propagation and Planting. New strawberry plantings are propagated from young plants produced on runners from older plantings. Because strawberries are subject to soil-borne root diseases, growers usually have a more productive planting if they use plants from a reliable local or mail-order nursery rather than from a neighbor's yard. In areas where winter temperatures are relatively cold strawberry plantings should be made in the early spring before the dormant plants have started to grow. In areas where the ground does not freeze through the winter they can be planted in the late fall or during the winter. They should be spaced about 18 inches (45 cm) apart in rows ranging from 2½ to 4 feet (¾ to 1¼ m) apart. Plants are usually permitted to spread across the bed, and only the center of the furrow is cultivated clean. When the plants are planted through plastic for weed and runner control, they can be spaced as close as a foot (30 cm) apart each way.

Strawberries can also be grown around the home as a ground cover or as window-box plants. Homeowners occasionally grow strawberries in an even more novel way by planting them in holes drilled through the sides of soil-filled wooden barrels, boxes, or hollow logs.

Culture and Management. During the season of planting all blossoms should be removed so that food reserves can be used for the development of large, vigorous plants. Runners that grow into the furrows should be cut off, and weeds must be controlled.

Fertilizer requirements vary depending on soil and location. For areas of relatively high rainfall 1 pound (500 g) of N and 2 pounds (1 kg) of P_2O_5/1,000 square feet (100 m^3) should be applied at the time plants are set into the field. An additional pound of N should be applied in July or August of that year. In drier areas about 1½ pounds of N should be applied at planting time. Other elements should be added, of course, if they are known to be required. Lime at the rate of 2 pounds/1,000 square feet applied a year before strawberries are planted is beneficial on some soils in the Atlantic Coast States. Boron is sometimes required in the Pacific Northwest and manganese in occasional plantings in the East. Other minor elements may be required in localized areas (see "Fertilizing Garden Crops," Chapter 14).

The initial fertilization can be a split application, part worked into the soil and part placed on the soil surface in a ring surrounding each plant. After plants have spread, fertilizer is most conveniently applied by broadcasting, followed by thorough irrigation with sprinklers to prevent burning where fertilizer is in contact with foliage.

Strawberry plants take most of their moisture from the top foot of soil, which will hold from ¾ to 1½ inches (2 to 4 cm) of water, depending on texture and organic matter content. The amount of water applied with each irrigation should be proportional to the amount the soil will hold and should be just enough to bring the top foot to field capacity. The two times when water is most critical to a planting of a June-bearing strawberry cultivar are just as berries are maturing and during August when blossom buds are forming for production the following season. Where spring rains do not supply 3 inches (8 cm) of moisture per month, some irrigation in April and May will be needed. Because moisture causes ripening fruit to rot, the planting should be thoroughly soaked just before the earliest berries begin to mature. Watering during the picking season is risky, but if the planting dries, irrigation may be worth the risk. Some moisture stress immediately after harvest does not seem to be damaging, but the planting should again be well watered during blossom-bud formation in August. Irrigation should be reduced in September and early October to force the plants into dormancy, but the planting should be brought back to field capacity before it is mulched for winter.

Most strawberry cultivars are subject to winter damage if grown east of the Cascade Mountains and the Sierra

Nevada and north of a line extending east from northern Arizona to northern Georgia and then northeast to southern Chesapeake Bay. Covering the plants with a mulch of waste hay, straw, shavings, sawdust, peat moss, or leaves is the usual method of cold protection (see Figure 7–17, Chapter 7). Mulch protects from cold or fluctuating temperatures and from heaving. By delaying bloom, mulching lessens the danger of blossom damage from early spring frost, a damage to which the low-growing, early-blooming strawberry is especially susceptible (Figure 7–11, p. 132). Mulch should not be placed over the plants until after the first hard autumn freeze. When growth starts in the early spring, surplus mulch should be raked from the rows into the alleyway. Where spring freezes are frequent, blossoms can be protected by temporarily raking the mulch back over the rows on the evenings when frost is anticipated. In areas where blossoms are frequently lost to spring frosts, the planting of everbearing cultivars is advisable.

Insects and Diseases. Although strawberries are subject to attack by a number of pests (see Table 13–3), a satisfactory crop can be produced without pest control in home gardens in many areas provided disease-free planting stock is used to establish the planting. Virus, root and leaf diseases, and insect eggs and larvae can be intro-

Table 13–3 Major insect and disease pests of strawberries

Insect pests	Symptoms and damage	Control
White grubs (with legs), crown borers, and root weevils (legless)	White grubs infect fields recently plowed from grass. White grubs and root weevils destroy or weaken plants by feeding on roots. Crown borers feed on and hollow crown.	Soil insecticide applied at time plants are set into the garden.
Cyclamen mites	Feed on and cause stunting and crinkling of new leaves, blasting of blooms, and distortion of fruit.	Planting stock from a good nursery will be treated to destroy mites and mite eggs. Prebloom and after-harvest miticide treatment.
Spider mites	Weaken plant by feeding on underside of older leaves. Usually most serious in arid regions during mid- and late summer.	Miticides after harvest. A jet of water strong enough to wash underside of leaves but applied so it does not splash mud on plant may help if washing is done several times after harvest.
Aphids	Cause damage mainly by spreading virus. Root aphids may require control in some locations.	Insecticides.
Slugs	Burrow into ripening fruit, leaving slimy trails.	Slug baits.
Symphylids	Centipedes that feed on feeder roots. Serious primarily in Pacific Northwest.	Soil insecticide at time of planting.
Diseases	**Symptoms and damage**	**Control**
Virus	Reduced runner formation and decreased vigor and yield. Often nonspecific problems.	Set only certified virus-free plants. Control aphids.
Leaf spots (various fungi)	Leaves show red or purple spots with or without white or grey centers, depending on species of fungus.	Plant resistant cultivars. Fungicidal spray.
Red stele	Plants wilt and die just before or during harvest. Roots decay, with red centers. Most severe in low-lying, poorly drained areas.	Plant disease-free, resistant cultivars on land that has not produced strawberries for several years.
Verticillium wilt	Decline and collapse of plants starting with roots and lower leaves about the time berries start to ripen. Most serious in West.	Do not plant where tomatoes, peppers, eggplant, black raspberries, or other wilt-susceptible crops have grown. Rotate planting.
Fruit rot	Most serious in humid areas.	Fungicidal sprays as fruits begin to ripen. Do not harvest any damaged or partially decayed fruit.

duced on planting stock. Considering that a healthy planting will produce for several years, it is usually profitable to purchase planting stock from a reliable nursery, or, better yet, to secure plants certified pest free by the state department of agriculture. County agents, experiment stations, or state departments of agriculture can supply addresses of certified plant growers.

Harvesting. Strawberries mature during a period of several weeks, and in order for all the berries to be harvested at the proper stage of maturity the patch will need to be picked every 2 to 5 days, depending on the temperature. Home-grown strawberries should not be harvested until they are at least 80% red. Although they will continue to soften and turn red if picked sooner, they will be lower in sugar content and have less flavor. Strawberries should be picked with the caps attached unless they are to be consumed or processed immediately. They should be harvested when the temperature is cool for best storage. Commercial growers often pick only between 4:00 and 10:00 AM. Decayed, overripe, or bird-damaged berries should be picked but discarded, because decay organisms that start in a single injured berry can quickly spread both rot and a disagreeable flavor throughout an entire container. The containers used for harvesting should be dry and shallow and constructed to allow air to circulate freely. The traditional small, thin wooden baskets (also called cups or hallocks) are ideal; mesh plastic and pulp paper baskets are also satisfactory (Figure 13–3).

Figure 13–3 A good berry container. Containers for raspberries and strawberries should be shallow and constructed so they permit free circulation of air. Strawberries should be picked with the caps attached.

Bramble Fruits

The bramble fruits (*Rubus* spp.) include red, black, and purple raspberries; erect blackberries; and trailing blackberries such as dewberries, loganberries, youngberries, and boysenberries. Except for some cultivars of trailing blackberries, production is mainly located in cooler areas. The general growth habit of bramble fruits is unique, the top being biennial and the roots perennial. Brambles bear a single crop of fruit on second-year canes except for a few everbearing or fall-bearing red raspberry cultivars.

Climatic Adaptation. The red raspberry is the most cold resistant of the bramble fruits, but long, hot summers, hot winds, and high soil temperatures prevent its successful culture in southern locations. Black raspberries are somewhat less winter-hardy than red raspberries. Erect blackberries are slightly less cold resistant than black raspberries, and trailing blackberries are grown extensively only where winters are relatively mild. Although they grow best when exposed to full sun, red raspberries are one of the very few garden fruit plants that produce a fair crop in partial shade. Brambles grow and yield better if they are given wind protection.

Brambles can be grown on a wide variety of soils if the soils are moderately fertile and have good water-holding capacity and drainage. The trailing types can be grown in somewhat unfertile, shallow soils if drainage is good. On heavier, more fertile soils trailing blackberries may make too much vegetative growth and produce few berries.

Propagation and Planting. Like strawberries, brambles are subject to many root and virus diseases, and plants secured from a reliable local or mail-order nursery will usually produce a more successful planting than those dug from a neighbor's yard. If they are obtainable, state-inspected certified plants will be most satisfactory. Growers sometimes purchase a few certified plants and propagate from them.

Red raspberries and most upright blackberry cultivars are propagated from root cuttings. Roots can be dug, cut into sections, and planted in a nursery bed. They should be covered with 3 inches (8 cm) of soil. A more common method of home propagation is to plant the suckers that emerge from the lateral roots some distance from the mother plant. Suckers with woody 1-year-old stems are more satisfactory than those with only soft current-season stems. Black and purple raspberries are propagated by tip layerage. Trailing blackberries can be propagated from tip cuttings or by tip or serpentine layerage. Because root cuttings of some trailing blackberry cultivars produce plants with thorny canes, root cuttings are not used to propagate thornless trailing blackberry cultivars (see "Layering to Renew or Multiply Plants," Chapter 14).

Both red and black raspberries and upright blackberries are planted in either a hill or hedgerow system. For the hill system plants are spaced 3 to 3½ feet (about 1 m) and for the hedgerow system, 2½ to 3 feet (⅔ to 1 m). Trailing blackberries should be 5 to 8 feet (1½ to 2½ m) apart in the row. Rows for all brambles should be 6 to 10 feet (2 to 3 m) apart depending on the method of cultivation and the space available. Planting is usually in the early spring, except in the mild winter areas of the South and Pacific Coast where fall or winter planting is often more practical.

Culture and Management. To facilitate cultural practices and harvest and to prevent the planting from becoming overcrowded, cultivate sucker plants from between rows and hills. After the planting becomes established, most small annual weeds will be shaded out, but the common practice of forgetting about weeds in the raspberry or blackberry planting is a mistake as competition by tall or climbing weeds can seriously retard the bramble planting and increase its susceptibility to disease.

Brambles respond to high organic matter, and the blackberry or raspberry planting should have high priority for available manure or compost. Immature (non-seed-bearing) weeds, grass clippings, and small twigs removed from the remainder of the yard can be profitably disposed of between rows of brambles to help control weeds, prevent evaporation, and supply organic matter.

As is true with most crops, fertilizer requirements of brambles vary with location and soil. Raspberries produce highest yields when new canes grow to 6 to 8 feet in height, so fertilizer should be adjusted to stimulate this kind of growth (see "Fertilizing Garden Crops," Chapter 14).

Because deep-rooted brambles can extract moisture from the upper 4 feet (1¼ m) of soil, they do not require as frequent irrigation as strawberries. Moisture requirements are greatest just before and during harvest to supply the dual needs of ripening fruit and rapidly developing new canes. Brambles planted in a medium-textured soil will require 2 inches (5 cm) of water every 2 weeks from either rain or irrigation during spring and early summer, 4 inches (10 cm) every 2 weeks during midsummer, and 1 to 2 inches during early fall. Watering should be tapered off after harvest to harden canes, but the soil should again be soaked to field capacity before hard freezes occur. Furrow irrigation is preferable to sprinkler irrigation during harvest as water from sprinklers can knock off ripening berries.

Brambles usually produce fruit buds on current-season canes during late summer or fall of the year preceding the crop year. In addition to their usual bloom a few so-called everbearing or fall-bearing cultivars of red raspberries form blossom buds during midsummer and set fruit that same fall on current-season canes (Figure 13–4).

Because brambles tend to bloom later in the spring, their blossoms are not as subject to freeze damage as are the blooms of many other fruits. All commonly produced cultivars are self-fruitful and do not require a pollinator.

Training and pruning of brambles is based on their biennial growth habit. All canes that have fruited, suckers growing in the furrow or away from the plant row, and the weakest canes should be removed from the row. The remaining strong canes should be cut back somewhat and/or tied to supporting wires or posts (see "Pruning and Training," Chapter 14).

Insects and Diseases. Because soil-borne root diseases and virus are the pest problems that destroy most bram-

Figure 13-4 Raspberry maturity. Raspberries ripen over a period of several weeks; (A) blossom stage, (B) ripe fruit. (A, courtesy M. M. Thompson, Oregon State University; B, courtesy D. K. Ourecky, New York Agricultural Experiment Station.)

ble plantings, gardeners should obtain disease-free planting stock and plant it in well-drained soil in which brambles have not previously grown. Furthermore, where virus disease is known to be a limiting factor in raspberry and blackberry production, it is usually best to grow only one kind of bramble—that is, raspberry, black raspberry, or blackberry—as each kind of bramble is often a masked carrier of viruses fatal to other kinds. In warm, moist areas where fungus disease is a problem old canes should be removed and destroyed as soon as they have fruited. Diseased and insect-infested canes or plants should be destroyed. Black and purple raspberries should not be planted where potatoes, tomatoes, eggplant, or peppers have been produced recently because of their susceptibility to verticillium wilt.

The common insects and disease pests of brambles are briefly described in Table 13-4.

Harvesting. Raspberries are ready to pick when they separate easily from the core that remains on the plant. Blackberries become fully colored and sweet before they are ripe. They can be picked when the depression in the tip of each drupelet or section of the berry is filled. The core of the blackberry is a part of the fruit. Boysenberries and youngberries do not develop good flavor or maximum size until they are fully ripe. Berries produced on

Table 13-4 Major insect and disease pests of bramble fruits

Insect pests	Symptoms and damage	Control
Aphids	Serious as vectors of virus disease.	Predators will often control aphids. If virus spread is a problem, periodic insecticidal sprays may be required.
Root weevils (several species)	White or brown legless grubs that feed on roots and cause plants to be unthrifty.	Preplant soil treatment with insecticide or treatment with insecticide washed in around crown of established plantings.
Raspberry crown borer	Swelling at base of cane from larvae tunnels. Breaking of fully grown year-old canes 2 in (5 cm) below soil surface. Larvae require 2 years to mature.	Soil insecticide applied in late fall or winter 2 years in succession to kill current-season larvae just under bark near crown.
Raspberry cane maggot	Adult fly lays eggs at tip of young canes. Maggot burrows into pith, girdles cane near tip, and continues downward in pith of cane.	Wilted canes should be cut off near ground and burned.
Fruit worms (various species)	Adult beetles feed on and distort flowers and fruit. The tiny worms burrow into fruit where they may remain until harvest.	Where numerous enough to be a problem, this pest is controlled by insecticides timed to kill adults before they lay eggs.

(continued)

Table 13–4 *(continued)*

Insect pests	Symptoms and damage	Control
Leaf mites (red spider and two-spotted)	These barely visible tiny mites form webs and feed on underside of leaves. Their webs and white flecking on leaves are diagnostic.	Damage is usually during a dry, postharvest period. Frequent sprinkling aids control. Except where severe defoliation occurs, pesticides should be avoided because they kill mite predators.
Fruit mites	Cause dry, mummified raspberries or hard red drupelets that "don't ripen" on blackberries.	Insecticide timed to kill wintering over mites just before they enter the fruit bud, just prior to flowering.
Sawfly (northern Great Plains)	Causes brown leaf blisters each containing $\frac{1}{3}$ in ($\frac{3}{4}$ cm) or shorter flattened active larvae.	Insecticide just before blossoms open.
Tree cricket (East)	Deposits curved orange eggs in closely spaced rows 2 in long. Egg-containing punctures may cause cane to break.	Damage is usually not serious enough to require control.
Orange tortrix (Northwest)	Larvae wriggle from rolled leaf shelters and fall into picking bucket.	Requires control only if berries are to be sold.

Diseases	Symptoms and damage	Control
Nematodes (various)	Unthrifty growth and gradual decline of plants over a period of years. Matted roots. Root knot nematode causes enlargements, and lesion nematode causes dead spots on small roots. Nematodes intensify other root disorders.	Prevention is only control. Use nematode-free planting stock. Have soil checked for nematodes before planting. Soil fumigation prior to planting or long rotation with grass, corn, or small grains will reduce nematodes.
Root rot	See Strawberry diseases, Table 13–3.	
Crown gall (East)	Cauliflowerlike swellings on roots, crown, or lower part of cane. Disease enters through wounds.	Avoid planting on infested sites. Disease can be spread with cultivating equipment.
Verticillium wilt (Northwest)	Most severe on black raspberries. Beginning at plant base, leaves turn yellow (often on one side). Canes turn blue and gradually die.	Don't plant where tomatoes, potatoes, or peppers have grown the previous 3 years. Plant disease-free plants and remove and burn infected plants.
Anthracnose	Most serious on black and purple raspberries in humid locations. Causes light-colored sunken spots, chiefly on canes. Canes may be girdled.	Plant on sites with good air drainage. Remove "handles" (old cane) from black raspberry at planting time. Thin planting to permit air circulation. Apply fungicide if necessary.
Cane blight (humid areas)	Enters canes through wounds, winter damage, pruning, etc. Dark brown cankers extend down from wounds to encircle cane.	Remove old canes as soon as they have fruited. Remove and burn infected canes.
Spur blight (humid North)	Brown or purple spots develop near buds on infected canes. Buds darken and shrivel, and leaves turn yellow and dry.	Thin planting to allow air circulation. Good soil drainage and weed control are important.
Viruses (mosaic, tomato ringspot, leaf curl, etc.)	Viruses produce various symptoms, including light and dark patches on leaves, distorted leaves, crumbly berries, reduced growth, sterile plants with no fruit, death of the plant. They are the agent primarily responsible for the decline of raspberry plantings in many areas.	Plant certified disease-free planting stock if available. Don't plant black and red raspberries in the same location. Don't plant healthy raspberries near diseased ones.
Fruit rot (various organisms)	Can occur on the plant or after harvest.	Pick only sound berries; pick when plants are dry and weather cool; refrigerate berries and keep them cool after harvest.
Crumbly berry	Drupelets on berries fail to stay together; generally results from failure of some drupelets to be fertilized and can be caused by a lack of bees during bloom or by plant diseases.	Don't apply insecticides near the bramble planting when it is blooming; control plant diseases.

any one plant mature during a period of several weeks, and the fruit that has ripened will need to be picked each 2 to 4 days during that period.

Raspberries and blackberries are even more fragile than strawberries. Recommendations concerning time of harvest and containers mentioned in the section on strawberries also apply to bramble fruits.

With reasonable care plantings of bramble fruits should produce profitable harvests for 6 to 10 years, though some have remained in production for as long as 20 years.

Currants and Gooseberries

Currants and gooseberries (*Ribes* spp.) are of minor commercial importance, but both are easy to grow and both produce large quantities of fruit that can be used for a variety of tasty dishes. In addition, both grow on attractive bushes, which makes them excellent crops for the home garden or dual-purpose landscape. Both are hardy, even into the prairie provinces of Canada and in southern Alaska, but do not grow well in the South or Southwest. Gooseberries are somewhat more resistant to heat than currants.

Currants and gooseberries can be propagated with hardwood cuttings or by layerage. Cuttings of previous-season wood 8 to 10 inches (20 to 25 cm) long are rooted during late winter or early spring in a propagating bed; mist chamber; or, in humid areas, in a shaded, protected location out-of-doors. They are inserted so that only two buds protrude above the propagation medium. If only a few plants are needed, they can be developed by layering the lower branches of established bushes. Often branches of currants and gooseberries layer spontaneously. It is fairly safe to obtain these layered currant and gooseberry plants from a neighbor provided the neighbor's plants are healthy and are standard cultivars that produce large fruit. Those transplanting from an older garden, however, run the risk of introducing soil-borne pests into their own gardens. Table 13–5 lists major pests of currants and gooseberries.

Table 13–5 Major insect and disease pests of currants and gooseberries

Insect pests	Symptoms and damage	Control
Currant aphids	Infested leaves curl downward, protecting the aphids.	Spray before leaves curl if required.
Currant borer	Feed in the pith of canes, sometimes causing them to be stunted or to break.	Remove and burn infested branches.
Scales	Soft-bodied insects covered with a hard scaly shell.	Treat with a suitable insecticide while buds are dormant.
Mites	See Bramble insect pests, Table 13–4.	
Currant fruit fly	Causes wormy currants and is perhaps the most difficult pest to control.	Treat with suitable insecticide at intervals while fruit is enlarging. Same control should control aphids.
Imported currant worm	Blue-green ¾-in (2 cm) worm with black head. May strip currants and gooseberries of almost all foliage.	Suitable insecticide.
Diseases	**Symptoms and damage**	**Control**
Leaf spot	Brown and, later, pale grey spots that cause defoliation.	Fungicidal treatment just before flowers open.
Anthracnose	Causes small brown spots on all above-ground parts of plant, resulting in early defoliation.	Keep plants pruned so the growth habit is open. Use fungicide if required.
White pine blister rust	Currants and gooseberries, especially black currants, are alternate hosts of this disease.	In white pine–growing areas home garden planting of some types of currants and gooseberries was often prohibited, but with the realization that wild currants and gooseberries cannot all be destroyed, restrictions have been relaxed in most places.

Each currant or gooseberry bush should have at least 16 square feet (2 m^2) of growing space. Bushes are best planted in the early spring and should be placed slightly deeper than they were in the nursery.

Mulching with a pest-free loose mulch is advantageous, especially in dry areas. Average fertilizer requirements are listed in Tables 14–16 and 14–17 (Chapter 14). Both commercial fertilizer and manure should be kept from touching the crowns of plants. Irrigation practices suggested for brambles should be satisfactory for currants and gooseberries.

Except for black currants, which are seldom grown in this country, planting two cultivars for pollination is not required. Pruning practices are simple. The fruit is produced on 1-, 2-, and 3-year-old wood; older wood produces inferior fruit. In pruning the older wood is removed, and it is replaced by 1-year-old wood. The average mature, well-pruned, fruit-bearing currant or gooseberry will consist of three to five shoots, each of 1-, 2-, and 3-year-old wood.

Both currants and gooseberries are used mainly for making jellies and jams. For jelly currants are usually picked when some of the fruit is not fully mature. Gooseberries are usually harvested when fully sized but not yet red in color (Figure 13–5). Many individuals enjoy eating these fruits out of hand or with sugar and milk or cream. Gooseberries also make excellent pies. Harvesting may extend over a period of a month or more. Fruit that remains will frequently dry on the bushes and is very attractive to birds in the late fall and early winter.

Figure 13–5 Fully ripe gooseberries. These berries are about the right stage to harvest for jelly but too mature for jam or pie.

Blueberries

Berries of a number of species of *Vaccinium* are harvested from the wild for fruit, but only two, the highbush blueberry and the rabbiteye blueberry, are propagated and grown in quantity under cultivation. Where adapted, the cultivated types produce fruit that is larger, although it may be less flavorful, than fruit of the wild types. The blueberry is a good home garden crop where its specific soil and climatic requirements can be met. The highbush blueberry grows best in soil of pH between 4.3 and 4.8 and has about the same temperature and chilling requirements as does peach. The rabbiteye blueberry is somewhat more susceptible to winter injury and requires less cold to overcome the rest period than does the highbush. Most cultivars are more flavorful when ripened where nights are cool. The highbush blueberry is adapted to the Pacific Coast valleys of Oregon, Washington, British Columbia, and northern California; the warmer parts of the Great Lakes region and Ohio Valley; and the East Coast south to South Carolina. Rabbiteye blueberries are adapted to the South Atlantic states.

Propagation and Planting. A reliable nursery is the best source of plants for the gardener who wants to grow a few blueberry bushes; however, blueberries are not difficult to propagate. New plants can be developed from softwood cuttings in a mist chamber, but rooting of hardwood cuttings is more popular. They should be gathered during late winter or early spring while plants are still completely dormant. Pencil-sized hardwood cuttings 4 to 6 inches (10 to 15 cm) long from healthy shoots of the previous season's growth are cut just below a bud at the base and above a bud at the top. These are placed in a rooting medium in partial shade with one or two buds above the surface. Propagation is usually accomplished in a cold frame, but cuttings should root satisfactorily in a mist chamber, a home propagating bed, or even a protected location out-of-doors. For outside propagation a dampened piece of coarse burlap should be placed over the cuttings until buds start to swell. When cuttings are well rooted, they are transplanted to nursery beds, where they have more room. They should be ready to plant in their permanent location the following spring. Each highbush blueberry bush requires about 30 square feet (3 m²), and each rabbiteye should have 40 to 60 square feet (4 to 6 m²). They should be planted in their permanent location slightly deeper than they were growing in the nursery.

Culture and Management. In their native habitat both rabbiteye and highbush blueberries grow on hummocks in and at the edges of swampy areas, so they will grow quite well in wet locations. They do, however, require at least 18 inches (45 cm) of soil above the water table. In most areas they grow best in full sun although, like raspberries, they will tolerate light shade.

In regions where blueberries are adapted, soils are usually low in all three major elements. Years ago there were numerous reports of manure being harmful to blueberry plantings, but more recent evidence suggests that these reports were exaggerated. If manure is applied, it should be kept from direct contact with the crown. J. S. Shoemaker, in *Small Fruit Culture*, suggests that the ammonium form of nitrogen is superior to the nitrate form and that the ammonium form may be essential for blueberries. Average fertilizer requirements are listed in Tables 14–16 and 14–17 (Chapter 14). With high pH, iron deficiency (stunted growth and yellowing) or magnesium deficiency (interveinal areas turn red and yellow beginning at leaf tip) are likely. Iron and magnesium chelates are the formulations of these elements most likely to correct deficiencies in most areas. Follow label directions on the purchased product. In some localities in the East less expensive formulations of these elements may be effective.

Blueberries are not drought tolerant. A loose mulch will help retain moisture, but where rainfall doesn't supply enough the planting should receive 1 to 2 inches (2½ to 5 cm) of moisture each 10 days during the harvest season and a little less, but still enough to keep the soil moist, before and after harvest.

Pests of blueberries are not numerous, and often common-sense sanitation and management practices will be all the control required (see Table 13–6).

Table 13–6 Major insect and disease pests of blueberries

Insect pests	Symptoms and damage	Control
Blueberry maggot	Causes wormy berries. Adult is a fly about size of housefly. Most serious insect pest of commercial blueberries.	Dust or spray when adult flies appear in large numbers but before they lay eggs (early to mid-July in most northern areas).
Cranberry and cherry fruit worm	Web and filth around some fruit clusters. Feeds on fruit.	Usually not numerous enough to require control in the garden.
Blossom weevil	Feeds on blossoms. May reduce blooms 50% in some areas.	Insecticidal spray.
Scale	Reduces vigor of canes (see Grape insect pests, Table 13–2).	Dormant scale sprays.
Diseases	**Symptoms and damage**	**Control**
Stem canker, stem blight (North Carolina)	Cankers girdle and weaken stems, prevent the growing of susceptible cultivars in the South. Blight causes dieback of stems.	Grow cultivars resistant to stem canker. Remove and destroy affected stems.
Stunt virus	Transmitted by sharp-nosed leafhopper. Internodes shortened; leaves reduced in size and chlorotic; premature red coloring.	Remove diseased plants. Don't propagate from any plant growing within ¼ mile (½ km) of diseased plants.
Necrotic ringspot virus	Spreads by nematodes. Leaves pucker, become spotted. Spots dry and crumble, leaving holes in leaves.	Plant virus-free plants on land not infested with nematodes.
Other viruses	Several viruses with various symptoms may affect plants.	Rogue diseased plants. Plant disease-free planting stock.
Mummy berry	A fungus that causes dieback of shoots, blighted blossoms, and mummified berries.	Remove and destroy mummy berries and hoe around plants to disturb sporulation on fallen berries.

Pollinating, Training, and Pruning. There is some evidence that cultivated blueberries produce better yields with cross-pollination, so homeowners should plant at least two cultivars.

The cultivated highbush blueberry fruits on wood of the previous season's growth, and the largest fruit is borne on the most vigorous wood. Most cultivars tend to overbear, and some flower buds must be removed by pruning to permit the plant to develop sufficient vigorous wood for the next year's crop and to prevent development of small berries. Heavy pruning definitely reduces the crop. Light pruning, although it may reduce the crop slightly for a given year, will stimulate more vigorous wood for the next year's crop. Under conditions of heavy cropping and no pruning some of the fruit may not mature. In contrast, light pruning will give earlier maturity of fruit as well as increased size.

When the plants are set in the field, about one-fourth of the top should be cut back along with bushy basal growth and damaged roots and shoots. During the first season all flower buds should be removed. No further pruning will be needed for several years other than removal of broken or interfering branches. After the plant matures, pruning will consist of removing dead, broken, and interfering branches and thinning bushy wood and old stems that are no longer vigorous. Eventually some thinning of the branches will be necessary. Erect-growing cultivars will need to have some center branches removed to permit better light penetration. Branches that droop and allow fruit to become dirty should be removed from spreading types.

Even with moderate pruning it may be desirable to rub off a few of the flower buds in each cluster to avoid overbearing and promote larger, more uniform berries.

Harvesting. Blueberries within a cluster mature over a period of several weeks, and for highest quality ripe berries should be picked each 4 to 5 days. Berries will not usually be mature-ripe when they turn blue, so other criteria—sweetness, ease of removal from the stem, and reduced resistance to pressure—must be considered in determining when to harvest.

Other Small Fruits

The cranberry, being a bog plant, is suitable for home gardens only in very localized special situations. Some cultivars of elderberries (*Sambucus* spp.) have been domesticated and are used for jelly and wine making. A number of small fruits, including elderberries and blueberries, are sometimes transplanted from the wild. The serviceberry (*Amelanchier*), sometimes called the Juneberry and in Canada known as the Saskatoon, is frequently grown around homes in the upper Midwest, the Rocky Mountain area, and the prairie provinces of Canada. It makes an excellent ornamental and produces delicious, although seedy, fruit. Wild currant (*Ribes* spp.), choke-cherry (*Prunus virginiana*), and hawthorn (*Crategus* spp.) are among the other berry-producing plants frequently transplanted from the wilds.

Tree Fruit Production in the Home Garden

During the last century virtually every yard contained a few fruit trees. As yards became smaller and Americans became more mobile during this century, the number of home gardens with fruit trees declined, probably because of what might be called the four Ws: Worms, Wood, Waste, and Wait. The spread of the codling moth, the cherry fruit fly, and certain other insect pests throughout the country made it impossible to grow good fruit without use of potent poisons that were the only available pest controls. Moreover, without proper pruning standard-sized fruit trees produced too much wood and grew to enormous sizes, their crowns filled with dead and intercrossing branches. These uncared-for trees produced small wormy fruit that fell to the ground and made a wasted mess. If new trees were planted, years of waiting were required before they eventually produced a crop, and, with the mobility of Americans, more often than not the individual who planted the tree had moved by the time it started to bear.

During the past few years increasing interest in home

gardening and scientific developments in the field of **pomology** (the science of fruit growing) have led to renewed interest in home production of tree fruits. Although it can hardly be said that tree fruits can be grown without effort, their production by the amateur is much more feasible today than it once was (see Figure 13–6). The introduction of dwarf trees and newer training systems has permitted the growing of a fruit tree in an area as small as 25 square feet (2 m^2). Also, one dwarf tree will produce only about as much fruit as a family is likely to consume during the course of the season. Thus a diversity of kinds of fruit maturing at different times can be produced in the area available. Furthermore, the development of relatively safe pesticides makes the production of tree fruits much easier today than it was a few years ago.

Tree fruits grown in the United States are mainly of two types, deciduous and evergreen. Of the deciduous fruits, the pomes—apple, pear, and the seldom-grown quince (Cydonia)—are most commonly planted. Peaches, apricots, cherries, and plums, the stone or drupe fruits, are the other major group of deciduous tree fruits.

Figure 13–6 Dwarf fruit trees in limited space. Two dwarf apple trees and a semi-dwarf plum tree have adequate growing room in the 18 x 30 foot (6 x 10 m) strip at the side of this home.

The Pome Fruits

Apples (*Malus* spp.) and pears (*Pyrus* spp.) are probably the most widely adapted tree fruits for the home garden. Cultivars are available that will grow in all but the extreme southern parts of the United States. Fruits of many cultivars store well for several months and can be utilized in a variety of ways.

Climatic Adaptation. Most cultivars of apples and pears can be hardened to withstand temperatures as low as −20°F (−29°C) (Figure 13–7). With optimum conditions for development of cold tolerance some of the hardier cultivars of apples can be produced where temperatures drop as low as −40°F (−40°C). Roots are not as hardy and may be injured if soil temperatures drop to 10° to 20°F (−12° to −7°C). Spring frosts of temperatures below 28°F (−3°C) occurring at the time of bloom or shortly thereafter can seriously damage blossoms and young developing fruits. Pear blossoms are more fre-

Figure 13–7 Cross section of a fruit tree trunk. A cross section of the trunk of a fruit tree tells a great deal about the growing conditions in the orchard. As mentioned earlier, differences in size of annual rings are correlated with seasonal growing conditions. The apple tree from which this section was taken was growing on a ridge having relatively shallow soil and subject to bright sun and prevailing southwest winds that reduced growth on the southwest (side adjacent to the split). The dark "heartwood" has resulted from fungal growth that invaded tissue damaged by a severe winter 6 years before the tree was cut.

quently damaged by early spring frosts than are apple blossoms because pears bloom earlier. Both apples and pears must be exposed to at least 6 weeks of average temperature below 40°F (4°C) to break rest.

Selecting and Planting Pome Trees. Today the only apple trees most home gardeners will consider planting are those grafted to dwarfing rootstocks. Dozens of rootstocks that dwarf to varying degrees have been developed, but only a few are widely used. The most dwarfing and popular for gardens of apple rootstocks is Malling IX. Other rootstocks commonly available that dwarf to varying degrees include E.M. 26, producing trees slightly larger and better anchored; E.M. VII and M.M. 106, producing trees that average 15 to 20 feet (5 to 6 m) high; and E.M. II, producing very vigorous trees about two-thirds standard size. Trees on Malling IX rootstock can be planted as close as 10 feet (3 m) apart, and they grow to a maximum height of about 10 feet. Their size can be reduced even further by pruning. The Malling IX rootstock does not give a tree very good anchor, and a stout steel or long-lasting wooden post should be driven next to the tree to provide support. Trees on Malling IX rootstock will often produce fruit 1 or 2 years following planting. Dwarf trees should be planted so that the graft union is above the soil line (see Figure 8–14, Chapter 8). If the graft union is buried, the scion part of the tree is likely to root, and a standard-sized tree will result. Pear trees are dwarfed by placing them on quince rootstock, as was mentioned in Chapter 8. Because quince is graft-incompatible with many of the better varieties of pears, an interstock is usually necessary. Thus at least 2 years are required for the development of a dwarf pear tree, and such a tree is likely to be fairly expensive. Better anchored dwarf apple trees are also sometimes developed by using a dwarfing interstock (Figure 13–8).

Figure 13–8 Apple tree with an interstock. To produce this tree a bud from a dwarf stock was budded to a seedling rootstock in August (lowest union). After the bud was well established the following spring, the seedling tree was cut off just above the inserted bud. The following summer a bud of the desired cultivar ('McIntosh' in this case) was budded onto the dwarfing interstock about 10 inches above the first union. The resulting tree will be dwarfed by the interstock but will still have a sturdy seedling root system. This photo was taken in late winter almost 30 months after the initial budding.

Pome trees are propagated by grafting or budding. Although most gardeners can quite easily master the art of grafting or budding, they usually purchase pome trees because dwarfing rootstocks are not readily available. Those who wish to use seedling rootstocks or to change a cultivar by topworking will find a description of grafting and budding in Chapter 14.

Trees should, of course, be purchased from a reliable

dealer. One-year-old trees are less expensive and easier to plant and train than older trees. If they have scaffolds at the right location and if they are vigorously growing and healthy, 2-year-old trees in a container may be a wise choice because they will produce earlier (see "Transplanting Woody Plants," Chapter 14).

There are literally hundreds of cultivars of apples and almost as many pears. Probably the most popular early summer apple is still the 'Yellow Transparent' or its improved and somewhat larger relative, 'Lodi.' Among the apple cultivars that mature a little later are 'Gravenstein,' 'Duchess,' and 'Wealthy.' 'McIntosh' is an early fall cultivar. The more popular cultivars that can be stored for winter use are 'Red Delicious,' 'Golden Delicious,' 'Jonathan,' 'Winesap,' 'Northern Spy,' 'Cortland,' and 'Idared.' The trees of 'McIntosh,' 'Wealthy,' 'Northern Spy,' and 'Winesap' are somewhat more winter hardy than are 'Red Delicious' and 'Golden Delicious.'

The most popular summer cultivar of pears is the 'Bartlett,' although it is not adapted to parts of the East because of fireblight. Later maturing pears that can be stored into the winter include 'D'Anjou,' 'Bosc,' and 'Winter Nellis.'

Culture and Management. Apples and pears can be grown either with a cover crop or with clean cultivation. More fertilizer and irrigation water are required if trees are being grown with cover crop; however, unless water is likely to be limiting, a cover crop should be considered because of the advantages of less erosion, easier access after rain, and the possibility of using the grassy home orchard as a recreation area. Sometimes the planting will be clean cultivated to reduce possible cover crop competition until the planting begins to produce, after which it is seeded to a cover crop. Grass that can be mowed as a lawn is the most popular cover crop for home orchards, but alfalfa, clover, or a mixture of grass and one of the legumes will reduce the amount of fertilizer required for the planting. Where the planting is to be cultivated during the summer, it is often desirable to plant a cold-tolerant annual cover crop such as rye or wheat that can be plowed under the next spring. These covers will protect the soil during the months it is most likely to erode and will provide organic matter for the planting.

Many factors affect the amount of fertilizer, especially nitrogen, needed for apple and pear trees. The amount of production and quality of fruit can be regulated to some extent by the amount of nitrogen. Too much nitrogen will cause the tree to be vegetative and produce few large, poorly colored fruits; too little will cause it to be highly reproductive and produce many small, low-quality fruits. Small, light-colored leaves are usually a good indicator of too little nitrogen (see Chapter 14).

Soil tests are not completely reliable guides to orchard tree fertilization because of the difficulty of obtaining representative soil samples for a crop having roots that may reach depths of 30 feet (9 m). Although a major portion of the feeding will be from the upper 2 feet (⅔ m) of soil, lower roots can pick up needed nutrients when they are deficient in the upper levels. In sandy soils in areas of high rainfall boron may be required. Iron and zinc deficiency are common in the West. Fertilizer is most effective if applied 3 to 4 weeks before buds begin to swell in the spring.

Apple and pear trees require 32 to 36 inches (80–90 cm) of water annually, and irrigation will be needed in areas where less than this amount falls in the form of rain or snow or when there are extensive periods of drought during the growing season. When trees are irrigated, the entire root zone should be brought to field capacity. Even in arid areas established trees growing in soil having good water-holding capacity should not require irrigation more often than once each 2 or 3 weeks. Water can be used to reduce damage from frost (see Chapter 7).

Pruning of pome trees should be kept to a minimum, because these trees grow slowly and new growth is slow to come into production. Because the trees may bear for 40 years, training must insure a strong framework. For this reason the modified central leader system of training is recommended (see "Pruning and Training," Chapter 14).

Fruiting Habit. Most cultivars of both apples and pears are self-unfruitful, and, unless close neighbors have trees of a different cultivar, the home planting should always consist of two or more cultivars (Figure 13–9).

Fruit of apple and pear trees is borne on small twigs called **spurs**, which are produced only on wood that is at least in its second year of growth. Some of the buds on wood that was developed during the previous season grow into short spurs in the spring. Later in the season mixed buds (buds from which both leaves and flowers will grow) develop at the tip of the spurs. The following spring these buds open to produce blossoms that grow into the fruit. Figure 13–10 shows that under ordinary circumstances the youngest wood to produce fruit is completing its third growing season as the fruit matures and also that an individual spur of the pomes produces fruit only every other year. An upset in nutritional balance or destruction of the crop is likely to cause all of the spurs to fruit during a single year and result in biennial bearing in apple or pear trees (see Chapter 3).

Insects and Diseases. The codling moth, which is responsible for most of the worms in apples and pears, is a universal pest. Fireblight, a bacterial disease that enters wounds and open blooms, causes progressive blistering and death of branches and eventually entire trees (Figure 13–11). It is especially prevalent in the East, where it literally prevents production of 'Bartlett,' the most popular pear cultivar. Major pests of pome fruits are described in Table 13–7.

Insect control is almost always essential for apples and pears, and the gardener who grows these fruits should plan on a systematic pest-control program. Insect and disease control usually require sprays in February (dormant), just as buds are beginning to swell (delayed dormant), just before blossoms open (prepink), 10 days later, and each 16 to 21 days thereafter until just before harvest. Each application may be a single pesticide or a combination of several and may control one or several pests.

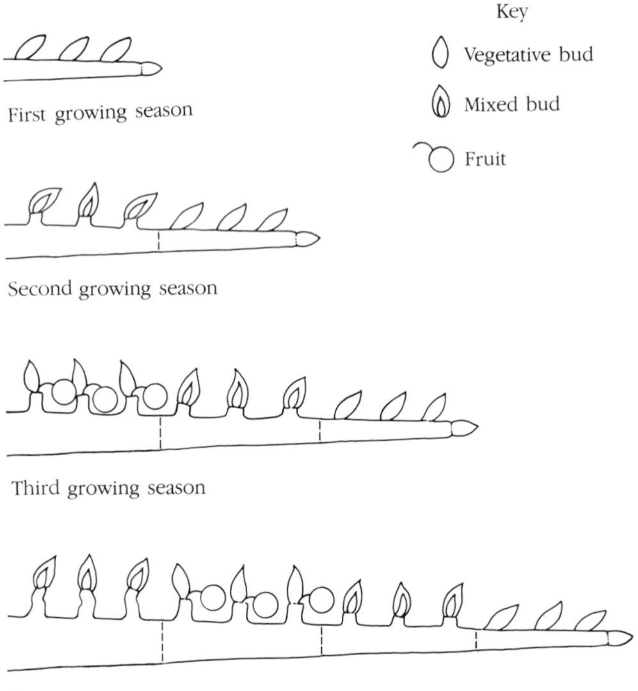

Harvesting and Storage. The time of pome fruit harvest is determined by type of fruit, variety, season, and use. Summer apples are picked for cooking when they have reached acceptable usable size. They have better flavor for eating fresh after they turn yellow or red, depending on the cultivar, and begin to soften slightly. Optimum harvest maturity of later cultivars is indicated by a number of factors. The stem of the fruit should detach relatively easily from the spur, and the flesh should be crisp, distinctly acid with little or no starchy taste. In apples the red color is not a good indicator of maturity, but the

Figure 13–9 Pollination of tree fruits by bees.

Key
Vegetative bud
Mixed bud
Fruit

First growing season

Second growing season

Third growing season

Fourth growing season

Figure 13–10 Growth and fruiting habit of apple and pear trees.

A

B

C

Figure 13-11 Pests of fruit trees. The codling moth larva (A) is responsible for most wormy pome fruits; the adult moth (B). Fireblight is a serious bacterial disease of apple and pear trees (C). (A and B courtesy USDA.)

Table 13–7 Major insect and disease pests of pome fruits

Insect pests	Symptoms and damage	Control
Scale (various kinds)	Several kinds cause damage by sucking juices from branches (see Grape insect pests, Table 13–2).	Dormant oil sprays before buds open.
Aphids	Several kinds of aphids infest apples and pears, including the woolly apple aphid (its name is diagnostic), which lives on all parts of the tree including roots and continues to infest aerial portions from root colonies.	Use rootstocks resistant to woolly aphid, dormant sprays to prevent eggs from hatching, and a periodic summer spray program.
Climbing cutworms	See Grape insect pests, Table 13–2.	
Tent caterpillar	Worms ½ to 1 in (1½ to 2½ cm) long emerge from eggs in early spring. Construct tentlike webs and defoliate whole limbs or entire tree.	Insecticide just before blossoms open.
Mites	Several kinds; very difficult to control (see Grape insect pests, Table 13–2).	Dormant oil spray for eggs. Periodic sprays through growing season.
Codling moth	The most serious pest of apples. Also infests pears. The white larvae usually enter fruit through calyx and leave, when mature, through a "worm hole" in the side. Moths lay eggs, which continue to hatch throughout the summer.	Insecticide spray 10 days after petal fall and each 16 to 21 days thereafter until 2 to 3 weeks before harvest. Clean up trash and scrape away loose bark to destroy adult moths.
Pear psylla	A 1/10 in (¼ cm) reddish-brown, four-winged, triangular insect found under pear bark during winter. Smaller yellow nymphs on fruit and leaves during summer. Causes leaves to turn brown and drop and fruit to be stunted and scarred, and spreads pear decline.	Dormant, delayed dormant, and prepink sprays with recommended insecticides.
Apple maggot	Most damaging east of the Great Plains and north of Arkansas and Ohio. Adult flies emerge from soil in midsummer, lay eggs in punctures in apple fruit. Eggs hatch into larvae that grow rapidly as fruit ripens.	Insecticide spray timed to kill adult flies as they emerge.
Curculios	Apple and plum curculios both damage pome fruits (see Stone fruit insect pests, Table 13–8).	Sprays used for codling moths will usually control curculios.

Diseases	Symptoms and damage	Control
Apple scab	Brownish spots on leaves in early spring, spread to fruit. Rough pustules on fruit enlarge, even in storage, to large rough scabs.	Fungicide, during spring where summer is dry, throughout growing season where summer rains are frequent.
Powdery mildew	Fungus winters over in infected buds. As weather warms during spring, it attacks leaves, buds, and sometimes fruit with white mealy coating. Causes russeting of pear fruit.	Fungicides during spring and early summer. Pruning away affected shoots is a supplemental control effective where spread is not excessive.
Fireblight	Bacteria enter blossom during pollination period, gradually move down the branch, causing darkening of bark. Leaves and small fruits die, blacken, and remain on twigs. Most serious on pears in the East, especially 'Bartlett' cultivar. Also serious on many apple cultivars.	Prune off infected branches 12 in (30 cm) below farthest visible symptoms. Disinfect pruning equipment between cuts with a solution of 1 part household bleach to 9 parts water. Streptomycin sprays at full bloom and 7 days later. During dormant season scrape and remove all dead bark from around cankers on larger limbs and disinfect with streptomycin.

green or yellow color is. Whenever the green area of the skin changes from a definite green to a yellowish green, the fruit is reaching maturity. The days from bloom have also been used as an index of maturity. Suggested times for a few cultivars are: 'Jonathan,' 135 to 145 days; 'Delicious,' 145 to 150 days; 'Golden Delicious,' 152 to 160 days; 'Winesap,' 165 to 175 days.

Fruit of many apple cultivars tends to fall from the tree before it is fully mature. This early drop is a serious problem when a heavy wind occurs just before harvest. It can be prevented by using a stop-drop spray. Naphthalene acetic acid at 10 parts per million is the most common active ingredient of such sprays. This material becomes effective approximately 3 days after application and holds the fruit for 10 to 14 days.

Pears are unlike most fruits and vegetables in that they reach highest eating quality only when they are picked at a slightly green stage. There are several commercial color and firmness tests used for pears. For the homeowner probably the best criterion is when full size has been reached and there is a slight change in color from green to yellowish-green, before fruit begins to soften. I usually harvest the bulk of my 'Bartletts' about the time the first two or three small fruits (usually wormy) become fully ripe. I leave the smaller, obviously immature fruit (about one-third of the crop) on the tree for 7 to 12 days longer.

Winter cultivars of apples and pears can be stored if they have been harvested at the proper stage of maturity and if there is a cool humid storage available, 32°F (0°C) and 95% humidity being ideal. In northern areas they can be stored in an insulated pit storage (see "Storing Horticultural Products," Chapter 14).

Neither apples nor pears should be placed in the same storage facility as vegetables because the ethylene gas they emit has an adverse effect on other products. For example, it causes carrots to become bitter, inhibits sprouting, and causes off-flavor of other root, bulb, and tuber crops.

Figure 13–12 Apricots. The fruit will grow large if it is thinned properly.

The Stone or Drupe Fruits

The stone fruits include peaches, nectarines, plums, sweet cherries, sour cherries, and apricots (Figure 13–12). Almonds are sometimes also classed with the stone fruits.

All of the group are closely related, belonging to the *Prunus* genus of the family Rosaceae. Most will intergraft, and there have been interspecific crosses between some of the different kinds. All grow on medium-sized trees.

Nectarines are perhaps the least well known of the group, because they don't ship well and are not as commercially available in supermarkets as the others. Nectarines are essentially fuzzless, somewhat fragile, peaches. Seeds from peaches will sometimes grow into nectarine trees and vice versa. Both originated in China and are usually listed under the botanical name of *Prunus persica* (Figure 13–13). Apricots (*Prunus armenica*) also are thought to have been domesticated in China.

Figure 13–13 Peaches just before harvest.

Three general groups of plums are recognized. The European plum, *Prunus domestica*, has been cultivated in Europe for many centuries and is thought to have originated from an interspecific cross between the myrobalan or cherry plum, *P. cerasifera* (now used as a rootstock for stone fruits), and the sloe plum, *P. spinosa*, of Western Europe. Most cultivars of prunes, the 'Damson' plum, and the 'Green Gage' or 'Reine Claude' plum are of the *P. domestica* group. Cultivars of the Japanese plum, *P. salicina*, are often larger and more tart than the European types. Trees of Japanese plum do not require as much cold to overcome their rest period, and they are less subject to disease in warm, humid climates. American plums include several *Prunus* species that are native to North America. Some have excellent cold-temperature and disease resistance, but they are not grown as extensively as are the other two types. Prunes are plum cultivars, such as 'Stanley' and 'Italian,' with a sugar content high enough that they can be dried without fermentation of the flesh around the pit.

Sweet cherries, *Prunus avium*, and sour cherries, *P. cerasus*, were both introduced from Europe, where there are many native related species (Figure 13–14). 'Duke' cherries are thought to have originated from a cross between sweet and sour cherries, because they range somewhere between the two in hardiness, tree size, and fruit flavor.

Figure 13–14 Sweet cherries.

Climatic Adaptation. Stone fruits vary somewhat as to temperature requirements, especially in cold tolerance. The peach is the least cold resistant; dormant peach flower buds will be injured at temperatures from −5° to −10°F (−21° to −23°C). Injury will occur at higher temperatures if the plants are not well hardened. Apricots generally are more cold resistant than peaches, but their early blooming habit exposes the blossoms to spring frost injury. Sweet cherries are somewhat more tolerant of low winter temperatures than peaches, and sour cherries are at least as cold tolerant as the more tender apple cultivars. Japanese plum cultivars vary widely in cold tolerance from those slightly less cold resistant than apples to those as tender as peaches. American-type plums are the hardiest of all stone fruits, some cultivars being as cold tolerant as the hardiest apples. European plums, including prunes, are somewhat less cold tolerant than apples but hardier than peaches or sweet cherries.

Propagation and Planting. Stone fruits are generally propagated by budding. Both sweet and sour cherries are propagated onto rootstocks of either 'Mazzard,' a sweet cherry having small, black, bitter fruits, or 'Mahaleb,' a wild European *Prunus* closely related to sour cherry. 'Mazzard' rootstock is preferred in the West and in areas of the East where moisture is sufficient and winter hardiness is not a problem. Trees with 'Mahaleb' rootstocks are usually slower growing, slightly smaller, and not as long-lived, but they are less subject to injury from drought and cold temperature.

Peach, nectarine, apricot, and plum will all intergraft. Peach and nectarine are usually budded to peach seedling rootstocks, although vegetatively propagated nematode-resistant peach rootstocks are used where nematodes are a problem, and 'Myrobalan' plum is used where the soil is heavy or poorly drained. Plums are mainly budded to 'Myrobalan' plum except where nematodes or droughty, sandy soils are a problem, in which case resistant peach rootstocks are used. Apricots are grafted to apricot seedling rootstocks, which have some nematode resistance, to peach rootstocks if soil is sandy or droughty, or to plum if the soil is fine textured or poorly drained. A good nursery will be able to tell the gardener what kind of rootstock has been used.

One- or 2-year-old trees can be purchased. The largest trees for their age are usually the most desirable as stunting is often a symptom of rootstock-transmitted virus. Planting arrangements and tree spacings should be based on the ultimate size of the trees and conditions of the area where they are grown. Although dwarf trees are occasionally advertised, dwarfing rootstocks of the stone fruits are not as available nor as reliable as those of apples and pears. Peach, sour cherry, and some plum cultivars are naturally small, and these can be pruned heavily enough to keep them down to reasonable size. Apricots will grow into large trees but will still produce fruit while being pruned heavily enough to maintain a reasonable size. Thus, with a certain amount of pruning, all of these trees can be planted as close together as 15 feet (5 m). Standard spacing recommendations if these trees are to be grown to their ultimate size are 18 to 25 feet (6 to 8 m) for peaches, plums, and sour cherries, and 25 to 40 feet (8 to 12 m) for sweet cherries and apricots. Spur-type sweet cherry trees that have shortened internodes and grow from about one-half to two-thirds the size of standard trees have been developed recently and may be available in the near future.

Culture and Management. Cherries and peaches generally grow best on land where there is no sod cover crop. In home gardens cultivating around trees is often impractical, which is perhaps one reason apricots and plums frequently grow better than peaches or cherries in home gardens. If trunks can be protected from rodents, mulching may be the answer to soil management. Satisfactory growth of peach trees is possible in sod if the trees are heavily fertilized with nitrogen.

Excess nitrogen does not affect fruiting of stone fruits to nearly the extent it affects pomes. Enough nitrogen should be applied to produce 12 to 24 inches (30 to 60 cm) of branch growth on nonbearing peach and sour cherry trees and 6 to 15 inches (15 to 38 cm) after they begin to bear at 4 to 6 years of age. Branch growth of sweet cherry should be somewhat greater and that of plum and apricot somewhat less. This amount of growth requires, on the average, an annual application of 1½ to 2 tablespoons of nitrogen per year of age until the tree is about 7 years old (see Chapter 14 for further recommendations). Zinc, manganese, and iron deficiencies frequently occur in stone fruits in the arid West.

In regions where annual rainfall is less than 30 inches (75 cm) or where trees are being grown with a cover crop and precipitation is less than 50 inches (127 cm), stone fruits will benefit from irrigation. The frequency with which irrigation is required will vary from each 10 to 14 days during mid-summer in areas having no summer rainfall and sandy soil, to three to four times each growing season in areas having good winter and spring rains and soil with high moisture-retaining capacity, to only an occasional irrigation during an extremely dry summer in some regions of the East and Midwest where summer rains normally supply enough moisture. Where water stress is likely to be a problem, peaches, nectarines, plums, and apricots should be grafted to peach and cherries to 'Mahaleb' rootstocks. The soil should be soaked to field capacity to at least 4 feet (1¼ m) of depth with each irrigation and then permitted to dry before water is again applied. Deep irrigation is essential if the trees are being grown with a permanent cover crop. The relative advantages and disadvantages of sprinkler vs. surface irrigation (Chapter 6) apply to stone fruits, except that cherry trees should not be sprinkled while fruit is maturing as wetting the fruit may cause it to split. Because they usually mature their crop prior to the time late summer drought occurs, cherries, apricots, and early peach cultivars can be more easily produced where water is limited than can late-maturing peach or plum cultivars or most apples or pears.

Apical dominance is not so pronounced in the peach as in the apple or pear tree, so the peach can be trained to the vase system, although in some areas the modified central leader system is preferred. Other stone fruits are usually trained to the modified central leader (see "Pruning and Training," Chapter 14).

Flowering and Fruiting. Unlike the apple tree, which produces fruit on wood that is at least 2 years old, the peach and the nectarine produce blossom buds only on current-season wood and blossoms and fruit only on wood that is 1 year old. Buds on all of the stone fruits are either blossom buds or vegetative buds, never mixed like those of apple and pear. New buds on peach trees are produced in the leaf axils starting about midsummer and continuing as long as the tree is producing new leaves so that blossoms will be produced to the tips of peach branches (Figure 13–15).

Plums, apricots, and sour cherries produce blossom buds on the older half of current-season growth and, to a

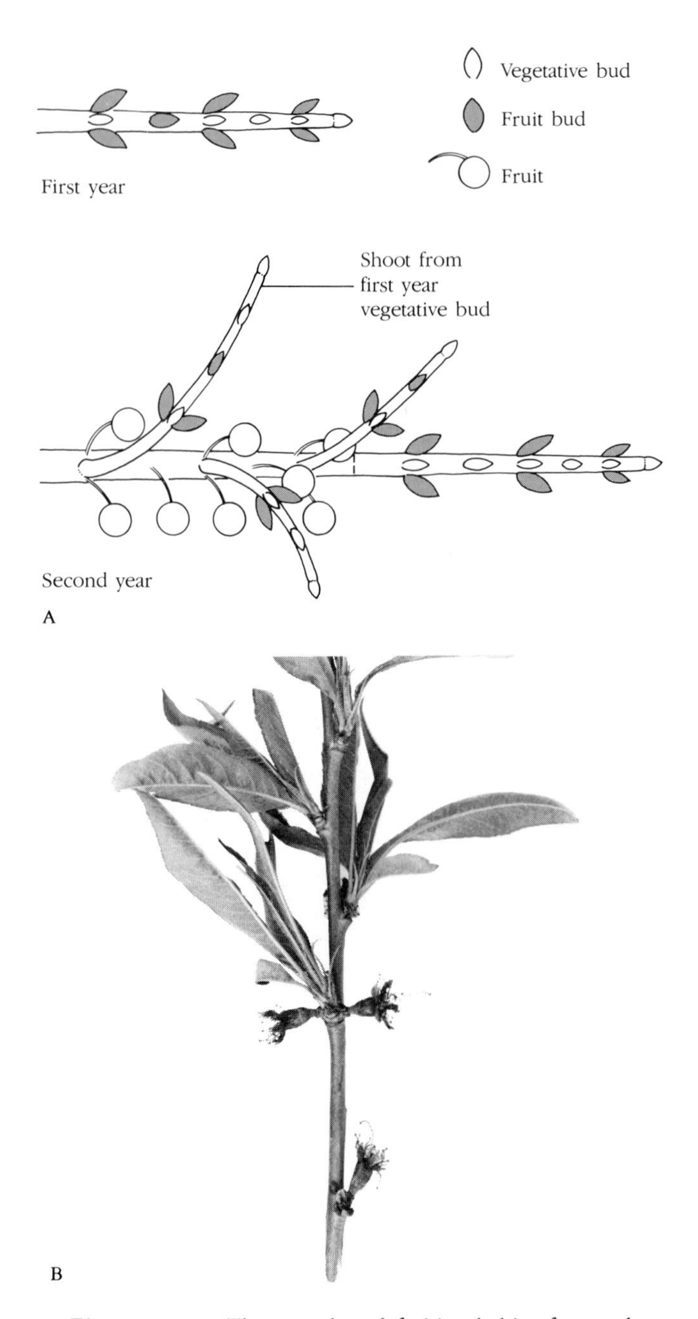

Figure 13–15 The growth and fruiting habit of a peach tree. (A) A healthy peach branch produces some buds in clusters of three. The outer two are fruit buds, and the center is a vegetative bud that develops into a new shoot. Only on an occasional node will both fruit buds develop into fruits, as is occurring in (B).

limited extent for some cultivars, on older wood. Sweet cherries produce them on older wood and in the first formed leaf axils of current-season wood. Thus blossoms and fruit will not be produced as far toward the ends of the branches on cherry, apricot, or plum as they are on peach trees.

As was mentioned in Chapter 3, most peach and apricot cultivars are self-fruitful, but some, such as the 'J. H. Hale,' peach produce poor pollen and must be cross-pollinated. Sour cherries are self-fruitful, but sweet cherries are not. The three sweet cherry cultivars popular in the West—'Bing,' 'Lambert,' and 'Royal Anne'—are also cross-incompatible and must have pollen from another cultivar if they are to set fruit.

Most Japanese and American plums and their hybrids require cross-pollination. Some cultivars of European plums are self-fruitful; others are not.

Figure 13–16 Cherry fruit fly maggot. The larva of the cherry fruit fly is a serious pest of cherries in almost all sections of America. (Courtesy M. T. Ali Niazee.)

Insects and Diseases. Peaches, apricots, sour cherries, and plums can be produced without insect or disease control in a few areas away from commercial production of stone fruits; but in most gardens a regular spray program will be necessary. (Insecticide applications at 10-day intervals from bloom until fruit maturity are almost always required to keep cherries free of cherry fruit fly maggots, Figure 13–16.) As with the pome fruits each application can be of a single or a combination of pesticides and can control one or several pests. The recommended timing of sprays for fruits in the garden or yard is early December and late January (two dormant sprays), just as buds are swelling (delayed dormant), just before blossoms open (prepink or pink), when blossom petals fall, late spring, summer, 10 to 14 days before harvest (preharvest) and postharvest. Generally it is not advisable to spray for insects unless they are present in damaging numbers (except for cherry fruit fly maggot), because the insecticide will also often kill predators that may otherwise keep that insect or another in check. Mite predators are frequently killed by pesticides used to control other insects, permitting disastrous buildup of mite populations. Table 13–8 lists major pests of stone fruits.

Cultivar Recommendations. Cultivar recommendations for stone fruits vary, of course, from one section of the country to another, and local recommendations are always best. However, a few standard cultivars that have proven to be highly adapted in many sections should be mentioned.

Peaches. Early peach cultivars of excellent quality include 'Red Haven,' 'Hale Haven,' and 'Golden Jubilee'; 'Early Elberta' is slightly later. 'Redskin' and 'Georgia Belle' are two later cultivars popular in the Southeast. In the West 'J. H. Hale' and 'Elberta' are standard freestone canning cultivars. They may not mature in cooler areas where peaches can be grown.

Apricots. 'Moorpak' and 'Tilton' are two commercial varieties frequently grown in home gardens in the West. Apricots have not been regularly productive in most of the eastern part of the United States. The New York experiment station recommends 'Alfred,' 'Veecot,' and 'Goldcot' as three cultivars worthy of home trials in the Northeast.

Sweet Cherries. 'Bing' and 'Lambert' have long been the standard cultivars of red or black sweet cherries in the western United States. 'Royal Anne' (sometimes called 'Napoleon') is the standard white variety. All three of these, as mentioned, are self- and cross-incompatible. 'Van,' a newer dark cultivar from Canada, or 'Ranier,' a

Table 13–8 Major insect and disease pests of stone fruits

Insect pests	Symptoms and damage	Control
Mites (various)	Mites hibernate under scales of bark (see Grape insect pests, Table 13–2).	Spraying before bloom, just after petals fall, and during the summer when mites are a problem.
Scale	See Grape insect pests, Table 13–2.	Dormant, delayed dormant, prebloom, petal-fall sprays. If serious a summer spray at crawler stage may be needed.
Lygus bug	Green or tan beetles that attack blossoms and fruit, causing blemish of fruit. They hide in cover crop.	Spray cover crop and trees at prebloom period.
Aphids	Several kinds attack stone fruits. They winter over as eggs and hatch in spring. They feed on young leaves, causing them to curl. Also transmit virus.	Delayed dormant and late-spring sprays as eggs are hatching and before leaves curl.
Peach tree borer	White worms with brown heads pass the winter in bark near base of tree. Worms become active in spring and feed on bark of tree. Can kill a tree in a few years. Gum, sawdust, and worm droppings exude from lower trunk area near ground level.	Fall treatment (October) with propylene dichloride emulsion in trench 2 in (5 cm) from tree trunk as follows: Age of tree (years): 1, 2, 3, 4 & up Water (parts): 8½, 7, 7, 6 Emulsion (parts): 1½, 3, 3, 4 Amount/tree (pints): ⅛, ¼, ½, ½–1 Amount/tree (liters): 1/16, ⅛, ¼, ¼–½ Also summer spray.
Peach twig borer	Larva passes winter as small brown worm that emerges about bloom time and feeds on tips of twigs, causing them to die back.	Sprays just as buds begin to swell in spring and just after the petals fall.
Oriental fruit moth	Kills twigs during spring and early summer. Bores into fruit much as codling moth does.	Early cultivars usually escape damage. Spring and summer sprays for late-maturing cultivars.
Plum curculio	Long-snouted beetle feeds on leaves and fruit. Lays eggs in fruit, causing it to be wormy. Main agent for spread of brown rot.	Periodic sprays starting 10 days after petal fall.
Earwig and Japanese beetle	Both insects are general feeders on leaves and fruits. Earwigs are relatively long insects with pincers on the rear. They may enter peaches and apricots through the stem end. Japanese beetles are metallic green beetles found mainly in the East.	Control program for other insects will usually control these two.
Cherry fruit fly	Adult fly lays eggs in fruit. These hatch into small white maggots about the time the fruit matures.	Insecticide 7 to 10 days after bloom and each 10 days thereafter to harvest.

Diseases	Symptoms and damage	Control
Brown rot	Most serious in humid areas or during wet seasons. Fruit shows soft brown areas as it ripens, becomes covered with tufts of grey mold.	Where brown rot is an annual problem, prepink, petal-fall, and preharvest fungicide sprays are required.
Powdery mildew	See Pome fruit diseases, Table 13–7.	Keep tree leaves dry if possible. Sulfur spray late spring and again 3 weeks later.
Peach leaf curl	Appears in early spring as leaves unfold. Leaves become thickened, puckered, and brittle. Fungus is carried from season to season in buds and on bark.	Spray trees with fungicide in November and again during late winter.
Coryneum blight	Causes gumming and death of buds and twigs; gumming and split bark on branches and trunk; and brown, red-bordered spots on leaves. Can cause heavy fruit losses.	Fungicide during October plus prebloom spray.
Bacterial canker	Enlarging lesions on trunks and larger limbs. Great amounts of gum are associated with active phase.	Scrape and disinfect cankers as described for fireblight of pome fruits (Table 13–7). Copper sprays.

light cultivar developed in Washington state, are normally planted now as pollinators. 'Emperor Francis,' a white cherry, 'Ulster,' and 'Windsor' are three cultivars recommended for the Northeast. 'Montmorency' is a standard cultivar of sour cherry. 'North Star,' developed by the University of Minnesota, is a sour cherry tree that does not grow over 8 feet (about 2½ m) in height and spread. It becomes dark and semisweet when fully ripe.

Plums. 'Stanley' and 'Italian' prune are the standard prune-type plums and can be grown in most areas where plums are productive. 'Green Gage,' a freestone, and 'Reine Claude' are two dessert-type plums that are excellent choices for the home gardener. 'Damson,' a small, blue, tart plum, is prized for jams and preserves. Japanese plums, which are not hardy in all locations, are commonly eaten fresh. 'Santa Rosa' and 'Burbank' are two widely adapted cultivars.

Harvesting, Storing, Preserving. Harvesting the stone fruits will depend to a large extent on their ultimate use. Peaches for canning or freezing are of a better quality if harvested before they become completely mature. Peach maturity is determined by changes in ground color. For yellow peaches this is when the green color starts to change to yellow. For home use the rest of the stone fruits should be fully mature but not soft when harvested.

None of the stone fruits can be stored for as long as apples, pears, and citrus, but most cultivars will keep for 10 to 14 days in a refrigerator. If storage is contemplated, fruit should be harvested when slightly immature and all damaged or imperfect fruits removed from the container before it is placed in storage.

All of the stone fruits can be easily processed by canning, freezing, or drying; directions are given in Chapter 14.

Citrus Fruit

The citrus are a subtropical group of evergreen fruits that include oranges, lemons, grapefruit, limes, and a number of lesser known kinds and hybrids. All types are injured by freezing, with critical temperatures ranging from 26° to 18°F (−3° to −8°C). Citrus trees have fragrant blossoms, attractive shapes, and waxy green leaves, and they start to bear fruit after only 2 to 3 years, all of which make them excellent dual-purpose ornamental fruit trees in areas where they are hardy. They are also grown to some extent in greenhouses farther north.

Climatic Adaptation. Citrus is limited to the very warmest locations of the United States. The sweet orange (*Citrus sinensis*), one of the earliest fruits to be domesticated, is also the most popular in the United States; it is consumed in larger quantities than any other fruit. Related to sweet orange are a group known as "kid glove" oranges, the skin of which separates easily from the flesh. This group includes the tangerines or mandarins (*C. reticulata*) and some recently developed hybrids among various citrus types, including citranges and tangelos.

Critical low temperatures are about 24°F (−4°C) for sweet orange cultivars and as low as 18°F for the satsuma. If temperatures drop lower, the crop will be destroyed and the trees badly injured.

The grapefruit (*Citrus paradisi*) grows especially well under desert conditions. Critical temperatures are between 24° and 26°F (−4° to −3°C). The lemon (*C. limon*) and lime (*C. aurantifolia*) are among the most tender of the citrus fruits, with severe injury occurring when temperatures drop to about 26°F.

Propagation and Planting. Although citrus trees can be easily rooted from cuttings, they are usually propagated by budding or grafting to rootstocks that are resistant to root diseases. Rootstocks that have been selected for specific conditions such as soil adaptation and disease resistance are available.

Most seed of citrus is produced without fertilization (apomictic), so, unlike most kinds of fruit trees, seedling citrus trees will usually produce fruit like that of the parent tree. Therefore, it is quite possible for a gardener to produce citrus trees from seed. However, seedling trees have more thorns than those propagated by grafting or from cuttings. Sour orange (*Citrus aurantium*), sweet orange (various cultivars), and 'Rough Lemon' are frequently used as rootstocks. Orange trees growing on 'Rough Lemon' rootstocks are reported to produce fruit with lower quality. Virus diseases create serious problems of graft incompatibility with some of the other rootstocks.

Nursery trees are available in containers, balled and burlapped, or, in some locations, bare-rooted. (Those sold bare-rooted usually have the leaves removed.) They are most easily planted during the cooler, more rainy season of the year. Each orange tree needs about 400 square feet (36 m^2) of space; grapefruit need a little less, lemons a little more.

Culture and Management. Citrus trees do not require soil as deep as most deciduous fruit trees, 3 feet (1 m) of depth being enough for maximum yield. Recommendations for fertilizing with the major elements are in Tables 14–16 and 14–17 (Chapter 14). In many Florida soils magnesium, zinc, copper, manganese, and calcium are often needed for citrus. These compounds can be most easily added with the general trace element package available at garden stores of that state. Zinc is often needed for citrus in the West. Nitrogen deficiency symptoms are light-green leaves, stunted growth, and excessive blossom production. Phosphorous deficiency causes excessively thick peeling on the fruit. Potassium deficiency causes reduced leaf size, reduced growth in the top of the tree, and small fruit with low acid content. Magnesium deficiency causes pale-colored leaves and increased leaf abscission, especially during heavy crop production.

Because they grow continuously, citrus trees use water throughout the year. Around 3 to 4 feet (about 1 m) annually or 1½ to 2½ inches (4 to 6 cm) each 2 to 3 weeks from rainfall or irrigation are required for a normal crop. It is essential that citrus not be subjected to water stress during periods of fruit set, development, and maturity. Fruiting occurs during winter and early spring with most kinds of citrus.

Citrus trees, vegetatively propagated, tend to bear some fruit 2 or 3 years after planting. Almost all kinds are self-fruitful. Blooms are not produced all at one time, as on deciduous fruit trees; however, where growth is slowed because of cool weather, as it is in much of the citrus-growing area of the United States, oranges and grapefruit tend to produce most of their blossoms during early spring.

Citrus are pruned very little except to thin out older trees to some extent and to remove crossing and dead branches (see "Pruning and Training," Chapter 14).

Disorders and Pests. Among the disorders that plague citrus fruit are (1) an extremely thick skin, sometimes caused by phosphorous deficiency but probably also by other factors; (2) dry, juiceless sections often caused by freezing temperatures during the ripening period; and (3) lack of color (especially in oranges) due to warm night temperatures and lack of sunlight during ripening.

Citrus is subject to attack by a great many diseases. Among the most serious is gummosis disease, caused by phytophthora and other fungi, which kills the bark around the graft union. It is controlled by using resistant rootstocks, by keeping the environment around the trunk as dry as possible, by pruning low-growing branches and controlling weeds, and by cutting out the bark and disinfecting the area around wounds. Scab, a condition that prevents the growing of lemons in the Southeast, can be controlled with fungicides on oranges, grapefruit, and other less susceptible kinds of citrus. Virus diseases and many unidentified disorders that may be caused by virus are controlled by using virus-free rootstock for propagating and by obtaining virus-resistant planting stock. Fruit mold, a serious postharvest disorder, can be controlled by a fungicidal dip.

A number of scale insects attack citrus, one of the most damaging being red scale. In Florida scale can be controlled with oil sprays during the rainy season; in California oil sprays damage citrus trees and don't give good control. Nematodes of several species also attack citrus.

Harvesting and Storage. Citrus fruits should not be harvested until they are fully ripe, which may be 10 to 13 months after bloom, depending on kind and cultivar. Unlike the temperate zone fruits, citrus can be "stored" by allowing the fruit to remain on the tree. The color of oranges stored on the tree will sometimes change from orange back to green, but this does not materially affect their quality. Citrus fruits do not deteriorate nearly as rapidly as apples and peaches, and they can be picked from the tree as needed over a considerable period of time—a decided advantage to the backyard gardener. Lemons are harvested over the entire year, with the heaviest pickings between December and March.

Cultivars. 'Washington Navel' and 'Valencia' are the two most popular sweet orange cultivars in the West, and 'Valencia' is probably the most important one in Florida. Both are relatively seedless. 'Texas Navel,' a bud mutation of 'Washington Navel,' is somewhat better adapted than its parent for growing in Texas. The navel oranges are very early, maturing before Christmas in the warmer areas, and the Valencias are late, maturing during the spring and summer. 'Pineapple,' an old cultivar that produces fruit with numerous seeds, is grown in Florida; 'Jaffa' is a new cultivar from Israel that is gaining in popularity. 'Temple,' an important cultivar in Florida, is probably a hybrid between sweet orange and mandarin. It peels easily and the segments separate more easily than other sweet oranges but not as easily as those of mandarin.

'Marsh Seedless' (yellow flesh) and its bud sports 'Thompson' (pink flesh), 'Ruby' (red flesh), and 'Webb' (red flesh) are the most important cultivars of grapefruit. 'Eureka,' 'Lisbon,' and 'Villa Franca' are important cultivars of lemon.

Other Subtropical Fruits

There are many kinds of fruits that can be produced in subtropical areas (Figure 13–17). Only a few of the more popular kinds will be mentioned in this text.

Figs. Although commercial production of figs (*Ficus* spp.) has been limited mainly to California and the adjacent southwestern states, figs are also grown for home use in the Gulf and southeastern states. Fig trees are grown mainly as ornamentals in warmer areas as far north as Trenton, New Jersey; Lewiston, Idaho; and Yakima, Washington. However, even the more cold-resistant sorts

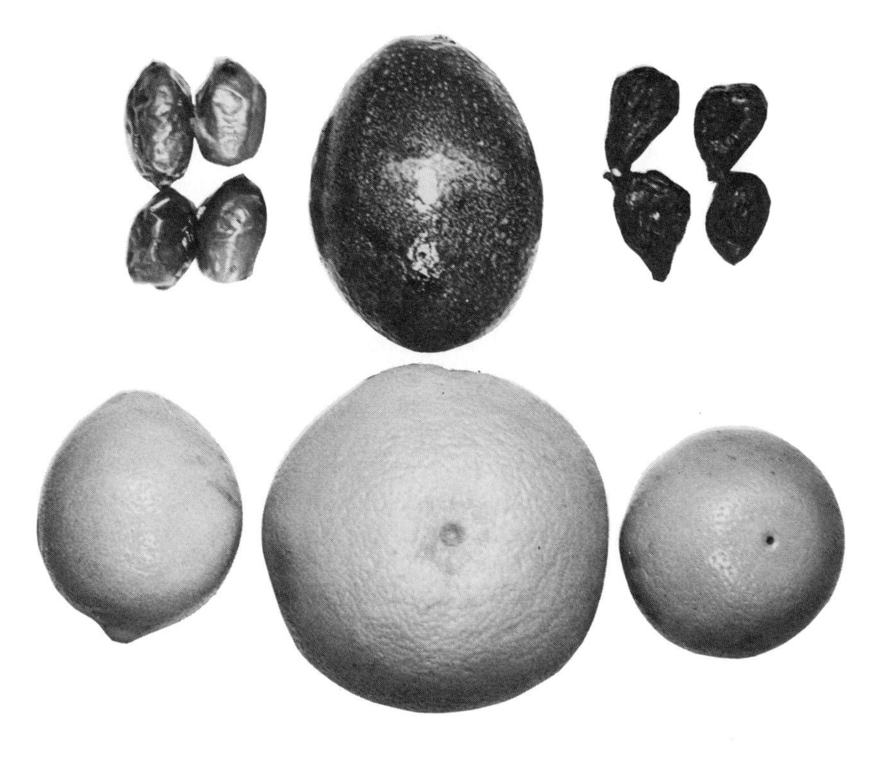

Figure 13–17 Various kinds of subtropical fruit—left to right (upper) dates, avocado, figs; (lower) lemon, grapefruit, orange.

will be injured by temperatures below 16°F (−9°C), and trees will be killed if temperatures drop below 5°F (−15°C). For satisfactory development growing seasons should be long, warm, and dry; cool summers prevent normal growth. Water requirements are not high, but at least 24 inches (60 cm) are needed. Fig plants generally are grown from cuttings, but they can be propagated by suckers, layers, buds, grafts, and seeds as well.

Flowers of edible figs produce no pollen, and, if none is brought into the orchard, common cultivars—'Adriatic,' 'Turkey,' 'Kadota,' 'Mission,' and 'Brunswick'—will set seedless fruits parthenocarpically (without pollination). Trees of the Smyrna type ('Calmyrna' is the main U.S. cultivar of this type) will not set fruit without pollination. When flowers are pollinated, pollen must be supplied from an almost inedible kind of fig known as caprifig. Insect pollination is required. Mature June-crop caprifigs in perforated bags are placed in Smyrna fig trees. The small fig wasps in the caprifigs become covered with pollen and seek another fig in which to lay their eggs. In moving about within the fruit, they brush pollen onto the pistils of the Smyrna fig. Because this type of pollination, caprification, is expensive, growth substances may be used to set fruit parthenocarpically.

Persimmons. There are two types of persimmons, the native American species (*Diospyros virginiana*) and the Chinese (Oriental or Kaki) persimmon (*D. kaki*); only the latter type is grown on a commerical basis. It is found as far north as New Jersey and southwestern Missouri and Utah. Some cultivars are quite cold resistant and have been known to withstand 0°F temperatures, but generally the Kaki is classed as a subtropical fruit. It will thrive wherever the fig is grown successfully. It is adapted to semiarid conditions, but it is also grown in regions of heavy rainfall. It thrives on rich, deep, friable soils. The trees are propagated by budding or grafting on native or Kaki seedlings.

Persimmon production in this country is limited. This may be partly because people are not educated about the necessity for ripening this fruit. One bite of a puckery, astringent, hard persimmon will discourage even the most dedicated connoisseur. Persimmons are harvested while still hard, but most cultivars do not become sweet and edible until they have softened.

The three main cultivars, 'Hachiya,' 'Tanenashi,' and 'Tamopan,' will set fruit without pollination when grown in California. When these cultivars are grown in the East or when other cultivars are planted, it is advisable to include a pollinator in the planting. The cultivar 'Gailey' is usually recommended as the pollinator.

Dates. The date palm (*Phoenix* spp.), a native of North Africa, is grown both for its fruit and as an ornamental in tropical and subtropical areas. In the United States commercial production is limited to the Imperial and Coachella Valleys in California and to southwestern Arizona. Date palms thrive best in a hot, dry climate; rain prevents fruit from maturing properly. Though some cultivars will withstand temperatures as low as 15° to 20°F (−9° to −7°C), commercial culture is limited to frost-free areas.

Because the date palm is dioecious, both a male and a female tree are needed if fruit is to be produced. For commercial production one staminate tree is planted to 25 pistillate trees. Dates are propagated from suckers that develop from the base of the palm or on its trunk. The 'Deglet Noor' and 'Maktoon' cultivars are grown extensively in California. For cooler areas, especially for ornamental purposes, the 'Hallaway' and 'Khodrawy' are used.

Olives. The olive (*Olea europea*) is native to the Mediterranean area and is grown widely in that region. Olive trees will withstand temperatures of 15° to 16°F (about −9°C) but may be severely injured if they drop as low as 13°F (−11°C). The tree will grow in both dry gravelly and poorly aerated soils and is very tolerant of drought, so it is a valuable ornamental in parts of the arid Southwest where water for irrigation is limited or nonexistent.

The olive is usually propagated from hardwood cuttings treated with a rooting hormone. Trees can also be budded or grafted on seedling rootstocks. Young trees for planting should have a trunk diameter of from ⅝ to ¾ inch (about 2 cm). If pruned heavily, trees can be spaced 20 x 20 feet (6 x 6 m), but they will need to be spaced 35 x 35 feet (11 x 11 m) if they are permitted to grow to full size. They are planted as far as 70 feet (21 m) apart where water is likely to be very limited. Olive trees require little pruning. Flowers are wind pollinated, and commonly grown cultivars do not require a pollinator. Olive trees continue to produce for many years after they are planted; some in the Middle East are claimed to be 2,000 years old.

Fruits have developed their maximum oil content 6 to 8 months after bloom, but they can be harvested a little earlier (for green olives) or can be left on the tree until later. For home consumption they are usually harvested when they turn from green to straw color or a little later, when they become reddish. (They turn black when processed.)

Bitter glucoside can be extracted by placing the olives in a solution made with 2 ounces sodium hydroxide in a gallon (60 g in 4 l) of water and leaving them for a few days until a color change shows that the lye has penetrated to the pit (or nearly to the pit if a slightly bitter flavor is desired). The olives must then be soaked for 3 days in water that is changed daily or more often to remove the lye. For table use the olives are then soaked for a day or two in a solution of 3 ounces of salt in a gallon (90 g in 4 l) of water. 'Mission,' a late small olive, 'Mansanillo,' a slightly larger olive excellent for canning, and 'Serillano,' a large-sized olive for fresh consumption or canning are three popular cultivars.

Avocado. The avocado (*Persea americana*) is a tropical evergreen fruit indigenous to Central and South America. In the United States it is grown in southern Florida, along the Gulf Coast, and in southern California. The trees may be large, reaching a height of 25 to 40 feet (8 to 12 m). They grow rapidly and come into bearing in 3 or 4 years.

Depending on the cultivar, the trees can withstand temperatures from 27° or 28°F (−2°C) down to 20°F (−7°C); however they cannot withstand extreme heat. The trees grow well on a wide range of deep, well-drained soils. They are propagated by budding or grafting on seedling stocks.

There are three types or races of avocado. The West Indian can withstand the least cold. The Guatemalan is moderately resistant to cold and is the type grown most extensively in California. The more cold-resistant Mexican avocados are grown commercially in both Florida and California. 'Fuerte,' a Mexican-Guatemalan hybrid, is the most popular cultivar in the West. 'Winter Mexican' and 'Lula,' Mexican-Guatemalan hybrids, and 'Nabal,' a Guatemalan type, are three cultivars popular in Florida.

Nut Trees

In popular terminology a **nut** is a hard vegetable product, usually a fruit, enclosing an edible or usable portion within a shell. There are, geographically or climatically, two types of nut trees—the nonhardy tropical evergreen and the hardy to semihardy deciduous. Nuts that grow on nonhardy evergreen trees—for example, the coconut, the Brazil nut, and the cashew—are not produced to any extent in the United States.

Vegetative propagation is necessary for all types of deciduous nut trees if high-quality, uniform nuts are to be produced. Budding and grafting are the methods used for all common nut trees except filberts, which are propagated from cuttings or suckers.

Trees of deciduous nuts, except for the almond, are wind pollinated. There is a wide range of self- and cross-fruitfulness. Almonds are self-unfruitful and require outside sources of pollen. Walnut cultivars appear to be both self- and cross-fruitful, and pollen from one species may function on pistils of another. Self-unfruitfulness, common in pecan cultivars, may be due to differences in the time when pistils are receptive and the staminate catkins shed pollen.

Nut trees usually are trained to the modified leader form, and do not require severe pruning. Pruning consists of the removal of superfluous and unsymmetrical branches and dead limbs. Lower branches of nut trees are removed up to 4 to 6 feet (1¼ to 2 m) from the ground to make cultivation and other operations easier. Where they can be grown, most kinds of nut trees make excellent shade and specimen trees and often are of more value as ornamentals than as food-producing plants.

Walnuts. Two types of walnuts are commonly grown—the American or native black walnut (*Juglans nigra*) and the Persian or English walnut (*J. regia*). Butternut (*J. cineria*), native to eastern North America, and the California black walnut (*J. hindsii*) are occasionally grown.

The black walnut is widely distributed, being found from north of the Canadian border to the Gulf of Mexico and from the Atlantic to the Pacific Ocean. It produces large crops of high-quality nuts and is a valuable timber tree as well. A number of selected cultivars with thinner shells and larger, more highly flavored kernels have been developed, including 'Thomas,' 'Ohio,' 'Stabler,' and 'Rohwer.'

The Persian or English walnut does not have a very wide range of distribution and production commercially is mainly in California, western Oregon, and southwestern Washington. Temperatures from 0° to −5°F (−18° to −21°C) are critical for most cultivars of these nuts, so production is limited to warmer areas. In addition, disease problems in humid climates are severe, so a moderately dry, cool growing season is desirable. 'Hartley' is the leading cultivar in California. 'Placentia' is popular in southern California and 'Fanguette,' because of late bloom, escapes spring frost and is popular in northern California and Oregon. In the Northeast and protected valleys of the Rocky Mountains Carpathian types from Poland, including 'Broadview,' 'Schafer,' 'Little Page,' and 'Colby,' can sometimes be produced in gardens.

Pecans. The pecan (*Carya illinoensis*) belongs to the hickory group of trees. In volume of nuts produced the pecan is surpassed only by the Persian walnut. The pecan tree is native to the southeastern and south-central United States as far west as New Mexico and as far north as southern Iowa and Indiana. Commercial plantings are confined largely to the Gulf Coast and adjacent states. Because pecans bloom late, they are not subject to spring frost.

The "paper shell" cultivars require a frost-free growing season of 240 to 250 days, whereas the northern types, which produce smaller nuts with harder shells, mature in 180 to 200 days. Improved hickory nuts hardy in all but the coldest regions are now available. Both pecan and hickory are good shade trees for the higher rainfall areas of the East, and, because they are native to the United States, there are dozens of cultivars. Many are adapted to a limited area, so local cultivar recommendations should be obtained.

Filberts. Primarily because of its unusual flowering habit, the filbert (*Corylus maximus*) or "hazelnut" is grown on a commercial scale mainly in mild-winter areas of Oregon and southwestern Washington. Its bloom period occurs during the winter. Ordinary frosts have no effect on either pistil or pollen, but pistils may be killed by temperatures lower than 10° to 12°F (−12° to −11°C). Most cultivars are self-unfruitful, so two should be planted. The filbert is an interesting ornamental that will survive in many areas of the country, but it does not produce consistently where winter temperatures drop to 5°F (−15°C) or lower (Figure 13–18). In Oregon and Washington 'Barcelona,' with 'Dariana' as pollinator, are the important cultivars. 'Cosford,' 'Medium Long,' 'Italian

Figure 13–18 A filbert tree. Because the filbert can be trained to grow into either a small tree or a large shrub, it has many uses in the home landscape. This shrubby specimen is about 9 feet (3 m) high.

Red,' and 'Royal' are cultivars also recommended for the Northwest. The cultivated filberts, as well as several more shrubby *Corylus* species, are important ornamentals.

Almonds. In many ways the almond (*Prunus amygdalus*) is similar to the peach, although the seeds or nuts have only a thin, hard layer of flesh over them. The flowers resemble peach blossoms, and the almond nut is the pit of the peachlike fruit. There are two types of almonds, the bitter and the sweet. The bitter almond, which is grown mainly in the Mediterranean countries, is used in the manufacture of flavoring extracts and prussic or hydrocyanic acid. The sweet almond is grown commercially mainly in central California.

Sweet almond flower buds start growth earlier in the spring than peach, so they are even more subject to late spring frost injury. They can be grown only in areas where late frosts are rare. Both trees and nuts are subject to rot where weather is cool or humid. Cultural practices are similar to those used in growing peaches. Almonds are insect pollinated and self-unfruitful, so a pollinator is required. 'Nonpareil,' 'Mission,' 'Ne Plus Ultra,' 'Peerless,' and 'Eureka' are important cultivars.

Selected References

Brooklyn Botanic Garden. *Fruit Trees and Shrubs.* Handbook 1967 (special printing of *Plants and Gardens*, vol. 27, no. 3). Brooklyn, NY. 1971.

Chandler, W. H. *Evergreen Orchards.* Lea and Febiger, Philadelphia. 1950.

Childers, N. F. *Fruit Science.* 6th ed. Rutgers University Press, New Brunswick, N.J. 1975.

Shoemaker, J. S. *Small Fruit Culture.* 4th ed. McGraw-Hill, New York. 1975.

Westwood, M. N. *Temperate Zone Pomology.* W. H. Freeman and Company, San Francisco. 1978.

14 The Handbook

Contents

Purpose and Organization

To a large extent the preceding 13 chapters tell the story of horticulture from a scientific viewpoint. Although they are not devoid of practical information, they do not present many of the how-to-do-it directions needed by gardeners, because including that information would have fragmented the presentation and lessened its readability.

This chapter, therefore, has been prepared to give illustrated instructions for many gardening practices. It is organized as much as possible in chronological sequence. Those tasks related to planting and propagating come first, followed by those that will need to be done while the garden is growing, followed, finally, by those associated with harvesting and storing.

Those who wish to gain maximum benefit from this handbook should initially read the entire chapter in order to ascertain the information it contains and to acquire the ideas that have application to situations already encountered. After that the handbook can be used as a reference for specific information.

Sources of Gardening Information

Most readers are aware that a textbook on gardening can cover the subject in only a general way. All gardeners will, from time to time, have questions that relate to their particular situation, location, or type of planting. In most areas there are several local sources of information.

Government Offices

Perhaps the most ubiquitous source of information and the one most likely to have the right answers for local gardening problems is the county extension staff. Counties where horticulture is important or where gardening is popular will usually have a horticultural specialist on the county extension staff.

In some urban areas experienced gardeners, often retired people, act as volunteers to supply information and assistance to those less experienced in horticultural practice. In some metropolitan areas experienced gardeners who have an interest and some free time are trained by extension horticultural specialists. When the extension staff is satisfied that the trainee has adequate knowledge, they may award him or her a title, such as (in Seattle) "Master Gardener." The trainees speak at garden club meetings, answer garden questions by telephone, and dispense information from booths at fairs and garden centers. They are able to answer most questions directed to them, and they can call on the county extension staff whenever more specialized information is required.

County extension offices also maintain a supply of information pamphlets compiled by state and federal specialists that are distributed free of charge or for a nominal

Table 14–1 Obtaining gardening information

State agricultural colleges and experiment stations			
State	University or experiment station	Post office	Zip code
Alabama	Auburn University	Auburn	36830
Alaska	University of Alaska	Fairbanks	99701
Arizona	University of Arizona	Tucson	85721
Arkansas	University of Arkansas	Fayetteville	72701
California	University of California	Berkeley	94720
	University of California	Davis	95616
	University of California	Riverside	92502
Colorado	Colorado State University	Fort Collins	80523
Connecticut	University of Connecticut	Storrs	06268
	University of Connecticut	New Haven	06504
Delaware	University of Delaware	Newark	19711
Florida	University of Florida	Gainesville	32611
Georgia	University of Georgia	Athens	30601
Hawaii	University of Hawaii	Honolulu	96822
Idaho	University of Idaho	Moscow	83843
Illinois	University of Illinois	Urbana	61801
Indiana	Purdue University	West Lafayette	47907
Iowa	Iowa State University	Ames	50010
Kansas	Kansas State University	Manhattan	66506
Kentucky	University of Kentucky	Lexington	40506
Louisiana	Louisiana State University	Baton Rouge	70893
Maine	University of Maine	Orono	04473
Maryland	University of Maryland	College Park	20742
Massachusetts	University of Massachusetts	Amherst	01002
Michigan	Michigan State University	East Lansing	48824
Minnesota	University of Minnesota	St. Paul	55101
Mississippi	Mississippi State University	Mississippi State	39762
Missouri	University of Missouri	Columbia	65201
Montana	Montana State University	Bozeman	59715
Nebraska	University of Nebraska	Lincoln	68503
Nevada	University of Nevada	Reno	89507
New Hampshire	University of New Hampshire	Durham	03824
New Jersey	Rutgers University	New Brunswick	08903
New Mexico	New Mexico State University	Las Cruces	88003

fee. These publications, along with many not stocked by the county extension office, are available from state and provincial agricultural information offices, which are usually a part of the agricultural college at the land-grant university of the state. Addresses of these offices are listed in Table 14–1. Federal publications are available from the Superintendent of Documents, U.S. Government Printing Office, Washington, DC, 20402, and from the Information Division, Canada Department of Agriculture, Ottawa K1A 0C7. Pamphlets listing both federal and state bulletins available will usually be found at the county extension office.

State agricultural colleges and experiment stations			
State	University or experiment station	Post office	Zip code
New York	Cornell University	Ithaca	14853
	Cornell University	Geneva	14456
North Carolina	North Carolina State University	Raleigh	27607
North Dakota	North Dakota State University	Fargo	58102
Ohio	Ohio State University	Columbus	43210
Oklahoma	Oklahoma State University	Stillwater	74074
Oregon	Oregon State University	Corvallis	97331
Pennsylvania	Pennsylvania State University	University Park	16802
Puerto Rico	University of Puerto Rico	Mayaguez	00708
Rhode Island	University of Rhode Island	Kingston	02881
South Carolina	Clemson University	Clemson	29631
South Dakota	South Dakota State University	Brookings	57006
Tennessee	University of Tennessee	Knoxville	37901
Texas	Texas A & M University	College Station	77843
Utah	Utah State University	Logan	84322
Vermont	University of Vermont	Burlington	05401
Virginia	Virginia Polytechnic Institute	Blacksburg	24061
Washington	Washington State University	Pullman	99163
West Virginia	West Virginia University	Morgantown	26506
Wisconsin	University of Wisconsin	Madison	53706
Wyoming	University of Wyoming	Laramie	82071

Canadian agricultural colleges and experiment stations			
Province	University	Post office	Code
Alberta	University of Alberta	Edmonton	T6G 2E1
British Columbia	University of British Columbia	Vancouver	V6T 1W5
Ontario	University of Guelph	Guelph	N1G 2W1
Quebec	Université Labal	Cité Universitaire Quebec, Quebec	G1K 7P4
Quebec	McGill University	P.O.B. 6070 Montreal	H3C 3G1
Manitoba	University of Manitoba	Winnipeg	R3T 2N2
Nova Scotia	Nova Scotia Agricultural Col.	Truro	
Saskatchewan	University of Saskatchewan	Saskatoon	S7N 0W0

Garden Clubs, Botanic Gardens, Commercial Garden Suppliers

Local garden clubs frequently can supply garden information. Nurseries and garden stores are also sources of up-to-date information relating to horticultural problems. Larger nurseries often hire a landscape architect to give advice on home landscaping problems. Local libraries maintain collections of garden publications, especially if gardeners let librarians know of their interest in such information.

Some of the larger public and private gardens disseminate gardening information. One of these, the Brooklyn Botanic Garden, distributes a quarterly publication, *Plants and Gardens*. Periodically this quarterly is issued as a handbook covering in depth a specialized horticultural topic; at last count, 59 illustrated handbooks were available, covering such diverse topics as *Plant Dyeing, Gardening in Containers, Plant Pests, Flowering Shrubs, Nursery Source Guide,* and *Terrariums*. A list of publications and prices are obtainable from Brooklyn Botanic Gardens, 1000 Washington Avenue, Brooklyn, NY, 11225.

Collectively, gardeners spend millions of dollars each year for equipment, supplies, and information, and the various companies catering to the needs and desires of gardeners dispense a great deal of useful information. This information ranges from advertising brochures and seed, nursery, and equipment catalogues to the gardening column or section in the local newspaper. Some of the better known periodicals catering wholly or partially to the gardening public are *Better Homes and Gardens, Sunset,* and *Organic Gardening and Farming*. The publisher of *Organic Gardening and Farming,* Rodale Press, has a series of publications dealing exclusively with the "natural," or "organic," way of gardening. *Good Housekeeping* and Greystone Press have each published an *Illustrated Encyclopedia of Gardening,* with 14 and 16 profusely colored volumes, respectively, on garden topics from A to Z. Time-Life has a series of books on gardening and *Sunset*'s Lane Publishers have produced an extensive how-to-do-it series on various aspects of gardening. Several companies that supply pesticides and fertilizers have published gardening literature, for example, the extensive series of books by the Ortho Division of the Chevron Chemical Company.

The Yellow Pages of the local telephone directory list various local businesses offering supplies and services, such as seed, nursery stock, turf, pesticides, fertilizer, tools and equipment for sale or rent, rototilling, landscape service, soil testing, and even medical care for sick plants.

Scientific and Trade Publications

Although most readers of this text are probably not vitally interested in the professional aspects of horticulture, they should be aware that there is a professional organization, called the American Society for Horticultural Science, that sponsors a national meeting each year where scientists interested in horticulture meet, discuss new developments and problems, and generally exchange ideas on many aspects of horticulture. This society also sponsors two bimonthly publications, *HortScience* and *The Journal of the American Society for Horticultural Science,* both of which publish papers reporting research findings of members. *HortScience* also publishes news of horticulture and horticulturists. Another organization, The American Horticultural Society, and its journal, *American Horticulturist,* cater to a national membership who are more interested in ornamental and hobby horticulture.

Trade publications, such as *American Vegetable Grower, Western Fruit Grower, Florist's Review,* and countless others, supply information primarily for commercial growers.

Soil Preparation

Sequential Preparation of Small Areas for Planting

1. Spread manure, compost, peat, and/or fertilizer evenly over the area to be dug. Determine that the soil has a moisture level suitable for working by pressing a quantity of soil in your hands. If it forms a ball that breaks easily when dropped, it is all right for working. If it forms a ball that doesn't break, it is too wet to work. Sometimes the soil may be so dry and hard that it is physically impossible to work.

2. Begin spading at the side of the garden that most needs to have the soil level raised. This may be the low corner or a side that needs to be raised for appearances or for irrigation and drainage.

3. Spading down 6 to 8 inches (15 to 20 cm), turn each shovelful of the first row directly over.

4. Pile the soil from the second row upside down on the first row, leaving a trench where the soil from the second row was removed.

5. Rake plant residue and added organic matter from the third row into the furrow and then dig the soil from

the third row, turning it upside down into the empty furrow.

6. Repeat Step 5 until the bed is completely worked.

7. Rake immediately to break up clods, remove trash from the surface, and prepare a uniform seedbed.

Sequential Preparation of Larger Areas for Planting

1. Broadcast organic matter and/or fertilizer and determine that soil moisture level is suitable as in Step 1 above.

2. Work the soil to a depth of at least 6 inches with a moleboard plow, disc, or rototiller.

3. Break up clods and level with a harrow or other equipment.

Commercially Prepared Media for Container-Grown Plants

Prepared growing media for starting seedlings, rooting cuttings, and growing plants in containers are available wherever plants are sold. These are generally composed of two of the following four materials:

1. Peat moss (plant residue, undecomposed because it accumulated in a swampy area).

2. Vermiculite (expanded mica).

3. Perlite (special type of expanded volcanic rock).

4. Sand (soil particles between 0.05 and 1 mm in diameter).

These mixes usually also contain a small quantity of fertilizer. Peat moss, vermiculite, perlite, and sand can be purchased separately for growers who want to synthesize their own growing mix. The first two are lightweight and retain water. Perlite and sand provide drainage to a mix. Perlite is somewhat more retentive of water and lighter in weight than sand. Vermiculite tends to lose its structure and become sticky if it is overworked or after prolonged use. If purchased in sealed bags, these products are usually sterile and do not require pasteurization. (Pasteurization is discussed below.)

Home-Prepared Mixes for Container-Grown Plants

Media for Rooting Cuttings. Sand, perlite, a mixture of sand and peat, and a mixture of sand and vermiculite are all used as rooting media. I prefer perlite when it is easily obtained and can be kept moist. The medium for rooting cuttings should not be fertilized.

Media for Starting Transplant Seedlings. Except for species with very small seeds, fertilizer is not needed if seeds are to be germinated in pots and pricked off into flats about the time the first true leaves are developing. Sand or perlite is often mixed with peat or vermiculite in equal portions. I am consistently most successful with sand and peat, but other individuals find other mixes more satisfactory. A soil mix or an artificial mix with fertilizer should be used if seedlings are to be grown to transplant size without being pricked off or if species like petunia or impatiens, having very tiny seed, are being germinated. The germination medium for small-seeded species should be sifted through a screen to provide a fine seed bed.

Media for Growing Transplants and Potting. A good potting soil can be made by mixing 1 part organic matter, 1 part sand, and 1 to 2 parts garden soil. Use a higher proportion of soil if the soil is sandy and a lower proportion if it is mostly clay. The organic matter can be peat moss, well-rotted manure, or well-decomposed compost.

Many experts advise pasteurizing all nonsterile potting mixes to prevent the fungal decay of seedling stems at ground level called damping-off (see Table 14–24). This may be advisable in many areas, especially if seedlings are to be grown to transplant size. When I use soil from uncultivated areas or from fields cropped to grains or forage and clean all containers and potting equipment, I find pasteurization to be unnecessary or even detrimental. Because there is little microbial competition in pasteurized soil, if root pathogens become established, they can spread and quickly destroy all plants in a container having pasteurized soil. For this reason it is doubly important to sterilize all equipment and containers when a sterile medium is used.

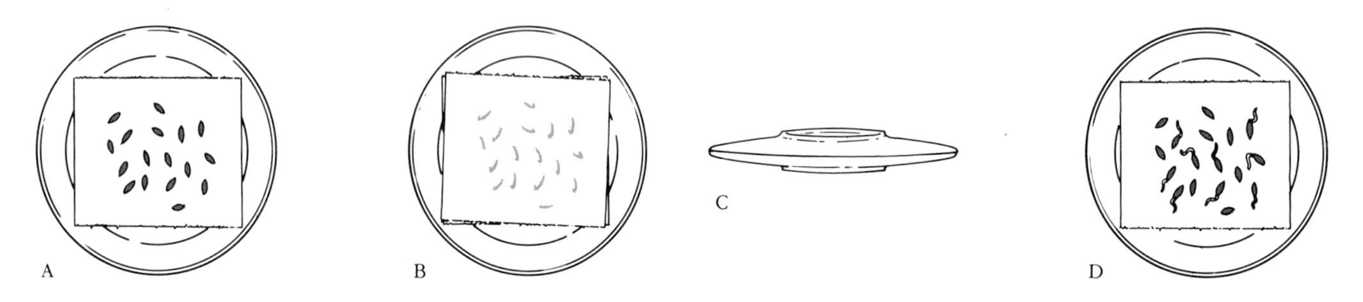

Figure 14–1 The "dinner plate" germination test. In this test of germinability of garden seeds absorbent paper or cotton cloth is moistened and placed on a plate. An appropriate number of seeds are counted onto it (A). The seed is covered with another piece of cloth or paper (B), and a second plate is inverted over the first (C). The germinator should be kept warm (between 70° and 80°F [21° to 27°C] is ideal) and moist but not too wet. Seed should not rest in water. Germinated seed (D) should be counted daily and discarded. After a week (or more, depending on the type of seed) you can calculate the percentage of seeds that germinated by dividing the number of seeds that germinated by the total number of seeds originally put on the plate.

Small amounts of soil can be pasteurized in shallow pans in an oven. If the soil layer is no thicker than 1½ inches (4 cm), most decay-causing organisms and most weed seeds should be destroyed when the soil has been baked 1 hour with the oven set at 220°F (105°C). If a wooden flat is being used, don't turn the oven higher than 250°F (120°C). The flat, as well as organic matter, can char and burn.

The growing medium should be moistened before it is used. Overwatering is not usually a problem with artificial mixes. Mixes using garden soil should be moistened and mixed by hand or with a spade to the point where a ball of earth sticks together when squeezed but crumbles apart when dropped.

Testing Seed Germinability

Vegetable and flower seed that has been stored from past seasons, has been subjected to unfavorable environmental conditions, or for any other reason has doubtful viability should be tested for its germination ability. This usually amounts to determining the percentage of seed that will germinate. Two simple home germination tests are illustrated in Figures 14–1 and 14–2. Several different kinds

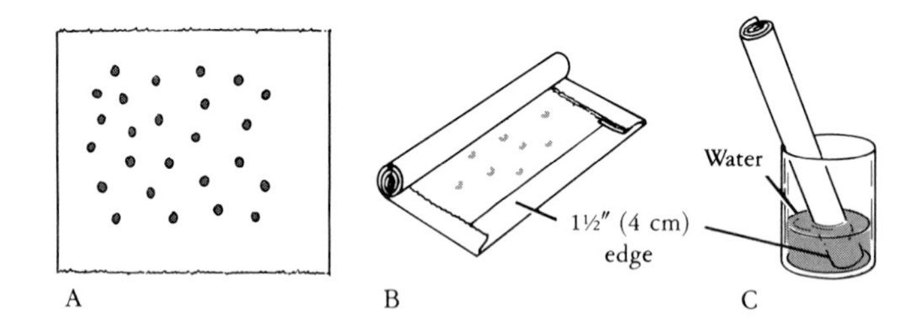

Figure 14–2 The "rag-doll" germination test.

(A) Fifteen to 25 seeds are counted onto a moistened paper towel. At least ¾ inch (2 cm) should be left free of seed on three edges for folding and 1½ inches (4 cm) or more on the other edge.

(B) A second moistened towel is laid over the first. All four edges are folded over and creased, and the towel is rolled so that the wider edge is at one end of the roll.

(C) The rolled "doll" is placed in a container with about 1 inch (2½ cm) of water, and the container is kept in a 70° to 80°F (21° to 27°C) room. The towel acts as a wick to keep the seeds moist but aerated. If the atmosphere is dry, additional water may be needed each day. Water should never reach above the level of the lowest seeds. No seeds should be covered by water.

of small seed can be tested simultaneously on different parts of the dinner plate. The "rag doll" test is most suitable for large seed. Germination time varies for various seed. Radish seeds will begin to germinate in two days and a majority of other vegetables and annual flowers in three to five, but peppers, New Zealand spinach, or asparagus may require several weeks.

Estimating Planting Date

Dates for Starting Transplants Indoors

A general guide on how soon to start plants indoors is to check the time required to grow them to transplanting size (Table 14–2). Then count back from one week after the average date of the last killing frost in your area (see Table 14–26). Use this date as the time to start warm-season plants, including annual flowers that require soil temperatures of 65°F (18°C) or above to germinate, tomatoes, peppers, eggplant, and melons. Members of the cabbage family (cabbage, broccoli, cauliflower, kale, collards), lettuce, celery, and onions, as well as those annual flowers that germinate at soil temperatures of 60°F (16°C) or below, can be started several weeks earlier as they will withstand considerable frost (see Table 14–3).

Dates for Spring and Fall Planting Outdoors

Vegetables that require or tolerate cool soil and flowers that germinate at 60°F or below (Tables 14–2 and 14–4) can be seeded in most areas as soon as the soil has dried and warmed sufficiently so that it can be worked. The exception might be in an arid region with sandy soil where soil can be worked whenever it isn't frozen. Soil temperature should have reached at least 45°F (7°C). Vegetables that require warm soil and flowers that germinate best at 65°F or above should not be seeded outdoors until danger of frost is over (Table 14–26).

Time of planting so that the crop will mature at a specific location before cold weather stops its growth can be determined from Table 14–2 and the climatic data further on in this chapter.

Table 14–2 Suggestions for successful propagation of common vegetables from seed

	Seeds		Distance				Soil temperature for seed[a]				
Vegetable	Depth to plant (inches)	No. to sow (per ft)	Between plants (inches)	Between rows (inches)	Days to germinate	Needs light to germinate	Needs cool soil	Tolerates cool soil	Needs warm soil	Weeks to grow to transplant size[b]	Days to maturity[c]
Asparagus	1½		18	36	7–21			X		1 year	3 years
Beans:											
Snap bush	1½–2	6–8	2–3	18–30	6–14				X		45–65
Snap pole	1½–2	4–6	4–6	36–48	6–14				X		60–70
Lima bush	1½–2	5–8	3–6	24–30	7–12				X		60–80
Lima pole	1½–2	4–5	6–10	30–36	7–12				X		85–90
Fava (Broadbean or Winsor bean)	2½	5–8	3–4	18–24	7–14			X			80–90
Garbanzo (Chick pea)	1½–2	5–8	3–4	24–30	6–12				X		105
Scarlet runner	1½–2	4–6	4–6	36–48	6–14				X		60–70
Soybean	1½–2	6–8	2–3	24–30	6–14				X		95–100

(continued)

Table 14–2 *(continued)*

Vegetable	Seeds: Depth to plant (inches)	Seeds: No. to sow (per ft)	Distance: Between plants (inches)	Distance: Between rows (inches)	Days to germinate	Needs light to germinate	Soil temperature for seed[a]: Needs cool soil	Soil temperature for seed[a]: Tolerates cool soil	Soil temperature for seed[a]: Needs warm soil	Weeks to grow to transplant size[b]	Days to maturity[c]
Beet	½–1	10–15	2	12–18	7–10			X			55–65
Black-eye cowpea (Southern pea)	½–1	5–8	3–4	24–30	7–10				X		65–80
Yardlong bean (Asparagus bean)	½–1	2–4	12–24	24–36	6–13				X		65–80
Broccoli, sprouting	½	10–15	18–24	24–30	3–10			X		5–7[d]	60–80T[e]
Brussels sprouts	½	10–15	18–24	24–30	3–10			X		4–6[d]	80–90T[e]
Cabbage	½	8–10	14–24	24–30	4–10			X		5–7[d]	65–95T[e]
Cabbage, Chinese	½	8–16	10–12	18–24	4–10			X		4–6	80–90
Cardoon	½	4–6	18	36	8–14			X		8	120–150
Carrot	¼	15–20	1–2	14–24	10–17			X			60–80
Cauliflower	½	8–10	18	30–36	4–10		X			5–7[d]	55–65T[e]
Celeriac	⅛	8–12	8	24–30	9–21		X			10–12[d]	90–120T[e]
Celery	⅛	8–12	8	24–30	9–21	X	X			10–12[d]	90–120T[e]
Celtuce (Asparagus lettuce)	½	8–10	12	18	4–10			X		4–6	80
Chard, Swiss	1	6–10	4–8	18–24	7–10			X			55–65
Chicory, witloof (Belgian endive)	¼	8–10	4–8	18–24	5–12	X		X			90–120
Chive	½	8–10	8	10–16	8–12			X			80–90
Chop Suey green (Shungiku)	½	6	2–3	10–12	5–14			X			42
Collard	¼	10–12	10–15	24–30	4–10			X		4–6[d]	65–85T[e]
Corn, sweet	2	4–6	10–14	30–36	6–10				X		60–90
Cornsalad	½	8–10	4–6	12–16	7–10			X			45–55
Cress, garden	¼	10–12	2–3	12–16	4–10	X		X			25–45
Cucumber	1	3–5	12	48–72	6–10				X	4	55–65
Dandelion	½	6–10	8–10	12–16	7–14	X		X			70–90
Eggplant	¼–½	8–12	18	36	7–14				X	6–9[d]	75–95T[e]
Endive	½	4–6	9–12	12–24	5–9			X		4–6	60–90
Fennel, Florence	½	8–12	6	18–24	6–17			X			120
Garlic (from sets)	1		2–4	12–18	6–10			X			90
Ground cherry husk tomato	½	6	24	36	6–13				X	6[d]	90–100T[e]
Horseradish (root divisions)	3		10–18	24				X			180–220
Jerusalem artichoke (tubers)	4		15–24	30–60				X			100–105
Kale	½	8–12	14–24	24–30	3–10			X		4–6	55–80
Kohlrabi	½	8–12	3–4	18–24	3–10			X		4–6	60–70
Leeks	½–1	8–12	2–4	12–18	7–12			X		10–12	80–90T[e]

Vegetable	Seeds: Depth to plant (inches)	Seeds: No. to sow (per ft)	Distance: Between plants (inches)	Distance: Between rows (inches)	Days to germinate	Needs light to germinate	Soil temperature for seed[a]: Needs cool soil	Soil temperature for seed[a]: Tolerates cool soil	Soil temperature for seed[a]: Needs warm soil	Weeks to grow to transplant size[b]	Days to maturity[c]
Lettuce:											
Head	¼–½	4–8	12–14	18–24	4–10	X	X			3–5	55–80
Leaf	¼–½	8–12	4–6	12–18	4–10	X	X			3–5	45–60
Muskmelon	1	3–6	18	48–72	4–8				X	3–4	75–100
Mustard	½	8–10	2–6	12–18	3–10	X		X			40–60
Nasturtium	½–1	4–8	4–10	18–36				X			50–60
Onion (sets)	1–2		2–3	12–24			X				95–120
(plants)	2–3		2–3	12–24			X			8	95–120T[e]
(seed)	½	10–15	2–3	12–24	7–12		X				100–165
Parsley	¼–½	10–15	3–6	12–20	14–28			X		8	85–90
Parsnip	½	8–12	3–4	16–24	15–25			X			100–120
Pea	2	6–7	2–3	18–30	6–15		X				65–85
Peanut	1½	2–3	6–10	30					X		110–120
Pepper	¼	6–8	18–24	24–36	10–20				X	6–8	60–80T[e]
Potato (tubers)	4	1	12	24–36	8–16			X			90–105
Pumpkin	1–1½	2	30	72–120	6–10				X		70–110
Radish	½	14–16	1–2	6–12	3–10		X				20–50
Rutabaga	½	4–6	8–12	18–24	3–10			X			80–90
Salsify	½	8–12	2–3	16–18			X				110–150
Salsify, black	½	8–12	2–3	16–18			X				110–150
Shallot (bulb)	1		2–4	12–18				X			60–75
Spinach	½	10–12	2–4	12–14	6–14		X				40–65
Malabar	½	4–6	12	12	10			X			70
New Zealand	1½	4–6	18	24	5–10			X			70–80
Tampala	¼–½	6–10	4–6	24–30				X			21–42
Squash (summer)	1	4–6	16–24	36–60	3–12				X		50–60
Squash (winter)	1	1–2	24–48	72–120	6–10				X		85–120
Sunflower	1	2–3	16–24	36–48	7–12				X		80–90
Sweet Potato (plants)			12–18	36–48					X		120T[e]
Tomato	½		18–36	36–60	6–14				X	5–7	55–90T[e]
Turnip	½		1–3	15–18	3–10		X				45–60
Watermelon	1	14–16	24–48	72	3–12				X	3–4	80–100

(Adapted and reprinted with permission from the Ortho book, *All About Vegetables*. Copyright 1973, Chevron Chemical Co. Information on light from Dr. Henry M. Cathey.)

[a]Seeds that "need cool soil" do best in a temperature range of 50°–65°F; those that "tolerate cool soil" in a 50°–85°F range; those that "need warm soil" in a 65°–85°F range.

[b]The variation of 4–6, 5–7, 10–12 weeks allows for hot-bed, greenhouse, window sill, and under grow-lamp conditions. Generally the warmer the growing conditions, the shorter the time to grow transplants. However, there must be allowance for a change from indoor to outdoor environment.

[c]The *relative* length of time needed to grow a crop from seed or transplant to table use. The time will vary by variety and season.

[d]Transplants preferred over seed.

[e]T = Number of days from setting out transplants; all others are from seeding.

Table 14-3 Latest dates, and range of dates, for safe fall planting of vegetables in the open (average dates of first fall frost shown in Table 14-26, p. 474)

Crop	Planting dates for localities in which average date of first freeze is—					
	Aug. 30	Sept. 10	Sept. 20	Sept. 30	Oct. 10	Oct. 20
Asparagus[a]	—	b	b	b	Oct. 20–Nov. 15	Nov. 1–Dec. 15
Bean, lima	—	—	—	June 1–15	June 1–15	June 15–30
Bean, snap	—	May 15–June 15	June 1–July 1	June 1–July 10	June 15–July 20	July 1–Aug. 1
Beet	May 15–June 15	May 15–June 15	June 1–July 1	June 1–July 10	June 15–July 25	July 1–Aug. 5
Broccoli, sprouting	May 1–June 1	May 1–June 1	May 1–June 15	June 1–30	June 15–July 15	July 1–Aug. 1
Brussels sprouts	May 1–June 1	May 1–June 1	May 1–June 15	June 1–30	June 15–July 15	July 1–Aug. 1
Cabbage[a]	May 1–June 1	May 1–June 1	May 1–June 15	June 1–July 10	June 1–July 15	July 1–20
Cabbage, Chinese	May 15–June 15	May 15–June 15	June 1–July 1	June 1–July 15	June 15–Aug. 1	July 15–Aug. 15
Carrot	May 15–June 15	May 15–June 15	June 1–July 1	June 1–July 10	June 1–July 20	June 15–Aug. 1
Cauliflower[a]	May 1–June 1	May 1–July 1	May 1–July 1	May 10–July 15	June 1–July 25	July 1–Aug. 5
Celery[a] and celeriac	May 1–June 1	May 15–June 15	May 15–July 1	June 1–July 5	June 1–July 15	June 1–Aug. 1
Chard	May 15–June 15	May 15–July 1	June 1–July 1	June 1–July 5	June 1–July 20	June 1–Aug. 1
Chervil and chive	May 10–June 10	May 1–June 15	May 15–June 15	b	b	b
Chicory, witloof	May 15–June 15	May 15–June 15	May 15–June 15	June 1–July 1	June 1–July 1	June 15–July 15
Collard[a]	May 15–June 15	May 15–June 15	May 15–June 15	June 15–July 15	July 1–Aug. 1	July 15–Aug. 15
Cornsalad	May 15–June 15	May 15–July 1	June 15–Aug. 1	July 15–Sept. 1	Aug. 15–Sept. 15	Sept. 1–Oct. 15
Corn, sweet	—	May 15–June 1	June 1–July 1	June 1–July 1	June 1–July 10	June 1–July 20
Cress, upland	May 15–June 15	May 15–July 1	June 15–Aug. 1	July 15–Sept. 1	Aug. 15–Sept. 15	Sept. 1–Oct. 15
Cucumber	—	—	June 1–15	June 1–July 1	June 1–July 1	June 1–July 15
Dandelion	June 1–15	June 1–July 1	June 1–July 1	June 1–Aug. 1	July 15–Sept. 1	Aug. 1–Sept. 15
Eggplant[a]	—	—	—	May 20–June 10	May 15–June 15	June 1–July 1
Endive	June 1–July 1	June 1–July 1	June 15–July 15	June 15–Aug. 1	July 1–Aug. 15	July 15–Sept. 1
Fennel, Florence	May 15–June 15	May 15–July 15	June 1–July 1	June 1–July 1	June 15–July 15	June 15–Aug. 1
Garlic	b	b	b	b	b	b
Horseradish[a]	b	b	b	b	b	b
Kale	May 15–June 15	May 15–June 15	June 1–July 1	June 15–July 15	July 1–Aug. 1	July 15–Aug. 15
Kohlrabi	May 15–June 15	June 1–July 1	June 1–July 15	June 15–July 15	July 1–Aug. 1	July 15–Aug. 15
Leek	May 1–June 1	May 1–June 1	b	b	b	b
Lettuce, head[a]	May 15–July 1	May 15–July 1	June 1–July 15	June 15–Aug. 1	July 15–Aug. 15	Aug. 1–30
Lettuce, leaf	May 15–July 15	May 15–July 15	June 1–Aug. 1	June 1–Aug. 1	July 15–Sept. 1	July 15–Sept. 1
Muskmelon	—	—	May 1–June 15	May 15–June 1	June 1–June 15	June 15–July 20

Planting dates for localities in which average date of first freeze is—					
Oct. 30	Nov. 10	Nov. 20	Nov. 30	Dec. 10	Dec. 20
Nov. 15–Jan. 1	Dec. 1–Jan. 1	—	—	—	—
July 1–Aug. 1	July 1–Aug. 15	July 15–Sept. 1	Aug. 1–Sept. 15	Sept. 1–30	Sept. 1–Oct. 1
July 1–Aug. 15	July 1–Sept. 1	July 1–Sept. 10	Aug. 15–Sept. 20	Sept. 1–30	Sept. 1–Nov. 1
Aug. 1–Sept. 1	Aug. 1–Oct. 1	Sept. 1–Dec. 1	Sept. 1–Dec. 15	Sept. 1–Dec. 31	Sept. 1–Dec. 31
July 1–Aug. 15	Aug. 1–Sept. 1	Aug. 1–Sept. 15	Aug. 1–Oct. 1	Aug. 1–Nov. 1	Sept. 1–Dec. 31
July 1–Aug. 15	Aug. 1–Sept. 1	Aug. 1–Sept. 15	Aug. 1–Oct. 1	Aug. 1–Nov. 1	Sept. 1–Dec. 31
Aug. 1–Sept. 1	Sept. 1–15	Sept. 1–Dec. 1	Sept. 1–Dec. 31	Sept. 1–Dec. 31	Sept. 1–Dec. 31
Aug. 1–Sept. 15	Aug. 15–Oct. 1	Sept. 1–Oct. 15	Sept. 1–Nov. 1	Sept. 1–Nov. 15	Sept. 1–Dec. 1
July 1–Aug. 15	Aug. 1–Sept. 1	Sept. 1–Nov. 1	Sept. 15–Dec. 1	Sept. 15–Dec. 1	Sept. 15–Dec. 1
July 15–Aug. 15	Aug. 1–Sept. 1	Aug. 1–Sept. 15	Aug. 15–Oct. 10	Sept. 1–Oct. 20	Sept. 15–Nov. 1
June 15–Aug. 15	July 1–Aug. 15	July 15–Sept. 1	Aug. 1–Dec. 1	Sept. 1–Dec. 31	Oct. 1–Dec. 31
June 1–Sept. 10	June 1–Sept. 15	June 1–Oct. 1	June 1–Nov. 1	June 1–Dec. 1	June 1–Dec. 31
b	b	Nov. 1–Dec. 31	Nov. 1–Dec. 31	Nov. 1–Dec. 31	Nov. 1–Dec. 31
July 1–Aug. 10	July 10–Aug. 20	July 20–Sept. 1	Aug. 15–Sept. 30	Aug. 15–Oct. 15	Aug. 15–Oct. 15
Aug. 1–Sept. 15	Aug. 15–Oct. 1	Aug. 25–Nov. 1	Sept. 1–Dec. 1	Sept. 1–Dec. 31	Sept. 1–Dec. 31
Sept. 15–Nov. 1	Oct. 1–Dec. 1	Oct. 1–Dec. 1	Oct. 1–Dec. 31	Oct. 1–Dec. 31	Oct. 1–Dec. 31
June 1–Aug. 1	June 1–Aug. 15	June 1–Sept. 1	—	—	—
Sept. 15–Nov. 1	Oct. 1–Dec. 1	Oct. 1–Dec. 1	Oct. 1–Dec. 31	Oct. 1–Dec. 31	Oct. 1–Dec. 31
June 1–Aug. 1	June 1–Aug. 15	June 1–Aug. 15	July 15–Sept. 15	Aug. 15–Oct. 1	Aug. 15–Oct. 1
Aug. 15–Oct. 1	Sept. 1–Oct. 15	Sept. 1–Nov. 1	Sept. 15–Dec. 15	Oct. 1–Dec. 31	Oct. 1–Dec. 31
June 1–July 1	June 1–July 15	June 1–Aug. 1	July 1–Sept. 1	Aug. 1–Sept. 30	Aug. 1–Sept. 30
July 15–Aug. 15	Aug. 1–Sept. 1	Sept. 1–Oct. 1	Sept. 1–Nov. 15	Sept. 1–Dec. 31	Sept. 1–Dec. 31
July 1–Aug. 1	July 15–Aug. 15	Aug. 15–Sept. 15	Sept. 1–Nov. 15	Sept. 1–Dec. 1	Sept. 1–Dec. 1
b	Aug. 1–Oct. 1	Aug. 15–Oct. 1	Sept. 1–Nov. 15	Sept. 15–Nov. 15	Sept. 15–Nov. 15
b	b	b	b	b	b
July 15–Sept. 1	Aug. 1–Sept. 15	Aug. 15–Oct. 15	Sept. 1–Dec. 1	Sept. 1–Dec. 31	Sept. 1–Dec. 31
Aug. 1–Sept. 1	Aug. 15–Sept. 15	Sept. 1–Oct. 15	Sept. 1–Dec. 1	Sept. 15–Dec. 31	Sept. 1–Dec. 31
b	b	Sept. 1–Nov. 1	Sept. 1–Nov. 1	Sept. 1–Nov. 1	Sept. 15–Nov. 1
Aug. 1–Sept. 15	Aug. 15–Oct. 15	Sept. 1–Nov. 1	Sept. 1–Dec. 1	Sept. 15–Dec. 31	Sept. 15–Dec. 31
Aug. 15–Oct. 1	Aug. 25–Oct. 1	Sept. 1–Nov. 1	Sept. 1–Dec. 1	Sept. 15–Dec. 31	Sept. 15–Dec. 31
July 1–July 15	July 15–July 30	—	—	—	—

(continued)

Table 14-3 *(continued)*

Crop	Planting dates for localities in which average date of first freeze is—					
	Aug. 30	Sept. 10	Sept. 20	Sept. 30	Oct. 10	Oct. 20
Mustard	May 15–July 15	May 15–July 15	June 1–Aug. 1	June 15–Aug. 1	July 15–Aug. 15	Aug. 1–Sept. 1
Okra	—	—	June 1–20	June 1–July 1	June 1–July 15	June 1–Aug. 1
Onion[a]	May 1–June 10	May 1–June 10	b	b	b	b
Onion, seed	May 1–June 1	May 1–June 10	b	b	b	b
Onion, sets	May 1–June 1	May 1–June 10	b	b	b	b
Parsley	May 15–June 15	May 1–June 15	June 1–July 1	June 1–July 15	June 15–Aug. 1	July 15–Aug. 15
Parsnip	May 15–June 1	May 1–June 15	May 15–June 15	June 1–July 1	June 1–July 10	b
Pea, garden	May 10–June 15	May 1–July 1	June 1–July 15	June 1–Aug. 1	b	b
Pea, black-eye	—	—	—	—	June 1–July 1	June 1–July 1
Pepper[a]	—	—	June 1–June 20	June 1–July 1	June 1–July 1	June 1–July 10
Potato	May 15–June 1	May 1–June 15	May 1–June 15	May 1–June 15	May 15–June 15	June 15–July 15
Radish	May 1–July 15	May 1–Aug. 1	June 1–Aug. 15	July 1–Sept. 1	July 15–Sept. 15	Aug. 1–Oct. 1
Rhubarb[a]	Sept. 1–Oct. 1	Sept. 15–Oct. 15	Sept. 15–Nov. 1	Oct. 1–Nov. 1	Oct. 15–Nov. 15	Oct. 15–Dec. 1
Rutabaga	May 15–June 15	May 1–June 15	June 1–July 1	June 1–July 1	June 15–July 15	July 10–20
Salsify	May 15–June 1	May 10–June 10	May 20–June 20	June 1–20	June 1–July 1	June 1–July 1
Shallot	b	b	b	b	b	b
Sorrel	May 15–June 15	May 1–June 15	June 1–July 1	June 1–July 15	July 1–Aug. 1	July 15–Aug. 15
Soybean	—	—	—	May 25–June 10	June 1–25	June 1–July 5
Spinach	May 15–July 1	June 1–July 15	June 1–Aug. 1	July 1–Aug. 15	Aug. 1–Sept. 1	Aug. 20–Sept. 10
Spinach, New Zealand	—	—	—	May 15–July 1	June 1–July 15	June 1–Aug. 1
Squash, summer	June 10–20	June 1–20	May 15–July 1	June 1–July 1	June 1–July 15	June 1–July 20
Squash, winter	—	—	May 20–June 10	June 1–15	June 1–July 1	June 1–July 1
Sweet potato	—	—	—	—	May 20–June 10	June 1–15
Tomato	June 20–30	June 10–20	June 1–20	June 1–20	June 1–20	June 1–July 1
Turnip	May 15–June 15	June 1–July 1	June 1–July 15	June 1–Aug. 1	July 1–Aug. 1	July 15–Aug. 15
Watermelon	—	—	May 1–June 15	May 15–June 1	June 1–June 15	June 15–July 20

(Adapted from ***Suburban and Farm Vegetable Gardens.*** USDA Home and Garden Bulletin 9. U.S. Government Printing Office, Washington, DC. 1967.)
[a]Plants.
[b]Generally spring-planted.

Planting dates for localities in which average date of first freeze is—					
Oct. 30	Nov. 10	Nov. 20	Nov. 30	Dec. 10	Dec. 20
Aug. 15–Oct. 15	Aug. 15–Nov. 1	Sept. 1–Dec. 1	Sept. 1–Dec. 1	Sept. 1–Dec. 1	Sept. 15–Dec. 1
June 1–Aug. 10	June 1–Aug. 20	June 1–Sept. 10	June 1–Sept. 20	Aug. 1–Oct. 1	Aug. 1–Oct. 1
b	Sept. 1–Oct. 15	Oct. 1–Dec. 31	Oct. 1–Dec. 31	Oct. 1–Dec. 31	Oct. 1–Dec. 31
b	b	Sept. 1–Nov. 1	Sept. 1–Nov. 1	Sept. 1–Nov. 1	Sept. 15–Nov. 1
b	Oct. 1–Dec. 1	Nov. 1–Dec. 31	Nov. 1–Dec. 31	Nov. 1–Dec. 31	Nov. 1–Dec. 31
Aug. 1–Sept. 15	Sept. 1–Nov. 15	Sept. 1–Dec. 31	Sept. 1–Dec. 31	Sept. 15–Dec. 31	Sept. 1–Dec. 31
b	b	Aug. 1–Sept. 1	Sept. 1–Nov. 15	Sept. 1–Dec. 1	Sept. 1–Dec. 1
Aug. 1–Sept. 15	Sept. 1–Nov. 1	Oct. 1–Dec. 1	Oct. 1–Dec. 31	Oct. 1–Dec. 31	Oct. 1–Dec. 31
June 1–Aug. 1	June 15–Aug. 15	July 1–Sept. 1	July 1–Sept. 10	July 1–Sept. 20	July 1–Sept. 20
June 1–July 20	June 1–Aug. 1	June 1–Aug. 15	June 15–Sept. 1	Aug. 15–Oct. 1	Aug. 15–Oct. 1
July 20–Aug. 10	July 25–Aug. 20	Aug. 10–Sept. 15	Aug. 1–Sept. 15	Aug. 1–Sept. 15	Aug. 1–Sept. 15
Aug. 15–Oct. 15	Sept. 1–Nov. 15	Sept. 1–Dec. 1	Sept. 1–Dec. 31	Aug. 1–Sept. 15	Oct. 1–Dec. 31
Nov. 1–Dec. 1	—	—	—	—	—
July 15–Aug. 1	July 15–Aug. 15	Aug. 1–Sept. 1	Sept. 1–Nov. 15	Oct. 1–Nov. 15	Oct. 15–Nov. 15
June 1–July 10	June 15–July 20	July 15–Aug. 15	Aug. 15–Sept. 30	Aug. 15–Oct. 15	Sept. 1–Oct. 31
b	Aug. 1–Oct. 1	Aug. 15–Oct. 1	Aug. 15–Oct. 15	Sept. 15–Nov. 1	Sept. 15–Nov. 1
Aug. 1–Sept. 15	Aug. 15–Oct. 1	Aug. 15–Oct. 15	Sept. 1–Nov. 15	Sept. 1–Dec. 15	Sept. 1–Dec. 31
June 1–July 15	June 1–July 25	June 1–July 30	June 1–July 30	June 1–July 30	June 1–July 30
Sept. 1–Oct. 1	Sept. 15–Nov. 1	Oct. 1–Dec. 1	Oct. 1–Dec. 31	Oct. 1–Dec. 31	Oct. 1–Dec. 31
June 1–Aug. 1	June 1–Aug. 15	June 1–Aug. 15	—	—	—
June 1–Aug. 1	June 1–Aug. 10	June 1–Aug. 20	June 1–Sept. 1	June 1–Sept. 15	June 1–Oct. 1
June 10–July 10	June 20–July 20	July 1–Aug. 1	July 15–Aug. 15	Aug. 1–Sept. 1	Aug. 1–Sept. 1
June 1–15	June 1–July 1	June 1–July 1	June 1–July 1	June 1–July 1	June 1–July 1
June 1–July 1	June 1–July 15	June 1–Aug. 1	Aug. 1–Sept. 1	Aug. 15–Oct. 1	Sept. 1–Nov. 1
Aug. 1–Sept. 15	Sept. 1–Oct. 15	Sept. 1–Nov. 15	Sept. 1–Nov. 15	Oct. 1–Dec. 1	Oct. 1–Dec. 31
July 1–July 15	July 15–July 30	—	—	—	—

Table 14–4 Guidelines for germination of annual, pot plant, and ornamental herb seeds

Common name and cultivar	Group[a]	Temp. for best germination (°F)	Contin. light or dark	Germination (days)
Ageratum 'Blue Mink'	VI	70	L	5
Ageratum 'Golden'	I	70	D	5
Alyssum 'Carpet of Snow'	I	70	DL	5
Amaranthus 'Molten Fire'	III	70	DL	10
Anise	IV	70	DL	10
Aster 'Ball White'	I	70	DL	8
Balsam 'Scarlet'	III	70	DL	8
Basil 'Dark Opal'	III	70	DL	10
Basil 'Lettuce Leaves'	III	70	DL	10
Begonia, fibrous-rooted 'Scandinavian Pink'	V	70	L	15
Begonia, tuberous-rooted 'Double Mix'	VI	65	L	15
Borage	VIII	70	D	8
Browallia 'Blue Bells' and 'Silver Bells'	VI	70	L	15
Browallia 'Sapphire'	V	70	L	15
Calceolaria multiflora nana	VI	70	L	15
Calendula 'Orange Coronet'	VIII	70	D	10
Campanula 'Annual Mix'	III	70	DL	20
Candytuft 'Giant White'	I	70	DL	8
Carnation 'Chaband's Giant' and 'Imp. Cardinal Red'	IV	70	DL	20
Celosia 'Toreador'	III	70	DL	10
Centaurea 'Blue Boy'	VIII	65	D	10
Centaurea 'Dusty Miller'	VIII	65	D	10
Centaurea, yellow	VIII	70	D	10
Chives (grass onion)	IV	60	DL	10
Christmas cherry 'Masterpiece'	III	70	DL	20
Cineraria 'Maritima Diamond'	VII	75	L	10
Cineraria 'Vivid'	III	70	DL	10
Clarkia 'Florist Mixture'	I	70	DL	5
Cobaea (cup-and-saucer vine), purple	I	70	DL	15
Coleus 'Red Rainbow'	VII	65	L	10
Coriander, annual	VIII	70	D	10
Cosmos 'Radiance'	II	70	DL	5
Cuphea 'Firefly'	VI	70	L	8
Cyclamen 'Pure White'	IX	60	D	50
Cynoglossum 'Firmament'	IV	60	D	5
Dahlia 'Unwins Dwarf Mix'	I	70	DL	5
Dianthus 'Bravo'	I	70	DL	5
Didiscus 'Blue Lace'	IV	65	D	15
Dill	IV	60	L	10
Dimorphotheca 'Orange Improved'	II	70	DL	10
Euphorbia, annual poinsettia	I	70	DL	15
Exacum 'Tiddly-Winks'	V	70	L	15
Fennel, sweet	IV	65	D	10
Feverdew 'Ball Double White Improved'	VII	70	L	15
Freesia 'White Giant'	IV	65	DL	25
Gaillardia 'Tetra Red Giant'	III	70	DL	20
Gazania 'Mix'	IV	60	D	8
Gloxinia 'Emperor Wilhelm'	V	65	L	15
Gomphrena 'Rubra'	III	65	D	15
Grevillea (Australian silk oak)	VI	80	L	20
Gypsophila 'Covent Garden'	I	70	DL	10
Helichrysum (everlasting)	VII	70	L	5
Heliotrope 'Marine'	IV	70	DL	25
Hollyhock 'Powderpuffs Mix'	IV	60	DL	10
Hunnemannia (bush escholtzia) 'Sunlite'	III	70	DL	15
Impatiens 'Holstii Scarlet'	VI	70	L	15
Kalanchoe 'Vulcan'	V	70	L	10
Kochia 'Bright green'	I	70	DL	15
Larkspur 'White Supreme'	IX	55	D	20
Lobelia 'Crystal Palace'	III	70	DL	20
Lupine 'Giant King' and 'Oxford Blue'	IV	55	DL	20
Marigold 'Doubloon'	I	70	DL	5
Marigold 'Spry'	I	70	DL	5
Marjoram, sweet	II	70	DL	8

Common name and cultivar	Group[a]	Temp. for best germination (°F)	Contin. light or dark	Germination (days)
Mesembryanthemum criniflorum	IX	65	D	15
Migonette 'Early White'	I	70	DL	5
Mimosa (sensitive plant)	VIII	80	D	8
Morning glory 'Heavenly Blue'	III	65	DL	5
Myosotis 'Ball Early'	IV	55	D	8
Naegelia 'Art Shades'	V	70	L	15
Nasturtium 'Golden Giant'	IV	65	D	8
Nemesia 'Fire King'	IX	65	D	5
Nicotiana 'Crimson Bedder'	VII	70	L	20
Nierembergia 'Purple Robe'	III	70	DL	15
Pansy 'Lake of Thun'	IX	65	D	10
Parsley 'Extra Triple Curled'	IX	75	D	15
Penstemon 'Sensation Mixture'	VIII	65	D	10
Perilla 'Burgundy'	VI	65	L	15
Petunia 'Maytime'	VI	70	L	10
Phlox 'Glamour'	VIII	65	D	10
Plumbago, blue	IV	75	DL	25
Poppy nudicaule 'Iceland'	I	70	D	10
Portulaca, yellow	IV	70	D	10
Primula 'Chinese Giant Fringed'	VIII	70	D	25
Primula malacoides 'White Giant'	VI	70	L	25
Primula 'Fasbender's Red'	VI	70	L	25
Rosemary, perennial	IV	60	DL	15
Rudbeckia, single 'Gloriosa Daisy'	III	70	DL	20
Sage, perennial	VIII	70	D	15
Saintpaulia 'Blue Fairy Tale'	V	70	L	25
Salpiglossis 'Emperor Mix'	III	70	D	15
Salvia (St. John's Fire)	VI	70	L	15
Savory, 'Bohnenkraut'	VI	65	L	15
Scabiosa 'Giant Blue'	III	70	DL	10
Schizanthus ball 'Giant Mix'	VIII	60	D	20
Shamrock 'True Irish'	IX	65	D	10
Smilax	VIII	75	D	30
Snapdragon 'Orchid Rocket'	VII	65	L	10
Statice 'Iceberg'	I	70	DL	15
Statice suworowii 'Russian'	VIII	70	D	15
Stock 'Lavender Column'	I	70	DL	10
Streptocarpus	V	70	L	15
Sweetpea 'Ruth Cuthbertson'	IV	55	D	15
Thunbergia gibsoni	III	70	DL	10
Thyme, perennial	IV	75	DL	10
Tithonia 'Torch'	VIII	70	D	20
Torenia	III	70	DL	15
Verbena 'Torrid'	VIII	65	D	20
Viola 'Blue Elf'	IX	65	D	10
Vinca, periwinkle (alba aculata)	VIII	70	D	15
Wallflower 'Golden Standard'	I	70	DL	5
Zinnia 'Isabellina'	III	70	DL	5

(Adapted from "Guidelines for germination of annual, pot plants, and ornamental herb seeds," *Florists Review* 144:26–29, 1975–1977.) By permission of Dr. Henry M. Cathey.

[a]Annuals divided into groups on basis of response to temperature and light:

Group I—Germination over a wide temperature range without a light requirement.
Group II—Germination only at cool temperatures without a light requirement.
Group III—Germination only at warm temperatures without a light requirement.
Group IV—Germination at a restricted range of temperatures without a light requirement.
Group V—Germination over a wide range of temperatures when exposed to light.
Group VI—Germination enhanced over a wide temperature range when exposed to light.
Group VII—Germination over a wide temperature range and enhanced at warm temperatures when exposed to light.
Group VIII—Germination over a wide temperature range when held in the dark.
Group IX—Germination over a wide temperature range and enhanced at warm temperatures when held in the dark.

Propagating Plants from Seed

by F. E. Larsen*

Starting Plants Indoors for Later Outdoor Planting

For time of planting see the previous section. Containers and media are described at the beginning of Chapter 4 and in the "Soil Preparation" section of this chapter. Preplanted trays of some crops are available at grocery and garden stores. All that is initially necessary to start plant growth in these trays is to punch holes in the top, water, place them in the proper environment, and wait.

Sowing Seeds. Place the germination mix in the selected container. Firm with the fingers at the container edges and corners, level even with the container top, and press down entire surface lightly but firmly to provide a uniform flat surface. For very small seeds, at least the top ¼ inch (½ cm) should be of the fine-screened mix (Figure 14–3).

For medium and large seeds make furrows about 1 inch (2½ cm) apart across the surface of the container. Individually space the large seeds. For medium-sized seeds, open the seed packet, hold in one hand, and lightly tap the packet with the index finger as you move the packet down the furrow; this will distribute the seeds fairly evenly. Do not plant too thick, as the seedlings will be crowded and spindly if they lack growing space. Cover the seeds lightly with the screened growing mix. A suitable planting depth is usually equal to about twice the diameter of the seed.

Broadcast small seed like petunias or begonias over the surface of the germinating medium rather than in rows or furrows. Do not cover with the growing mix.

If you use peat pots, strips, cell paks, or other individual plant containers (Figure 14–4), they must be filled with planting mix and firmed, as with larger containers. Before seeds are sown in them, pellets must be expanded with water. Plant the seeds in the center of each small container or cell. For medium and large seeds punch a small hole in the center of the growing mix of each container or cell, place two or three seeds in each hole, and cover the seeds as with larger containers. Place small seeds on the surface.

Watering. After you sow the seed, wet the planting mix. Place the containers in a pan, tray, or tub of water that has about 1 inch of water in the bottom. When the water has seeped upward through the container to the surface, remove the container and set aside to drain for an hour or two. Then slip the container in a clear plastic bag and tie shut (Figure 14–5). A pane of glass can also be used to cover the containers to hold in moisture.

Place the containers in a warm or cool place as required (see discussion on temperature) but not in direct sunlight. Some types of seed must be placed in the dark (see discussion on lighting). Check each day to be sure the mix is still moist and to watch for emerging plants. When plants are emerging, remove the plastic or glass covering and place in full natural or artificial light (if placed in sunlight, be sure it does not get too hot). Continue to check the moisture level of the mix. If water is needed, apply as previously described for small seeded plants or add to the top of the container with medium- and large-seeded plants. Be careful not to overwater or wash out the plants.

Lighting. The germination of some seed is held back by light, some require light for germination, and many will germinate in either light or dark. Of the common garden vegetable crops, only a few are affected significantly by light. Consult Tables 14–2 and 14–4 for light requirements for seed germination.

When seeds have germinated, they will require adequate lighting for good growth. If you have a window that is well lighted through the day, this will often be adequate. Since the plants will bend toward the window, turn the containers daily to prevent permanent curves in the stems and to allow sturdier plants to develop.

If a place with good natural light is not available, artificial light must be provided. Even where good natural

*Adapted from *Propagating Plants from Seed,* by F. E. Larsen. Pacific Northwest Cooperative Extension Publication 170. Extension Services of Washington State University, Oregon State University, and University of Idaho. Pullman, WA. 1977.

Figure 14–3 Preparing the seeding mix. (Courtesy Washington State University.) (A) Screening the mix. (B) Leveling the mix in the flat. (C) Making furrows for the larger seed.

Figure 14–4 Plants seeded into individual containers. (Courtesy Washington State University.)

Figure 14–5 Placing flat in plastic bag after seeding. (Courtesy Washington State University.)

light is found, it is usually of benefit to supplement it with artificial light to extend the day length and increase the intensity.

A fluorescent fixture with 40-watt cool or warm light or gro-type tubes provides a good source of light. Suspend about 4 to 6 inches (10 to 15 cm) above the small plants. Provide light 16 to 18 hours per day. Turning on and off can be controlled automatically by commercially available timing devices.

Transplants can, of course, be grown in hotbeds or greenhouses where there is ample natural light.

Temperature. Rapid germination of most flower and vegetable seeds will take place at 75° to 80°F (24° to 27°C). At temperatures of 60°F (16°C) or less, germination is slow, and damage from damping-off diseases can be a major problem. There are some plants, however, that germinate better at 50° to 65°F (10° to 18°C). See Tables 14–2 and 14–4 for favorable germination temperatures of a variety of garden plants.

The temperature variations within a home can often be taken advantage of to give the right temperature conditions for seed germination. Temperatures can be more precisely controlled by placing the containers in a cool area on top of thermostatically controlled heating cables to raise the temperature of the germinating medium.

Pricking Off. If plants have not been seeded in individual containers, they must be transplanted soon after they germinate to give greater growing space and to allow proper plant development; this is called **pricking off.** Transplant when plants develop their first true leaves (Figure 14–6). Do not confuse the cotyledons (seed leaves) with true leaves, which have quite a different appearance. Failure to transplant on time results in hardened, spindly, or overdeveloped plants that seldom grow properly.

Fill growing containers as described for germinating seeds. In flats, mark rows about 2 inches (5 cm) apart. Using a sharpened pencil, make holes about 2 inches apart along the row for individual plants.

Several approaches can be used to extract plants from their germinating containers. First water the seedlings. If they were germinated in a very loose mix, you can usually lift them out gently with the tip of a pencil with little loss of roots. However, if the mix contains sand and garden soil, it will be more compact. In this case cut the mix in a flat into sections as if you were cutting a cake. Lift out a section at a time. If pots were used, remove the entire contents at once.

Figure 14–6 Plants ready to transplant from trays. (Courtesy Washington State University.)

Gently break up each section by separating the plant roots, and remove one plant at a time. Where a heavy intertwining root mass is present, place the sections in water in a clean bucket and gently agitate the mass. The soil will be washed away, and you can separate the plants with no loss of roots. Place the plants on moist newspaper while transplanting to prevent damage from drying roots. Discard spindly or poorly developed plants.

Hold each separated plant carefully by the top and tuck the roots in the holes previously prepared in flats or pots. Attempt to keep roots pointed downward. Do not jam roots in the hole in a ball or use a hole that is too shallow, so that the roots form a "U" going down into the soil and back out. Plant the seedlings slightly deeper than they were in the germinating containers. Place soil around each plant as you set it in place and gently firm the soil around the stem and roots with your fingers. Figures 14–7 and 14–8 illustrate pricking off.

Figure 14-7 Pricking off into a flat. (Courtesy Washington State University.)

(A) Separate plants grown in a loose mix.

(B) Punch holes for transplants.

(C) Firm planting medium about the roots.

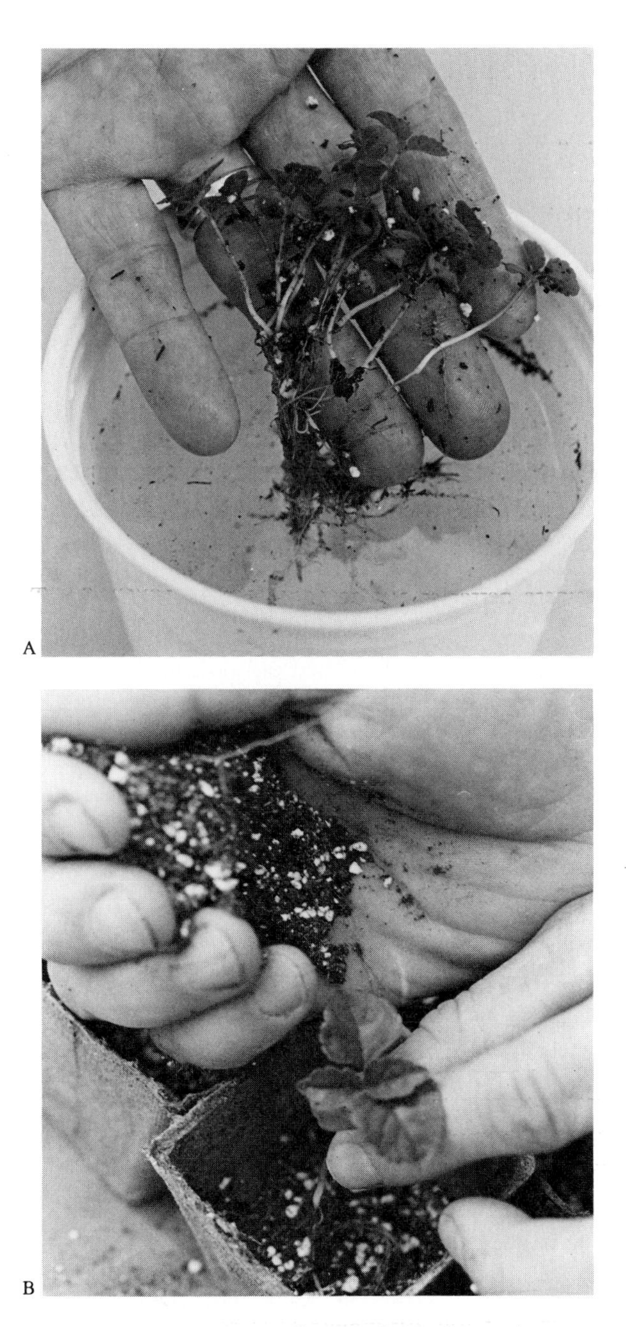

Figure 14-8 Pricking off into individual containers. (Courtesy Washington State University.)

(A) Separate entwined roots by agitating in water.

(B) Position plant in container, fill with mix, and firm about roots.

After transplanting, water the plants as described for seedlings or gently water from the top, taking care not to wash out or flatten the plants with the water.

For a few days after transplanting give plants special care to help them overcome transplanting shock. Syringing tops with water or covering the plants with a plastic bag to maintain high humidity will help prevent wilting. Keep away from drafts or high temperatures.

Growing Transplants or Plants Started in Individual Containers. Plants started in individual containers must be thinned. Remove all but the most vigorous, healthy plant in each container or cell. After transplanting, or with plants of the same age started in individual containers, keep plants growing at a steady, rapid rate to be in good condition for outdoor planting.

Adequate light insures adequate plant development. Poor lighting results in spindly, weak plants. If natural light is used, be sure it is bright throughout the day. Artificial lighting, as described earlier, is usually of benefit to provide adequate light intensity and day length. For most plants 14 to 16 hours of light daily is adequate. Keep the lights 4 to 6 inches above the plants.

The information in Tables 14–2 and 14–4 will provide a clue to growing temperatures for various plants. However, optimum growing temperatures may differ from optimum seed germination temperatures. In general, most plants grow best between 60° and 80°F. Day temperatures should usually be between 70° and 80°F and night temperatures between 60° and 70°F. Do not subject the plants to cold drafts.

At each watering apply enough water to wet the entire volume of growing mix so that some drains through the bottom. Apply water early enough in the day so plant leaves dry before nightfall. This prevents damping-off diseases. Apply water carefully to avoid washing the plants out or splashing soil on the leaves. Keep the soil slightly moist to the touch but not waterlogged.

If the plants have a pale green or otherwise discolored look, apply a water-soluble house plant fertilizer at half strength. They should not need fertilizer more than once or twice before you plant them outdoors.

Hardening Plants for Moving Outdoors. For best results do not abruptly transfer seedlings from indoors to the uncertain outdoor climate. Gradually harden or toughen plants for about two weeks before you plant them in the garden. Do this by slowing down their rate of growth to prepare them to withstand such conditions as cold, drying winds, water shortage, or high temperatures. Reducing water and lowering the temperature are methods usually used to harden plants.

A suitable way to harden plants is to place them out-of-doors in the daytime in partial shade. Over a period of a few days, gradually move them into direct sunlight. One or two days before planting outdoors leave them outside during the night on the sidewalk or steps but not directly on the ground. During the hardening period reduce water application but do not allow them to wilt. Transplants can be hardened in a cold frame.

Transplanting to the Garden. Prepare the planting area. If plants are in flats, cut the mix into cubes as if cutting cake. The plants can then be lifted out individually. If individual plastic or clay pots are used, place one hand on top of the pot with the fingers around the plant. Tip the pot up-side-down and tap the edge of the pot on a table edge. The contents of the pot should fall out intact in your hand.

When setting the plants in the garden soil, use the same procedures and the precautions about root placement as described earlier for indoor transplanting. Set plants slightly deeper in the soil than they were previously growing. You can place containers such as peat pots, expanded pellets, and cubes directly in the garden soil. Be sure to plant these containers deep enough to cover the upper edges and surfaces with soil. If exposed, these surfaces sometimes act as a wick, resulting in rapid loss of moisture from around the plant's roots.

After the plant is set, leave a slight dish-shaped depression around the stem. Fill this with water or starter solution (see "Fertilizing Garden Crops," this chapter).

Choose a cloudy day for transplanting or wait until evening. Avoid windy or very cold days. The more moderate the conditions, the better the survival. If conditions the day following transplanting are very hot or windy, protect with shingles pushed into the soil on the sunny or windy side of each plant.

You can plant outdoors one or two weeks earlier with some plants by covering them with waxed paper "hotcaps" after transplanting. Plastic tunnels supported by wire frames can be used to cover entire rows.

Hotcaps and plastic tunnels will trap heat from the soil to protect against light frosts. Check carefully during warm days to be sure inside temperatures are not too high. Venting slots may be cut on the side away from the sun to reduce this problem.

The transplanting sequence is illustrated in Figure 14–9.

Figure 14–9 Transplanting to the garden. (Courtesy Washington State University.)

(A) Place plant slightly deeper than it was indoors.
(B) Firm soil about roots.
(C) Water generously.
(D) Apply hotcap for earlier outdoor planting.
(E) Anchor hotcap with soil.

Planting Seeds Outdoors

Directions for seedbed preparation are given earlier in this chapter. For suggestions regarding timing, irrigation, and depth of planting, see Chapter 4 and Tables 14–2 and 14–4.

Seeds are planted outdoors after the soil has warmed sufficiently and the seedbed is raked smooth. Measure the correct between-row distance and mark the ends of the rows with small stakes. Using either a string attached to stakes or a straight board as the row guide, dig a shallow trench for the seeds with the corner of a hoe. Straight rows make cultivation easier. Place the seeds in the furrow, cover to the proper depth, and tamp the soil firmly with the hoe or a rake (Figure 14–10). Packing with a rake is recommended with heavy soil that tends to

Figure 14–10 Direct seeding without furrow irrigation. (Courtesy Washington State University.)
(A) Marking planting furrows using a string for a guide.
(B) Marking planting furrows with a straight-edge guide.
(C) Placing seed in furrow.
(D) Packing soil firmly over seed.

A

B

D

C

crust after rain or sprinkling if the surface is left smooth. Planting a few more seeds than are necessary to insure adequate stand in case of poor germination is all right if the planting is thinned soon after seedlings emerge. Generally, however, amateurs tend to plant far too many seeds, with the result of crowding and stunted growth.

Crops that require wide between-plant spacing—squash, watermelon, muskmelon, cucumbers, pole beans, sweet corn—can be planted in hills (Figure 14–11) and thinned to two or three plants per hill. The total number of plants remaining after thinning should be the same as the number that would remain if the planting were thinned to the suggested between-plant distance in Table 14–2.

Where crusting is a problem, it can be prevented with a narrow, shallow mulch of peat moss, vermiculite, bark-dust, or sawdust over the row.

If the garden is to be furrow-irrigated or if drainage after heavy rains is required, it is necessary to furrow the garden area and plant the garden on beds. Large gardens can be furrowed with power equipment, but smaller areas can be furrowed with the corner of a hoe as the garden is being planted. Crops should be seeded near the furrow rather than in the center of the bed so that irrigation water can soak to the planting. Furrows should slope gradually so that water can flow slowly from top to bottom. On a sloping site this may require planting on the contour. Beds are seldom made less than 2½ feet (about ¾ m) wide from center to center, and closely spaced crops are planted two rows per bed (Figure 14–12).

Begin thinning as soon as germination is complete. Do not thin all at once but go over the rows several times. The first plants thinned out, if carefully removed, can be transplanted to areas where germination was poor or can be used to plant other areas. Edible plants removed later can be eaten. Consult Table 14–2 or follow directions on seed packets for thinning.

Figure 14–11 Planting (A) and covering (B) a hill of seed. (Courtesy Washington State University.)

Starting Woody Plants from Seed

As most woody plants are heterozygous, those grown from seed will usually differ somewhat from the parent plant and from each other. Many will be inferior to the parent plant in horticultural qualities.

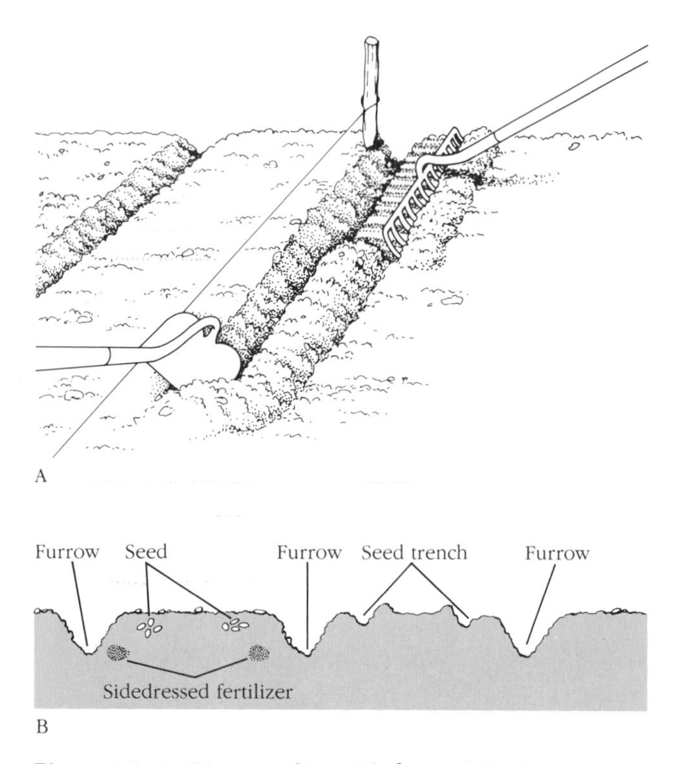

Figure 14–12 Direct seeding with furrow irrigation.
(A) Making furrows. With a taut string as a guide, use the corner of a hoe to dig a furrow 3 to 4 inches (8 to 10 cm) deep. The soil from the furrow should be pulled away from the completed bed and piled. The pile of soil is then raked over the area of the next bed. At the same time humps are leveled and depressions filled in that area.
(B) Cross section of furrowed garden. Shallow trenches for planting the seed are dug a few inches from the furrow. Seed should be distributed and covered immediately after the seed furrow is dug in order to limit drying of the soil in the seed trench. Fertilizer sidedressing, if needed, is most effective if the fertilizer is placed slightly below and on the furrow side of the seed or seedling.

Treating the Seed. Seed from woody plants is frequently affected by one or more types of dormancy, which must be overcome before it will germinate. Scarification and stratification to overcome seed dormancy are described in Chapter 4; see also Figures 4–4 (Chapter 4) and 14–13. Seeds should be planted immediately after they are treated.

Planting the Seed. After dormancy has been overcome, the seed can be grown as described for vegetables and herbaceous ornamentals. They can be started indoors for later transplanting outside, or they can be planted outdoors at the proper time. Table 14–5 lists suggestions for handling seed of selected woody plants.

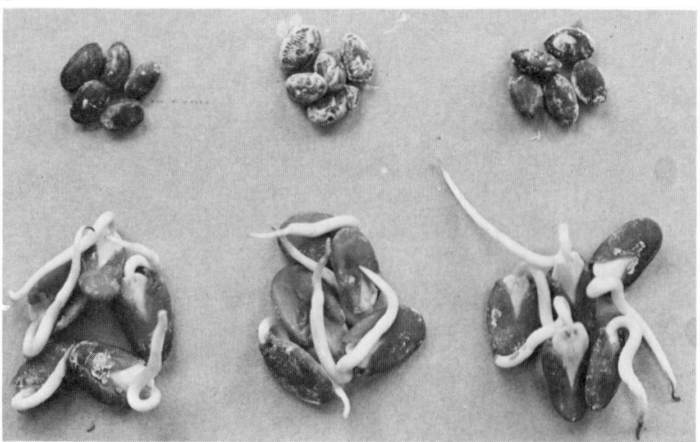

Figure 14–13 Overcoming seed dormancy. Seeds can be stratified in a plastic bag (A). Seeds scarified with acid germinate; untreated seeds remain dormant (B). (Courtesy Washington State University.)

Table 14–5 Treatments necessary to stimulate seed germination of selected woody plants

Plant	Outdoor planting time	Stratification conditions necessary
Apple	Fall	None
	Spring	30–90 days; 32–50°F; sand or peat
Arborvitae	Fall	None
	Spring	30–60 days; 32–50°F; sand or peat
Ash	Fall	None
	Spring	60–90 days; 35–41°F; sand or peat
Barberry	Fall	None
	Spring	15–40 days; 32–41°F; sand or peat
Birch	Fall	None
	Spring	30–60 days; 32–50°F; sand or peat
Blackberry	Plant immediately from the berry	None
Cherry	Fall	None
	Spring	60–170 days; 35–45°F; sand or peat
Chestnut (American)	Fall	None
	Spring	90–120 days; 41°F; sand or peat
Cotoneaster[a]	Fall	None
	Spring	90–120 days at 60–75°F followed by 90–120 days at 41°F; sand or peat
Dogwood[b]	Fall	None
	Spring	60 days at 70–85°F followed by 90–100 days at 41°F; sand or peat
Elm	Fall	None
	Spring	60–90 days; 41°F; sand or peat
Euonymus	Fall	None
	Spring	90–120 days; 32–50°F; sand or peat
Filbert	Fall	None
	Spring	90 days; 41°F; sand
Fir	Fall	None
	Spring	30–90 days; 41°F; sand or peat
Hickory	Fall	None
Hickory	Spring	90–150 days; 32–45°F; sand or peat
Holly	Fall	None (may require 2 years or more for complete germination)
	Spring	60 days at 68–86°F followed by 60 days at 41°F; sand or peat
Honeylocust[c]	Spring	None
Horse chestnut	Fall	None
	Spring	120 days; 41°F; sand
Lilac	Fall	None
	Spring	30–90 days; 41°F; sand
Maple	Fall	None
	Spring	60–120 days; 41°F; sand or peat
Mountain ash	Fall	None
	Spring	60–150 days; 41–50°F; sand or peat
Pear	Fall	None
	Spring	60–90 days; 32–45°F; sand or peat
Pine	Fall	None
	Spring	30–90 days; 32–41°F; sand or peat
Plum	Fall	None
	Spring	60–120 days; 41°F; sand or peat
Poplar	Spring	No treatment necessary
Raspberry	Plant immediately from the berry	
Rose	Plant immediately from rose hip	
Spruce	Fall	None
	Spring	Some may require 30–90 days; 41°F; sand or peat
Walnut	Fall	None
	Spring	90–120 days; 34–41°F; sand or peat
Willow	Sow immediately after collection	
Yew	Fall	None (may require 2 years for complete germination)
	Spring	90–270 days; 37–41°F; sand or peat

(Adapted from *Propagating Plants from Seed*, by F. E. Larsen. Pacific Northwest Cooperative Extension Publication 170. Extension Services of Washington State University, Oregon State University, and University of Idaho. Pullman, WA. 1977.)

[a]Requires scarification treatment: 1½ hours in concentrated sulfuric acid.

[b]Requires mechanical scarification or 1–3 hours in concentrated sulfuric acid.

[c]Requires scarification treatment: 1–2 hours in concentrated sulfuric acid.

Propagating with Cuttings

by F. E. Larsen*

Cuttings are detached vegetative plant parts that will develop into complete new plants by reproducing their missing parts. In accordance with what is likely to work best for each plant, cuttings might be made from stems, roots, or leaves (see Figure 14–14).

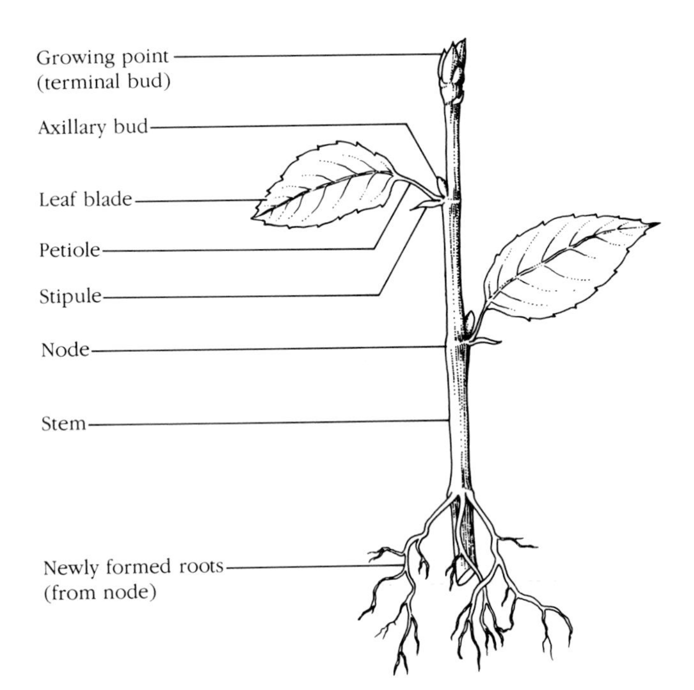

Figure 14–14 Rooted stem cutting. Both leaves and new roots normally arise from the nodes (enlarged areas on the stem). (Courtesy Washington State University.)

Types of Woody Plant Cuttings

Cuttings of deciduous plants are usually made from stem sections or tips one year or less in age, evergreens usually just from stem tips. The basal part of a cutting is sometimes older wood.

Tip cuttings are probably the most common type for use with deciduous plants during the growing season; they generally do not give the best results any other part of the year. The tip section of a shoot is more subject to winter cold damage, may have flower buds rather than shoot buds, and may not have the proper internal nutritional and hormonal balance for good rooting during the dormant season. Simple or straight cuttings, starting 8 to 10 inches (20 to 25 cm) from the shoot tip, are usually more satisfactory for dormant cuttings.

For evergreen plants tip cuttings are the most common type and generally give satisfactory results. They are cut about 4 to 10 inches long from stem tips, using stems 1 year or less in age. Tip cuttings can be made from the main shoot or long side branches. Large cuttings produce a usable plant in shorter time than small cuttings but may require more care while rooting.

Simple or straight cuttings are made from long, 1-year-old shoots that can be cut into sections. This is the most common type of cutting for propagating dormant cuttings of deciduous plants. It might occasionally be used for broadleaved evergreens.

Heel cuttings are made from side shoots produced on stems 2 or more years in age. To make the cuttings, pull the side shoots from the main stem. Pull directly away from the tip end of the main stem. This usually leaves a "heel" of older, main-stem tissue attached to the basal end of the side shoot. The heel cutting can also be made by cutting the heel portion from the main stem with a knife.

Mallet cuttings are similar to heel cuttings, but they include a complete cross-section of the older, main stem at the base of the side shoot. This requires the use of a knife or, better yet, a pair of small pruning shears.

Some evergreens may root better from heel or mallet cuttings because these plants normally develop root **primordia** (specialized cells that will develop into roots) in older stems. These root primordia remain dormant until the stem bends naturally to the moist soil or until the stem is cut from the plant and placed in a rooting

---

*Adapted from *Herbaceous Plants from Cuttings,* by W. E. Guse and F. E. Larsen. Pacific Northwest Cooperative Extension Publication 151; and from *Propagating Deciduous and Evergreen Shrubs, Trees and Vines with Stem Cuttings,* by F. E. Larsen and W. E. Guse. Pacific Northwest Cooperative Extension Publication 152. Extension Services of Washington State University, Oregon State University, and the University of Idaho. Pullman, WA. 1975.

medium. Some deciduous plants also produce root primordia.

Figure 14–15 illustrates woody plant cuttings.

Types of Herbaceous Plant Cuttings

Stem cuttings are the most frequently used type of cutting for herbaceous plants. Leaves or parts of leaves are also used for propagating herbaceous plants. Leaves of some plants form roots but will not produce new shoots. To propagate these, you must use a leaf-bud cutting, which includes the leaf plus an axillary bud and a portion of the stem; the shoots of the new plant will then form from the axillary bud.

Factors Affecting Rooting

Time of year cuttings are taken may affect rooting of woody plants. Best results can be expected from cuttings of many deciduous plants that are taken from late fall to early winter before there has been enough cold weather to complete the rest requirements of the leaf buds. This allows the cutting to be rooted under warm conditions without the development of leaves. After rooting has started, however, the cuttings must be subjected to cold temperatures according to the individual plant's requirements. Some deciduous plants can only be rooted from leafy softwood or semihardwood cuttings taken during the growing season. Others root readily almost any time of the year.

Best results can be expected of cuttings of many narrow-leaved evergreens that are taken from late fall to late winter. Exposure of the mother or stock plant to cold temperatures prior to taking the cuttings stimulates rooting. But cuttings of broadleaved evergreen plants usually root best if you take them during the growing season after a flush of growth, when the wood is partially matured. Some plants root readily almost any time of the year.

Softwood cuttings are taken during the growing season from new growth that has not matured or hardened significantly. When the wood is partially matured, they are called **semihardwood** cuttings. Those taken during

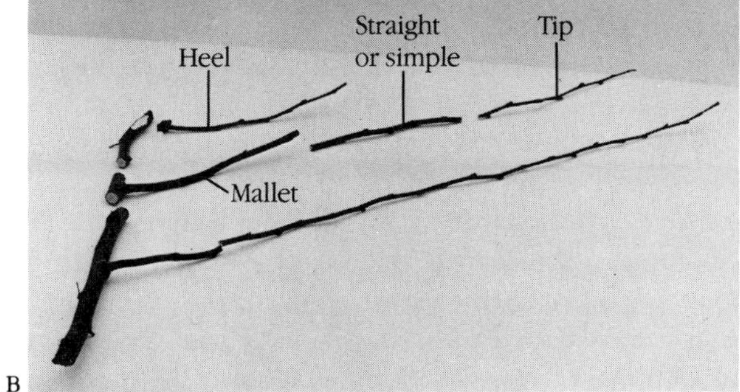

Figure 14–15 Woody plant cuttings. (Courtesy Washington State University.)
(A) Several types of hardwood cuttings made from a narrow-leaved evergreen plant. Tip cuttings might also be made from very long side shoots.
(B) Several types of hardwood cuttings made from a deciduous plant.

the subsequent dormant season when the wood is mature and hardened or from wood older than one year are called **hardwood cuttings.**

Herbaceous plants are most easily propagated in the spring when a natural increase in the rate of growth is occurring. They can, however, be propagated successfully throughout the year as long as succulent plant tissue is available and proper environmental conditions are provided.

Age of stock plant may be an important factor with hard-to-root plants. Cuttings from young seedling plants may root better than those from older plants. Chances of rooting cuttings from large old trees or shrubs may not be very good unless they are easy-to-root types.

Wounding the basal end of the cutting often stimulates rooting of such evergreen plants as rhododendrons and junipers, especially if the cutting has older wood at its base. Slight wounding is done by using the tip of a sharp knife to make a 1- to 2-inch (2½ to 5 cm) vertical cut down each side of the base of the cutting. Stripping off the lower side branches of the cutting during its preparation can also be considered as a slight wounding (see Figure 14–16). For more severe wounding on difficult-to-root types or larger diameter cuttings make several vertical cuts. Wounding can also be done by removing a thin slice of bark down one or both sides of the base of the cutting. Expose the cambium but avoid cutting very deeply into the wood.

Wounding may stimulate rooting by promoting cell division and more absorption of water or applied root-promoting chemicals, or it may remove tough tissue that is a barrier to outward root growth from the cutting. Wounding is most often used on evergreen plants, but it may at times be useful on deciduous plants.

Physical condition of the stock will affect the rooting of cuttings. Cuttings taken during the growing season often root poorly if they are from rapidly growing, succulent shoots. Instead, take them after growth has stopped and the wood has begun to harden; otherwise, many may rot. Shoots that have grown very little also root poorly. Neither type of shoot has the optimum physical condition and nutritional balance for the best rooting.

Figure 14–16 Wounding. A narrow-leaved evergreen cutting can sometimes be induced to root more easily when a thin layer of bark is sliced or scraped from its base. (Courtesy Washington State University.)

Too much or too little fertilizer on the stock plants may hinder rooting ability of the cuttings because of its effect on growth and internal nutritional balance. Because of better physical condition and nutritional balance lateral shoots may root better than terminal shoots from the same plant. Similarly, where a very long shoot can be made into several cuttings, sections from the central part of the shoot may root better than those from either end.

Shoots with flower buds or flowers may not root as well as shoots that are strictly vegetative in some hard-to-root plants. In such plants removal of flower buds sometimes helps rooting.

Take both woody and herbaceous cuttings only from healthy plants that are free of insects, disease, and nutritional disorders.

Plant species may influence rooting. Many deciduous and evergreen plants are propagated from hardwood cuttings, but they vary considerably in ease of rooting. Honeysuckle, currant, grape, and willow root readily. Apple and pear are more difficult, and cherry and lilac are usually very difficult to root using hardwood cuttings.

Among evergreen plants, false cypress, arborvitae, and low-growing juniper generally root readily. Yew roots fairly well. The upright junipers, spruces, and hemlocks are difficult to root. Cuttings of firs and pines are usually very difficult to root.

There is also considerable variation among species within these groups. Even genetic variability from plant to plant may affect ease of rooting.

Environmental conditions, number of leaves remaining on the cutting, internal nutritional and chemical factors, and synthetic hormones all affect rooting (see Chapter 4).

Procedures for Making and Handling Cuttings

Cuttings of Woody Evergreen and Leafy Deciduous Plants. Remove from the parent plant a portion of stem 4 to 8 inches (10 to 20 cm) long with the leaves attached. For most deciduous plants, a tip, simple, or straight cutting will suffice. For most evergreen plants use tip or heel cuttings. Snip off leaves (or needles) that would be in contact with the rooting medium (the bottom 1½ to 2 inches of stem) to prevent rotting of these leaves. The remaining leaves will continue to produce substances that aid in root formation on the cutting. If hardwood cuttings of evergreen plants are used, wound the base of the cutting by one of the methods described. Use the more severe methods of wounding for hard-to-root types.

Spread a small amount of auxin compound on waxed paper or in a clean dish. Dip the base (cut end) of the cutting in the powder so that some adheres to the cut surface and wounded areas. Discard leftover powder to prevent contamination. Talc preparations lose their effectiveness after about eight months, even if kept in a closed container and refrigerated.

Make a hole in the rooting medium so that the powder is not scraped off when you insert the cutting. Insert the base of the cutting into the prepared hole in the rooting medium. If bottom heat is used, insert the base of the cutting nearly to the bottom of the container so that it is close to the heat source. Firm the rooting medium around the base of each cutting. After all cuttings are inserted and firmed in place, apply sufficient water to the rooting medium to settle it around the cuttings. This "watering-in" procedure will leave the rooting medium in close contact with the base of each cutting.

Place the cover over the propagation box (Figure 14–17) or container and keep in a moderately warm room. Most leafy cuttings do best at temperatures of 60–70°F (16° to 21°C). Inspect the cuttings daily and remove any leaves that have fallen. Syringe the tops of the cuttings and keep the rooting medium moist. When the cuttings resist a slight tug and begin to feel anchored, they are beginning to root. Some types may require two to three months or more to form sufficient roots to allow removal from the rooting medium.

When the cuttings have two or three roots about one-half inch long, place them in pots about 4 inches in diameter. Use a good potting soil. Because the cuttings have been accustomed to the humid atmosphere of the propagating box (or mist, if used), accustom them to the "outside" atmosphere by gradually aerating the propagation box (or reducing the mist) before potting. Another way is to cover the potted cutting with perforated plastic film for about a week after potting; this is called hardening off.

After potting do not expose the cuttings to direct sun-

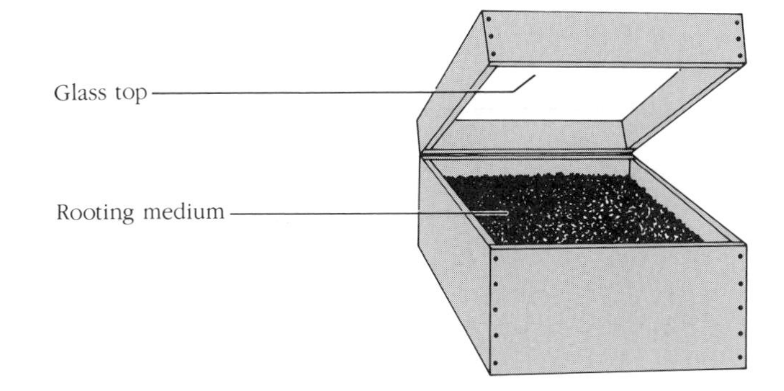

Figure 14–17 Propagation chamber. A simple propagation chamber can be made from two boxes. Top box has glass cover to admit light (cover could also be of plexiglass). When closed, interior of box becomes humid, giving effect similar to mist chamber. Provide drainage in bottom box to prevent waterlogging. For small number of cuttings same effect can be obtained by putting plastic bag over a flower pot. (Courtesy Washington State University.)

light or temperature extremes until they have had several weeks to become accustomed to outdoor conditions.

Hardwood Cuttings of Deciduous Plants. Many of the techniques for rooting hardwood cuttings (Figure 14–18) are similar to those just described for rooting leafy deciduous and evergreen cuttings. However, these cuttings may not need a covered propagation box or mist unless they require a long time to root, during which leaves develop from dormant buds. Easy-to-root types can be taken in the fall and rooted outdoors in the soil in mild climates, or they can be taken in the spring if winters are cold. Take difficult-to-root types in late fall. They will require treatment in a moist, warm (60°–70°F) rooting medium until rooting begins. This is followed by holding in cool (40°–45°F; 4°–7°C) moist storage until spring weather allows outdoor planting. Wounding and treatment with root-promoting chemicals may be of value.

Cuttings handled in this way are often tied in bundles with the basal ends all in the same direction. The bundle is plunged into the rooting or storage medium. When the root initiation and storage period are over, the bundles are untied and the cuttings are individually planted, usually outside in a closely spaced nursery row.

A

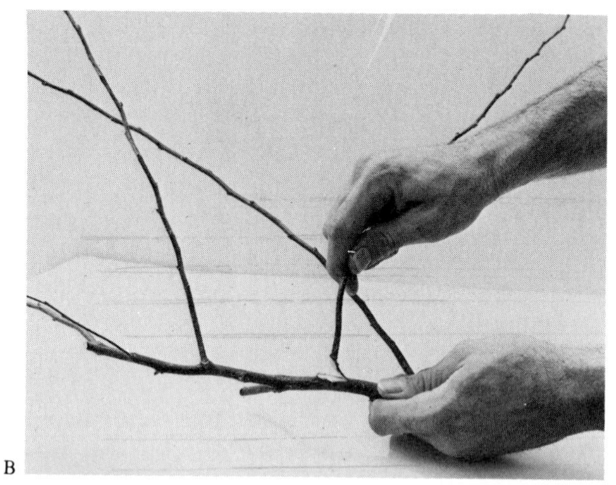

B

C

D

E

Figure 14–18 Procedures for making and handling cuttings. (Courtesy Washington State University.)

(A) Making tip and straight cuttings from a one-year whip of a deciduous plant.

(B) Making heel cuttings from a deciduous plant.

(C) Treating the basal end of a simple or straight cutting with a root-promoting chemical.

(D) Inserting a cutting into the rooting medium.

(E) Rooted hardwood cuttings of deciduous plants. Both shoots and roots developed while the cutting was in the rooting medium.

Figure 14–19 Making an herbaceous stem cutting. (Courtesy Washington State University.)

(A) Cuttings are taken below a node from a healthy, vigorous plant.

(B) A small amount of rooting hormone is applied to the cut surface of the cutting. Excessive amounts of hormone may inhibit rooting.

(C) A furrow or hole is made in the rooting medium, and the cutting is inserted. The medium is then pressed firmly about the cutting.

Herbaceous Plant Cuttings. Remove from the parent plant a portion of stem 3 to 5 inches (8 to 13 cm) long with the leaves attached. Make a clean cut or break just below a node of the donor plant; nodes contain actively dividing cells and are the areas where root formation is likely to occur most readily.

Snip off leaves and stipules from the bottom 1½ inches of stem, as they would be in contact with the rooting medium and would rot. The remaining leaves will continue to produce substances that aid in root formation on the cutting.

Rooting hormone may be applied to the base of the cutting. Then place the base of the cutting in firm contact with a moist, warm rooting medium (see Figure 14–19). After roots form, transplant the cutting to a permanent pot (Figure 14–20).

Figure 14–20 Rooted cuttings ready for transplanting—(left to right) peperomia, carnation, coleus.

Cuttings of plants that exude a sticky sap—such as geraniums and cacti—will do better if the cut ends are allowed to dry a few hours before being placed in the rooting medium. This allows the wounded tissues to dry and helps prevent the entrance of disease organisms.

Some plants can be propagated from a single leaf (Figure 14–21). Generally, you can place the petiole of the leaf into the rooting medium just like the stem of a stem cutting. Roots and shoots form at the base of the petiole.

Leaves of such plants as begonia, bryophyllum, and jade plant are laid flat with their lower surface in firm contact with the rooting medium. Roots and shoots form from the leaf, which eventually decays.

The swordlike leaves of sansevieria can be cut into cross sections. Place the base of each section in the rooting medium. In a similar way leaves of many begonias can be cut into pie-shaped sections. Shoots and roots form from the basal end of the cutting. Figure 14–22 shows some leaf cuttings.

Tables 14–6, 14–7, and 14–8 list selected plants that can be propagated from cuttings.

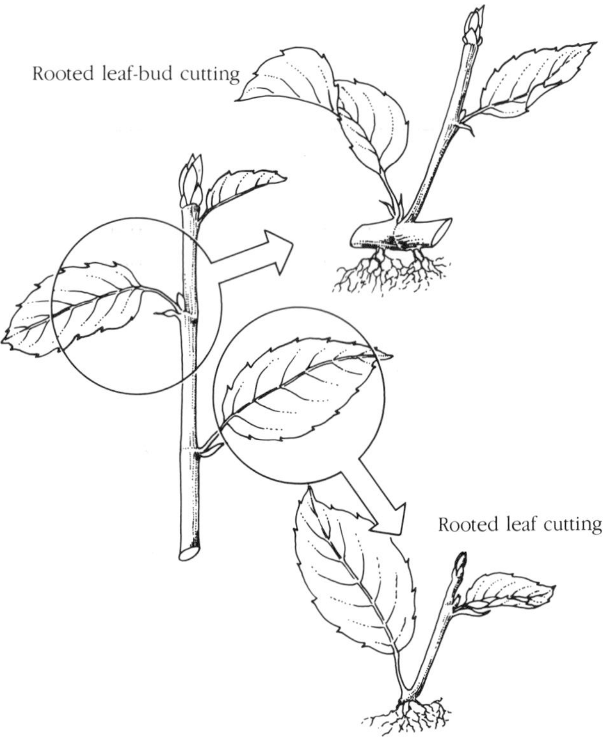

Figure 14–21 Propagation from leaf cuttings. A single leaf is used for leaf cuttings. A leaf-bud cutting includes a portion of the stem and an axillary bud; the new shoots arise from this bud. (Courtesy Washington State University.)

Figure 14–22 Leaf cuttings. (A, left to right) Peperomia leaf with roots formed from the petiole, bryophyllum leaf with new plants forming in the notches of the leaf edge, unrooted jade leaf. (B) Whole-leaf cutting of begonia showing root formation from both the petiole and at cuts through the large veins on the underside. (A, courtesy Washington State University.)

Table 14-6 Selected woody evergreen plants that might be propagated from stem cuttings[a]

Plant		Favorable time to take cuttings	Kind of cutting[b]	Rooting success expected
Common name	Scientific name			
Arborvitae	*Thuja occidentalis* (American)	Summer Winter	Semihardwood Hardwood	High
	Thuja orientalis (Oriental)	Late spring	Softwood	Low
Azalea	*Rhododendron* spp.	Summer	Semihardwood	Moderate to high
Barberry	*Berberis* spp.	Summer Fall	Softwood Hardwood	Moderate to high
Boxwood	*Buxus* spp.	Spring, summer, fall	Softwood or semihardwood	High
Camellia	*Camellia* spp.	Summer	Semihardwood	Moderate to high
Cedar	*Cedrus* spp.	Late summer or fall	Semihardwood	Low
Chamaecyparis (false-cypress)	*Chamaecyparis* spp.	Late fall or winter	Hardwood	Moderate to high
Cryptomeria	*Cryptomeria japonica*	Summer	Semihardwood	Slow to root
Daphne	*Daphne* spp.	Summer	Semihardwood	Moderate
Euonymus	*Euonymus* spp.	Summer	Semihardwood	High
Fir	*Abies* spp.	Winter	Hardwood	Low
Heath	*Erica* spp.	Summer Winter	Semihardwood Hardwood	High
Heather	*Calluna vulgaris*	Summer Winter	Semihardwood Hardwood	High
Hemlock	*Tsuga* spp.	Fall Winter	Semihardwood Hardwood	Low
Holly	*Ilex* spp.	Summer Winter	Semihardwood Hardwood	Moderate to high
Ivy	*Hedera helix*	Summer	Semihardwood	High
Juniper	*Juniperus* spp.	Summer Winter	Semihardwood Hardwood	Upright types may root poorly
Laurel (cherry)	*Prunus laurocerasus*	Summer Winter	Semihardwood Hardwood	High
Laurel (mountain)	*Kalmia latifolia*	Winter	Hardwood	Low
Madrone (Pacific)	*Arbutus menziesii*	Fall	Semihardwood	Moderate
Magnolia	*Magnolia* spp.	Summer	Softwood or semihardwood	Moderate to high
Oleander	*Nerium oleander*	Summer	Semihardwood	High
Oregon grape	*Mahonia aquifolium*	Summer	Semihardwood	Moderate to high
Pachistima	*Pachistima canbyi*	Summer	Softwood	Moderate to high
Pachysandra (spurge)	*Pachysandra terminalis*	Summer	Semihardwood	High
Pieris	*Pieris* spp.	Summer	Semihardwood	Moderate to high
Pine	*Pinus* spp.	Winter	Hardwood	Low
Privet	*Ligustrum* spp.	Summer Winter	Softwood Hardwood	Low to high
Pyracantha	*Pyracantha* spp.	Summer	Semihardwood	High
Rhododendron	*Rhododendron* spp.	Summer	Softwood or semihardwood	Low to high
Spruce	*Picea* spp.	Winter	Hardwood	Low
Viburnum	*Viburnum* spp.	Summer	Semihardwood	Moderate to high
Yew	*Taxus* spp.	Fall Winter	Semihardwood Hardwood	Moderate

(Adapted from *Propagating Deciduous and Evergreen Shrubs, Trees and Vines with Stem Cuttings*, by F. E. Larsen and W. E. Guse. Pacific Northwest Cooperative Extension Publication 152. Extension Services of Washington State University, Oregon State University, and University of Idaho. Pullman, WA. 1975.)

[a]Some plants listed also have closely related deciduous types that might be handled differently.

[b]Good results can usually be obtained with tip cuttings of most evergreen plants. The use of heel or mallet cuttings might be desirable with juniper and yew.

Table 14–7 Selected woody deciduous plants that might be propagated from stem cuttings[a]

Plant		Favorable time to take cuttings	Kind of cutting[b]	Genera in which some species are difficult to root
Common name	Scientific name			
Alder	*Alnus* spp.	Winter	Hardwood	
Azalea	*Rhododendron* spp.	Summer	Softwood	X
Barberry	*Berberis* spp.	Summer Winter	Softwood or semihardwood Hardwood	
Bittersweet	*Celastrus* spp.	Summer Winter	Softwood or semihardwood Hardwood	
Blueberry	*Vaccinium* spp.	Summer Winter	Softwood Hardwood	
Boston ivy	*Parthenocissus tricuspidata*	Summer Winter	Softwood Hardwood	
Bottlebrush	*Callistemon* spp.	Summer	Semihardwood	
Boxwood	*Buxus* spp.	Summer Winter	Softwood or semihardwood Hardwood	
Broom	*Cytisus* spp.	Summer Winter	Semihardwood Hardwood	
Butterfly bush	*Buddleia* spp.	Summer	Softwood or semihardwood	
Catalpa	*Catalpa* spp.	Summer	Softwood	
Ceanothus	*Ceanothus* spp.	Summer Winter	Softwood or semihardwood Hardwood	
Cherry	*Prunus* spp.	Summer	Softwood or semihardwood	X
Clematis	*Clematis* spp.	Summer	Softwood or semihardwood	
Cotoneaster	*Cotoneaster* spp.	Summer	Softwood or semihardwood	
Crabapple	*Malus* spp.	Summer Late fall	Softwood or semihardwood Hardwood	X
Currant	*Ribes* spp.	Summer Winter	Softwood Hardwood	
Deutzia	*Deutzia* spp.	Summer Winter	Softwood Hardwood	
Dogwood	*Cornus* spp.	Summer	Softwood or semihardwood	X
Elderberry	*Sambucus* spp.	Summer	Softwood	
Elm	*Ulmus* spp.	Summer	Softwood	
Euonymus (spindle tree)	*Euonymus* spp.	Winter	Hardwood	
Forsythia	*Forsythia* spp.	Summer Winter	Softwood Hardwood	
Fringe tree	*Chionanthus* spp.	Summer	Softwood	X
Ginkgo (maidenhair)	*Ginkgo biloba*	Summer	Softwood	
Goldenrain tree	*Koelreuteria* spp.	Summer	Softwood	
Grape	*Vitis* spp.	Summer Winter	Softwood Hardwood	
Hawthorn	*Crataegus* spp.	Summer Winter	Softwood Hardwood	
Hibiscus (rose mallow)	*Hibiscus* spp.	Summer Winter	Softwood or semihardwood Hardwood	
Honeylocust	*Gleditsia triacanthos*	Winter	Hardwood	
Honeysuckle	*Lonicera* spp.	Summer Winter	Softwood Hardwood	

Plant		Favorable time to take cuttings	Kind of cutting[b]	Genera in which some species are difficult to root
Common name	Scientific name			
Hydrangea	*Hydrangea* spp.	Summer Winter	Softwood Hardwood	
Jasmine	*Jasminum* spp.	Summer Winter	Semihardwood Hardwood	
Lilac	*Syringa vulgaris*	Summer	Softwood	X
Locust (black)	*Robinia pseudoacacia*	Summer	Semihardwood	
Maple	*Acer* spp.	Summer	Softwood	X
Mock orange	*Philadelphus* spp.	Summer Winter	Softwood Hardwood	
Magnolia	*Magnolia* spp.	Summer	Softwood or semihardwood	
Mulberry	*Morus alba*	Summer	Softwood	
Peach	*Prunus* spp.	Summer	Softwood or semihardwood	
Pear	*Pyrus* spp.	Late fall	Hardwood	X
Plum	*Prunus* spp.	Summer	Softwood or semihardwood	
Poplar	*Populus* spp.	Summer Winter	Softwood Hardwood	
Quince (flowering)	*Chaenomeles* spp.	Summer Winter	Semihardwood Hardwood	
Redbud	*Cercis* spp.	Summer	Softwood	X
Rose	*Rosa* spp.	Summer Winter	Softwood or semihardwood Hardwood	
Russian olive	*Elaeagnus angustifolia*	Winter	Hardwood	
St. John'swort	*Hypericum* spp.	Summer	Semihardwood	
Serviceberry	*Amelanchier alnifolia*	Summer	Softwood	
Smoke tree	*Cotinus coggygria*	Summer	Softwood	
Snowberry	*Symphoricarpos* spp.	Summer Winter	Softwood Hardwood	
Spiraea	*Spiraea* spp.	Summer Winter	Softwood or semihardwood Hardwood	X
Sumac	*Rhus* spp.	Summer	Softwood	X
Sweetgum	*Liquidambar* spp.	Summer	Softwood	
Tulip tree	*Liriodendron tulipifera*	Summer	Softwood	
Viburnum	*Viburnum* spp.	Summer Winter	Softwood or semihardwood Hardwood	X
Virginia creeper	*Parthenocissus quinquefolia*	Summer Winter	Softwood Hardwood	
Weigela	*Weigela* spp.	Summer Winter	Softwood or semihardwood Hardwood	
Willow	*Salix* spp.	Summer Winter	Softwood or semihardwood Hardwood	
Wisteria	*Wisteria* spp.	Summer Winter	Semihardwood Hardwood	

(Adapted from *Propagating Deciduous and Evergreen Shrubs, Trees and Vines with Stem Cuttings*, by F. E. Larsen and W. E. Guse. Pacific Northwest Cooperative Extension Publication 152. Extension Services of Washington State University, Oregon State University, and University of Idaho. Pullman, WA. 1975.)

[a]Some plants listed also have closely related evergreen types that may be handled differently.

[b]In general, use tip cuttings for those taken during the growing season (softwood or semihardwood) and simple (straight) cuttings for the leafless dormant type of cutting. Heel or mallet cuttings might be used for quince, which may have preformed root initials in 2-year-old wood.

Table 14–8 Selected herbaceous plants that can be propagated from cuttings

Plant		Type of cutting	Approximate time to root (weeks)[a]
Common name	Scientific name		
African violet	*Saintpaulia* spp.	Leaf	3–4
Aluminum plant	*Pilea* spp.	Stem	2–3
Aloe	*Aloe* spp.	Leaf	4–6
Aphelandra	*Aphelandra* sp.	Stem	2–3
Arrowhead plant	*Syngonium albolineatum*	Stem	2–3
Begonia	*Begonia* spp.	Stem (fibrous rooted); whole leaf or leaf section (Rex)	4–5
Cactus	*Cephalocereus senilis*	Stem	3–4
	Opuntia microdasys	Stem	3–4
Chrysanthemum	*Chrysanthemum* spp.	Stem	1–2
Carnation	*Dianthus* spp.	Stem	2–3
Coleus	*Coleus blumei*	Stem	1–2
Crown of thorns	*Euphorbia splendens*	Stem	4–5
Dahlia	*Dahlia* spp.	Stem or leaf-bud	3–4
Dieffenbachia (dumbcane)	*Dieffenbachia* spp.	Stem	4–6
Dracaena	*Dracaena* spp.	Stem	3–4
Echeveria	*Echeveria* spp.	Leaf or stem	4–6
Euphorbia	*Euphorbia* spp.	Stem	4–6
Fittonia	*Fittonia* spp.	Stem	2–3
Fuchsia	*Fuchsia* spp. (also hybrids)	Stem	1–2
Geranium	*Pelargonium* spp.	Stem	1–2
Hoya	*Hoya* spp.	Stem	3–4
Hydrangea	*Hydrangea* spp.	Stem	2–3
Impatiens	*Impatiens* spp.	Stem	2–3
Ivy	Several genera and species	Stem	2–3
Jade	*Crassula* spp.	Stem or leaf	4–5
Kalanchoe (bryophyllum)	*Kalanchoe* spp.	Stem or leaf	4–5
Lantana	*Lantana* sp.	Stem	3–4
Monstera (split-leaf philodendron)	*Monstera deliciosa*	Stem	4–5
Mint	*Mentha* spp.	Stem	2–3
Peperomia	*Peperomia* spp.	Leaf, leaf-bud, or stem	4–6
	P. obtusifolia	Leaf-bud or stem works best	4–6
	P. obtusifolia variegata	Leaf-bud or stem works best	4–6
Periwinkle (myrtle)	*Vinca* spp.	Stem	3–4
Petunia	Petunia hybrids	Stem	2–3
Philodendron	*Philodendron* spp.	Stem	2–4
Piggyback plant	*Tolmiea menziesii*	Leaf with plantlet	3–4
Pothos	*Scindapsus aureus*	Stem	2–3
Poinsettia	*Euphorbia pulcherrima*	Stem	2–3
Sansevieria (snake plant)	*Sansevieria* spp.	Leaf, leaf section	4–6
Velvet plant	*Gynura* spp.	Stem	1–2
Wandering Jew	*Tradescantia* spp.	Stem	2–3
	Zebrina spp.		

(Adapted from *Herbaceous Plants from Cuttings*, by W. E. Guse and F. E. Larsen. Pacific Northwest Cooperative Extension Publication 151. Extension Services of Washington State University, Oregon State University, and University of Idaho. Pullman, WA. 1975.)

[a]The indicated time for rooting is only approximate and may be longer under some conditions. Where new shoots must develop in addition to roots, the time required for shoot development is often longer.

Layering to Renew or Multiply Plants

by F. E. Larsen*

Layering, or layerage, is the process of forming roots on a stem that is still attached to a plant. Some plants reproduce themselves naturally by layering.

Many plants can be propagated in limited numbers by layering. Layering does not require the skilled techniques necessary for grafting or the close attention to environment necessary for rooting cuttings. It is an ideal method for the home gardener to renew an old plant or to produce a limited number of new plants from an existing one. A layer (the stem on which roots are formed) is supported by the parent plant, from which it draws water and nutrients during root formation.

Tip Layering

Tip layering is not adapted to as many kinds of plants as are some other methods of layering. It is used primarily for black and purple raspberries and some blackberries that are commercially propagated and may reproduce naturally by tip layering.

In early summer when new canes are 24 to 30 inches (60 to 75 cm) high, remove the cane (shoot) tips. This stimulates growth of lateral shoots from the main shoot. The laterals are used for tip layering when they are long enough to reach the ground (about late July or early August) and the tips have small, curled leaves giving them a rat-tailed appearance.

Insert the lateral shoot tips vertically upside-down into a 4- to 6-inch (10 to 15 cm) hole in the soil next to the parent plant. Firm the soil around the tip. Keep the soil moist (see Figure 14–23).

By fall the tips will form a good root system. Cut the lateral cane from the parent plant 6 to 8 inches (15–20 cm) above the soil where the tip was buried. The rooted tip can be dug and replanted the same fall or the following spring. The new plant is composed of the rooted tip

Figure 14–23 Tip layering. Some raspberries and blackberries can be tip layered by inserting the stem tip into a hole 6 or 8 inches (15 to 20 cm) deep. The stem should be cut about 6 inches above the ground when the tip is rooted. (Courtesy Washington State University.)

*Adapted from *Layering to Renew or Multiply Plants,* by F. E. Larsen. Pacific Northwest Cooperative Extension Publication 165. Extension Services of Washington State University, Oregon State University, and University of Idaho. Pullman, WA. 1977.

with a terminal shoot bud, a good root system, and the section of cane from the parent plant. When growth of the layer resumes, the terminal bud or another bud turns up, pushes out of the soil, and produces a new shoot oriented in the proper direction.

Simple Layering

This method of layering can be done with many deciduous and evergreen plants. With deciduous plants simple layering is usually done in the spring, using long, low branches that were produced during the previous growing season; however, layering later in the season with current-season shoots is also done.

One-year-old branches or shoots can also be used for broadleaved evergreen plants such as magnolia or rhododendron. Layering of broadleaved plants is more often done during the growing season, using shoots of current-season growth when they are long enough and after they have matured to the point where they will break when bent sharply. Simple layering can also be done on low-growing, narrow-leaved evergreens such as yew and juniper.

To layer deciduous and broadleaved evergreen plants, bend a branch to the ground into a hole or trench about 6 inches deep. Bend the branch up sharply about 12 inches (30 cm) from the tip and cover the bent portion with soil, leaving about 6 inches of the tip exposed. Remove leaves that will be covered with soil. In some cases, you will need to place a wire loop or wooden peg over the lowest point of the bend to hold the branch in the soil. You may have to stake the protruding shoot tip to hold it upright (Figure 14–24).

Rooting may be stimulated by a shallow cut or notch in the underside of the shoot at the point of the bend. This also makes bending easier on very stiff stems. A root-promoting chemical such as Rootone might be applied to the cut. Rooting may also be improved by twisting the stem in the region of the bend to loosen the bark slightly or by placing a tight wire around the stem just behind the bend, between the bend and the parent plant. Rooting is also encouraged by keeping the soil moist where the stem is covered (see Figure 14–25).

Figure 14–24 Simple layering. Simple layering is accomplished by bending a long stem (still attached to the mother plant) in a **U**-shape in the bottom of a hole. Use a wire loop or forked wooden stake to secure the bent stem to the ground. (Courtesy Washington State University.)

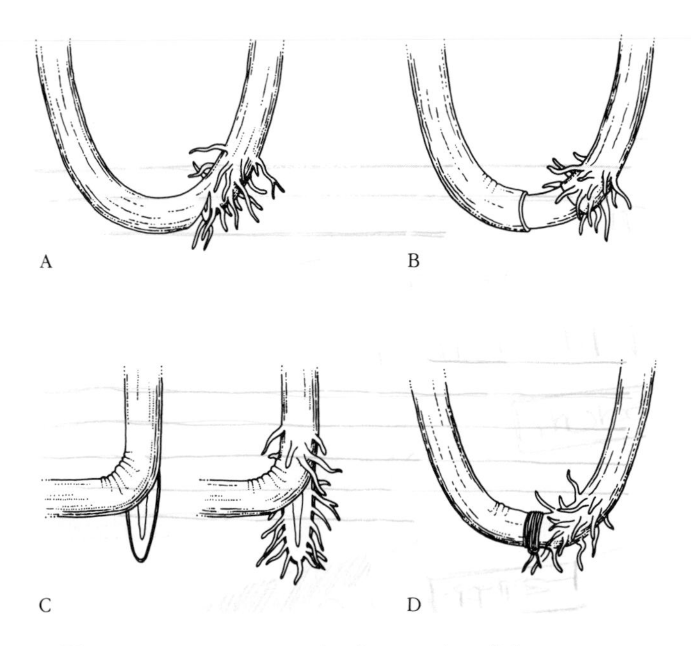

Figure 14–25 Ways to stimulate rooting. (A) stem cut on underside, (B) stem girdled by bark removal, (C) stem cut and twisted, (D) stem girdled by tight wire. (Courtesy Washington State University.)

You can often layer low-growing, narrow-leaved evergreens by covering a branch with 3 or 4 inches (8–10 cm) of soil about 10 to 12 inches (25 to 30 cm) back from the tip. Some, such as 'Tam' juniper, will often layer themselves.

Branches that were layered in the spring should be well rooted by fall. You can sever them from the parent plant after they are dormant. They can be transplanted either that fall or next spring before growth begins. Do not dig those layered during the summer until the following spring.

If the top of the new plant (layer) is large in proportion to the root system, prune the top to reduce its size so that the water demands of the top can be met by the roots. Dig evergreen layers with a ball of soil (about 6 inches in diameter) attached to the roots. Broadleaved evergreens will do best if you pot them after digging and hold them for three or four weeks in a shaded area before planting in a permanent location.

Compound (Serpentine, Multiple) Layering

Compound layering is used on plants with long, flexible vine-type stems, such as grape or clematis. The techniques are similar to simple layering except that the long, flexible stems of these plants allow alternate coverage and exposure along their length.

Treatments such as notching and chemical root promoters applied at the lowest point of each buried part of the stem may aid rooting. See that each exposed part of the stem has one or more leaves attached. This assures that a bud will be present to form a shoot for the new plant.

The stem should root at each buried location. After it has rooted, cut the stem into sections so that each portion contains roots and a node (point of leaf attachment) with a shoot bud (Figure 14–26). Time of layering, digging, and methods of handling are the same as for simple layers.

Mound (Stool) Layering

Many well-established, vigorous shrubs and plants that have stiff branches that do not bend easily to the soil can be propagated by mound layering. This method may be

Figure 14–26 Compound layering. This long flexible stem to be compound layered is alternately covered and exposed. When roots develop, several new plants can be obtained by cutting the stem into sections. (Courtesy Washington State University.)

used commercially to produce rootstocks for fruit and nut trees. It might also be used to propagate trees that produce suckers at their base.

If a well-established shrub can be sacrificed for a year, many new plants can be propagated from it by mound layerage. The entire plant is cut to within 2 to 3 inches (5 to 8 cm) of the ground in early spring before growth starts. When growth begins, the remaining branch stubs of the plant should produce numerous shoots.

When the new shoots are 3 or 4 inches long, mound soil, compost, or shavings around the plant so that the bases of the new shoots are covered to about one-half their length. Add soil two or three more times throughout the growing season as the shoots grow. The final depth of soil should be 8 to 10 inches.

Shoots will be rooted at their base by fall and can be removed and replanted in the fall or the following spring (Figure 14–27). If the shoots are removed in the fall, the original plant should be recovered with soil until spring to protect it from winter damage. More plants can be produced each year for several years after the mounding process is started.

An alternate method for plants that root poorly is to cover completely the cut stubs of the plant in the spring before growth starts so that the stub ends are buried ½ inch deep. The new shoots will be forced to grow through the soil; gradually add soil to the base of the shoots as they grow. Final depth is the same as indicated above. By this method the base of the shoots will never develop a green color (chlorophyll), and the stem tissue will produce roots more easily.

Girdling with wire at the base of the new shoots about 1½ to 2 months after they begin to grow may stimulate rooting of hard-to-root types. Notching or partial cutting of the stem or removal of a ring of bark may also aid rooting (see Figure 14–25).

If established plants are not available to mound layer, a new plant can be planted, allowed to grow for a year, and then handled as outlined above. Even if the plant initially has only one stem, usually it will produce a half dozen or more shoots that can be layered.

Soil to be used for this type of layering should be porous and well drained; sawdust or wood shavings are sometimes used for mounding instead of soil. The

A

B

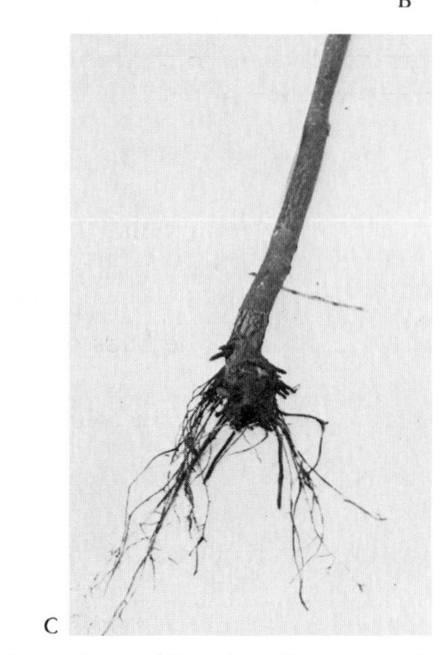
C

Figure 14–27 Mound layering. For mound layering a rooted plant is allowed to grow in place for a year, then is headed back close to the ground (A). The stub is covered with moist soil, shavings, or similar material. Shoots develop from mound-layered plants, and rooting at the base of the shoots results (B). The rooted shoots can be removed and planted (C). (A and B courtesy Washington State University.)

mound must be kept moist, as drying will slow rooting or destroy roots already formed. Do not waterlog the area, however.

Trench Layering

Trench layering is an alternate method to mound layering and is often used by nurseries to propagate rootstocks for fruit and nut trees.

With this method rooted cuttings or liners are planted at an angle of about 30° from horizontal and allowed to grow 1 year. The spring following planting, before growth starts, a 2-inch-deep trench is opened in the direction the plant is leaning. The plant is bent into the bottom of the trench and held against the soil with wire loops or wooden pegs. Weak, thin branches are removed or cut back severely.

Before the buds begin to grow, cover the plant with about 1 inch (2½ cm) of soil, sawdust, or wood shavings. New shoots from buds on the plant will push up through the covering. Add more covering as the shoots elongate. The process from this point is like that for mound layering (Figure 14–28).

A similar method, called continuous layering, is often used on ornamental shrubs or other plants with long stems that can be bent to the ground. In this case the long stem is placed in the trench and covered over its entire length except for the tip, which is left exposed. New shoots will develop from the buds along the stem. They should usually be mounded as above to develop a greater area for root formation at the stem bases.

Figure 14–28 Trench layering. To trench layer, set plant at an angle in a shallow trench (A). Remove small, weak branches and head back the others. Wire hoops can be used to hold the plant close to the ground (B). Branch stubs are mounded with a moist medium. New rooted shoots will develop from the stubs. (Courtesy Washington State University.)

Air Layering

This method is used on plants that have stiff or large stems that cannot be bent to the soil or for plants that do not readily produce shoots at their base. It is a good method to use on such ornamental house plants as fig, rubber plant, or dieffenbachia, but it is adapted to many other plants as well.

The layer should usually be made on a portion of a branch that is 1 year old. This means that the layer will be made within about 1 foot of the end of a branch.

First, either girdle the stem at this point by removing a ½-inch-wide ring of bark or make a slanting cut about half way through the stem on the bottom side. Apply a root-promoting chemical to the girdled area or force the root promoter into the knife cut. It may also help to force a toothpick into the knife cut to hold it open and prevent healing of the cut.

The final step is to place a handful of moist (not saturated) sphagnum or peat moss around the treated area. Hold this in place by a wrap or two of clear polyethylene film. Tie the polyethylene to the stem above and below the treated area. Add moisture periodically as needed at the top of the film after loosening the tie. Do not overwater, as excess moisture will promote decay (Figure 14–29).

Figure 14–29 Completed air layer on a rubber plant. (Courtesy Washington State University.)

Figure 14–30 Uncovered air layer, showing several large roots that developed from wounded area. (Courtesy Washington State University.)

With some plants under some conditions producing a good root system on an air layer may require as long as a year, although usually it will be more rapid. When you see a mass of roots through the polyethylene film, remove the ties and film. The stem can be cut off below the rooted area (Figure 14–30). If possible, do this when the plant is not in active growth.

Pot the new plant immediately. Handle carefully, because there is usually a large top in relation to the size of the root system. Where possible, prune the top to balance more evenly the size of the top to that of the roots. Where this is not possible, particularly with leafy foliage plants, place the newly potted plant in a cool, shaded, humid area until it has time to recover and establish a larger root system. Frequent syringing or misting, or covering with a plastic bag (left open at the bottom) will help aid recovery.

Table 14–9 lists selected plants that might be propagated by layering.

Table 14–9 Selected plants that might be propagated by layering

Plant			
Common name	Scientific name	Type of layerage	Timing
Apple	*Malus* spp.	Mound or trench	Spring
Ash	*Fraxinus* spp.	Simple	Spring, summer
Azalea	*Rhododendron* spp.	Simple, mound	Spring, summer
Barberry	*Berberis* spp.	Simple, mound	Spring
Blueberry	*Vaccinium* spp.	Mound	Spring
Blackberry	*Rubus* spp.	Tip	Summer
Beech	*Fagus* spp.	Simple	Spring
Buckthorn	*Rhamnus* spp.	Simple	Spring
Birch	*Betula* spp.	Simple	Spring, summer
Bittersweet	*Celastrus* spp.	Simple	Spring, summer
Boxwood	*Buxus* spp.	Simple	Summer
Camellia	*Camellia* spp.	Simple	Spring
Ceanothus	*Ceanothus* spp.	Simple	Spring, summer
Chestnut	*Castanea* spp.	Simple	Spring, summer
Clematis	*Clematis* spp.	Simple, compound	Spring, summer
Cotoneaster	*Cotoneaster* spp.	Simple	Summer
Currant	*Ribes* spp.	Mound, simple	Spring
Daphne	*Daphne cneorum*	Simple	Spring, summer
Dogwood	*Cornus* spp.	Simple, continuous	Spring, summer
Dumbcane	*Dieffenbachia* spp.	Air	Anytime
Dracaena	*Dracaena* spp.	Air	Anytime
Euonymus	*Euonymus* spp.	Simple	Spring, summer
Elder	*Sambucus* spp.	Simple	Spring
Fiddle-leaf fig	*Ficus lyrata*	Air	Anytime
Filbert	*Corylus* spp.	Simple	Summer
Forsythia	*Forsythia* spp.	Tip, simple	Summer
Grape	*Vitis* spp.	Simple, compound	Spring
Gooseberry	*Ribes* spp.	Mound	Spring
Heather	*Calluna vulgaris*	Simple	Summer
Hemlock	*Tsuga* spp.	Simple	Spring
Hibiscus, Chinese	*Hibiscus rosa-sinensis*	Air	Spring, summer
Holly	*Ilex* spp.	Air, simple	Summer, fall
Honeysuckle	*Lonicera* spp.	Simple	Spring, summer
Horse chestnut	*Aesculus* spp.	Simple	Spring
Hydrangea	*Hydrangea* spp.	Mound	Spring, summer
Ivy	*Hedera helix*	Simple	Spring
Juniper	*Juniperus* spp.	Simple	Summer
Jasmine	*Jasminium* spp.	Simple	Spring
Laburnum	*Laburnum* spp.	Simple	Summer
Laurel (mountain)	*Kalmia latifolia*	Simple	Summer
Laurel (sweet bay)	*Lauris nobilis*	Air	Anytime
Lilac	*Syringa vulgaris*	Air	Spring

(continued)

Table 14-9 *(continued)*

Plant			
Common name	Scientific name	Type of layerage	Timing
Linden	*Tilia americana*	Simple, mound (suckers)	Spring
Magnolia	*Magnolia* spp.	Simple	Spring
Oleander	*Nerium oleander*	Simple	Spring, summer
Philodendron	*Philodendron* spp.	Air	Anytime
Pittosporum	*Pittosporum tobira*	Air	Anytime
Quince	*Chaenomeles* spp.	Mound	Spring
Rhododendron	*Rhododendron* spp.	Trench, simple	Spring, summer
Raspberry, black, purple	*Rubus* spp.	Tip	Summer
Redbud	*Cercis* spp.	Simple, mound	Spring
Rose	*Rosa* spp.	Simple, tip	Spring, summer
Rubber plant	*Ficus elastica decora*	Air	Anytime
Split-leaf philodendron	*Monstera deliciosa*	Air	Anytime
Tree ivy	*Fatshedera lizei*	Air	Anytime
Trumpet creeper	*Campris* spp.	Simple	Summer
Viburnum	*Viburnum* spp.	Simple	Spring, summer
Virginia creeper	*Parthenocissus quinquefolia*	Compound	Summer
Weeping fig	*Ficus benjamina*	Air	Anytime
Weigela	*Weigela* sp.	Simple	Spring
Willow	*Salix* sp.	Simple	Spring
Wisteria	*Wisteria* sp.	Simple	Spring

(Adapted from *Layering to Renew or Multiply Plants,* by F. E. Larsen. Pacific Northwest Cooperative Extension Publication 165. Extension Services of Washington State University, Oregon State University, and University of Idaho. Pullman, WA. 1977.)

Propagating from Fleshy Storage Organs

by F. E. Larsen*

Tunicate Bulbs

There are a number of ways in which tunicate bulbs can be used for propagation.

Offsets. The simplest way to produce more plants from tunicate bulbs is by using **offsets,** small bulbs that form naturally from the parent bulb (Figure 14–31). When the bulb is dug, these can be separated and planted to produce a new plant. Small offsets will usually produce only leaves the first year and will not flower until the second year or later.

In some plants, such as hyacinths, offsets do not readily form, and artificial stimulation is used to induce new bulb formation. Such methods are scooping, scoring, coring, and sectioning (Figure 14–32).

Scooping. Scooping removes the entire basal plate. When properly done, this will remove the main shoot and flower bud at the center of the bulb and expose the bases of the rings of leaves (bulb scales). Small bulblets will then form at the base of these modified leaves.

Dip the bulb in a fungicide after cutting to protect the cut surface. Keep the bulbs in a warm (about 70°F;

*Adapted from *Propagation from Bulbs, Corms, Tubers, Rhizomes, and Tuberous Roots and Stems,* by F. E. Larsen. Pacific Northwest Cooperative Extension Publication 164. Extension Services of Washington State University, Oregon State University, and University of Idaho. Pullman, WA. 1977.

Figure 14–31 Propagation of tunicate bulbs. (A) The tunicate-type onion bulb has formed an offset. (B) A longitudinal section of an onion bulb, showing an early stage in formation of an offset. (Courtesy Washington State University.)

21°C), dark place for about two weeks to dry and form wound tissue on the cut surface. They are usually kept in a shallow box with a wire mesh bottom.

When the scales begin to swell (by the third week), increase the temperature from 70°F to about 85°F (30°C) and hold the relative humidity at about 85%. The bulblets are ready for planting when new roots begin to form.

The first year plant them still attached to the mother bulb. Plant in loose soil with the tip of the mother bulb about 1 inch (2½ cm) deep. They will produce only foliage the first year. When the foliage is yellow and dry, they can be dug and separated. Although scooping will produce 25 to 50 bulblets per mother bulb, these will be small and may require 4 or 5 years of growth before flowering.

Scoring. In scoring three V-shaped cuts are made through the basal plate of the bulb so that there are six pie-shaped sections. Make cuts deep enough to destroy the main growing shoot and reach to just below the widest point of the bulb. Bury cut bulbs upside down in

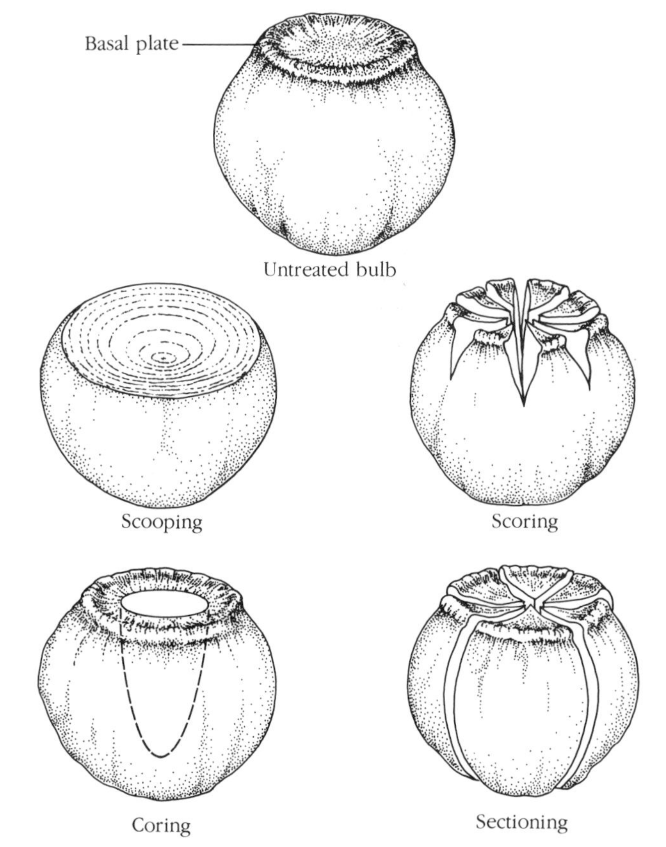

Figure 14–32 Several methods for propagating hyacinth bulbs. (Courtesy Washington State University.)

clean, dry sand about 2 inches (5 cm) below the surface. After wound tissue has formed in the cuts, remove the bulbs from the sand and treat the same as scooped bulbs. Scoring will produce about half as many bulbs as scooping, but they will be larger and should flower in 3 or 4 years.

Coring. Coring removes the center portion of the basal plate and the main growing point of the bulb. Cored bulbs are treated like scored bulbs. They will produce still fewer bulblets, but these will be larger and should flower in 2 or 3 years.

Sectioning. Some bulbs can be cut into sections with a portion of the basal plate attached to each section. Bulblets then form from the basal plate of each cutting. The bulb is cut into five to ten pie-shaped vertical sections. These sections can be further divided by slipping a knife down between each third or fourth pair of leaf-scale rings and cutting through the basal plate. Each bulb section will include a portion of basal plate with segments of three or four leaf scales attached. The length of time required for bulblets formed from sections to flower will depend on the size of the sections and may vary from 2 to 4 or 5 years.

Scaly Bulbs

Offsets. Like tunicate bulbs, scaly bulbs are most easily propagated by offsets. Offset production is often too slow for commercial purposes.

Scaling. Scaly bulbs are often propagated by removing the outer scales from the bulb and planting them one-half their length in a rooting medium. Bulbs for this purpose are best dug after flowering. Plant the scales about 1 inch apart in rows 6 inches (15 cm) apart. New bulblets will form at the base of the scales by fall, when they can be planted out. The mother bulb can be replanted and will regenerate to provide more scales in one or two years. If the bulbs are not dug until fall, the scales can be stored in moist sand until spring and then planted out.

Aerial Bulbs (Bulbils). Some lilies produce tiny aerial bulbs called bulbils where the leaves join the stem. These can be used to produce new plants by separating and growing them for two or three years to flowering size. Aerial bulbs can be induced to form more readily by removing flower buds from the plant.

Underground Stem Bulblets. Bulbs will also form below ground on the flower stalks of some lilies. These can sometimes be induced by pulling the stalk from the plant after it flowers and covering the basal portion with the moist rooting medium. Bulbs formed in this way can be separated and planted out to form new plants. They require 1 to 2 years to reach flowering size.

Figures 14–33 and 14–34 illustrate methods of scaly bulb propagation.

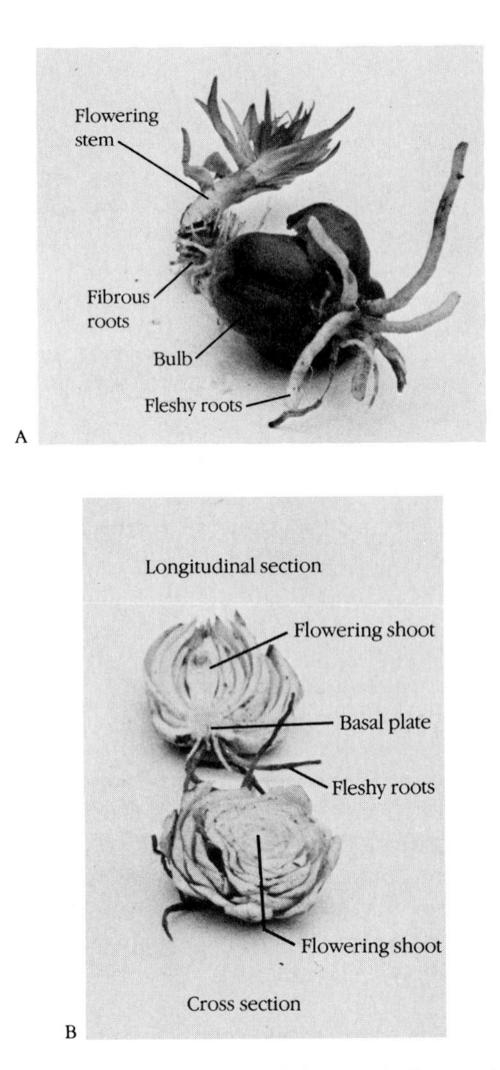

Figure 14–33 Scaly bulbs. (A) A lily bulb in early stage of growth of a flowering shoot. (B) Longitudinal and cross sections of a lily bulb. (Courtesy Washington State University.)

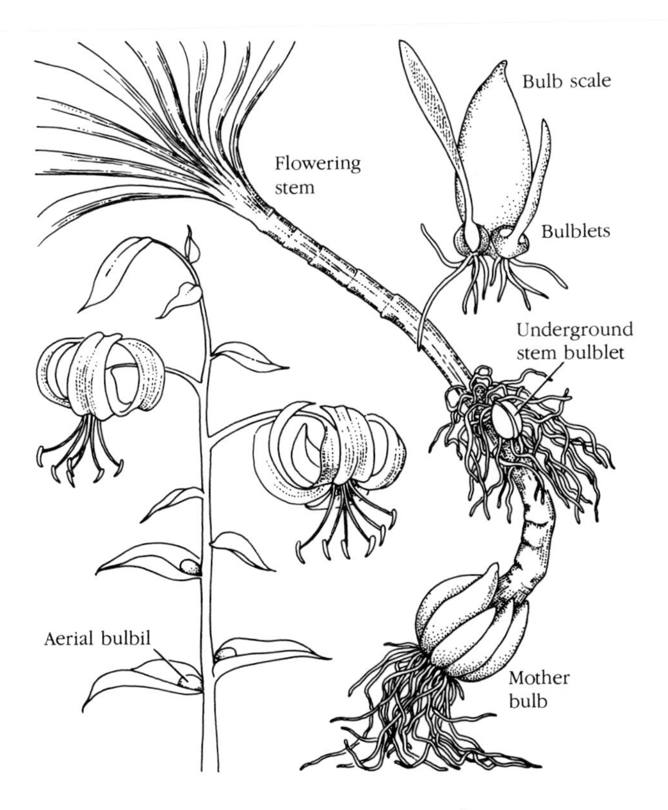

Figure 14–34 Several methods of lily propagation. (Courtesy Washington State University.)

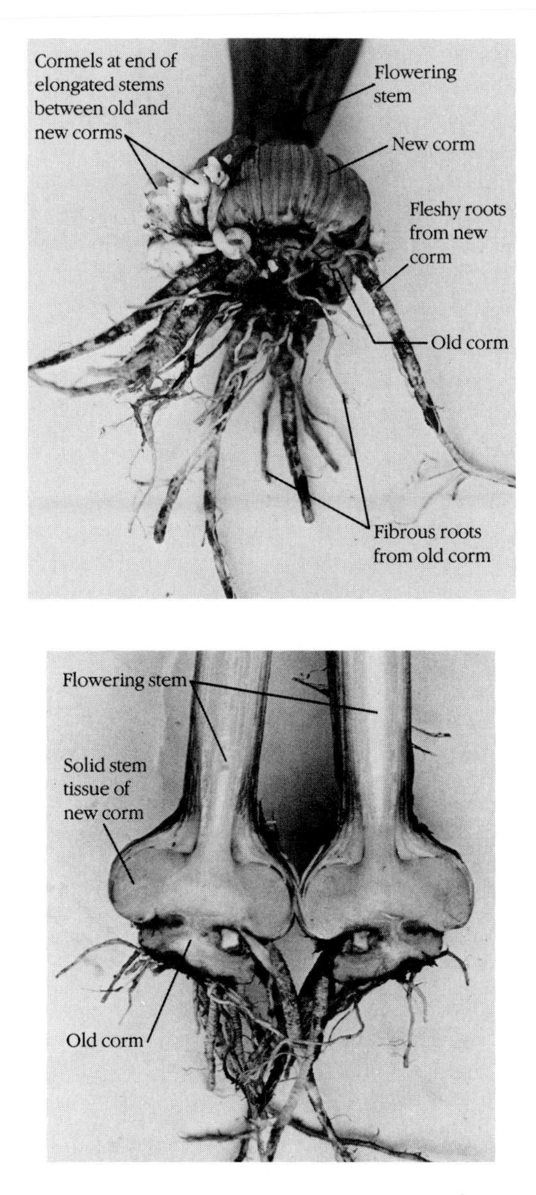

Figure 14–35 Structure of a gladiolus corm. (Courtesy Washington State University.)

Corms

Natural Corm Production. Plants that grow from corms produce a new corm at the base of each shoot every year. The old corm is used up producing the plant, and the plant then produces the new corm. More than one flowering shoot may grow from a large corm. As a result two (not usually more) corms of smaller size will form from the original corm (Figure 14–35).

Cormels. Small corms called cormels are formed at the base of the new corm. Shallow planting of corms encourages their formation. These can be separated and planted and will flower in 1 to 3 years. Plant cormels only about 1 inch deep because of their small size.

Corms are dug in the fall while the leaves are still somewhat green or after a light frost. They may be dried either outdoors or indoors, but rapid drying is best. Only after the corms and leaves have thoroughly dried should the corms be cleaned and the cormels separated. Treat with a fungicide for disease control and insecticide to control thrips. Store with adequate ventilation and at temperatures below 60°F (16°C) and about 70% relative humidity. If corm diseases are a problem, storage below 40°F (4°C) is recommended.

Hardy corms, such as spring-flowering crocus, are fall-planted. Tender corms, such as gladiolus, are usually spring-planted.

Tubers, Tuberous Roots, and Tuberous Stems

Tuberous plants can be propagated by cutting the tuber into sections so that each section has a bud. Allow the cut surface to dry. Treatment with a fungicide will help prevent the section from decaying before a new plant forms.

Tuberous roots are sometimes confused with tubers. However, tuberous roots are actually root tissue and don't have "eyes" or buds on them as tubers do. When propagating with tuberous roots, it is usually necessary to include a section of the crown or stem with a bud on it to produce the shoots of the new plant. Tuberous roots, such as the dahlia, are usually divided so that each root has one bud from the crown of the plant with it (Figure 14–36). Divide the clumps before the buds begin to grow. This is usually done in February or March just before planting.

Sweet potatoes do not need a section of crown with a shoot bud for propagation. They produce shoots directly from the tuberous root, and these shoots then form roots. These young plants, called slips, can then be separated and planted.

Tuberous roots dug in the fall can be stored on shelves in a cellar where temperatures range from 40° to 50°F (4° to 10°C). If the storage area is rather dry, the roots can be packed in dry peat or dry sand. Where loss from drying is severe, the roots can be dipped in melted paraffin wax that is floated on hot water.

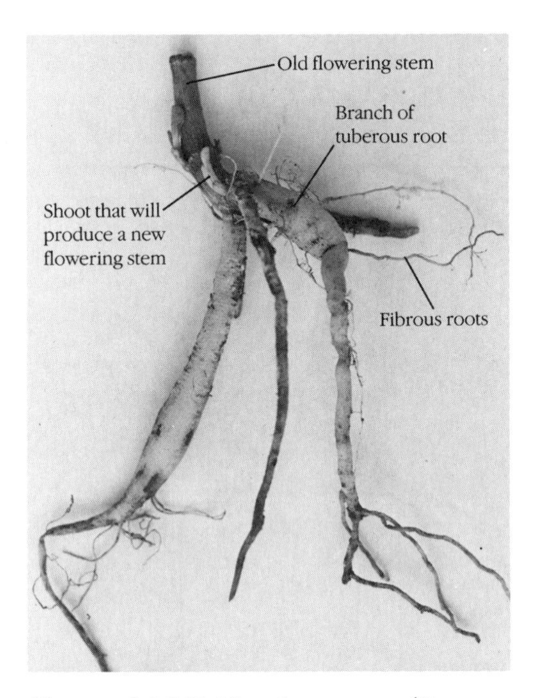

Figure 14–36 Dahlia tuberous root. (Courtesy Washington State University.)

Some plants, such as the tuberous begonia and gloxinia, have enlarged stems located between the regular stem and the roots. These stems, called tuberous stems, enlarge each year. The plants can be propagated by cutting the tuberous stem into sections so that each section includes a bud (Figure 14–37). Apply fungicide to the cut surface to combat decay and dry each section for several days after cutting before you plant it. Propagation of these plants can often be done better by using stem, leaf, or leaf-bud cuttings.

Rhizomes

A rhizome is a modified stem that runs horizontally just under or partially under the ground. Rhizomes contain shoot buds and can be propagated by being cut into sections so that each section contains at least one bud (Figure 14–38). Plant the sections horizontally, the same

Figure 14–37 Tuberous stem of tuberous begonia. (Courtesy Washington State University.)

as the parent plant. Lily of the valley and German (bearded) iris are well-known examples of this type of plant.

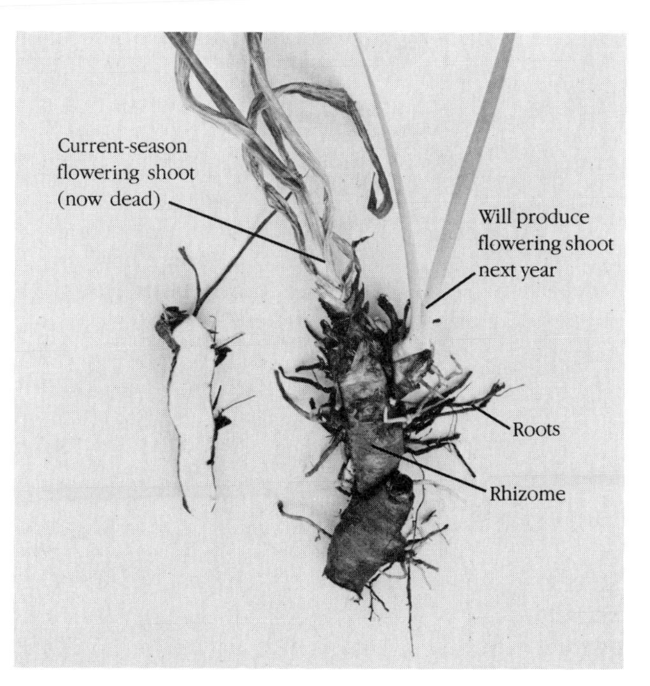

Figure 14–38 Iris rhizome. (Courtesy Washington State University.)

Pseudobulbs

Orchids produce both rhizomes and another type of modified stem called a pseudobulb. The exact structure of these organs varies among different species of orchids. Rhizome cuttings of new growth that include four or five nodes can be used to propagate orchids. The pseudobulb itself can sometimes be used, a new shoot forming at the base. Either back bulbs (those without leaves) or green bulbs (those with leaves) can be used to propagate the species *Cymbidium*.

Treating pseudobulbs with the rooting hormone IBA (indolebutyric acid) helps. Place the pseudobulbs in a rooting medium to about half the depth of the "bulb." When one new shoot is formed, separate it with its roots from the pseudobulb. Additional shoots will usually be produced from the pseudobulb (Figure 14–39).

Table 14–10 lists common flowering plants that are propagated from fleshy storage organs.

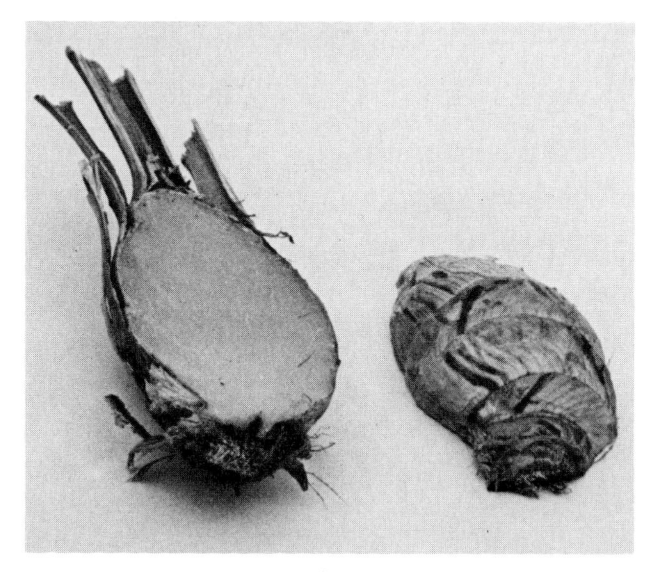

Figure 14–39 Pseudobulb of orchid.

Table 14–10 Common flowering plants grown from specialized structures

HARDY fall-planted, spring-flowering	SEMIHARDY fall-planted, summer- and fall-flowering	TENDER spring-planted, summer- and fall-flowering; dug and stored in winter	TENDER greenhouse or houseplants
		Bulbs	
Glory of the snow (*Chionodoxa*) Snowdrop (*Galanthus*) Hyacinth (*Hyacinthus*) Bulbous iris (*Iris*) Grape hyacinth (*Muscari*) Daffodil (*Narcissus*) Squill (*Scilla*) Tulip (*Tulipa*)	Lily (*Lilium*) Hardy amaryllis (*Lycoris*)	Tuberose (*Polianthes*) Summer hyacinth (*Galtonia*) *Amaryllis belladonna*	Amaryllis (*Hippeastrum*) Oxalis (bulbous) Amazon lily (*Eucharis grandiflora*)
		Corms	
Crocus (*Crocus*)	Autumn crocus (*Colchicum*) Crocus (some species)	Freesia (*Freesia*) Gladiolus (*Gladiolus*)	
		Tubers	
	Aerial begonia (*Begonia evansiana*)	Caladium (*Caladium*)	Caladium (*Caladium*)
		Tuberous roots and stems	
Wind flower (*Anemone*) Jack in the pulpit (*Arisaema*) Bleeding heart (*Dicentra*)		Tuberous begonia (*Begonia*) Dahlia (*Dahlia*) Gloriosa (*Gloriosa*)	Elephant's ear (*Colocasia*) Cyclamen (*Cyclamen*) Ranunculus (*Ranunculus*) Black calla (*Arum*) Gloxinia (*Sinningia*)
		Rhizomes	
Lily of the valley (*Convallaria*) Bearded iris (*Iris*)		Canna (*Canna*)	Calla (*Zantedeschia*) Achimenes (*Achimenes*)

(Adapted from *Plant Propagation: Principles and Practices*, 3rd ed., by Dale E. Kester and Hudson T. Hartmann. Prentice-Hall, Englewood Cliffs, NJ, 1975. With permission of the publisher.)

Grafting and Budding

by F. E. Larsen*

Graftage refers to any process of inserting a part of one plant into or on another in such a way that the two will unite and continue growth as a single unit. The term includes two similar processes—grafting and budding. Specific methods are described in the sections that follow; see also Chapter 4.

Factors That Influence Grafting and Budding Success

Time of Year. Grafting is usually done during the dormant season, whereas budding is done during the growing season or just as growth is beginning in the spring (spring budding). Budding done in early summer in areas with long growing seasons (e.g., parts of California and the South) is called June budding. Budding done in August or early September is called fall budding.

Growth Condition of Plant Parts. Scions for grafting should be dormant. Spring budding requires dormant budsticks (shoots from which buds are taken). For other budding, buds are taken from actively growing shoots. For budding (except chip budding) and bark grafting the cambium of the stock must be in active growth, allowing the bark to "slip" (separate) readily from the wood. Grafting potted evergreens requires stock in active growth. For other forms of budding and grafting the stock is usually dormant or just beginning active growth.

Age of Plant Parts. Budsticks are usually 1 year or less in age. Scions for grafting are usually 1 or not more than 2 years old. Rootstocks for producing new plants are usually 2 years or less in age.

Temperature. Grafting is usually done during the dormant season when temperatures are cool. Graft union depends on the uniting of new cells formed from the cambium of stock and scion. These new cells, which are not differentiated into a specific plant tissue, are called callus. Callus formation is slow at 40°F (4°C) or lower. However, if warm temperatures occur too soon after grafting, shoot buds may grow and produce a leaf surface that depletes moisture reserves in the scion before a graft union is formed. For this reason temperatures should not exceed 60°F (16°C) for 2 to 3 weeks following grafting unless scion buds are still in their rest period.

Union formation following summer budding is favored by temperatures around 70°F (21°C) because callus formation is rapid at these temperatures. Temperatures above 90°F (32°C) slow or stop callus formation.

Protective Measures. The surfaces of stock and scion where the union is to form must be protected from drying. This is usually done by covering the exposed surfaces after scion and stock are fitted together with grafting wax or other protective materials. Any material that will prevent tissue drying without itself causing tissue damage or excessive oxygen reduction is satisfactory. Following bench grafting, completed grafts are often protected from drying by covering with moist peat moss. In the field grafts can sometimes be protected by mounding with moist soil.

For some kinds of budding no special protective measures except tying are necessary.

Compatibility. Only plant parts with relatively close botanical relationships can be joined by graftage. Unrelated plants have chemical and physiological differences that prevent a union. Viruses may also cause incompatibility.

Scion Orientation. Scions (or stocks) for grafting can only form a permanently successful union if they are oriented as they normally grow. Those fitted together upside down will not grow properly. Budding scions often can grow normally regardless of orientation.

Soil Moisture. For maximum cambial activity the soil moisture supply must be ample. This is particularly important during the time of summer budding. Active growth of rootstocks during the summer must be maintained by keeping soil moisture high prior to budding and for a time afterward.

*Adapted from *Grafting and Budding to Propagate, Topwork, Repair*, by F. E. Larsen. Washington State Extension Bulletin 683. Extension Service, Washington State University. Pullman, WA. 1977.

Applied Pressure. Graft and bud unions are promoted by a good, snug fit and intimate contact of stock and scion. In some cases this is provided by the nature of the graft itself; in others it is promoted by the use of tying materials.

Follow-up Attention. Grafts may require rewaxing, tying for support, or light pruning to direct growth after shoot growth commences. (Waxing is explained below.)

Tools and Materials (see Figure 14–40). Although one type of knife can be used for budding and grafting, there are special knives best suited for each operation. They differ primarily in blade shape. Grafting knives have a straight blade, whereas the budding knife blade curves upward at the end. Some budding knives also have an attachment of the handle opposite the blade for opening the bark on the stock prior to insertion of the bud. A cutting blade with high-quality steel that will take and hold a good edge is desirable.

A good-quality, fine-grained sharpening stone is important for developing a good edge on your knife. A sharp knife is essential for ease in making smooth, straight grafting and budding cuts. Good cuts are essential for a high degree of success.

Several kinds of protective and tying materials are used. Waxes are the most common material to protect grafts from drying. Some must be heated and applied with a brush; others are soft and pliable enough to be applied cold by hand. Cold liquid wax emulsions and latex-base materials to be brushed on are also available.

Adhesive cloth or masking-type paper tape are frequently used to protect grafts. If applied under tension, tapes will hold the joined parts snugly and promote union formation. If tapes are properly applied, waxes or similar coverings are not needed. However, tapes are not readily adapted for application to all types of grafts. Waxed string or rubber budding strips are sometimes used to tie grafts; they must be covered with a layer of wax to adequately protect the graft.

For budding strips of thin rubber about 3⁄16 inch (2⁄3 cm) wide are most commonly used to tie buds securely to the stock. String, tape, and strips of plastic may be used.

Figure 14–40 Tools and materials used in grafting and budding. Clockwise from top center: latex-base grafting compound, wooden mallet, cleft graft tool, grafting knife (left), budding knife (right), hand clippers, pruning saw, sharpening stone, brush (for applying grafting wax or compound), grafting wax, lantern (for keeping grafting wax liquid). Center: nails (for fastening bark graft), tape (for fastening cleft and splice grafts), budding rubbers. (Courtesy Washington State University.)

Other tools are sometimes needed. For cleft grafting a wooden mallet and clefting tool are useful. The clefting tool is used to split the stock and hold it open for insertion of the scions. Pruning shears and saws are also handy tools.

Handling of Scion Wood

If freezing damage is likely to occur, collect grafting scion wood in the fall after normal leaf drop but before severe winter temperatures. Otherwise, wait to collect until late winter. Store the wood in a plastic bag. Enclose a moist cloth, but no free water should be in the bag. Store the wood in a refrigerator between 35° and 40°F (2° to 4°C). If refrigeration is not available, store the wood outdoors in moist sand in a well-drained, protected location where the soil will not freeze.

For spring budding the same scion wood as for grafting is used. For budding during the growing season new shoots of the current season's growth with mature, plump buds are used. The leaves are removed by snipping through the petiole (the stalk of the leaf) and leaving a petiole stub of about ¼ inch (⅔ cm) attached to the budstick. Store scion wood budsticks in a refrigerator like grafting wood. It will not keep as long as dormant wood and should be used in a few days. It is best if it can be collected and used immediately.

For either budding or grafting select only healthy wood free of insect, disease, and winter damage and from plants of known quality or performance. For fruit trees collect wood only from those in production to be sure that the kind and quality of fruit will be what you expect. The terminal ends of long shoots collected for scion wood should normally be discarded, because the buds and wood are usually poorly developed. Extreme basal buds may also be undesirable.

Kinds of Grafts

Splice Grafting. This method is the simplest way to join stock and scion. It is best suited for herbaceous plants in a protected location. Stock and scion should be less than 1 inch (2½ cm) in diameter and of equal thickness. Make long, diagonal cuts of equal length on stock and scion. Fit cut surfaces of stock and scion together and use tape or other tying materials to hold the parts together. Additional protection with wax or similar materials is usually advisable.

Whip-and-Tongue Grafting (Whip Grafting). Whip-and-tongue is one of the most commonly used and useful grafts for woody plants (Figure 14–41). It is used for topworking and producing new plants, primarily deciduous trees. It works best with stock and scion of equal diameter and less than 1 inch (preferably ¼ to ½ inch; ⅔ to 1¼ cm) thick.

Two identical cuts are made on stock and scion. The first is a diagonal cut like that of a splice graft. The length of the cuts varies with the diameter of stock and scion, increasing in length with increased diameter. In general, the length of cut should be four to five times the stock or scion diameter. Make this cut with a single knife stroke. Wavy cuts may prevent a satisfactory union. Make the second, or tongue, cut on stock and scion by placing the knife on the surface of the first cut about half the distance between the pith (the center core of stem) and outer bark, or epidermis, on the upper part of the cut. Bring the knife down through the pith until it is opposite the base of the first cut. This cut should not follow the grain of the wood but should tend to parallel the first cut.

When the tongues are cut, insert into each other until they are interlocked. Then secure the parts by wrapping tightly with tape or other tying materials. If tape is properly applied, additional protection with wax may not be necessary.

If the scion is smaller than the stock, fit the tongues together so that the outside surfaces of stock and scion are aligned on one side only.

Cleft Grafting. Cleft grafting is used for topworking and should be done before active growth of the stock. Scions are usually about ¼ inch in diameter and two to three buds long. Stocks should be 1 to 4 inches (2½ to 10 cm) in diameter and straight-grained. Saw off the stock at a right angle in relation to its main axis. Make the cut so there are 4 to 6 inches (10 to 15 cm) below, with no knots or side branches. Use a clefting tool or heavy knife to split the stock down the center for 2 to 3 inches (5 to 8 cm). Drive the tool in with a wooden mallet. Remove the cutting edge of the clefting tool and drive the wedge part of the tool in the center of the stock to open the split to receive the scions.

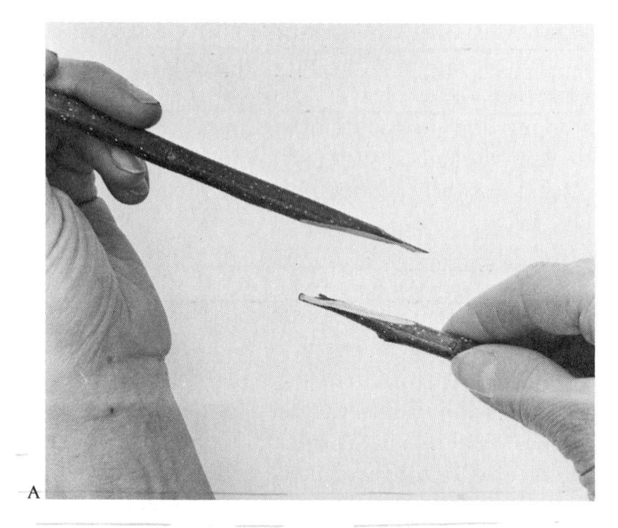

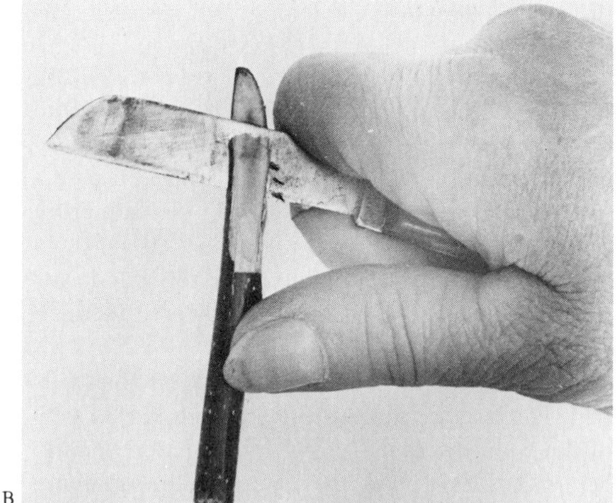

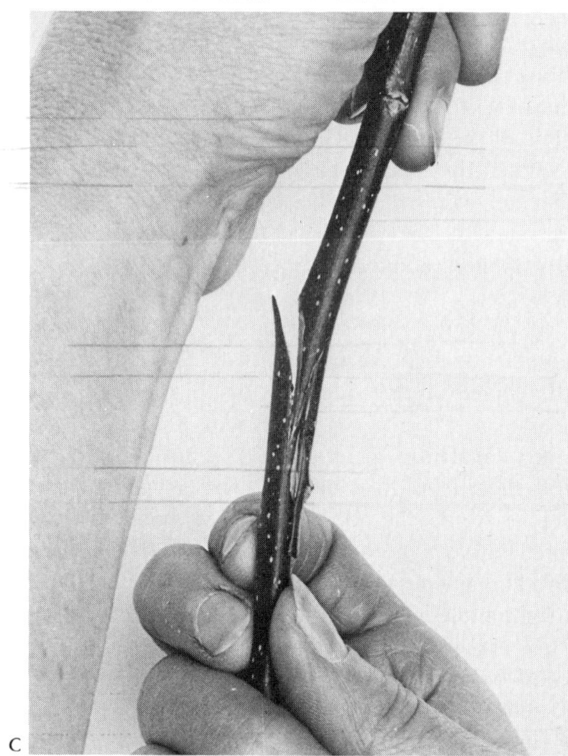

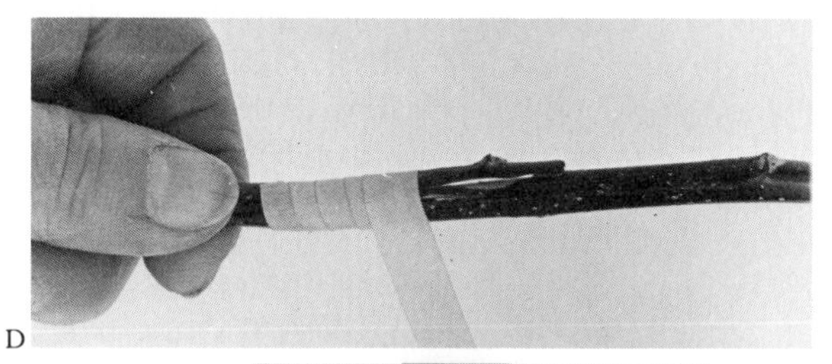

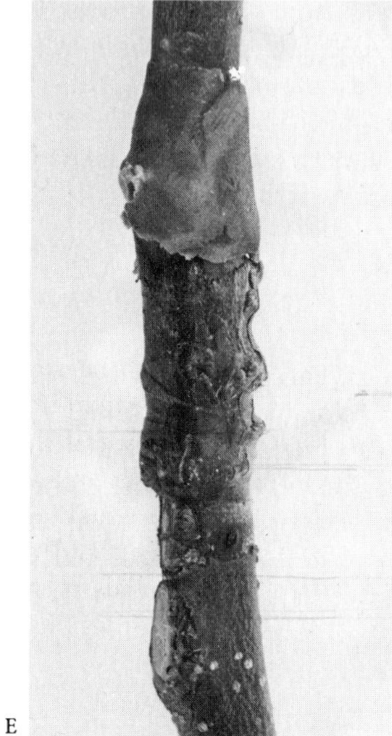

Figure 14–41 The whip-and-tongue graft. (A) Splice graft, or initial cut for whip graft, (B) final cut on whip graft, (C) fitting scion and stock together, (D) wrapping the whip graft, (E) established whip graft. (E, courtesy Washington State University.)

Prepare two scions. Cut the basal end of each into about a 2-inch long tapered wedge. One side of the wedge should be slightly thicker. These long cuts on the scion should be smooth and each made with a single sweep of the knife. The wedge should not be too short, otherwise stock-scion contact will be only at one point.

Insert the cut wedge of the scions, one on each side of the stock, with the narrow part of the wedge toward the center of the stock. Align cambium layers of stock and scion without regard to outside surfaces. Remove the metal wedge of the clefting tool from the stock, which leaves the scions held snugly in place. Completely cover the cut surfaces and the splits down the side of the stock with wax. Also cover the tip of the scion.

If both scions grow, the healing of the large stock stub will be more rapid than if only one scion grows. However, eventually one scion must be removed. From the beginning one scion should be dominant; keep the other small by pruning. After 2 or 3 years prune out the smaller scion (see Figure 14–42).

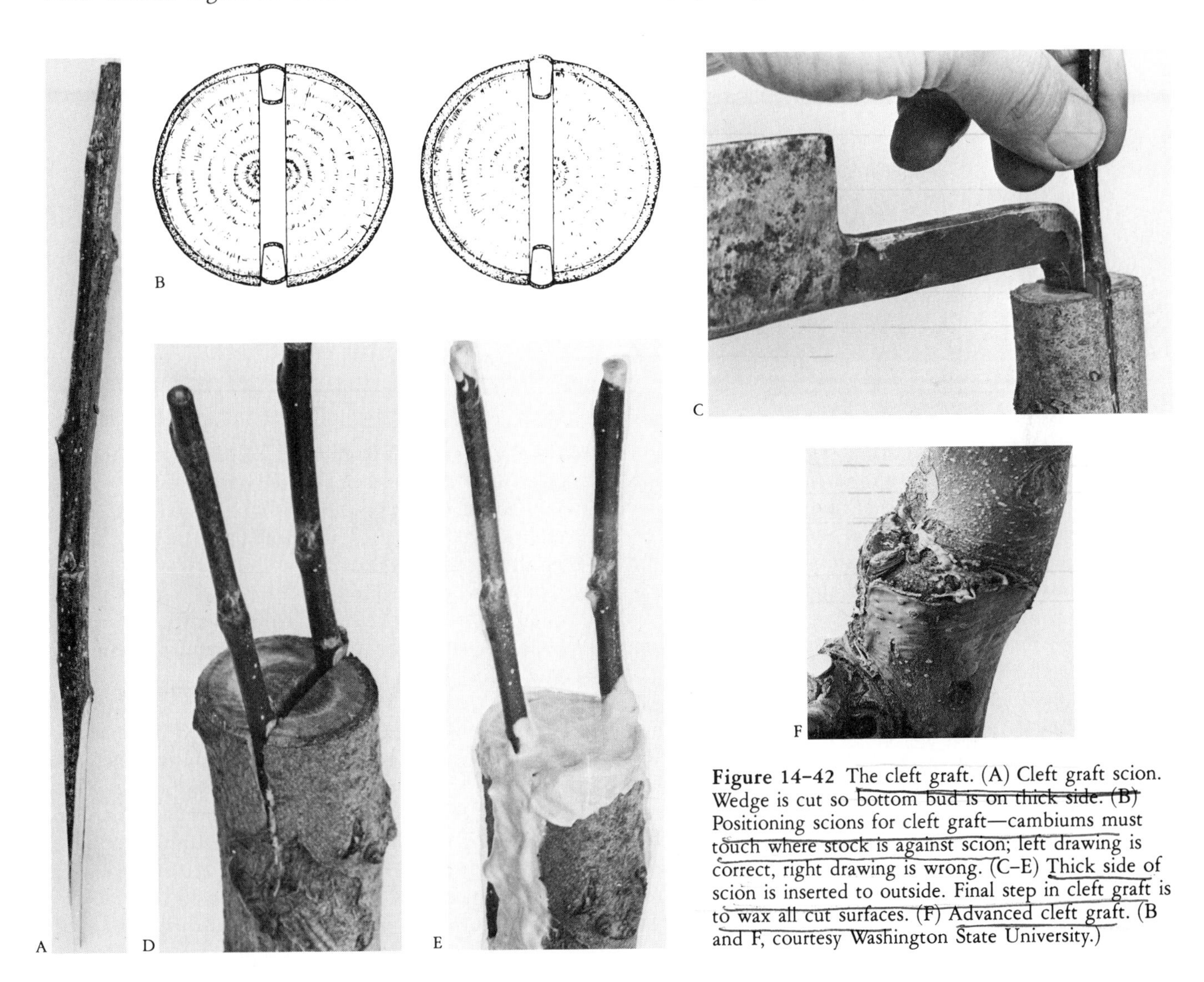

Figure 14–42 The cleft graft. (A) Cleft graft scion. Wedge is cut so bottom bud is on thick side. (B) Positioning scions for cleft graft—cambiums must touch where stock is against scion; left drawing is correct, right drawing is wrong. (C–E) Thick side of scion is inserted to outside. Final step in cleft graft is to wax all cut surfaces. (F) Advanced cleft graft. (B and F, courtesy Washington State University.)

Figure 14–43 The bark graft. (A) Bark graft scion is inserted after a short slit is made in bark. (B) Bark graft scion can also be inserted after cutting bark flap on stock the width of scion. (C) The bark graft scion is held in place with nails. The final step is to wax all cut surfaces. (D) Bark grafts after becoming established. (D, courtesy Washington State University.)

Bark Grafting. Bark grafting is used for topworking (Figure 14–43). It can be used with larger stocks (up to 12 inches [30 cm] in diameter) than for cleft grafting, but the scions are similar in size. Several scions can be inserted around the stock. The stock is cut off as for cleft grafting, except it is not split through the center. It must be done when cell division in the stock has begun in early spring, allowing the bark to separate readily from the wood.

Cut the base of the scion on one side with a long, smooth, sloping cut about 1½ inches (4 cm) long, going completely through the scion so that it comes to a point at the base. Make a vertical cut about 1½ inches long going through the bark in the stub of the stock. Slightly loosen the bark at the top of the cut and insert the wood surface of the scion base next to the wood of the stock. Push the scion down in behind the bark to the extent of the cut on the scion base. Secure in place by driving one or two wire brads through the base of the scion into the stock. Cover all cuts and exposed surfaces with grafting wax. Because the union with this type of graft is weak for a year or two, the scions may need to be tied up for support after growth begins.

For topworking place scions every 2 to 4 inches around the stock stub. As for cleft grafting, the intent is usually for only one to eventually remain. Train and prune as for cleft grafting.

Side Grafting. Side grafting can be used for topworking or producing new plants. Several forms of side grafting are used. The stub-side graft is used primarily for topworking fruit trees with branches too small for cleft or bark grafting and too large for whip grafting. The other forms are used mostly for producing new evergreen plants by grafting on small seedling stocks.

Make **stub-side** grafts on stock branches that are between ½ and 1 inch in diameter. Make a cut in the stock at a 45° angle going about halfway through the stock. Cut the scion as for a cleft graft except with a shorter wedge. Open the cut in the stock by pulling down on the stock branch beyond the cut. Insert the scion into the stock cut with the thick side of the wedge out. Little or no cut surface should show on the scion wedge. Let the stock branch spring back into place. The natural tension of the branch will hold the scion in place. Cut the stock off about 6 inches beyond the graft. Cover the graft area, stock stub, and end of the scion with a protective compound (Figure 14–44).

For **side-veneer** grafting make a rather shallow cut about 1½ inches long at the base of the stock, cutting slightly inward as the cut is made. At the base of this cut make a short inward, downward cut to intersect the first cut, thus allowing removal of a piece of wood and bark. Prepare the scion with a long cut the same length and width as that of the first cut on the stock. Make a short cut on the opposite side of the base of the scion.

Insert the scion in the stock with the long cut of the stock next to the long cut on the scion. Secure the scion by wrapping with tape or rubber budding strips. Cover the graft region with a protective material. When the graft union forms, cut off the stock just above the union (Figure 14–45).

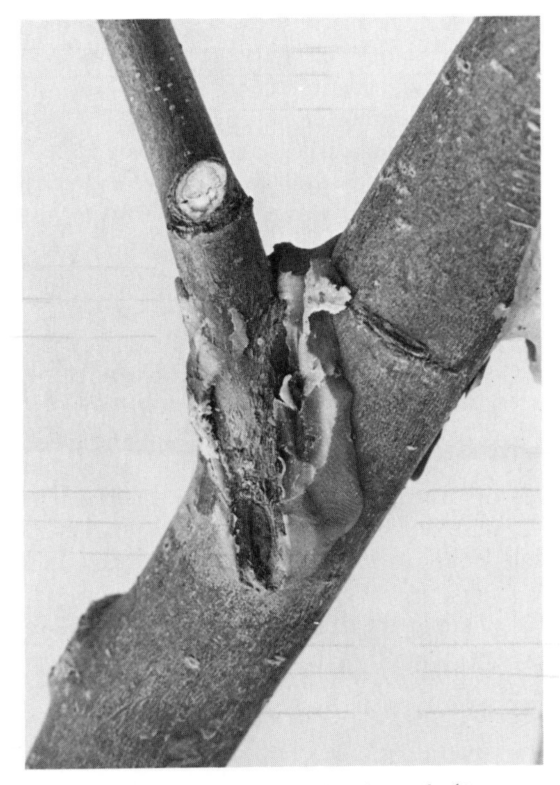

Figure 14–44 Established stub-side graft. (Courtesy Washington State University.)

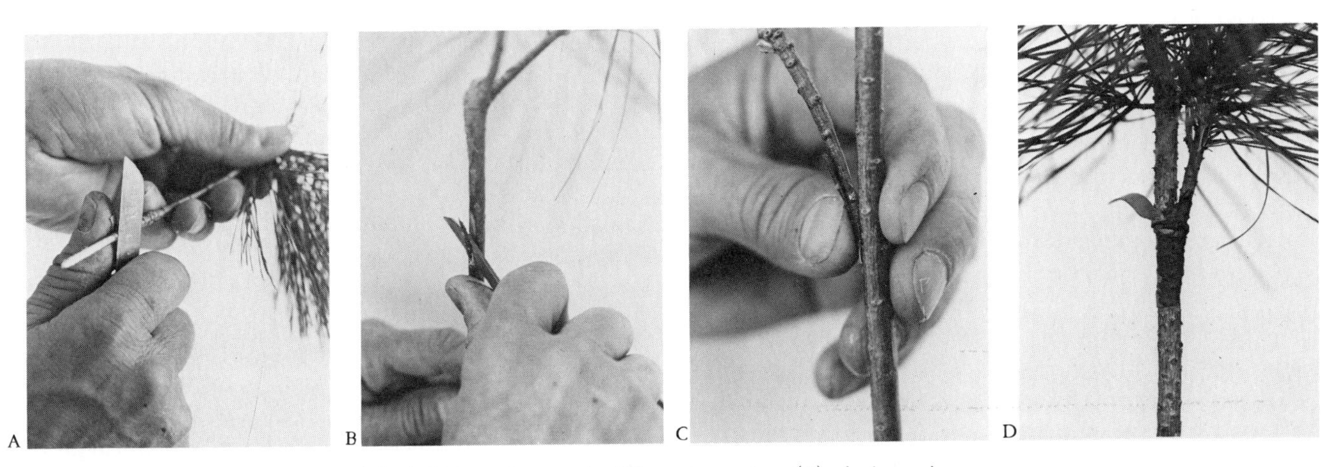

Figure 14–45 The side-veneer graft. (A) Preparing stock, (B) cutting scion, (C) placing scion on stock, (D) wrapped and waxed side-veneer graft. (Courtesy Washington State University.)

For **side-tongue** grafting cut the scion as for whip grafting. On a smooth place on the stock remove a thin slice of wood and bark about 1 to 2 inches long. Make a tongue cut by starting about one-third of the distance from the top of the first cut and progressing to the base of the first cut. Fit the tongues of stock and scion together; wrap the graft with tape or other tying material and add a protective covering to the graft region. Cut off the stock above the graft after union occurs (see Figure 14–46).

Bridge Grafting. Bridge grafting is used to repair damaged bark areas at the base of a tree. Bark damage caused by cold, rodents, or implements may kill a tree if it is severe enough. If the trunk is not completely girdled, the tree can usually be saved by bridge grafting.

Bridge grafting must be done when the bark "slips"; it is usually done just as active growth is becoming apparent in the spring. Select dormant 1-year-old scion wood long enough to bridge the damaged area. Make cuts like those for bark grafting on both ends of the scions, being sure that both cuts are in the same plane. Make a flap cut the width of the scion through the bark of the stock below the injured area. Pull the end of the bark flap loose from the wood, insert the base of the scion under the flap, and push the scion under the flap until the cut surface on the base of the scion is covered. Drive nails through the base of the scion as for bark grafting. In a similar fashion attach the top of the scion to the stock above the injured area.

Attach scions to the stock about 3 inches apart across the injured area. If the tree is young enough to allow the trunk to bend in a strong wind, the scions must retain an outward bow after completing the graft. This allows the trunk to bend without pulling the ends of the scions loose from the trunk. Cover all cut surfaces with protective waxes, being sure to seal the areas where the scion is inserted into the stock, particularly under the scion. Remove shoots that grow from scion buds (see Figure 14–47).

If trunk damage extends below ground level, the lower graft must be made on a large root. Uncover the root and proceed as if you were working on the trunk. Recover the root with soil.

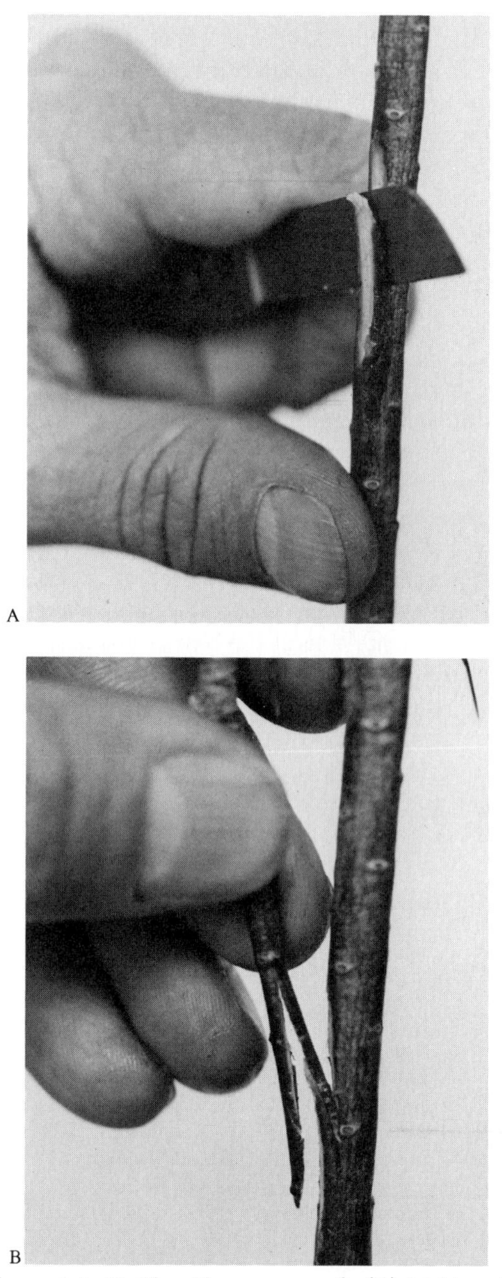

Figure 14–46 The side-tongue graft. (A) Making the second cut on stock for the side-tongue graft. The first cut is similar to the cut made for the side-veneer graft. (B) Inserting the scion (cut as for a whip graft) on stock. (Courtesy Washington State University.)

Figure 14–47 The bridge graft. (A) Single bridge graft scion held in place with nails. The exposed areas are coated with wax or other protective material. (B) An established bridge graft. (B, courtesy Washington State University.)

Inarching. Inarching is a repair technique similar to bridge grafting except that suckers growing at convenient locations next to the injured area or seedlings planted next to the damaged trunk are used as scions. These are bark grafted into the trunk above the injured area. Cover the graft area with a protective material (Figure 14–48).

Figure 14–48 Inarching. (A) The tip of a sucker from tree base can be bark grafted above wounded area to form an inarch. (B) Established inarch grafts. (Courtesy Washington State University.)

Remove shoots that develop on the inarches. If the tree is on a size-controlling rootstock, the inarch should be of the same variety as the rootstock, otherwise the size-controlling benefits of the original rootstock will be reduced or eliminated.

Approach Grafting. The approach graft method is used to support a weak crotch in a tree or to graft together two plants while both remain on their own roots. For giving support, two adjacent branches are joined together. For two plants on their own roots the main stems are joined together, because the objective is to eventually remove one top and the opposite root system. This method has the advantage of an uninterrupted flow of water to the scion from its own roots until the union is formed and its roots are removed. Similarly, the rootstock receives manufactured food from its top during the healing process.

Stock and scion or adjacent branches within a tree may be joined by the **spliced** method, which uses single, long, smooth cuts on adjacent surfaces. Bring the cut together and use wrapping material to hold tightly in place. Cover the area with a protective material. If opposing tongues are cut on the face of the splice cuts of stock and scion as for side-tongue grafting, the method is referred to as the **tongued-approach** graft. In this case the tongues are slipped together for added stability while healing. The other procedures are the same as for the spliced approach (see Figure 14–49).

Figure 14–49 A spliced-approach graft. The first step in making a spliced-approach graft is to remove thin slices of bark from adjacent surfaces of both stock and scion (A). Cut surfaces are then brought together and wrapped (B and C) and all surfaces are covered with grafting wax. (Courtesy Washington State University.)

Budding Methods

T- or Shield Budding. The T-budding method is used to topwork or produce new plants. It is the most common budding method for producing fruit and ornamental plants. It works best on rootstocks of ¼ to 1 inch (⅔ to 2½ cm) in diameter with thin bark. It must be done when the bark slips.

Prepare the stock by removing side branches to provide a smooth area of at least 6 inches (15 cm) in which to work. For summer and fall budding, this preparation begins by early season removal of shoots while they are still soft and succulent (Figure 14–50). Scions are selected from dormant 1-year-old wood for spring budding and current-season growing shoots for summer budding.

To begin the budding process, select a smooth, branch-free area on the stock for insertion of the bud. If you are topworking, this area should be on the side of a lateral branch, 12 to 18 inches (30 to 45 cm) from the main stem or trunk. If you are budding on seedlings to produce new plants, select an area about 6 inches from the ground. On similar size clonal rootstocks buds are often inserted 12 to 18 inches from the ground.

Make an upward vertical cut through the bark for about 1½ inches (4 cm) on the selected area. Next, make a horizontal cut to form a "T" with the vertical cut. As the last cut is made, hold the knife at an acute angle in relation to the upper part of the rootstock. This technique will open the bark to allow easy start of the scion bud.

Remove scion buds from the budstick in one smooth stroke. Start the cut about ½ to ¾ inch (1¼ to 2 cm) below the bud; pass just under the bud, taking a sliver of wood with the bud; and extend about the same distance above the bud. As the cut is completed, pinch the bud shield against the knife blade to force the knife through the bark. This positions the bud in your hand so it can be immediately inserted into the stock.

To insert the bud, place the lower tip of the bud shield in the opening at the top of the T-cut. When the opening is properly made in cutting the top of the T, there should be no difficulty in starting the bud. When the bud is started, push it down into the opening with the tip of the budding knife inserted in the shield above the bud. Slide the bud down so that the top of the shield is even with the top of the T in the stock.

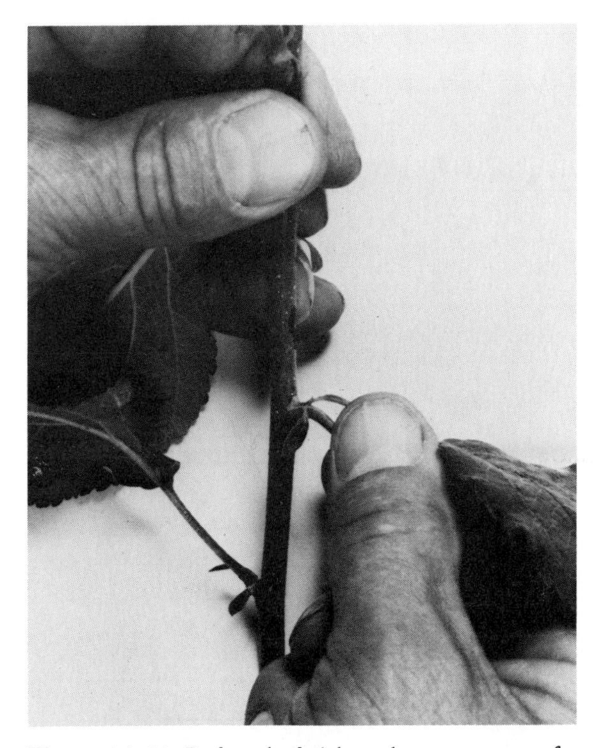

Figure 14–50 Bud on budstick at the proper stage for summer and fall T-budding. Leaves should be clipped from budsticks to reduce transpiration. (Courtesy Washington State University.)

Wrap the bud in place with rubber budding strips by starting a self-binding loop just below the lower end of the bud shield. While stretching the rubber strip, make three or four loops below and above the bud, being sure to cover the top of the T. Insert the end of the rubber strip under the last wrap and pull tight (see Figure 14–51).

Cut off the top of the rootstock at a point just about even with the top of the T-cut. For fall budding this is not done until the following spring, just as growth starts. For spring budding it is done 2 to 4 weeks after budding. With June budding it is done in two steps starting 4 to 5 days after budding. Cut 5 or 6 inches (13 to 15 cm) above

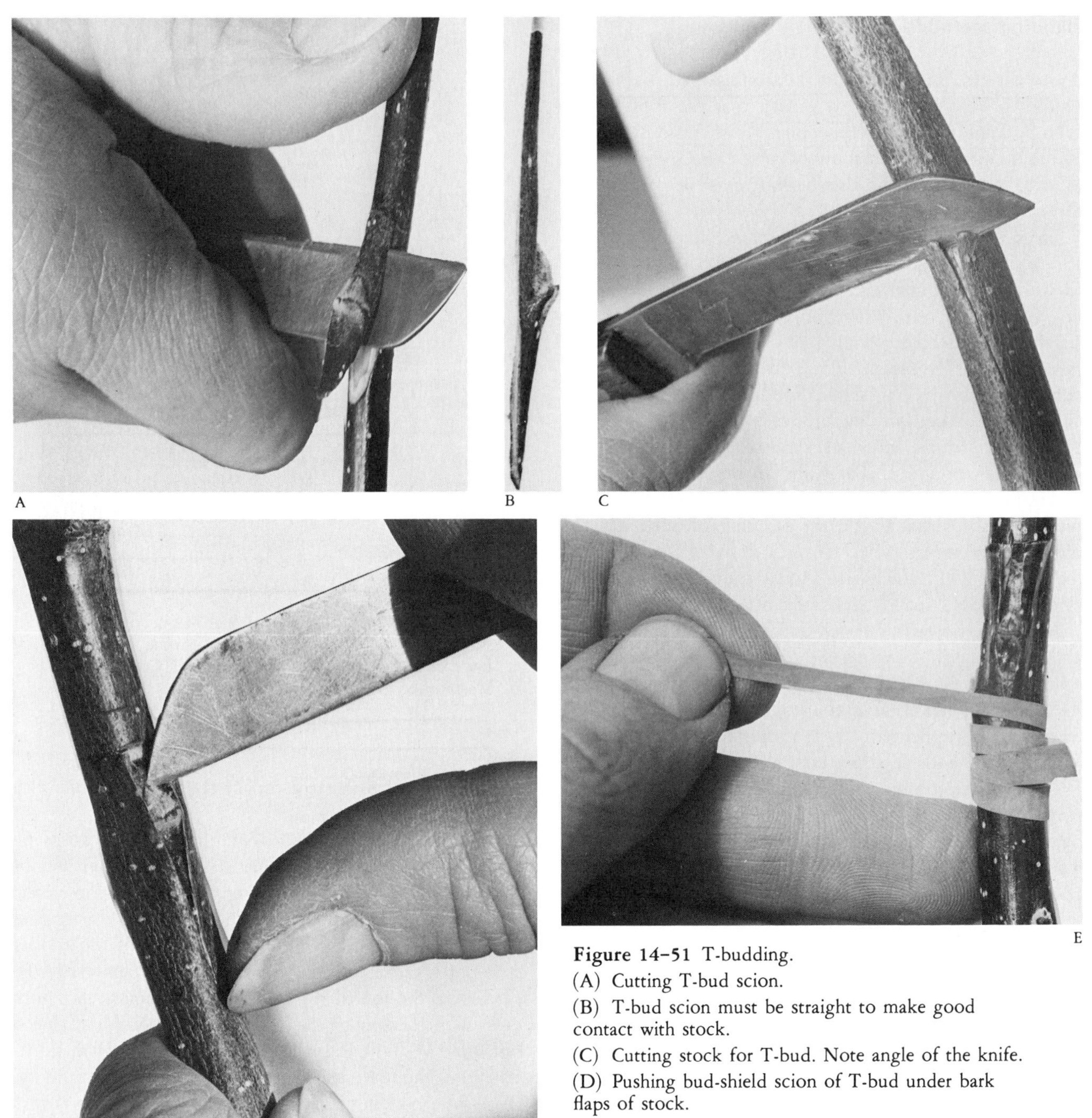

Figure 14–51 T-budding.
(A) Cutting T-bud scion.
(B) T-bud scion must be straight to make good contact with stock.
(C) Cutting stock for T-bud. Note angle of the knife.
(D) Pushing bud-shield scion of T-bud under bark flaps of stock.
(E) Securing scion to stock with rubber strips.

the bud, allowing at least one leaf of the rootstock to remain above the scion bud. About 2 weeks later, when the scion bud has started, cut the rootstock just above the scion shoot.

After removal of its top, the rootstock may produce a number of shoots around and below the bud. Break these off as soon as they appear, except in June budding, where these shoots should remain until the scion shoot is 10 or 12 inches (25 to 30 cm) long.

Patch Budding and Related Forms. Patch budding is usually done in the summer and includes several related forms used on plants with thick bark that give poor results with T-budding. The scions used with these methods include a bud on a patch of bark varying from about 1 x 1½ inches to a complete ring of bark removed from the budstick. The scion contains no wood behind the bark patch. Budsticks and stocks should be approximately equal in diameter; the best size is between ½ and 1 inch. The bark of both stock and budstick must slip readily.

The scion patch is best removed from the budstick with a special four-bladed cutting tool with the blades fixed in a square or rectangle. With this tool an exact duplicate of the scion patch can be removed from the stock, and the fit will be perfect. Another common approach is to use a double-bladed knife to make the horizontal cuts on budsticks and stock while using a single-bladed knife to make the vertical cuts. It is especially important that the scion-stock fit be perfect at the horizontal cuts.

When removing the scion patch from the budstick, lift only the edges; remove the remainder of the patch by sliding sideways to break the woody connection between the bud and the budstick (Figure 14–52). If the entire patch is lifted off, this connection will be torn from the bud, and the bud will fail to unite properly with the stock.

Fit the scion bud patch into the prepared area of the stock and wrap the patch in place. If the stock bark is thicker than the scion, this bark around the edge of the patch may require trimming so that the patch can be held snugly in place. Wrap the bud in place with tape, being

Figure 14–52 Removing scion for patch budding. (Courtesy Washington State University.)

careful to seal all cuts, but do not cover the bud. If wrapping is carefully done, no other protection may be necessary. An added precaution, however, is to use wax or other protective materials. Cut the stock back as for T-budding.

Chip Budding. Chip budding does not require bark that slips on either stock or budstick. It is usually done in the spring just as growth begins. However, it may be done in the summer at the same time as other techniques. Stocks and budsticks are usually ½ to 1 inch in diameter.

The chip from the rootstock is removed by making two cuts. The first is a downward cut at a 45° angle going about one-fourth through the stem. The second starts about 1 inch higher than the first, going downward and inward until it connects with the first cut. A similar cut is made on the budstick to remove the scion (Figure 14–53). The first cut starts about ¼ inch below the bud, the second about ½ inch above the bud. Fit the scion to the stock so that the cambium layers of stock and scion match on at least one side, but preferably on both. Wrap the bud in place with tape to cover all cut edges but not the bud. An added precaution is to cover the area with grafting wax. Cut the stock back as for T-budding.

Table 14–11 lists selected plants that can be propagated by budding or grafting.

Topworking and Repair

Topworking can be done by a variety of grafting or budding techniques. Trees of less than 3 or 4 years of age can be topworked rapidly by T-budding or whip grafting. The branches should be less than ½ inch in diameter. For larger branches side, cleft, or bark grafts must be used.

Graft or bud scions should usually be placed within about 18 inches (45 cm) of the main trunk of a tree. This means that large cuts are required on older trees. Insert buds on the side of a horizontal branch and the outside of a vertical branch to encourage outward growth.

When the buds or grafts begin to grow, they may require support or pruning.

A

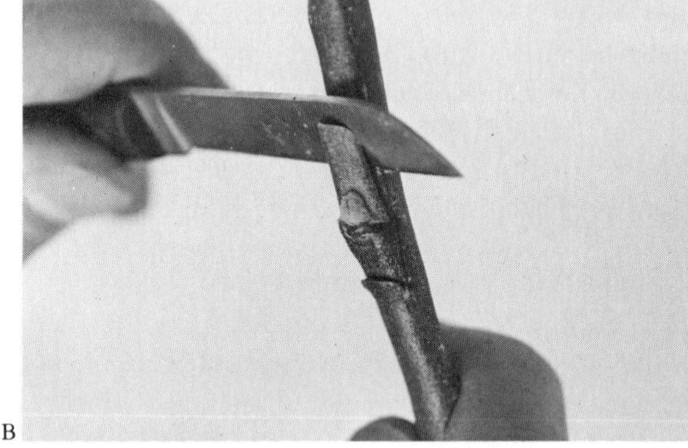
B

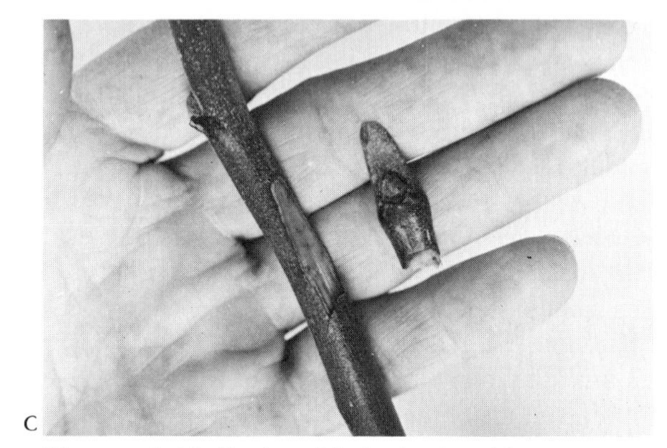
C

Figure 14–53 Cutting chip bud scion. (Courtesy Washington State University.)
(A) Making the first cut.
(B) Making the second cut.
(C) Scion of chip bud ready to be inserted into prepared cut on stock.

Table 14–11 Selected plants that can be propagated by budding or grafting and common methods and rootstocks for each

Plant	Method[a]	Time[b]	Rootstock
Almond	T	S, Su, F	Almond, peach
Apple	T Whip	S, F W	Apple
Apricot	T	S, Su, F	Apricot, peach
Arborvitae	Side	W	Arborvitae
Avocado	T Whip, side	S, Su, F W	Avocado
Azalea	Side	F	Azalea
Beech	Whip, side	W	Beech
Birch	T Side	F W	Birch
Butternut	Bark	S	Black walnut
Camellia	Side	W	Camellia
Carob	T	S	Carob
Cedar	Side	W	Cedar
Cherry	T	S, F	Cherry
Chestnut	Whip	W	Chestnut
Citrus species	T	S, F	Citrus
Clematis	Side	W	Clematis
Dogwood	T Whip	F W	Dogwood
Fir	Side	W	Fir
Fringe tree	T Side	F W	Fringe tree
Ginkgo	Patch Side	F W	Ginkgo
Grape	Chip Whip	Su, F W	Grape
Hackberry	Chip Side	S S	Hackberry
Hawthorn	T Whip, side	F W	Hawthorn
Hibiscus	T Whip, side	F W	Hibiscus
Hickory	Patch Side	F W	Hickory
Holly	T Whip, side	F W	Holly
Honeylocust	T Whip, side	F W	Honeylocust
Horsechestnut	T Whip, side	F W	Horsechestnut
Juniper	Side	W	Juniper
Kiwi	T Whip	F W	Kiwi
Larch	Side	W	Larch
Lilac	T Whip, side	F W	Lilac, privet, ash
Magnolia	Side	W	Magnolia
Maple	T Side	F W	Maple
Mountain ash	T Whip	F W	Mountain ash
Mulberry	T Whip, side	F W	Mulberry
Nectarine	T	Su, F	Peach, apricot, some plums
Oak	Bark, side, whip	W	Oak
Olive	T, patch Whip, side	F W	Olive
Pawpaw	Side	W	Pawpaw
Peach	T	Su, F	Peach, apricot, some plums
Pear	T Whip	F W	Pear, quince
Persimmon	T Whip	F W	Persimmon
Pine	Side	W	Pine
Pistache	T	F	Pistache
Plum, prune	T	F	Plum, peach, apricot, almond
Quince	T	F	Quince
Redbud	T	Su	Redbud
Rhododendron	Side	W	Rhododendron
Rose	T	F	Rose
Spruce	Side	W	Spruce
Viburnum	Side	W	Viburnum
Walnut	Patch Whip	F W	Walnut
Wisteria	Whip, side	W	Wisteria
Witch hazel	Whip	W	Witch hazel
Yew	Side	W	Yew

(Adapted from *Grafting and Budding to Propagate, Topwork, Repair,* by F. E. Larsen. Washington State Extension Bulletin 683. Extension Service, Washington State University. Pullman, WA. 1977.)

[a]The grafting and budding methods indicated are primarily for producing new plants. For topworking other methods might be used.

[b]S = spring, Su = summer, F= fall, W = winter.

Figure 14–54 Flowering crabapple topworked to a fruit-producing apple variety. (Courtesy Washington State University.)

Shoots will develop from the old or original parts of the tree (Figure 14–54). Unless you wish to have more than one variety on the tree, these shoots must all be removed eventually. This should not be done all at one time, but removal should be gradual during the first year. Some shoots will continue to appear for 2 to 3 years.

Tree repair is done by bridge grafting, inarching, or possibly by approach grafting. See the discussions related to these specific methods.

Growing Garden Seed at Home

Caution. *Don't save or attempt to grow seed of F_1 hybrid cultivars* (see Chapter 3).

Isolation and Pollination

Except for the few self-pollinated species (Table 14–12), garden plants must be isolated from other cultivars of the same species if they are to produce seed true to type. Commercial seed producers are required to plant seed fields from ¼ to 1 mile (about ½ to ⅔ km) from the nearest planting of a different cultivar of the same species. Although an occasional stray pollen grain may reach the planting, seed satisfactory for home gardening usually can be produced with an isolation of 200 yards (about 180 m). Where this kind of distance is not feasible, individual plants can be isolated with a screened cage, and individual florets with a piece of cheesecloth (Figure 14–55). Isolation must occur before the flowers from which seed is to be saved have opened.

Table 14–12 Classification of garden plants according to reproduction

Asexually propagated garden plants				
Fruits and nuts (almost all)			Ornamental shrubs and trees	Vegetables
Almond	Currant	Peach	Almost all	Garlic
Apple	Date	Pear		Globe artichoke
Apricot	Fig	Pecan		Horseradish
Avocado	Gooseberry	Pineapple		Jerusalem artichoke
Banana	Grape	Plum		Potato
Blackberry	Grapefruit	Raspberry		Rhubarb
Blueberry	Lemon	Strawberry		Sweet potato
Cherry	Lime	Walnut		
Cranberry	Orange			

Sexually propagated garden plants
(including some normally propagated asexually
that can be propagated sexually)

Naturally self-pollinated crops		
Vegetables		Flowers
Bean	Pepper	Larkspur
Eggplant	Potato	Nicotiana
Lima bean	Tomato	Sweet pea
Pea		

Cross-pollinated crops			
Vegetables			Other plants
Asparagus	Cucumber	Pumpkin	Hops
Beet	Endive	Radish	Sugar beets
Broccoli	Jerusalem artichoke	Rhubarb	Sunflowers
Brussels sprouts	Leek	Rutabaga	Most flowers
Cabbage	Lettuce	Spinach	Most forest trees
Cantaloupe	New Zealand spinach	Squash	Most ornamental trees and shrubs
Carrot	Okra	Sweet corn	
Cauliflower	Onion	Sweet potato	
Celery	Parsley	Turnip	
Chard	Parsnip	Watermelon	

Figure 14–55 Isolation cage to prevent random pollination of an individual plant.

Nature will usually take care of the pollination of uncaged plants. Caged flowers should be hand pollinated when they become receptive, which for most species is shortly after they open. Pollen from another plant of the same cultivar should be used to assure fertilization in case the cultivar is self-incompatible. Pollen is generally most viable shortly after it matures and begins to be shed from the anthers. Pollen can be transferred by brushing the pollen-shedding stamens onto the stigma or by touching the stigma with the point of a knife blade containing a bit of pollen scraped from an anther (Figure 14–56).

Collecting the Seed

The seed of most plants that produce a dry fruit or pod should be allowed to fully mature and dry on the plant. Seed pods of species that burst and scatter their seed before they are fully dry need to be gathered and dried artificially before they burst. Because the gardener will usually produce only small quantities of seed of any one cultivar, chaff and other extraneous material can be blown away by pouring the seed from one container to another in a slight breeze or the air stream from a small fan.

Seed produced in a moist fruit like tomato or muskmelon should be squeezed or scraped carefully into a container with as little extraneous pulp as possible. If the liquid extracted with the seed is not adequate to keep the seed from drying out, a small amount of water should be added. A loose cover is placed over the container and the mixture is allowed to ferment for 4 to 7 days; it is then rinsed several times with water to remove the fermented liquid and pulp. The seed is then placed on paper toweling or other absorbent material and allowed to dry thoroughly (Figure 14–57).

Seed Production of Biennial Crops

Biennial crops must be subjected to a cool rest period after their first season of growth before they will grow a seedstalk. This cold period can be provided by allowing them to winter over in the garden. Where freezing damage is likely, root and bulb crops can be protected with a heavy loose mulch of straw or leaves. Root and bulb crops and cabbage also can be dug and wintered over in a cool, moist root cellar or pit.

Transplanting Woody Plants

The technique required for successful transplanting of woody plants depends on the kind of plant; its age and size; whether it is dormant or actively growing; its nutri-

Figure 14-56 Artificial pollination of a squash flower. Stray pollen can be prevented from entering the female cucurbit flower by fastening it shut with a piece of paper and a paper clip (A). The petals of a pollen shedding male flower are stripped away (B), and the anther cone is brushed against the stigma (C) to complete pollination. After being pollinated, the female flower is again clipped shut to prevent stray pollen from entering.

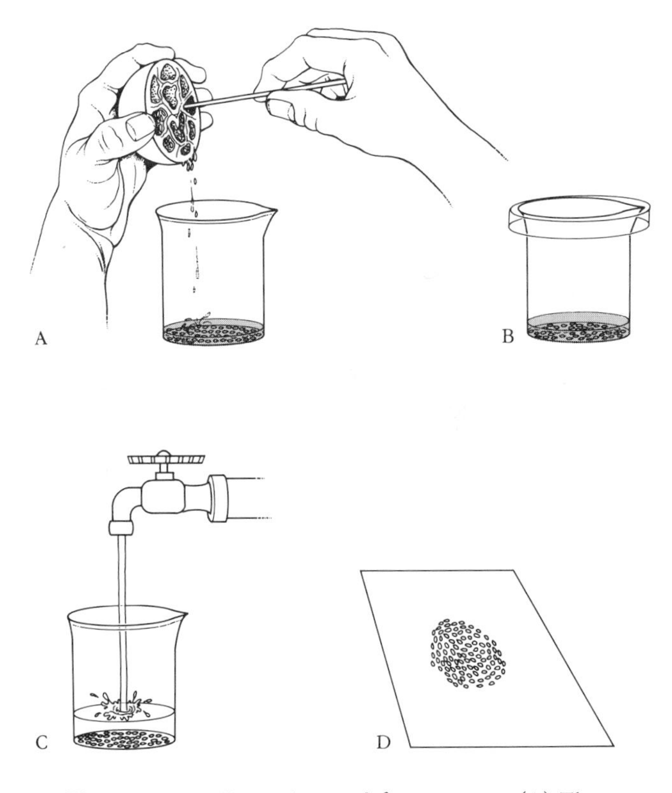

Figure 14-57 Extracting seed from tomato. (A) The fruit is cut open and the seed with as little pulp as possible is scraped and squeezed into a container (B). A loose cover is placed over the container and the pulp and juice are allowed to ferment for 4 to 7 days. The seed is rinsed several times (C) and allowed to dry on a paper towel (D).

tional status; whether it is nursery grown, growing around a home, or a native plant; and the weather conditions and climate of the area.

Securing and Planting Nursery-Grown Woody Plants

Woody plants are sold in three ways: bare-root, balled and burlapped, and in containers.

Sources. Nursery stock can be purchased from local nurseries and garden stores, from mail-order nurseries, and from supermarkets and department stores. Good buys on trees and shrubs are often available from department stores and supermarkets, and stores of this kind that have plant and garden departments managed by plant specialists are reliable sources of nursery stock. Those that sell plants seasonally as an adjunct to produce or hobby departments may be a less reliable source as they may not have facilities or personnel trained to properly care for plant materials. Occasionally plants not salable elsewhere are sold at bargain prices to these outlets. Mail-order nurseries are a good source of bare-root planting stock, especially of uncommon and unusual trees and shrubs, during the dormant season. A local nursery or garden store is usually the most reliable, albeit the most expensive, year-round source of woody plants, and sometimes the only source during the summer.

Method of Planting. Select a well-drained site that provides the environment required by the tree or shrub—sunny, shady, wind-free, and so on. Dig a hole that allows plenty of room for root growth—a minimum of 3 to 4 feet (about 1 m) in diameter and 2 feet (⅔ m) deep for a tree and 2 feet in diameter and 1½ feet (½ m) deep for shrubs. If the soil dug from the hole is not good topsoil, it should be mixed with 20–25% well-decomposed organic matter (peat moss, compost, aged manure). Don't place unmixed organic matter in the bottom of the hole where it will be in contact with roots.

Prune away at least 50% of the leaf-bearing surface of the bare-root plants and about 20–25% of balled and burlapped plants. Because leaves are borne primarily on smaller twigs, the pruning can be accomplished without extensive reduction in size. Overpruning is preferable to underpruning. Leaving too many leaves to dissipate the limited moisture absorbed by the drastically reduced root system is the most common cause of death due to transplanting. Damaged and excessively long roots also should be removed. Container-grown plants generally do not require pruning.

Roots of bare-root transplants should be spread around the hole. The string around the trunk or crown of balled and burlapped stock should be cut after the plant has been properly placed in the hole. The burlap should be pulled back from the top of the ball so that it will be covered with soil, but it need not be removed from the plant. Container-grown plants can often be knocked from wider mouthed pots by tipping the container upside down and tapping lightly. The sides of metal cans should be cut from top to bottom in three places and pulled back so that the plant with its ball of earth can be removed without disturbing the root system (Figure 14–58). Planting of the tree is shown in Figure 14–59.

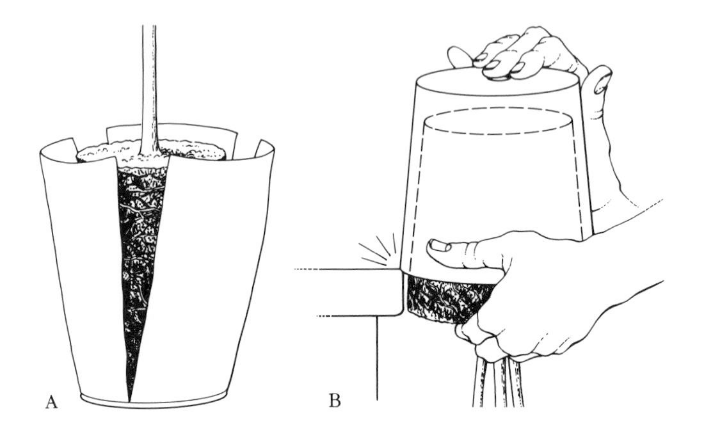

Figure 14–58 Removing a nursery plant from its container. The sides of a metal straight-sided can should be cut from top to bottom in three places (A). A tapered container with an opening wider than its base can be turned upside down and tapped against the edge of a table or bench (B).

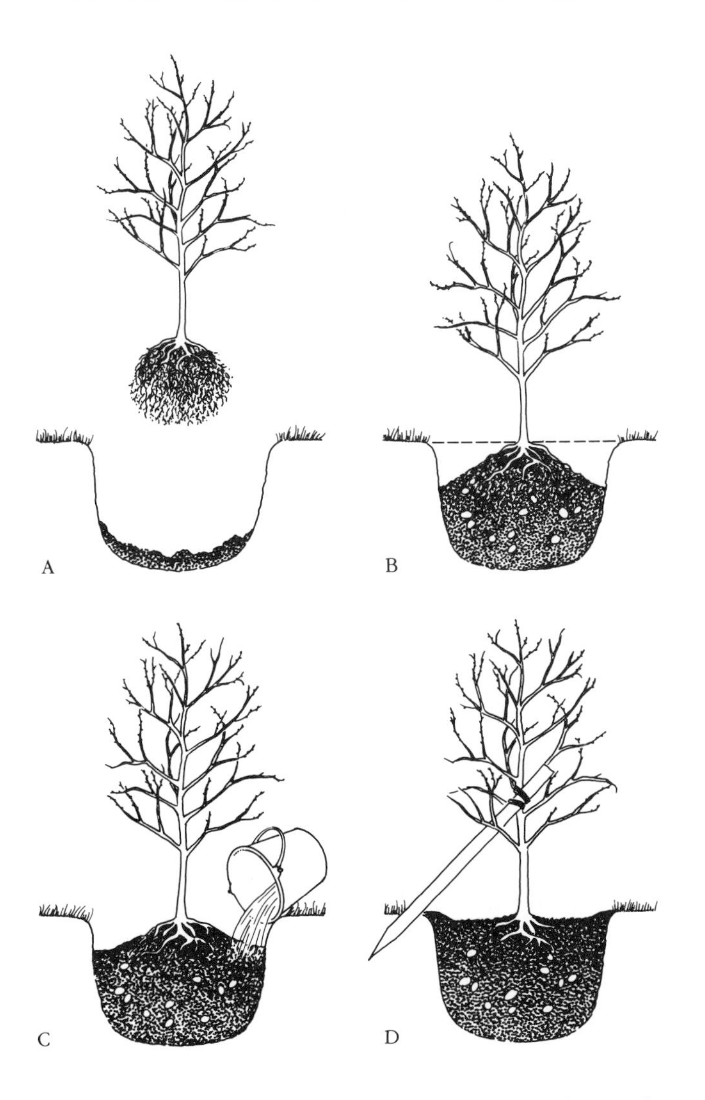

Figure 14–59 Planting a tree. (From *Transplanting Woody Plants,* by B. Adams. Washington State University Extension Bulletin 675. Pullman, WA. 1976.)

(A) Dig a hole a foot deeper than height of roots and twice as wide as root span or rootball. Loosen several inches of soil at bottom of hole to aid drainage.

(B) Add soil to hole and build it up in a mound beneath and among the roots so plant is at same ground level as before it was moved.

(C) Fill three-fourths of hole with soil, then water. Complete filling the hole with soil and water again. In arid regions leaving a slight depression around the tree will aid irrigation during the first summer.

(D) Drive a stake into the soil at a 45° angle to brace small trees. Pad stake at point of intersection and secure tree with soft cord or rag strips.

Transplanted trees and shrubs require a steady supply of water until roots are regenerated; however, more transplants are killed by overwatering than by underwatering. In most areas a thorough soaking each 10 to 14 days is enough. It is a good idea to dig down 4 to 6 inches (10 to 15 cm) near the plant occasionally to check soil moisture.

Don't permit fertilizer to come in contact with the root system. If the plant needs fertilizer it can be watered with a gallon or two of nutrient solution (see "Fertilizing Garden Crops," this chapter).

Transplanting Established Plants

The main difference between nursery-grown plants and native or cultivated plants that need to be moved from their present location is that the nursery plants have been root pruned to confine and concentrate their roots. Digging around a plant to cut wide-spreading roots 6 months and/or a year before it is to be transplanted will greatly increase the chances of its being successfully moved.

As was mentioned in Chapter 2, plants with fibrous roots are easier to move with a ball of earth intact than are those with tap roots. Determining which plants are easily moved is mostly trial and error; I know of no reliable root-system classification. For instance, when moving trees in the woods, I have had no problem keeping a good ball of earth around the roots of Douglas fir (*Pseudotsuga menziesii*) and Engelmann spruce (*Picea engelmannii*), but a great deal of difficulty keeping any soil whatsoever on the less concentrated root system of grand fir (*Abies grandis*).

Size and Transplantability. Smaller plants are obviously easier to move, and the smaller the plant, the greater the probability of successful reestablishment. However, with some experience two strong people can move (tug or drag may be more appropriate terms) shrubs and coniferous trees up to 6 to 8 feet (2 to 2½ m) with a ball of earth that can be encased in the cloth made by opening three sides of two 100-pound burlap bags and splicing the two together with nails.

Method of Transplanting. If the plant cannot be transplanted with bare roots, several options may be possible (Table 14–13). Often, especially when plants are small, roots are fibrous, and soil is heavy, the ball of earth will adhere without being encased in a sack. In some cold winter areas it is common practice to move larger shrubs and trees with their roots in a frozen ball of earth. Before the ground freezes, a trench is dug around and under the root system just as for balling and burlapping. The planting hole is also prepared. After the ball is frozen, it is broken from the soil below with a digging bar and moved to its new location. The roots should be protected from extreme cold and dehydration with straw, peat moss, or similar material until the ground thaws enough to complete the planting.

Balling and Burlapping. The standard method of moving medium-sized and large trees and shrubs without exposing their root system is balling and burlapping, illustrated in Figure 14–60.

Repotting

As houseplants and other potted plants increase in size, they will eventually need to be transferred into larger containers.

There are several conditions that suggest that a plant needs to be repotted. These include:

1. The plant looks overgrown for the size of its pot. The balance between plant and container is no longer aesthetically pleasing.

2. The top part of the plant has become so large that its container tends to be unstable and tips easily.

3. The plant has become rootbound—roots are either compacted along the top and sides of the container or are growing through the drainhole.

Table 14–13 Season and method of transplanting that will give reasonable success for most woody plants

Class of plants	Season	Balled and burlapped	Containerized	Bare root
Evergreens under 3 years	Dormant	X	X	X
Evergreens over 3 years	Dormant	X	X	
Evergreens over 3 years	Summer		X	
Deciduous	Dormant	X	X	X
Deciduous	Summer		X	

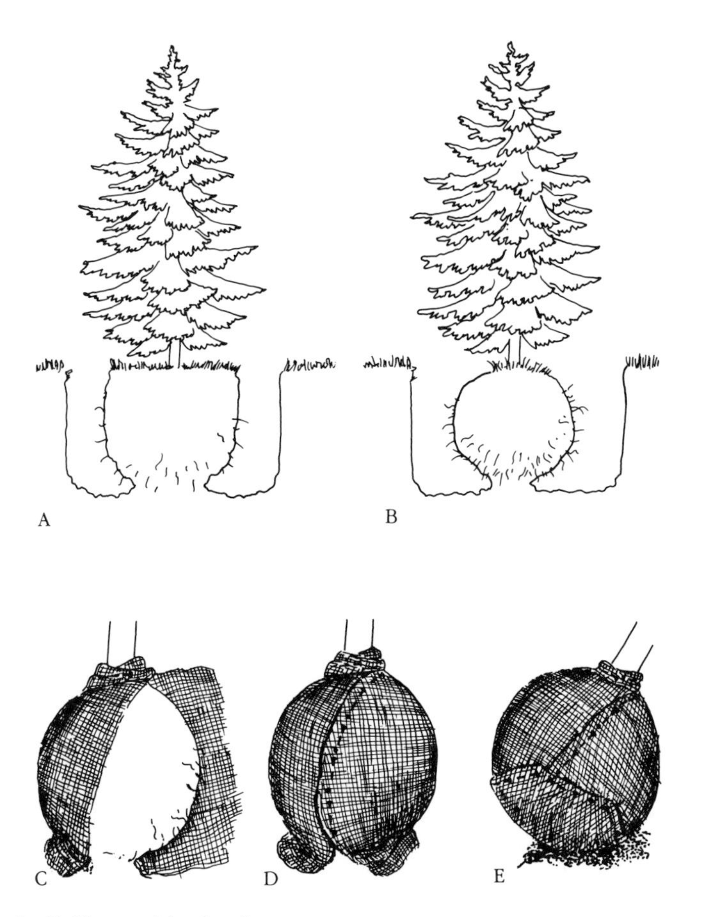

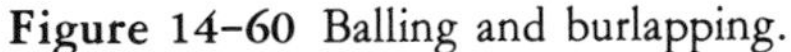

Figure 14–60 Balling and burlapping.

(A) Using a sharpened, straight-handled, round-mouth spade, dig a trench about half as deep as the plant is high all the way around the tree or shrub at about the outer perimeter of the leaf canopy. Diameter of the ball relative to the canopy spread and height of the plant will vary somewhat for cultivars of unusual shape. The initial cuts should be made with the back of the spade toward the tree so that the ball of earth is kept intact.

(B) Round off the ball of earth and dig under it until only a small part of it remains attached to the soil beneath.

(C) Wrap one corner of a piece of burlap of adequate size around the trunk or crown. Fasten using a 6- or 8-penny nail as a pin.

(D) After the sack is securely fastened around the trunk or crown, it can be pulled tightly around the earth ball and pinned with nails. With some kinds of soil nails will hold (and a tighter fit is possible) if they are pushed straight into the soil ball through both overlapping layers of burlap. The excess burlap should be positioned at one side of the base of the ball.

(E) With its upper soil tightly bound by the burlap, the ball can be cut from the soil supporting its base and rolled onto one side against the edge of the hole. The loose end of the unpinned burlap is pulled under and pinned to enclose the basal soil and complete the operation.

4. The soil in the container dries rapidly. It has to be watered more frequently than other plants in the home.

5. The plant is doing poorly, and there is no other readily discernible cause.

Figure 14–61 explains the method of repotting a houseplant.

Figure 14–61 Repotting a houseplant.
(A) Remove the roots and rooting medium from the pot by carefully tipping the plant upside down and tapping the container lightly against the edge of a bench or table.
(B) Use your fingers to thin out and spread the root system matted around the edge and base of the pot.
(C) Place a piece of broken clay pot or other loose-fitting material over the drain hole of the new pot. Add ½ to 1 inch (about 2 cm) of fine gravel to the bottom for drainage and enough potting medium so that the plant and its old soil ball will be at the desired level. (Setting the plant somewhat deeper than it originally was to hide a stemmy base usually will not be detrimental.) Finally, fill the remaining space in the pot with the potting medium and water.

Mulching with Plastic

Benefits of mulching are discussed in Chapter 5. Plastic mulch can be conveniently used for crops that are normally spaced a foot or further apart and either transplanted or direct-seeded in hills (Figure 14–62, p. 398).

Controlling Plant Growth with a Permanent Mulch

Polyethylene can be made to last for many years by preventing it from being exposed to the sun's rays. The usual way to preserve it is to completely bury it with several inches of sand, gravel, or river rock. A thick layer of cedar or Douglas fir bark or other loose material can be used as a covering if holes are punched in the plastic so that rainwater doesn't accumulate above it and float the covering away.

Following is a brief description of the application of a semipermanent mulch to a small front yard that the owner wanted maintenance free:

The area to be mulched was cleared of weed growth, excavated to 3½ inches (9 cm) below the level of the sidewalk and drives, and leveled to eliminate humps and depressions. Four mil black polyethylene was spread over the area, stretched taut, and held in place with cinder blocks as weight and large nails as pins. Plants were planted through the plastic so that they would be at the proper depth when the loose mulch was added. Holes were punched through the plastic 2 feet (30 cm) apart each way to allow water to drain. The plastic was covered with 2 inches of coarse sand topped with an overlapping layer of round (4 to 8 inch [10 to 20 cm] diameter), flat (1½ inch [4 cm] thick) river rocks common in the area. The cinder blocks were removed, and the area under them was covered with sand and rock.

Composting

Materials to Use

Straw Sawdust Leaves Manure
Vegetable and fruit peelings, tops, cores
Garden refuse Lawn clippings Immature weeds

Materials Not to Use

Woody branches Bones Diseased plant material
Weeds that have formed seed
Weeds or garden refuse that have been killed with herbicides or have been recently treated with other pesticides

The Composting Structure

Organic matter can be composted by piling it in a corner, but generally the composting process will occur more rapidly and satisfactorily in an enclosure. The enclosure can be a pit or a plastic bag, but more commonly it is built with mesh wire, wood, or other structural material. A common composting bin has three enclosed sides each 4 to 6 feet (about 1¼ to 2 m) long (a size convenient for handling) and a height of 4 or 5 feet. (Aeration is reduced by compaction resulting in slowed decomposition if the pile is over 6 feet high.) One side is often open to permit easy mixing of the compost.

The compost pile is made by alternating layers of 6 to 8 inches (15 to 20 cm) of organic matter with about an inch of garden soil (Figure 14–63). One-fourth pound (112 g) of nitrogen should be mixed with each cubic foot or bushel of composted dry matter (about half that amount if the compost is mainly kitchen wastes or lawn clippings). The pile should have 60–70% moisture. During dry weather the pile should be checked each week (or each time it is turned) and water should be added if the material feels dry. It should not remain wet enough so that water can be squeezed from it. The process of decomposition will keep the composting material warm, and it should be turned a few times during the summer (or every 3–4 days for rapid decomposition). Decomposition rate will vary and can occur in as short a time as 3 weeks. It requires less effort to let the process proceed for most of one growing season. The volume of the pile will decrease as decomposition occurs, but it is generally more convenient to start a new pile each year and to spread the compost from the previous season around the garden as the need arises.

Fertilizing Garden Crops

Determining the Amount of Fertilizer

Fertilization of the garden should be based on:

1. Your own or neighbor's experience during previous seasons.

2. Soil test from your state university soil-testing laboratory (see Table 14–1), from a reliable private soil-testing laboratory or, as a rough guide, from a do-it-yourself soil-test kit.

3. If better estimates are not available, use Tables 14–16 and 14–17 as a general guide.

4. In humid areas soil pH will often be too low for good growth of most garden crops. Quick pH test kits, available at garden stores, can be used if a more accurate determination of pH from an official soil test is not available. If the soil has a pH below 6.2, lime should be added. The amount initially required to bring soil pH to the desired level between 6.2 and 6.8 varies (Table 14–14). After the pH has been raised to the desired level,

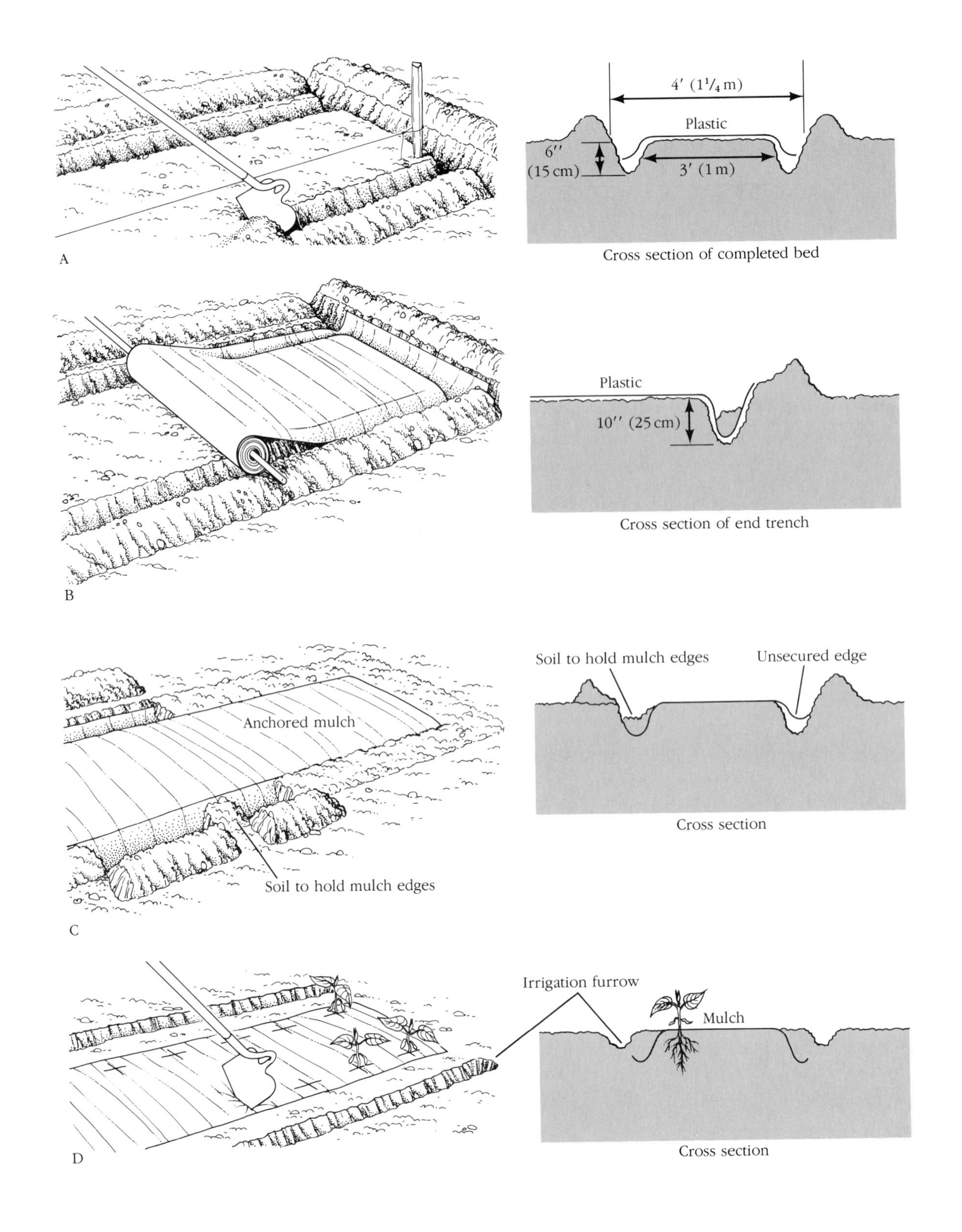
4′ (1¼ m)
Plastic
6″
(15 cm)
3′ (1 m)
A
Cross section of completed bed
Plastic
10″ (25 cm)
B
Cross section of end trench
Soil to hold mulch edges
Unsecured edge
Anchored mulch
Soil to hold mulch edges
C
Cross section
Irrigation furrow
Mulch
D
Cross section

Figure 14–62 Covering a small garden bed with plastic mulch.

(A) With the corner of a hoe dig a furrow at least 6 inches (15 cm) deep on each side of the bed that is to be mulched, and a furrow 10 inches (25 cm) deep at one end, leaving the soil mounded beside the furrows. The completed bed should be 1 foot (30 cm) narrower than the width of the plastic.

(B) Begin unrolling the roll of plastic by placing its end across the end furrow and tucking it all the way to the bottom. While holding the plastic in place, fill the trench with soil and tamp it to secure the end of the plastic.

If two people are working together, they should place a hoe or rake handle through the roll of plastic and, keeping the mulch stretched tightly, unroll it over the bed. One person can unroll a roll as wide as 4 feet (1¼ m) by holding it upright over his open hands. With the tips of his fingers at the point where the sheet is leaving the roll, he can cause the roll to turn over his wrists by walking backward along the center of the bed. (Don't attempt to lay plastic mulch by yourself if the wind is blowing.)

(C) After you have unrolled about 50 feet (15 m) of plastic sheet, hold its edges in place in the furrow by scraping a small quantity of soil onto them. Do this each 5 to 6 feet (about 2 m) on both sides of the bed. Then bury both edges by shoveling soil into the furrow. Make certain that the mulch is stretched tightly across the bed. The weight of soil will usually do this, but some manipulation by hand occasionally may be necessary. Tamp the soil well to seal down the edges of the mulch.

(D) Using the corner of the hoe or a small garden trowel, cut through the plastic at intervals to set transplants or plant hills of seed at their proper spacing. If the plot is to be furrow-irrigated, plant near one edge of the bed. Moisture will soak under the edge of the mulch. If the plot is to be sprinkler-irrigated, punch holes in the plastic over a few low places in the bed. It is necessary to cover only a part of the plot for a wide-spaced crop like melons or cantaloupe, e.g., 3 feet of their 9-foot bed.

If the plastic used is polyethylene, its exposed part will deteriorate by the end of the growing season, but the soil-covered strips along the edge will remain intact and will have to be removed before the soil can be reworked.

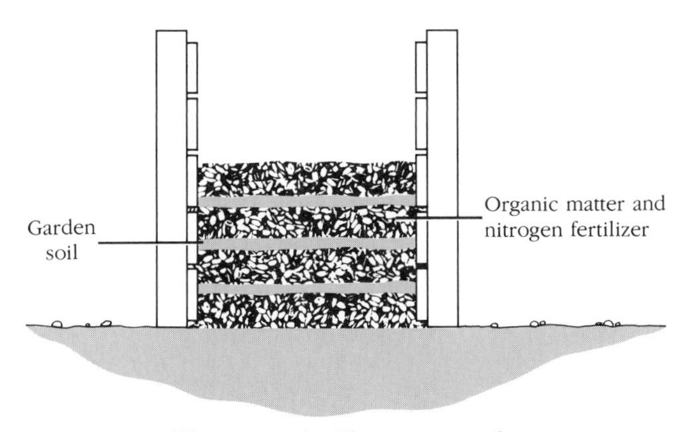

Figure 14–63 The compost pile.

an application of 60 to 80 pounds (30 to 40 kg) of ground limestone, or its equivalent, each 4 to 5 years will usually keep pH at the optimum level. Blueberries, rhododendrons, azaleas, heathers, and heaths require a low pH and should not be limed.

5. An additional ⅛–¼ pound (75–125 g) of nitrogen should be added for each 100 pounds (45 kg) of organic mulch or crop residue that has less than 1% nitrogen (Table 14–15).

Fertilizing Vegetables and Flowers

See Tables 14–16, 14–17, and 14–18. Most fertilizer for vegetables and annual flowers should be applied in the

Table 14–14 Approximate amounts of finely ground limestone needed to raise the pH of a 7-inch layer of soil as indicated[a]

Soil regions and textural classes	Limestone requirements: From pH 3.5 to pH 4.5 (tons per acre)[c]	From pH 4.5 to pH 5.5 (tons per acre)[c]	From pH 5.5 to pH 6.5 (tons per acre)[c]
Soils of warm-temperate and tropical regions:			
Sand and loamy sand	0.3	0.3	0.4
Sandy loam	—	0.5	0.7
Loam	—	0.8	1.0
Silt loam	—	1.2	1.4
Clay loam	—	1.5	2.0
Muck	2.5[b]	3.3	3.8
Soils of cool-temperate and temperate regions:			
Sand and loamy sand	0.4	0.5	0.6
Sandy loam	—	0.8	1.3
Loam	—	1.2	1.7
Silt loam	—	1.5	2.0
Clay loam	—	1.9	2.3
Muck	2.9[b]	3.8	4.3

(Adapted from *Soil Survey Manual.* USDA Agricultural Handbook 18. U.S. Government Printing Office, Washington, DC. 1951.)

[a]All limestone goes through a 2 mm mesh screen and at least half through a 0.15 mm mesh screen. With coarser materials applications need to be greater. For burned lime about half the amounts given are used; for hydrated lime about three-quarters.

[b]The suggestions for muck soils are for those essentially free of sand and clay. For those containing much sand or clay the amounts should be reduced to values midway between those given for muck and the corresponding class of mineral soil. If the mineral soils are unusually low in organic matter, the recommendations should be reduced about 25%; if unusually high, increased by about 25% or even more.

[c]Multiply the tons per acre required by 4.6 to obtain pounds per 100 square feet. Multiply tons per acre by 2.24 to obtain metric tons per hectare.

Table 14–15 The nitrogen, phosphorus, and potassium content, as a percentage of fresh weight, of several soil-improving crops and organic wastes used frequently in gardening

Crop or product	N (%)	P_2O_5 (%)	K_2O (%)
Alfalfa, vetch	2.5	0.5	2.0
Clovers	2.0	0.5	1.5
Bean and pea vines	0.5	0.1	0.5
Small grains, grass clippings	0.4	0.1	0.3
Dairy manure (medium amount bedding)	0.7	0.3	0.7
Sawdust, shavings	0.2	0.1	0.2
Grain straw	0.6	0.2	0.1
Peat moss (woody or sedge)	0.2	<0.1	<0.1
Peat moss (sphagnum)	0.1	<0.1	<0.1
Vegetable and fruit peelings	0.4	0.1	0.5
Dry leaves	0.8	0.2	0.6

(Information for compiling this table came from various sources, including: *Western Fertilizer Handbook*, 5th ed. Calif. Fertilizer Assoc., Sacramento. Copyright © 1975. *Composts for the Garden*, by A. A. Duncan. Fact Sheet 134. Extension Service, Oregon State University, Corvalis, OR. 1971. *Soil*, U.S. Department of Agriculture Yearbook of Agriculture, 1957. U.S. Government Printing Office, Washington, DC. 1957.)

Table 14-16 Average pounds of N, P, and K needed for 1,000 square feet of garden crops in humid and subhumid areas.[a,b,h] (Divide by 2 to determine kg/100 m^2.)

Unit	Method of application[c]	Low organic matter sandy soil				High organic matter clay soil			
		When[c,d]	N	P	K	When[c,d]	N	P	K
Temp. zone tree fruits, nuts	B[e]	E. Sp & Su	3 & 1	3	3	E. Sp	2	2	2
Citrus fruits	B[e]	Sept., Jan.	3 & 2	3	3	Sept., Jan.	2,2	2	2
Strawberries	B	E. Sp	2	4	0	E. Sp	1½	3	0
Grapes	B & Si[e]	E. Sp & Su	3 & 1	3	3	E. Sp	2	0	2
Bushberries	Si	E. Sp	3	6	3	E. Sp	2	4	2
Asparagus	Si	E. Sp & Su	1 & 1	3	4	Su	1½	2	3
Other perennial vegetables	Si	E. Sp & Su	1 & 1	3	4	E. Sp	1½	2	3
Lettuce, radishes, spinach, turnips[f]	B & Si	PT	3	3	3	PT	2½	2½	2½
Peas, beans, lima beans	B	PT	1	2	2	PT	1	2	2
Carrots, cucurbits, tomatoes	B & Si	PT & Su	2 & 1	4	2	PT	1½	3	1½
Other long-season vegetables	B & Si	PT & Su	3 & 2	6	3	PT & Su	2½ & 1½	5	2½
Bulbs and corms	B	E. Sp	2½	5	2½	E. Sp	2	4	2
Annual flowers	B & Si	PT & Su	2½ & 2	5	2½	E. Sp & early Su	2 & 1½	4	2
Perennial flowers spring and summer blooming	B	E. Sp	3	6	3	E. Sp	2	4	2
Perennial flowers fall blooming	B & Si	Sp & Su	3 & 1½	6	3	Sp & Su	2 & 1	4	2
Roses	B	E. Sp & Su	2 & 2	4	2	E. Sp & Su	1½ & 1½	3	1½
Other shrubs and lianas	B[e]	E. Sp & Su	2 & 1	4	2	E. Sp	2	4	2
Spreading trees	B[e]	E. Sp & Su	2 & 1	4	2	E. Sp	2	4	2
Tall trees	B[e]	E. Sp & Su	3 & 1½	2	2	E. Sp	3	2	2
Turf[g]	B	March, June, August	2,2,2	2	2	March, June, August	1½,1½,1½	1½	1½

[a]Humid and subhumid areas are those areas east of a line drawn from Winnipeg, Manitoba, to Corpus Christi, Texas; plus coastal Alaska and British Columbia, Oregon, and Washington west of the Cascades and California north of San Francisco and north and west of the Central Valley.

[b]Where annual rainfall is over 60 inches, the growing season is more than 200 days, and the fertilizer used is not slow-release, N, P, and K additional to the one or two applications recommended in the table should be applied during mid-summer in an amount equal to half the initial spring recommendation.

[c]B = broadcast; Si = sidedressed; E. Sp = early spring; Su = summer; PT = planting time.

[d]If the main source of fertilizer is slow-release, a small amount of sidedressed, fast-release N and P will often enhance growth of early-planted crops. With slow-release fertilizers the later summer application may not be necessary.

[e]The area to be fertilized around a tree or shrub is a square or rectangle just beyond the periphery of the canopy. If there is no cover crop of if the cover is sparse, the fertilizer can be broadcast on the soil surface. If the woody plant is growing in a lawn, fertilizer should be placed in holes punched to a depth of 14 to 18 inches and spaced 2½ feet each way across the entire area. Turf growing under a tree or shrub should also receive its allotted broadcast of fertilizer. Other suggestions for woody plant fertilization are found in the text.

[f]Broadcast 1 or 2 pounds of N, and other elements if required, on each 1,000 square feet of vegetable garden before the soil is worked. Apply the extra amount as a sidedressing at planting time to those vegetables that require more.

[g]Lawns composed of Bermuda grass will need more N, 5–10 pounds of actual N per 1,000 sq ft per year applied three to five times during the growing season.

[h]Lime is needed periodically on many soils in humid areas. See Item 4 under "Determining the Amount of Fertilizer" in the text; also Table 14–14.

spring. If manure is available, it should be incorporated when the soil is worked, preferably by digging or plowing it under. In areas where phosphorus is deficient, the application of manure should be supplemented with a fertilizer high in phosphorus. Tomatoes, the cucurbits, carrots, peas, beans, and some of the short-season vegetables do not require as much fertilizer as do most other annual vegetables. The amount required by these crops can be broadcast onto the garden before the soil is dug in the spring. More fertilizer can be **banded**—applied in a narrow band parallel to the crop row—at planting time for crops requiring it. Ideally, fertilizer should be banded 2

Table 14–17 Average pounds of N, P, and K needed for 1,000 square feet of garden crops in arid and semiarid areas.[a] (Divide by 2 to determine kg/100 m².)

Unit	Method of application[b]	Low organic matter sandy soil				High organic matter clay or silt loam			
		When[c,d]	N	P	K	When[b,c]	N	P	K
Temp. zone tree fruits	B[d]	E. Sp	3	3	0	E. Sp	2	0	0
Citrus trees	B[d]	Sept., Jan.	3,2	3	0	Sept., Jan.	2,2	0	0
Strawberries	B	E. Sp	2	2	0	E. Sp	1½	0	0
Grapes	B[d]	E. Sp	3	2	2	E. Sp	2	0	0
Bushberries	Si	E. Sp	2½	2½	0	E. Sp	2	0	0
Asparagus	Si	E. Sp	2	2	0	Su	2	0	0
Other perennial vegetables	Si	Su	2	2	0	E. Sp	2	0	0
Lettuce, radishes, spinach, turnips[e]	B & Si	PT	3	3	0	PT	2	0	0
Peas, beans, lima beans	B	PT	1	1	0	PT	1	0	0
Carrots, cucurbits, tomatoes	B	PT	2	2	0	PT	1½	0	0
Other long-season vegetables	B & Si	PT & Su	3 & 2	3	0	PT	3	0	0
Bulbs and corms	B	E. Sp	2½	2	0	E. Sp	2	0	0
Annual flowers	B & Si	PT & Su	2 & 1	2	0	PT	2	0	0
Perennial flowers									
spring and summer blooming	B	E. Sp	3	3	0	E. Sp	2	0	0
fall blooming	B & Si	Sp & Su	3 & 1½	3	0	Sp	2	0	0
Roses	B	E. Sp & Su	2 & 1½	2	0	E. Sp	2	0	0
Other shrubs and lianas	B[d]	E. Sp	2	2	0	E. Sp	2	0	0
Spreading trees	B[d]	E. Sp	2½	2½	0	E. Sp	2	0	0
Tall trees	B[d]	E. Sp & Su	3½ & 3½	3½	0	E. Sp	3	0	0
Turf[f]	B	March, June, August	1½,1½,1½	1½	0	March, June, August	1½,1½,1½	0	0

[a]Arid and semiarid areas would include the region west of a line drawn from Winnipeg, Manitoba, to Corpus Christi, Texas, except for the area west of the Cascades in Washington and Oregon, the coastal valleys of northern California, coastal British Columbia, and coastal Alaska.
[b]B = broadcast; Si = sidedressed; E. Sp = early spring; Su = summer; PT = planting time.
[c]See footnote [d], Table 14–16.
[d]See footnote [e], Table 14–16.
[e]See footnote [f], Table 14–16.
[f]See footnote [g], Table 14–16.

inches (5 cm) to the side and 1 inch (2½ cm) below the seed, especially if the planting is to be furrow irrigated or if rain is not anticipated. The band should be between the plant row and the irrigation furrow so that moisture can dissolve and carry the fertilizer to the plant roots. Soluble nitrate-nitrogen fertilizer can be banded on the surface if the plot is watered by sprinkler or rainfall.

For short-season vegetables and flowers and on soils with high organic matter content and high cation-exchange capacity, a spring application of fertilizer may be all that is required. For long-season crops growing on soils low in organic matter, one or more additional fertilizer applications may be needed. Fertilizer applications banded after the crop is growing are called **side-dressings.** These later applications are usually banded a few inches from the plant 6–10 weeks after it is planted.

Perennial flowers and perennial vegetables, except asparagus, are usually fertilized in the early spring. Asparagus responds best to being fertilized right after the cutting season. The fertilizer for perennial flowers and vegetables can be either sidedressed or spread in a circle a few inches away from the crown of each plant.

Table 14–18 Approximate number of pounds of various analyses of fertilizer to apply if recommendation is given in pounds of actual N, P, or K per 1,000 square feet or per acre. (Divide by 2 to determine kg/100m².)

Recommendation[a] →	40 lb/acre equivalent to 1 lb/1,000 sq ft		60 lb/acre equivalent to 1½ lb/1,000 sq ft		80 lb/acre equivalent to 2 lb/1,000 sq ft		100 lb/acre equivalent to 2½ lb/1,000 sq ft		120 lb/acre equivalent to 3 lb/1,000 sq ft	
	Pounds required		Pounds required		Pounds required		Pounds required		Pounds required	
Analysis	100 sq ft	1 acre	100 sq ft	1 acre	100 sq ft	1 acre	100 sq ft	1 acre	100 sq ft	1 acre
0.3	35.0	14,000	50.0	20,000	70.0	26,000	80.0	35,000	100.0	42,000
0.5	20.0	8,000	30.0	12,000	40.0	16,000	50.0	20,000	60.0	24,000
0.7	15.0	6,000	22.5	9,000	30.0	12,000	37.0	15,000	45.0	18,000
1.0	10.0	4,000	15.0	6,000	20.0	8,000	25.0	10,000	30.0	12,000
1.5	6.7	2,700	10.0	4,000	13.3	5,300	16.7	6,700	20.0	8,000
2	5.0	2,000	7.5	3,000	10.0	4,000	12.5	5,000	15.0	6,000
3	3.3	1,300	5.0	2,000	6.7	2,700	8.3	3,300	10.0	4,000
5	2.0	800	3.0	1,200	4.0	1,600	5.0	2,000	6.0	2,400
7–8	1.3	500	1.9	750	1.5	1,000	3.3	1,250	3.7	1,500
9–10	1.0	400	1.5	600	2.0	800	2.5	1,000	3.0	1,200
11–12	0.8	330	1.3	500	1.7	670	2.1	830	2.7	1,000
15–16	0.6	240	0.9	360	1.2	480	1.5	600	1.8	720
18–20	0.5	200	0.75	300	1.0	400	1.2	500	1.5	600
25–27	0.4	160	0.60	240	0.8	320	1.0	400	1.2	480
33–36	0.3	120	0.45	180	0.6	240	0.8	330	0.9	360
46–50	0.2	80	0.30	120	0.4	160	0.5	200	0.6	240

[a]For higher application rates—160, 200, or 240 lb/acre or 4, 5, or 6 lb/1,000 sq ft—apply twice the amount recommended for 80, 100, or 120 lb/acre, respectively. For lower application rates, 20 or 30 lb/acre or ½ or ¾ lb/1,000 sq ft, apply half the amount recommended for 40 or 60 lb/acre.

Fertilizing Woody Plants

General directions for fertilizing most woody plants are given in Tables 14–16 and 14–17. Amount of fertilizer for plants like grape, where the location of branches may not be directly above the root system, can be calculated on the basis of normal spacing of the plant (Chapter 13 and Table 14–25). Fertilizer should be spread around the trunk over the soil area allotted to the plant.

General Precautions for Woody Plant Fertilization

1. Use less N if vegetative growth is excessive and/or if flowering and fruit set are not adequate.

2. With plants and at locations where winter damage is common, don't fertilize after mid-summer.

3. Don't place large amounts of fertilizer close to the crown of a plant.

4. When fertilizer is broadcast on low-growing shrubs, be sure it is washed from their leaves and soaked into the soil.

Recommendations for Some Specific Plants

1. Grapes growing on medium and high organic-matter soils require about ⅕ lb (90 g) elemental nitrogen per plant per year; and currants, gooseberries, and blueberries require about a tablespoon or ½ oz (14 g) per plant.

2. Another common method of calculating fertilizer requirements for apple and pear trees is

$$\frac{\text{trunk diameter}}{9} = \frac{\text{pounds of N}}{\text{required per tree}}$$

3. In arid regions both tree and bush fruit plants often exhibit a yellow interveinal chlorosis as a result of calcium-

induced iron deficiency. Applications of a chelated iron is usually the most satisfactory way of overcoming this problem.

Fertilizing Lawns

In areas where they stay green year-round, lawns should be fertilized approximately every 3 months. In northern areas they should generally receive an application in March, in June, and in August. Fertilizer requirements vary, and recommendations for specific locations are available from the county agent or state experiment stations. For lawns composed entirely of grass, from 1 to 1½ lb of elemental nitrogen will be required for every 1,000 sq ft (500–700 g/100 m^2) every three months (Table 14–16 and 14–17).

It is important that fertilizer be spread uniformly over the lawn area. This can be accomplished with a fertilizer spreader, which can be purchased or rented from a local hardware or garden store. Directions for calibrating should be followed carefully, and fertilizer should be spread over the entire area. Often an operator allows too little overlap to fertilize wheel tracks, which results in strips of light green grass. Fertilizer can also be spread by hand broadcasting. When this is done, the amount needed should be divided into two equal portions so that the area can be covered twice, the second application being at right angles to the first.

Fertilizing Houseplants

Houseplants should be fertilized approximately once each month. Various houseplant fertilizer formulations are available at nurseries and garden stores. For the gardener who already has commercial fertilizer on hand, approximately 3 tablespoons (1½ oz or 40 g) of 10:10:10 analysis fertilizer mixed in a gallon of water will provide about the right nutrient concentration and can be used in place of a usual watering once every 3 or 4 weeks. A tablespoon of treble superphosphate, a tablespoon of ammonium nitrate, and a tablespoon of potassium sulfate in a gallon of water will also provide about the same balance.

Examples to Illustrate Use of Fertilizer Tables

In each example, if the fertilizer recommendation is not given, it must be determined using either Table 14–16 or Table 14–17. Find the plant and its soil type in the appropriate table. With the figures on the amount of N, P, and K needed per 1,000 sq ft, use Table 14–18 to find how many pounds of fertilizer to apply, given the analysis of the fertilizer. If the fertilizer analysis is not given, check Table 14–19, which lists the analysis of many common fertilizers. In Table 14–18, select either "100 sq ft" or "1 acre," depending on the size of the area to be fertilized. This "pounds required" figure from Table 14–18 must then be multiplied by (square footage to be fertilized/100) or by the number of acres to be fertilized. Table 14–19 gives weight/volume equivalents for various units of measure, to convert pounds to the appropriate unit for dispensing the fertilizer. This final number is the amount (in desired units) of a fertilizer with given analysis to be broadcast or sidedressed over a given area. Tables 14–16 and 14–17 give the method and times for application.

I. A 40 x 50 ft lawn growing in a silt loam soil in the upper Midwest; no soil test; fertilizer available is ammonium nitrate and 10-10-10.

   1. Required fertilizer for this soil and climate are 1½ lb of N, 1½ lb of P_2O_5, and 1½ lb of K_2O for each 1,000 square feet for the March application; and 1½ lb of N in June and August (Table 14–16).

   2. From Table 14–18: 1½ lb of a fertilizer with an analysis of 10 are required for each 100 sq ft. Because this lawn (2,000 sq ft) is 20 times as large, it will require 20 x 1½ or 30 lb of 10-10-10 fertilizer.

   3. The 30 pounds can be weighed out and spread on the lawn, or it can be measured. Table 14–19 shows that 1 pound of 10-10-10 fertilizer equals approximately 1 pint, so 30 pints (15 quarts or 3¾ gal) are needed.

   4. In June $0.45 \times 20 = 9$ lb of ammonium nitrate (analysis 33-0-0) will be required, and the same amount will be needed in August (Tables 14–18 and 14–19).

II. An apple tree growing in a yard at Greeley, Colorado, in a relatively light sandy soil. An abundance of

Table 14–19 Composition, effect on soil pH, and weight/volume equivalent of common fertilizers

Formula	Analysis	Effect on soil pH[a]	Pounds	Cu ft or bushels	Pints	Tbsp	Kg	Liters
Ammonium nitrate	33-0-0	A						
Ammonium sulfate	20-0-0	A						
Potassium chloride	0-0-55	N						
Sodium nitrate	16-0-0	B	1	1/60	1	32	0.45	0.45
Ammonium phosphate sulfate	16-20-0	A	60	1	60		27	27
Mixed lower analysis fertilizers, including	5-10-5	A	2.2	1/27	2.2		1	1
	10-10-10	A						
	10-6-4	A						
Mixed higher analysis fertilizers, including	19-9-0	A						
	27-12-0	A						
Treble superphosphate	0-48-0	N						
Potassium sulfate	0-0-50	N						
Ground limestone	(Ca)	B						
Ground dolomitic limestone	(Ca & Mg)	B						
			1	1/70	0.8	27	0.45	0.4
Superphosphate	0-18-0	N	1.2	1/60	1	32	0.55	0.45
Ammonium nitrate (high density)	33-0-0	A	70	1	60		32	27
Mixed fertilizers, including	12-12-12	A	2.2	1/32	1.8		1	0.8
	16-16-16	A						
	11-48-0	A						
	15-5-25	A						
	18-46-0	A						
Bonemeal (steamed)	0-20-0	N						
Urea	48-0-0	A	1	1/45	1.3	42	0.45	
Borax	(B)	N	3/4	1/60	1	32	0.34	0.45
Digested sewage sludge[b]	2-3-0	N	45	1	60		20	27
Activated sewage sludge[b]	6-3-0.5	N	2.2	1/20	2.9		1	1.3
Urea-form[b]	36-0-0	A						
Dairy manure[b]	0.7-0.3-0.7	N						
Hog manure[b]	1-0.7-0.7	N						
Poultry droppings[b]	4-3-2	N						
Poultry manure[b]	1.5-1-1	N						
Sulfur	(S)	A						
Cottonseed meal[b]	6-3-1.5	N	1	1/30	2	64	0.45	0.9
Tankage (dry)[b]	8-9-1.5	N	1/2	1/60	1	32	0.22	0.45
Fish scraps (dry)[b]	8-6-1	N	30	1	60		13	27
Cattle feedlot manure[b]	2-0.5-2	N	2.2	1/14	4.4		1	2
Horse manure[b]	0.7-0.3-0.5	N						
Sheep manure[b]	2-1-2	N						
Rabbit manure[b]	2-1-1	N						
Goat manure[b]	3-1.5-3	N						
Compact wood ashes	0-2-5	B						

[a] A = acid, B = basic, N = neutral.
[b] Slow-release fertilizers to which plants will usually not show a response for at least 14 days after application. Other fertilizers are quick- to moderate-release, and plants will respond to them in 3–14 days.

manure is available from cattle feeding operations nearby.

1. Measure a square just beyond the outer spread of branches (Table 14–16, note e). For this medium-sized tree the area is 18 × 18 = 324 sq ft. An apple tree requires about 3 lb each of N and P_2O_5/1,000 sq ft in this climate and soil (Table 14–17).

2. Manure from a feedlot has an N analysis of about 2 (Table 14–19), and about 15 lb of this analysis are required to supply N for each 100 sq ft (Table 14–18). The 325 sq ft under this tree will require 3¼ × 15 = 49 lb of manure. Because a bushel of feedlot manure weighs 30 lb (Table 14–19), about 1⅔ bushels would be required.

3. The manure supplies only about one-fourth of the needed P and will need to be supplemented with about 2¼ lb of P_2O_5 from a phosphate-containing fertilizer. There is considerable latitude in the amount of all elements except N, so the extra K supplied by manure usually will not cause a problem. If manure is used year after year over a period of many years, potassium and some other salts can accumulate in arid soils to a point where they could be detrimental.

III. A 5 x 10 ft bed of annual flowers growing in sandy soil along the Virginia Coastal Plain.

1. From Table 14–16: Annuals in this soil require a complete fertilizer with 2½ lb N, 5 lb P_2O_5 and 2½ lb K/1,000 sq ft (fertilizer with 1:2:1 ratio). This suggests that a fertilizer of an analysis 5-10-5 might be used.

2. Table 14–18 shows that this application rate requires 5 lb of a fertilizer with an analysis of 5 for each 100 sq ft. The 50 sq ft garden would require half that much, or 2½ lb, to supply the needed quantiy of all three major fertilizer elements.

IV. A soil analysis shows your garden soil to have a pH of 5.2, to be low in organic matter and potassium, and to be slightly low in phosphorus. It recommends 1,200 lb of ground limestone, 160 lb of nitrogen, and 160 lb of potassium per acre. You want to fertilize a 6-foot wide 50-foot row of black raspberries with a pile of rabbit manure that your neighbor wants to get rid of.

1. From Table 14–19, rabbit manure has an analysis of 2-1-1. Table 14–18 does not list 160 lb/acre, but it does show 10 lb of 2% fertilizer/each 100 sq ft for a rate of 80 lb/acre. For 160 lb/acre, apply double the 80 lb rate, or 20 lb for each 100 sq ft. This would be 60 lb to supply the N for the 300 sq ft of raspberries. Each bushel of rabbit manure weighs 30 lb, so you need about 2 bushels.

2. The 2 bushels of manure will supply only half the potassium required. The remainder (80 lb/acre) could come from potassium chloride. From Table 14–18, 0.4 lb of a 46–50% fertilizer would supply 100 sq ft. Because potassium chloride has an analysis of 55, you will need a little less than 3 × 0.4 = 1.2 lb. About 1 lb would supply the additional potassium.

3. None of the tables lists values as high as 1,200 lb/acre, the amount of limestone required; however, Table 14–18 shows that 120 lb/acre is approximately equivalent to 3 lb/1,000 sq ft. That means that 1,200 lb/acre would be equivalent to 30 lb/1,000 sq ft or 3 lb/100 sq ft. The raspberries need about 9 lb or (from Table 14–19) 9 × 0.8 = about 7 pints.

4. Because rabbit manure is organic and thus a slow-release kind of fertilizer, the total required amount could probably be applied at one time during the early spring. Applying 160 lb of N in a rapid-release form, for example, ammonium nitrate or ammonium sulfate, could be injurious; if these forms are used, a split application, 80 to 100 lb during the early spring and 60 to 80 lb 6 weeks later, would be desirable.

V. Fertilizer recommendation is 1,200 lb of 5-10-5 for your acre of garden, but you have ammonium nitrate (33-0-0), treble superphosphate (0-48-0), and potassium sulfate (0-0-50) available.

1. Multiply 1,200 by the concentration of each element (0.05 N, 0.10 P, and 0.05 K) to find that you need 60 lb of N, 120 lb of P_2O_5, and 60 lb of K_2O.

2. Table 14–18 shows that you need 180 lb of 33-0-0, 240 lb of 0-48-0, and 120 lb of 0-0-50 to supply the N, P, and K for your acre.

VI. Estimate the amount of horse manure to supply enough nitrogen to decompose 3 cu ft of dry leaf compost. According to Table 14–19, horse manure has about .7% or 0.007 N. Directions in the "Composting" section of this chapter recommend ¼ (0.25) lb nitrogen for each cu ft, which would be 0.75 for 3 cu ft. Manure needed would be $\frac{0.75}{0.007}$ or $\frac{750}{7}$, or a little over 100 lb!

Irrigating Garden Crops

Irrigating Vegetables, Flowers, and Strawberries

If the soil is dry at seeding time, it is better to irrigate before annual crops are seeded and then, if possible, not to water again until after seedlings have emerged. If rain does not fall, annual garden plants will require frequent light irrigation while they are becoming established in the early spring. Because small plants use and transpire less moisture than large ones, irrigation can be reduced as soon as plants have become well established but should be increased gradually as plants increase in size and the season becomes warmer and drier. Most garden soils will hold 1 to 2 inches (1½ to 2½ cm) of available moisture in their upper 2 feet (60 cm) (see Table 6–1) and, despite variation in water requirement from crop to crop, most garden plants during the warmer part of the season will use, on the average, 1 inch of water per week (up to 1½ inches [4 cm] in some desert areas). Thus an average irrigation recommendation for annual crops would be 1 inch per rainless week. There are some exceptions. Celery and onions, for example, require frequent lighter irrigation. In its native environment celery is a swamp plant that has evolved a very small root system; the root system of onion is also restricted. Plants of both crops can quickly deplete the small amount of moisture available in their limited root zones. Although onions require frequent irrigation early in the season, their moisture must be limited later or the bulbs will not mature properly.

Deep-rooted crops, such as melons, cucumbers, tomatoes, and hollyhocks, are quite drought tolerant and even in semiarid climates can frequently be grown with very little added water. Sweet corn is tolerant of droughty conditions until the period just before it starts to silk. At that time it must have moisture, or no crop will be produced.

Frequency of irrigation of herbaceous crops is somewhat dependent on the type of soil and the amount of moisture it will hold. Light, sandy soils may need their inch each 5 to 7 days, whereas heavy soils and those high in organic matter may be better supplied with 1½ to 2 inches every 10 to 14 days.

Irrigation amount and frequency for these crops should be reduced as the weather becomes cooler and the crop matures.

Perennial flowers, vegetables, and strawberries require a good moisture supply until after they have bloomed or the crop is harvested. After bloom or harvest they can be irrigated less frequently but they still require enough water to keep them producing vegetative growth and manufacturing carbohydrates for the following season's crop. Some, like rhubarb, peony, and bearded iris, become semidormant during late summer. Extreme care is required in watering dormant plants during hot weather. Because little water is being transpired by them, heavy irrigation can result in prolonged soil saturation, reducing the oxygen supply essential for root and crown respiration. Respiration may be quite rapid due to the heat, and a lack of oxygen under these conditions can result in severe injury or death of the root system.

Irrigating Woody Plants

Trees and shrubs require an average of about an inch of water each week during the growing season. It is important that their entire rooting area be moistened, so irrigations should be heavier and less frequent than for herbaceous crops, as much as 3 or 4 inches (8 to 10 cm) and as infrequent as each 3 to 4 weeks for large trees in soil with good moisture retention.

Peaches are somewhat more drought tolerant than are other tree fruits, so peaches, apricots, plums, and cherries are grafted onto peach rootstock when drought-resistant fruit trees are needed. Orchard or forest management also will determine the amount of water required. A tree grown with clean cultivation will require less water than one grown with grass or another cover crop, because enough water must be applied to the latter to supply the tree as well as the cover crop. Fruit trees will require more water when fruit is expanding rapidly.

The most frequent mistake in irrigating home plantings of shrubs, including shrubby small fruits, is applying too little water too often. A thorough irrigation every two weeks will supply the moisture required by most flowering shrubs, although roses and berries may need water each week while they are flowering or fruiting.

In most situations woody plants can be either surface or sprinkler irrigated unless height and foliage interfere with the sprinkler pattern. Roses, however, are subject to leaf diseases and generally should not be sprinkled. Juniper and arborvitae are subject to damage by mites, which are partially controlled by sprinkling. In dry locations and along unpaved roadways sprinkling may be necessary to remove dust that not only detracts from appearance but also reduces photosynthesis and growth.

Irrigation of woody plants growing in areas where they could be winter damaged needs to be carefully monitored as winter approaches. The soil in which the plants are growing should be permitted to become relatively dry in September and early October in order to allow the plants to enter their rest period. After the plants become dormant but before the soil freezes soil moisture should be brought to field capacity, either by autumn rains or by irrigation.

Irrigating Potted Plants

Houseplants should not be watered too frequently and should receive a good soaking when they are watered. They will usually need more frequent watering during winter than during summer, because the atmosphere of a heated home is often very dry. Those growing in porous containers require more frequent irrigation than those growing in nonporous containers, and different kinds of plants require different amounts of water. Thus the temptation to water all plants at once with equal amounts should be avoided. Each plant should be watered according to its individual needs, which can be determined by feeling the soil in each container with a finger two or three times each week. Water should not be added to any container in which the soil feels moist. It is essential that the soil below peat moss or other mulch be tested, as most materials used to mulch the tops of houseplant containers can remain dry even when floating on water.

Cacti and some succulents should not be watered until after they have been dry for 7 to 10 days, but most other plants should be watered well as soon as the soil in their containers feels dry. Because most houseplants are tropical, lukewarm water should be used. Generally the water is poured into the container from the top, but it can also be added from the bottom, a practice advisable for plants such as African violet and gloxinia that develop dead spots wherever cold water touches their leaves. Containers of plants to be watered from the bottom are usually kept in a flat pan or dish into which the water is poured as the plant requires it. Sometimes a cloth or fiber wick is used to conduct water from the drainhole up into the soil medium, but this is usually not necessary as water is conducted through most potting media almost as well as along a wick. Newly potted plants should be carefully watched the first few days to make certain that no obstruction is preventing water from being conducted up through the pot.

Bottom watering is also desirable for ferns, because the pan of water supplies needed humidity as well as irrigation. During the winter when humidity is low, spraying foliage each week will benefit ferns and other humidity-loving plants.

Irrigating Lawns

A lawn uses more water per unit area than does any other part of the garden if it is to be kept growing vigorously. Most well-established grass lawns will, however, survive prolonged dry periods if irrigation is impossible.

An area in which a new lawn is seeded must be kept wet until the seedlings have germinated and become firmly rooted. This may require two or three irrigations every day if the weather is warm and dry. The frequency of irrigation can be gradually reduced as the seedling roots penetrate the soil.

Established lawns should be irrigated thoroughly and no oftener than necessary. **Pan evaporation** is sometimes used to determine the frequency of lawn irrigation and the amount of water needed. This procedure involves placing a pan under the sprinkler at the time of lawn irrigation. When the pan has approximately ¾ inch of water in it, dig along one edge of the lawn to see how far down the water has soaked. Most soils require from about ¾ to 1¼ inches (2 to 4 cm) of moisture to bring the top foot (30 cm) of soil to field capacity (Table 6–1), and as grass roots do not extend much below a foot, there is no need to soak the soil below this depth. If the top foot of soil is wet, stop irrigating; if not, continue until that depth is reached. Leave the pan containing the amount of water required to soak 1 foot in the yard; when the water has nearly all evaporated, it is time to irrigate again. For example, if you live in an area where 1 inch of water will evaporate from a pan each week, and it requires 1 inch to soak the soil to 1 foot, the recommended irrigation for your lawn during the warmer part of the season would be 1 inch of water each week. After observing the amount of evaporation that occurs each week for a few weeks, you can usually judge when to irrigate and how much water to apply.

Pruning and Training

Pruning Tools

Strong, well-sharpened tools are essential for good pruning. Purchasing cheap, poorly constructed pruning tools not only wastes money but, more importantly, may waste considerable time and be harmful to plants. A home gardener needs three basic tools: a pair of loppers, a pair of one-hand pruning shears, and a pruning saw. Many kinds of pruning saws are available. Some gardeners attempt to use a carpenter's handsaw or a bowsaw, but most handsaws are not designed with either the type of teeth or set (the pattern of outward bending of the teeth that aids the removal of sawdust and determines the width of the cut) to conveniently cut green, wet wood. A bowsaw is difficult to manipulate among branches and to guide through a large cut. I prefer a nonfolding curved speed saw with raker teeth because it cuts rapidly and is easy to grasp.

Depending on the amount and the type of planting being trimmed, a gardener may also need a pair of hand or power hedge clippers, grass clippers, a pole saw (for reaching limbs to about 12 feet [4 m] from the ground), a pole pruner, and/or a small power chain saw (Figure 14–64).

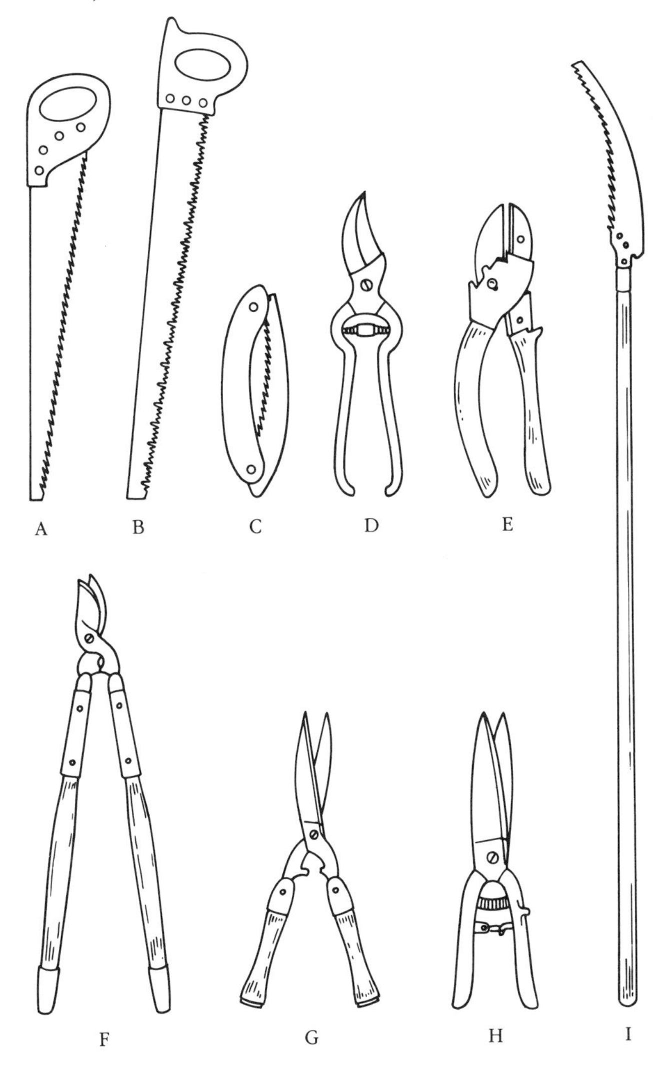

Figure 14–64 Pruning tools. (A) Speed saw with lance teeth. (B) Speed saw with raker teeth. (C) Folding saw. (D) One-hand hook and blade shears. (E) One-hand anvil shears. (F) Hook and blade lopping shears. (G) Hedge shears. (H) Grass clippers. (I) Pole saw.

Pruning equipment will last for many years if it is properly cared for. Plant sap should be washed from the equipment and the cutting blades, and the assembly joint should be oiled lightly after the tool has been used. Dull or slightly nicked loppers and shears with a simple nut and bolt assembly can be disassembled and sharpened with a mill file or oil stone. The sharpening of all saws and of shears badly nicked or with complicated assembly mechanisms requires special tools; they should be taken to experts who have the proper equipment to repair them.

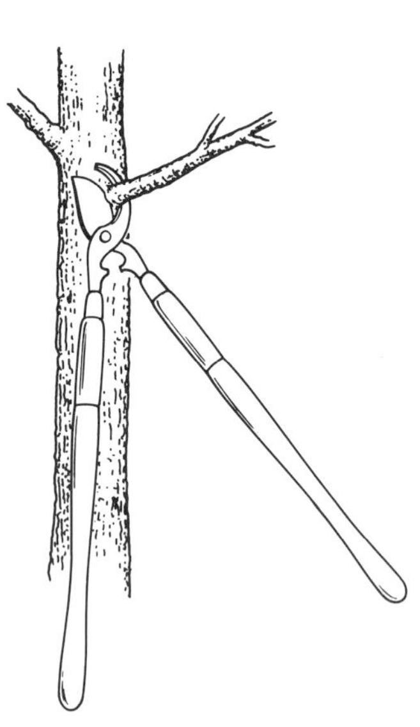

Figure 14–65 Pruning a small branch. When a branch is being pruned from a larger limb or trunk, the sharp blade of the loppers or shears should be held as close to the larger limb or trunk as possible so that a stub will not remain. Even more important, the blade should be positioned so that it cuts with the grain, which usually means cutting from the lower side. This lessens splitting and the development of small cracks that permit entry of pathogenic organisms.

How to Make a Pruning Cut

Canes or branches originating at soil level that are to be completely removed should be cut as near the soil as possible to eliminate unsightly stubs that interfere with new growth and future management of the planting. For thinning out cuts of trees and branching shrubs, the limb should be removed flush with the main trunk or a larger limb. With smaller branches closer cuts are possible if the sharp blade of the loppers or shears is placed next to the trunk away from the branch being removed. The cut should be made with the grain of the wood (Figure 14–65). Figure 14–66 shows the correct method for removing larger branches. Wounds larger than about 1½ inches (4 cm) in diameter should be sealed with tree paint to prevent them from drying or decaying. Ordinary paint should not be used as it may contain additives harmful to the plant.

Training and Pruning Small Fruits

The following general principles apply to pruning small fruits:

1. Unless the number of fruit buds is reduced by pruning, small fruit plants set more fruit than they can mature to optimum size and quality.

2. The best fruit is produced on young wood.

3. The best fruit is produced on the most vigorous branches.

Grapes. Grape vines should be pruned while dormant. If they are pruned during the growing season, plants may "bleed" or lose sap. Bleeding does not, however, cause serious damage.

Grapes should be pruned so that only enough previous-season wood to contain 40 to 80 buds remains. These buds can be distributed on numerous short canes (Figure 14–67) or on a few long canes (Figure 14–68). When the grape plant is trained to cover an arbor, the new wood may be at the end of fairly long, older branches, or it may consist of a few buds on each of a number of shortened canes. In the East growers leave a minimum of 40 buds on a moderately growing plant. In the West, where plants are generally grown larger due, perhaps, to higher light intensity, 60 buds may be left. If

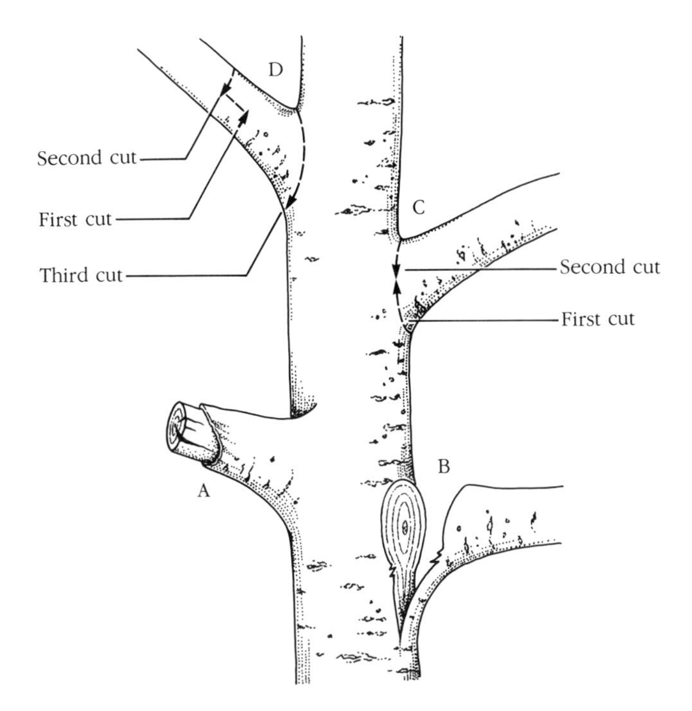

Figure 14–66 Pruning larger branches. When a large limb is pruned, the main concerns are that the cut be made so that it doesn't leave a stub (A) that will die back and serve as an entrance point for decay organisms, and that it doesn't tear the bark (B), leaving a larger wound than necessary. There are several correct ways of removing a large limb. I prefer (C) because it requires a minimum of cutting. If it is difficult to make the upper and lower cuts meet, three cuts can be made (D).

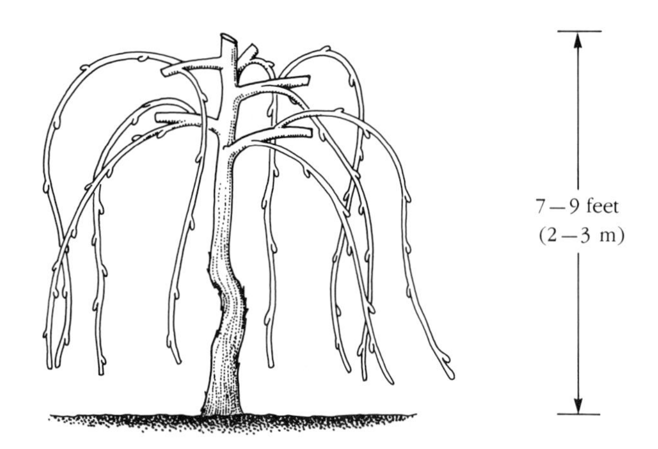

Figure 14–67 Grape plant trained on a self-supporting trunk. Canes that have fruited are pruned away each spring, and canes grown the previous year, having 40 to 80 buds, are left to replace them. Clematis can be trained in the same manner.

the plant is vigorous it will support 10 to 20 additional buds.

Because it can be used with almost all cultivars in almost all areas, the four-cane Kniffen is a popular home garden system of pruning grapes (see Figure 14–68). The principles described for this system can be used as a pruning guide no matter how the grape vine is to be trained. For the Kniffen system the trellis to support the vines is two wires, one 5 to 6 feet (about 2 m) and the other 2 to 3 feet (about 1 m) from the ground, with supporting posts. When plants are first set, they are cut back to two or three buds on the most vigorous cane. If sufficient growth has been made at the end of the first year in the field, the cane is tied vertically to the trellis wires to form the trunk. If growth is too short, the cane is again cut back to two or three buds, and the best cane is tied up to the wires at the end of the second season. Next, four canes or branches are selected for fruiting wood each year with one on each side of the trunk at each wire.

From the fourth year on pruning is a replacement operation. The canes that have produced fruit are removed and are replaced by new 1-year-old canes. In addition to the four fruiting canes short two-or-three bud branches, or spurs, are left near the trunk to grow fruiting canes for the following year so that the newly pruned plant has four fruiting canes and four to six short spurs (see Figure 14–68).

Brambles. Training and pruning of bramble fruits are based on the biennial-bearing habit of these crops. After canes have fruited, they should be removed. This should be done promptly in areas where fungus diseases of the cane are prevalent. In areas free of disease and having heavy winter snow the old canes should remain until the following spring as they afford a degree of protection from breaking by wind and snow.

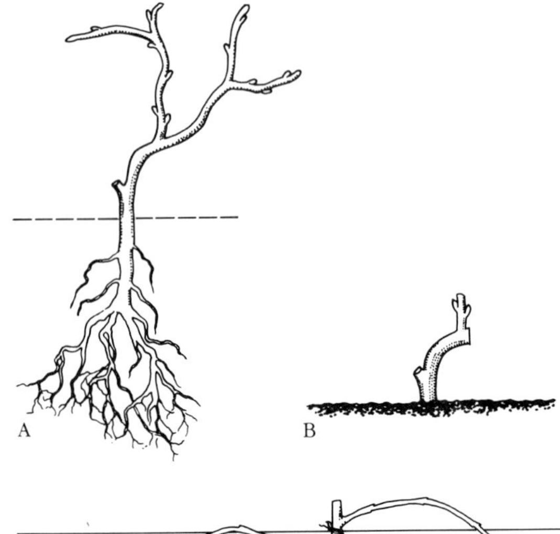
A
B

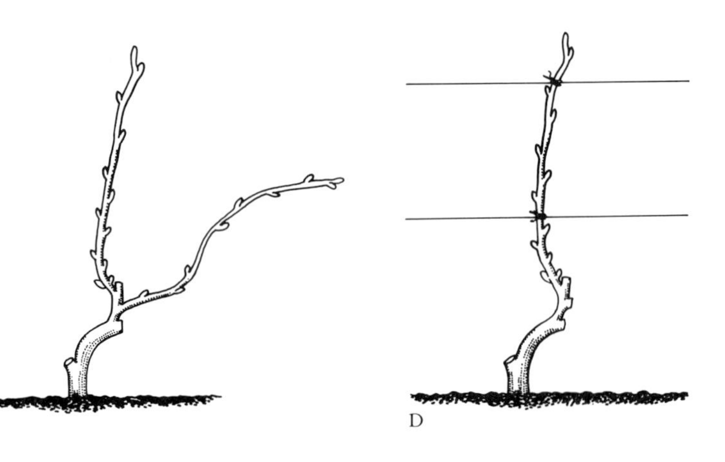
C
D

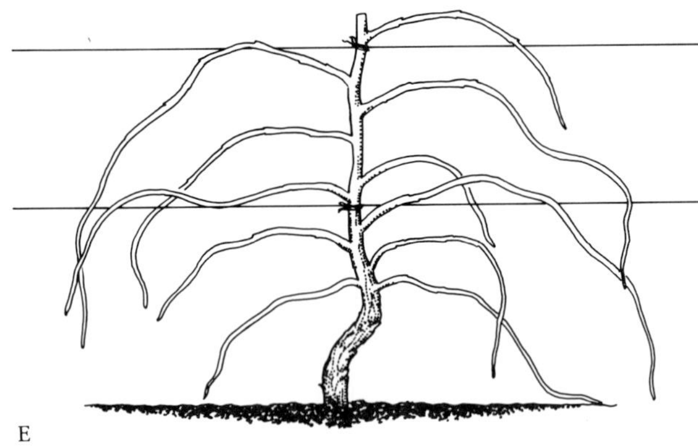
E

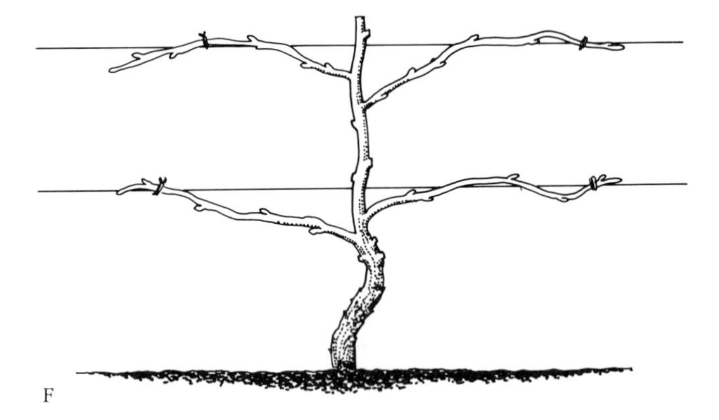
F

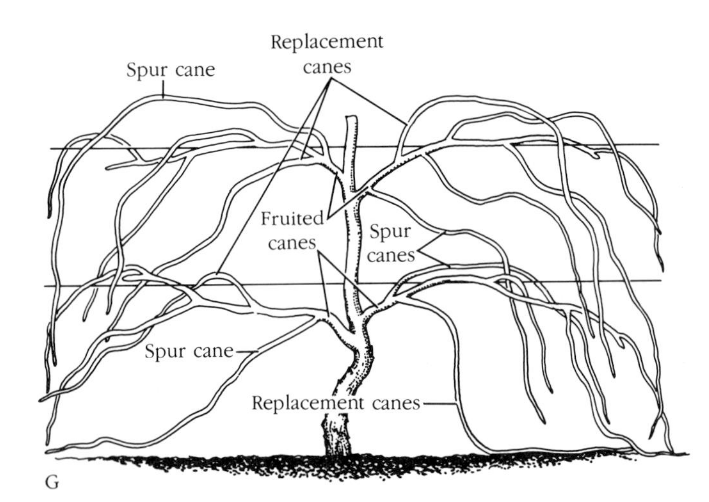
Replacement canes
Spur cane
Fruited canes
Spur canes
Spur cane
Replacement canes
G

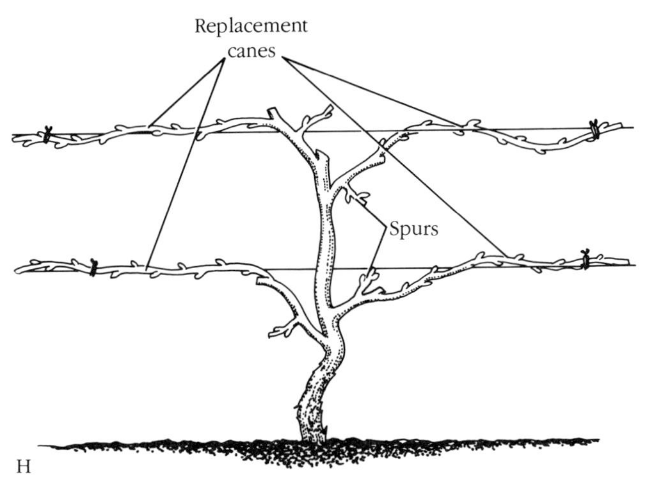
Replacement canes
Spurs
H

Figure 14–68 The four-cane Kniffen system.

The newly planted grape plant (A) is cut back to two or three buds (B). At the end of the first growing season (C) one branch is selected for a permanent trunk and tied to the supporting wires (D); other growth is pruned away. At the end of the second or third and each succeeding season (E), four canes of wood grown that season are tied to wires to produce the crop the following summer (F).

From the third or fourth season the pruning of grapes is a replacement operation. The canes that have fruited are replaced by four canes from the past season's growth. In addition, four to six spurs each containing one or two buds are left near the trunk to produce new canes for the following year. A mature grape plant before (G) and after (H) it is pruned.

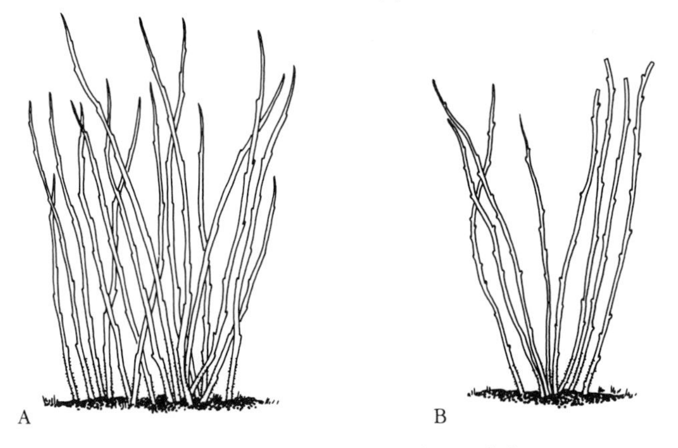

Figure 14–69 Pruning bramble fruits. (A) Red raspberry plant before pruning. (B) The same plant after pruning. (After USDA.)

Red raspberries send up canes from buds initiated on the roots, and these canes may be handled in a number of ways. In the **hedgerow** system canes are allowed to develop freely, but the row is restricted to a width of from 1½ to 3 feet (½ to 1 m). Generally, the taller canes are more productive than the shorter ones, so tall canes should be cut back only a short distance and short ones should be removed. Where wire supports are used, longer canes may be left. The distances between canes also should be controlled. A suggested guide to cane spacing is to thin the canes in the row so that they stand from 5 to 7 inches (13 to 18 cm) apart each way, leaving only vigorous, well-matured canes. If the canes are left in hills or clumps, there should be about 7 to 14 per hill, depending on the vigor of the plant—more for a plant that produces numerous large canes (Figures 14–69 and 14–70).

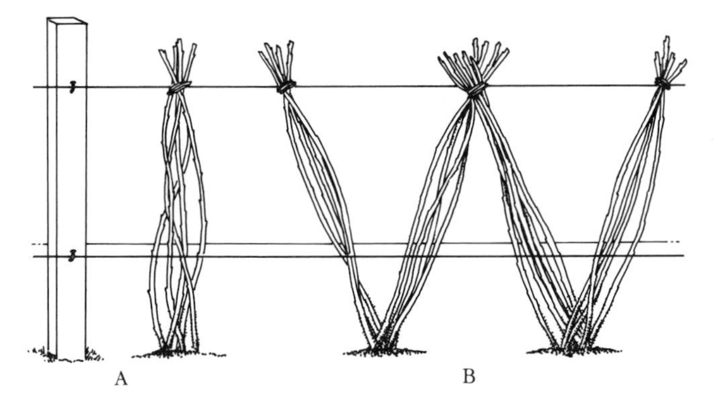

Figure 14–70 Three-wire trellis for longer raspberry canes. Wires are fastened directly to posts. The single top wire is stapled to the posts, and the two bottom wires are attached on hooks or bent nails so they can be lifted and swung out over new canes to pull them in. Hills with up to six or seven canes can be brought up and tied in a single bundle (A). Split hills are possible for more canes (B), with part going half way to the next plant in the row. (Courtesy Washington State University.)

Black and purple raspberries grow from a single-stem base or crown, so they always are grown as individual plants. As the new canes develop, they are pinched back to a height of 18 to 30 inches (about 45 to 75 cm) to force the development of fruiting lateral branches (these are the laterals used for layerage). In the spring before growth starts these laterals are headed back to 8 to 12 buds each (Figure 14–71).

With blackberries annual pruning and frequent removal of suckers is essential to prevent a few plants from becoming an inpenetrable briar patch. Upright blackberries are pruned in the same manner as black raspberries except that the canes are headed from 30 to 48 inches (about 75 to 120 cm) from the ground, and in the spring the lateral branches are cut back to 10 to 24 inches (25 to 60 cm). Different cultivars of blackberries fruit on different sections of the lateral branches, so severity of heading back is dependent on the fruiting habit of the cultivar in question.

Trailing blackberries such as the boysenberry or loganberry are vines and must be given support. Most of their fruit is produced near the base of vigorous canes and on spurs 15 inches (40 cm) or more in length, so canes of moderate length are sufficient. Boysenberry, youngberry, and loganberry canes may be left from 9 to 15 feet (3 to 5 m) long. Training systems vary considerably. Most often these fruits are trained to a single-wire or two-wire trellis, a horizontal trellis, or stakes. During the summer the new canes are grown on the ground in the line of the row or on the lower wire of a two-wire trellis. Later, after the old canes have fruited and have been removed, the young canes are raised to the upper wires where they are to fruit.

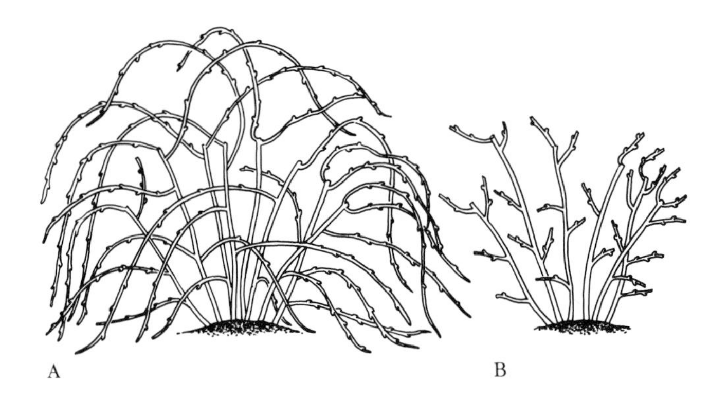

Figure 14–71 Mature black raspberry plant before and after pruning. New shoots were headed during the growing season by pinching out the tips. (A) Before dormant pruning. Old fruited-out canes and some of the weaker ones have been removed. Note length of laterals resulting from heading. (B) After pruning. Weak canes (less than ⅓ inch [¾ cm] in diameter at 1 foot above ground) were removed. Laterals were cut five to eight buds long. The stronger the cane, the lighter the pruning. (After USDA.)

Training and Pruning the Home Orchard

Pruning of fruit trees may be done at any time during the dormant season in areas with mild winters, but it should be delayed until February or early March in areas where winter damage can occur.

Apple and Pear Trees. The modified central leader system of training (Chapter 8) is the most satisfactory for both dwarf and standard apple and pear trees.

First year. When a tree is planted, it will ordinarily have been growing for one season after having been grafted. It is usually unbranched and is referred to as a whip. The whip is headed back to about 30 inches and little more is necessary the first year (see Figure 14–72).

Second and third years. Three or four well-spaced scaffold branches are selected during the second and third years. If possible, the lowest scaffold should be about 18 inches from the ground. In northern areas where winter sun is intense and winds are from the southwest it is best to have the lowest scaffold on the southwest side of the tree. This helps reduce winter damage, which frequently results from dehydration of the southwest side of the trunk. The other laterals should be spaced 6 to 10 inches (15 to 25 cm) vertical distance apart and should also be evenly spaced around the tree on all sides. The central leader is usually allowed to grow during these years because its growth causes the laterals to produce wider, stronger crotch angles (Figure 14–73). The extra laterals are stubbed back rather than being cut off completely.

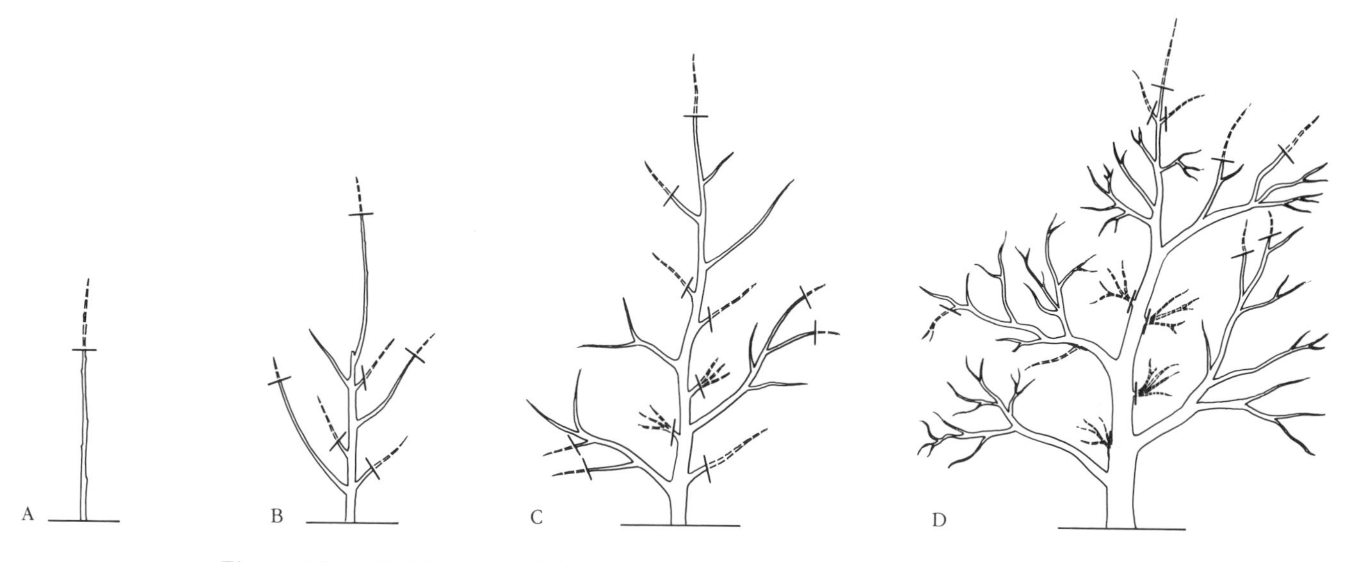

Figure 14–72 Training pome fruits. To train trees to a modified central leader, cut the newly planted whip (A) back to about 30 inches (75 cm). During the first 2 years, three or four main scaffold branches are selected. Other branches from the main trunk are cut back but not entirely removed, and the central leader and other secondary branches on the scaffolds tending to grow out of bounds are headed back slightly (B and C). The central leader and extra side branches are removed after the fruit load starts to spread the tree, in 4 to 7 years (D).

They produce leaves for food manufacture and afford some competition, which seems to aid in developing wider crotch angles and a stronger framework.

Fourth through seventh years. During these years the tree should be pruned as little as possible, because pruning delays production. The central leader can be headed back somewhat; with standard and semidwarf trees it may be entirely removed during the sixth or seventh year. It is usually not necessary to remove the central leader of fully dwarf trees. The brushy interior caused by the growth of the stubbed laterals must be kept in bounds, and this extra wood will be removed during the fifth to seventh years.

Fruiting period. With apples and pears only light pruning is necessary during the fruiting period. Weak, unfruitful, and dead branches are removed, and some heading back of wood that tends to become too high may be

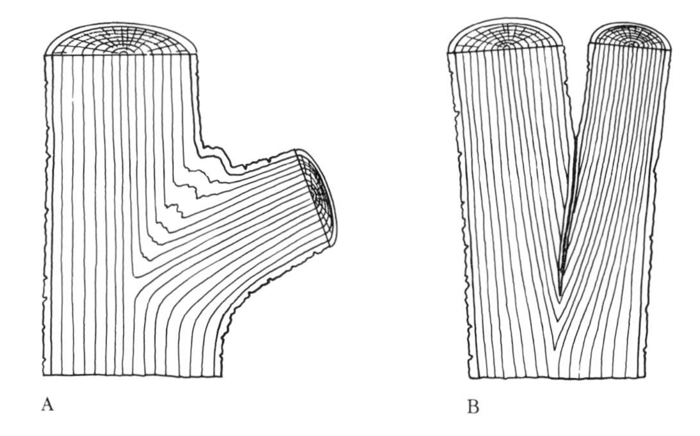

Figure 14–73 Crotch angles. A wide crotch angle (A) provides strength to the framework of a tree. A narrow crotch angle (B) is weak and often splits as the branches become larger and heavier or when they are loaded with fruit.

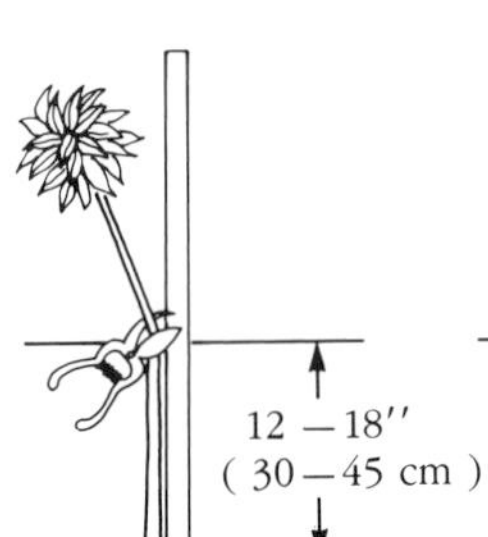

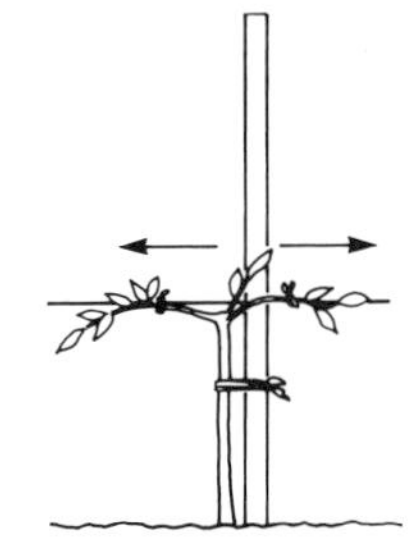

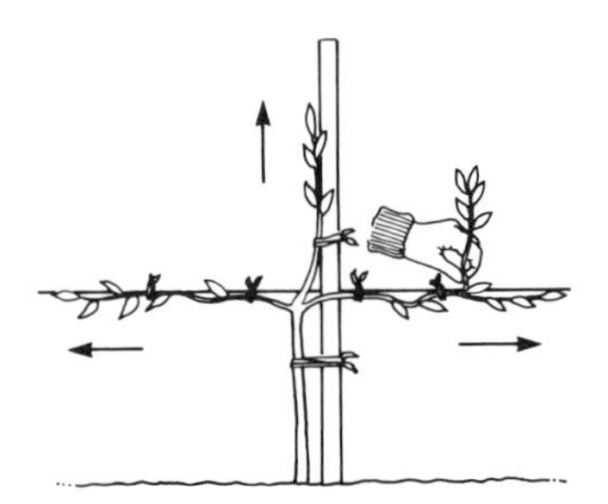

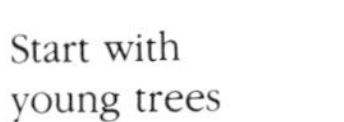

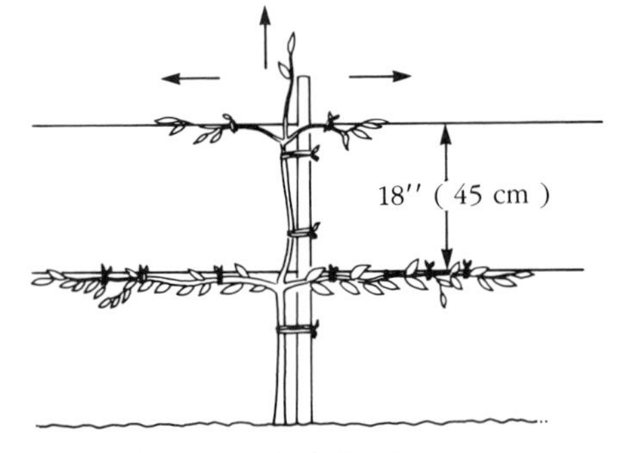

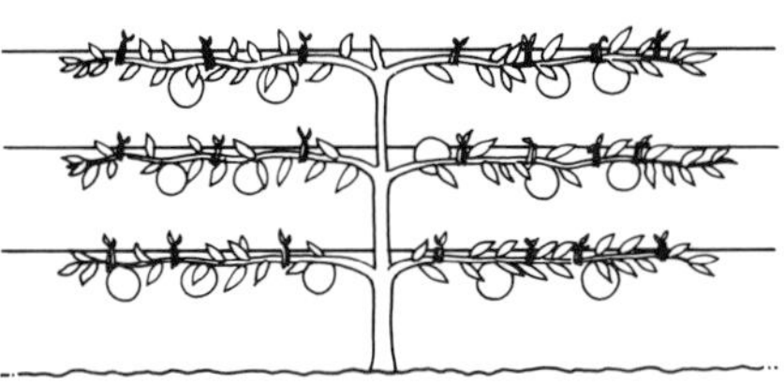

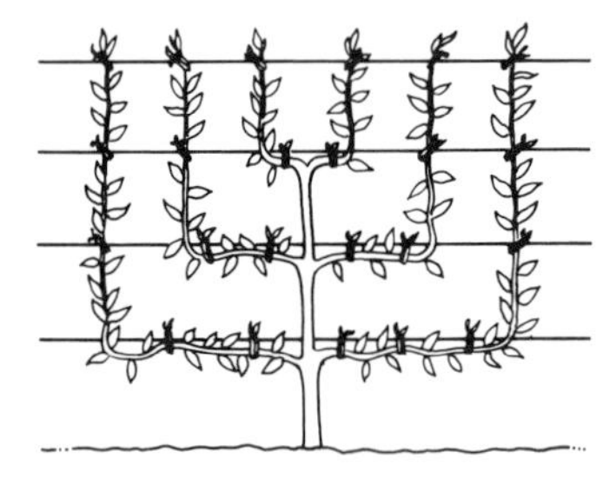

Other forms

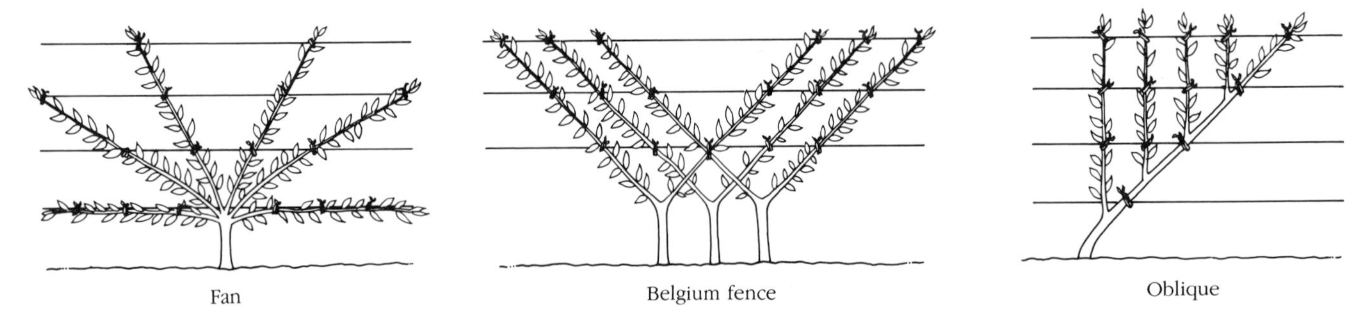

Figure 14–74 The creation of an espalier. (From *Horticultural Science,* 3rd ed., by Jules Janick, W. H. Freeman and Company, San Francisco. Copyright © 1979.)

necessary. As the trees reach 25 to 30 years of age, somewhat heavier pruning to stimulate more vigorous shoot growth may be desirable.

Apples, pears, citrus, and many shrubs can be trained flat against a wall or on wires (similar to the Kniffen system of training grapes). The two-dimensional plant produced by this system is called an espalier. The advantage of the espalier for the homeowner is a saving of space and the assurance that sunlight reaches all parts of the plant. The most popular form of espalier for fruit trees is a vertical stem with horizontal branches (Figure 14–74), although the branches can be angled, curved into a crescent, turned upward halfway to the tip to form an L, or trained to other forms or combinations of forms.

The pillar system of training common in commercial European orchards also has possibilities around a home where space is limited. Dwarf or semidwarf trees are permitted to grow with their central leader intact. Stiff upright side branches are removed as they begin to develop, which encourages the growth of more horizontal willowy laterals. Bearing trees are pruned in the late winter or early spring. All laterals that have produced fruit are completely removed or cut back to a single bud. Buds originating on branches near the trunk or on the trunk are left to produce wood for future crops. Excess bushy growth from the trunk and the past season's growth on laterals beyond the fruiting spurs is also removed (Figure 14–75). With this system trees can be planted 5 feet apart and still have adequate light for quality fruit.

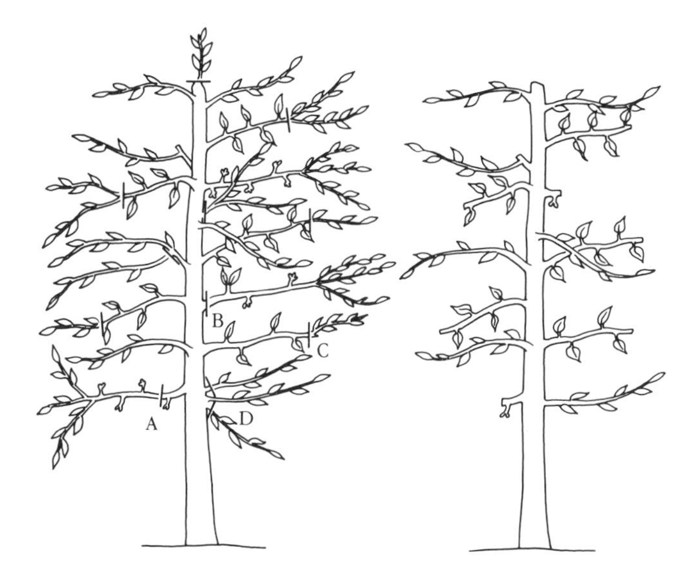

Figure 14–75 Diagrammatic sketch of a pillar-trained tree before (left) and after (right) pruning. Laterals that have fruited are stubbed back to a single bud (A) or entirely removed (B). The past season's growth on laterals beyond fruiting spurs (C) and excess brushy growth from the trunk (D) are removed.

Stone Fruits. Peach trees grow rapidly and normally are branched when they come from the nursery. Often it is possible to cut back the top when the peach tree is first planted, leaving only the three or four laterals that will become the scaffolds as the tree begins to grow. Ideally the scaffolds should originate about 20 to 24 inches (50 to 60 cm) above the ground. Stubbing back but not entirely removing the central leader will help to spread the scaffolds. Little pruning is necessary during the next 2 or 3 years. At the end of the third year the central leader is removed to give an open tree for ease of spraying and to allow sunlight to reach the fruit (Figure 14–76).

After the third year when the tree begins to bear, pruning must be moderately heavy to keep the new fruiting wood near the central axis of the tree and to stimulate new wood production. Because the peach tree tends to produce many more fruits than it can support, pruning becomes a way of thinning the crop. Enough wood is removed to eliminate at least 50% of the bloom. Moderately vigorous branches with a high proportion of buds in groups of three produce the best fruit, so pruning should remove both the weak, slow-growing and the extremely vigorous shoots.

Because the fruiting habits and rates of growth of plum, sour cherry, and apricot are intermediate between those of apple and peach, their training and pruning are also intermediate. They are usually trained to the modified central leader system as described for apples and pears. Pruning of mature trees should be somewhat heavier than for apple or pear but somewhat less than for peach. If it is necessary because of space limitations, the stone fruits can be pruned heavily as they do not tend toward alternate bearing.

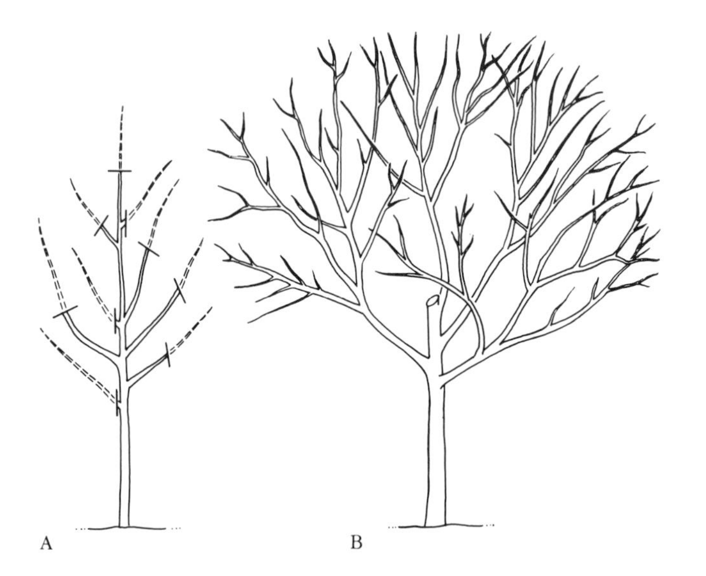

Figure 14–76 Training stone fruits. To be trained to an open center a peach tree should be cut back to three or four selected scaffolds at the time of planting (if it has sufficient branches then) or at the end of the first growing season (A). Only minimal further pruning will be required until the tree begins to bear fruit two growing seasons later. The central leader should be headed back each spring and entirely removed after 2 years (B). Even with proper training peach trees may need to be propped or braced to prevent their being broken down by a heavy load of fruit (C).

The sweet cherry grows slowly, and pruning practices would be similar to those described for apple.

Citrus Trees. With the exception of lemon, citrus trees do not require much pruning. They produce fruit at the ends of current-season wood, and most kinds tend to grow slowly. Pruning at any time of year is possible. At planting time young trees should be headed back to about 3 feet (1 m). During the first 2 years, four to six strong laterals are selected for scaffolds, and the remaining branches are pinched back somewhat but not removed. The lower branches and surplus twiggy interior growth are removed gradually during the first 3 or 4 years to obtain a clean trunk to the desired height. The gardener should attempt to develop a tree with a strong framework of upright branches because the weight of foliage and fruit will bend the branches downward. Shade is not as detrimental to citrus as to stone and pome fruits, so gardeners usually cut back branches that start to grow out of bounds in order to develop a pleasingly compact tree that will be ornamental as well as produce fruit. If the terminals of unwanted growth are pinched back several times a year, almost no actual pruning will be required.

The lemon tree has a more vigorous and unpredictable growth habit. It should be headed back to 36 inches (90 cm) at planting time, and the three to four laterals selected for scaffolds should be headed quite severely. Several more scaffolds should be selected high up the trunk from second-year growth. If there are gaps in the tree, new scaffolds can be formed later by gradually arching one or more upright suckers toward the ground. If the sucker is of the right age, it will stay in the arched position; if it is too old, it may require tying. There will usually be considerable sucker growth each year on lemon trees, and this should be removed along with the ends of long pendulous branches that often grow beyond the remainder of the tree.

Citrus can also be trained as espaliers (see Figure 14–74), and the sour orange is often clipped to form a hedge.

Training and Pruning Woody Ornamentals

When to Prune. Ornamentals can be lightly pruned whenever they require cutting back; however, there is often a season when a plant is physiologically and structurally best conditioned for major pruning and this season varies depending on the type of ornamental. General guidelines for some woody ornamentals follow:

1. Spring-flowering shrubs: right after they bloom.
2. Pines: spring, when the new growth is succulent (at the "candle" stage) and can be broken off easily.
3. Rhododendrons and azaleas: mid-spring, as the blooms fade and while the new growth is succulent enough to be pinched back.
4. Formal hedges: whenever they begin to look ragged or to grow out of bounds, several times a growing season.
5. Birches, walnuts, maples: late fall or winter. These "bleed" profusely if pruned in the spring.
6. Most other trees, lianas, and shrubs: late winter and very early spring.

Deciduous Trees. If the homeowner has been careful to select trees to fit the location, there will be little need to prune shade trees. Some trees such as the American elm or the Norway maple have no definite leaders like those found in conifers, birches, most oaks, and poplars; however, a tree's natural growth habit usually produces its most desirable form and is certainly easiest to maintain. Except for removing crossing and dead or broken branches, cutting back side branches that occasionally tend to outgrow the central leader, and eliminating balanced crotches, very little pruning of young shade trees is required. Balanced crotches, when two branches of the same size form a crotch, are always narrow and weak (see Figures 8–11 and 14–73) and will usually split when the tree becomes older. A balanced crotch is especially undesirable when it produces two central leaders of equal size. When shade trees have reached a height of 15 to 19 feet (5 to 6 m), lower branches should be removed to give a basal clearance of 7 to 10 feet (2 to 3 m).

Drastic amounts of pruning to rejuvenate or reduce the size of an older tree should be accomplished gradually, a few large limbs and some smaller ones being removed or headed back each year over a period of 3 to 6 years until the desired size and shape is developed. The common practice of stubbing back a mature tree to its basic scaffolds is deleterious to its health and beauty and will frequently result in the decline and death of the tree.

Pruning Deciduous and Broadleaved Evergreen Shrubs

Some shrubs are slow growing and produce most of their growth at the tips of canes or branches each year. If they have plenty of room for full development, they need pruning or pinching back only to maintain good form and shape. These shrubs include azalea, many of the viburnums, flowering dogwood (*Cornus florida*), redbud (*Cercis canadensis*), mountain laurel (*Kalmia latifolia*), and Rhododendron. Another group can be left alone to grow naturally into large shrubs or can be kept smaller by heavy pruning. These include bush arbutus (*Abelia grandiflora*), beauty bush (*Kolkwitzia amabilis*), African tamarisk (*Tamarix africana*), winged euonymus (*Euonymus alata*), English laurel (*Prunus laurocerasus*), Oregon grape (*Mahonia aquifolium*), and pyracantha (*Pyracantha* spp.).

Other shrubs send up new unbranched canes each year. These canes branch the second and third years but do not grow from the tip. A few of these kinds of shrubs that are late-blooming produce better flowers or fruits if the whole shrub is cut back to within 6 to 12 inches of the ground each spring. These include rose of Sharon (*Hibiscus syriacus*), beautyberry (*Callicarpa dichotoma, C. ja-*

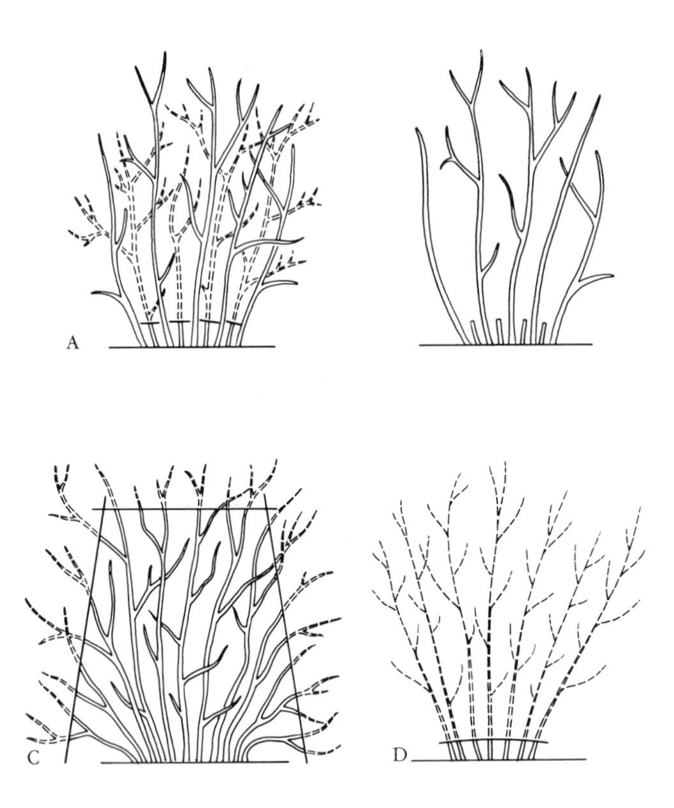

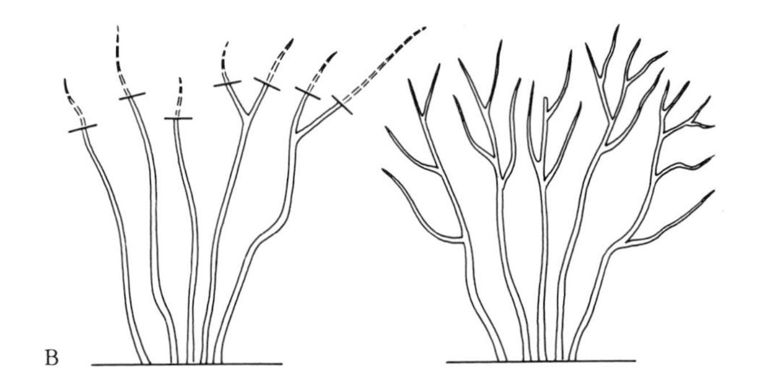

Figure 14–77 Pruning shrubs.
(A) Thinning, before and after. Thinning removes old, twiggy, weak, crossing, or excess branches and produces a softer texture and more desirable flowers on plants that bloom.
(B) Heading, before and after a season's growth. Heading produces a more compact and symmetrically shaped shrub.
(C) Shearing or clipping shapes shrubs to regular, predetermined lines. It is used mainly for formal hedges.
(D) Rejuvenation is used to reinvigorate shrubs. Part of the canes can be cut back each season for 2 or 3 years.

ponica), butterfly bush (*Buddleia* spp.) hydrangea, hills of snow (*Hydrangea arborescens* 'Grandiflora'), various kinds of summer flowering *Spiraea*, umbrella catalpa (*Catalpa bignonioides*), and the tamarisk species that bloom on current-season wood. Older unthrifty or overgrown shrubs of kinds like mock orange (*Philadelphus*), lilac (*Syringa*), many *Prunus* species, and rhododendron can also be cut back severely to rejuvenate them. Like trees mentioned earlier, these shrubs respond better if a few old canes are removed each year over a period of 2 to 4 years rather than all at once.

Many common flowering shrubs respond best to a moderate amount of annual pruning. Removal of 10 to 30% of their wood is enough to stimulate them to produce a supply of young vigorous wood. The first step is to cut out dead wood and then remove weak and dead flowers and seed pods. With most of these kinds of shrubs all branches more than 3 years old should be removed. Those, like forsythia, that grow rapidly and tend to become leggy may have to be pruned several times each year to keep them within their allotted confines.

Various methods of pruning shrubs are illustrated in Figure 14–77.

Hedges. Informal hedges or background plantings require no more pruning than the same plants growing under other circumstances, but a hedge that is to be trained to a specific form must be given periodic attention. Formal hedges must be clipped one or more times per year according to their growth habit. This trimming is to keep the hedge thick, neat, and under control.

In development of a hedge, the base should be wider than the top. This permits growth to remain dense and leaf covered to the ground, because light can reach all parts of the hedge. There also will be less injury from snow to a hedge with a base wider than the top (Figure 14–78).

Roses. Pruning rose bushes depends on the type of plant and the blooms desired. Roses in the hybrid perpetual, the hybrid tea, and the grandiflora groups should be pruned toward the end of the dormant season but before the buds break in the spring. All broken, winter-damaged, borer-infested, weak, and crossing canes should be removed. If the plant is to produce exhibition-quality blooms, pruning must be severe. Only two or three buds on three or four of the strongest canes should be left. Where greater quantities of blooms are desired for general garden growing, more buds should be left. The more vigorously growing cultivars can be cut back to five or six buds on the four to seven heaviest canes. On less vigorous plants three to five buds per cane should be left (Figure 14–79). Wherever possible the upper bud left should face outward so that new growth will be away from the center, keeping the plant open.

Polyantha, old fashioned, hybrid perpetual, shrub, and miniature roses require little pruning other than the removal of dead wood and a few old branches each spring. Old branched canes and weak new canes should be removed from climbing and trailing roses when the plants have finished blooming in the fall or early spring. Four to six of the strongest unbranched canes should be tied to a trellis or other support. Long unwieldy branches can be cut back. Vigorous older branches will bloom, but if they are left, their lateral side shoots should be shortened to spurs of two to four buds (Figure 14–80).

Figure 14–78 Good shapes and a poorly shaped formal hedge (end views).

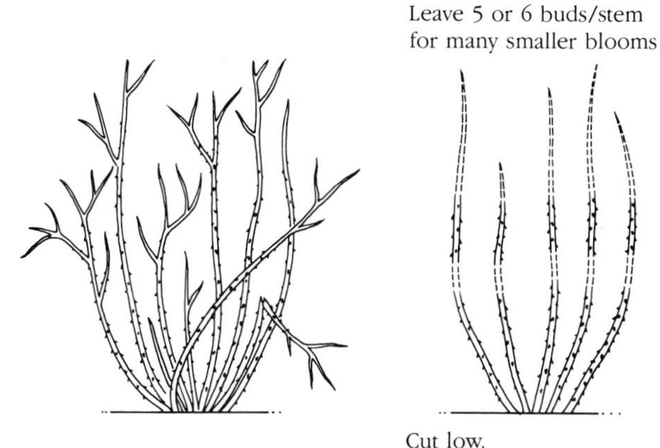

Figure 14–79 Pruning hybrid tea and other similar rose bushes.

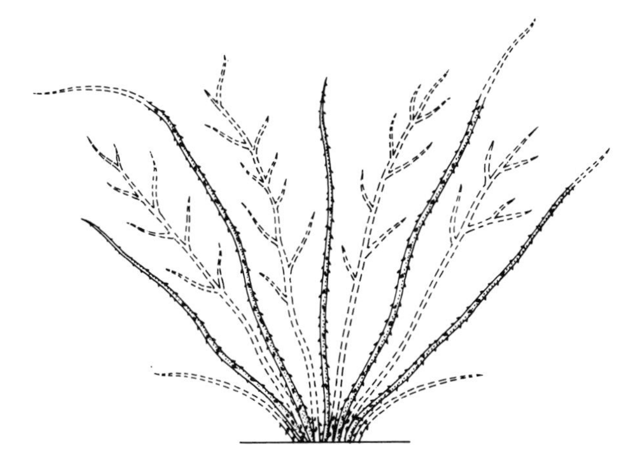

Figure 14–80 Pruning climbing roses. Four to six of the youngest strongest canes are left to produce the blooms.

Rambler roses should be pruned after they bloom. Much of the wood that flowered can be cut off, as the ramblers produce flowers the next season on new canes growing from the base of the plants.

Figure 14–81 Pruning conifers. Growth of pine can be regulated by the amount of "candle" removed. Removing part of the candle (A) will result in moderate growth. Removing the entire candle (B) will result in branched compact growth. The branch tips of fir and spruce can be cut back slightly each spring (C). This will produce trees with more compact growth habit.

Conifers. Pruning of conifers will vary with type and use. Pines produce new growth only at the terminal end of branches and are best pruned by pinching back the candles in the spring before the wood has hardened. The amount of length growth permitted will be determined by the percentage of the candle removed. Pinching back the candle will cause new buds to be initiated and make the pine more compact. The various fir and spruce species possess several buds toward the end of each branch; these are usually pruned by clipping back the tips of branches to give them a more compact growth habit (Figure 14–81). If the leader of pine, spruce, or fir is lost, it is wise to form a new one by using a stick to tie a branch from the terminal branch whorl into an upright position

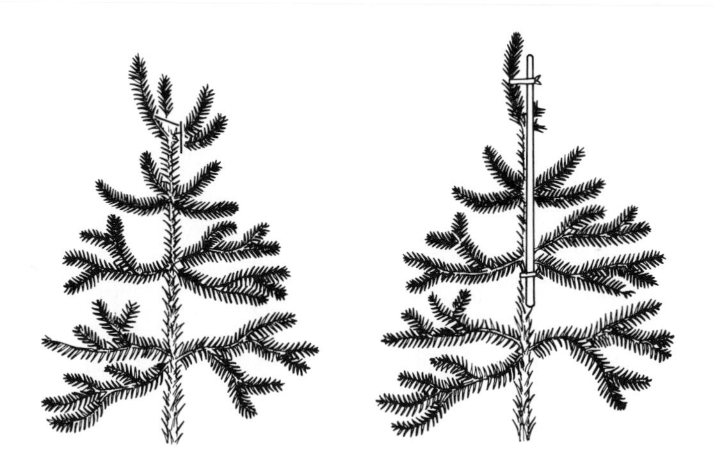

Figure 14–82 Renewing the leader on a conifer.

(Figure 14–82). Otherwise the tree is likely to form a number of leaders simultaneously.

Yew, arborvitae, cypress, and juniper will form new growth from any part of the plant and consequently can be pruned in almost any way the owner desires. In restricting the growth of spreading evergreens, it is well to cut the branches back so that unsightly stubs are hidden by other branches. Cutting out complete branches produces a more naturally shaped spreading evergreen than does heading back or shearing. Because they are so tolerant of various pruning practices, these conifers along with various species of boxwood (*Buxus*) were used for the ornate topiary of the formal gardens of earlier years.

Vines and Lianas. Some pruning is needed in training a vine to climb or cover an object. The method of handling will depend on the way the plant "climbs" or attaches itself to its support. English ivy climbs by aerial roots that penetrate into small cracks. Woodbine and Boston ivy have adhesion disks, clematis and grapes have twining petioles, and bittersweet has a twining stem growth.

Most vines and lianas require little pruning. Dead or injured wood and growth that has developed where it is not wanted should be cut off. Lianas grown for their bloom, such as clematis and wisteria, require more extensive pruning. Clematis types that bloom only during summer are blooming on current-season wood. They should be pruned back to within 2 feet (60 cm) from the ground during each dormant season. Those that bloom during both spring and summer need light corrective pruning during the dormant season plus heavy pruning of spring-blooming portions immediately after they have bloomed.

Wisteria can be trained as a tree, a shrub, or a liana. During the first and second year it is trained to its desired form by staking, tying, and pinching back unwanted growth. During subsequent years long streamers are pinched back, and unwanted growth is removed to keep it in bounds (Figure 14–83). Rampant-growing older

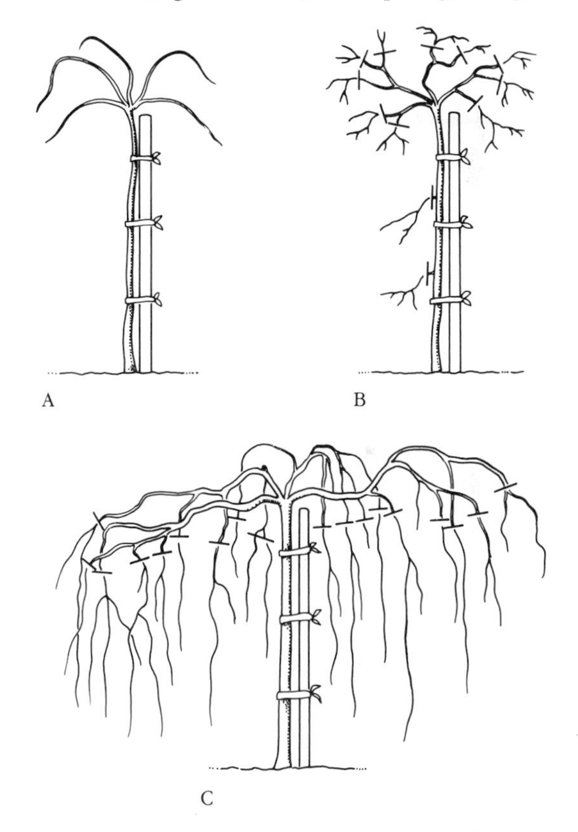

Figure 14–83 Training a wisteria to tree form. The first year the plant is staked and tied, and the top is pinched back at the height it is to head (A). The second year suckers are removed and the branches shortened (B). The third and subsequent years streamers are pinched or cut back to shape the head and keep the plant in bounds (C).

plants that fail to bloom can often be made more reproductive by digging vertically with a sharp spade to prune the roots.

Pruning Herbaceous Plants

Most herbaceous plants do not require pruning, but a few are benefited by pinching back, disbudding, and/or staking (Figures 14–84 to 14–87).

Landscape Construction

Legal Considerations

Most incorporated municipalities have ordinances that specify minimum standards for sidewalks, curbs, gutters, and driveways. The specifications will vary depending on

Figure 14–84 Training tomato plants. Plants of indeterminate tomato cultivars can be trained to a single stem by pinching out the vegetative buds that form in the leaf axils and tying the main stem to a stake. Fruit buds are formed between the nodes. Note on the soil the size of axillary shoots that have been removed. (Courtesy Washington State University.)

Figure 14–85 Pruning peonies. Most peony cultivars produce flower stems having a larger center bud and two small side buds (A). The main flower will be stronger if the side buds are removed (B).

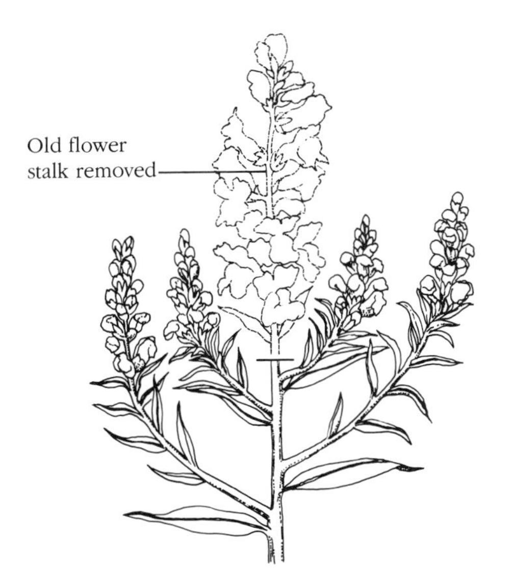

Figure 14–86 Pruning faded flowers. Removing the flower stalk from snapdragons as the florets fade will force new flower shoots to elongate from the flower leaf axils and eventually flower. Removing the spent flower stalk from delphinium will produce the same response. Pansies and violas will continue to bloom only if the flowers are picked before they form seed. In fact, most annual and many perennial flowering plants will bloom longer and more profusely if faded blooms are removed promptly.

the climate and soil of the region as well as on municipal traditions. Although these ordinances often apply only to the sidewalks and portions of drives that are along the street and used by the public, they require standards that assure reasonable endurance of the structures in the local environment. Thus, the specifications are just as essential for all walks and drives in the area. Following is a brief summary of essential points of a representative sidewalk standards code for a small city (the full text of the code requires six printed pages).

1. Sidewalks shall be constructed adjacent to the curb.

2. They shall be 4.5 feet wide exclusive of curbs.

3. Side slope shall be ¼ inch to 1 foot of sidewalk width.

4. Sidewalk slab shall be not less than 3⅜ inches thick except at driveway crossing sections, where thickness shall be not less than 5.5 inches.

5. Sidewalks containing curbs and gutters shall be made of Portland cement concrete mix, not less than six bags of Portland cement per cubic yard, and shall have ultimate strength of 3,600 psi, minimum, at 28 days after mixing.

6. Expansion joints shall be not less than 20 or more than 25 feet apart.

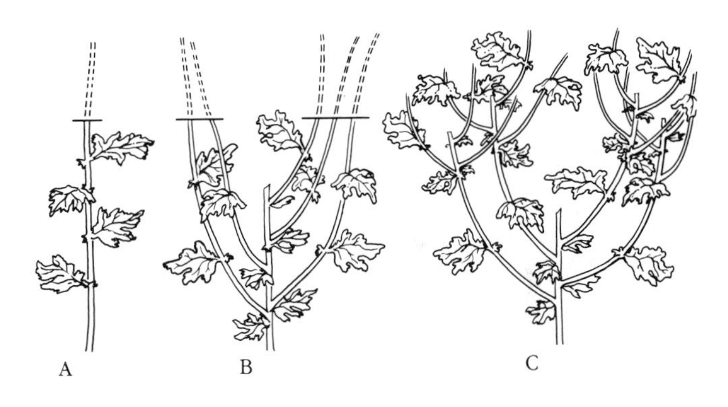

Figure 14–87 Pruning for compact growth habit. Chrysanthemum plants will produce more flowers and become more bushy and compact if their new growth is pinched back. From a few inches to about one-third of the growth should be removed in early to mid-July (A). This will force new branches from the leaf axils, which, in turn, can be pinched back in early August (B) in time for new growth, on which flower buds will be initiated and blooms produced in September and October (C). Similarly, rooted cuttings of many houseplants, including coleus, bergenia, and velvet plant, need to be pinched back for a compact growth habit.

Construction of Walks, Drives, Patios

Gravel, Cinders, Bark, and Turf. These materials are inexpensive to purchase and easy to lay. Bark or turf paths have appeal in naturalistic gardens. However, these materials also have disadvantages. Gravel and cinder paths and drives can become dusty in areas with a dry climate. Turf paths will not withstand heavy traffic. Bark breaks up and must be renewed frequently. Gravel tends to get scattered into lawn areas, where it can be a hazard with rotary mowers.

Drives, walks, and patios made of these materials require considerable upkeep. Turf must be trimmed, and weeds and grass tend to come through the other materials. A layer of 4 mil perforated polyethylene or a thorough treatment with a nonselective, persistent herbicide applied before the gravel, cinders, or bark are laid will help control weeds.

Loose Bricks, Cement Blocks, Field Stone, Cedar Rounds. When these materials are laid singly as stepping stones, they should be centered at about 28 to 32 inches (70 to 80 cm) apart, the distance of a normal step. They should be anchored well, with the surface about ¼ inch (⅔ cm) above the surrounding turf so that they don't teeter when walked upon and don't collect moisture when it rains.

In the South and along the Pacific Coast, where frost seldom penetrates the soil, walks and patios can be made with loose blocks laid close together in well-settled soil from which the plant cover has been removed. Herbicides and/or perforated polyethylene can be used to limit weeds. Wooden 2 x 4s or 2 x 6s treated to resist decay should be used as a border to keep bricks or smaller stones from working their way out of place. In colder locations, where freezing and thawing of soil water tends to displace individual blocks, walks and patios must be laid on a well-drained foundation or else be mortared into a unified structure (Figure 14–88).

Concrete or Masonry. Where soil is porous and well drained, concrete or solid masonry walks, drives, and patios can be placed directly on a well-packed soil surface. Where drainage is less adequate, several inches of gravel should be spread before the structure is laid.

Walks should be at least 2 feet (60 cm) wide for each person who is to walk abreast. Walks and patios should be constructed slightly higher than the surrounding area and tilted so that water drains from them. They should be at least 4 inches (10 cm) thick.

Drives should be at least 8 feet (about 2⅓ m) wide for each automobile, have a minimum thickness of 6 inches (15 cm) of concrete reinforced with steel, and be constructed so that water drains from them. Monolith drives and patios should have expansion joints every 8 to 12 feet (2⅓ to 4 m) in each direction. Expansion joints are ½- to ¾-inch (1¼ to 2 cm) spaces that are filled with asphalt or other material to allow concrete to shrink and expand with changes in temperature.

Concrete walks, drives, and patios are constructed by placing forms around the excavated area they are to occupy. Metal forms can be rented, or unwarped 2 x 4 or 2 x 6 planks can be used as forms for straight segments. Thinner bendable boards or metal can be used as forms for curved sections. After the forms are adjusted to the correct position and firmly fixed in place with stakes and soil, the excavation is filled with concrete that has been mixed with correct proportions of sand, gravel, Portland cement, and water. Today concrete already mixed is usually purchased and delivered from a premix plant. After the concrete is poured, the slab is leveled with a plank that is pushed back and forth to force excess cement into areas not yet filled. The poured mixture should be worked with a shovel and trowel so that air pockets are eliminated and coarse aggregates are worked beneath the surface, but not to the point that the surface becomes completely smooth. A smooth surface will be slippery when wet. The surface can be brushed lightly with a broom after it has begun to set but before it has hardened to provide better traction. Concrete that is to be the base for a brick or flagstone and masonry patio walk should receive a minimum of troweling so that its surface will remain rough, thereby permitting the masonry to bond more tightly.

Newly finished concrete should be protected from rain until it has hardened. After hardening, concrete must un-

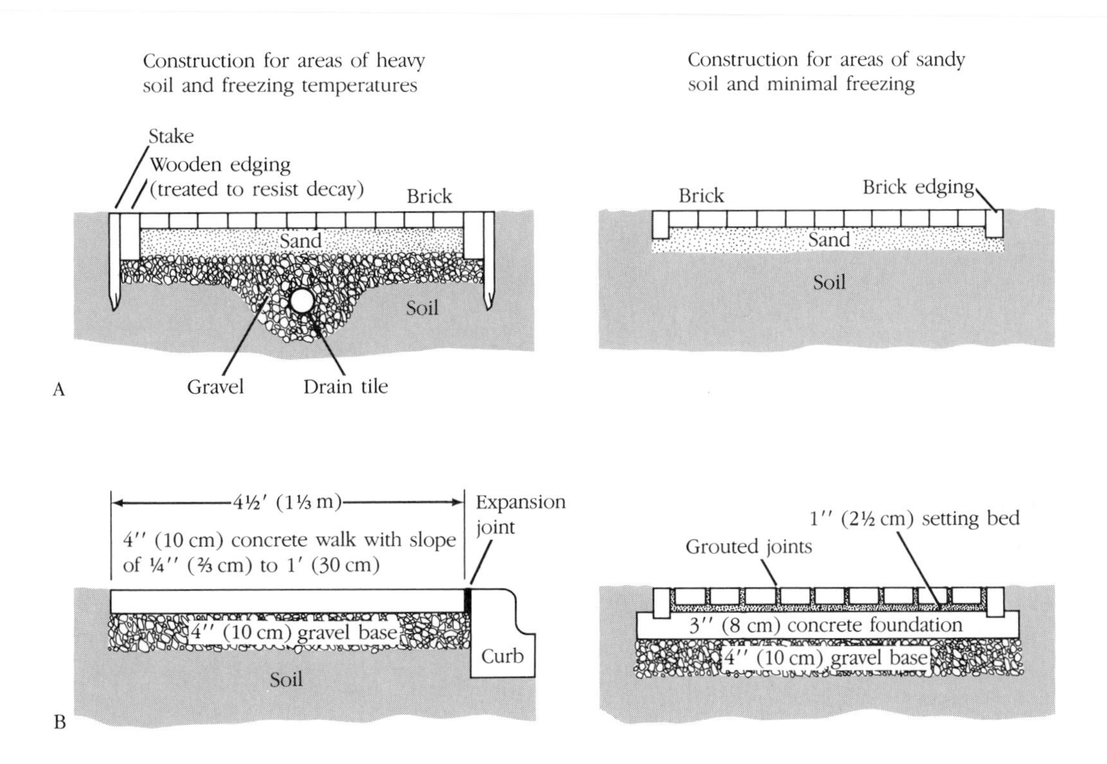

Figure 14–88 Construction detail of unmortared brick walks (A) and of cement and mortared brick walks (B). The same construction principles apply to flagstone walks and patios. Even with adequate drainage loose bricks or stones are likely to be displaced by frequent freezing and thawing and require leveling every few years (see A, left).

dergo a series of chemical and physical changes, called curing, that require 4–5 days and are essential for durability. It must be protected from freezing temperatures and not be allowed to dry out until it is cured. During hot dry weather concrete that is curing may need to be sprinkled as many as eight or ten times each day.

Patios and walks of brick or flagstone with mortared joints are not difficult to construct (see Figure 14–88). The area should be excavated to a depth of 10 inches (25 cm). A 4-inch layer of fine gravel is placed in the bottom, wet thoroughly, and rolled or tamped. A 3-inch (8 cm) base layer of concrete is spread over the gravel and leveled. After the concrete foundation has set for 24 hours, the brick or stones are laid in the desired pattern on a thin layer of mortar. When the mortar has set, the joints should be filled. If the joints are over ¼ inch wide, they can be filled carefully with a mixture of wet mortar. If joints are less than ¼ inch wide, a dry mixture of 1 part cement to 3 parts sand can be dusted onto the surface and swept into the joints. When the joints are filled and the stones or bricks have been swept clean, the surface can be watered with a fine gentle mist until the mortar in the joints has become thoroughly wet.

Construction of Retaining Walls

The construction of retaining walls made of loose stone (dry wall) is illustrated in Figure 14–89. They are subject to movement from frost and erosion and, especially in

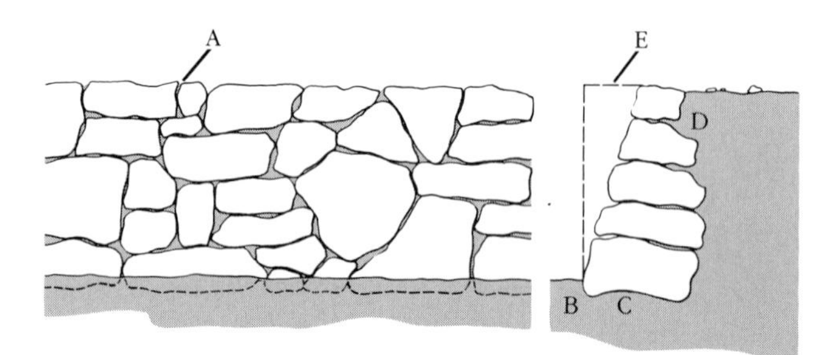

Figure 14–89 Dry wall. A retaining wall made of loose stones with no mortar (dry wall) should have the stones fit together as well as possible (A), should have a solid base (B), should have the longest dimension of the stones fit back into the retained soil (C), should have soil packed solidly into the area behind and crevices between the stones (D), and should slope strongly toward the soil it is to retain (E).

areas that have heavy or poorly drained soil, will require some maintenance and repair each spring.

Solid retaining walls are of two types: mass concrete or gravity type, which rely on the weight of the wall to prevent movement (Figure 14–90), and cantilever type, which are prevented from sliding by the weight of soil on a toe or footing (Figure 14–91). Figure 14–92 shows the average depth of frost penetration in the United States. (The specifications in Figures 14–90 and 14–91 come from *Architectural Graphic Standards*, 6th ed., by C. G. Ramsey and H. R. Sleeper, Wiley, New York, 1970; *Time Saver Standards for Architectural Design Data*, 5th ed., ed. J. H. Callender, McGraw-Hill, New York, 1974; and *Use of Concrete on the Farm*, Farmer's Bulletin 2203, U.S. Department of Agriculture, U.S. Government Printing Office, Washington, DC, 1970.)

Grading Around a Tree

Woody plants already partly grown are an asset to a building lot and should be protected. Before any construction begins, the owner should establish a written agreement with construction contractors concerning the value of plant materials that are to be saved. Placing barriers around trees or shrubs during excavation will help prevent damage to the trunks.

Figure 14–93 illustrates ways of protecting trees from various grading procedures.

Harvesting

As was mentioned in previous chapters, the quality of perishable vegetables and fruits deteriorates rapidly once they are harvested. Wherever possible, harvesting of crops such as strawberries, peas, and sweet corn should be done when the temperature is cool, and the harvested product should be prepared for the table, processed, or placed in storage immediately. Carelessness during the harvest operation can cause serious problems. Spurs and branches broken from fruit trees and berry plants will not produce fruit in succeeding years. Throwing or dropping fruit,

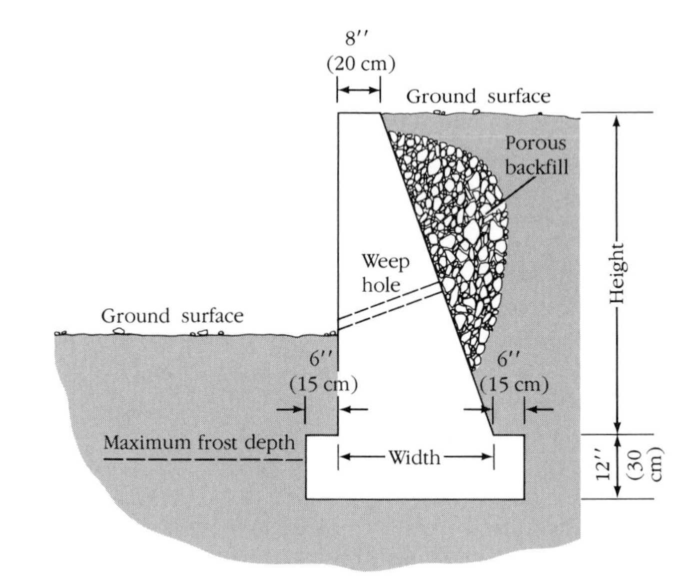

Figure 14–90 Solid concrete gravity-type retaining wall. The base width should be a minimum of half the height, the top should be at least 8 inches (20 cm) wide, and the footing should extend below the maximum frost penetration for the area. Other minimum specifications are shown in the diagram.

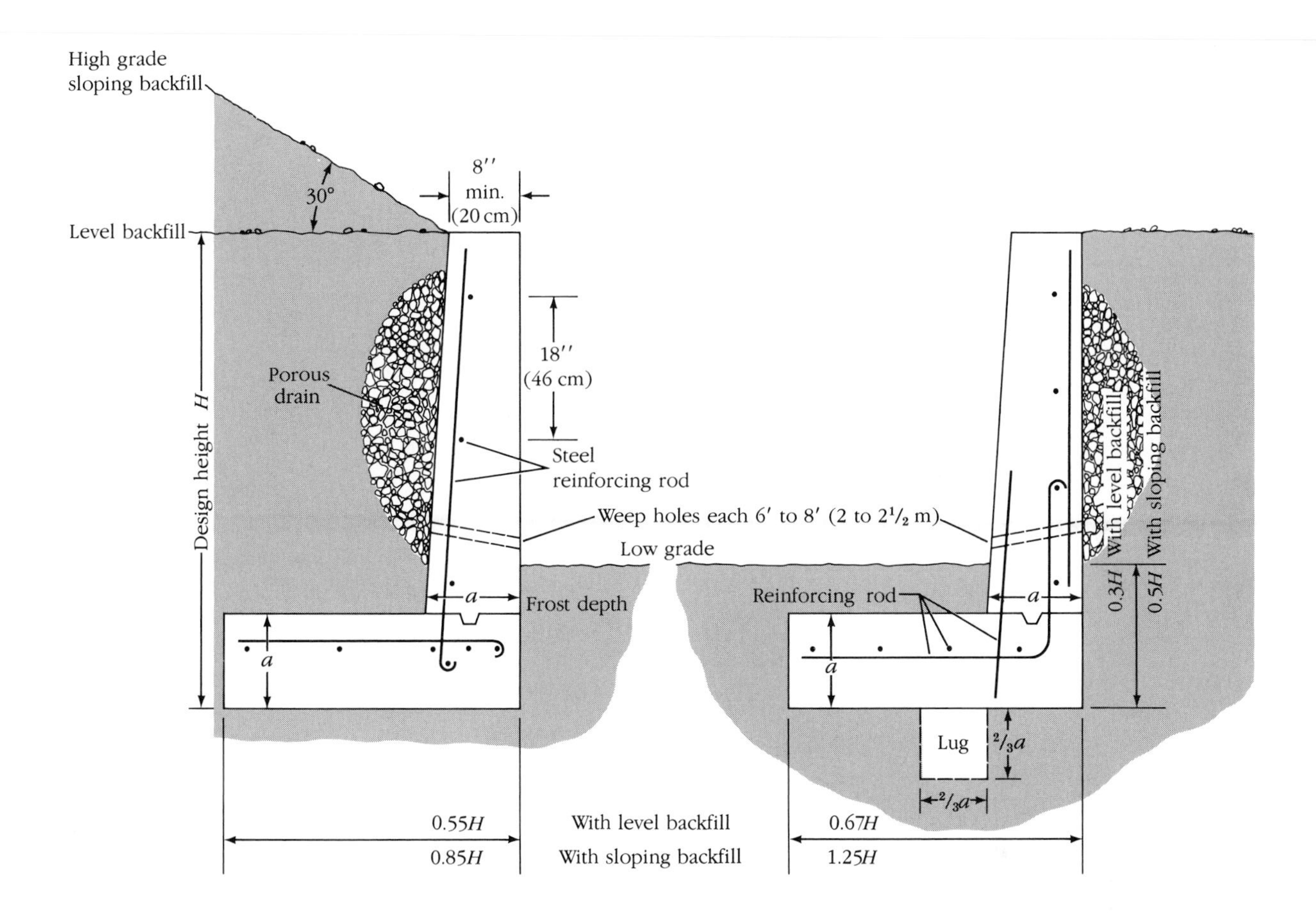

Figure 14–91 L-type reinforced concrete cantilever retaining wall. Reinforcing rods should be 18 inches (46 cm) apart each way and near to, but at least 2 inches (5 cm) from, the "fill" side of the wall. The thickness of the wall base and the footing or toe, *a*, is 11 inches (28 cm) for a wall 3 feet (1 m) high. Add ½ inch (1¼ cm) to the thickness of *a* for each foot of added wall height (*H*) if the backfill is level, and 1 inch (2½ cm) for each added foot of *H* if the backfill slopes 30°. A lug (lower right) will help prevent sliding of walls built on moist clay. The specifications of a T-type wall are the same as for the L-type, with toe extending in the direction of the backfill (left), provided no more than one-third of the T-type footing extends toward the low-grade side of the wall. An engineer should design walls over 10 feet (3 m) high or 25 feet (8 m) long. Solid brick or stone masonry walls up to 6 feet (2 m) high can be constructed with similar specifications. They should be at least 9 inches (23 cm) wide with a masonry cap 2 inches thick. The fill side of a masonry wall should be waterproofed before the wall is backfilled to prevent discoloration from moisture and salts.

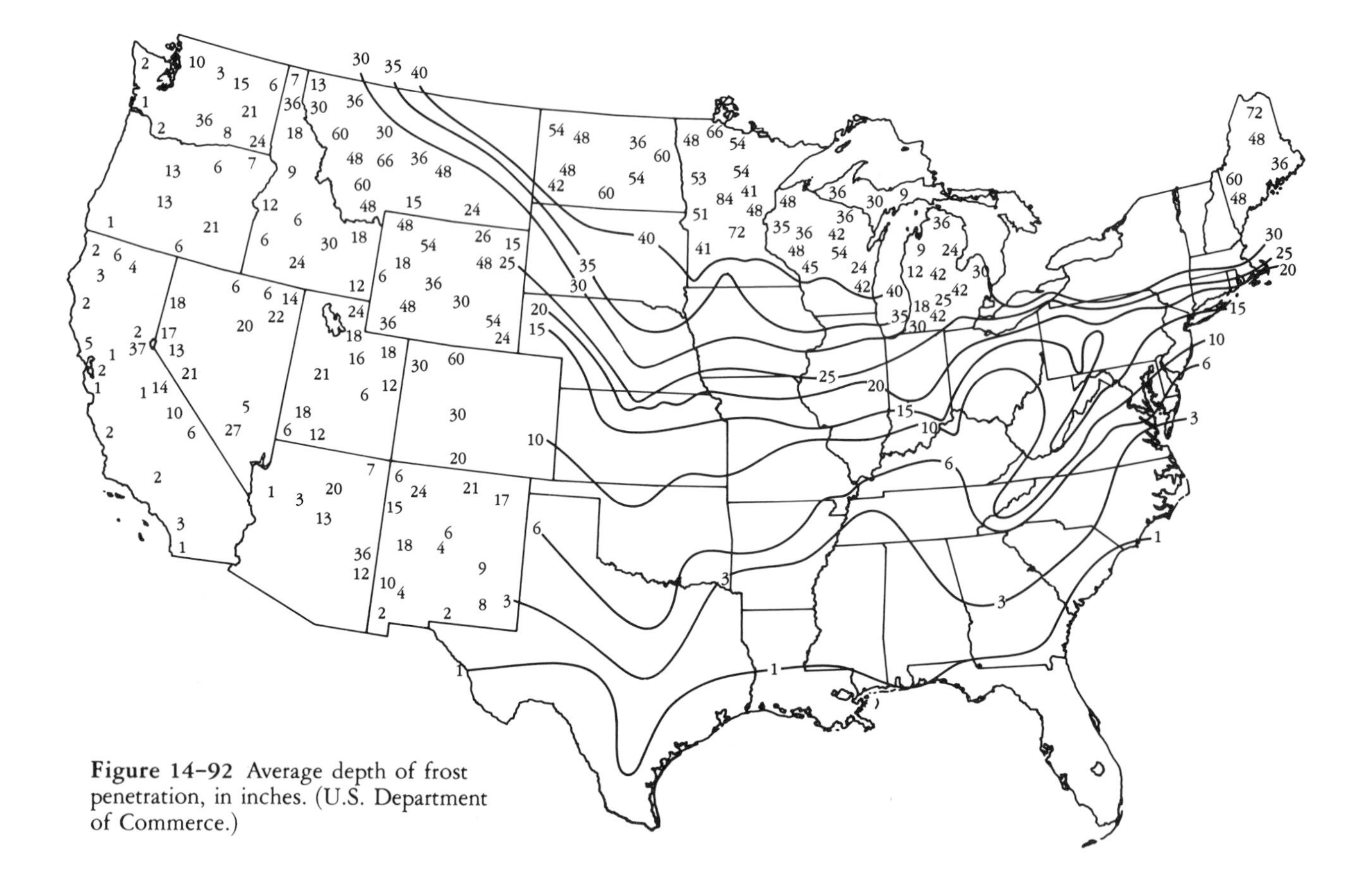

Figure 14–92 Average depth of frost penetration, in inches. (U.S. Department of Commerce.)

roots, bulbs, tubers, or even dry seed, or other careless handling greatly reduces quality and storage life. All horticultural products are easily bruised, and the damage from careless handling at harvest may not be obvious at the time the damage occurs. Even so, deterioration and disease, which get their start on bruised or damaged tissue, can, in a short time, spread not only through the damaged fruit or vegetable but through other items stored in the same container or area.

Stage of Maturity and Method of Harvest

Cutting Flowers. Cut flowers for arranging should be gathered before the florets have fully expanded to insure the longest possible life. Roses are cut when they are in the bud stage. Gladioli, snapdragons, and other spiked flowers should be harvested soon after the lowest florets have opened. Mums, zinnias, marigolds, daisies, and other composites should be cut when the outer petals are fully colored but before the inner parts of the flower have matured or shed pollen.

Wilted flowers will not last, so cutting beds should be kept well-watered, and flowers should be cut during late afternoon or early morning when the temperature is cool. If possible, take a bucket of lukewarm water to the garden and immerse the stems immediately after they are cut. The cut ends of flowers such as dahlias and poppies that give off a milky exudate should be seared by immersion in boiling water, by burning in a flame for a few seconds, or by immersion in very hot tap water for 10 minutes. Some woody-stemmed flowers such as lilac are

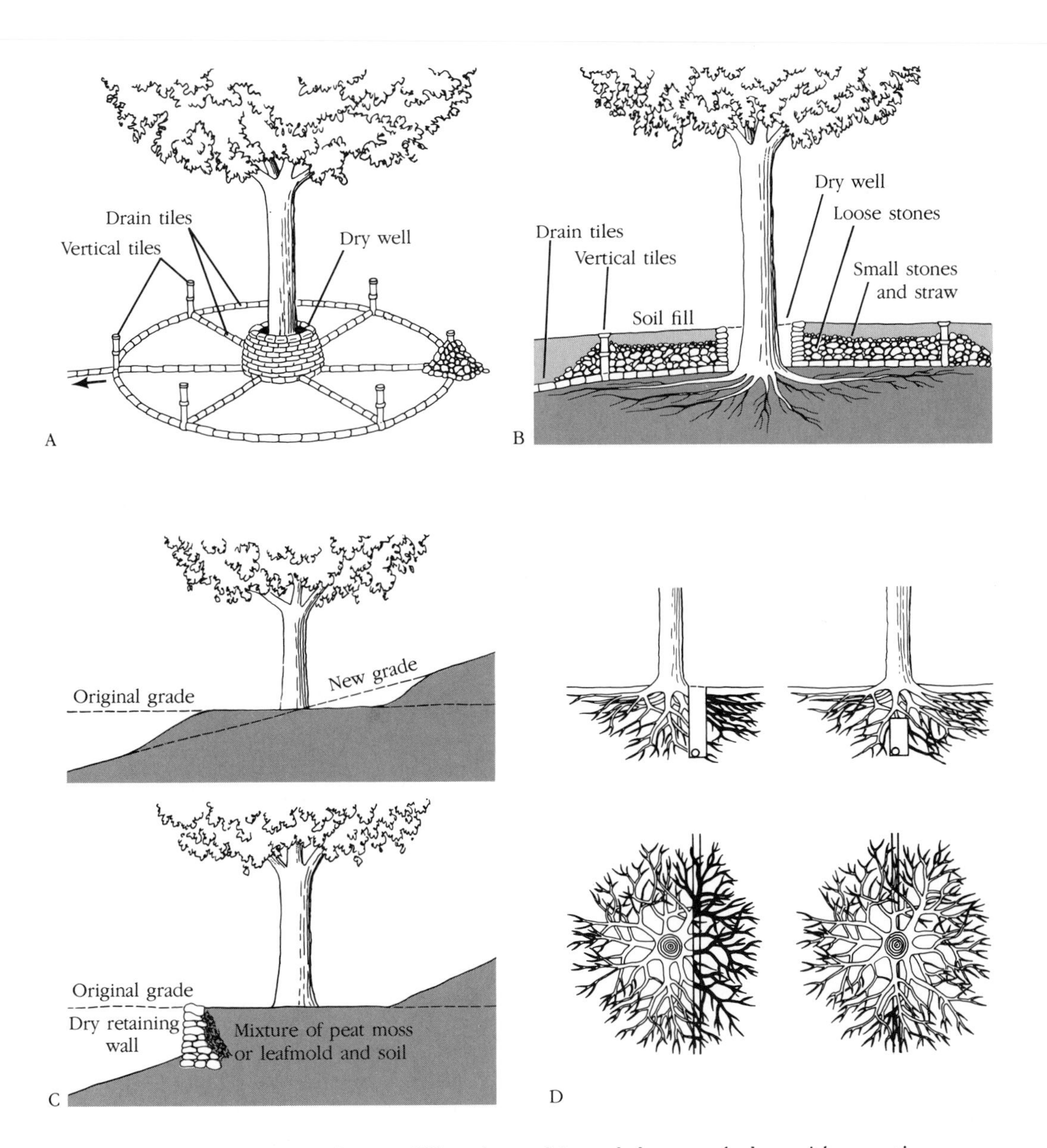

Figure 14–93 Grading around a tree. When the yard is graded or trenched, special precautions are necessary to protect and provide aeration for tree roots. A tile system protects a tree from a raised grade. (A) The tile is laid out on the original grade, leading from a dry well around the tree trunk. (B) The tile system is covered with small stones to allow air to circulate over the root area. (C) A retaining wall protects a tree from a lowered grade. (D) Tunnel beneath root systems, don't trench through them. Left: these trenches will probably kill the tree. Right: tunneling under the tree will preserve many of the important feeder roots. (From *Protecting Shade Trees During Home Construction*. USDA Home and Garden Bulletin 104. U.S. Government Printing Office, Washington, DC. 1965.)

better able to take up water if the ends of their cut stems are crushed with a hammer. The flowers, with their container of warm water, should be kept in a cool environment for at least 3 hours and preferably overnight.

All containers, vases, frogs, and other material used for gathering and arranging flowers should be sterilized with heat or chemicals after they have held or been in contact with cut flowers. Shredded styrofoam, floral blocks, and other disposable material used to support flower arrangements should be discarded or sterilized in boiling water before being used a second time.

Harvesting Vegetables and Fruits. The timing of fruit and vegetable harvest, especially of those crops produced for their fruit or immature seed, is critical. Timing is usually less important for crops where the leaf, stem, root, or petiole is the edible portion.

The time and method of harvest of selected fruits and vegetables are summarized in Table 14–20.

Storing Horticultural Products

In most parts of the United States the season during which a horticultural product is available directly from the garden is rather short. Proper storage and/or processing can prolong the season during which the table is supplied from the garden.

Causes of Deterioration

You will recall from Chapter 2 that, as long as a plant remains living, respiration continues to utilize carbohydrates and oxygen to produce plant energy, releasing carbon dioxide, water, and heat energy as by-products. Most living products also continue to transpire and thus are gradually losing moisture.

In addition to the reduction of sugars and moisture, which occurs as a result of respiration and transpiration,

Table 14–20 Time and method of harvesting vegetables and fruits

Vegetable	Stage for harvest	Methods of harvest
Artichoke (globe)	When bud is fully expanded but still tight.	Cut ¾ in (2 cm) below bud.
Asparagus	When spears are 6–10 in (15–25 cm) long; harvest all spears for 8–10 weeks in spring; don't harvest after July 1; don't harvest first year after transplanting.	Cut below soil surface or grasp the tip and bend; all of the spear harvested will be useable with the latter method.
Rhubarb	As soon as petioles are of sufficient size; never harvest more than half the petioles from a plant at one time; don't harvest after July 1; don't harvest first year after transplanting.	Pull petiole from stalk, don't cut it off; pull and discard seed stalks.
Radish	As soon as roots have expanded to sufficient size but before seed stalks form or roots become pithy.	Pull roots; first harvest can thin plants so nearby roots can enlarge.
Spinach, chard	As soon as plants reach sufficient size; before spinach bolts. Early harvest can thin plants to provide more room for those that remain.	Cut chard above crown so regrowth can occur. Spinach can be cut above or below crown.
Lettuce	Thinnings can be used for early salad; heads of Iceberg should be solid but not overmature; butterhead and Cos can be harvested as soon as heads form, leaf lettuce, anytime before it bolts.	Pull or cut at ground level.
Celery	When fully grown; usually requires the full season.	Cut below crown.
Cabbage	When head is solid but before it cracks. Twisting the mature head part way around to sever half the roots will permit cabbage to remain in the garden longer.	Cut below the head.
Cauliflower	Tie leaves above the head to shade it when head reaches golf ball size; cut head 4–6 days later depending on temperature but before curd starts to separate.	Cut below first whorl of wrapper leaves.

Vegetable	Stage for harvest	Methods of harvest
Broccoli	Cut center head before buds begin to separate; continue to cut side shoots as they form.	Cut 3 in (8 cm) below the flower buds.
Brussels sprouts	Remove lower leaves as sprouts start to enlarge. Harvest sprouts as they become solid; the lowest will mature first.	Break sprouts from stalk.
Beets, carrots, turnips, rutabagas	As soon as roots are large enough to use. Early harvests can thin to provide more room for plants remaining. Small beets with tops can be used for greens; harvest for storage before they become woody.	Pull and cut off tops *above* the crown.
Onions	Thinnings can be used as green onions or scallions. For storage bulbs should be pulled when two-thirds of the tops have fallen over.	Pull and cut off tops *above* the crown. Cutting below the crown will create an entrance wound for decay organisms.
Potatoes	For immediate use dig as soon as tubers are large enough to use; small tubers will have formed about two weeks after bloom. For storage dig a few days after vines have died, been frozen, or artificially killed.	Cut stem of fruit from plant. Leave some stem on fruit.
Peas and edible-podded peas	When pods are nearly full but before pods begin to wrinkle; edible podded peas should be harvested while pods are still flat.	Pick from vine during cool part of the day.
Snap beans	When pods are 3 in (8 cm) long but before seed is much larger in diameter than a pencil lead.	Pick from vine each 3–4 days (disturbing wet vines spreads rust and other diseases).
Lima beans	When pods are fully expanded but before pods turn yellow or seeds white.	Pick from vines and shell.
Sweet corn	When liquid squeezed from a kernel with the thumb nail is milky (watery, immature; creamy or solid, overmature); when individual kernels can be felt through the husk.	Pull cob from stalk with a quick downward motion.
Sweet potatoes	Whenever "potatoes" are large enough; before soil temperature drops below 50°F (10°C).	Dig carefully; injured roots rot.
Tomatoes	After red color first shows they will ripen to high quality indoors. For fall storage pick all green fruit that is nearly full size.	Pick from vine and remove stem if it stays with tomato.
Eggplant	After fruit reaches egg size but before it matures; quality of young fruit is better than that of older fruit.	Cut stem of fruit from plant.
Peppers	Anytime after they are large enough to use.	Break stem from plant.
Summer squash and cucumbers	As soon as fruits have reached desired size but before cucumbers turn yellow or summer squash form hard seeds or rinds.	Break fruit from vine; leave a piece of stem with the fruit.
Winter squash	When fully mature, rind is hard.	Leave stem on fruit.
Pumpkins	When they turn gold in color.	Leave stem on fruit.
Muskmelon	When a crack forms around the area where fruit attaches to stem; at this time melon will slip easily from the vine.	Pull to detach stem from fruit.
Watermelon	When vine tendril next to fruit yellows and dries; when thumping with fingers causes a dull rather than a ringing sound. If melon is to be used immediately it can be tested by applying pressure with the palm of hand; a splitting sound indicates a ripe melon, but pressure will cause an inside crack.	Cut to leave some stem with the fruit. Pulling the stem from the fruit may damage the fruit.

(continued)

Table 14–20 *(continued)*

Fruit	Stage for harvest	Methods of harvest
Apples—summer and fall	'Yellow Transparent' and 'Lodi' can be picked whenever they have reached size desired; with others ground color (color between stripes or where they are not red) turns from green to light green or yellow; flesh softens somewhat.	Turn fruits back on stem so that stem but not spur remains with fruit.
Apples—winter	Ground color turns green to yellow; seeds turn brown; flesh softens slightly.	Same as above.
Pears	Picked when full size but still hard; pears are of low quality if ripened on trees. Commercially they are picked according to pressure required to push a plunger into their flesh.	Same as apple.
Sweet and sour cherries	For home use when fully ripe if birds will permit them to ripen. Dark sweet cherries should be sweet and almost black; pie cherries will be bright or dark red depending on variety.	Pick with stems on fruit unless they are to be canned or otherwise preserved immediately.
Apricots	When fully grown and orange colored but before they soften. Some varieties may require picking the same tree several times in order to harvest fruit at the right stage.	Twist carefully to prevent damage to nearby spurs or the fruit.
Peaches	When fully grown but at least 2 days before they become soft; peaches are better quality, especially for canning, if they are allowed to ripen after being picked.	Same as above.
Plums	When mature but before they begin to shrivel.	Same as above.
Gooseberries	Gooseberries full size but still firm and green in color are best for jam and pies; can be ripe or green for jelly.	Pick with stem and blossom; remove stem and blossom for jam or pie but not for jelly.
Currants	For the best jelly currants should be picked when 90% are ripe and 10% are still green. Green ones supply more pectin.	Pick entire cluster with stem; stem need not be removed for jelly. Pick all fruit from each cane before moving to the next.
Raspberries and blackberries	Pick when berries slip easily from stem.	Pick berries without stem.
Blueberries	When fully ripe.	Harvest individual berries without stem.
Strawberries	When red but still firm.	Harvest with cap and stem.
Citrus	When fully mature and sweet; citrus can be left on the tree for some time after it is mature without deteriorating. With certain weather conditions fully ripe oranges may be green in color.	Twist to avoid tree or fruit injury.
Nuts	When they drop to the ground in the fall.	Shake from the tree and gather before squirrels do. Filberts must be separated from husks. Black walnut husks are allowed to decay for removal.

many plant products undergo gradual chemical changes in storage. Most fruits, for example, give off ethylene, a growth-regulating substance that may have profound effects on other products stored nearby. Stored carrots turn bitter, and potatoes may fail to sprout and may undergo other changes if there is ethylene in the atmosphere where they are stored. Fruits are likely to absorb an earthy flavor if they remain very long in close contact with root crops.

Another major cause of fruit and vegetable deterioration in storage is decay or spoilage due to the action of microorganisms. A horticultural product usually possesses considerable resistance to microorganism attack unless the product has been damaged by rough handling or unless it is beginning to deteriorate as a result of age or improper or too lengthy storage. Spoilage microorganisms generally enter through wounds or through dead or dying cells.

The length of time a plant product will store depends partly on the nature of the product. The relation of low respiration rate to ability to store has been mentioned in earlier chapters. A waxy epidermis or thick periderm provide protective coverings that prevent entry of microorganisms and retard moisture loss, enabling products like apples and potatoes to be stored for relatively long periods. Physical structure and chemical composition are also related to storageability.

The optimum storage temperature and humidity vary depending on the product being stored. The majority of fruits, vegetables, and flowers that originated in the temperate and subtropical zones store best at 32°F (0°C) and 90–100% relative humidity. Crops of tropical origin, such as tomatoes, bananas, eggplant, and avocado, undergo physical deterioration if stored at temperatures lower than about 40°F (4°C). Optimum storage environment for a number of fruits and vegetables is listed in Table 14–21.

How to Store

Short-Term Storage. For the person who raises a large garden, a used refrigerator that can be reserved entirely for fruit and vegetable storage and can be placed in some out-of-the-way corner of the basement is a wise investment. Lettuce, spinach, cauliflower, broccoli, radishes, chard, summer squash, and a host of fruits can be kept from 1 to 3 weeks in this kind of storage. One precaution—if fruits and vegetables are to be kept in the same refrigerator, be sure that one or the other is in a tightly closed container.

Table 14–21 Freezing points, recommended storage conditions, and length of storage period of vegetables and fruits

Commodity	Freezing point (°F)	Place to store	Storage conditions		Length of storage period
			Temperature (°F)	Humidity	
Vegetables:					
Dry beans and peas		Any cool, dry place	32° to 40°	Dry	As long as desired
Late cabbage	30.4	Pit, trench, or outdoor cellar	Near 32° as possible	Moderately moist	Through late fall and winter
Cauliflower	30.3	Storage cellar	Near 32° as possible	Moderately moist	6 to 8 weeks
Late celery	31.6	Pit or trench; roots in soil in storage cellar	Near 32° as possible	Moderately moist	Through late fall and winter
Endive	31.9	Roots in soil in storage cellar	Near 32° as possible	Moderately moist	2 to 3 months
Onions	30.6	Any cool, dry place	Near 32° as possible	Dry	Through fall and winter
Parsnips	30.4	Where they grew, or in storage cellar	Near 32° as possible	Moist	Through fall and winter
Peppers	30.7	Unheated basement or room	45° to 50°	Moderately moist	2 to 3 weeks
Potatoes	30.9	Pit or in storage cellar	35° to 40°	Moderately moist	Through fall and winter
Pumpkins and squashes	30.5	Home cellar or basement	55°	Moderately dry	Through fall and winter
Root crops (miscellaneous)		Pit or in storage cellar	Near 32° as possible	Moist	Through fall and winter
Sweet potatoes	29.7	Home cellar or basement	55° to 60°	Moderately dry	Through fall and winter
Tomatoes (mature green)	31.0	Home cellar or basement	55° to 70°	Moderately dry	4 to 6 weeks
Fruits:					
Apples	29.0	Fruit storage cellar	Near 32° as possible	Moderately moist	Through fall and winter
Grapefruit	29.8	Fruit storage cellar	Near 32° as possible	Moderately moist	4 to 6 weeks
Grapes	28.1	Fruit storage cellar	Near 32° as possible	Moderately moist	1 to 2 months
Oranges	30.5	Fruit storage cellar	Near 32° as possible	Moderately moist	4 to 6 weeks
Pears	29.2	Fruit storage cellar	Near 32° as possible	Moderately moist	See text

(Adapted from *Storing Vegetables and Fruits in Basements Cellars, Outbuildings and Pits.* USDA Home and Garden Bulletin 119. U.S. Government Printing Office, Washington, DC. 1970.)

Winter pears and cultivars of late fall apples, such as 'Red' and 'Golden Delicious,' 'Yellow Newtown,' 'Winesap,' 'Idared,' and 'York Imperial,' can be kept during the fall almost until Christmas in an unheated garage or shed in many areas where the temperature during this period remains between 20° and 40°F (−7° and +4°C). A heavy quilt or other cover to provide insulation on colder nights, a thermometer to measure the temperature of the area around the fruit, and a careful watch of weather so that the stored produce can be moved inside during excessively cold periods are necessary if fruit is to be stored this way. The respiration of the fruit will provide considerable heat if the containers are insulated. Plastic box liners perforated with ten to twelve ¼-inch holes will help keep apples of cultivars such as 'Golden Delicious' and 'Yellow Newtown' from shriveling in this kind of storage.

Root vegetables in light plastic bags can be stored in the same way, although in many areas they can be left in the garden during the fall. In fact, parsnips and salsify can be stored all winter where they are growing, because they are not damaged by freezing. In some areas beets, carrots, turnips, gladioli corms, and the storage organs of other flowers normally dug in the fall and kept indoors through the winter can also be left all winter where they are growing. After the first light frost but before heavy freezes occur, a thick cover of straw is raked over the row of plants. Leaves can be used to cover flowers, but many kinds of leaves impart an undesirable flavor to root vegetables.

Winter Storage in a Cellar. A number of horticultural products can be stored through the winter if proper storage conditions can be provided. A refrigerated storage, where temperature and humidity are automatically controlled, would be ideal; however, it is not economically practical for most homeowners. A version of the old cellar storage works well in areas where winters are cold and may be practical for the homeowner who grows a large garden and has the right location. This kind of structure is most easily constructed by digging into a sloping site. The walls can be constructed of concrete, cinder block, or mortared stone and can be insulated by soil piled against their outside. If the storage is dug into a slope, the floor and sometimes the back wall are exposed soil, which can be watered down occasionally to increase humidity in case natural moisture is insufficient. Only the front and roof are exposed (Figure 14–94). The U.S. and Canadian Departments of Agriculture have publications and working plans of storages. These are available at many county extension offices for a small fee or from the two national departments.

An unheated basement with a dirt floor under a home may be an ideal storage location, or sometimes an unheated area of a concrete basement can be insulated from the rest of the home for storage. The basement storage area should have a window that can be opened for ventilation. Where basement humidity is lower than 90%, root crops can be kept from dehydrating by being covered with sand that is kept moist.

Storage Pits. I usually store root vegetables, gladioli corms, and dahlia tubers in a pit that provides about the same storage environment as the cellar storage. I dig a hole in a well-drained location and line it with straw and a section of a hollow cedar log. A barrel or wooden box would do as well. A solid liner is convenient but not essential. Vegetables are washed and placed in gallon-sized perforated plastic bags, and other products are cleaned and placed in bags. The bags are placed in the pit. The pit is covered with a wooden lid and a layer of straw or leaves; soil is piled on the straw so that water drains away from the pit. The straw provides enough insulation to protect the pit from freezing. Throughout most of the winter it is possible to reach down through the straw and pull out a bag of carrots that are ready to be popped into the refrigerator. In areas where soil temperatures remain below 45°F (7°C) most of the winter carrots can be kept until May in this kind of pit (Figure 14–95).

Where the temperature doesn't get too cold or where enough insulation is used, a barrel or above-ground pit storage is feasible (Figure 14–96). Cabbage and celery also can be stored in outdoor pits (Figure 14–97). Apples and pears will store in a pit but are likely to pick up an earthy flavor.

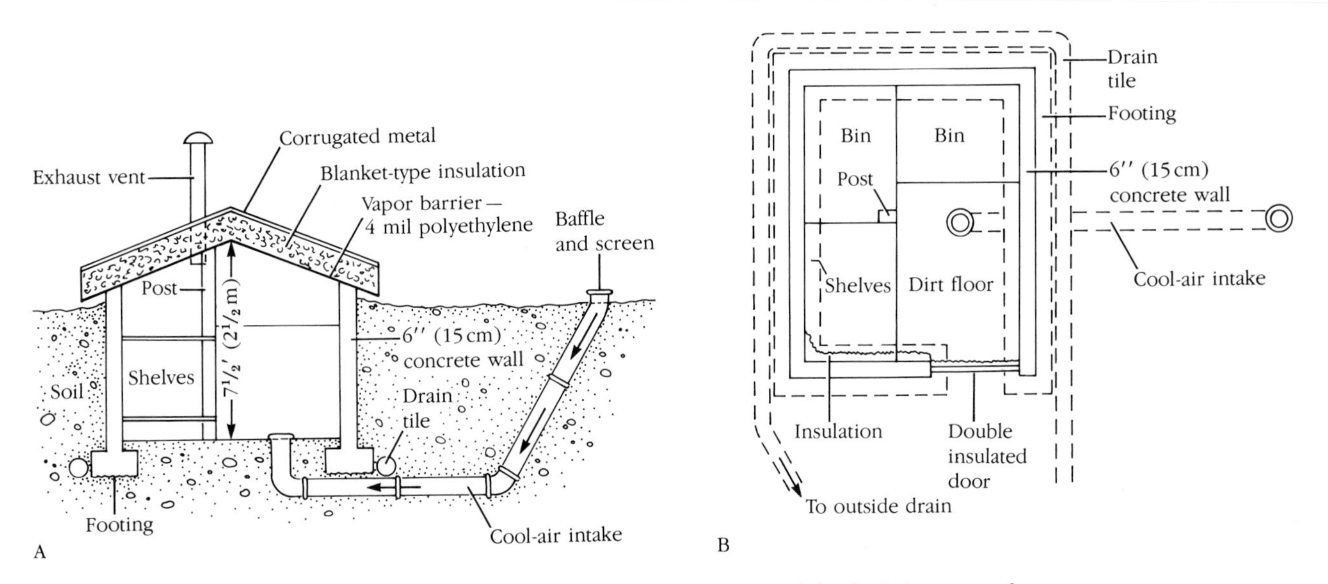

Figure 14–94 Vertical cross section (A) and floor plan (B) of a below-ground storage cellar with three sides of undisturbed earth and an open front. Vents that can be opened and closed are necessary to permit heavier cold air to enter at the bottom and warm air to escape through the roof. At least 6 inches (15 cm) of rock wool batting or similar insulation are required to keep up the temperature during cold weather and to prevent condensation on the roof. The exposed front also must be insulated and drainage provided.

Underground storage can be made with roofs of reinforced concrete covered by soil for insulation. Some cellars are still being built with supports made entirely with timbers. These usually have three sides of undisturbed earth with a roof supported by upright posts set in concrete. The roof poles are covered with brush and baled straw topped with a layer of fine soil low in organic matter.

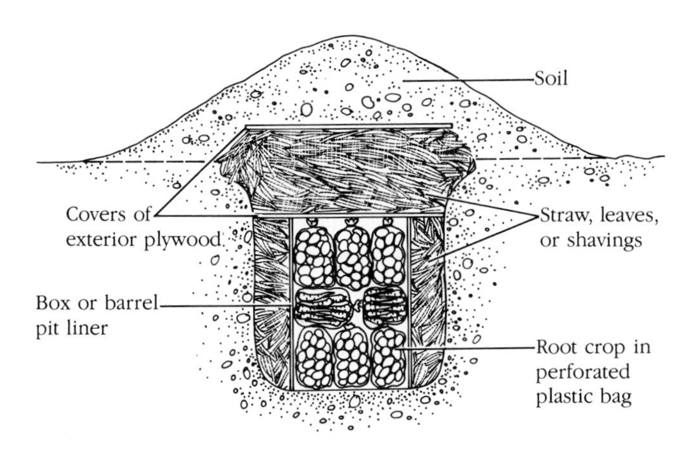

Figure 14–95 Cross section of a below-ground storage pit.

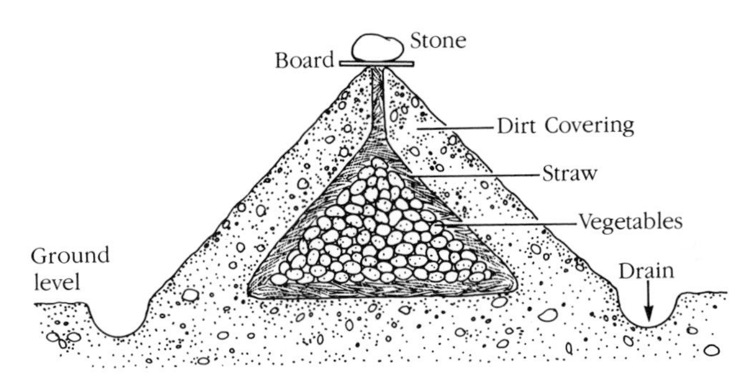

Figure 14–96 Cross section of an above-ground storage pit for areas with a moderate climate or drainage problems. (From *Storing Vegetables and Fruits*. USDA Home and Garden Bulletin 119. U.S. Government Printing Office, Washington, DC. 1970.)

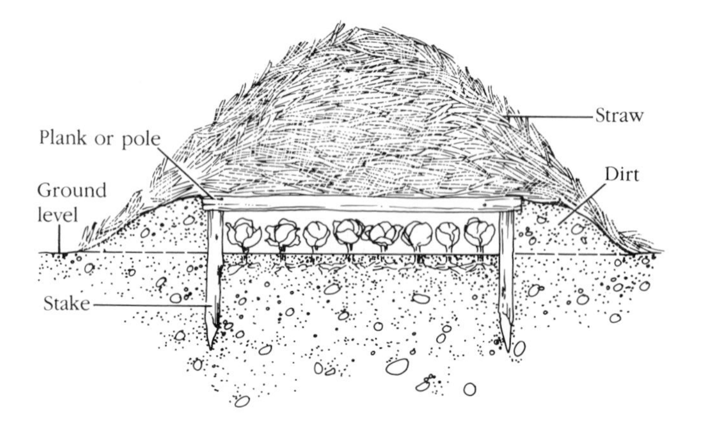

Figure 14–97 Cross section of a cabbage storage pit made of stakes and poles covered with straw. (From *Storing Vegetables and Fruits.* USDA Home and Garden Bulletin 119. U.S. Government Printing Office, Washington, DC. 1970.)

Washing Products to Be Stored. Generally home storage instructions advise against washing vegetables before they are stored on the grounds that washing increases the likelihood of decay. My own experience with carrots and beets has been that washing and then spreading the roots to permit them to barely air dry before they are placed in storage does not increase the incidence of decay; in fact, it may actually reduce it. There are undoubtedly situations where this would not be true, and the reader's own experience and inclination should be the determining factor. Potatoes to be stored and fruits and vegetables that grow above ground should ordinarily not be washed before they are stored.

Storage Requirements for Specific Crops. If very much fruit is to be stored through the winter, a special storage should be built for it. Storage temperature for temperate zone fruits and citrus should be as near 31°F (−1°C) as possible. Winter pears are harvested while still green and held at 32°F (0°C). A week or two before they are to be used, winter pears should be transferred to a room where the temperature is 60° to 65°F (16° to 18°C), which will allow them to ripen.

Potatoes require a high humidity for storage, and once their rest period is broken, usually in December or January, they will sprout if temperatures rise much higher than 40°F (4°C). Moreover, the starch in potato tubers rapidly changes to sugar at temperatures below 38°F (3°C). As a result, the ideal temperature range for potato storage, at least during the spring, is 38° to 40°F, with 90–100% humidity.

If potatoes have been stored at too low a temperature, the resulting undesirable sweetness can be eliminated by keeping them where the temperature is above 55°F (13°C) for a week or 10 days. The potato storage should be kept dark as potato tubers develop green chlorophyll and associated bitter alkaloids when exposed simultaneously to light and temperatures higher than 40°F.

If possible, after they are dug and before they are stored, onions should be left out-of-doors in the sun for a week to 10 days until the top and scales are completely dry. If the weather is rainy or cloudy they can be cured indoors in a warm, dry location. Dirt and loose scales should be removed before they are stored. Onions with large necks or decay around the neck should not be stored. Onions store best at 32°F with a relatively low humidity and can be kept on a shelf in a corner of the root cellar. Onions and well-matured pumpkins and squash will all keep for several months in a cool, dry basement or on a shelf in an unheated bedroom where temperatures are 55°F or lower.

Because tomatoes are very subject to freezing and are injured by long-term exposure to temperatures below 45°F (7°C), mature green fruits should be harvested before frost or before daytime temperatures start to remain in the low 40s. They can be spread on a shelf after harvest and ripened in a room where the temperature is 55° to 65°F. I have kept tomatoes harvested in early October ripening from then until Thanksgiving both in New York and Washington states.

After being harvested, sweet potatoes should be cured in a warm moist atmosphere for from 10 days at 85°F (29°C) to 2 to 3 weeks at 65° to 75°F (18° to 24°C). They should then be moved to a dry 55° to 65°F (13° to 18°C) area. They are subject to cold temperature injury at temperatures below 50°F (10°C).

A summary of storage requirements for various vegetables and fruits is given in Table 14–21, p. 435.

Home Processing

Processing of fruits and vegetables is subjecting them to treatments that preserve their edibility for from one to several years. Changes that bring about spoilage of fruits and vegetables are attributable to three factors: (1) continuation of life processes, including respiration and transpiration; (2) enzymatic changes; and (3) action of microorganisms, either by their direct feeding or from their chemical by-products. Thus stopping life processes, inactivating enzymes, and preventing the multiplication of microorganisms are necessary to preserve fruits and vegetables.

Canning

When foods are processed by canning, their life processes and enzymes are inactivated, and microorganisms associated with them are killed by heat. The microorganisms are prevented from reentering by the product's being sealed in a container impervious to them.

Botulism. One problem associated with home canning that has received considerable publicity is poisoning by botulism, a frequently fatal condition caused by ingestion of even minute quantities of a toxin produced by the spore-forming bacteria *Clostridium botulinum*. These bacteria are ubiquitous; however, they are poor competitors and will not multiply in the presence of other microorganisms. The spores they produce, being heat tolerant, are a problem in canned products because other microorganisms are destroyed with less heat, leaving the spores of *C. botulinum* free to multiply without competition. Fortunately, *C. botulinum* will not grow in an acid medium, and so there is no danger in eating home-canned fruits or pickles. Home canning of nonacid vegetable and meat products should be done only in a pressure cooker that has been recently tested to see that it is operating properly. In addition, home-canned meats and vegetables should be boiled for ten minutes before they are eaten as a further precaution.

Because of the strict adherence to quality control by commercial processing companies, there is almost no danger of botulism from products canned commercially in North America.

Canning Fruits and Juices. Most home-canned fruits are packed in screw-cap Mason jars. The jars should be washed with soap and water and scalded, or washed in an electric dishwasher. The metal rings can be reused, but used caps should be discarded. Manufacturer's directions for heating the caps and placing them on the filled jars should be followed.

In preparation of fruit for canning peaches and tomatoes are peeled by dipping the fruit in hot water for a few minutes, just long enough to loosen the skin. Pears and apples are peeled and cored with a knife or small hand peeler made specifically for that purpose. Peeled peaches, pears, and apples can be kept from browning if they are peeled into a solution of 1 tablespoon (15 g) salt and 1 tablespoon (15 ml) vinegar dissolved in a gallon (4 liters) of water. We (canning is a family enterprise at my house) seldom use the salt/vinegar solution with peaches and pears, preferring to pack the peeled fruit directly into jars so that it can be processed as quickly as possible.

Plums should have their skins pricked with a fork if they are to be canned whole. Rhubarb is cut into very small (about ½-inch) pieces. Apricots (which usually are not peeled), peaches, and sometimes plums are halved to remove the stone. Most berries and sweet cherries are canned whole. Sour cherries are often pitted before they are processed, especially if they are to be used for pies. My own family enjoys unpitted sour cherries as breakfast fruit, and to save time we hose down our dwarf cherry tree and pick the cherries without stems directly into the quart jars in which they are to be canned, being careful to exclude all foreign matter. Don't attempt to wash sweet cherries on the tree, as water will cause them to split. If raspberries and blackberries have been grown without pesticides in an insect- and dust-free, unpolluted atmosphere, and if hands and picking containers are scrupulously clean, they may not require washing. Sprinkling will knock ripe raspberries from the bushes and washing can crumble them. All purchased fruit should, of course, be washed.

Fruit may be canned by either the cold- or the hot-pack method. Cold packing involves fewer operations, so we usually prefer that method for all fruits but apples and a few plum cultivars. These two fruits contain considerable air. That air is replaced by liquid during processing, and cold-packed apples and certain plums often have a solid mass of fruit floating on liquid in jars that are one-third empty. When fruit is exposed to air in the top of the jar, it is likely to turn brown unless the jars are stored upside down part of the time. For cold-pack canning the prepared fruit is packed in a jar and covered with hot water or a syrup to within ½ inch (1¼ cm) of the jar top. A light syrup is made by heating 2 parts sugar with 4 parts water until sugar is dissolved. Medium syrup is 3 parts sugar and 4 parts water and heavy syrup is equal measures of sugar and water (all volume measures). Canning fruit can also be sweetened with corn syrup or honey. A sweetener is not needed to prevent spoilage, so fruit can be covered with water if that is preferred. For most tree-ripened fruit a light syrup provides ample sweetening. The liquid used to cover the fruit should be hot to lessen the likelihood that the jars will break when they are lowered into the hot-water bath. As a further precaution the stove unit should be turned off while the canner is being unloaded and reloaded to allow the water bath to cool slightly. After the jars are filled, the lids are screwed on tightly, and the jars are completely submerged in a hot-water bath.

Almost any container deep enough to completely submerge the jars can be used for the water bath. A low perforated platform should be placed in the bottom of the water bath container to keep the jars from close contact with the heat unit of the stove, and a loose-fitting lid should be provided. When we have a single jar to be processed, we have occasionally used an empty and thoroughly cleaned 1-gallon (4 liter) latex-base paint can with a piece of folded and flattened chicken wire in the bottom and the lid from a kitchen saucepan. A commercially made home canner, however, is more convenient. These are available in various sizes, the most common being an enameled kettle that holds 7 quart (or pint) jars. A lid and a wire rack with handles for lowering and raising the jars into and from the water bath are included.

In canning by the hot-pack method, fruit (other than apples), along with the desired amounts of sweetener and water, is heated to boiling before it is packed into jars. Apples should be boiled about 5 minutes to eliminate gases. After the jars are filled, the lids are screwed on tightly, and the jars are processed in a water bath.

Canning guides do not entirely agree on processing times. We've processed thousands of quart jars of cold-packed fruit for a standard 20 minutes after the water bath starts to boil (15 minutes for pints, 30 minutes for half-gallons) and have never had a problem with spoilage. Slightly longer times—22, 16, and 33 minutes—would be required for liter, half-liter, and 2-liter containers, respectively. Hot-packed fruit is processed 15 minutes for quarts and 10 minutes for pints. Processing time should be increased 5% for each 1,000 foot (300 m) rise in elevation.

To can fruit juices, tomato juice, and purees including apple sauce, the raw product is washed and blemishes are removed. The calyx but not the core or seeds of apples is removed. Fruit for purees and tomatoes for juice are strained through a colander or sieve with holes small enough to exclude seeds. Fruit for juice is filtered through a cloth bag. The strained product is sweetened if desired (1 teaspoon of salt/quart is added to tomato juice), poured into hot jars to within ½ inch of the top, and processed in a water bath for 10 minutes (20 minutes for tomato juice).

Peeled tomatoes are packed in jars and pushed down with a spoon or fork to eliminate air pockets and allow them to become covered by their own juice. A teaspoon of salt per quart is added and they are processed in a water bath for 35 minutes.

Canning Vegetables. Many home-canned vegetables become mushy during the heating necessary to assure their being free of *Clostridium botulinum*. For this reason and because freezing is a much more convenient method of processing them, vegetables are usually processed by freezing if a home freezer or rented locker is available.

Vegetables, except tomatoes and pickles, should be canned only in a steam-pressure canner. Pressure canning raises temperatures high enough to eliminate danger of botulism. A pressure saucepan with a weighted gauge for accurately controlling pressure at 10 lb (510 mm) (240°F; 115°C) is satisfactory for processing pint jars, but 20 minutes should be added to the processing times listed in

Table 14–22 Canning directions and processing times at 240°F (115°C) recommended for common vegetables

Vegetable	Directions[a]	Minutes of processing time, 10 lb (510 mm) pressure	
		Pints	Quarts
Asparagus	Break tender portions into 1-inch lengths.	25	30
Beans, fresh lima	Leave 1½-inch space at top of quart jar and 1 inch at top of pints for expansion.	40	50
Beans, snap	Trim ends and cut into 1-inch lengths.	20	25
Beets	Cook with 1 inch of stem and entire root attached. Peel and pack.	30	35
Carrots	Slice, dice, or pack small carrots whole.	25	30
Corn (cream style)	Cut from cob at center of kernel and scrape cob.	90	—
Corn (whole kernel)	Cut at about ⅔ depth of kernel.	55	85
Mushrooms	Steam 4 minutes or heat gently for 15 minutes. Pack hot.	30	—
Okra	Pack only young pods. Cook for 1 minute. Pack hot.	25	35
Peas (blackeye)	Shell. Leave 1½ inch at top of quarts and 1 inch at top of pints.	35	40
Peas (fresh green)	Pack to 1 inch of top.	30	35
Potatoes	Cook cubes or small tubers for 2 minutes. Pack hot.	35	40
Spinach	Steam for 10 minutes. Pack loosely and hot.	70	90
Squash, summer	Cut into ½-inch slices.	25	30
Squash, winter	Cube and bring to boil. Pack hot.	55	90
Sweet potatoes	Boil 20 to 30 minutes. Remove skin. Pack with or without liquid.	65	90

(Prepared from information in *Home Canning of Fruits and Vegetables*. U.S. Department of Agriculture Home and Garden Bulletin 8. U.S. Government Printing Office, Washington, DC. 1972.)

[a]Except where otherwise noted, all vegetables can be packed cold or brought to a boil and packed hot. Jars should be filled to and covered with liquid to ½ inch from the top.

Table 14–22 if such a container is used. Pressure canners of various sizes with racks for holding jars are more convenient, and one should be acquired if more than a few jars of vegetables are to be canned. The gauge should be thoroughly cleaned. The gauge on a pressure canner should be checked after the first year of use and every few years thereafter to determine its accuracy; consult the manufacturer or the county extension office to see where this can be done.

Most vegetables can be packed by either the raw-pack or the hot-pack method. Exceptions are spinach and other greens that have to be cooked for a few minutes in order to concentrate their bulk, and beets that are usually cooked for a few minutes to loosen their peeling. Vegetables for canning are prepared as they would be for the table and packed lightly into jars. One teaspoon of salt is added for each quart, boiling water is added to fill the jar to ½ inch from the top.

For processing jars are placed on a rack in the canner containing 2 or 3 inches (5–8 cm) of water. The canner cover should be fastened securely so that steam can escape only through the petcock or vent opening. After the steam has been coming through the opening for 10 minutes or more, the air will have been driven from the canner and then the petcock can be closed or the weighted gauge placed on the opening. Counting of processing time should be begun as soon as the pressure has risen to 10 pounds. When processing time is up, the canner should be removed from the heat. With glass jars the canner should be left undisturbed until the pressure returns to zero. Then the petcock can be slowly opened or the weighted gauge removed, and after a couple of minutes more the lid can be taken off and the jars removed. Table 14–22 lists processing directions for common vegetables.

Freezing

In processing by freezing, microorganism growth and most chemical changes are prevented by the lack of heat. Certain enzymes that bring about undesirable flavors in vegetables are inactivated by blanching. Blanching is heating in steam or boiling water for a short period. Once the desired amount of heat has been applied, the product is immediately cooled, usually in cold water to stop the softening action of high temperatures. Enzymatic activity is stopped in some fruits by the addition of sugar or an acid, usually ascorbic acid.

Products processed by freezing should be sealed in airtight containers so that they aren't exposed to air, which causes freezer burn, and so that escaping volatiles do not flavor other products in the freezer. Never fill a freezing jar to more than 85% capacity, especially if the product is packed in liquid; some head space is necessary to allow for the expansion of freezing. Newly packaged containers should be separated so that they will freeze more rapidly. After they are frozen, vegetables packaged in plastic bags without a rigid cover should be placed together in a box or heavy bag. Loose packages not protected are often punctured during the daily use of the freezer. The temperature of the freezer should be kept at 0°F (−18°C) or lower.

The general recommendation for freezing cherries, berries, and most other fruits is to pack them with dry sugar or in a light syrup. Peaches, fruit cocktail, and sliced strawberries keep best in a light syrup, but my own experience suggests that most other fruit and berries do not juice out as much if packed without sugar. Sugar can be added when they are consumed, if desired.

Blanching is done with a blancher, which has a blanching basket and lid, or with a large covered kettle and wire basket. At least 1 gallon of water is used for each pound of produce, and blanching time should start as soon as the vegetable is plunged into the boiling water. Blanching time in boiling water for vegetables is given in Table 14–23. We usually blanch snap beans somewhat longer than recommended so that they will require less time to cook when they are prepared for the table.

Dehydrating

Because of the ready availability of supplies and equipment for home canning and freezing and because of the lengthy cooking time required to reconstitute most home-dried products, home drying of fruits and vegetables has not been as popular as home canning or freezing. Moreover, it is difficult to remove enough moisture so that the product will store for many months without spoilage. In parts of the West where humidity is low and summer rainfall is infrequent it is possible to spread fruit on trays covered with a screen in direct sun for a week to 10 days. Some kinds of fruits and vegetables can be dehydrated satisfactorily by cutting them into thin strips and spreading them on a rack in a partly open oven turned to the lowest heat possible. A small fan to circulate out the air humidified by the moisture from the fruit will speed oven drying. The oven can be used in conjunction with sun drying in cases of unexpected wet weather. If completely dried, fruits and vegetables will keep indefinitely. Fruit that is dried just to the rubbery stage will keep without refrigeration for several weeks, and, if stored in a home freezer, is an excellent supplement for backpack trips.

Today renewed interest in almost forgotten rural arts has encouraged several companies to manufacture small home fruit and vegetable driers, and plans are available for driers that can be built at home with materials as common as a light bulb, a small fan, and plywood. Check

Table 14–23 Blanching time in hot water for selected vegetables

Vegetable	Blanching time (minutes)	Vegetable	Blanching time (minutes)
Asparagus	3	Kohlrabi, whole	3
Beans, lima	3	Kohlrabi, cubed	1
Beans, green shell	1	Okra	3
Beans, snap	3	Parsnips, cubed	2
Broccoli	3	Peas, blackeye	2
Brussels sprouts	4	Peas, green	1½
Cabbage, shredded	1½	Peppers	No blanching
Carrots, whole	5	Pumpkin, squash, sweet potato	Cook until tender
Carrots, diced	2	Rutabagas and turnips, cubed	2
Cauliflower	3	Spinach, chard, kale, collards, and New Zealand spinach	2
Celery	3	Beet greens and mustard greens	2
Corn, sweet, whole kernel (blanch on cob)	4		
Corn, frozen on cob	9		

(Prepared from information in *Home Freezing of Fruits and Vegetables.* U.S. Department of Agriculture Home and Garden Bulletin 10, revised. U.S. Government Printing Office, Washington, DC. 1971.)

with the county extension office, a local natural food store, or some of the "back-to-the-earth" movement books if you are interested in home dehydrating.

Other Methods of Processing

Some of the oldest known methods of preserving foods are pickling; fermenting; and preserving with sugar, salt, and/or some other chemical. These techniques all involve a change in the chemistry of the food and are sometimes classed in a group as chemistry alteration. Often chemistry alteration is not permitted to continue to the point where the product will keep without further treatment, so such foods as chili sauce, catsup, and even some pickles are canned to prevent spoilage.

Vinegar pickles are made by soaking the product in acetic acid and combinations of sugars, herbs, and spices. Cucumbers are the most common pickled product, but such items as crabapples, apple slices, beets, peppers, cauliflower, and carrots also are sometimes pickled.

Common table salt in precise but not too concentrated amounts is added to cucumbers and cabbage to bring about lactic acid fermentation and produce "fermented" pickles and sauerkraut. Yeast acting on sugars or starches of various kinds produces ethyl alcohol, which is the basis of beer and wine making.

Jelly and jam are made by combining fruit juice, high concentrations of sugar, and pectin. Pectin is found naturally in many fruits. Boiling the juice or puree of those fruits for a long period of time evaporates off enough moisture to concentrate the pectin and cause the product to "set" into jelly or jam. Jelly making used to be a real art; today commercially manufactured pectin, available everywhere, greatly simplifies the making of jelly and jam. Recipes for making jelly, jam, and conserves from most kinds of fruit are found on each package or bottle of pectin.

Six Processing Recipes for Laboratory or Home

The following six recipes illustrate the various methods of fruit and vegetable processing. The ingredients required are readily available for purchase during the entire year, and college and university laboratories will already have most of the utensils and equipment needed. I divide a laboratory section into groups of three or four students, each group responsible for processing one product. During a later laboratory, the same group prepares its product for consumption (or tasting) by the class and gives a brief report on procedures and the scientific reasons for those procedures. This is an effective and enjoyable method for teaching the rudiments of food processing. Each of these recipes could also provide the basis for an interesting family activity at home.

Sauerkraut (Pickling by Fermentation). Using a knife, hand shredder, or commercial cabbage cutter, shred 5 pounds of cabbage into pieces no thicker than a dime. Sprinkle 3½ tablespoons plain salt (do not use iodized) over cabbage and mix thoroughly. The exact measure of salt is important for proper fermentation. Pack cabbage into clean glass jars, pressing it down thoroughly with a wooden spoon until juice is emitted from the cabbage. Fill jars to within 1½ to 2 inches from the top. Be sure the released juice covers the cabbage. Lay pads of cheesecloth over cabbage, tucking edges down against inside of jar. Hold cabbage down by crisscrossing two dry wood strips and placing them under the neck of jar (Figure 14–98). Wipe off the outside of jar and put the lid on loosely to prevent fungus spores from entering. If sealed, the jar may be burst by the gases created by the fermenting process. Set jars in shallow pans or on folded newspapers and keep at 70°F (21°C) for 15 days. The juice usually overflows the jar during the first few days and then recedes. Add more brine (made by dissolving 1½ tablespoons salt in 1 quart of water) if this is necessary to keep kraut covered. After the kraut has fermented, jars can be sealed and processed for 15 minutes in a water bath for long-term storage.

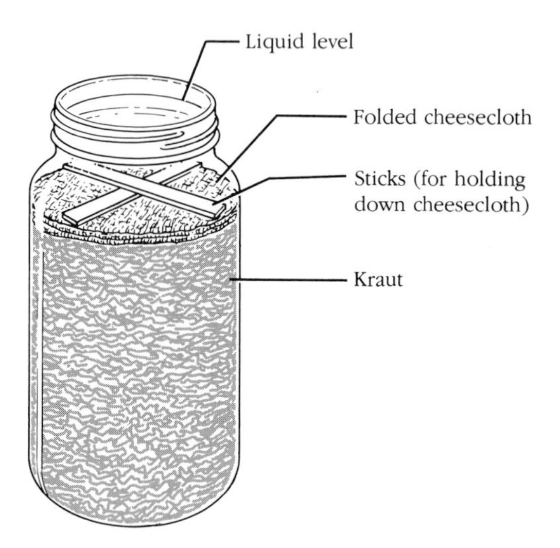

Figure 14–98 Half-gallon (2 liter) jar prepared for fermenting cabbage to sauerkraut.

Winter Apple or Grape Jelly (Concentrated Sugar to Prevent Spoilage).

1 quart bottled or canned apple or grape juice

5 drops (approximately) red food color (for apple)

1 package powdered fruit pectin

5½ cups sugar

Combine juice, color, and pectin in large saucepan; bring to a full boil. Add sugar; stir until dissolved. Return to boil; boil 2 minutes. Remove from heat; skim. Pour into hot jars; seal. Makes about 7 half-pints. (For cinnamon-apple jelly add ⅓ cup red cinnamon candies with the sugar to the apple juice and omit the red food color.) We have found that two or three layers of plastic wrap are as effective and much more convenient than melted paraffin in preventing mold on jelly. The wrap should be fastened over the mouth of the jar with a string or rubber band immediately after the hot liquid is poured into the glasses.

Apple Sauce (Canning). Place unpeeled quartered apples in a saucepan and barely cover with water. Boil until soft. Force the sauce through a sieve to remove peels and cores. Add sugar to taste. Pack hot apple sauce to within ½ inch (1¼ cm) of the top of clean Mason jars. Screw on caps and process for 15 minutes in a boiling-water bath.

Root Beer (Fermentation). (The following half-batch recipe is adapted from the "Home Recipe" label that comes with Hires Root Beer Extract, Hires Division, 2201 Main Street, Evanston, Illinois 60202. Use it or the more complete directions for varying environmental conditions on the extract label.)

Shake extract well before using. Pour 1½ ounces (½ bottle) of root beer extract over 2 pounds refined sugar and mix well. Dissolve sugar and extract mixture in 9½ quarts lukewarm water and mix well. Dissolve a level ¼ teaspoon dried yeast in 1 pint lukewarm water and let stand 5 minutes. Add yeast mixture to sugar and extract solution. Mix well and pour into sterilized beverage bottles immediately. Fill bottles to within ½ inch of top. Cap bottles with crown caps and capper. (Tight fitting corks can also be used.) Place bottles on their sides in a warm (70°–80°F; 21°–27°C) location. The beverage should be ready to drink in about 5 days. *Note:* Only returnable-type beverage bottles should be used. Other jars cannot withstand the fermentation pressure.

Fruit Leather (Dehydration). Wash fully ripe fruit and remove pits (if necessary), cut away blemishes, and measure (up to 5 pints for any one batch). Crush the fruit while heating it. Bring to a boil but don't cook longer than necessary to soften the fruit. Whirl fruit in a blender or put it through a food mill (or strainer if there are objectionable seeds to be removed). Add ½ cup of sugar for each 5 pints of puree, or 1 tablespoon for each cup. The amount of sugar added and cooking time will vary with kind of fruit and taste preference.

Stretch clear plastic film on a perfectly level surface (table top) and fasten with cellophane tape. Spread the puree over the film in a thin layer and place it where it will have full sun all day. The puree can be protected by cheesecloth stretched over boards (to keep it above the fruit). In a dry climate it will require 20 to 24 hours of hot sunlight to dry. The leather is sufficiently dry when it will separate from the film in a sheet with no puree adhering to the plastic. In humid climates the puree may need to be partially or completely dried in an oven. Cookie sheets can be covered with plastic film and the oven set at 140° to 150°F. The oven door should be left slightly open.

The leather can be stored by rolling up the sheet with the plastic film attached and sealing it tightly in more plastic film.

Frozen Diced Carrots (Freezing). Peel carrots; dice or slice into small chunks (¼ inch thick). Blanch in boiling water for 2 minutes. Remove from heat and immediately cool in ice water. Place in containers and seal containers shut. Place filled containers in a freezer where the temperature is 0°F (−18°C) or lower.

Arranging Flowers

by E. W. Kalin*

Flower arranging is an art based on natural forms. Nature designs and colors the flowers; you select the flowers and design the arrangement.

You may have natural talent for arranging flowers, but a few basic principles of design and color will help you. And, like other artists, you must be willing to practice.

When you have flowers in the garden, make one or more arrangements every day. Get acquainted with your flowers and flower types and learn the kinds of arrangements for which they are best suited.

Forms and Balance for Arrangements

You use the lines and types of balance found in drawings, paintings, and sculpture in planning the basic shape of your arrangement. Because lines form the skeleton of the arrangement, they are even more important than color and mass.

Lines may show strength and vitality or gentleness and delicacy. Lines can produce a rhythmic quality in an arrangement and give a feeling of motion. The way lines are arranged creates balance.

Most arrangements come under four main types—vertical, horizontal, triangular, and radial. You may hear of other forms, such as ovals, diagonals, crescents, and Hogarth curves, but these are slight variations of the four main types.

Vertical. Vertical lines are the most natural to use because most plants and flowers grow vertically. These lines suggest growth and vigor. Putting long, narrow flowers in the middle of the vase gives a vertical feeling (Figure 14–99). The arrangement may be like an exclamation point demanding attention, or it may have gentle curves expressing dignity.

---

*Adapted from *Arranging Flowers*, by E. W. Kalin. Pacific Northwest Bulletin 13, revised. Extension Services of Washington State University, Oregon State University, and University of Idaho. Pullman, WA. 1972.

Figure 14–99 Vertical line arrangement of snapdragons and carnations in a vertical container. (Courtesy Washington State University.)

Horizontal. Horizontal lines are popular for table centerpieces and arrangements on coffee tables, mantels, or wherever the background demands a long, low design (Figure 14–100). These arrangements have a restful effect. Long, narrow flowers emphasize the horizontal lines and round flowers create the focal point or center of interest.

Triangular. Triangular designs are a combination of horizontal and vertical lines to give a "two-line" composition. These designs fit many places and can be made to be seen from all sides or from only the front.

When the vertical line comes up from the center of the horizontal line, the triangle is the same on each side of the center. This is conventional balance in triangular design (Figure 14–101). This type of balance is also called formal or symmetrical balance.

Figure 14–100 Horizontal line arrangement of callas, passion flower vine, and calla leaves in a low container for a church altar. (Courtesy Washington State University.)

Figure 14–101 Conventional triangular arrangement of red roses and a few sprays of cedar in a bowl. (Courtesy Washington State University.)

Placing a long, narrow flower along the vertical line and two long, narrow flowers along the horizontal line makes the skeleton or frame of the design. By changing the length of the horizontal and vertical lines you may make a tall, narrow triangle or a low, flat one.

When the vertical line is not in the exact center of the horizontal line, the two sides of the triangle are different. This is called naturalistic balance (Figure 14–102). It is also known as informal or asymmetrical balance.

This type of balance is often more interesting than conventional or formal balance. You may be more imaginative and also leave something for the imagination of the viewer. Often with only a few flowers you can create a pleasant effect.

Radial. Radial lines form part of a circle. The construction is like a wheel, with the spokes made by long, narrow flowers. The focal point is the hub of the wheel (Figure 14–103). Round flowers give contrast and create interest within the arrangement. This design has conventional balance.

Figure 14–102 Naturalistic triangular arrangement of snapdragons and carnations. (Courtesy Washington State University.)

Figure 14–103 Radial arrangement of gladiolus and roses in a shallow round container. (Courtesy Washington State University.)

Flowers and Foliages

Before you gather flowers and foliages, have a clear mental picture of the type of arrangement you want to make. Consider the occasion and where the arrangement will be used as well as the season and kinds of flowers available. Try to use no more than three or four different kinds of flowers and foliages in one arrangement. Carefully select each flower or flower cluster for color, stem length, size, and suitability.

Types of Flowers. Three main types of flowers are used in arrangements: spikes, buttons, and background flowers.

Spikes have a narrow, straight look and pointed tips. Examples are gladiolus, snapdragon, stock, delphinium, rose buds, iris leaves, and grain heads. These flowers are put in the arrangement first and form the outline.

Buttons are round or round looking. Zinnias, marigolds, irises, open roses, petunias, carnations, and chrysanthemums are examples. These flowers are used to create an accent or focus and to break space into interesting patterns. They are put into the arrangement last.

Background flowers have fine textures and irregular outlines. Baby's breath, perennial statice, stevia, asparagus fern, and love-in-a-mist are examples. Background flowers add lightness and airiness to an arrangement. They are added after the spikes but before the buttons. They must be used sparingly. Background flowers may be used if available, but they are not as essential as the other two types.

Containers and Flower Holders

Select the bowl or vase and the flower holder with the same care you use in selecting the flowers.

Selecting Bowls and Vases. The lines of the bowl or vase influence the lines of the arrangement. Except for arrangements to go with period settings, the container is usually less important than the flowers (Figure 14–104).

Here are some points to consider when you select containers:

Shape. Containers should have simple lines and good proportions. They should stand firmly without tipping. Often the shapes of the flowers and foliage will help in choosing a container with the proper lines.

Texture. Use light, fragile, smooth-textured containers for small, delicate flowers. Choose bold, sturdy, heavy-looking containers for large, coarse flowers.

Color. Plain glass, copper, burnished silver, or neutral-colored containers are always suitable. If you use a colored container, choose one with a subdued color and a dull finish. The color may be the same as the flowers or a contrasting or complementary color.

Size. Make sure the container is big enough to hold the flowers without crowding the stems and deep enough to hold an ample supply of water.

Selecting Flower Holders or Frogs. Flower holders are available in many types (Figure 14–105). When you buy, look for holders that:

A

B

C

Figure 14–104 Bowls and vases for flower arrangements. (Courtesy Washington State University.)

(A) Round and free-form bowls suited for low designs.

(B) Rectangular and square bowls for low and angular types of designs.

(C) Vases for vertical and tall triangular arrangements (top row), period and modern designs (center row), and tall, spreading arrangements (bottom row).

Figure 14–105 Flower holders. Hairpin and needlepoint holders (top row); bird-cage holders, chicken wire (center row); floral foam, shredded styrofoam (bottom row). (Courtesy Washington State University.)

1. hold flowers rigidly at the needed angles with minimum injury to stems.
2. weigh enough to keep from tipping over when they hold heavy flowers.
3. hold both large and small stems firmly.
4. resist rust and do not discolor water.
5. are not too difficult to conceal.

The common types of holders have both good and bad points. Hairpin holders have most of the desirable qualities. They are made of brass wire shaped like hairpins and set into a lead base. They are especially good for bowl arrangements.

Needlepoint holders are available in many sizes and shapes. However, it is difficult to put flowers in at extreme angles, particularly flowers with hollow stems.

Bird-cage holders do not injure the flowers and hold them at several angles but not always at the exact angle desired.

Chicken wire may be used in vase and basket designs alone or along with shredded styrofoam. It is available at hardware stores.

Floral foam is a moisture-holding material that is strong enough to support flower stems inserted into it. It comes in cylinder or brick forms that can be easily tailored with a knife to fit the container.

Shredded styrofoam is a porous plastic material that will not absorb water. It is packed into the vase to hold the flowers in place.

Use floral clay to fasten metallic-base holders to the bottom of the bowl. To be certain the clay holds, make sure the container, frog, and clay are all dry.

Principles of Design

All the parts of a design must be related to each other and to the whole. Here are some basic principles you will want to keep in mind when you create any flower arrangement.

Scale. The apparent size and visual weight of each part should be in scale with the rest of the arrangement. This includes the flowers, the foliage, and the container. In addition, the arrangement must be in scale with its surroundings.

As a general guide, the flowers and foliage should be 1½ times as high as the average width of a low container or bowl or 1½ to 2 times as high as the height of an upright container or vase.

Although the flowers must be in scale with each other, it is possible to use small flowers with large ones if you group the small flowers together.

Unity. Every flower and stem should appear as though it actually belongs to the arrangement. There should be no straying elements or parts.

Unity is easiest to achieve if you limit yourself to a small number of flowers and only a few different kinds. This also keeps the flowers from being crowded and allows them to retain their individuality.

The flowers should have something in common with each other. Avoid using exotic or unusual flowers with ordinary garden flowers. Leaves that grow naturally with a flower will blend better than leaves from another species.

Each arrangement should have one main color. You may use more than one color, but a single hue should always be dominant.

Harmony. Most of us notice lack of color harmony, but harmony of texture, shape, and design also contribute to the success of an arrangement. When all the parts harmonize, each complements the other and shows it off to best advantage.

The container should harmonize with the flowers and plant materials in design, weight, and feeling. The flowers and foliage should have something in common with each other in size, shape, texture, idea, time of flowering, or color. In addition, the arrangement should harmonize with its immediate setting.

Balance. Balance makes a design look stable. The highest point of the arrangement should be directly above the visual center or focal point, or as close to it as possible.

Dark-colored flowers are visually heavier than light-colored ones and should usually be kept low and near the center of the arrangement. Texture and shape affect visual weight too. Coarse-textured flowers seem heavier than fine-textured ones, and round flowers look heavier than long, narrow ones.

Rhythm. Rhythm is a visual effect that suggests motion. The eye should move from one point to another in a natural and rhythmic order. This effect can be created by using several flowers of the same color or shape or by using the same flower in different sizes.

All the lines in the arrangement should contribute to the rhythmic effect. There should be no crossed stems to interfere with the sweeping motion of the eye.

Repetition. Form, texture, color, and kinds of plant material need to be repeated to assist in achieving unity. Without repetition there will be an undesirable feeling of isolation and a lack of coherence in the design.

Focus. Every arrangement needs a focal point—a center of interest where all the lines in the design come together. This point should usually be low in the design. In a conventional or formal arrangement it is low and in the center. In a naturalistic or informal arrangement it is where the imaginary vertical and horizontal lines cross.

The focal point can be accented by using a sharp contrast in form, size, or color. A large, rich-colored flower will often provide the necessary emphasis.

Color in Arrangements

The Language of Color. Color has a language of its own. Here are some terms you will want to know as you work with color:

Hue is the quality or difference between one color and another, as red and yellow.

Primary hues are red, yellow, and blue. All other hues can be made by mixing these in various proportions.

Secondary hues are orange, green, and violet. These result when two primary hues are mixed in equal amounts.

Intermediate hues result when a primary hue and a secondary hue are mixed in equal proportions. Yellow-green and blue-violet are examples.

Warm hues are colors made up mainly of red or yellow, such as red-orange.

Cool hues are colors that are mostly blue, green, or violet.

Neutral hues are black, white, and gray. When neutrals are mixed with colors, they dilute or dull the colors, but when they are put near other colors, they make them brighter by contrast.

Value is the lightness or darkness of a hue. Value depends upon the amount of white or black in a hue. If white is added to red, the result is pink. Pink is a light value of red.

Intensity is the purity or brightness of a hue. When a hue is mixed with gray or with a small amount of a hue from the opposite side of the color wheel (Figure 14–106), it becomes duller and less intense. Pure yellow is clear and bright. Adding gray or a small amount of violet makes it softer and less intense.

Color Harmony. Color harmony will be important in every arrangement you make. You will probably find harmonies of related hues easier to master at first.

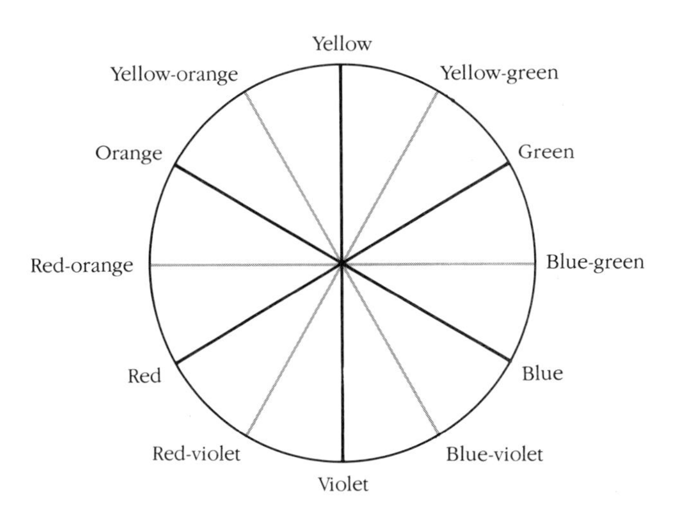

Figure 14–106 The color wheel. Complementary colors are at opposite ends of each straight line.

Single-hue harmony is the use of one color in different values and intensities, such as dark blue and light blue. These are always safe combinations.

Analogous harmony is a combination of colors that lie near each other on the color wheel. Orange, yellow-orange, and yellow are analogous colors.

Harmonies of **contrasting** colors are more striking, but they must be done carefully, with each color in the right amount. One of the colors should generally predominate.

Complementary harmony is a combination of two colors that are directly opposite on the color wheel. Yellow and violet are complementary colors. So are blue and orange.

Double complementary harmony is the use of two colors that lie side by side on the color wheel with their complements. For example, orange and yellow-orange with their complements of blue and blue-violet would be double complementary harmony.

Split complementary harmony is a combination of a color with the two colors on either side of its complement. Yellow with red-violet and blue-violet is an example.

Triads are three colors that are an equal distance apart on the color wheel. Red, blue, and yellow are a triad. So are violet, green, and orange.

The Finishing Touch

Good design and color harmony alone do not guarantee a successful arrangement. You must also give some attention to the final details or "finish" of the arrangement.

The way the frog or holder is hidden is a part of finish. If the water surface in the bowl shows, keep it clean and free from pieces of leaves, petals, and other debris.

Keep the arrangement in a cool room at night and change the water daily. You may use a meat baster or syringe to change the water without disturbing the arrangement.

To keep the flowers fresh as long as possible, never expose the arrangement to direct sunlight or drafts. Commercial flower foods are satisfactory only for certain flowers—roses and carnations, for instance—and then only if the water is soft.

Table 14–24 *(continued)*

Pest	Description and damage	Control
Insects and insectlike pests (many pictured in Figure 9–9)		
Thrips	Tiny, slim insects that damage by rasping and distorting leaves. Often distort and/or prevent flower buds from opening. Gladiolus and onions seem especially susceptible.	Encourage predators by avoiding unnecessary pesticide applications. Spray with insecticide when damage is noted and repeat as recommended on label or as needed.
Weevils	Long-snouted insects with hard protective wing covers. Adults feed on leaves and fruit during night. Larvae destroy roots or hatch in grain products. Pea and bean weevils lay eggs in blossoms. Larvae grow and reach adulthood in seed.	Treat soil with insecticide before planting or treat after and water in. Use insecticide for adults when damage is noted. Spray peas and beans during bloom where dried seed is being grown in areas where weevils are a problem.
Whiteflies	Small, pure white adults flutter erratically when disturbed. Scalelike young attach to and feed on underside of leaves. Common on house and greenhouse plants as well as outdoors.	Apply insecticide when you see the pest and repeat to destroy those newly hatched. Difficult to control. A systemic can be effective with ornamentals.
Wireworms	Larvae of "click" beetle. Wavy yellow hard grubs from ½ to 2 in (1¼ to 5 cm) long. Kill roots, burrow into bulbs, fleshy roots, tubers. Attack germinating seed.	Treat soil with recommended insecticide before planting or treat after planting and water in.
Diseases		
Root knot nematodes	Nematodes are microscopic wormlike entities. In a field of susceptible plants infection starts with circular patches of dead or unthrifty plants that enlarge each year until the entire field is affected. Root symptoms are swellings of various shapes and sizes on major roots, causing death of feeder roots and areas for entry of other pathogenic organisms.	Prevent garden infestation by planting only nematode-free stock from a reliable nursery. Clean thoroughly garden tools that have been borrowed or used in another location. Fumigate infested areas before planting susceptible crops. Grow grasses, grains, and sweet corn, which are resistant, for several years to reduce the nematode population. Plant other monocots, most of which are highly tolerant or resistant.
Other nematodes	Cause unthrifty plants. Many affect only one or two kinds of plants. Some cause root lesions, and nematodes and organisms for which they provide points of entry may cause decay.	Same control as for root knot nematodes except that resistant species of garden plants will be different for each nematode.
Root diseases	Unthrifty plants that wilt readily. Infection by some pathogens result in sudden death. Roots will often be rotted. A longitudinal cut will often reveal a discolored vascular system.	Same sanitation, rotation, and fumigation practices as for nematodes. Don't plant stone fruits where potatoes or tomatoes have recently grown. Dig onions before cold, wet weather. Treatment with fungicide will often control organisms that are seed-borne. Dig and burn infected plants; don't use them for compost. Don't overwater.
Damping-off	A number of organisms kill plants by destroying their stems at ground level. This is called damping-off. Young seedlings are especially susceptible.	Treat seed with fungicide. Germinate seed and prick off seedlings into a pasteurized growing medium if damping-off has been a problem. Make certain that all containers for growing seedlings are free of soil and thoroughly clean. A light dusting of sulfur or other fungicide will help control these diseases.
Rots and decays	These are caused by both bacteria and fungi or sometimes a combination of pathogens. Rotting organisms usually enter through a break in the protective layer. Breaks are usually caused by rough handling or insects or other pathogens.	Keep plants and plant parts from becoming crowded, which prevents free circulation of air. Don't permit free moisture to remain on plant parts for long periods. Don't allow fruits and vegetables to become overmature. Harvest and handle all perishable products carefully to avoid bruising. Provide optimum storage conditions (Table 14–21), and cool products as quickly as possible. Sort out and destroy plants, plant parts, or stored products that have begun to decay.

Pest	Description and damage	Control
	Diseases	
Leaf-spot diseases	Leaf spots in center or along margins of leaves. Spots may spread and leaves yellow and fall. Usually more serious after heavy rains or sprinkler irrigation.	Avoid frequent light sprinkling where leaf diseases are a problem. Water during early morning so leaves can dry before evening. Sulfur and other fungicidal dusts will prevent some leaf diseases.
Mildews	Weblike or powdery white fungal growth (mycelium), usually beginning on underside of leaves or lower crowded stems, often spreading to remainder of plant. Mildewed plants may be attacked by other decay-causing organisms.	Same as for leaf-spot diseases. Avoid crowding of plants.
Sclerotinia	Sclerotinia is almost ubiquitous in garden soils. Affects flowers, fruits, stems, and leaves that are in contact with soil, usually during cool weather of fall. Infection results in water-soaked areas that soon begin to rot and in white cottony mycelial growth often containing hard black fruiting bodies.	Avoid overwatering. Stake tomato plants. Pumpkins and melons can be placed on platforms before they mature. Don't leave fruit in contact with soil after it is mature. Give plenty of space to fall flowers and keep blooms from touching the soil.
Virus diseases	Virus diseases are difficult to diagnose because symptoms are so variable. A combination of two or more of the following usually suggests a virus: stunted growth or year-by-year decline of perennials; yellowed or markedly light-colored leaves; cupping up or down of leaves; brittle leaves, i.e., leaves that break with a crackling sound when crushed; areas of light and dark green (mosaic pattern) in leaves and stems; dead spots on leaves or around leaf margins; dark colored or dead streaks in leaves and/or on petioles and stems, flowers having streaks of two colors; dark streaks in the vascular system; pitting of stems, leaves, or fruit; dying back of new shoots; distorted flowers, leaves, or other plant parts; excessive branching; growth habit different from that of other plants of the species.	Once a plant is infected with a virus, it is impractical to attempt a cure. Plant a resistant cultivar. Use certified virus-free planting stock whenever available. Rogue out and destroy plants that become infected. Wash hands and pruning tools frequently and avoid use of tobacco products when handling plants; especially potato, tomato, pepper, eggplant, petunia, and nicotiana. Keep plants free of aphids and leafhoppers. Keep plants vigorously growing and healthy.

Plants for the Landscape

Characteristics of many of the woody plants most popular for beautification of home grounds are included in Table 14–25. Selected flowers, turf plants, rock garden plants, cacti and related plants, ferns, vines, and roses are briefly described in the tables in Chapter 11. These lists are not meant to be complete, and they exclude many fine and popular ornamentals.

Climate, Hardiness, and Maturity

The mean temperatures, hardiness zones, record temperatures, precipitation, and degree days listed in Table 14–26 coupled with Figure 14–107 will help gardeners determine which kinds of plants are likely to mature or survive and do well in the climate where they live. Spring and fall frosts are listed, wherever available, to help determine planting date and length of growing season, and precipitation amounts will suggest to a degree whether or not irrigation is likely to be essential. It is important to remember that many kinds of cool-season plants do not do well where temperatures are hot. Thus the range of zones to which each plant is adapted is given in Table 14–25 as a rough guide to both cold hardiness and heat tolerance.

Cold Hardiness

Traditionally, woody plants for North American gardens have been classified into winter hardiness categories based on the lowest temperature they are likely to survive.

(text continues on page 479)

Table 14–25 Selected woody plants for the landscape

Name	Hardiness zone[a]	Growth habit	Height (feet)	Soil and exposure[b]	Remarks
		Trees—conifers			
Abies amabilis Pacific silver fir	3–(9)	Columnar	60	B	Grows best in cool humid areas.
Abies concolor White fir	3–8	Columnar	75	A	Better specimens are bluish-green and very symmetrical.
Abies pinsapo Spanish fir	7–10	Pyramidal	30	A	Slow growing; rigid needles make this an interesting specimen.
Araucaria araucana Monkey puzzle tree	8–10	Oval	50	A	Twisted, ropelike branches; a unique specimen.
Calocedrus decurrens Incense cedar	5–10	Columnar	80	A	Widely adapted; good as background.
Cedrus atlantica Atlantic cedar	7–9	Pyramidal	60	B	Blue-green needles; spreading to 30 ft.
Cedrus deodara Deodar cedar	6–9	Pyramidal	80	A	Fast growing; spreading to 40 ft; many variations.
Cedrus libani Cedar of Lebanon	3–9	Dense; pyramidal	60	A	Slow growing, scarce; wide, flat crown at maturity.
Chamaecyparis lawsoniana Lawson false-cypress	6–8	Variable	to 100	A	200 forms known from small shrubs to large trees.
Chamaecyparis nootkatensis Alaska yellow cedar	4–8	Pyramidal	60	A	Slow growing with striking coarse, pendulous branches.
Cupressus glabra Arizona cypress	7–10	Pyramidal	40	D	Adapted to hot interior climates; drought tolerant; fast growing.
Cupressus sempervirens Italian cypress	7–9	Narrow; columnar	60	A	Normally only the narrow upright varieties are sold.
Juniperus spp. Junipers	3–10	Variable	to 40	A	Many species and types; most are extremely adaptable.
Larix decidua European larch	3–9	Pyramidal	75	A	Larch are deciduous conifers. Lacy foliage; pendulous branches.
Metasequoia glyptostroboides Dawn redwood	5–9	Pyramidal	80	A	Can grow 60 ft in 20 years; deciduous; good reddish-brown autumn color.
Picea abies Norway spruce	2–9	Pyramidal	100	A	Fast growing; extremely hardy and wind resistant.
Picea glauca White spruce	2–7	Pyramidal	60	A	Tolerates extreme cold. Alberta spruce is a slow growing, compact form.
Picea pungens Colorado blue spruce	3–9	Pyramidal	90	A	Foliage varies from green to blue; one of the most popular conifers.
Pinus cembra Swiss stone pine	3–9	Dense; pyramidal	30	A	Slow growing; very hardy.
Pinus contorta latifolia Lodgepole pine	4–9	Bushy	40	A	Tolerant of adverse environment; variety of shore pine.

Name	Hardiness zone[a]	Growth habit	Height (feet)	Soil and exposure[b]	Remarks
		Trees—conifers			
Pinus densiflora Japanese red pine	5–9	Irregular	50	A	Damaged by cold winds and desert heat; good informal effect.
Pinus mugo Swiss mountain pine	3–9	Irregular; variable	(20) Variable	A	Hardy; damaged by desert heat; bushy, twisted, open.
Pinus nigra Austrian pine	4–9	Pyramidal	40	A	Attractive rough bark; dense when young; open when old.
Pinus ponderosa Ponderosa pine	5–9	Oval	100	A	Fast-growing western native; long needles; drought tolerant.
Pinus strobus Eastern white pine	3–9	Oval	100	A	Fine-textured, handsome; subject to blister rust.
Pinus sylvestris Scotch pine	2–8	Pyramidal	70	A	Regular and compact when young; irregular and open when older.
Platycladus orientalis Oriental arborvitae	5–9	Varied	to 25	A	Mostly used in its various shrubby forms.
Pseudotsuga menziesii Douglas fir	4–9	Pyramidal	100	A	In cold areas the hardy form 'Glauca' should be grown.
Taxodium distichum Bald-cypress	4–10	Pyramidal; columnar	80	C	Deciduous; feathery foliage and attractive shredding bark; native to southern swamps.
Thuja occidentalis American arborvitae	3–9	Pyramidal	60	A	Tolerant of most environments; many types.
Thuja plicata Western red cedar	4–9	Pyramidal	100	A	Fast-growing massive tree; can be pruned to hedge.
Tsuga canadensis Canadian hemlock	4–9	Pyramidal	90	B	Graceful dense tree; tolerant of shade.
		Trees—palms			
Jubaea chilensis Chilean wine palm	(9)–10	Feather leaves	60	A	Slow growing; hardy to 20°F.
Phoenix spp. Date palms	10	Feather leaves	80	D	Tall, stately; grow and fruit best in arid regions.
Roystonea spp. Royal palms	10	Feather leaves	80	A	Magnificent palms of the more rainy tropics.
Sabal mexicana Mexican palmetto	9–10	Fan leaves	40	A	Sabals are some of the hardiest of palmlike plants.
Sabal palmetto Cabbage palm	9–10	Fan leaves	40	A	Dense globular head with big green leaves.
Washingtonia filifera California fan palm	(9)–10	Fan leaves	60	A/D	Fast growing; old leaves bend to form thatch; open crown.
Washingtonia robusta Mexican fan palm	(9)–10	Fan leaves	100	A/D	More slender and more compact crown than *W. filifera*.

(continued)

Table 14–25 *(continued)*

Name	Hardiness zone[a]	Growth habit	Height (feet)	Soil and exposure[b]	Remarks
		Trees—broadleaved deciduous (planted for foliage)			
Acer circinatum Vine maple	5–(9)	Irregular	12	B	Orange fall color; maples don't grow well in warmer areas of zones 8–10.
Acer ginnala Amur maple	2–(9)	Shrubby tree	18	A	Fragrant flowers; attractive winged fruit; red fall color.
Acer palmatum Japanese maple	5–(9)	V-shaped	20	A	Refined, deeply cut foliage; several cultivars; red fall color; slow growing.
Acer platanoides Norway maple	3–(9)	Round; dense	90	A	Casts heavy shade; many types; lacks fall color; seedy; honeydew from aphids.
Acer rubrum Red maple	3–(9)	Round	100	A	Rapid growth; attractive spring flowers; early autumn color.
Acer saccharinum Silver maple	3–(9)	Oval	80	A	Rapid growth; fairly open form; subject to breakage and aphids.
Betula papyrifera Paper birch	2–8	Pyramidal	60	A	Cold tolerant; open habit; white bark.
Betula verrucosa European white birch	3–8	Oval; weeping	40	A	Several forms; dramatic in effect; lacy and open.
Carpinus betulus European hornbeam	5–(9)	Round	70	A	Dark grey bark; medium growth rate; easily grown.
Carya illinoensis Pecan	6–9	Oval	120	A	Massive and beautiful; no nuts in cool areas; won't tolerate alkalinity.
Celtis occidentalis Common hackberry	4–9	Oval	50	A	Similar to elm; deep rooted; doesn't heave sidewalks.
Cercidiphyllum japonicum Katsura tree	4–9	Spreading	45	B	Multiple trunks; clean appearance; protect from hot sun, dry wind.
Elaeagnus angustifolia Russian olive	3–9	Irregular	20	D	Hardy and adaptable; fine grey foliage is landscape contrast.
Fagus sylvatica European beech	5–9	Oval	80	A	Many forms; pest free, handsome large tree.
Fraxinus americana White ash	3–8	Oval	80	A	Rapid growing, cold-climate tree; twigs litter; use male form.
Fraxinus pennsylvanica Green ash	3–8	Oval	50	A	Good for Plains states; choose male or seedless form.
Ginkgo biloba Maidenhair tree	4–9	Pyramidal or V-shaped	40	A	Plant only male type; graceful; yellow fall color; slow growing; pest free.
Gleditsia triacanthos inermis Thornless honeylocust	4–9	Round	50	A	Rapid growing; open habit makes this a good lawn tree.

Name	Hardiness zone[a]	Growth habit	Height (feet)	Soil and exposure[b]	Remarks
Trees—broadleaved deciduous (planted for foliage)					
Gymnocladus dioica Kentucky coffee tree	4–9	Contorted	50	A	Winter form is effective in landscape; tolerant of poor soil.
Juglans nigra Black walnut	4–9	V-shaped; open	100	A	Big, hardy shade tree; inhibits growth of nearby plants.
Juglans regia English walnut	7–9	V-shaped	60	A	Many forms; messy as a shade tree.
Liquidambar styraciflua American sweet gum	5–9	Narrow oval	80	B	Nice fall color and winter branch pattern; shows iron chlorosis in alkaline soils.
Morus alba White mulberry	4–9	Round	35	D	Birds are attracted to fruit; fruitless form good for desert and polluted areas.
Nyssa sylvatica Sour gum	4–9	Pyramidal; irregular	40	A	Slow growth; dependable fall color even in warm climates.
Platanus acerifolia London plane tree	5–9	Round	80	A	Colorful bark; withstands smog; subject to anthracnose.
Platanus occidentalis Sycamore	5–9	Round	100	A	Massive tree, large leaves; not as adaptable as *P. acerifolia*.
Populus nigra 'Italica' Lombardy poplar	4–9	Narrow; columnar	80	A	Graceful upright sentinel; fast growing but short lived.
Populus tremuloides Quaking aspen	1–9	Oval	40	A	Widely adapted; bark white; leaves flutter; good for naturalizing.
Quercus alba White oak	5–9	Oval	80	A	Stately; slow growing, long lived; red fall color; nice winter effect.
Quercus palustris Pin oak	4–9	Pyramidal	70	A	Branches drooping; autumn color variable; needs ample water.
Quercus rubra Red oak	4–9	Pyramidal	90	A	Needs deep irrigation and fertile soil; fast growing; red fall color.
Salix alba 'Tristis' Golden weeping willow	2–7	Weeping	60	C	The most colorful weeping willow; fast growth; willows drop litter.
Tilia americana American linden	3–9	Oval	50	A	Straight trunk; narrow crown; slow growth; long lived.
Tilia cordata Littleleaf linden	3–9	Pyramidal	40	A	Good lawn tree; pest free; July flowers attract bees.
Ulmus americana American elm	2–9	Vase	120	A	Subject to Dutch elm disease; not recommended for East.
Ulmus parvifolia Chinese elm	6–9	Round	60	A	Resistant to Dutch elm disease; fast growing; best elm where hardy.

(continued)

Table 14–25 *(continued)*

Name	Hardiness zone[a]	Growth habit	Height (feet)	Soil and exposure[b]	Remarks
		Trees—broadleaved evergreen			
Arbutus menziesii Pacific madrone	(7)–10	Round	60	B	Large spring flowers, red berries; exacting in requirements.
Ceratonia siliqua Carob	10	Round or shrubby	30	A	Rapid growing; subject to root rot if watered heavily.
Cinnamomum camphora Camphor tree	9–10	Oval	50	A	Needs hot summers; roots competitive.
Citrus spp. various kinds	9–10	Round	20	A	Good fruit and ornamental trees in warm climates; potted plants elsewhere.
Eucalyptus camalduensis Red gum	8–10	Vase	80	A	Many species available; rapid growth; pest free.
Ficus macrophylla Moreton Bay fig	10	Massive; spreading	80	A	Large leathery leaves; massive trunk.
Ficus retusa India laurel fig	(9)–10	Pendulous	30	A	New rose-colored leaves produce two-toned effect.
Ilex aquifolium English holly	7–10	Columnar	25	B	Dioecious; male and female plants necessary for berries.
Ilex opaca American holly	7–10	Columnar	25	B	Better than English holly for the East.
Jacaranda acutifolia Jacaranda	9–10	Round	35	A	Lavender flowers; may lose leaves in February.
Magnolia grandiflora Southern magnolia	(7)–10	Round	80	A	Large blooms through summer.
Myrica californica Pacific wax myrtle	(7)–10	Round	20	B	Tolerant of infertile soil; neat appearing.
Olea europaea Olive	8–10	Round	30	A	Grey narrow-leaved foliage; tolerant of drought.
Prunus caroliniana Carolina laurel cherry	7–10	Round	35	A	Single or multistemmed tree; drought tolerant.
Prunus laurocerasus English laurel cherry	7–10	Shrubby	25	A	Grows rapidly; can be trained to small tree or hedge.
Quercus agrifolia Coast live oak	8–10	Round	50	A	Rapid-growing, handsome shade tree.
Quercus ilex Holly oak	9–10	Round	60	A	Moderate growth rate; pest tolerant; less graceful than coast oak.
Quercus virginiana Southern live oak	7–10	Spreading crown	60	A	Better than other live oaks for hot interior climates.
Spathodea campanulata African tulip tree	10	Round	50	A	Fiery-red flowers year round.
		Trees—flowering			
Aesculus carnea Ruby horsechestnut	4–9	Oval	40	A	Bright-red flowers, May; coarse appearance; fall color.
Albizia julibrissin Silk tree	6–(10)	Round	25	A	Fluffy pincushion flowers, July; rapid, open growth.

Name	Hardiness zone[a]	Growth habit	Height (feet)	Soil and exposure[b]	Remarks
		Trees—flowering			
Amelanchier canadensis Serviceberry	3–9	Columnar	25	A	White flowers, April; tree or shrub; attractive all year.
Cassia spp. Senna	9–10	Flat; round	20	A	Pea-type flowers over a long season; many types.
Castanea mollissima Chinese chestnut	5–9	Round	60	B	Long white catkins, June; plant two trees if nuts are desired.
Catalpa speciosa Catalpa	4–10	Pyramidal	80	A	Very adaptable, but large and coarse; white panicles, June.
Cercis canadensis Eastern redbud	4–9	Round	25	A	Cultivars with several flower colors; not well adapted to Northeast or coast.
Cornus florida Flowering dogwood	(5)–9	Oval	25	A	Several varieties; white or pink flowers before leaves; widely adapted.
Cornus nuttallii Pacific dogwood	(6)–9	Oval	30	A	Grown mostly in West; white flowers before leaves.
Crataegus oxyacantha English hawthorn	4–9	Round	15	A	Dense foliage; white or pink flowers, May; red berries; several varieties.
Crataegus phanenopyrum Washington hawthorn	4–9	Round	20	A	White flowers, May; red berries; autumn foliage color.
Franklinia alatamaha Franklinia	5–9	Oval	25	B	Large white foliage, August, September; leaves turn scarlet, fall.
Koelreuteria paniculata Goldenrain tree	5–9	Oval	30	A	Yellow flowers, July; pods, late summer; good winter form.
Laburnum watereri Goldenchain tree	5–9	Oval	20	A	Yellow flowers, mid-May; needs neutral pH.
Liriodendron tulipifera Tulip tree	4–9	Pyramidal	100	A	Yellow cup-shaped blooms, late spring; yellow leaves, fall.
Magnolia grandiflora Southern magnolia	8–10	Round	80	A	Evergreen; many types; huge white flowers, summer through fall.
Magnolia kobus Kobus magnolia	5–9	Oval	40	A	Creamy-white flowers, April; several types.
Magnolia soulangiana Saucer magnolia	5–9	Oval	20	A	Many forms; large rose or white flowers before leaves.
Malus spp. Flowering crabapples	2–9	Various	10–30	A	Many types; beautiful year round; white to pink flowers, May; some have colorful fruit and leaves.
Oxydendrum arboreum Sourwood	4–9	Narrow; oval	35	B	White panicles, July; attractive seed pods; fall color.
Prunus persica 'Alboplena' Double white-flowering peach	5–9	Rounded; flat	15	A	White flowers, mid-April; also pink- and white-flowered forms.

(continued)

Table 14–25 *(continued)*

Name	Hardiness zone[a]	Growth habit	Height (feet)	Soil and exposure[b]	Remarks
		Trees—flowering			
Prunus serrulata Japanese flowering cherry	5–9	Rounded; upright	15–30	A	Year-round interest; white to pink flowers, April; many types.
Prunus subhirtella 'Pendula' Weeping higan cherry	5–9	Weeping	30	A	Several forms; pink flowers, April.
Pyrus calleryana 'Bradford' Bradford pear	4–9	Pyramidal	40	A	White flowers, spring; red foliage, fall; resistant to fireblight.
Sophora japonica Pagoda tree	4–9	V-shaped	40	A	Shape resembles American elm; pods are messy; flowers, mid-summer.
Sorbus aucuparia European mountain ash	3–9	Round	35	A	White flowers, spring; clustered red berries, fall.
Styrax japonica Japanese snowdrop	5–9	Rounded	30	A	Pendulous bell-shaped flowers, late spring; dainty, refined.

Name	Hardiness zone[a]	Growth habit	Flowers	Soil and exposure[b]	Remarks
		Tall shrubs and shrubby trees—8–20 feet high			
Caragana arborescens Siberian peashrub	3–9	V-shaped or oval	Yellow, March–May	A	Deciduous; hedge, screen or windbreak; cut back for dense form.
Ceanothus thyrsiflorus Blueblossom ceanothus	7–9	Oval	Blue, May–July	A	Broadleaved evergreen native of West; to 15 ft.
Ceanothus velutinus Snowbrush ceanothus	7–9	Rounded	White, June–July	A	Glossy green broadleaved evergreen native of West; to 15 ft.
Corylus avrellana Hazelnut	6–9	Rounded	Catkins, late winter	A	Grown for nuts and as an ornamental.
Cotinus coggygria Common smoketree	(5)–9	Rounded	See remarks; June–August	A	Deciduous; yellowish flowers; large plumes of smoky-appearing flower parts.
Hibiscus syriacus Rose of Sharon	5–9	V-shaped or oval	See remarks; August–September		Deciduous shrub; tolerant of urban settings; white, red, or purple to violet flowers.
Holodiscus discolor Creambush rockspirea (Oceanspray)	4–9	Oval	Creamy white, June–July	A	Deciduous; to 12 ft; panicles of flowers; flowerlike seed pods; persistent.
Ilex crenata Japanese holly	7–9	Rounded	Unimportant	B	A fine-textured broadleaved evergreen foliage plant; useful as screen.
Juniperus chinensis 'Reeves' Reeves Chinese juniper	4–9	Pyramidal	Unimportant	A	Vigorous conifer; light blue-green color.
Ligustrum amurense Amur privet	4–9	Oval	Unimportant	A	Hardiest of privets; screen or hedge plant.
Ligustrum ovalifolium California privet	6–9	Oval	White, July	A	Deciduous or half-evergreen hedge plant; to 15 ft.

Name	Hardiness zone[a]	Growth habit	Flowers	Soil and exposure[b]	Remarks
			Tall shrubs and shrubby trees—8–20 feet high		
Lonicera tatarica Tatarian honeysuckle	3–9	Rounded	Pink to white, April–May	A	Deciduous; to about 10 ft; pink or white flowers.
Magnolia stellata Star magnolia	6–9	Oval to round	White, March–April	B	Deciduous shrub; to 10 or 12 ft.
Photinia serrulata Chinese photinia	7–9	Oval	White, April–May	A	Broadleaved evergreen; 20 ft; leaves reddish when young, turning glossy green; red berries.
Prunus laurocerasus Common laurel cherry	7–9	Round	White, June–July	A	Vigorous, coarse-textured broadleaved evergreen.
Prunus triloba Flowering almond	4–9	Oval	Pink, April–May	A	Deciduous; magnificent show of double flowers.
Pyracantha coccinea lalandi Laland firethorn	5–9	Irregular	White, May–June	A	Hardy broadleaved evergreen; to 20 ft; orange-red berries.
Rhododendron macrophyllum Coast rhododendron	7–9	Round	Rosy-pink, May–June	B	Broadleaved evergreen; West Coast native; to 12 ft.
Rhododendron vaseyi Pinkshell azalea	6–9	Oval; irregular	Light rose, April–May	B	Deciduous; to 10 or 15 ft.
Rhus typhina Staghorn sumac	4–9	Round or irregular	Greenish, June–July	D	Deciduous; to 20 ft; greenish flowers, crimson fruiting bodies; fall color bright red.
Salix discolor Pussy willow	2–8	Irregular	Catkins, late winter	C	Vigorous shrub; moist or well-drained soil.
Sambucus canadensis American elder	3–9	Loose	White, June	C	Coarse-leaved suckering plant; black edible berries.
Syringa chinensis Chinese lilac	4–9	Rounded	Purplish-rose, May	A	Free-flowering lilac; grows to 15 ft; similar to Persian lilac.
Syringa vulgaris varieties Common lilac	3–9	Oval to rounded	Various, April–May	A	Many cultivars with blooms from white, pink, blue to purple in single- or double-flower forms.
Tamarix tetrandra Spring-flowering tamarisk	4–9	Upright; irregular	Pink, April–May	D	Deciduous; 15 ft; tolerant of alkaline soil and drought.
Tamarix pentandra Salt cedar	3–9	Oval to round	Rosy-pink, August–September	D	Deciduous to 25 ft; tolerant; heavy pruning encourages bloom.
Thuja occidentalis 'Pyramidal' Pyramidal eastern arborvitae	4–9	Narrowly columnar	Cones	A	Slender conifer useful as narrow hedge; to 25 ft ultimately.
Thuja occidentalis 'Ware Gold' Ware Gold eastern arborvitae	4–9	Narrowly pyramidal	Cones	A	Confier with golden-yellow foliage; to 25 ft.
Viburnum lantana Wayfairing tree viburnum	4–9	Rounded	White, April–May	A	Deciduous shrub; to 15 ft; fruit red in July, black later.
Viburnum opulus European cranberrybush viburnum	4–9	Round	White, May–June	A	Deciduous plant to 12 ft; red fall color and red berries.
Viburnum rhytidophyllum Leatherleaf viburnum	(7)–9	Oval	Light-yellow, May–June	A	Broadleaved evergreen; to 15 ft; large, leathery leaves; coarse.

(continued)

Table 14–25 *(continued)*

Name	Hardiness zone[a]	Growth habit	Flowers	Soil and exposure[b]	Remarks
		Large shrubs—5–8 feet high			
Abelia gaucheri Gaucher abelia	7–9	Round	Lavender-pink, June–November	A	Evergreen with bronzy foliage in winter.
Abelia grandiflora Glossy abelia	7–9	Round; dense	Flushed-pink, June–November	A	Broadleaved evergreen; glossy foliage; shorter in North.
Aucuba japonica Japanese aucuba	7–9	Round; dense	Purplish-green, March–April	B	Broadleaved evergreen; red berries, winter; needs pollenizer.
Berberis darwini Darwin barberry	7–9	Oval; dense	Golden-yellow, April	A	Broadleaved evergreen plant; most useful at 4–5 ft, has been known to grow 8 ft; foliage small, hollylike, and dark green.
Berberis julianae Wintergreen barberry	6–9	Oval	Yellow, May	A	Broadleaved evergreen; makes a dense, spiny hedge.
Berberis mentorensis Mentor barberry	6–9	Round	Yellow, May	A	Dense thorny shrub with leaves persistant into winter.
Buddleia davidii Orange-eye butterfly bush	5–9	Lanky	Various, summer	A	Long flower spikes; cut back to ground each winter in North.
Buxus sempervirens Common box	6–9	Round	Unimportant	A	Broadleaved evergreen hedge or specimen; grows slowly to 10 ft.
Calycanthus floridus Common sweetshrub	5–9	Round	Reddish-brown, June–July	A	Deciduous shrub with dark-green leaves; dark reddish-brown fragrant flowers.
Camellia japonica Common camellia	7–9	Oval	Various	B	Many varieties of this broadleaved evergreen shrub available.
Camellia sasanqua Sasanqua camellia	7–9	Oval	Various	B	Many varieties of this broadleaved evergreen shrub available.
Ceanothus impressus Santa Barbara ceanothus	(7)–9	Irregular; arching	Dark-blue, April–May	A	Fine-textured broadleaved evergreen. *Ceanothus* spp. do best in West.
Chaenomeles Flowering quince	4–9	Round	White to red, early spring	A	Deciduous; scarlet or white to dark-red flowers, depending on variety.
Chamaecyparis obtusa 'Nana gracilis' Dwarf ninoki cypress	5–9	Conical	Unimportant	A	Very slow growth to 7 ft.
Cornus sericea (stolonifera) Red-osier dogwood	2–9	Round	White, May	A	White berries, summer; red twigs, winter; hardy.
Cortaderia selloana Pampas grass	7–9	Round	Silvery-yellow, September	A	Giant ornamental grass; arching, razor-sharp blades.
Cotoneaster divaricata Spreading cotoneaster	5–9	Vase; arching	Pink, profuse	A	Refined shrub; small leaves; red berries.
Cotoneaster franchetii Franchet cotoneaster	(7)–9	Round	Pinkish-white, June	A	A broadleaved evergreen; slender arching stems; grey leaves, orange-red fruit.

Name	Hardiness zone[a]	Growth habit	Flowers	Soil and exposure[b]	Remarks
		Large shrubs—5–8 feet high			
Elaeagnus umbellata Autumn elaeagnus	4–9	Tall; spreading		A	Silver foliage; brown berries turn red in fall.
Enkianthus perulatus White enkianthus	(5,6)–9	Erect; oval	White, May	B	Deciduous shrub; fall color scarlet.
Escallonia 'Appleblossom' Appleblossom escallonia	7–9	Round; arching	Pinkish-white, summer	A	Broadleaved evergreen; arching branches; many cultivars.
Euonymus alata Winged euonymus	4–9	Rounded	Unimportant	A	Deciduous; to 10 ft; crimson to scarlet fall color.
Euonymus japonica Evergreen euonymus	7–9	Oval	Unimportant	A	Broadleaved evergreen hedge plant; tolerates pruning.
Exochorda racemosa Common pearlbush	5–9	Round	White, April–May	A	Deciduous shrub to 10 ft; flowers pearllike in bud.
Fatsia japonica Japan fatsia	8–9	Oval to round	White, October–January	B	Bold tropical-appearing broadleaved evergreen; leaves to 16 in across; flowers and fruit, October to January.
Forsythia intermedia Showy forsythia	5–9	Round; upright	Yellow, early	A	Several named cultivars; the most showy forsythia.
Forsythia viridissima Greenstem forsythia	6–9	Round	Yellow, March–April	A	This forsythia develops a purple-green fall color.
Hydrangea arborescens Smooth hydrangea	4–9	Round to oval	White, June–July	B	Deciduous; appears to grow best in zones 6, 5, and 4.
Hydrangea macrophylla Bigleaf hydrangea	6–9	Round	Blue or pink, June–July	B	Coarse-textured, vigorous; several varieties available.
Hydrangea paniculata 'Grandiflora' Peegee hydrangea	4–9	V-shaped or round	White, purplish, August–September	B	Deciduous; to 10 ft; huge flower clusters.
Ilex crenata 'Convexa' Box-leaf holly	5–9	Round; dense	Inconspicuous	A	Broadleaved evergreen; widely adapted; several forms of *I. crenata*.
Juniperus chinensis 'Hetz' Hetz Chinese juniper	5–9	Oval; ascending	Unimportant	A	Broad conifer; fountain effect with age.
Juniperus squamata meyeri Meyer singleseed juniper	5–9	V-shaped; irregular	Unimportant	A	Unique-shaped conifer; blue color.
Kalmia latifolia Mountain laurel	5–9	Round	White, pink, May–June	B	Dense evergreen; moist acid soil; same care as rhododendron.
Kerria japonica 'Pleniflora' Double Japanese kerria	4–9	V-shaped to oval	Yellow, April–May	A	Green stems in winter; may spread if not confined.
Kolkwitzia amabilis Beautybush	4–9	Round	Pinkish-lavender, May–June	A	Vigorous deciduous shrub; grows in infertile soil with minimum moisture.
Lagerstroemia indica Crape myrtle	7–10	Oval	Red to white, summer	A	Deciduous to 10 ft; beautiful in flower in summer; numerous cultivars.
Ligustrum obtusifolium regelianum Regel privet	5–9	Round	White, July	A	Reliable screen plant; horizontal character; deciduous.

(continued)

Table 14-25 *(continued)*

Name	Hardiness zone[a]	Growth habit	Flowers	Soil and exposure[b]	Remarks
		Large shrubs—5–8 feet high			
Lonicera morrowi Morrow honeysuckle	4–9	Round	White, April–May	A	Hardy old-timer; most frequently seen in zones 5 and 4.
Magnolia liliflora Lily magnolia	6–9	Oval	Purple, white, March–April	A	Deciduous shrub; grows to 10 or 12 ft; petals deep-purple outside, white inside.
Myrica pensylvanica Northern bayberry	2–8	Oval	Inconspicuous	A	Deciduous; aromatic leaves; silver berries; likes sandy soil.
Nandina domestica Nandina	7–9	Oval	White, June–July	A	Broadleaved evergreen; erect habit gives distinction; grows in most soils.
Osmanthus heterophyllus Holly osmanthus	7–9	Round	White, September–October	A	Broadleaved evergreen with hollylike leaves; good screen.
Pieris japonica Japanese pieris	6–9	Oval	White, April–May	A	Refined evergreen to 7 ft; clusters of bell-shaped flowers.
Pinus aristata Bristlecone pine	5–9	Irregular	Unimportant	A	Slow-growing shrub in East to small tree in West.
Pinus mugo mughus Mugho Swiss mountain pine	3–8	Round; irregular	Unimportant	A	Hardy conifer; compact habit; useful as foundation plant.
Prunus laurocerasus schipkaensis Schipka laurel cherry	5–9	Round	Yellowish-white, May	A	Hardiest of laurels; compact; lustrous dark-green leaves.
Prunus tomentosa Nanking cherry	3–8	Rounded; upright	White, April	A	Hardy; attractive edible red fruit.
Pyracantha 'Government Red' Government red firethorn	7–9	Rounded; irregular	White, May–June	A	Bright-red fruit; broadleaved evergreen; some winter injury in zone 7.
Pyracantha 'Rosedale' Rosedale firethorn	7–9	Rounded; irregular	White, May–June	A	Glossy evergreen foliage; bright-red berries; screen, espalier, or hedge.
Rhododendron calendulaceum Flame azalea	5–9	Oval	Yellow-orange to scarlet, May–June	B	Deciduous; several brillant color variations.
Rhododendron nudiflorum Pinxterbloom azalea	(4)–9	Oval	Light-pink, white, April–May	B	Deciduous; very hardy.
Rhododendron occidentale Western azalea	5–9	Upright to round	Pinkish with yellow, May–June	B	Flowers fragrant; deciduous.
Rhododendron schlippenbachii Royal azalea	(4)–9	Oval to round	Pink, April–May	B	Deciduous azalea with clear-pink blossom.
Rhododendron named hybrids	7–9	Round; upright	Various	B	Dozens of named hybrids, 5–8 ft, evergreen and deciduous, are available.
Rhus typhina 'Laciniata' Cutleaf staghorn sumac	3–9	Irregular	Greenish, June–July	A	Foliage crimson in fall; suckers freely; attractive seed head.
Salix purpurea Purple osier willow	4–9	Rounded	Unimportant	A	Deciduous hedge plant; grey-green foliage.
Sorbaria sorbifolia Ural false-spirea	2–8	Oval	Cream, June–July	A	Deciduous screen plant; ultimate height 6 ft; suckers freely.
Spiraea prunifolia Bridalwreath spirea	5–9	Round	White, April–May	A	Brilliant red and yellow fall color.

Name	Hardiness zone[a]	Growth habit	Flowers	Soil and exposure[b]	Remarks
			Large shrubs—5–8 feet high		
Spiraea vanhouttei Vanhoutte spirea	4–9	Round	White, May	A	Arching white cascades of bloom.
Syringa persica Persian lilac	4–9	Rounded	Pale lilac, May	A	Hardy, free-flowering lilac.
Viburnum burkwoodii Burkwood viburnum	(5)–9	Rounded	Pinkish-white, March–April	A	Semievergreen; vigorous; fragrant flowers; dark-green foliage.
Viburnum odoratissimum Sweet viburnum	7–9	Oval	White, May–June	B	Large glossy leaves; to 10 ft; fragrant flowers.
Viburnum plicatum Japanese Snowball viburnum	(5)–9	Rounded	White, May–June	A	Similar to 'Doublefile' except sterile; flowers in double file of round clusters.
Viburnum plicatum tomentosum Doublefile viburnum	(5)–9	Rounded	White, May–June	A	Deciduous; grows to about 10 ft; bright-red fall color.
Viburnum tinus Laurestinus viburnum	(7)–9	Rounded	White, March–April	A	Broadleaved evergreen; useful as screen.
Weigela florida Old-fashioned weigela	(4)–9	Rounded	Various, May–June	A	Deciduous plant; several color forms, white, pink, or rose.
			Shrubs 3–5 feet high		
Berberis darwinii Darwin barberry	7–9	Oval; dense	Yellow, early spring	A	Broadleaved evergreen; foliage small, hollylike, and dark-green.
Berberis thunbergii Japanese barberry	4–9	Rounded; dense	Yellow, April	A	Spiny, deciduous shrub; brilliant scarlet and yellow fall color; red berries.
Berberis thunbergii 'Atropurpurea' Redleaf Japanese barberry	4–9	Rounded; dense	Yellow, early spring	A	Same as above; leaves dark reddish-purple, spring and summer.
Berberis verruculosa Warty barberry	4–9	Round; dense	Yellow, April	A	Spiny broadleaved evergreen; fine texture, dark-green leaves.
Buxus sempervirens Common box	6–9	Round	Unimportant	A	Broadleaved evergreen hedge or specimen; grows slowly to height of 10 ft or more, can be kept lower with pruning; occasionally seen in zone 5.
Chamaecyparis pisifera 'Filifera Nana' Dwarfthread false-cypress	6–9	Round; ascending	Unimportant	A	This conifer forms a dense, rounded mass with weeping effect.
Chamaecyparis pisifera 'Nana Aurea' Yellowdwarf false-cypress	6–9	Round; ascending	Unimportant	A	Same as above except for golden-yellow color of foliage.
Cotoneaster microphylla Small-leaf cotoneaster	7–9	Spreading	White, early spring	A	Spreads quickly to 3–4 ft.
Cytisus praecox Warminister broom	6–9	Round; dense	Cream, April	A	Green branches; useful in sunny, dry locations or poor soil.

(continued)

Table 14-25 *(continued)*

Name	Hardiness zone[a]	Growth habit	Flowers	Soil and exposure[b]	Remarks
		Shrubs—3–5 feet high			
Daphne mezereum February daphne	5–9	Erect; oval	Rosy-purple, February–March	A	Early flowering; deciduous; flowers fragrant.
Daphne odora Winter daphne	7–9	Dense mound	White, rose, March–April	B	Broadleaved evergreen; dark-green in partial shade, yellowish in full sun.
Erica stricta Corsican heath	7–9	Round; ascending	Purple, late summer	A	Can be kept low by pruning; attractive brown seed capsules.
Euonymus fortunei 'Vegeta' Bigleaf wintercreeper	6–9	Viny or shrubby	Unimportant	A	Broadleaved evergreen; can be trained to shrub, liana, espalier.
Hydrangea arborescens Smooth hydrangea	4–9	Erect; oval	White, July–fall	A	Deciduous; flowers, large snowballs.
Hydrangea quercifolia Oakleaf hydrangea	5–9	Loose; spreading	White, early summer	A	Coarse-textured; deciduous; dark-green summer, red fall color.
Hypericum prolificum Shrubby St. Johnswort	6–9	Dense	Bright-yellow, July	A	Brown stems; should be pruned to the ground in spring.
Juniperus chinensis 'Pfitzeriana' Pfitzer juniper	4–9	Spreading	Unimportant	A	Conifer; tolerant of dry soil.
Leucothoe fontanesiana Drooping leucothoe	6–9	Round; arching	White, late spring	B	Broadleaved evergreen; leaves bronze in winter.
Ligustrum vicaryi Vicary golden privet	5–9	Round	Unimportant	A	Golden foliage; likes full sun; deciduous.
Ligustrum vulgare 'Lodense' Lodense privet	4–9	Mound; dense	Unimportant	A	Suitable for untrimmed hedge.
Mahonia aquifolium Oregon grape	5–9	Open	Golden, spring	A	Evergreen leaves resemble holly; variable in type.
Picea glauca 'Conica' Dwarf Alberta spruce	3–9	Pyramidal; dense	Unimportant	A	Grows slowly; very formal and full.
Pieris floribunda Mountain pieris	5–9	Round	White, early spring	A	Broadleaved evergreen; several planted 4 ft apart will form a single mass.
Pieris japonica Japanese pieris	6–9	Oval	White, early spring	A	Broadleaved evergreen; can be kept lower if higher branches are removed.
Pinus mugo mughus Swiss mountain pine	3–9	Round to spreading	Unimportant	A	Hardy conifer; compact growth; climatic adaptability.
Pinus strobus 'Nana' Dwarf white pine	3–9	Round	Unimportant	A	Bluish coniferous shrub.
Prunus glandulosa 'Sinensis' Dwarf pink flowering almond	(5)–9	Rounded	Pink, spring	A	Profuse display of double flowers; foliage somewhat coarse.
Prunus laurocerasus zabeliana Zabel laurel cherry	7–9	Rounded to round	White, spring	A	Broadleaved evergreen; broader than high; foundation plant or hedge.

Name	Hardiness zone[a]	Growth habit	Flowers	Soil and exposure[b]	Remarks
			Shrubs—3–5 feet high		
Punica granatum 'Dwarf' Dwarf pomegranate	(7)–9	Round; ascending	Orange-red, June–September	A	Deciduous shrub; rich soil required; protect from cold winds.
Raphiolepis umbellata ovata Yeddo hawthorn	8–10	Round	White, May–June	A	Broadleaved evergreen plant; grows slowly; mainly for South and West.
Rhododendron (Azalea mollis hybrids) Mollis azalea	(5)–9	Oval	Various, April–May	B	Loosely classified group of azaleas of good garden quality.
Rhododendron mucronulatum Korean rhododendron	(4)–9	Oval	Rosy-purple, March–April	B	Deciduous; grows to 7 ft; fall color yellow and crimson.
Rhododendron 'Blue Diamond'	6–9	Oval to round	Lavender-blue, April	B	Broadleaved evergreen; fertilizer not beneficial.
Rhododendron 'Bowbells'	6–9	Rounded; compact	Shell-pink, May	B	Broadleaved evergreen; 3 or 4 ft tall; small, coinlike bronzed leaves.
Rhododendron 'Bric-a-Brac'	6–9	Round	White with pink, February–March	B	Broadleaved evergreen; grows to about 30 in.
Rhododendron 'Brittania'	6–9	Round	Crimson-red, June		Broadleaved evergreen; to 5 ft or slightly higher; compact, broad mass; flowers excellent.
Rhododendron 'Broughtoni Aureum'	6–9	Oval	Yellow, May	B	Cross between deciduous and broadleaved evergreen; persistent leaves.
Rhododendron 'Gomer Waterer'	(7)–9	Rounded; compact	Apple-blossom pink, June	B	Broadleaved evergreen; to about 4 ft.
Rhododendron 'Mars'	(6)–9	Rounded; compact	Blood-red, May–June	B	Broadleaved evergreen plant; hardy.
Rhododendron PJM-hybrids	(5)–9	Round; compact	Lavender, April	B	Tight mound; small bronze-tipped leaves.
Rhododendron 'Unique'	6–9	Round; compact	Pale-yellow, April–May	B	Broadleaved evergreen; flowers pink in bud, pale yellow in full bloom.
Ribes alpinum Alpine currant	(2)–7	Round; compact	Greenish, April	A	Maplelike leaves; hedge for cold climates; red autumn foliage.
Senecio greyii Grey's groundsel	8–9	Rounded	Yellow, summer	D	Broadleaved evergreen; 3 ft high; grey foliage.
Skimmia japonica Japanese skimmia	7–9	Rounded	Light-yellow, April–May	B	Evergreen with bright-red fruit in winter; male plant needed for fruit.
Spiraea bumalda 'Froebel' Froebel spirea	5–9	Rounded	Crimson, July	A	Deciduous; similar to Anthony Waterer spirea but somewhat taller with darker foliage.
Spiraea thunbergii Thunberg spirea	5–9	Round	White, February–May	A	Deciduous shrub with light-green leaves; good fall color.
Symphoricarpos orbiculatus Indiancurrant coralberry	3–9	Oval to round	Yellow, August	A	Deciduous; purplish-red fruit ornamental during winter.

(continued)

Table 14–25 *(continued)*

Name	Hardiness zone[a]	Growth habit	Flowers	Soil and exposure[b]	Remarks
		Shrubs—3–5 feet high			
Taxus cuspidata 'Densa'	4–8	Cushion	Unimportant	A	Slow growing conifer; 4 ft high, 8 ft broad.
Thuja orientalis 'Berckmanns' Berckmanns Oriental arborvitae	6–9	Oval	Unimportant	A	Slow growing, golden-foliaged conifer.
Vaccinium ovatum Box blueberry	7–9	Rounded; stems ascending	Light-pink, April–May	B	West Coast broadleaved evergreen native; bronze in spring and darker bronze in winter; edible fruit.
Viburnum carlesii Korean spice viburnum	5–9	Oval; spreading branches	Light-pink, April–May	A	Fragrant-flowered, deciduous viburnum.
		Shrubs 18 inches to 3 feet			
Abies balsamea 'Nana' Dwarf balsam fir	3–8	Round	Unimportant	A	Hardy dwarf conifer; good for cold locations.
Berberis thunbergii 'Crimson pygmy' Crimson pygmy barberry	4–9	Round	Unimportant	A	Red-leaved barberry that doesn't grow over 2 ft high.
Buxus microphylla koreana Korean box	5–9	Spreading	Unimportant	A	Evergreen; loose-growing; selected types are more compact.
Calluna vulgaris Scotch heather	(6)–9	Spreading	White-pink, summer–fall	B	Many types; all low-growing evergreens.
Chamaecyparis obtusa 'Nana' Dwarf false cypress	5–9	Mound	Unimportant	B	Dwarf conifer; squat habit; dark green foliage.
Cotoneaster horizontalis Rock cotoneaster	6–9	Spreading	Pale-pink, late spring	A	Bright-red berries; good bank cover; can espalier.
Cotoneaster microphylla Rockspray cotoneaster	7–9	Spreading	White, late spring	A	Broadleaved evergreen; red berries.
Cryptomeria japonica 'Dwarf' Dwarf Japanese cryptomeria	6–9	Dense; round	Unimportant	A	Dwarf conifer; green in summer, reddish in winter.
Cycas revoluta Sago palm	9–10	Spreading	Unimportant	A	Primitive plant; looks like dwarf palm; taller with age.
Deutzia gracilis Slender deutzia	5–9	Round; stems ascending	White, spring	A	Needs pruning after bloom.
Euonymous fortunei	5–9	Varies	Unimportant	A	Evergreen with broad waxy leaves; many kinds with varied growth.
Gaultheria shallon Salal	7–9	Spreading; stems ascending	Light-pink, late spring	B	Broadleaved evergreen.
Ilex crenata 'Hetzii' Hetz Japanese holly	(6)–9	Round; dense	Unimportant	B	Evergreen; other dwarf hollies with various forms.
Juniperus chinensis sargentii Sargent juniper	4–9	Spreading	Unimportant	A	More dense and slower growing than most low junipers.
Juniperus sabina 'Tamariscifolia' Tamarix juniper	5–9	Spreading	Unimportant	A	A most useful form; does not exceed height of 2 ft.

Name	Hardiness zone[a]	Growth habit	Flowers	Soil and exposure[b]	Remarks
		Shrubs 18 inches to 3 feet			
Lavandula officinalis True lavender	6–9	Round; stems ascending	Lavender, late summer	D	Tolerant of dry, alkaline soil; flowers fragrant.
Lonicera pileata Privet honeysuckle	6–9	Spreading horizontal	White, spring	A	Semievergreen; 2 to 3 ft tall with 6-ft spread.
Mahonia nervosa Cascades mahonia	7–9	Spreading; stems ascending	Yellow, spring	A	A 2- to 3-ft broadleaved evergreen; western native; similar to Oregon grape but much smaller.
Nandina domestica 'Nana' Dwarf nandina	7–9	Round; stems ascending	White, summer	B	Resembles a diminutive bamboo; thrives in moist soil.
Picea abies 'Nidiformis' Nest spruce	3–9	Spreading	Unimportant	A	Slow-growing conifer; to 18 in high; spreads to 3 ft.
Potentilla fruticosa Bush cinquefoil	2–8	Round; stems ascending	Yellow, white, all summer	A	Deciduous; several varieties; tolerant of wet, dry, acid, or alkaline soil.
Raphiolepis indica 'Rosea' Pink India raphiolepis	7–9	Spreading loose	Pink, late spring	A	Broadleaved evergreen; may reach 4 or 5 ft when old.
Rhododendron 'Bluetit' Bluetit rhododendron	6–9	Round; compact	Blue, early spring	B	2 to 3 ft high.
Rhododendron impeditum	5–9	Compact mound	Lavender, early spring	B	Small scaly evergreen leaves; 18 in high.
Rhododendron 'Macrantha' Macrantha azalea	6–9	Spreading	Deep-rose, early summer	A	Double flowers; broadleaved evergreen; 24 to 30 in high.
Rhododendron mucronatum Snow azalea	6–9	Spreading; stems ascending	White, spring	A	Large white flowers and dull-green leaves.
Rhododendron 'Ramapo'	5–9	Compact mound	Violet, spring	A	2-ft evergreen mound; light blue-green foliage.
Salix purpurea 'Gracilis' Dwarf Arctic willow	2–(9)	Compact	Unimportant	A	Grey foliage; dwarf hedges.
Skimmia japonica Dwarf skimmia	7–9	Round	White, spring	B	A compact broadleaved evergreen shrub for a shady place; bright-red berries.
Spiraea bumalda 'Anthony Waterer' Anthony Waterer spirea	3–9	Stems ascending	Lavender, late summer	A	Deciduous shrub to about 3 ft; desirable for summer color.
Symphoricarpos albus Common snowberry	3–9	Round	Pink, all summer	A	Deciduous; 3 ft high; tolerant of most conditions; white berries.
Taxus baccata 'Repandens' Weeping English yew	5–9	Weeping	Unimportant	A	Dwarf slow-growing conifer; tolerates shade.
Thuja occidentalis 'Hetz Midget' Hetz midget arborvitae	3–9	Globe	Unimportant	A	Slow growing; to 2½ ft.
Viburnum davidi David viburnum	7–9	Spreading; dense	White, June	A	Broad, dark-green evergreen leaves; 2 to 3 ft high; spreads 5 to 6 ft.

(continued)

Table 14–25 *(continued)*

Name	Hardiness zone[a]	Growth habit	Flowers	Soil and exposure[b]	Remarks
		Shrubs 18 inches to 3 feet			
Viburnum opulus 'Nanum' Dwarf European cranberry bush	4–9	Round; compact	Unimportant	A	Slow growing; deciduous; ornamental foliage.
Yucca filamentosa Adam's needle yucca	4–9	Round; ascending	Cream, late summer	D	Hardy desert plant; flower spikes 3 to 5 ft; leaves spear-shaped.
		Small shrubs to 18 inches high (see also Table 11–9)			
Andromeda polifolia Dwarf Bog-rosemary	2–9	Dense mound	Pink, May	C	Narrow, grey-green evergreen leaves.
Berberis buxifolia 'Nana' Dwarf Magellan barberry	6–10	Dense mound	Unimportant	A	Evergreen; small leaves.
Buxus sempervirens 'Suffruticosa' Truedwarf common box	6–9	Round; dense	Inconspicuous	A	Slow-growing broadleaved evergreen; to 3 ft high, usually 6 to 18 in.
Calluna vulgaris 'Aurea' Goldleaf Scotch heather	6–9	Mound; stems ascending	Pink, July–September	B	Golden leaves in summer; red in winter.
Calluna vulgaris 'County Wicklow' County Wicklow Scotch heather	7–9	Spreading; ascending	Shell-pink, August–September	B	Double-flowered form; ground cover.
Ceanothus gloriosus Point Reyes ceanothus	7–9	Spreading	Blue, April	A	Broadleaved evergreen plant; to about 12 in with spread of 3 to 4 ft.
Daboecia cantabrica 'Alba' White bell Irishheath	7–9	Spreading; ascending	White, May–November	B	Broadleaved evergreen; heathlike; masses well.
Daboecia cantabrica 'Atropurpurea' Purple bell Irishheath	7–9	Spreading; ascending	Purple, June–November	B	Same comment as above; space 2 ft.
Erica spp. Heath	7–9	Spreading; ascending	Various, summer	B	Colorful spreading ground covers.
Euonymous fortunei Purple leaf wintercreeper	4–9	Spreading	Unimportant	A	Broadleaved evergreen; winter protection in zones 4 and 5.
Hedera canariensis Algerian ivy	7–9	Spreading or climbing	Unimportant	A	Vigorous ground cover or climbing vine; tolerates dense shade.
Hedera helix English ivy	6–9	Spreading or climbing	Unimportant	A	See above; plant ivy 4 ft apart.
Helianthemum nummularium Sunrose	5–9	Spreading	Various, May–June	D	To 12 in; drought tolerant; red, white, yellow.
Hypericum calycinum Aaron's beard St. Johnswort	6–9	Spreading; stoloniferous	Yellow, July–September	A	Vigorous and invasive ground cover; keep in bounds.
Hypericum moserianum St. Johnswort	7–9	Spreading; stems ascending	Gold, July–September	A	Same as above.
Juniperus horizontalis 'Bar Harbor' Bar Harbor creeping juniper	4–9	Spreading	Unimportant	A	Red foliage in winter; too low to discourage weeds.

Name	Hardiness zone[a]	Growth habit	Flowers	Soil and exposure[b]	Remarks
Small shrubs to 18 inches high (see also Table 11–9)					
Juniperus horizontalis 'Douglasii' Waukegan creeping juniper	4–9	Trailing	Unimportant	A	Prostrate conifer 12 to 18 in; 4- to 5-ft spread.
Rhododendron impeditum Cloudland rhododendron	7–9	Rounded; ascending	Bluish-purple, April	B	Broadleaved evergreen to 20 in; foliage slightly grey.
Lianas and perennial vines					
Actinidia polygama Silver vine	6–9	15 ft; spreading	White, early summer	A	Green fruit in September; sun or shade; needs support.
Akebia quinata Fireleaf akebia	5–9	15 ft; spreading	Purple, spring	A	Sun or shade; flowers not showy; dainty five-parted leaves.
Campsis radicans Common trumpet creeper	5–10	30 ft; ascending	Orange, July–September	A	Clings to walls or trees; bright foliage; spectacular flowers.
Celastrus scandens American bittersweet	4–9	15 ft; spreading	White, June	A	Brilliant orange-red berries persist all winter; need male and female plants for fruit.
Clematis spp.	5–9	15 ft; trailing	Various	A	Large blooms, profuse; cultivars of many colors.
Euonymus fortunei radicans Evergreen euonymus	5–9	20 ft; ascending	Greenish-white, June	A	One of the hardiest evergreen lianas; sun to full shade; bank or wall cover.
Hedera (see "Small shrubs," above)					
Lonicera henryi	5–9	20 ft; spreading	Red-yellow, June–August	A	Semievergreen; showy flowers; purple berries.
Lonicera japonica 'Halliana' Hall's honeysuckle	5–9	30 ft; spreading	White, June–September	A	Evergreen; rampant growth; can be a weed.
Lonicera sempervirens Trumpet honeysuckle	5–9	30 ft; spreading	Orange, May–August	A	Evergreen; very showy flowers.
Parthenocissus quinquefolia Virginia creeper	5–9	50 ft; ascending	Unimportant	A	Rapid-growing wall or bank cover; sun to shade.
Parthenocissus tricuspidata Boston ivy	5–9	75 ft; ascending	Unimportant	A	Wall or bank cover; only on north or east walls in warm climates.
Wisteria spp.	(5)–9	20 ft; spreading	White-violet, June	A	Deciduous; many types; can be trained as tree, shrub, vine; showy flowers.

[a]See Figure 14–107. Parentheses around a zone number indicate marginal tolerance in that zone.
[b]A—General garden loam; sun to light shade tolerant.
B—Needs acid, well-drained soil; usually shade tolerant.
C—Tolerates wet marshy situations.
D—Needs perfect drainage, full sun; usually alkaline tolerant.

Table 14–26 Climatic data for selected municipalities in the United States and Canada

City and State	Temperature (mean) (Fahrenheit)		Average annual	Killing frost		Hardiness zone[a]	Record temperatures (Fahrenheit)		Precipitation average annual (inches)	Degree days Average Annual	
	January	July		Spring	Fall		High	Low		(Base 50°F)	(Base 40°F)
ALABAMA											
Birmingham	45.5	80.0	63.2	3-19	11-14	7	107	−10	53.25	5,191	8,509
Mobile	51.7	81.8	67.5	2-17	12-12	9	104	− 1	63.11	6,412	10,062
Montgomery	49.2	81.2	65.4	2-27	12- 3	8	107	− 5	53.66	5,694	9,301
ALASKA											
Anchorage	11.4	57.6	34.7	5-18	9-13	4	86	−38	14.83	532	1,863
Fairbanks	−12.0	60.4	25.6	5-24	8-29	2	99	−66	11.57	783	2,004
Juneau	22.8	55.5	39.9	4-27	10-19	6	89	−21	54.18	391	1,832
ARIZONA											
Flagstaff	27.8	65.7	45.5	6- 8	10- 2	5	96	−30	20.16	937	3,238
Phoenix	51.6	90.7	70.2	1-27	12-11	8	118	16	7.41	7,412	11,062
Tucson	50.3	86.1	67.4	3- 6	12-23	8	112	6	11.22	6,382	10,032
ARKANSAS											
Little Rock	41.7	81.3	62.2	3-16	11-15	7	110	−13	47.87	5,067	8,125
CALIFORNIA											
Bakersfield	47.4	84.3	65.1	2-14	11-28	9	118	13	6.36	5,644	9,176
Los Angeles	55.9	71.3	63.6	b	b	10	110	28	14.87	4,953	8,603
Red Bluff	45.5	83.8	63.5	2-25	11-29	7	115	17	22.05	5,200	8,612
Sacramento	44.5	75.9	60.5	1-24	12-11	9	115	17	17.32	4,180	7,499
San Diego	55.0	68.3	61.9	b	b	10	111	25	9.76	4,359	8,009
San Francisco	50.2	58.8	56.5	b	b	10	101	27	21.51	2,391	6,041
COLORADO											
Colorado Springs	29.5	70.7	48.7	5- 7	10- 8	5	100	−32	14.81	2,200	4,208
Denver	30.2	72.7	50.2	5- 2	10-14	5	105	−30	14.47	2,500	4,565
Grand Junction	26.0	78.3	52.6	4-20	10-22	5	105	−23	8.51	3,391	5,611
CONNECTICUT											
Hartford	27.0	73.0	50.0	4-22	10-19	6	102	−26	42.38	2,684	4,824
New Haven	29.1	71.2	49.7	4-15	10-27	7	101	−15	44.99	2,463	4,573
DELAWARE											
Dover	36.8	77.2	56.3	4-15	10-26	7	104	−11	46.40	3,724	6,231
DISTRICT OF COLUMBIA	35.6	78.4	57.1	4-10	10-28	7	106	−15	39.54	4,013	6,564
FLORIDA											
Jacksonville	55.9	82.6	69.5	2- 6	12-16	9	105	12	53.36	7,141	10,791
Miami	66.9	82.2	75.3	b	b	10	100	28	60.19	9,255	12,905
Orlando	60.5	82.3	72.4	1-31	12-17	9	103	20	50.90	8,186	11,836
Tallahassee	53.9	81.3	68.0	2-26	12- 3	9	100	10	58.86	6,597	10,247
Tampa	60.9	81.8	72.2	1-10	12-26	10	98	18	49.30	8,128	11,778
GEORGIA											
Atlanta	43.3	78.5	61.5	3-20	11-19	8	103	− 9	48.40	4,714	7,872
Columbus	47.8	81.1	64.4	3-10	11-20	8	104	3	48.67	5,439	8,941
Savannah	51.7	81.5	66.8	2-21	12- 9	9	105	8	48.47	6,164	9,814
HAWAII											
Hilo	71.0	75.1	73.2	Never		>10	94	51	133.27	8,467	12,117
Honolulu	72.4	78.8	76.0	Never		>10	92	54	24.30	9,506	13,156
IDAHO											
Boise	29.9	74.5	51.3	4-29	12-16	6	112	−28	11.97	2,613	4,833
Lewiston	32.7	73.8	51.6	4-21	10-17	6	117	−23	13.24	2,606	4,827
Pocatello	22.3	72.4	47.0	5- 8	9-30	5	105	−31	10.85	2,169	4,116

City and State	Temperature (mean) (Fahrenheit)		Average annual	Killing frost		Hardiness zone[a]	Record temperatures (Fahrenheit)		Precipitation average annual (inches)	Degree days Average Annual	
	January	July		Spring	Fall		High	Low		(Base 50°F)	(Base 40°F)
ILLINOIS											
Chicago	24.7	73.7	49.9	4-19	10-28	6	105	−23	33.23	2,840	4,926
Peoria	24.4	75.6	51.2	4-22	10-16	5	113	−27	34.80	3,204	5,356
Springfield	27.3	77.3	53.3	4- 8	10-30	6	112	−24	25.33	3,598	5,828
INDIANA											
Evansville	34.7	78.2	56.9	4- 2	11- 4	6	105	−23	41.37	4,024	6,239
Ft. Wayne	26.3	73.5	49.9	4-24	10-20	5	103	−17	34.21	2,778	4,852
Indianapolis	28.4	75.7	52.6	4-17	10-27	6	107	−25	39.90	3,310	5,507
South Bend	24.5	73.1	49.4	5- 3	10-16	5	109	−22	35.19	2,778	4,858
IOWA											
Burlington	24.4	76.6	51.2	4-22	10-17	5	111	−27	34.60	3,287	5,427
Des Moines	20.8	76.0	49.9	4-20	10-19	5	110	−30	31.24	3,128	5,268
Sioux City	18.7	77.4	49.1	4-28	10-12	4	111	−37	24.77	3,212	5,319
Waterloo	17.9	73.8	47.2	4-28	10- 4	4	112	−34	31.48	2,760	3,946
KANSAS											
Dodge City	30.3	79.9	55.0	4-22	10-24	5	109	−26	20.58	3,874	6,176
Topeka	28.6	79.0	54.8	4- 9	10-26	6	114	−25	33.18	3,901	6,227
Wichita	31.7	80.2	56.6	4- 5	11- 1	6	114	−22	29.99	4,210	6,651
KENTUCKY											
Lexington	33.5	76.2	55.1	4-13	10-28	6	108	−21	43.14	3,640	6,036
Louisville	34.6	78.4	56.8	4- 1	11- 7	6	107	−20	42.60	4,025	6,525
LOUISIANA											
Baton Rouge	52.4	81.1	67.5	2-22	12- 1	9	103	19	59.13	6,407	10,057
New Orleans	54.0	82.4	69.0	2-13	12-12	9	102	7	57.49	6,953	10,603
Shreveport	47.4	83.2	66.0	3- 1	11-27	8	110	− 5	44.33	5,975	9,523
MAINE											
Caribou	10.5	64.9	38.7	5-19	9-21	3	103	−41	35.76	1,251	2,887
Portland	22.4	68.2	45.5	4-29	10-15	6	103	−39	41.89	1,841	3,756
MARYLAND											
Baltimore	32.5	76.8	54.9	4-28	11-17	7	107	− 7	40.61	3,585	5,965
MASSACHUSETTS											
Boston	38.7	72.5	50.3	4-16	10-25	6	104	−18	41.40	2,635	4,781
Worcester	23.8	69.9	47.0	5- 7	10- 2	5	102	−24	45.09	2,157	4,150
MICHIGAN											
Detroit	25.2	73.0	49.1	4-25	10-23	6	105	−24	31.52	2,680	4,721
Grand Rapids	24.4	72.8	48.5	4-25	10-27	5	108	−24	33.02	2,610	4,651
Marquette	17.4	65.8	41.7	5-14	10-17	5	108	−27	31.66	1,446	3,178
MINNESOTA											
Duluth	8.6	65.1	38.6	5-22	9-24	4	106	−41	28.19	1,310	2,949
Minneapolis	13.2	73.0	44.9	4-30	10-13	4	108	−34	26.72	2,482	4,499
MISSISSIPPI											
Jackson	48.2	81.8	65.5	3-10	11-13	8	107	− 5	50.46	5,754	9,333
Meridian	47.5	80.9	64.5	3-13	11-14	8	105	− 7	55.07	5,449	8,965
MISSOURI											
Kansas City	29.8	79.5	55.6	4- 5	10-31	6	113	−22	36.47	4,054	6,437
St. Louis	31.7	79.4	56.2	4- 2	11- 8	6	115	−23	36.65	4,084	6,513
Springfield	33.6	78.8	56.5	4-10	10-31	6	113	−29	41.08	4,017	6,431

(continued)

Table 14–26 *(continued)*

City and State	Temperature (mean) (Fahrenheit)		Average annual	Killing frost		Hardiness zone[a]	Record temperatures (Fahrenheit)		Precipitation average annual (inches)	Degree days Average Annual	
	January	July		Spring	Fall		High	Low		(Base 50°F)	(Base 40°F)
MONTANA											
Billings	23.2	74.7	47.5	5-15	9-24	5	112	−49	13.23	2,420	4,425
Great Falls	22.1	69.4	44.7	5-14	9-26	5	107	−49	14.07	1,734	3,605
Kalispell	19.8	65.7	42.8	5-12	9-23	5	105	−38	15.42	1,360	3,122
Miles City	16.5	75.3	45.8	5- 5	10- 3	4	111	−49	12.17	2,512	4,492
NEBRASKA											
North Platte	23.7	74.8	49.3	4-30	10- 7	5	112	−35	18.61	2,767	4,868
Omaha	22.0	77.4	51.1	4-14	10-20	5	114	−32	28.44	3,391	5,531
NEVADA											
Las Vegas	44.0	89.4	65.8	3-13	11-13	8	117	8	3.85	6,161	9,455
Reno	31.2	69.6	49.4	5-14	10- 2	6	104	−16	6.96	1,993	4,089
NEW HAMPSHIRE											
Concord	21.3	70.1	46.0	5-11	9-30	5	102	−37	38.08	2,093	4,046
NEW JERSEY											
Atlantic City	30.8	74.4	53.0	3-31	11-11	7	106	− 9	41.27	3,121	5,438
Newark	31.6	76.2	53.7	4- 3	11- 8	7	105	−14	41.32	3,342	5,666
NEW MEXICO											
Albuquerque	34.6	77.1	55.7	4-16	10-29	6	104	−17	8.42	3,652	6,101
Clayton	33.2	74.5	53.1	5- 2	10-15	5	105	−21	14.51	3,057	5,276
Roswell	37.9	78.6	58.5	4- 9	10- 2	6	110	−29	11.62	4,222	6,893
NEW YORK											
Albany	22.9	72.5	48.1	4-27	10-13	5	104	−28	36.29	2,555	4,590
Buffalo	25.0	70.2	47.3	4-30	10-25	6	99	−21	35.09	2,226	4,165
New York	32.1	76.1	54.0	4- 7	11-12	7	106	−15	43.00	3,388	5,745
Syracuse	24.1	71.1	47.7	4-30	10-15	5	102	−26	35.63	2,368	4,373
Watertown	20.4	70.5	46.3	5- 7	10- 4	5	99	−39	38.85	2,279	4,254
NORTH CAROLINA											
Asheville	34.9	72.6	54.6	4-12	10-24	7	99	− 7	45.39	3,165	5,619
Charlotte	42.7	79.2	60.8	3-21	11-15	8	104	− 5	43.38	4,598	7,627
Raleigh	41.6	78.3	60.0	3-24	11-16	8	105	− 2	45.23	4,348	7,321
Wilmington	47.9	80.0	63.8	3- 8	11-24	8	105	5	51.29	5,209	8,701
NORTH DAKOTA											
Bismarck	9.2	72.1	41.7	5-11	9-24	3	114	−45	15.40	2,116	3,916
Fargo	5.0	69.9	40.1	5-13	9-27	3	114	−48	21.00	1,996	3,768
Williston	10.0	70.9	41.3	5-14	9-23	3	110	−50	14.66	1,958	3,745
OHIO											
Cincinnati	31.7	76.3	54.6	4-15	10-25	6	109	−17	39.40	3,576	5,938
Cleveland	27.4	72.2	49.9	4-21	11- 2	6	103	−19	33.82	2,680	4,778
Columbus	29.3	74.8	52.3	4-17	10-30	6	106	−20	36.79	3,167	5,373
Toledo	26.0	73.3	49.8	4-24	10-25	6	105	−17	31.43	2,763	4,843
Youngstown	27.5	72.3	49.8	4-23	10-20	5	100	−12	41.33	2,683	4,757
OKLAHOMA											
Oklahoma City	37.3	81.5	60.1	3-28	11- 7	7	113	−17	31.47	4,738	7,486
Tulsa	37.3	82.5	60.7	3-31	11- 2	7	115	−16	37.37	4,854	7,658
OREGON											
Burns	23.9	69.8	46.8	5-20	9-21	5	103	−25	10.25	1,838	3,789
Eugene	38.2	66.6	52.4	4- 9	10-31	8	105	− 4	37.51	1,986	4,587
Medford	37.2	71.8	54.5	4-25	10-20	8	115	−10	18.15	2,687	5,249
Pendleton	32.5	73.3	52.4	4-27	10- 8	6	119	−28	12.97	2,633	4,970
Portland	38.4	65.3	52.3	2-25	12- 1	8	107	− 3	37.72	1,964	4,564

City and State	Temperature (mean) (Fahrenheit) January	July	Average annual	Killing frost Spring	Fall	Hardiness zone[a]	Record temperatures (Fahrenheit) High	Low	Precipitation average annual (inches)	Degree days Average Annual (Base 50°F)	(Base 40°F)
PENNSYLVANIA											
Harrisburg	30.3	75.2	52.6	4-10	10-28	6	107	−14	37.38	3,177	5,425
Philadelphia	33.0	76.6	54.6	3-30	10-17	7	106	−11	40.97	3,498	5,873
Pittsburgh	30.6	74.6	52.7	4-20	10-23	6	103	−20	35.99	3,170	5,406
Scranton	27.1	72.2	49.6	4-24	10-14	5	103	−19	36.33	2,595	4,705
RHODE ISLAND											
Providence	20.2	72.7	50.5	4-13	10-27	6	102	−17	40.30	2,645	4,809
SOUTH CAROLINA											
Charleston	50.0	81.1	65.8	2-19	12-10	9	104	7	48.51	5,786	9,436
Columbia	46.9	81.6	64.0	3-14	11-21	8	107	− 2	46.82	5,374	8,772
SOUTH DAKOTA											
Rapid City	21.1	72.3	46.1	5- 7	10- 4	4	109	−33	17.10	2,211	4,158
Sioux Falls	15.2	74.1	46.3	5- 5	10- 3	4	110	−42	25.29	2,632	4,685
TENNESSEE											
Knoxville	39.3	77.7	58.9	3-31	11- 6	7	104	−16	47.68	4,186	6,926
Memphis	41.2	81.2	61.9	3-20	11-12	7	106	−13	48.23	4,999	8,010
Nashville	38.9	79.4	59.5	3-28	11- 7	7	107	−15	46.12	4,434	7,191
TEXAS											
Amarillo	36.6	77.9	57.2	4-20	10-28	6	108	−16	20.54	3,946	6,498
Corpus Christi	57.4	84.1	71.8	2- 9	12-12	9	105	−11	28.34	7,973	11,623
Dallas	45.5	85.0	65.9	3-18	11-22	8	111	− 3	35.00	6,063	9,498
El Paso	42.9	81.9	63.3	3-14	11-12	7	109	− 8	7.89	5,312	8,533
Houston	51.7	83.2	68.0	2- 5	12-11	8	108	5	46.42	6,603	10,253
Midland	44.0	82.9	64.3	4- 3	11- 6	7	107	−11	14.24	5,618	8,907
San Antonio	52.1	84.0	69.1	3- 3	11-26	9	107	0	27.14	6,995	10,645
UTAH											
Milford	23.8	74.0	49.0	5-21	9-26	5	104	−34	8.44	2,483	4,557
Salt Lake City	28.1	77.2	51.7	4-12	11- 1	5	107	−30	15.58	2,974	5,121
St. George	38.7	83.5	60.8	4- 3	10-30	6	116	−11	8.22	4,825	7,648
Vernal	15.4	69.5	44.3	5-29	9-25	4	106	−38	8.22	1,919	3,871
VERMONT											
Burlington	17.9	69.6	44.5	5- 8	10- 3	5	98	−27	32.30	2,037	3,924
VIRGINIA											
Norfolk	41.4	78.6	59.8	3-18	11-27	8	105	2	15.13	4,333	7,259
Richmond	37.7	77.7	57.6	4- 2	11- 8	7	107	−12	43.11	3,973	6,573
Roanoke	36.3	76.2	56.4	4-20	10-24	7	105	−12	39.09	3,719	6,218
WASHINGTON											
Aberdeen	39.7	60.1	50.3	4-16	10-30	9	105	6	84.54	1,302	3,795
Seattle	38.2	64.2	50.9	2-23	12- 1	8	100	0	40.15	1,665	4,053
Spokane	26.9	70.2	48.2	4-20	10-12	5	108	−30	16.17	2,049	4,083
Walla Walla	33.2	76.0	54.2	3-28	11- 1	7	113	−16	15.50	3,153	5,548
Yakima	27.5	71.0	49.8	4-19	10-15	6	111	−25	7.86	2,296	4,495
WEST VIRGINIA											
Charleston	36.5	76.1	56.4	4-18	10-28	6	108	−17	43.35	3,737	6,264
Parkersburgh	34.4	75.7	54.9	4-16	10-21	6	106	−27	39.11	3,511	5,913
WISCONSIN											
Green Bay	16.1	69.9	43.6	5- 6	10-13	5	104	−36	26.51	2,052	3,897
La Crosse	15.7	74.0	46.2	5- 1	10- 8	4	108	−43	28.92	2,638	4,676
Madison	17.3	71.3	46.0	4-26	10-19	5	107	−37	30.13	2,359	4,388
Milwaukee	20.9	70.7	46.5	4-20	10-25	5	105	−25	30.04	2,243	4,220

(continued)

Table 14-26 *(continued)*

City and State	Temperature (mean) (Fahrenheit) January	July	Average annual	Killing frost Spring	Fall	Hardiness zone[a]	Record temperatures (Fahrenheit) High	Low	Precipitation average annual (inches)	Degree days Average Annual (Base 50°F)	(Base 40°F)
WYOMING											
Casper	22.3	71.1	45.1	5-18	9-25	5	104	−40	14.09	1,939	3,758
Cheyenne	26.2	67.8	45.1	5-20	9-27	5	100	−38	14.55	1,640	3,411
Lander	16.8	70.4	43.2	5-15	9-20	4	102	−40	14.18	1,847	3,583
Sheridan	20.1	70.6	44.5	5-21	9-21	4	106	−41	16.75	1,884	3,724

City and Province	Temperature (mean) (Celsius) January	July	Average annual	Killing frost Spring	Fall	Hardiness zone[a]	Record temperatures (Celsius) High	Low	Precipitation average annual (mm)	Degree days Average Annual (Base 10°C)	(Base 5°C)
ALBERTA											
Calgary	− 9.9	16.7	3.6	5-26	9-10	3	36	−43	444	483	1,254
Edmonton	−14.1	17.3	2.7	5-15	9-17	3	37	−48	473	590	1,358
Lethbridge	− 8.2	18.9	5.4	5-21	9-17	4	39	−43	439	765	1,616
BRITISH COLUMBIA											
Pr. George	−11.3	14.9	3.3	6-10	8-28	2	34	−50	626	357	1,112
Pr. Rupert	1.8	13.4	7.6	4-28	10-28	6	32	−19	2,399	340	1,250
Vancouver	2.9	17.7	10.2	3-22	11-11	8	33	−18	1,044	870	1,987
MANITOBA											
Churchill	−27.5	12.0	− 7.2	6-25	9-11	2	33	−45	407	112	467
The Pas	−21.7	18.2	− 0.4	5-27	9-19	2	37	−45	451	588	1,313
Winnipeg	−17.7	20.2	2.5	5-25	9-21	3	42	−47	517	911	1,714
NEW BRUNSWICK											
St. John	− 6.9	17.2	5.4	5-10	10-12	5	34	−33	1,362	662	1,511
NEWFOUNDLAND											
Goose	−16.6	16.3	0.2	6- 6	9-17	5	38	−39	837	401	1,014
St. John's	− 4.3	15.4	4.7	6- 2	10- 3	6	30	−23	1,551	404	1,085
NORTHWEST TERRITORIES											
Ft. Smith	−24.5	16.2	− 3.2	6-15	8-19	1	34	−54	337	409	1,037
Frobisher Bay	−26.5	7.9	− 8.9	6-30	8-29	1	24	−45	380	0	149
NOVA SCOTIA											
Halifax	− 3.3	18.5	7.4	5- 8	10-23	5	34	−25	1,384	843	1,769
ONTARIO											
Kapuskasing	−17.8	17.3	0.8	6-13	9- 5	2	36	−42	858	556	1,256
Moosonee	−20.6	15.6	− 1.1	6-21	9- 1	2	36	−47	789	386	1,002
North Bay	−12.2	18.7	3.8	5-18	9-23	3	33	−40	1,034	782	1,596
Ottawa	−10.8	20.7	5.7	5- 8	10- 4	4	38	−36	850	1,100	1,973
Toronto	− 3.9	21.9	8.7	5- 7	10-14	6	41	−30	776	1,299	2,279
QUEBEC											
Ft. Chimo	−23.9	11.8	− 5.2	6-27	8-30	2	32	−46	417	71	468
Montreal	− 8.7	21.6	6.9	5- 4	10-12	5	36	−34	1,048	1,232	2,169
Quebec	−11.5	19.3	4.4	5-12	10- 6	5	36	−36	1,058	858	1,682
SASKATCHEWAN											
Regina	−16.9	19.3	2.2	6- 1	9-10	3	41	−46	394	783	1,551
Saskatoon	−17.6	19.3	2.0	5-27	9-15	3	40	−46	352	861	1,626
YUKON											
White Horse	−18.1	14.2	− 0.7	6- 5	9- 1	1	33	−52	257	283	907

[a]See Figure 14-107.
[b]Occurs less than 1 year in 10.

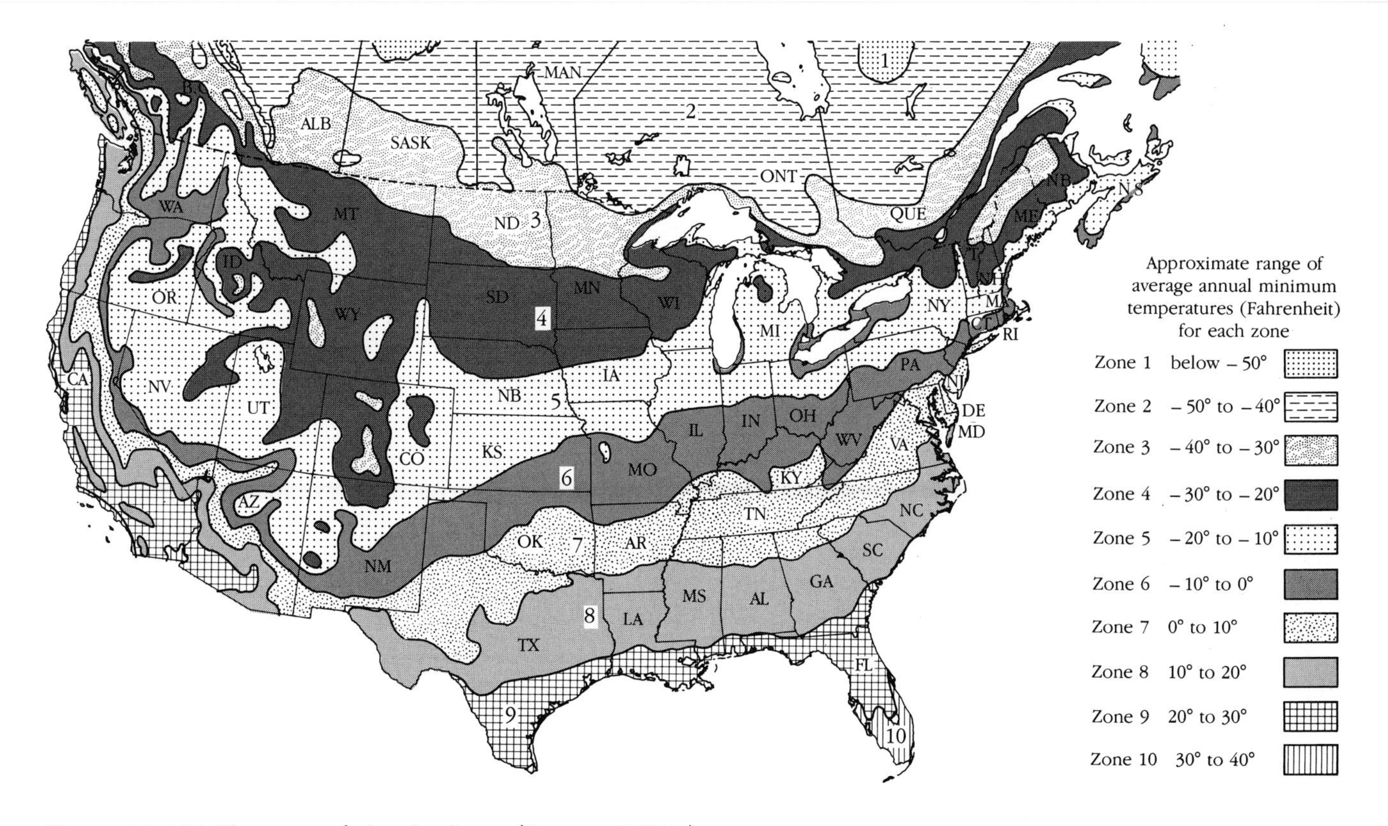

Figure 14–107 The zones of plant hardiness. (Courtesy USDA.)

The classification having the widest use is the one established by the USDA, which classifies into 10 hardiness groups based on 10°F increments, from those that will survive 40° to 30°F (4° to −1°C) (hardiness group 10) to those that will survive temperatures below −50°F (−46°C) (hardiness group 1). North America has been divided into 10 zones based on the average minimum winter temperature (see Figure 14–107).

To determine whether or not a plant will survive, look at the map to find the zone in which you live and then at the hardiness zone listed for the plant in Table 14–25. As a further check, I have listed the hardiness zones for a few weather station locations in each state and province (Table 14–26). Those living in the vicinity of these weather stations will generally be aware as to whether their gardens are in a warmer or colder location.

The minimum temperature recorded for each location listed in Table 14–26 will enable growers to determine which kinds of plants may require occasional winter protection in their climate.

Weather stations are sometimes located where conditions deviate from the average of the surrounding areas and, as mentioned in Chapter 7, many factors affect cold temperature survival. Consequently, although the plant hardiness zone designation is useful as a rough guide of winter survival, plants listed as hardy in a specific zone cannot always be counted on to do well there. As is true with so many facets of gardening, the experience of local growers is likely to be the most accurate predictor of plant winter survival.

Predicting Maturity

The ability of warm-season crops to mature in cold-climate areas depends largely on the amount of heat energy accumulating during the frost-free period. The accumulation of heat is usually measured in heat units, also called degree days, described in Chapter 7.

The utilization of degree days is based on the fact that most cool-season crops do not grow until the temperature is above about 40°F (5°C) and that most warm-season crops do not grow until the temperature reaches 50°F (10°C). Also, within the growing range, the rapidity of growth increases as the temperature rises. Degree days for any one day are calculated by subtracting the base temperature (40° or 50°F) from the mean temperature for that day. Thus if on June 15 the maximum temperature is 80°F and the minimum is 60°F, the mean would be 70°F and the degree days for June 15 would be 70 minus 50 or 20 for warm-season crops, and 70 minus 40 or 30 for cool-season crops.

The long-term average degree days likely to accumulate on any one day are determined by subtracting the base temperature from the long-term mean temperature for that day. The average annual degree days for a geographic location (Table 14–26) are determined by adding together average degree days for each day of the growing season.

The degree days (base 50°F) required for various grape cultivars are listed in Table 13–1. The 'Red Delicious' apple and 'Red Haven' peach cultivars require about 4,000 degree days (base 40°F) to mature, and earlier maturing cultivars should be planted in locations where there are lower heat accumulations. Mid-season sweet corn and early pepper and tomato cultivars require 1,900 to 2,100 degree days, and standard mid-season watermelon and muskmelon cultivars will not mature with less than 2,500 to 2,700 (base 50°F).

Glossary

Abscission Dropping off of a leaf, fruit, or flower; shedding.

Adventitious Organ Plant organ, such as a bud, shoot, or root, produced in an abnormal position or at an unusual time of development.

Aeration (soil) Air in the spaces around soil particles.

Aerial Bulb See Bulbil.

Aggregate Fruit Fruit developed from a flower having a number of pistils, all of which ripen together and are more or less cohesive at maturity.

Aggregates (soil) Masses or clusters of soil particles variable in shape, size, and degree of coherence; as granules, nuciform aggregates, clods, prisms.

Agronomy Science of soil management and production of crops grown on large acreages.

Air Layering Method of propagation by which plant parts are rooted while they remain attached to the mother plant.

Allele One of the contrasting genes that may exist at a particular location on a pair of chromosomes. The gene that produces pink flower color in sweet peas is an allele of the one that produces white flowers and also of the one that produces lavender flowers.

Allelopathy Detrimental influence of one plant on another due to chemical interactions.

Alternate Bearing Bearing of heavier and lighter crops of flowers or fruits, or both, in successive years.

Alternate Buds Leaf buds (or leaves) that occur singly at a node.

Analysis Plan Preliminary drawing of an area to be landscaped, showing the environmental, topographical, and construction features that must be considered in designing the landscape.

Angiosperms Flowering plants; plants having their seeds enclosed in an ovary (the most advanced class of plants).

Annual Plant in which the entire life cycle is completed in a single growing season.

Anther Upper portion of a stamen, containing the pollen grains.

Anthocyanin Pigment in sap responsible for scarlet to purple or blue coloration in plants.

Apical Dominance Influence exerted by a terminal bud in suppressing the growth of lateral buds.

Apical Meristem Meristematic cells of the apex of the root and shoot.

Apomictic Seed Seed developed from an unfertilized egg or other ovarian cell without sexual fusion.

Approach Grafting Method of grafting by which two branches are joined while each remains attached to its own root system.

Arthropods Phylum or division of the animal kingdom that includes insects, spiders, and Crustacea; characterized by a coating that serves as an external skeleton and by legs with distinct movable segments or joints.

Asexual (or vegetative) Propagation Propagation by utilizing a part of the body tissue of the mother plant as opposed to sexual union.

Auxin Organic compound active at low concentrations that promotes plant growth by cell enlargement and affects other aspects of plant development. Sometimes used loosely as synonymous with growth substance.

Available Moisture Amount of water in soil that can be absorbed by the roots of plants. Technically, it is the difference in the weight of moisture held in a soil at field capacity and that held at the wilting point.

Axil Angle on upper side between leaf and stem.

Bacteria Microscopic, one-celled organisms that lack chlorophyll, multiply by fission, and do not form noticeable fungus threads. May be parasites on other plants or animals, or may live saprophytically on nonliving, organic matter.

Bactericide Substance that destroys bacteria; germicide.

Balanced Crotch An undesirable situation that occurs when two branches grow from the same point on a tree at the same rate. Balanced crotches tend to split easily.

Balled and Burlapped Plant Plant with a compact mass of earth, covered with sacking, left on the roots for transplanting.

Banding Fertilizing close to a row of seed or seedlings.

Bare-root Transplanting Method of transplanting in which plants are taken from the ground with little soil left on the roots.

Bark Grafting Method of joining plants in which the scion is inserted between the bark and the xylem of the stock.

Bark Inversion Type of phloem disruption in which a strip of bark is removed, turned upside down, and tacked back to the area from which it was removed. Used as a means of hastening reproduction of a nonfruiting tree or shrub.

Basal Pertaining to the base or lowest part of an organ or part.

Basal Plate Short, fleshy stem axis within a bulb.

Bedding Plants Flowers appropriate for growing in flower beds for massed decorative effect.

Berry Simple fleshy fruit, as a grape or tomato.

Biennial Plant that normally requires two growing seasons to complete the life cycle. Only vegetative growth occurs the first year; flowering and fruiting occur the second year.

Biennial Bearing Bearing fruit once every two years.

Binomial System of Nomenclature System in which the scientific name for any plant is composed of two Latin terms that designate genus and species.

Biological Pest Control (biocontrol) Destruction or suppression of undesirable insects, plants, or animals by the introduction or propagation and dissemination of predators, parasites, and diseases.

Blanching 1. Bleaching or whitening a vegetable as it is growing by wrapping the stalk and leaves with paper or outer leaves, or by throwing soil around the portion to be whitened, as celery is blanched. Also called etiolating. 2. Heating vegetables in water, live steam, or dry heat, to inactivate enzymes preparatory to processing.

Blastula Hollow sphere formed during tissue development that is reminiscent of certain stages of lower life forms; also one of the stages of animal embryonic development.

Blossom-end Rot Disorder of tomato fruits in which a sunken dry rot develops on the distal end; associated with calcium deficiency and water stress.

Bolting Premature flower and seedstalk formation; usually refers to seed formation in biennial crops during their first year of growth.

Bonsai Culture of miniature potted trees that have been dwarfed by pruning and controlled nutrition.

Border Masses Masses of shrubs and trees grown as plant walls.

Bract Modified and reduced leaflike structure.

Bramble Any plant of the genus *Rubus,* family Rosaceae; as the blackberry, raspberry, and dewberry.

Bridge Grafting Method of preserving trees that have suffered bark damage of the lower trunk. The damaged area is bridged by long suckerlike scions grafted into healthy bark above and below the wound.

Bud Protuberance on a plant stem, leaf, or root that gives rise to vegetative shoots, flowers, and/or leaves.

Budding Grafting by inserting a single bud under the bark.

Bud Mutation Genetic change in a bud that causes it to develop a shoot, flower, or fruit different from other shoots, flowers, or fruits of that plant.

Bud Shield Bud with attached segment of bark and wood cut in the shape of a shield. (See T-budding.)

Bud Sport Cultivar originating from bud mutation.

Budstick Shoot from which buds are removed for propagation.

Bulb Subterranean budlike storage organ produced by some plants; has a short stem surrounded by overlapping, fleshy leaf bases, as in onions and tulips.

Bulbil (aerial bulb, brood bulb, bulbel) Small bulb produced above the ground, among the flowers, or in the axil of a leaf.

Bulblet Small bulb borne in the axil of a bulb scale.

Bulb Scale Fleshy sheathing leaf base of a bulb.

Bundle Sheath Group of cells surrounding or associated with the vascular bundles in stems and leaves of some plants.

Callus Protective covering that forms over a wounded plant surface.

Calyx Outer or lowest of the four series of floral parts composed of the sepals. Usually green and leaflike, but may be colored like the petals.

Cambium Zone or cylinder of meristematic cells, lateral in position, which gives rise to secondary xylem and secondary phloem. Derived from provascular tissue and located between the xylem and phloem.

Cane 1. Woody stem of any small fruits, such as grape or raspberry. 2. Stem of reeds and large grasses, such as bamboos and sorghums. Also sometimes applied to the stem of rosebushes and some small palms.

Capillary Moisture Water held by the soil against the force of gravity and available for plant absorption. The amount of water a soil will hold between wilting point and field capacity.

Carbohydrate/Nitrogen Balance Relative proportion of accumulated carbohydrates and nitrogen in stems and leaves of plants, important because it influences flower bud initiation and fruit set.

Catface Healing or healed wound on the trunk of a tree, frequently occurring on the southwest side of trees growing in cold winter areas and caused by physiological drought.

Cation Ion carrying a positive charge of electricity.

Cation Exchange Interchange between a cation in solution and another cation on the surface of a colloidal or other surface-active material such as a clay or organic matter particle.

Cation Exchange Capacity Measure of a soil's ability to retain fertility, the sum total of exchangeable cations absorbed by a soil, expressed in milliequivalents per 100 g of soil equivalent to the milligrams of H^+ that will combine with 100 g of dry soil.

Cell Structural unit composing the bodies of plants and animals; an organized unit of protoplasm, in plants usually surrounded by a cell wall.

Cell Membrane Structure inside the cell wall that appears to have the function of regulating the flow of nutrients and other materials into and out of the cell.

Cellulose A carbohydrate, the chief component of the cell wall in most plants.

Cell Wall Membranous covering of a cell secreted by the cytoplasm in growing plants; consists largely of cellulose but may contain chitin in some fungi and silica in some algae.

Central Leader System of tree training in which the trunk is encouraged to form a central axis with branches distributed laterally around it.

Chelate Metal ion bonded on to an organic molecule from which it can be released. For example, iron, only slightly available to plants growing in calcarious soils in its usual ferric hydroxide form, is readily absorbed when applied as chelate sequestrene 138 Fe.

Chemical Pest Control Use of chemical pesticides to control diseases, weeds, insects, and other pests that reduce crop yields.

Chemistry Alteration Methods of preserving foods that involve a change in the chemistry of the food, including pickling; fermenting; and preserving with sugar, salt, and/or some other chemical.

Chlorophyll Green pigment located in plastids; necessary to the process of photosynthesis.

Chloroplast Specialized body in the cytoplasm that contains chlorophyll.

Chlorosis Interveinal yellowing of foliage that results from chlorophyll deficiency.

Chromosome Threadlike structural unit in the nucleus that preserves its individuality from one cell generation to the next and is the site of the hereditary determiners, the genes.

Class Taxonomic grouping of plants more comprehensive than an order and more specialized than a division (phylum).

Clay 1. Soil particle less than 0.002 mm in diameter. 2. Textural class of soil; one that contains 40% or more of particles of clay size.

Clean Cultivation Intensive cultivation of a field so as to remove all plant growth except the crop.

Cleft Grafting Method of grafting in which large trees are used for stock. The branch is sawed squarely across and split lengthwise, and two scions are inserted into the cleft.

Climacteric Peak (climacteric) Maximum point of the respiration rate of mature fruit. The respiration rate rises dramatically just prior to reaching the climacteric.

Clone Group of plants derived by asexual propagation from a single mother plant.

Colchicine Poisonous alkaloid extracted from the common autumn crocus and used in plant breeding to block anaphase separation during cellular division, thereby producing polyploidy and doubling the number of chromosomes.

Cold Frame Bottomless box with a removable glazed top; used to protect, propagate, or harden off plants. No heating device is used.

Cole Crop Any plant of the genus *Brassica,* family Cruciferae (e.g., cabbage, cauliflower, broccoli).

Coleoptile Sheathlike pointed structure covering the shoot of grass seedlings; commonly interpreted as the first leaf of the plant above the cotyledon.

Colloid 1. Insoluble substance consisting of particles small enough to remain suspended indefinitely in a medium. 2. Mineral particle less than 0.002 mm in diameter. 3. Substance that does not form a true solution in water and does not diffuse readily through animal or vegetable membranes. Its presence does not affect the freezing point or vapor tension of the solution.

Companion Cropping Form of intercropping in which specific kinds of plants are supposed to be mutually benefited by close association in the garden.

Complete Flower Flower having all four series of floral parts—stamen, pistil, corolla, calyx.

Compost Material formed when organic residues, such as peat, manure, or discarded plant material, are aerobically decomposed.

Compound (serpentine) Layering Propagation, usually of vines, in which several portions of a branch are covered with soil and the intervening portions left above the soil until rooting takes place.

Contact Herbicide Herbicide that kills a plant primarily by contact with tissue rather than by internal absorption.

Cool-season Crop Crop that thrives best in cool weather; as apple, dahlia, and radish.

Core 1. Innermost part of pome and certain other fruits that contains the seeds. 2. Receptacle in certain plants, as the raspberry.

Cork Cambium Lateral meristem producing cork in woody and some herbaceous plants.

Corm Short, thickened underground storage organ formed usually by enlargement of the base of the main plant stem.

Cormel Small corm, usually produced from a bud of the parent or major corm.

Corolla Second (beginning from below) of the series of floral organs composed of petals.

Cortex Outer primary tissues of the stem or root, extending from the primary phloem (or endodermis, if present) to the epidermis composed chiefly of parenchyma cells.

Cotyledon Leaves (seed leaves) of the embryo, one or more in number.

Cover Crop Crop grown to add organic matter to soil and/or protect against erosion.

Cross-compatible (cross-fruitful) Condition in which each of two cultivars is capable of fertilizing and/or inducing fruit set on the other.

Cross-incompatible (cross-unfruitful) Condition based on genetic factors in which two different cultivars cannot fertilize or induce fruit set on each other.

Cross-pollination Transfer of pollen from the anther of one plant to the stigma of another plant.

Crown 1. Upper part of a tree, which bears branches and leaves. 2. Place at which the stem and root join in a seed plant; top of the root.

Crown Division Method of reproduction of perennial plants, in which the crown is divided to form several separate plants.

Crumb (soil) A natural structure in soil; small, spheroidal, very porous, easily crushed aggregate.

Cucurbit Any plant of the family Cucurbitaceae, as cucumber, squash, watermelon.

Cultivar Official name for all cultivated variants of plants; also called horticultural or agricultural varieties, but distinguished from the botanical use of the term variety.

Cultivation 1. Planting, tending, and harvesting of plants. 2. Tillage of the soil to promote crop growth after the plant has germinated and appeared above ground. 3. Loosening of the soil and removal of weeds from among desirable plants.

Cultural (pest) Control Control of pests through cultural practices; e.g., cultivating to destroy pests, planting at a time when pests are not likely to be destructive, regulating irrigation.

Cuticle Thin layer of cutin that covers the epidermis of above-ground plant parts.

Cuttage Method of propagation by which stem, leaf, or root tissue is removed from the plant and caused to form new roots and shoots.

Cutting Any part that can be severed from a plant and be capable of regeneration.

Cytoplasm Protoplasm of the cell exclusive of the nucleus.

Damping-off Rotting of seedlings and cuttings caused by any of several fungi. Usually refers to fungal attack near the soil line that results in falling over and death of cuttings or emerged seedlings, although preemergence damping-off is the killing of seedlings after germination but before the seedlings appear above soil.

Day-neutral Plant Plant in which the flowering period or some other process is not influenced by length of daily exposure to light.

Deciduous 1. Parts of a plant that fall at the end of the growing period, such as leaves in autumn or fruits or flower parts at maturity. 2. Broadleaved trees or shrubs that drop their leaves at the end of each growing season; contrasted with evergreen plants.

Delayed Dormant Spray Pesticide applied to fruit trees when the fruit buds show green leaf tips about ¼ in (½ cm) long.

Deoxyribonucleic Acid (DNA) Hereditary material of the cell; molecules composed of a double helix of sugar-phosphate linkages (forming the sides of a twisted ladder) with purine and pyrimidine bases joined across (to form the rungs).

Design Elements One of the three areas of consideration in planning the landscape, including cubic space, topography, plants, rock, water, and building materials.

Determinant A type of growth habit that eventually terminates in a flower cluster and the end of shoot elongation.

Dicot Plant whose embryo has two cotyledons or seed leaves.

Dioecious Bearing staminate and pistillate flowers (or pollen and seed cones of conifers) on different individuals of the same species.

Diploid Having two sets of chromosomes; the $2n$ number characteristic of the sporophyte generation.

Disbudding Removal of flower buds and/or shoot buds from a plant.

Diurnal Plant whose blossoms open during the day and close at night. Occurring daily.

Division 1. See Crown Division. 2. The highest category of classification of plants according to rules of nomenclature; an aggregation of classes; synonymous with phylum as used by zoologists.

Dormancy Period of inactivity in bulbs, buds, seeds, and other plant organs.

Dormant Spray Pesticide applied to a plant during its period of physiological inactivity.

Double Working Rebudding or regrafting on a previously established graft on a plant. This system is used when two varieties or species of stock and scion do not unite except through an intermediary.

Drupe Simple fleshy fruit in which the inner part of the ovary wall develops into a hard stony or woody endocarp, as in the peach.

Drupelet A small drupe, often one of many in a cluster or group; e.g., the pulpy grain of bramble fruits.

Dwarf Plant, especially one that has been grafted, that is much smaller when mature than others of its species.

Ecosystem Interacting system of one to many living organisms and their nonliving environment.

Egg Cell Female reproductive cell of animals and plants.

Embryo Rudimentary plant formed in a seed resulting from fusion of the egg cell with a sperm.

Embryo Sac (ovule) Female gametophyte of angiosperms, containing at maturity typically eight cells, three of which, the egg and two polar nuclei, are important in the formation of the embryo and endosperm tissue of the seed.

Endocarp Inner leathery, woody, or stony part of the wall of a fruit, as in a drupe (peach) or pome (apple).

Endodermis One-celled layer of specialized cells, frequently absent in stems but usually present in young roots, which separates the pericycle from the cortex. Often acts as a barrier to loss of moisture from vascular system of older roots.

Endogenous Growing from within or developing internally.

Endoplasmic Reticulum Much-folded submicroscopic, double-layered membranes found in the cytoplasm; associated with the major biosynthesis of the cell.

Endosperm Nutritive portion of seeds formed by fusion of the two polar bodies of the embryo sac with a sperm. In many plants the endosperm is absorbed as food by the embryo before the seed matures, but in some, such as cereal grains, a large part of the mature seed is endosperm tissue.

Enology Art and science of wine making.

Epicotyl Part of a seedling stem above the cotyledons but below the first foliage leaves.

Epidermis Outermost layer of cells of the leaf and of young stems and roots.

Espalier System Method of training a woody plant in which the tree is usually planted against a wall and the main branches trained in a plane parallel to the wall in a geometric design.

Ethylene C_2H_4, a colorless, flammable, unsaturated hydrocarbon gas manufactured at several locations in the plant and considered a growth regulator. Ripening fruit and damaged tissues give off large quantities. Used artificially for many purposes, including ripening and coloring fruit.

Everbearing (everblooming) Producing fruit or bloom throughout most of the season.

Evergreen Plant that retains its leaves or needles longer than one growing season so that leaves are present throughout the year.

Exocarp Outermost layer of the pericarp or fruit wall; often the skin of the fruit.

F_1 Hybrid First generation following cross-pollination; F_2 and F_3 are the second and third generations.

Family Taxonomic grouping of plants more comprehensive than a genus and more specialized than an order; composed of one or (usually) a number of genera.

Fasciation Flattening and enlargement of a branch as if several stems were fused, often accompanied by curving. Believed to be caused by injury to the cells of the bud or by multiple terminal buds arranged in a single plane.

Fertilizer Analysis Statement, usually on the label of a fertilizer container, of the percentages of nitrogen, phosphoric acid, and potash contained.

Fertilizer Formula Quantity and grade of crude stock materials used in making a fertilizer mixture.

Fibrous Root Root system in which the roots branch near the crown and become finely divided.

Field Capacity Amount of water held in the soil after the excess or gravitational water has drained away.

Filament Stalk of the stamen, supporting the anther.

Flat Shallow box containing soil in which seeds are sown or to which seedlings are transplanted from the seedbed.

Floret Single individual flower that is a part of a flower head.

Flower Reproductive structure of the angiosperms.

Foliar Analysis Detailed laboratory procedure that analyzes leaf tissue to furnish information on the minerals or other compounds within the plant.

Foot-candle Density of light striking the inner surface of a sphere with all surface area being one foot away from a one candle-power source.

Forma Group of individuals within a population that differs from the rest of the population in a regular but trivial way. Usually there is less difference between two forma of a population than between two varieties or subspecies. In horticulture the term is often used with ornamentals to distinguish groups with different growth characteristics not reproducible with seed.

Friable Soil Soil with aggregates that can be readily ruptured and crushed with application of moderate force; one easily pulverized or reduced to crumb or granular structure.

Fruit 1. (botanical) Matured ovary of a flower and its contents, including any external part that is an integral portion of it. 2. (horticultural) Fleshy, ripened ovary of a woody plant, tree, shrub, or vine, used as a cooked or raw food.

Full Slip In harvesting of melons, the easy separation of the fruit from the vine.

Fungi (Kingdom Fungi) A lower order of plant organisms, excluding bacteria, that contain no chlorophyll, have no vascular system, and are not differentiated into roots, stems, or leaves. Many cause diseases of horticultural crops; others, the mushrooms, are grown as food.

Fungicide Chemical that kills or inhibits fungi.

Furrow Depression in the ground surface dug along a prescribed line for planting seed, irrigating, controlling surface water, or reducing soil loss.

Gamete Reproductive body capable of fusion with another; most frequently refers to the sperm from the pollen grain and the egg from the ovule.

Gametophyte Plant during the haploid part of its life cycle. In higher plants, the sperm and egg and the haploid cells from which they develop.

Gene Unit of inheritance; located on chromosomes.

Generative Nucleus One of the two nuclei of the pollen grain, responsible for fertilization of the ovule.

Genetic Resistance Ability of an organism, because of its genetic makeup, to tolerate a condition (e.g., pesticide application, insect, disease) that would ordinarily kill other members of the same species.

Genus Group of closely related species clearly differentiated from other groups.

Germination Resumption of growth of an embryo or spore, including pollen grain on a stigma; the sprouting of a seed.

Germ Plasm Hereditary materials (chromosomes, genes, and any other self-propagating particles) transmitted to the offspring through the reproductive cells; often used to denote the total genetic resources available in the entire population of a crop or species.

Girdling Encircling a living plant with a wound involving the absence of tissues as deep or deeper than the cambium layer; often done to reduce the downward flow of carbohydrates through the phloem in order to encourage reproduction.

Graftage (grafting) Process of inserting a part of one plant into or on another in a way that the two will unite and continue growth as a single unit.

Graft-compatibility Ability of parts of two different plants, when grafted together, to produce a successful union, and of the resulting single plant to develop satisfactorily. This condition is usually met when scion and stock are relatively closely related botanically.

Graft-incompatibility Inability of parts of two different plants, when grafted together, to produce a successful union, and of the resulting single plant to develop satisfactorily.

Graft Union Healed wound, with an additional, foreign, piece of tissue (scion) incorporated into it. The union is accomplished entirely by cells that develop after the actual grafting operation has been made.

Granule 1. Rounded or subangular, relatively dense soil aggregate. 2. Particle containing volatile forms of a pesticide designed to be scattered among plants.

Gravel Rounded or angular particles of rocks or minerals greater in size than coarse sand and up to 75 mm in diameter.

Gravitational Water Water that moves through soil under the influence of gravity; the source of spring and well water.

Green Manure Crop plowed under when green for its beneficial effect on the soil.

Ground Cover 1. Any vegetation that grows close to the ground, producing protection for the soil. 2. Any of many different plants, usually perennials, that grow well on sites on which grass does not thrive. Often used as a substitute for grass.

Groundwater Water contained in reservoirs in the soil.

Growth Substance (plant growth regulator, plant hormone) Organic chemical that circulates in the plant in minute quantities and is a primary regulator of growth and other plant activities.

Guard Cells Specialized crescent-shaped epidermal cells surrounding a stomate.

Gymnosperms Seed plants having ovules borne on open scales; mostly the needle evergreens, like pine, fir, and cedar.

Half-slip In harvesting of melons, a stage of ripeness in which, as the fruit is pulled from the vine, only a portion of the stem separates easily from the base of the fruit.

Hardening Treating plants to make them more resistant to adverse environmental conditions, usually by increasing their cold resistance.

Hard Seed Seed that is unable to take in water because of an impervious seed coat.

Hardwood Stem Cutting Mature shoot of the last season's growth that is removed from the plant after the leaves have fallen to be used in propagating new plants.

Heading Back Pruning the end of a branch or stem by cutting back to a bud or side branch.

Heaving Upward movement of soil caused by freezing and thawing of free water in the soil, thus involving expansion and contraction. Damage, and sometimes destruction, of plants may result from the lifting action.

Hedgerow 1. Method of training strawberries by definite placement in the row of the runners from each mother plant. 2. Widened row used for some bramble fruits in which sucker plants are permitted to grow between the mother plants in the row. 3. Close within-row planting of dwarf fruit trees so that they resemble a hedge.

Heel Cutting Basal end of a plant stem cutting along with a piece of the older stem.

Herbaceous Plant or portion of a plant that lacks a pronounced woody structure.

Herbicide Plant killer; chemical used for weed control. A nonresidual herbicide kills only at the time of application, whereas a residual herbicide remains active in the soil for a few days to a number of years. A nonselective herbicide kills all vegetation to which it is applied, whereas a selective herbicide kills weeds without injuring the surrounding plants.

Hermaphroditic Bearing both male and female sex organs, e.g., a flower having anthers and ovaries.

Hesperidium A berry fruit, primarily from citrus, in which the exocarp and mesocarp are the rind and the endocarp is fleshy.

Heterozygous Condition in which the genes for a given character on the homologous chromosomes are unlike.

High-analysis Fertilizer Fertilizer containing more than 30 pounds total N, P, K per 100-pound bag.

Homologous Chromosomes Chromosomes that associate in pairs in the first stage of meiosis; each member of the pair is derived from a different parent.

Homozygous Condition in which the genes for a given character on homologous chromosomes are alike.

Horizon Stratum of the soil. Horizons start with "A" at the surface and usually end with "C," which is the parent material from which the soil formed.

Horticulture 1. Department of the science of agriculture that relates to intensively cultivated crops. 2. Cultivation of gardens or orchards, including the growing of vegetables, fruits, flowers, and ornamental shrubs and trees.

Host Any animal or plant upon or in which another organism lives as a parasite.

Host Range Various kinds of plants that may be affected by a given pathogen.

Hotbed Small enclosed garden bed, having a transparent covering, in which the soil is heated.

Hybrid Progeny of a cross between two individuals differing in one or more genes.

Hybridization Process of crossing two individuals that differ in genetic makeup.

Hygroscopic (unavailable) Water Moisture a soil contains after it has dried or been depleted to the wilting point. Such water does not move in the soil and cannot be used by plants.

Hypocotyl Part of the stem of an embryo or seedling below the cotyledons and above the radicle or embryonic root.

Imperfect Flower Flower containing either stamens or pistil but not both.

Inarching Method of propagation in which a plant, still attached to its own roots, is grafted to another; frequently done with root suckers or seedling trees to bridge lower trunk damage of fruit trees.

Inbred Line Homozygous line of plants or animals produced by inbreeding and selection.

Inbreeding Breeding of closely related plants or animals; in plants it is usually brought about by self-pollination.

Incomplete Flower Flower lacking one or more of the four kinds of floral organs: sepals, petals, stamens, or pistil.

Indeterminant Growth habit that can continue indefinitely.

Indole-3-acetic Acid (IAA) Plant hormone that causes elongation of cells when it is present in suitable concentrations; usually considered the natural plant auxin.

Indole-3-butyric Acid (IBA) Synthetic auxinlike hormone used to stimulate root formation and other plant growth responses.

Inorganic Not made up of or derived from plant or animal materials.

Insecticide Substance that kills insects by chemical action.

Integrated (pest) Control Type of pest control in which the pest-host relationship is monitored carefully, and every conceivable control measure, including cultural practices, environmental manipulation, and the use of natural predators, is utilized to control the pest. Chemicals are used mainly as the last resort and then with as low a concentration and as minimal a coverage as possible.

Intercropping Growing two or more crops simultaneously, as in alternate rows in the same field or single tract of land.

Intermittent Mist Propagating system in which water sprays over the rooting bed at intervals to prevent the cuttings from dehydrating while roots are being initiated.

Internode Region of the stem between any two nodes.

Interspecific Cross Cross between two different species.

Interstock Intermediate stock grafted between a rootstock and a scion.

Ion Atom or group of atoms carrying an electrical charge, which may be positive (cation) or negative (anion). Usually formed when salts, acids, or bases are dissolved in water.

June Drop Shedding of tree fruits during the early summer; believed to be caused most frequently by, or associated with, embryo abortion.

Juvenile Type of growth in young plants that is not found in older plants of the same kinds.

Kingdom One of the five main areas of biological classification now recognized by most biologists. The five kingdoms include the traditional Kingdom Animalia; Kingdom Monera, or the bacteria, including blue-green algae; Kingdom Protista, or the algae and slime molds; Kingdom Fungi; and Kingdom Plantae.

Kniffen System Method of training grape vines, in which one or more trunks are carried to wires of a trellis along which fruiting canes are renewed and tied annually.

Landscape 1. In soil geography the total natural and human-made characteristics that distinguish a certain area of the earth's surface from other areas; as soil types, vegetation, rock formations, hills, valleys, streams, cultivated fields, roads, and buildings. 2. To beautify terrain, as with plantings or ornamental features.

Larvicide Chemical used to kill the larval or preadult stages of insects.

Lateral Bud Bud attached to the side of a branch or spur.

Layering Method of propagating woody plants by covering portions of their stems or branches with moist soil or sphagnum moss so that they take root while still attached to the parent plant.

Leaching 1. Removing alkali and/or salt from soil by abundant irrigation combined with drainage. 2. Removing soluble materials in downward percolating water.

Leaf More or less flattened outgrowth from a plant stem, varying in size and shape and usually green in color, that is concerned primarily with the manufacture of carbohydrates by photosynthesis.

Leaf Bud Cutting Method of propagation in which parts of a stem are split lengthwise so that each cutting has one leaf and its axillary bud. The bud is then buried under a shallow covering of soil to root.

Legal (pest) Control Exclusion of pests from an area by prohibiting importation of plant materials from areas where the pest is endemic.

Legume Plant of the family Leguminosae, as alfalfa, clovers, peas, beans, and others; characterized botanically by a fruit called a legume or pod that opens along two sutures when ripe.

Lenticel Opening through the bark or outer covering of fruits, etc., that permits exchange of gases from the inner tissues with the surrounding air.

Liana Woody climbing plant.

Lignification Process in which plant cells become woody by conversion of certain constituents of the cell wall into lignin; generally considered to include the hardening, strengthening, and cementing of the cell walls in the formation of wood.

Lignin Principal noncellulosic constituent of wood. Tends to harden and preserve the cellulose.

Loam Soil not definitely sandy or claylike; one not strongly coherent, mellow, and well supplied with organic matter. Technically, a soil that contains 7 to 27% clay, 28 to 50% silt, and less than 52% sand.

Locus Position on a chromosome occupied by a particular gene.

Lodging Pertaining to plants that break, bend over, or lie flat on the ground, sometimes forming a tangle.

Long-day Plant Plant in which the flowering period or other process is accelerated by daily exposure to light longer than a certain minimum number of hours.

Low-analysis Fertilizer Fertilizer containing less than 30 pounds of total N, P, K per 100-pound bag.

Macroclimate Long-term weather pattern of a fairly broad geographical area.

Macronutrient Mineral required in relatively large amounts for the healthy growth of plants.

Megaspore Large spore that germinates to form a female gametophyte.

Megaspore Mother Cell Cell that undergoes two meiotic divisions to produce four megaspores.

Meiosis Process of two divisions during which the chromosomes are reduced from the diploid to the haploid number. In higher plants it occurs only in the reproductive cells.

Meristem Undifferentiated, thin-walled tissue of a plant from which, as development proceeds, the permanent tissues are produced.

Mesocarp Middle layer (usually fleshy) of the fruit wall of a drupe or other fruit.

Metabolism Total of the chemical processes in the plant body.

Microclimate Purely local variations from the general or regional climate due to slight differences in elevation, direction of slope exposure, soil, density of vegetation, etc.

Micronutrient Nutrient essential to plants in only very small amounts.

Micropylar End Part of the embryo sac that is attached to the remainder of the ovary. Has a thin wall from which there is living cellular tissue leading to the main stylar tissue of the flower.

Microspore Small spore that germinates to form the male gametophyte.

Microspore Mother Cell Cell that divides to produce microspores.

Middle Lamella (intercellular layer) The wall layer common to two adjoining cells and lying between the primary walls.

Mineral Soil Common soil of land surfaces, consisting of broken-down minerals and rocks. The solid matter is preponderantly inorganic (as contrasted to organic soil).

"Mini-til" (minimum tillage) Method of cultivation utilizing the least amount of tillage possible. Usually residue from the previous crop remains between the rows until the current crop is established.

Miticide Any chemical mixture or compound used to kill mites.

Mitochondria Minute protoplasmic bodies in the cytoplasm, believed to be the site of the enzymes responsible for the oxygen-requiring steps in respiration.

Mitosis Process during which the chromosomes become doubled longitudinally, the daughter chromosomes then separating to form two genetically identical daughter nuclei. Mitosis is usually accompanied by cell division, in which each daughter cell has the same number of chromosomes as the mother cell.

Mixed Bud Bud that produces both leaves and flowers.

Modified Central Leader System A training system extensively utilized with apples, pears, and some stone fruits, in which the central leader is headed back slightly but not completely removed until after the tree is five or six years old and has borne a crop of fruit, at which time the main framework will have been established.

Moisture Cycle Circulation of the earth's moisture, involving evaporation from oceans; falling to earth in rain, snow, or hail; various pathways through the soil and in streams back to the ocean; and evaporation from soil and transpiration from plants back to the atmosphere.

Mold and Hold System System of training in which trees are permitted to grow to any convenient size desired and are then kept at that size by heavy pruning. The downward orientation of side branches overcomes the inhibition of fruiting that would otherwise occur as a result of the heavy pruning.

Monocot Plant whose embryo has a single cotyledon.

Monocropping (monoculture) Cultivation of a single crop, such as wheat or cotton, to the exclusion of other possible uses of the land.

Monoecious Bearing both staminate and pistillate flowers (or pollen and seed cones of conifers) at different locations on the same plant.

Morphology Form, structure, and development of plants.

Mound Layering Rooting of branches of woody plants or shrubs by leaving the branches upright and mounding soil about the basal portions of the stems.

Muck Organic soil derived mainly from plant matter that is more highly decomposed than in peat and has lost its botanical identity. May be granular or amorphous but is humified and blackish.

Mulch 1. Soil, straw, peat, plastic, or any other loose or sheet material placed on the ground to conserve soil moisture, promote early maturity, or prevent undesirable plant growth or soil erosion. 2. Material, such as straw, placed over plants or plant parts to protect them from cold or heat.

Multiple Cropping Growing of two or more crops consecutively on the same field in a single year.

Multiple Fruit Fruit composed of a number of closely associated ovaries derived from different flowers that, with the fleshy tissue surrounding them, forms one body at maturity; as a pineapple.

Mutation Change in a gene that results in a change in descendents produced from the cell containing that gene.

Mycelium Hyphae or filaments (plant body) of a fungus.

α-Naphthalene-acetic Acid (NAA) One of the synthetic organic acids of hormone mixtures used to stimulate root development.

Necrosis Death of a cell or group of cells, usually while a part of the plant is still living.

Nematocide Agent that kills nematodes.

Nematode Any of the round, threadlike, unsegmented animal worms of the phylum Nematoda, ranging in size from microscopic to one meter long. May be saprophytes or parasites of plants and animals. Responsible for important animal and plant diseases resulting in much economic loss.

Nitrogen-fixing Bacteria Species of the genus *Rhizobium*, family Rhizobiaceae, which live symbiotically in the root nodules of leguminous plants, upon which they are dependent. They are capable of extracting nitrogen from the air and converting it to a form utilizable by the plant.

Node Region of the stem where one or more leaves are attached. Buds are commonly borne at the node, in the axils of the leaves.

Nucleus Specialized body within the protoplasm that contains the chromosomes.

Nut Hard, dry, one-seeded fruit.

Obligate Parasite Any parasite that cannot exist independently of its living host.

Offset Short, prostrate, many-noded branch growing from the crown of a plant and having a somewhat fleshy, scaly bud or a rosette of leaves located terminally. Offsets often form roots and are used to propagate some kinds of plants.

Offshoot Lateral shoot or branch that rises from one of the main stems of a plant; often used for propagation.

Olericulture Vegetable culture.

Open-pollinated Cultivars of cross-pollinated crops, seed of which is produced by allowing plants in seed fields to intercross freely; contrasted to inbred, F_1 hybrid, and other cultivars in which pollination is controlled.

Opposite Buds Buds (and leaves) occurring in pairs at a node.

Order Category of classification more comprehensive than a family and more specialized than a class; composed of one or more families.

Organelle Part of the protoplasm of a cell having a particular function.

Organic 1. Of plant or animal origin. 2. More inclusively, chemical compounds that contain carbon.

Organic Soil Any soil whose solid part is predominantly organic matter, as contrasted to mineral soil; muck or peat.

Ornamental Horticulture Production and utilization of flowers, shrubs, and trees.

Osmosis Flow of a fluid through a semipermeable membrane separating two solutions that permits passage of the solvent but not of the dissolved substance. The liquid will flow from a weaker to a more concentrated solution, thus tending to equalize concentrations.

Ovary Swollen basal portion of a pistil; the part containing the ovules or seeds.

Ovicide Substance that kills parasites in the egg stage.

Ovule (embryo sac) Part of the ovary containing one female gametophyte. Following fertilization, the ovule develops into the seed.

Palisade Layer Cells of the mesophyll lying next to the upper epidermis of leaves. The process of food manufacture, or photosynthesis, is most active in these cells.

Parasite Organism that lives at least for a time on or in and at the expense of living animals or plants.

Parenchyma Unspecialized, simple cell or tissue, usually thin-walled, living at maturity and retaining a capacity for renewal of cell division.

Parthenocarpic Fruit Fruit produced without fertilization.

Peat Geologic deposit, consisting predominantly of plant remains only very slightly decomposed, that accumulates in lakes, marshes, and some swamps. Various kinds are recognized according to origin, texture, and plant composition.

Peat Pellet Type of seedling container made of compressed peat moss surrounded by coarse netting. Usually contains fertilizer and expands to several times its size when soaked in water.

Pectin Any of the fruit juice substances that form a colloidal solution with water and are derived from pectose (protopectin) in ripening processes or other forms of hydrolysis. Pectin is derived from citrus fruits and apple wastes and is used in jelly making to firm the body of the product.

Pepo Fleshy or succulent fruit, often of large size, formed from an inferior syncarpous ovary, and containing many seeds.

Perennial Woody or herbaceous plant living from year to year not dying after flowering once.

Perfect Flower Flower containing both stamens and pistil; also called a bisexual flower.

Pericycle Cylinder of vascular tissue, three to six cells thick, lying immediately inside the endodermis of a root from which branch roots are initiated.

Periderm Secondary protective tissue formed in secondarily thickened stems and roots. It consists of the cork cambium, cork, and secondary cortex.

Perlite White and very porous volcanic mineral that is sometimes used as a medium for rooting cuttings.

Pest Cycle Sequential changes and interactions that occur in the host/pathogen relationship through their respective life cycles.

Petal One of the units of the corolla of a flower.

Petiole Thin stem supporting the blade of a leaf.

pH Index designating relatively weak acidity and alkalinity, as encountered in soils and biological systems. A pH of 7.0 indicates neutrality; higher values indicate alkalinity, lower values, acidity.

Phloem Vascular tissue that conducts synthesized foods in vascular plants. Characterized by the presence of sieve tubes and in some plants companion cells, fibers, and parenchyma.

Phloem Disruption (as growth regulator) Disruption of phloem by scoring, girdling, or bark inversion for the purpose of hastening the reproduction of a nonfruiting tree or shrub.

Photoperiodism Reaction of plants to periods of daily exposure to light; generally expressed in formation of blossoms, tubers, fleshy roots, runners, etc.

Photosynthesis Production of carbohydrate from carbon dioxide and water in the presence of chlorophyll, using light energy and releasing oxygen.

Phototropism Response of a plant to the stimulus of sunlight, in which parts of the plant receiving the direct rays grow more slowly and the plant appears to turn toward the light.

Phylum Major division of the plant or animal kingdom. The term division is used more frequently in the plant kingdom.

Physical (pest) Control Control of pests through physical force; e.g., physical removal of infested and infesting entities, barriers against infestation.

Physical Seed Dormancy Failure of a seed to germinate for a physical reason, such as a hard seed coat that is impervious to water.

Physiological Drought Inability of a plant to obtain water from soil although the water may be present in it, as when a soil is frozen or by reason of weak osmotic force of plant roots.

Physiological Seed Dormancy Failure of a seed to germinate for physiological reasons, such as an immature embryo when the seed is otherwise ready to harvest, or chemical inhibitors within or outside the seed.

Physiological Self-incompatibility Condition in which self-pollination frequently occurs but self-fertilization does not. Pollen may fail to germinate on the stigma, or the pollen tube may grow only part way through the style.

Pillar System A system of training in which a permanent trunk is developed and new side branches are permitted to grow each year. The following year's crop comes from newer side branches. Side branches oriented downward are kept to encourage fruiting.

Pistil Central or female organ of the flower, composed of one or more carpels and enclosing the ovules.

Pistillate Designating a flower that has a pistil or pistils but lacks stamens; an imperfect flower.

Pith Tissue occupying the center of the stem within the vascular cylinder. Usually consists of parenchyma, but other types of cells may also occur.

Plant Band Short strip of heavy paper that may be folded or rolled into a potlike unit to replace clay pots for growing young plants, which are later transplanted directly to the field without removal of the plant band.

Plant Pathology Branch of botanical science that deals with the diseases and disorders of plants.

Plant Protector Any device used to protect plants from nuisances such as rabbits and birds or from excessive sunlight, winter injury, or cold temperatures.

Plant Volatile Aromatic substance emitted by plants in gaseous form. Some repel pests; for example, a cedar closet repels moths.

Plasmodesmata Streams of protoplasm that extend through pores in the cell wall and connect the protoplasts of adjacent living cells.

Plug Sodding Method of planting a new lawn with grasses that reproduce asexually, in which plugs of the grass are set at specified intervals and allowed to spread.

Plumule Bud of the embryo. May consist of a shoot apex alone or of the apex and one or more embryonic leaves. Also applied to the primary bud of a seedling.

Polar Bodies Two of the eight cells of the embryo sac. They stay in the approximate center of the embryo sac, and during fertilization they fuse with one of the nuclei of the pollen grain to form the initial endosperm cell.

Pollen (grain) Mature, usually two-nucleated, microspore of seed plants.

Pollen Tube Tube formed in the style following germination of the pollen grain.

Pollination Transfer of pollen from the anther to the stigma of the same or another flower.

Pollinator Cultivar grown primarily as a source of pollen for other cultivars.

Pome Fleshy fruit with a leathery endocarp, as produced by apple, pear, and quince.

Pomology Art and science of growing and handling fruits, especially tree fruits.

Pore In plant and animal membranous tissue, minute opening for absorption and transpiration of matter.

Postbloom Drop Period of time right after petals fall from the blooms of a fruit tree when small fruits are normally lost.

Postemergence Herbicide Herbicide that is applied after the crop has emerged.

Potherb Greens; any plant yielding foliage that is edible when cooked, such as spinach, kale, chard.

Predator Any animal, including insects, that preys upon and devours other animals. Distinguished from a parasite, which lives on only one host at a time and usually does not destroy the host.

Preemergence Herbicide Herbicide that is applied after the crop has been seeded but before it comes through the ground.

Prepink Spray Spray applied to apples in the stage just before the flower buds turn pink.

Preplant Herbicide Herbicide that is applied before the crop is seeded.

Pricking Off Lifting very small seedling plants from seed beds and transferring them to transplant flats.

Primordia Organ of a plant in its earliest condition.

Propagating Causing to generate or to multiply by sexual or asexual means.

Propagule Newly propagated plant.

Protectant Chemical containing heavy metals, sulfur, or organic compounds used as sprays, dusts, or dips on seeds, stems, leaves, or wounds of living plants to prevent entrance of fungi or bacteria.

Protoplasm Living material of a cell.

Pruning Removal of live or dead branches, roots, and other parts from trees, shrubs, vines, etc., for purposes of improvement.

Pseudobulb Swollen, stemlike base of many orchids, or an elongated above-ground, fleshy plant stem in which food and moisture are held.

Qualitative Characteristics Plant characteristics in which differences of expression, such as color, leaf form, presence of hairs, can be easily distinguished.

Quantitative Characteristics Characteristics in which differences of expression, such as size, shape, yield, and quality, intergrade. Usually under the control of large numbers of genes, each of which adds to or modifies the characteristic.

Radicle Basal end of the embryonic axis, which grows into the primary root.

Random Segregation Capacity of members of each pair of alleles to randomly separate into different sex cells or gametes and thence into different offspring.

Receptacle Part of the axis of a flower stalk that surrounds the floral organs.

Region of Cell Division Area of the root tip between the root cap and the region of cell elongation.

Region of Cell Elongation Area just behind the region of cell division at the tip of stems or roots or at nodes of plants having growth in the nodal area.

Repellent Substance obnoxious to insects (or chordates) that prevents them from injuring their hosts or laying eggs.

Reproductive Phase Stage in the growth of a plant when it changes from purely vegetative growth to production of flowers, fruit, and seeds.

Respiration Intracellular process in which food is oxidized with release of energy.

Rest Period Stage during which a plant is physiologically unable to initiate growth, even though environmental conditions may be favorable.

Rhizome Underground stem, usually horizontal in position. Distinguished from a root by the presence of nodes and internodes and sometimes buds and scalelike leaves at the nodes.

Ribonucleic Acid (RNA) Nucleic acid found in some nuclei and in the cytoplasm. Believed to carry a copy of the genetic information contained in the genes and to apply it in the synthesis of specific protein molecules.

Ribosome Submicroscopic granule in the protoplasm that contains RNA and protein and is associated with protein formulation.

Rill Furrow.

Roguing Removing and destroying undesirable plants.

Root Lower portion of a plant that usually develops underground and anchors the plant in the soil; bears the root hairs, which absorb water and mineral nutrients.

Root Cap Thimble-shaped mass of parenchyma cells over the root apex, protecting it from mechanical injury.

Root Hair Zone Area of the root tip behind the region of cell elongation. In the root hair zone and younger root tissue the endodermis is permeable enough to permit passage of water and nutrients.

Root Primordia Specialized cells that will develop into roots.

Rootstock Underground stock upon which a desirable variety may be grafted.

Root Sucker Sprout that rises from a root.

Rotation Growing of two or more crops on the same piece of land in different years in sequence and according to a definite plan.

Runner Horizontal shoot or branch forming roots at the tip or nodes.

Russeting Brownish, roughened areas on the skins of fruit, tubers of potatoes, etc., resulting from abnormal production of cork tissue. May be caused by disease, insects, or injury; or may be a natural varietal characteristic.

Salad Crop Food plant grown primarily for its edible leaves, stalk, or other vegetative parts used in salads.

Samara Single-seeded, dry fruit having a winglike extension of the pericarp.

Sand Group of textural classes of soil in which the particles are finer than gravel but coarser than silt, ranging in size from 2.00 to 0.5 mm in diameter. Class of any soil that contains 85% or more sand and not more than 10% clay.

Scaling Method for propagating scaly bulbs, in which individual bulb scales are separated from the mother bulb and placed in growing conditions so that adventitious bulblets form at the base of each scale.

Scaly Bulb Bulb, such as lily, with fleshy overlapping leaves resembling scales.

Scarification Abrasion, scratching, or modification of a surface for increasing water absorption; as scarification of an impervious seed coat.

Scion Branch or portion of a branch having one or more buds. Detached from a woody plant, it is then used in grafting or budding.

Sclerenchyma Fiber Elongated cells with tapering ends and thick secondary walls; usually nonliving at maturity; supporting tissue.

Scooping Method of basal cuttage in which the entire basal plate of a bulb is scooped out. Adventitious bulblets develop from the base of the exposed bulb scales.

Scoring 1. Method of bulb propagation in which three straight knife cuts are made across the base of a bulb, each deep enough to go through the basal plate and the growing point. Growing points in the axils of the bulb scales grow into bulblets. 2. Type of phloem disruption that consists of running a knife blade around the tree or branch to cut through the phloem.

Sectioning Method of propagation in which a bulb is cut into sections with a portion of the basal plate attached to each section. Bulblets then form from the basal plate of each cutting.

Seed Organ formed by seed plants following fertilization. Embryonic plant with food supply and protective covering.

Seed Coat Hard outer layer of a seed; the protective covering, or integument.

Self-compatible (self-fruitful) Able to fertilize and/or mature fruit without the aid of pollen from another cultivar.

Self-incompatible (self-unfruitful) Unable to set viable seed and/or fruit from self-pollination although pollen and ovules may be normal; may result from the arrest of pollen tube growth in the style.

Self-pollination Transfer of pollen from the anther to the stigma of the same flower or of another flower on the same plant or within a clone.

Senescence Stage in the life of an individual plant or plant part when the rate of metabolic activities declines and there is a change in the physiology prior to death.

Sepal One of the units of the calyx.

Serpentine Layering See Compound Layering.

Set Small propagative part—a bulb, shoot, tuber, etc.—suitable for setting out or planting.

Sexual (seed) Propagation Propagation utilizing the fusion of male and female gametes, or their nuclei, to form a zygote, which develops into a new individual.

Sheath Leaf base when it forms a tubular casing around the stem.

Short-day Plant Plant in which flowering period or some other process is accelerated by daily exposure to light shorter than a specified maximum.

Short-season Crop Crop that reaches maturity in a short period of time.

Sidedressing Applying fertilizer to the soil at the side of a plant row, usually after the crop has started to grow.

Side Graft A type of graft used for topworking or for producing new plants. The scion is inserted into the side of the stock, which is generally larger in diameter than the scion.

Silt 1. Small, mineral soil particles ranging from 0.05 to 0.002 mm in diameter. 2. Textural class of soils that contains 80% or more silt and less than 12% clay.

Simple Fruit Fruit derived from a single pistil, simple or compound; ovary superior or inferior.

Slip Softwood or herbaceous cutting from a plant, used for propagation or grafting.

Sludge Residual solids after sewage treatment; solids deposited by sedimentation in sewage treatment. Often used as an organic fertilizer.

Sodding Removing sod from one area and placing it on a bare soil area in another location.

Softwood Immature, succulent stem of a woody plant.

Soil Profile Vertical section of a soil. The section, or face of an exposure made by a cut, may exhibit with depth a succession of separate layers, although these may not be separated by sharp lines of demarcation.

Solanaceous Crop Crop that is a member of the family Solanaceae, including potato, tomato, pepper, eggplant.

Species Taxonomic grouping of plants more specialized than genus. Species is the basic unit in which each plant is classified, even though some species may be further divided into subspecies, varieties, or other groupings. Plants within a species will usually intercross.

Sperm Mature male germ cell.

Spheroidal Aggregate Soil aggregate that is built around a central core and has rounded or irregular surfaces; the most desirable structural form of soil aggregate.

Spine Stiff pointed protuberance from a plant, especially one that is a modified leaf or leaf part.

Spongy Parenchyma Leaf tissue composed of loosely arranged chloroplast-bearing cells; also called spongy tissue. Found usually toward the lower side of the leaf.

Sporophyte Plant during the diploid part of its life cycle. In higher plants, all parts of the cycle but egg and sperm and the cells from which they develop.

Sprigging Method of planting a lawn with grasses that reproduce asexually, in which individual plants, cuttings, or stolons are planted at spaced intervals. Sprigs are obtained by tearing apart established lawns.

Spur Short, stubby shoot, as in some fruit trees.

Spur-type Tree Fruit tree (primarily apple and cherry) that has shortened internodes and consequently buds and spurs much closer together; about two-thirds the height of normal trees.

Stamen Male organ of the flower producing the pollen; usually composed of anther and filament.

Staminate Designating a flower that has stamens but no pistil and hence is imperfect.

Stem Axis of a plant bearing leaves with buds in their axils. It may be above or below ground, and the leaves may be functional or scales.

Stem Cutting Any part of a stem used for plant propagation by cuttage.

Stigma Summit of the pistil; receives the pollen grains.

Stipule Basal appendage of a leaf or petiole. May photosynthesize or be scales, and may protect the axillary buds.

Stock Plant or plant part upon which a scion is inserted in propagation by graftage.

Stolon Trailing and rooting shoot (in higher plants); also called runner.

Stolonizing Method of planting a lawn with grasses that reproduce asexually. Shredded stolons are spread on the soil surface and raked lightly to firm them into the soil.

Stoma (Stomate) Opening or pore, mainly in leaves, through which CO_2 for photosynthesis is absorbed and transpired moisture lost.

Stone Fruit Fruit with the seed or kernel surrounded by a hard endocarp (or stone) within the pulp or flesh; as a plum, peach, or cherry.

Storage Organ Any plant part in which elaborated food materials are stored.

Strain Group of plants of common lineage that, although not taxonomically distinct from others of the species or variety, are distinguishable on the basis of ecological or physiological characteristics.

Stratification Method of storing seeds at a temperature between 35° and 45°F (2° and 7°C) in alternate layers or mixed in moist sand, peat moss, or other medium, as a means of overcoming physiological dormancy.

Strip Sodding Laying of sod in strips separated by unsodded spaces.

Structural Incompatibility Evolutionary adaptation to insure cross-pollination, in which the stigma and anthers of a plant are located in such a way that self-pollination does not occur.

Structure Natural arrangement of individual particles in soil into separate aggregates, various in form and size.

Style Part of the pistil that connects the ovary and stigma, through which the pollen tube grows to the ovule.

Suberization Healing of wounded plant tissue by formation of a corky layer.

Subspecies Group of individuals within a species, distinguished by certain common geographical or varietal characters.

Subtropical Crop Crop that will take some freezing temperatures but will not survive in areas with a cold winter climate.

Succession Cropping Growing of two or more crops, one after the other on the same land, in one growing season.

Succulent Juicy; plant having a high percentage of water.

Sucker Secondary shoot that develops from the root, crown, or stem of a plant; a rapid-growing upright vegetative shoot.

Summer Fallowing System of farming where a crop is produced every other year; used primarily in dry regions where the moisture that falls on the soil during two seasons is needed to produce one crop.

Supercooling Phenomenon by which water kept absolutely motionless does not form ice crystals until the temperature falls several degrees below the freezing point.

Taxa Categories in a plant classification system.

Tap-root Stout, tapering main root from which arise smaller, lateral branches.

T (shield) Budding Bud grafting by insertion of a bud of a specified variety, with an attached segment of bark and wood in the shape of a shield, into an opening in the bark of a stem or branch of a different stock. The opening in the bark of the stem is a T-shaped cut.

Temperate Zone Crop Crop able to adapt so that it survives temperatures considerably below the freezing point.

Temperature-induced Hardiness Cold hardiness of plants native to the temperate zones that develops only if the plant is subjected to below-freezing temperatures.

Temperature Inversion Condition in the atmosphere in which the temperature rises with increased elevation.

Terminal Bud Bud that develops at the end of a branch or stem.

Terracing Construction of a raised, level area of earth supported on one side by a wall, bank, etc.; as a terraced yard.

Terrarium Tightly fitted, glass-enclosed, indoor garden, resembling an aquarium, in which plants are grown in earth.

Tetraploid Organism whose cells contain four haploid (monoploid) sets of chromosomes.

Texture Relative proportion in a soil of the various size groups of individual soil grains. The coarseness or fineness of the soil depends on the predominance of one or the other of these groups, which are silt, clay, and sand.

Thatch Dry layer of organic matter at the soil surface in a lawn.

Thinning Out Removal of an entire shoot or branch.

Tilling Cultivating land.

Tissue Group of organized plant cells that perform a specific function.

Tissue Culture Growth of detached pieces of tissue in nutrient solutions under sterile conditions; an important research tool and method of propagation for some plants.

Topiary Shrubs that have been clipped and trained into ornamental but unnatural shapes.

Topworking Top grafting; changing the cultivar of an old tree by graftage.

Tracheophyta Division of the plant kingdom containing plants with vascular (xylem and phloem) tissue, such as ferns and seed plants.

Training Directing the growth of a plant to a desired shape by pruning while young or fastening the stem and branches to a support.

Translocation Movement of a substance, such as water or an herbicide, from one part of a plant to another.

Transpiration Loss of water from plant tissues in the form of vapor.

Trashy Fallowing Soil management procedure in which the soil is stirred but dried weeds, stubble, and other debris are left as a mulch to protect fine soil particles from erosion.

Trickle Irrigation System of irrigation in which a constant drip of water is supplied.

Triple Fusion Union in the embryo sac of the two polar nuclei and a male nucleus; the starting point for the development of the endosperm.

Triploid Having three times the haploid number of chromosomes.

Tropical Crop Crop that originated in tropical areas of the earth; subject to cold injury at temperatures considerably above the freezing point.

Tube Nucleus The nongenerative nucleus in a pollen-tube; probably plays a part in regulating the development and behavior of that organ.

Tuber Enlarged, fleshy, usually terminal portion of a rhizome, bearing "eyes" or buds.

Tuberous Root Enlarged root that tapers toward both ends, as in dahlia and sweet potato.

Tuberous Stem Enlarged stem located between the regular stem and the roots; enlarges each year.

Tunica Dry, protective cover surrounding a tunicate bulb.

Tunicate Bulb Bulb in which the fleshy scales are in continuous, concentric layers. Has a dry and membranous outer covering that protects the bulb. Examples are onion and tulip bulbs.

Ultraviolet Portion of the spectrum composed of light waves just shorter than violet light. It is used in irradiation, disinfection, and sterilization.

Unavailable (hygroscopic) Water Water held by the soil below the wilting point.

Vacuole Cavity within the protoplasm containing a solution of sugars, salts, pigments, etc., together with colloidal materials.

Variety Subdivision of a species. Horticultural variety is synonymous with cultivar.

Vascular Bundle Strandlike portion of the vascular tissue of a plant, composed of xylem and phloem.

Vase (open center) System System of training in which the central leader is cut off 18 to 30 inches (45 to 75 cm) from the ground, and two or three side branches become the scaffolds and spread to form the framework of the tree.

Vegetative Growth, tissues, or processes concerned with the maintenance of the plant body; contrasted with tissues or activities involved in sexual reproduction.

Vegetative Phase Stage in the growth of a plant when food resources are directed primarily to the growth of leaves, stems, and roots.

Vegetative (asexual) Propagation Reproduction by plant parts other than seed of the parent plant.

Vein Vascular bundle forming a part of the framework of the conducting and supporting tissue of a leaf or other expanded organ.

Vermiculite Mineral or minerals, classified with the micas, which, with treatment at high temperatures, expand into wormlike scales and become a loose, absorbent mass. Commercial vermiculite is used as a mulch for seed beds, as a medium for rooting plant cuttings, and in potting plants.

Vine Crop Plant of the family Cucurbitaceae, such as cucumber, squash, pumpkin.

Virus Self-reproducing agent that is considerably smaller than a bacterium and can multiply only within the living cells of a suitable host. Often severely damages or kills the host.

Viticulture Art and science of growing grapes.

Warm-season Crop Crop, usually of tropical origin, that is killed as soon as temperatures drop slightly below freezing.

Water Stress Physiological disorder of plants (usually manifested by wilting) due to a lack of water.

Whorl 1. Three or more leaves or branches at a node. 2. Circle of floral organs, such as a whorl of sepals or stamens.

Wilted Lacking turgidity, drooping, or shriveling of plant tissue usually due to a deficiency of water.

Wilting Point Stage in soil moisture depletion where a plant is unable to take additional moisture from the soil and, as a consequence, becomes wilted.

Windbreak Object that serves as an obstacle to free movement of surface winds; most frequently refers to rows of trees that serve that purpose.

Wounding Practice of making wounds in the basal end of a cutting to stimulate rooting.

Xylem One of the two component tissues of vascular tissue; primarily responsible for transporting water and mineral nutrients from the roots.

Conversion Tables

Temperature

Fahrenheit (°F)	to Celsius (°C)
−40	−40
−35	−37
−30	−34
−25	−32
−20	−29
−15	−26
−10	−23
−5	−21
0	−18
5	−15
10	−12
15	−9
20	−7
25	−4
30	−1
32	0
35	2
40	4
45	7
50	10
55	13
60	16
65	18
68	20
70	21
75	24
80	27
85	29
90	32
95	35
100	38
120	49
140	60
160	71
180	82
200	93
212	100

Celsius (°C)	to Fahrenheit (°F)
−40	−40
−35	−31
−30	−22
−25	−13
−20	−4
−15	5
−10	14
−5	23
0	32
5	41
10	50
15	59
20	68
25	77
30	86
35	95
40	104
50	122
60	140
70	158
80	176
90	194
100	212

Conversion formulas:
$C° = 5/9\,(F° - 32)$
$F° = (9/5\,C°) + 32$

Length

1 micrometer (micron)	=	0.001 millimeter
1 millimeter	=	0.001 meter
	=	0.0394 inch
1 centimeter	=	10 millimeters
	=	0.3937 inch
	=	0.01 meter
1 meter	=	39.37 inches
	=	3.281 feet
	=	1,000 millimeters
	=	100 centimeters
1 kilometer	=	3,281 feet
	=	1,094 yards
	=	0.621 mile
	=	1,000 meters
1 inch	=	25.4 millimeters
	=	2.54 centimeters
1 foot	=	30.48 centimeters
	=	0.3048 meter
	=	12 inches
1 yard	=	0.9144 meter
	=	91.44 centimeters
	=	3 feet
1 mile	=	1,609.347 meters
	=	1.609 kilometers
	=	5,280 feet
	=	1,760 yards

Area

1 square centimeter	=	0.155 sq inch
	=	100 sq millimeters
1 square meter	=	1,550 sq inches
	=	10.764 sq feet
	=	1.196 sq yards
	=	10,000 sq centimeters
1 square kilometer	=	0.3861 sq mile
	=	1,000,000 sq meters
1 hectare	=	2.471 acres
	=	10,000 sq meters
1 square inch	=	6.452 sq centimeters
	=	1/144 sq foot
	=	1/1296 sq yard
1 square foot	=	929.088 sq centimeters
	=	0.0929 sq meter
1 square yard	=	8,361.3 sq centimeters
	=	0.8361 sq meter
	=	1,296 sq inches
	=	9 sq feet
1 square mile	=	2.59 sq kilometers
	=	640 acres
1 acre	=	0.4047 hectare
	=	43,560 sq feet
	=	4,840 sq yards
	=	4,046.87 sq meters

Weight

1 milligram	=	0.001 gram
	=	0.0154 grain
1 centigram	=	0.01 gram
	=	0.1543 grain
1 gram	=	0.0353 avoirdupois ounce
	=	15.4324 grains
1 kilogram	=	1,000 grams
	=	353 avoirdupois ounces
	=	2.2046 avoirdupois pounds
1 metric ton	=	1,000 kilograms
	=	2,204.6 pounds
	=	1.102 short tons (U.S.)
	=	0.984 long ton (British)
1 grain	=	1/7000 avoirdupois pound
	=	0.064799 gram
1 ounce (avoirdupois)	=	28.3496 grams
	=	437.5 grains
	=	1/16 pound
1 pound (avoirdupois)	=	453.593 grams
	=	0.45369 kilograms
	=	16 ounces
1 short ton	=	907.184 kilograms
	=	0.9072 metric ton
	=	2,000 pounds

Yield

1 kilogram per 100 sq meters	=	2.05 pounds per 1,000 sq ft
1 kilogram per hectare	=	0.89 pound per acre
1 cubic meter per hectare	=	14.2916 cubic feet per acre
1 pound per 1,000 sq ft	=	.488 kilograms per 100 sq meters
1 pound per acre	=	1.121 kilograms per hectare
1 ton (2,000 lb) per acre	=	2.242 metric tons per hectare
1 cubic foot per acre	=	0.0699 cubic meter per hectare
1 bushel (60 lb) per acre	=	67.26 kilograms per hectare

Volume

1 liter	=	1.057 U.S. quarts liquid
	=	0.9081 quart, dry
	=	0.2642 U.S. gallon
	=	0.221 Imperial gallon
	=	1,000 milliliters or cc
	=	0.0353 cubic foot
	=	61.02 cubic inches
	=	0.001 cubic meter
1 cubic meter	=	61,023.38 cubic inches
	=	35.314 cubic feet
	=	1.308 cubic yards
	=	264.17 U.S. gallons
	=	1,000 liters
	=	28.38 U.S. bushels
	=	1,000,000 cu centimeters
	=	1,000,000,000 cu millimeters
1 fluid ounce	=	6 teaspoons
	=	2 tablespoons
	=	1/128 gallon
	=	29.57 cubic centimeters
	=	29.562 milliliters
	=	1.805 cubic inches
	=	0.0625 U.S. pint (liquid)
1 U.S. quart liquid	=	946.3 milliliters
	=	57.75 cubic inches
	=	32 fluid ounces
	=	4 cups
	=	1/4 gallon
	=	2 U.S. pints (liquid)
	=	0.946 liter
1 quart dry	=	1.1012 liters
	=	67.20 cubic inches
	=	2 pints (dry)
	=	0.125 peck
	=	1/32 bushel
1 cubic inch	=	16.387 cubic centimeters
1 cubic foot	=	28,317 cubic centimeters
	=	0.0283 cubic meter
	=	28.316 liters
	=	7.481 U.S. gallons
	=	1,728 cubic inches
1 U.S. gallon	=	16 cups
	=	3.785 liters
	=	231 cubic inches
	=	4 U.S. quarts liquid
	=	8 U.S. pints liquid
	=	8.3453 pounds of water
	=	128 fluid ounces
	=	0.8327 British Imperial gallon
1 British Imperial gallon	=	4.546 liters
	=	1.201 U.S. gallons
	=	277.42 cubic inches
1 U.S. bushel	=	35.24 liters
	=	2,150.42 cubic inches
	=	1.2444 cubic feet
	=	0.03524 cubic meter
	=	2 pecks
	=	32 quarts (dry)
	=	64 pints (dry)

(From Janick, J., Schery, R. W., Woods, F. W., and Ruttan, V. W. *Plant Science*, 2nd ed. W. H. Freeman and Company, San Francisco. Copyright © 1974.)